Cabling
The Complete Guide to
Copper and Fiber-Optic Networking

Fourth Edition

Andrew Oliviero
Bill Woodward

WILEY *Custom*
LEARNING SOLUTIONS

Printed in the United States of America

ISBN   978-0-470-878095

Printed and bound by Victor Graphics, Inc.

10 9 8 7 6 5 4 3 2

Cabling

The Complete Guide to Copper and Fiber-Optic Networking

Fourth Edition

Andrew Oliviero

Bill Woodward

WILEY

Wiley Publishing, Inc.

Acquisitions Editor: Agatha Kim

Development Editor: Stef Jones

Technical Editors: Charlie Husson, Paul Neveux

Production Editor: Elizabeth Campbell

Copy Editor: Liz Welch

Production Manager: Tim Tate

Vice President and Executive Group Publisher: Richard Swadley

Vice President and Publisher: Neil Edde

Book Designers: Maureen Forys and Judy Fung

Compositor: Craig W. Johnson, Happenstance Type-O-Rama

Proofreader: Josh Chase, WordOne

Indexer: Nancy Guenther

Cover Designer and Illustrator: Richard Miller, Calyx Design

Library of Congress Cataloging-in-Publication Data

Oliviero, Andrew, 1966-
 Cabling : the complete guide to copper and fiber-optic networking / Andrew Oliviero, Bill Woodward. — 4th ed.
   p. cm.
 ISBN 978-0-470-47707-6 (pbk.)
 1. Telecommunication cables. 2. Copper wire. 3. Fiber optic cables. 4. Computer networks. I. Woodward, Bill. II. Title.
 TK5103.15.O43 2009
 621.382'3—dc22
          2009016551

Dear Reader

Thank you for choosing *Cabling: The Complete Guide to Copper and Fiber-Optic Networking, Fourth Edition*. This book is part of a family of premium quality Sybex books, all written by outstanding authors who combine practical experience with a gift for teaching.

Sybex was founded in 1976. More than thirty years later, we're still committed to producing consistently exceptional books. With each of our titles we're working hard to set a new standard for the industry. From the paper we print on, to the authors we work with, our goal is to bring you the best books available.

I hope you see all that reflected in these pages. I'd be very interested to hear your comments and get your feedback on how we're doing. Feel free to let me know what you think about this or any other Sybex book by sending me an email at nedde@wiley.com. Or, if you think you've found a technical error in this book, please visit http://sybex.custhelp.com. Customer feedback is critical to our efforts at Sybex.

Best regards,

Neil Edde
Vice President and Publisher
Sybex, an Imprint of Wiley

In loving memory of my brother Maurice. I miss you.

And to my parents, Mario and Colomba, and my brother Dominick. Thank you for all of your support, encouragement and great memories throughout the years. Although we are miles apart, you are with me every step of the way.

—*AO*

To Marc K. Boustany, Biswa Bhattacharyya, Matthew M. Tignor, and Nasrollah Fatehi for making this project possible!

—*BW*

Acknowledgments

First, I would like to thank David Barnett, David Groth, and Jim McBee, the authors of the last edition of *Cabling—The Complete Guide to Network Wiring*. They provided a strong foundation for this new edition. Thank you to the staff at John Wiley & Sons and Sybex for the opportunity to work on this book. Specifically, I would like to thank Agatha Kim, Pete Gaughan, Stef Jones, and Elizabeth Campbell for their support and patience. I thank our technical editor, Paul Neveux of Superior Essex, for doing an excellent job on the editing process and making this book well balanced. Thanks to Bill Woodward for helpful discussions before embarking on this work. Applause goes to Herb Congdon of Tyco-AMP, and chair of TIA's TR-42 committee, in managing the successful revision leading to the ANSI/TIA-568-C series standard. I think anyone involved with the revision of this standard would agree. I also thank John Kamino and Andy Ingles of OFS for providing some of the figures used in Part 1. I would like to thank my employer OFS for their support and encouragement on this project; specifically, my manager Pete Roberts. And finally, thanks to all of my friends for their unfailing loyalty and interest in this project.

—Andrew Oliviero

Writing a book is a team effort that takes a dedicated group of professionals. I am very fortunate to have been able to work with this team of talented and dedicated individuals.

First, I would like to especially thank Sybex for giving me the opportunity to write this book. Special thanks to Acquisitions Editor Agatha Kim, Production Editor Elizabeth Campbell, Developmental Editor Stef Jones, Editorial Manager Pete Gaughan, and Connor O'Brien and Jenni Housh of the Editorial department for the outstanding job you did guiding me through this project from start to finish.

Thanks to Chuck Schue, Randy Hall, Jim Moore, and Pat McGillvray at UrsaNav, Inc. for all your support with this project.

Thanks, Charlie Husson, for the outstanding job with the technical edits. You are an exceptional engineer, great mentor, and friend. I have learned so much from you over the years and look forward to working with you on future projects.

Many companies also provided technical information, equipment, and photographs. Special thanks to Donald Stone from KITCO Fiber Optics, Harvey Stone and Mike Jones from MicroCare, Mark Messer from Carlisle Interconnect Technologies, Manish Sharma from Corning Optical Fiber, Ryan Spillane from Corning Cable Systems, Bob Scharf from Protokraft, Mark Roehm from Amphenol Fiber Systems International, Chris Pegge from Stran Technologies, Scott Kale from Norfolk Wire, Christine Pons from OptiConcepts, and Dave Edwards from W.R. Systems.

Dick Glass has been a friend, mentor, and co-worker for many years; he has spent many hours guiding me through various writing projects. I feel very blessed to have met Dick and greatly appreciate his guidance over the years and assistance with this project.

Thanks to the host of people behind the scenes who I did not mention for all your efforts to make this book the best that it can be.

Last but not least, thank you to my family—to the love of my life, my beautiful wife Susan, for making this possible; and to my children, Mike, Brandon, Eric, Nathan, and Kathryn, for your patience, inspiration, encouragement, and prayers. I am the luckiest man alive to have all of you in my life.

—Bill Woodward

About the Authors

Andrew Oliviero is Senior Product Manager of optical fiber products at OFS (formerly Lucent Technologies), a supplier of optical fiber, cable apparatus and specialty products. He is responsible for developing global product strategies for the Enterprise, FTTX, Metro, Long Haul, and Submarine markets. Andrew is presently focused on new product development for single-mode fiber and is leading OFS's product development teams in developing advanced bend-insensitive single-mode fiber for Central Office and FTTX applications

Andrew began his career in Research and Development and has worked in single-mode and multimode optical fiber Product Management, Research & Development, Engineering and Manufacturing Operations for the last fourteen years with OFS, Lucent Technologies, and SpecTran Corp.

Andrew has presented world-wide at a variety of seminars, conferences and symposiums and is involved in the development of optical fiber and cabling architecture standards in TIA, IEC, ISO and IEEE. He is an active member of TIA's TR-42 committee where he was involved in the development of the ANSI/TIA-568-C standard, 10Gbps Multimode fiber specifications and measurement processes. He has held the Chair position of TIA's Fiber Optic LAN Section in 2007.

Andrew holds a B.S.E and Ph.D. in Chemical Engineering from University of Pennsylvania and University of Massachusetts.

Bill Woodward is a principal electrical engineer with UrsaNav, Inc. an engineering services company. Bill has been teaching fiber optics and other technical courses since 1992. He has more than 25 years of experience in the operation, maintenance, troubleshooting and repair of electronic and electrical systems.

Bill is licensed in the Commonwealth of Virginia as a professional electrical engineer. He is a member of SAE International and serves on three committees within the Aerospace Avionic Systems Division. He is vice-chairman of the AS-3 Fiber Optics and Applied Photonics Committee, co-chairman of the AS-3B Fiber Optic Supportability Subcommittee, and chairman of the AS-3B2 Education and Design Working Group. He recently received SAE International's Technical Standards Board Outstanding Contribution Award. He is also a member of the Electronics Technicians Association (ETA) International; he has served three tours as chairman of the ETA and currently chairs the Fiber Optic Committee.

Contents at a Glance

Contents

Part I

LAN Networks and Cabling Systems

Chapter 1

Introduction to Data Cabling

"Data cabling! It's just wire. What is there to plan?" the newly promoted programmer-turned-MIS-director commented to Jim. The MIS director had been contracted to help the company move its 750-node network to a new location. During the initial conversation, the director had a couple of other "insights":

He said that the walls were not even up in the new location, so it was too early to be talking about data cabling.

To save money, he wanted to pull the old Category 3 cabling and move it to the new location. ("We can run 100Base-TX on the old cable.")

He said not to worry about the voice cabling and the cabling for the photocopier tracking system; someone else would coordinate that.

Jim shouldn't have been too surprised by the ridiculous nature of these comments. Too few people understand the importance of a reliable, standards-based, flexible cabling system. Fewer still understand the challenges of building a high-speed network. Some of the technical problems associated with building a cabling system to support a high-speed network are comprehended only by electrical engineers. And many believe that a separate type of cable should be in the wall for each application (PCs, printers, terminals, copiers, etc.).

Data cabling has come a long way in the past 25 years.

You are probably thinking right now that all you really want to know is how to install cable to support a few 10Base-T workstations. Words and phrases such as *attenuation, crosstalk, twisted-pair, modular connectors*, and *multimode optical-fiber cable* may be completely foreign to you. Just as the world of PC LANs and WANs has its own industry buzzwords, so does the cabling business. In fact, you may hear such an endless stream of buzzwords and foreign terminology that you'll wish you had majored in electrical engineering in college. But it's not really that mysterious and, armed with the background and information we'll provide, you'll soon be using "cable-speak" like a cabling professional.

In this chapter, you will learn to:

◆ Identify the key industry standards necessary to specify, install, and test network cabling

◆ Understand the different types of unshielded twisted-pair (UTP) cabling

◆ Understand the different types of shielded twisted-pair cabling

◆ Determine the uses of plenum- and riser-rated cabling

◆ Identify the key test parameters for communications cables

The Golden Rules of Data Cabling

Listing our own golden rules of data cabling is a great way to start this chapter and the book. If your cabling is not designed and installed properly, you will have problems that you can't even imagine. Using our experience, we've become cabling evangelists, spreading the good news of proper cabling. What follows is our list of rules to consider when planning structured-cabling systems:

♦ Networks never get smaller or less complicated.

♦ Build one cabling system that will accommodate voice and data.

♦ Always install more cabling than you currently require. Those extra outlets will come in handy someday.

♦ Use structured-cabling standards when building a new cabling system. Avoid anything proprietary!

♦ Quality counts! Use high-quality cabling and cabling components. Cabling is the foundation of your network; if the cabling fails, nothing else will matter. For a given grade or category of cabling, you'll see a range of pricing, but the highest prices don't necessarily mean the highest quality. Buy based on the manufacturer's reputation and proven performance, not the price.

♦ Don't scrimp on installation costs. Even quality components and cable must be installed correctly; poor workmanship has trashed more than one cabling installation.

♦ Plan for higher-speed technologies than are commonly available today. Just because 1000Base-T Ethernet seems unnecessary today does not mean it won't be a requirement in five years.

♦ Documentation, although dull, is a necessary evil that should be taken care of while you're setting up the cabling system. If you wait, more pressing concerns may cause you to ignore it.

The Importance of Reliable Cabling

We cannot stress enough the importance of reliable cabling. Two recent studies vindicated our evangelical approach to data cabling. The studies showed:

♦ Data cabling typically accounts for less than 10 percent of the total cost of the network infrastructure.

♦ The life span of the typical cabling system is upward of 16 years. Cabling is likely the second most long-lived asset you have (the first being the shell of the building).

♦ Nearly 70 percent of all network-related problems are due to poor cabling techniques and cable-component problems.

TIP   If you have installed the proper category or grade of cable, the majority of cabling problems will usually be related to patch cables, connectors, and termination techniques. The permanent portion of the cable (the part in the wall) will not likely be a problem unless it was damaged during installation.

Of course, these were facts that we already knew from our own experiences. We have spent countless hours troubleshooting cabling systems that were nonstandard, badly designed, poorly documented, and shoddily installed. We have seen many dollars wasted on the installation of additional cabling and cabling infrastructure support that should have been part of the original installation.

Regardless of how you look at it, cabling is the foundation of your network. It must be reliable!

The Cost of Poor Cabling

The costs that result from poorly planned and poorly implemented cabling systems can be staggering. One company that moved into a new datacenter space used the existing cabling, which was supposed to be Category 5e cable. Almost immediately, 10 Gigabit Ethernet network users reported intermittent problems.

These problems included exceptionally slow access times when reading email, saving documents, and using the sales database. Other users reported that applications running under Windows XP and Windows Vista were locking up, which often caused them to have to reboot their PC.

After many months of network annoyances, the company finally had the cable runs tested. Many cables did not even meet the minimum requirements of a Category 5e installation, and other cabling runs were installed and terminated poorly.

WARNING   Often, network managers mistakenly assume that data cabling either works or it does not, with no in-between. Cabling *can* cause intermittent problems.

Is the Cabling to Blame?

Can faulty cabling cause the type of intermittent problems that the aforementioned company experienced? Contrary to popular opinion, it certainly can. In addition to being vulnerable to outside interference from electric motors, fluorescent lighting, elevators, cell phones, copiers, and microwave ovens, faulty cabling can lead to intermittent problems for other reasons.

These reasons usually pertain to substandard components (patch panels, connectors, and cable) and poor installation techniques, and they can subtly cause dropped or incomplete packets. These lost packets cause the network adapters to have to time out and retransmit the data.

Robert Metcalfe (inventor of Ethernet, founder of 3Com, columnist for *InfoWorld*, and industry pundit) helped coin the term *drop-rate magnification*. Drop-rate magnification describes the high degree of network problems caused by dropping a few packets. Metcalfe estimates that a 1 percent drop in Ethernet packets can correlate to an 80 percent drop in throughput. Modern network protocols that send multiple packets and expect only a single acknowledgement are especially susceptible to drop-rate magnification, as a single dropped packet may cause an entire stream of packets to be retransmitted.

Dropped packets (as opposed to packet collisions) are more difficult to detect because they are "lost" on the wire. When data is lost on the wire, the data is transmitted properly but, due to problems with the cabling, the data never arrives at the destination or it arrives in an incomplete format.

You've Come a Long Way, Baby: The Legacy of Proprietary Cabling Systems

Early cabling systems were unstructured, proprietary, and often worked only with a specific vendor's equipment. They were designed and installed for mainframes and were a combination of thicknet cable, twinax cable, and terminal cable (RS-232). Because no cabling standards existed, an MIS director simply had to ask the vendor which cable type should be run for a specific type of host or terminal. Frequently, though, vendor-specific cabling caused problems due to lack of flexibility. Unfortunately, the legacy of early cabling still lingers in many places.

PC LANs came on the scene in the mid-1980s; these systems usually consisted of thicknet cable, thinnet cable, or some combination of the two. These cabling systems were also limited to only certain types of hosts and network nodes.

As PC LANs became popular, some companies demonstrated the very extremes of data cabling. Looking back, it's surprising to think that the ceilings, walls, and floor trenches could hold all the cable necessary to provide connectivity to each system. As one company prepared to install a 1,000-node PC LAN, they were shocked to find all the different types of cabling systems needed. Each system was wired to a different wiring closet or computer room and included the following:

◆ Wang dual coaxial cable for Wang word processing terminals

◆ IBM twinax cable for IBM 5250 terminals

◆ Twisted-pair cable containing one or two pairs, used by the digital phone system

◆ Thick Ethernet from the DEC VAX to terminal servers

◆ RS-232 cable to wiring closets connecting to DEC VAX terminal servers

◆ RS-232 cable from certain secretarial workstations to a proprietary NBI word processing system

◆ Coaxial cables connecting a handful of PCs to a single Novell NetWare server

Some users had two or three different types of terminals sitting on their desks and, consequently, two or three different types of wall plates in their offices or cubicles. Due to the cost of cabling each location, the locations that needed certain terminal types were the only ones that had cables that supported those terminals. If users moved—and they frequently did—new cables often had to be pulled.

The new LAN was based on a twisted-pair Ethernet system that used unshielded twisted-pair cabling called SynOptics LattisNet, which was a precursor to the 10Base-T standards. Due to budget considerations, when the LAN cabling was installed, this company often used spare pairs in the existing phone cables. When extra pairs were not available, additional cable was installed. Networking standards such as 10Base-T were but a twinkle in the IEEE's (Institute of Electrical and Electronics Engineers) eye, and guidelines such as the ANSI/TIA/EIA-568 series released, called ANSI/TIA-568-C; it is discussed at length in Chapter 2, "Cabling Specifications and Standards."

The IEEE maintains the industry standards for Ethernet protocols (or applications). This is part of the 802.3 series of standards and includes applications such as 1000Base-T, 1000Base-SX, 10GBase-T, and 10GBase-SR.

The structured cabling market is estimated to be worth approximately $5 billion worldwide (according to the Building Services Research and Information Association [BSRIA]), due in part to the effective implementation of nationally recognized standards.

Cabling and the Need for Speed

The past few years have seen some tremendous advances not only in networking technologies but also in the demands placed on them. In the past 25 years, we have seen the emergence of standards for 10Mb Ethernet, 16Mb Token Ring, 100Mb FDDI (Fiber-Distributed Data Interface), 100Mb Ethernet, 155Mb ATM (Asynchronous Transfer Mode), 655Mb ATM, 1Gb Ethernet, 2.5Gb ATM, and 10Gb Ethernet. Network technology designers are already planning technologies to support data rates of up to 100Gbps.

The average number of nodes on a network segment has decreased dramatically, while the number of applications and the size of the data transferred have increased dramatically. Applications are becoming more complex, and the amount of network bandwidth required by the typical user is increasing. Is the bandwidth provided by some of the new ultra-high-speed network applications (such as 1Gb Ethernet) required today? Maybe not to the desktop, but network backbones already take advantage of them.

Does the fact that software applications and data are putting increasing demands on the network have anything to do with data cabling? You might think that the issue is related more to network interface cards, hubs, switches, and routers but, as data rates increase, the need for higher levels of performance on the cable also increases.

Types of Communications Media

Four major types of communications media (cabling) are available for data networking today: unshielded twisted-pair (UTP), shielded or screened twisted-pair (STP or ScTP), coaxial, and fiber-optic (FO). It is important to distinguish between backbone cables and horizontal cables. Backbone cables connect network equipment such as servers, switches, and routers and connect equipment rooms and telecommunications rooms. Horizontal cables run from the telecommunications rooms to the wall outlets. For new installations, multistrand fiber-optic cable is essentially universal as backbone cable. For the horizontal, UTP accounts for 85 percent of the market for typical applications. Much of the focus of this book is on UTP cable; however, newer fiber-optic–based network topologies are covered as well, as they are providing more and more advantages over UTP.

TWISTED-PAIR CABLE

In traditional installations, the most economical and widely installed cabling today is twisted-pair wiring. Not only is twisted-pair wiring less expensive than other media, installation is also simpler, and the tools required to install it are not as costly. Unshielded twisted-pair (UTP) and shielded twisted-pair (STP) are the two primary varieties of twisted-pair on the market today. Screened twisted-pair (ScTP) is a variant of STP.

CABLING @ WORK: THE INCREASING DEMANDS OF MODERN APPLICATIONS

A perfect example of the increasing demands put on networks by applications is a law firm that a few years ago was running typical office-automation software applications on its LAN. The average document worked on was about four pages in length and 12KB in size. This firm also used email; a typical email size was no more than 500 bytes. Other applications included dBASE III and a couple of small corresponding databases, a terminal-emulation application that connected to the firm's IBM minicomputer, and a few Lotus 1-2-3 programs. The size of transferred data files was relatively small, and the average 10Base-T network-segment size was about 100 nodes per segment.

Today, the same law firm is still using its 10Base-T and finding it increasingly insufficient for their ever-growing data processing and office-automation needs. The average document length is still around four pages but, thanks to the increasing complexity of modern word processing software and templates, the average document is nearly 50KB in size!

Even simple email messages have grown in size and complexity. An average simple email message size is now about 1.5KB, and, with the new message technologies that allow the integration of inbound/outbound faxing, an email message with a six-page fax attached has an average size of 550KB. Further, the firm integrated the voice mail system with the email system so that inbound voice mail is automatically routed to the user's mailbox. The average 30-second voice mail message is about 150KB.

The firm also implemented an imaging system that scans and stores many documents that previously would have taken up physical file space. Included in this imaging system are litigation support documents, accounting information, and older client documentation. A single-page TIFF file can vary in size (depending on the resolution of the image) from 40 to 125KB.

Additional software applications include a client/server document-management system, a client/server accounting system, and several other networked programs that the firm only dreamed about 2 years before. Most of the firm's attorneys make heavy use of the Internet, often visiting sites that provide streaming audio and video.

Today, the firm's average switched segment size is less than 36 nodes per segment, and the segments are switched to a 100Mbps backbone. Even with these small segment sizes, many segments are congested. Although the firm would like to begin running 100Base-TX Ethernet to the desktop, it is finding that its Category 3 cabling does not support 100Base-TX networking.

When this firm installs its new cabling system to support the next-generation network applications, you can be sure that it will want to choose the cabling infrastructure and network application carefully to ensure that its needs for the next 10 to 15 years will be accommodated.

Unshielded Twisted-Pair (UTP)

Though it has been used for many years for telephone systems, unshielded twisted-pair (UTP) for LANs first became common in the late 1980s with the advent of Ethernet over twisted-pair wiring and the 10Base-T standard. UTP is cost effective and simple to install, and its bandwidth capabilities are continually being improved.

NOTE An interesting historical note: Alexander Graham Bell invented and patented twisted-pair cabling and an optical telephone in the 1880s. During that time, Bell offered to sell his company to Western Union for $100,000, but it refused to buy.

UTP cabling typically has only an outer covering (jacket) consisting of some type of nonconducting material. This jacket covers one or more pairs of wire that are twisted together. In this chapter, as well as throughout much of the rest of the book, you should assume unless specified otherwise that UTP cable is a four-pair cable. Four-pair cable is the most commonly used horizontal cable in network installations today. The characteristic impedance of UTP cable is 100 \ ohms plus or minus 15 percent, though 120-ohm UTP cable is sometimes used in Europe and is allowed by the ISO/IEC 11801 Ed. 2 cabling standard.

A typical UTP cable is shown in Figure 1.1. This simple cable consists of a jacket that surrounds four twisted pairs. Each wire is covered by an insulation material with good *dielectric* properties. For data cables, this means that in addition to being electrically nonconductive, it must also have certain properties that allow good signal propagation.

FIGURE 1.1
UTP cable

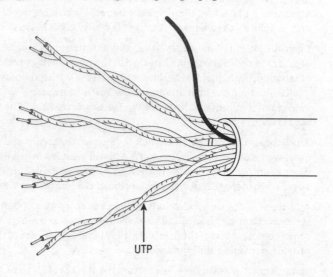

UTP

UTP cabling seems to generate the lowest expectations of twisted-pair cable. Its great popularity is mostly due to the low cost and ease of installation. With every new generation of UTP cable, network engineers think they have reached the limits of the UTP cable's bandwidth and capabilities. However, cable manufacturers continue to extend its capabilities. During the development of 10Base-T and a number of pre–10Base-T proprietary UTP Ethernet systems, critics said that UTP would never support data speeds of 10Mbps. Later, the skeptics said that UTP would never support data rates at 100Mbps. After that, the IEEE approved the 1000Base-T (1 Gb/s) standard in July 1999, which allows Gigabit Ethernet to run over Category 5 cable. Just when we thought this was the end of copper UTP-based applications, in 2006 the IEEE approved the 10GBase-T standard, which allows 10 Gigabit Ethernet over unshielded Category 6 and 6A cable!

Shielded Twisted-Pair (STP)

Shielded twisted-pair (STP) cabling was first made popular by IBM when it introduced type classification for data cabling. Though more expensive to purchase and install than UTP, STP offers some distinct advantages. The current ANSI/TIA-568-C cabling standard recognizes IBM Type 1A horizontal cable, which supports frequency rates of up to 300MHz, but does not recommend it for new installations. STP cable is less susceptible to outside electromagnetic interference (EMI) than UTP cabling because all cable pairs are well shielded.

NOT ALL UTP IS CREATED EQUAL!

Though two cables may look identical, their supported data rates can be dramatically different. Older UTP cables that were installed to support telephone systems may not even support 10Base-T Ethernet. The ANSI/TIA-568-C standard helps consumers choose the right cable (and components) for their application. The standard has been updated over the years and currently defines four categories of UTP cable: Categories 3, 5e, 6, and 6A. Here is a brief rundown of categories past and present:

Category 1 (not defined by ANSI/TIA-568-C) This type of cable usually supports frequencies of less than 1MHz. Common applications include analog voice telephone systems. It was never included in any version of the 568 standard.

Category 2 (not defined by ANSI/TIA-568-C) This cable type supports frequencies of up to 4MHz. It's not commonly installed, except in installations that use twisted-pair ARCnet and Apple LocalTalk networks. Its requirements are based on the original, proprietary IBM Cabling System specification. It was never included in any version of the 568 standard.

Category 3 (recognized cable type in ANSI/TIA-568-C) This type of cable supports data rates up to 16MHz. This cable was the most common variety of UTP for a number of years starting in the late 1980s. Common applications include 4Mbps UTP Token Ring, 10Base-T Ethernet, 100Base-T4, and digital and analog telephone systems. Its inclusion in the ANSI/TIA-568-C standard is for voice applications.

Category 4 (not defined by ANSI/TIA-568-C) Cable belonging to Category 4 was designed to support frequencies of up to 20MHz, specifically in response to a need for a UTP solution for 16Mbps Token Ring LANs. It was quickly replaced in the market when Category 5 was developed, as Category 5 gives five times the bandwidth with only a small increment in price. Category 4 was a recognized cable in the 568-A Standard, but was dropped from ANSI/TIA/EIA-568-B and also does not appear in ANSI/TIA-568-C.

Category 5 (was included in ANSI/TIA/EIA-568-B for informative purposes only) Category 5 was the most common cable installed, until new installations began to use an enhanced version. It may still be the cable type most in use because it was the cable of choice during the huge infrastructure boom of the 1990s. It was designed to support frequencies of up to 100MHz. Applications include 100Base-TX, FDDI over copper, 155Mbps ATM over UTP, and, thanks to sophisticated encoding techniques, 1000Base-T Ethernet. To support 1000Base-T applications, the installed cabling system had to pass performance tests specified by TSB-95 (TSB-95 was a Telecommunications Systems Bulletin issued in support of ANSI/TIA/EIA-568-A, which defines additional test parameters). It is no longer a recognized cable type per the ANSI/TIA-568-C standard, but for historical reference purposes, Category 5 requirements, including those taken from TSB-95, are specified in ANSI/TIA-568-C.2.

Category 5e (recognized cable type in ANSI/TIA-568-C) Category 5e (enhanced Category 5) was introduced with the TIA/EIA-568-A-5 addendum of the cabling standard. Even though it has the same rated bandwidth as Category 5, that is, 100MHz, additional performance criteria and a tighter transmission test requirement make it more suitable for high-speed applications such as Gigabit Ethernet. Applications are the same as those for Category 5 cabling. It is now the minimum recognized cable category for data transmission in ANSI/TIA-568-C.

Category 6 (recognized cable type in ANSI/TIA-568-C) Category 6 cabling was officially recognized with the publication of an addition to ANSI/TIA/EIA-568-B in June 2002. In addition to more stringent performance requirements as compared to Category 5e, it extends the usable bandwidth to 250MHz. Its intended use is for Gigabit Ethernet and other future high-speed transmission rates. Successful application of Category 6 cabling requires closely matched components in all parts of the transmission channel, that is, patch cords, connectors, and cable.

Category 6A or Augmented Category 6 (recognized cable type in ANSI/TIA-568-C) Category 6A cabling was officially recognized with the publication of ANSI/TIA/EIA-568-B.2-10 in February 2008. In addition to more stringent performance requirements as compared to Category 6, it extends the usable bandwidth to 500MHz. Its intended use is for 10 Gigabit Ethernet. Like Category 6, successful application of Category 6A cabling requires closely matched components in all parts of the transmission channel, that is, patch cords, connectors, and cable. The cabling standards are discussed in more detail in Chapter 2. Additional information on copper media can be found in Chapter 7, "Copper Cable Media," and Chapter 10, "Connectors."

Category 7 (recognized cable type in ISO 11801) Category 7 is an ISO/IEC category suitable for transmission frequencies up to 1GHz. It is widely used in Europe and is gaining some popularity in the United States. It is not presently recognized in ANSI/TIA-568-C.

Some STP cabling, such as IBM types 1 and 1A cable, uses a woven copper-braided shield, which provides considerable protection against EMI. Inside the woven copper shield, STP consists of twisted pairs of wire (usually two pairs) wrapped in a foil shield. Some STP cables have only the foil shield around the wire pairs.

NEW NOMENCLATURE FOR TWISTED-PAIR CABLES

TIA is addressing the potentially confusing nomenclature for different types of twisted-pair cables:

◆ Shielded twisted-pair (STP) will be called U/FTP.

◆ Screened twisted-pair (ScTP or FTP) will be called F/UTP.

◆ Category 7 screened shielded twisted-pair (S/STP or S/FTP) will be called ScFTP.

Figure 1.2 shows a typical STP cable. In the IBM design, the wire used in STP cable is 22 AWG (just a little larger than the 24 AWG wire used by typical UTP LAN cables) and has a nominal impedance of 150 ohms, but category versions can have a nominal impedance of 100 ohms.

Constructions of STP in 24 AWG, identical in copper conductor size to UTP cables, are more commonly used today.

Simply installing STP cabling does not guarantee you will improve a cable's immunity to EMI or reduce the emissions from the cable. Several critical conditions must be met to achieve good shield performance:

◆ The shield must be electrically continuous along the whole link.

◆ All components in the link must be shielded. No UTP patch cords can be used.

◆ The shield must fully enclose the pair, and the overall shield must fully enclose the core. Any gap in the shield covering is a source of EMI leakage.

◆ The shield must be grounded at both ends of the link, and the building grounding system must conform to grounding standards (such as J-STD-607-A).

FIGURE 1.2
STP cable

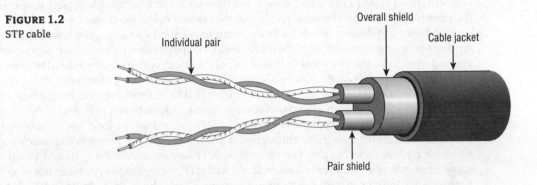

Overall shield

Cable jacket

Individual pair

Pair shield

If even one of these conditions is not satisfied, shield performance will be badly degraded. For example, tests have shown that if the shield continuity is broken, the emissions from a shielded cabling system increase by 20dB on the average.

Screened Twisted-Pair (ScTP)

A recognized cable type in the ANSI/TIA-568-C standard is screened twisted-pair (ScTP) cabling, a hybrid of STP and UTP cable. ScTP cable contains four pairs of unshielded 24 AWG, 100 ohm wire (see Figure 1.3) surrounded by a foil shield or wrapper and a drain wire for grounding purposes. Therefore, ScTP is also sometimes called foil twisted-pair (FTP) cable because the foil shield surrounds all four conductors. This foil shield is not as large as the woven copper-braided jacket used by some STP cabling systems, such as IBM types 1 and 1A. ScTP cable is essentially STP cabling that does not shield the individual pairs; the shield may also be smaller than some varieties of STP cabling.

FIGURE 1.3
ScTP cable

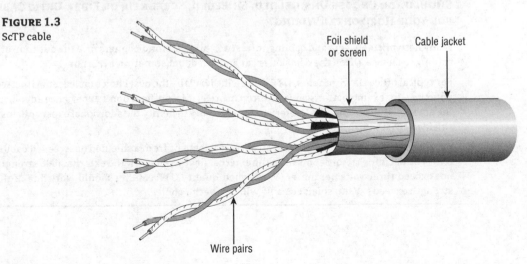

Foil shield
or screen

Cable jacket

Wire pairs

The foil shield is the reason ScTP is less susceptible to noise. To implement a completely effective ScTP system, however, the shield continuity must be maintained throughout the entire channel—including patch panels, wall plates, and patch cords. Yes, you read this correctly; the continuity of not only the wires but also the shield must be maintained through connections. Like STP cabling, the entire system must be bonded to ground at both ends of each cable run, or you will have created a massive antenna, the frequencies of which are inversely proportional to the length of the shield. The net effect is that the noise is out of band.

Standard eight-position modular jacks (commonly called RJ-45s) do not have the ability to ensure a proper ground through the cable shield. So special mating hardware, jacks, patch panels, and even tools must be used to install an ScTP cabling system. Many manufacturers of ScTP cable and components exist—just be sure to follow all installation guidelines.

ScTP is recommended for use in environments that have abnormally high ambient electromagnetic interference, such as industrial work spaces, hospitals, airports, and government/military communications centers. For example, ScTP is used in fast-food restaurants that use wireless headsets for their drive-through-window workers; some wireless frequencies can interfere with Ethernet over copper. The value of an ScTP system in relation to its additional cost is sometimes questioned, as some tests indicate that UTP noise immunity and emissions characteristics are comparable with ScTP cabling systems. Often, the decision to use ScTP simply boils down to whether you want the warm and fuzzy feeling of knowing an extra shield is in place.

Screened Shielded Twisted-Pair (S/STP or S/FTP)

S/STP cabling, also known as screened fully shielded twisted-pair (S/FTP), contains four individually shielded pairs of 24 AWG, 100 ohm wire surrounded by an outer metal shielding covering the entire group of shielded copper pairs. This type of cabling offers the best protection from interference from external sources, and also eliminates alien crosstalk (discussed later), allowing the greatest potential for higher speeds.

Category 7 is an S/STP cable standardized in ISO 11801 Ed. 2, which offers a usable bandwidth to 600MHz. Its intended use is for the 10 Gigabit Ethernet, 10GBase-T application. S/STP cable looks similar to the cable in Figure 1.2, but has four individually shielded conductor pairs.

SHOULD YOU CHOOSE UNSHIELDED, SHIELDED, SCREENED, OR FIBER-OPTIC CABLE FOR YOUR HORIZONTAL WIRING?

Many network managers and cabling-infrastructure systems designers face the question of which cabling to choose. Often the decision is cut and dried, but sometimes it is not.

For typical office environments, UTP cable will always be the best choice (at least until active components—for example, transceivers—drop in price). Most offices don't experience anywhere near the amount of electromagnetic interference necessary to justify the additional expense of installing shielded twisted-pair cabling.

Environments such as hospitals and airports may benefit from a shielded or screened cabling system. The deciding factor seems to be the external field strength. If the external field strength does not exceed three volts per meter (V/m), good-quality UTP cabling should work fine. If the field strength exceeds 3V/m, shielded cable will be a better choice.

However, many cabling designers think that if the field strength exceeds 3V/m, fiber-optic cable is a better choice. Further, these designers will point out the additional bandwidth and security of fiber-optic cable.

Although everyone has an opinion on the type of cable you should install, it is true that the only cable type that won't be outgrown quickly is optical fiber. Fiber-optic cables are already the media of choice for the backbone. As hubs, routers, and workstation network interface cards for fiber-optic cables come down in price, fiber will move more quickly into the horizontal cabling space.

FIBER-OPTIC CABLE

As late as 1993, it seemed that in order to move toward the future of desktop computing, businesses would have to install fiber-optic cabling directly to the desktop. It's surprising that copper cable (UTP) performance continues to be better than expected. Fiber-optic cable is discussed in more detail in Chapter 8, "Fiber-Optic Media."

NOTE   *Fiber* versus *fibre*: Are these the same? Yes, just as *color* (U.S. spelling) and *colour* (British spelling) are the same. Your U.S. English spell checker will probably question your use of *fibre*, however.

Although for most of us fiber to the desktop is not yet cost-effective for traditional LAN networks, fiber-optic cable is touted as the ultimate answer to all our voice, video, and data transmission needs since it has virtually unlimited bandwidth and continues to make inroads in the LAN market. Some distinct advantages of fiber-optic cable include:

◆   Transmission distances are much greater than with copper cable.

◆   Bandwidth is dramatically higher than with copper.

◆   Fiber optic is not susceptible to outside EMI or crosstalk interference, nor does it generate EMI or crosstalk.

◆   Fiber-optic cable is much more secure than copper cable because it is extremely difficult to monitor, "eavesdrop on," or tap a fiber cable.

NOTE   Fiber-optic cable can easily handle data at speeds above 10Gbps; in fact, it has been demonstrated to handle data rates exceeding 500Gbps!

Since the late 1980s, LAN solutions have used fiber-optic cable in some capacity. Recently, a number of ingenious solutions that allow voice, data, and video to use the same fiber-optic cable have emerged.

Fiber-optic cable uses a strand of glass or plastic to transmit data signals using light; the data is carried in light pulses. Unlike the transmission techniques used by its copper cousins, optical fibers are not electrical in nature.

Plastic-core cable is easier to install than traditional glass core, but plastic cannot carry data as far as glass. In addition, graded-index plastic optical fiber (POF) has yet to make a widespread appearance on the market, and the cost-to-bandwidth value proposition for POF is poor and may doom it to obscurity.

Light is transmitted through a fiber-optic cable by light-emitting diodes (LEDs) or lasers. With newer LAN equipment designed to operate over longer distances, such as with 1000Base-LX, lasers are commonly being used.

A fiber-optic cable (shown in Figure 1.4) consists of a jacket (sheath), protective material, and the optical-fiber portion of the cable. The optical fiber consists of a core (8.3, 50, or 62.5 microns in diameter, depending on the type) that is smaller than a human hair, which is surrounded by a cladding. The cladding (typically 125 micrometers in diameter) is surrounded by a coating, buffering material, and, finally, a jacket. The cladding provides a lower refractive index to cause reflection within the core so that light waves can be transmitted through the fiber.

FIGURE 1.4
A dual fiber-optic cable

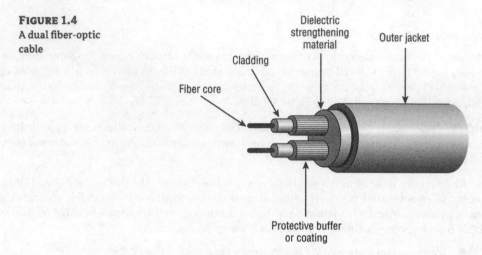

Dielectric strengthening material

Outer jacket

Cladding

Fiber core

Protective buffer or coating

FIBER-OPTIC CABLING COMES OF AGE AFFORDABLY

Fiber-optic cable used to be much harder to install than copper cable, requiring precise installation practices. However, in the past few years, the cost of an installed fiber-optic link (just the cable and connectors) has dropped and is now often the same as the cost of a UTP link. Better fiber-optic connectors and installation techniques have made fiber-optic systems easier to install. In fact, some installers who are experienced with both fiber-optic systems and copper systems will tell you that with the newest fiber-optic connectors and installation techniques, fiber-optic cable is easier to install than UTP.

The main hindrance to using fiber optics all the way to the desktop in lieu of UTP or ScTP is that the electronics (workstation network interface cards and hubs) are still significantly more expensive, and the total cost of a full to-the-desktop FO installation (Centralized Cabling, per ANSI/TIA-568-C) is estimated at 30 percent greater than UTP. However, the Fiber-to-the-Telecommunications-Enclosure topology can bring fiber closer to the desk, while still using UTP for the final connection, and actually lower the cost compared to traditional topologies. This will be discussed in more detail in Chapter 3, "Choosing the Correct Cabling."

Two varieties of fiber-optic cable are commonly used in LANs and WANs today: single-mode and multimode. The mode can be thought of as bundles of light rays entering the fiber; these light rays enter at certain angles.

KEY TERM   *dark fiber*   No, *dark fiber* is not a special, new type of fiber cable. When telecommunications companies and private businesses run fiber-optic cable, they never run the exact number of strands of fiber they need. That would be foolish. Instead, they run two or three times the amount of fiber they require. The spare strands of fiber are often called *dark fiber* because they are not then in use—that is, they don't have light passing through them. Telecommunications companies often lease out these extra strands to other companies.

Single-Mode Fiber-Optic Cable

Single-mode fiber-optic cable is most commonly used by telephone companies in trans-continental links and in data installations as backbone cable interconnecting buildings. Single-mode fiber-optic cable is not used as horizontal cable to connect computers to hubs and is not often used as a cable to interconnect telecommunications rooms to the main equipment room. The light in a single-mode cable travels straight down the fiber (as shown in Figure 1.5) and does not bounce off the surrounding cladding as it travels. Typical single-mode wavelengths are 1,310 and 1,550 nanometers.

FIGURE 1.5
Single-mode fiber-optic cable

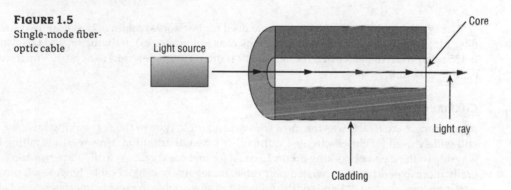

Before you install single-mode fiber-optic cable, make sure the equipment you are using supports it. The equipment that uses single-mode fiber typically uses expensive lasers to transmit light through the cable because a laser is the only light source capable of inserting light into the very small (8- to 10-micron) core of a single-mode fiber.

Multimode Fiber-Optic Cable

Multimode fiber (MMF) optic cable is usually the fiber-optic cable used with networking applications such as 10Base-FL, 100Base-F, FDDI, ATM, Gigabit Ethernet, and 10 Gigabit Ethernet that require fiber optics for both horizontal and backbone cable. Multimode cable allows more than one mode (a portion of the light pulse) of light to propagate through the cable. Typical wavelengths of light used in multimode cable are 850 and 1,300 nanometers.

There are two types of multimode fiber-optic cable: step index or graded index. Step-index multimode fiber-optic cable indicates that the refractive index between the core and the cladding is very distinctive. The graded-index fiber-optic cable is the most common type of multimode fiber. The core of a graded-index fiber contains many layers of glass; each has a lower index of refraction going outward from the core of the fiber. Both types of multimode fiber permit multiple modes of light to travel through the fiber simultaneously (see Figure 1.6). Graded-index fiber is preferred because less light is lost as the signal travels around bends in the cable; therefore, the cable offers much greater bandwidth.

FIGURE 1.6

Multimode fiber-optic cable (graded-index multimode)

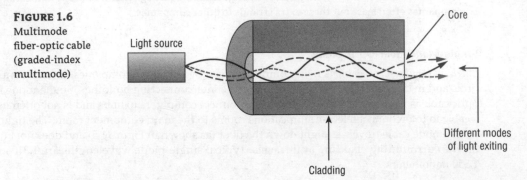

Light source

Core

Different modes of light exiting

Cladding

The typical multimode fiber-optic cable used for horizontal cabling consists of two strands of fiber (duplex); the core is either 50 or 62.5 microns (micrometers) in diameter, and the cladding is 125 microns in diameter (the measurement is often simply referred to as 50/125 micron or 62.5/125 micron).

COAXIAL CABLE

At one time, *coaxial cable* was the most widely used cable type in the networking business. It is still widely used for closed-circuit TV and other video distribution. However, it is falling by the wayside in the data networking arena. Coaxial (or just coax) cable is difficult to run and is generally more expensive than twisted-pair cable. In defense of coaxial cable, however, it provides a tremendous amount of bandwidth and is not as susceptible to outside interference as is UTP. Overall installation costs might also be lower than for other cable types because the connectors take less time to apply. Although we commonly use coaxial cable to connect our televisions to our VCRs, we will probably soon see fiber-optic or twisted-pair interfaces to television set top boxes.

Coaxial cable comes in many different flavors, but the basic design is the same for all types. Figure 1.7 shows a typical coaxial cable; at the center is a solid (or sometimes stranded) copper core. Some type of low dielectric insulation material, like fluorinated ethylene-propylene (FEP), polyethylene (PE), or polypropylene (PP) is used to surround the core. Either a sleeve or a braided-wire mesh shields the insulation, and a jacket covers the entire cable.

The shielding shown in Figure 1.7 protects the data transmitted through the core from outside electrical noise and keeps the data from generating significant amounts of interference. Coaxial cable works well in environments where high amounts of interference are common.

A number of varieties of coaxial cable are available on the market. You pick the coaxial cable required for the application; unfortunately, coaxial cable installed for Ethernet cannot be used for an application such an ArcNet. Some common types of coaxial cable are listed in Table 1.1.

FIGURE 1.7
Typical coaxial cable

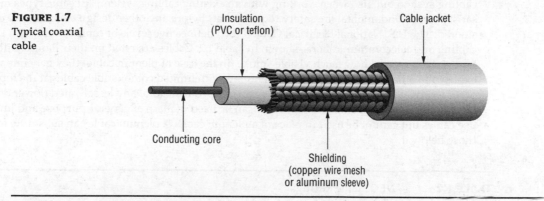

TABLE 1.1 Common Coaxial-Cable Types

CABLE	DESCRIPTION
RG-58/U	A 50 ohm coaxial cable with a solid core. Commonly called thinnet and used with 10Base-2 Ethernet and some cable TV applications.
RG-58 A/U	A 50 ohm coaxial cable with a stranded core. Also known as thinnet. Used by 10Base-2 Ethernet and some cable TV applications.
RG-58 C/U	A military-specification version of RG-58 A/U.
RG-59U	A 75 ohm coaxial cable. Used with Wang systems and some cable TV applications.
RG-6U	A 75 ohm coaxial cable. The current minimum grade to install in residences because it will handle the full frequency range of satellite service, plus high-definition TV and cable-modem service.
RG-6 Quad Shield	Same as RG-6U, but with additional shielding for enhanced noise immunity. Currently the recommended cable to use in residences.
RG-62U	A 93 ohm coaxial cable. Used with IBM cabling systems and ArcNet.

Cable Design

Whether you are a network engineer, cable installer, or network manager, a good understanding of the design and components of data cabling is important. Do you know what types of cable can be run above the ceiling? What do all those markings on the cable mean? Can you safely untwist a twisted-pair cable? What is the difference between shielded and unshielded twisted-pair cable? What is the difference between single-mode and multimode fiber-optic cable?

You need to know the answer to these questions—not only when designing or installing a cabling system but also when working with an existing cabling system. All cable types must satisfy some fundamental fire safety requirements before any other design elements are considered. The U.S. National Electrical Code (NEC) defines five levels of cable for use with LAN cabling and telecommunications, shown in Table 1.2. Cables are rated on their flammability, heat resistance, and how much visible smoke (in the case of plenum cable) they generate when exposed to a flame. The ratings are a hierarchy, with limited combustible cables at the top. In other words, a cable with a higher rating can be used instead of any lesser-rated (lower down in the table) cable. For example, a riser cable can be used in place of general-purpose and limited-use cables but cannot be used in place of a plenum cable. A plenum cable can substitute for all those below it.

TABLE 1.2   NEC Flame Ratings

Optical Fiber Article 770	Twisted-Pair Article 800	Coaxial Cable Article 820	Common Term	Notes
OFNP[1] OFCP[2]	CMP[3] MPP[4]	CAVTP	Plenum	Most stringent rating. Must limit the spread of flame and the generation of visible smoke. Intended for use in HVAC (heating, ventilation, and air conditioning) plenum areas; can be substituted for all subsequent lesser ratings.
OFNR	CMR MPR	CATVR	Riser	When placed vertically in a building riser shaft going from floor to floor, cable must not transmit flame between floors.
OFCR			Riser	This is a conductive version of OFNR.
OFNG OFCG	CMG MPG	CATVG	General purpose	Flame spread limited to 4″–11″ during test. Cable may not penetrate floors or ceilings, i.e., may only be used within a single floor. This designation was added as a part of the harmonization efforts between U.S. and Canadian standards.

TABLE 1.2   NEC Flame Ratings   *(CONTINUED)*

OPTICAL FIBER ARTICLE 770	TWISTED-PAIR ARTICLE 800	COAXIAL CABLE ARTICLE 820	COMMON TERM	NOTES
OFNOFC	CM	CATV	General purpose	Flame spread limited to 4″–11″ during test. Cable may not penetrate floors or ceilings, i.e., may only be used within a single floor.
Not applicable	CMX	CATVX	Limited use	For residential use, but can only be installed in one- and two-family (duplex) housing units. Often co-rated with optional UL requirements for limited outdoor use.

[1] *OFN = Optical fiber, nonconductive (no metallic elements in the cable)*

[2] *OFC = Optical fiber, conductive (contains a metallic shield for mechanical protection)*

[3] *CM = Communications cable*

[4] *MP = Multipurpose cable (can be used as a communications cable or a low-voltage signaling cable per NEC Article 725)*

WARNING   The 2008 edition of the NEC requires that the accessible portion of all abandoned communications cables in plenums and risers be removed when installing new cabling. The cost of doing so could be significant, and your cabling request for quote (RFQ) should clearly state both the requirement and who is responsible for the cost of removal.

NOTE   More details on the National Electrical Code are given in Chapter 4, "Cable System and Infrastructure Constraints."

Plenum

According to building engineers, construction contractors, and air-conditioning people, the *plenum* (shown in Figure 1.8) is the space between the false ceiling (a.k.a. drop-down ceiling) and the structural ceiling, *when that space is used for air circulation, heating ventilation, and air-conditioning (HVAC).* Occasionally, the space between a false floor (such as a raised computer room floor) and the structural floor is also referred to as the plenum. Typically, the plenum is used for returning air to the HVAC equipment.

Raised ceilings and floors are convenient spaces in which to run data and voice cable, but national code requires that plenum cable be used in plenum spaces. Be aware that some people use the word *plenum* too casually to refer to all ceiling and floor spaces, whether or not they are plenums. This can be expensive because plenum cables can cost more than twice their nonplenum equivalent. (See the sidebar "Plenum Cables: Debunking the Myths.")

Cable-design engineers refer to *plenum* as a type of cable that is rated for use in the plenum spaces of a building. Those of us who work with building engineers, cabling professionals, and contractors must be aware of when the term applies to the air space and when it applies to cable.

of cabling standards were not yet formulated (see the next section for more information on ANSI/TIA-568-C). Companies deploying twisted-pair LANs had little guidance, to say the least.

Much of the cable that was used at this company was sub–Category 3, meaning that it did not meet minimum Category 3 performance requirements. Unfortunately, because the cabling was not even Category 3, once the 10Base-T specification was approved, many of the installed cables would not support 10Base-T cards on most of the network. So three years into this company's network deployments, it had to rewire much of its building.

KEY TERM   *application* Often you will see the term *application* used when referring to cabling. If you are like us, you think of an application as a software program that runs on your computer. However, when discussing cabling infrastructures, an application is the technology that will take advantage of the cabling system. Applications include telephone systems (analog voice and digital voice), Ethernet, Token Ring, ATM, ISDN, and RS-232.

Proprietary Cabling Is a Thing of the Past

The company discussed in the previous section had at least seven different types of cables running through the walls, floors, and ceilings. Each cable met only the standards dictated by the vendor that required that particular cable type.

As early as 1988, the computer and telecommunications industry yearned for a versatile standard that would define cabling systems and make the practices used to build these cable systems consistent. Many vendors defined their own standards for various components of a cabling system.

The Need for a Comprehensive Standard

Twisted-pair cabling in the late 1980s and early 1990s was often installed to support digital or analog telephone systems. Early twisted-pair cabling (Level 1 or Level 2) often proved marginal or insufficient for supporting the higher frequencies and data rates required for network applications such as Ethernet and Token Ring. Even when the cabling did marginally support higher speeds of data transfer (10Mbps), the connecting hardware and installation methods were often still stuck in the "voice" age, which meant that connectors, wall plates, and patch panels were designed to support voice applications only.

The original Anixter Cables Performance Levels document only described performance standards for cables. A more comprehensive standard had to be developed to outline not only the types of cables that should be used but also the standards for deployment, connectors, patch panels, and more.

A consortium of telecommunications vendors and consultants worked in conjunction with the American National Standards Institute (ANSI), Electronic Industries Alliance (EIA), and the Telecommunications Industry Association (TIA) to create a Standard originally known as the Commercial Building Telecommunications Cabling Standard, or ANSI/TIA/EIA-568-1991. This standard has been revised and updated several times. In 1995, it was published as ANSI/TIA/EIA-568-A, or just TIA/EIA-568-A. In subsequent years, TIA/EIA-568-A was updated with a series of addendums. For example, TIA/EIA-568-A-5 covered requirements for enhanced Category 5 (Category 5e), which had evolved in the marketplace before a full revision of the standard could be published. A completely updated version of this standard was released as ANSI/TIA/EIA-568-B in May 2001. At the time of this writing, a new standard is about to be

Some local authorities and building management may also require plenum-rated cable in nonplenum spaces. Know the requirements in your locale.

FIGURE 1.8
The ceiling space
and a riser

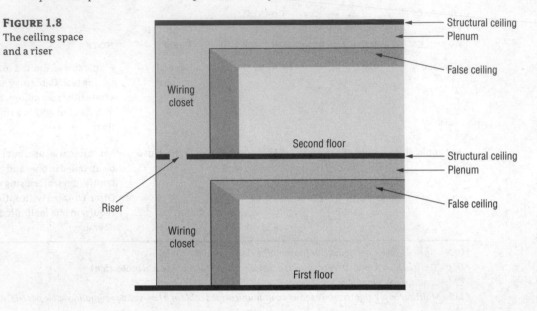

PLENUM CABLES: DEBUNKING THE MYTHS

It's time to set the record straight about several commonly held, but incorrect, beliefs about plenum-rated cable. These misconceptions get in the way of most discussions about LAN cabling but are especially bothersome in relation to UTP.

Myth #1: Any false or drop-ceiling area or space beneath a raised floor is a plenum, and I must use plenum-rated cables there.   Not true. Although many people call all such spaces the plenum, they aren't necessarily. A plenum has a very specific definition. It is a duct, raceway, or air space that is part of the HVAC air-handling system. *Sometimes*, or even *often*, the drop-ceiling or raised-floor spaces are used as return air passageways in commercial buildings, but not always. Your building maintenance folks should know for sure, as will the company that installed the HVAC. If it isn't a plenum space, then you don't have to spend the extra for plenum-rated cable.

Myth #2: There are plenum cables and PVC cables.   The wording here is nothing but sloppy use of terminology, but it results in the widespread notion that plenum cables don't use poly-vinyl chloride (PVC) in their construction and that nonplenum cables are all PVC. In fact, vir-tually all four-pair UTP cables in the United States use a PVC jacket, plenum cables included. And guess what? No Category 5e or better cables on the market use any PVC as an insulation material for the conductors, no matter what the flame rating. So a plenum-rated cable actu-ally has just as much PVC in it as does a so-called PVC nonplenum cable. Unless you have to be specific about one of the lesser flame ratings, you are more accurate when you generalize about cable flame ratings if you say *plenum* and *nonplenum* instead of *plenum* and *PVC*.

Myth #3: Plenum cables don't produce toxic or corrosive gasses when they burn. In Europe and in the United States (regarding specialized installations), much emphasis is placed on "clean" smoke. Many tests, therefore, measure the levels of toxic or corrosive elements in the smoke. But for general commercial and residential use, the U.S. philosophy toward fire safety as it relates to cables is based on two fundamentals: first, give people time to evacuate a building and, second, don't obscure exits and signs that direct people to exits. NEC flame-test requirements relate to tests that measure resistance to spreading a fire, to varying degrees and under varying conditions based on intended use of the cable. The requirements satisfy part one of the philosophy—it delays the spread of the fire. Because all but plenum cables are intended for installation behind walls or in areas inaccessible to the public, the second part doesn't apply. However, because a plenum cable is installed in an air-handling space where smoke from the burning cable could spread via HVAC fans to the populated part of the building, the plenum test measures the generation of *visible* smoke. Visible smoke can keep people from recognizing exits or suffocate them (which actually happened in some major hotel fires before plenum cables were defined in the code).

Myth #4: I should buy plenum cable if I want good transmission performance. Although FEP (fluorinated ethylene-propylene, the conductor insulation material used in plenum-rated Category 5e and higher cables) has excellent transmission properties, its use in plenum cables is due more to its equally superb resistance to flame propagation and relatively low level of visible-smoke generation. In Category 5e and higher nonplenum cables, HDPE (high-density polyethylene) is commonly used as conductor insulation. It has almost as good transmission properties as FEP and has the added benefit of being several times lower in cost than FEP (and this explains the primary difference in price between plenum and nonplenum UTP cables). HDPE does, however, burn like a candle and generate copious visible smoke. Cable manufacturers can adjust the PVC jacket of a four-pair construction to allow an HDPE-insulated cable to pass all flame tests except the plenum test. They also compensate for differences in transmission properties between FEP and HDPE (or whatever materials they select) by altering the dimensions of the insulated conductor. End result: No matter what the flame rating, if the cable jacket says Category 5e or better, you get Category 5e or better.

Myth #5: To really protect my family, I should specify plenum cable be installed in my home. This is unnecessary for two reasons. First, communications cables are almost never the source of ignition or flame spread in a residential fire. It's not impossible, but it's extremely rare. Second, to what should the "fireproof" cable be attached? It is going to be fastened to wooden studs, most likely—wooden studs that burn fast, hot, and with much black, poisonous smoke. While the studs are burning, the flooring, roofing, electrical wiring, plastic water pipes, carpets, curtains, furniture, cabinets, and woodwork are also blazing away merrily, also generating much smoke. A plenum cable's potential to mitigate such a conflagration is essentially nil. Install a CMX-rated cable, and you'll comply with the National Electric Code. Install CM, CMG, or CMR, and you'll be exceeding NEC requirements. Leave the CMP cable to the commercial environments for which it's intended and don't worry about needing it at home.

Riser

The *riser* is a vertical shaft used to route cable between two floors. Often, it is nothing more complicated than a hole (core) that is drilled in the floor and allows cables to pass through. However, a hole between two floors with cable in it introduces a new problem. In the fire disaster movie *The Towering Inferno* the fire spread from floor to floor through the building's cabling. That should not happen nowadays because building codes require that riser cable be rated properly. So the riser cable must have certain fire-resistant qualities.

TIP The National Electrical Code permits plenum cable to be used in the riser, but it does not allow riser cable to be used in the plenum.

The Towering Inferno had a basis in reality, not only because cables at the time burned relatively easily but also because of the chimney effect. A chimney works by drawing air upward, through the fire, invigorating the flames with oxygen flow. In a multistory building, the riser shafts can act as chimneys, accelerating the spread and intensity of the fire. Therefore, building codes usually require that the riser be firestopped in some way. That's accomplished by placing special blocking material in the riser at each penetration of walls or ceilings after the cables have been put in place. Techniques for firestopping are discussed in Chapter 13, "Cabling System Design and Installation."

General Purpose

The general-purpose rating is for the classic horizontal cable that runs from the wiring closet to the wall outlet. It is rated for use within a floor and cannot penetrate a structural floor or ceiling. It is also the rating most commonly used for patch cords because, in theory, a patch cord will never go through a floor or ceiling. You should be aware that riser-rated cable is most commonly used for horizontal runs, simply because the price difference between riser and general-purpose cables is typically small and contractors don't want to haul more cable types than they have to.

Limited Use

The limited-use rating is for single and duplex (two-family) residences only. Some exceptions in the code allow its use in other environments, as in multitenant spaces such as apartments. However, the exceptions impose requirements that are typically either impractical or aesthetically unpleasant, and so it is better to consider limited-use cables as just for single- and two-family residences.

Cable Jackets

Because UTP is virtually ubiquitous in the LAN environment, the rest of this chapter will focus on design criteria and transmission-performance characteristics related to UTP cable.

The best place to start looking at cable design is on the outside. Each type of cable (twisted-pair, fiber optic, or coaxial) will have a different design with respect to the cable covering or the jacket.

KEY TERM *jacket and sheath* The cable's *jacket* is the plastic outer covering of the cable. *Sheath* is sometimes synonymous with jacket but not always. The sheath includes not only the jacket of the cable but also any outside shielding (such as braided copper or foil) that may surround the inner wire pairs. With UTP and most fiber-optic cables, the sheath and the jacket are the same. With ScTP and STP cables, the sheath includes the outer layer of shielding on the inner wires.

One of the most common materials used for the cable jacket is *polyvinyl chloride* (PVC); UTP cables in the United States are almost exclusively jacketed with PVC, regardless of the flame rating of the cable. PVC was commonly used in early LAN cables (Category 3 and lower) as an insulation and as material for jackets, but the dielectric properties of PVC are not as desirable as those of other thermoplastics, such as FEP or PP, that can be used for higher-frequency transmission. Figure 1.9 shows a cutaway drawing of a UTP cable.

FIGURE 1.9
Cutaway drawing of a UTP cable showing insulated wire pairs, slitting cord, and jacket

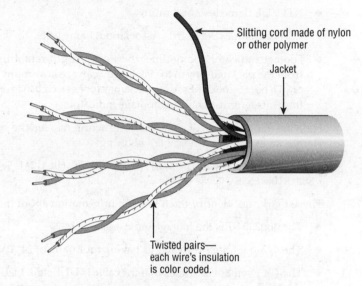

Slitting cord made of nylon or other polymer

Jacket

Twisted pairs—each wire's insulation is color coded.

Other substances commonly used in cable jackets of indoor cables include ECTFE (HALAR), PVDF (KYNAR), and FEP (Teflon or NEOFLON). These materials have enhanced flame-retardant qualities as compared to PVC but are much more costly. Where PVC can do the job, it's the jacket material of choice.

KEY TERM *rip cord* Inside some UTP cable jackets is a polyester or nylon string called the *rip cord*, also known as the *slitting cord* or *slitting string*. The purpose of this cord is to assist with slicing the jacket open when more than an inch or two of jacket needs to be removed. Some cable installers love them; many find them a nuisance, as they get in the way during termination.

NOTE No standard exists for the jacket color, so manufacturers can make the jacket any color they care to. You can order Category 5e or 6 cables in at least a dozen different colors. Colors like hot pink and bright yellow don't function any differently than plain gray cables, but they sure are easier to spot when you are in the ceiling! Many cable installers will pick a different color cable based on which jack position or patch panel the cable is going to so that it is easier to identify quickly.

CABLE MARKINGS

Have you examined the outside jacket of a twisted-pair or fiber-optic cable? If so, you noticed many markings on the cable that may have made sense. UL has requirements on how their designations are applied and the FCC requires that the category be placed every foot. For cables manufactured for use in the United States and Canada, these markings may identify the following:

- Cable manufacturer and manufacturer part number.

- Category of cable (e.g., UTP).

- NEC/UL flame tests and ratings.

- CSA (Canadian Standards Association) flame tests.

- Footage indicators. Sometimes these are "length-remaining markers" that count down from the package length to zero so you can see how many feet of cable remains on a spool or in a box. Superior Essex (www.superioressex.com) is one cable manufacturer that imprints length-remaining footage indicators.

For a list of definitions of some marking acronyms, see the section "Common Abbreviations." Here is an example of one cable's markings:

```
000750 FT 4/24 (UL) c(UL) CMP/MPP VERIFIED (UL) CAT 5e
 SUPERIOR ESSEX COBRA 2313H
```

These markings identify the following information about the cable:

- The 000750 FT is the footage indicator.

- The 4/24 identifies the cable as having four pairs of 24 AWG wire.

- The (UL) symbol indicates that the cable is UL listed. Listing is a legal requirement of the NEC.

- The symbol c(UL) indicates that the cable is UL listed to Canadian requirements in addition to U.S. requirements. Listing is a legal requirement of the CSA.

- The CMP/MPP code stands for communications plenum (CMP) and multipurpose plenum (MPP) and indicates that the cable can be used in plenum spaces. This is the NEC flame/smoke rating.

- The term VERIFIED (UL) CAT 5e means that the cable has been verified by the UL as being Category 5e compliant (and TIA/EIA-568-B compliant). Verification to transmission properties is optional.

- SUPERIOR ESSEX is the manufacturer of the cable.

- COBRA is the cable brand (in this case, a Category 5e–plus cable, which means it exceeds the requirements for Category 5e).

- The numbers 2319 indicate the date of manufacture in Julian format. In this case, it is the 231st day of 2009.

- H indicates the Superior Essex manufacturing plant.

Some manufacturers may also include their "E-file" number instead of the company name. This number can be used when calling the listing agency (such as the UL) to trace the manufacturer of a cable. In the case of UL, you can look up the E-file numbers online at www.ul.com.

WARNING Cables marked with *CMR* (communications riser) and *CMG* (communications general) must *not* be used in the plenum spaces.

COMMON ABBREVIATIONS

So that you can better decipher the markings on cables, here is a list of common acronyms and what they mean:

NFPA The National Fire Protection Association

NEC The National Electrical Code that is published by the NFPA once every three years

UL The Underwriters Laboratories

CSA The Canadian Standards Association

PCC The Premise Communication Cord standards for physical wire tests defined by the CSA

Often, you will see cables marked with NFPA 262, FT-4, or FT-6. The NFPA 262 (formerly UL-910) is the test used for plenum cables. The FT-4 is the CSA equivalent of UL 1666 or the riser test, and FT-6 is the CSA equivalent of NFPA 262.

Wire Insulation

Inside the cable jacket are the wire pairs. The material used to insulate these wires must have a very low dielectric constant and low dissipation factor. Refer back to Figure 1.9 for a diagram of the wire insulation.

KEY TERM *dielectric and dissipation factor* A material that has good *dielectric* properties is a poor conductor of electricity. Dielectric materials are insulators. In the case of LAN cables, a good dielectric material also has characteristics conducive to the transmission of high-frequency signals along the conductors. The dissipation factor is the measure of loss-rate of power and becomes the dominant factor for signal loss at high frequencies.

MATERIALS USED FOR WIRE INSULATION

A variety of insulating materials exists, including polyolefin (polyethylene and polypropylene), fluorocarbon polymers, and PVC.

The manufacturer chooses the materials based on the material cost, flame-test ratings, and desired transmission properties. Materials such as polyolefin are inexpensive and have great transmission properties, but they burn like crazy, so they must be used in combination with material that has better flame ratings. That's an important point to keep in mind: don't focus on a particular material. It is the material *system* selected by the manufacturer that counts. A manufacturer will choose insulating and jacketing materials that work together according to the delicate balance of fire resistance, transmission performance, and economics.

The most common materials used to insulate the wire pairs in Category 5e and greater plenum-rated cables are fluorocarbon polymers. The two varieties of fluorocarbon polymers used are fluorinated ethylene-propylene (FEP) and perfluoroalkoxy (PFA).

These polymers were originally developed by DuPont and are also sometimes called by their trademark, Teflon. The most commonly used and most desirable of these materials is FEP. Over the past few years, the demand for plenum-grade cables exceeded the supply of available FEP. During periods of FEP shortage, Category 5e plenum designs emerged that substituted another material for one or more of the pairs of wire. In addition, some instances of marginal performance occurred in the UL-910 burn test for plenum cables. These concerns, coupled with increases in the supply of FEP and substitutes like MFA, have driven these designs away.

TIP When purchasing Category 5e and higher plenum cables, ask whether other insulation material has been used in combination with FEP for wire insulation.

In non-plenum Category 5e and higher and in the lower categories of cable, much less expensive and more readily available materials, such as HDPE (high-density polyethylene), are used. You won't sacrifice transmission performance; the less stringent flame tests just allow less expensive materials.

Insulation Colors

The insulation around each wire in a UTP cable is color-coded. The standardized color codes help the cable installer make sure each wire is connected correctly with the hardware. In the United States, the color code is based on 10 colors. Five of these are used on the *tip* conductors, and five are used on the *ring* conductors. Combining the tip colors with the ring colors results in 25 possible unique pair combinations. Thus, 25 pair groups have been used for telephone cables for decades.

NOTE The words *tip* and *ring* hark back to the days of manual switchboards. Phono-type plugs (like the ones on your stereo headset cord) were plugged into a socket to connect one extension or number to another. The plug had a tip, then an insulating disk, and then the shaft of the plug. One conductor of a pair was soldered into the tip and the other soldered to the shaft, or ring. Remnants of this 100-year-old technology are still with us today.

Table 1.3 lists the color codes found in a binder group (a group of 25 pairs of wires) in larger-capacity cables. The 25-pair cable is not often used in data cabling, but it is frequently used for voice cabling for backbone and cross-connect cable.

TABLE 1.3   Color Codes for 25-Pair UTP Binder Groups

Pair Number	Tip Color	Ring Color
1	White	Blue
2	White	Orange
3	White	Green
4	White	Brown

TABLE 1.3 Color Codes for 25-Pair UTP Binder Groups   *(CONTINUED)*

PAIR NUMBER	TIP COLOR	RING COLOR
5	White	Slate
6	Red	Blue
7	Red	Orange
8	Red	Green
9	Red	Brown
10	Red	Slate
11	Black	Blue
12	Black	Orange
13	Black	Green
14	Black	Brown
15	Black	Slate
16	Yellow	Blue
17	Yellow	Orange
18	Yellow	Green
19	Yellow	Brown
20	Yellow	Slate
21	Violet	Blue
22	Violet	Orange
23	Violet	Green
24	Violet	Brown
25	Violet	Slate

With LAN cables, it is common to use a modification to this system known as *positive identification*. PI, as it is sometimes called, involves putting either a longitudinal stripe or circumferential band on the conductor in the color of its pair mate. In the case of most four-pair UTP cables, this is usually done only to the tip conductor because each tip conductor is white, whereas the ring conductors are each a unique color.

Table 1.4 lists the color codes for a four-pair UTP cable. The PI color is indicated after the tip color.

TABLE 1.4   Color Codes for Four-Pair UTP Cable

Pair Number	Tip Color	Ring Color
1	White/Blue	Blue
2	White/Orange	Orange
3	White/Green	Green
4	White/Brown	Brown

WAITER! THERE'S HALOGEN IN MY CABLE!

Much of the cable currently in use in the United States and elsewhere in the world contains halogens. A *halogen* is a nonmetallic element, such as fluorine, chlorine, iodine, or bromine. When exposed to flames, substances made with halogens give off toxic fumes that quickly harm the eyes, nose, lungs, and throat. Did you notice that fluorine and chlorine are commonly found in cable insulation and jackets? Even when cables are designed to be flame-resistant, any cable when exposed to high enough temperatures will melt and burn. PVC cables contain chlorine, which emits toxic fumes when burned.

Many different manufacturers are now making low-smoke, zero-halogen (LSZH or LS0H) cables. These cables are designed to emit no toxic fumes and produce little or no smoke when exposed to flames. Tunnels, enclosed rooms, aircraft, and other minimum-ventilation areas are prime spots for the use of LSZH cables because those areas are more difficult to escape from quickly.

LSZH cables are popular outside the United States. Some safety advocates are calling for the use of LSZH cables in the United States, specifically for the plenum space. Review your local building codes to determine if you must use LSZH cable. Non-LSZH cables will produce corrosive acids if they are exposed to water (such as from a sprinkler system) when burned; such acids may theoretically further endanger equipment. But many opponents of LSZH cable reason that if an area of the building is on fire, the equipment will be damaged by flames before it is damaged by corrosives from a burning cable.

Why, you might ask, would anyone in his or her right mind argue against the installation of LSZH cables everywhere? First, reducing toxic fumes doesn't necessarily mean the cable is more fireproof. The flame-spread properties are worse than for cables in use today. Numerous studies by Bell Labs showed that cables composed of LSZH will not pass the plenum test, not because of smoke generation but because of flame spread. Most LSZH cable designs will only pass the riser test where the allowable flame spread is greater. Second, consider practicality. LSZH is an expensive solution to a problem that doesn't seem to really exist in the United States.

Twists

When you slice open a UTP communications cable, you will notice that the individual conductors of a pair of wire are twisted around one another. At first, you may not realize how important these twists are.

TIP   Did you know that in Category 5e cables a wire pair untwisted more than half of an inch can adversely affect the performance of the entire cable?

Twisted-pair cable is any cable that contains a pair of wires that are wrapped or twisted around one another between 2 and 12 times per foot—and sometimes even more than 12 times per foot (as with Category 5e and higher). The twists help to cancel out the electromagnetic interference (EMI) generated by voltage used to send a signal over the wire. The interference can cause problems, called crosstalk, for adjacent wire pairs. Crosstalk and its effects are discussed in the "Speed Bumps" section later in this chapter.

Cables commonly used for patch cables and for horizontal cabling (patch panel to wall plate) typically contain four pairs of wires. The order in which the wires are crimped or punched down are very important.

TIP   Companies such as PANDUIT (www.panduit.com) and Leviton (www.leviton.com) have developed termination tools and patch cables that all but eliminate the need to untwist cables more than a tiny amount.

Wire Gauge

Copper-wire diameter is most often measured by a unit called AWG (American Wire Gauge). Contrary to many other measuring systems, as the AWG number gets smaller, the wire diameter actually gets larger; thus, AWG 24 wire is smaller than AWG 22 wire. Larger wires are useful because they have more physical strength and lower resistance. However, the larger the wire diameter, the more copper is required to make the cable. This makes the cable heavier, harder to install, and more expensive.

NOTE   The reason the AWG number increases as the wire diameter decreases has to do with how wire is made. You don't dump copper ore into a machine at one end and get 24 AWG wire out the other end. A multistep process is involved—converting the ore to metal, the metal to ingots, the ingots to large bars or rods. Rods are then fed into a machine that makes them into smaller-diameter rods. To reach a final diameter, the rod is pulled through a series of holes, or dies, of decreasing size. Going through each die causes the wire to stretch out a little bit, reducing its diameter. Historically, the AWG number represented the exact number of dies the wire had to go through to get to its finished size. So, the smaller the wire, the more dies involved and the higher the AWG number.

The cable designer's challenge is to use the lowest possible diameter wire (reducing costs and installation complexity) while at the same time maximizing the wire's capabilities to support the necessary power levels and frequencies.

Category 5e UTP is always 24 AWG; IBM Type 1A is typically 22 AWG. Patch cords may be 26 AWG, especially Category 3 patch cords. The evolution of higher-performance cables such as Category 6 and Category 6A has resulted in 23 AWG often being substituted for 24 AWG.

Table 1.5 shows 22, 23, 24, and 26 AWG sizes along with the corresponding diameter, area, and weight per kilometer.

TABLE 1.5 American Wire Gauge Diameter, Area, and Weight Values

AWG	NOMINAL DIAMETER (INCHES)	NOMINAL DIAMETER (MM)	CIRCULAR-MIL AREA1	AREA (SQ. MM)	WEIGHT (KG/KM)
22	0.0253	0.643	640.4	0.3256	2.895
23	0.0226	0.574	511.5	0.2581	2.295
24	0.0201	0.511	404.0	0.2047	1.820
26	0.0159	0.404	253.0	0.1288	1.145

The dimensions in Table 1.5 were developed more than 100 years ago. Since then, the purity and, therefore, the conductive properties of copper have improved due to better copper-processing techniques. Specifications that cover the design of communications cables have a waiver on the actual dimensions of a wire. The real concern is not the dimensions of the wire, but how it performs, specifically with regard to resistance in ohms. The AWG standard indicates that a 24 AWG wire will have a diameter of 0.0201″, but based on the performance of the material, the actual diameter of the wire may be slightly less or slightly more (but usually less).

Solid Conductors vs. Stranded Conductors

UTP cable used as horizontal cable (permanent cable or cable in the walls) has a solid conductor, as opposed to patch cable and cable that is run over short distances, which usually have stranded conductors. Stranded-conductor wire consists of many smaller wires interwoven together to form a single conductor.

TIP Connector types (such as patch panels and modular jacks) for solid-conductor cable are different than those for stranded-conductor cable. Stranded-conductor cables will not work with insulated displacement connector (IDC)-style connectors found on patch panels and 66-style punch-down blocks.

Though stranded-conductor wire is more flexible, solid-conductor cable has much better electrical properties. Stranded-conductor wire is subject to as much as 20 percent more attenuation (loss of signal) due to a phenomenon called *skin effect*. At higher frequencies (the frequencies used in LAN cables), the signal current concentrates on the outer circumference of the overall conductor. Since stranded-conductor wire has a less-defined overall circumference (due to the multiple strands involved), attenuation is increased.

KEY TERM *core* The *core* of the cable is anything found inside the sheath. The core is usually just the insulated twisted pairs, but it may also include a slitting cord and the shielding over individual twisted pairs in an STP cable. People incorrectly refer to the core of the cable when they mean the *conductor* (the element that conducts the electrical signal).

Most cabling standards recommend using solid-conductor wire in the horizontal or permanent portion of the link, but the standards allow for stranded-conductor wire in patch cables where flexibility is more important. We know of several UTP installations that have used stranded-conductor wires for their horizontal links. Although we consider this a poor practice, here are some important points to keep in mind if you choose to use a mixture of these cables:

◆ Stranded-conductor wire requires different connectors.

◆ Stranded-conductor wires don't work as well in punch-down blocks designed for solid-conductor cables.

◆ You must account for reduced horizontal-link distances.

Cable Length

The longer the cable, the less likely the signal will be carried completely to the end of the cable, because of noise and signal attenuation. Realize, though, that for LAN systems the time it takes for a signal to get to the end is also critical. Cable design engineers are now measuring two additional performance parameters of cable: the *propagation delay* and the *delay skew*. Both parameters are related to the speed at which the electrons can pass through the cable and the length of the wire pairs in the cable. The variables are discussed in the "Speed Bumps" section later in this chapter.

Cable Length vs. Conductor Length

A Category 5e, 6, or 6A cable has four pairs of conductors. By design, each of the four pairs is twisted in such a fashion that the pairs are slightly different lengths. Varying twist lengths from pair to pair improves crosstalk performance. Therefore, signals transmitted simultaneously on two different pairs of wire will arrive at slightly different times. The *conductor length* is the length of the individual pair of conductors, whereas the *cable length* is the length of the cable jacket.

Part of a modern cable tester's feature set is the ability to perform conductor-length tests. Here is a list of the conductor lengths of a cable whose cable length is 139′ from the wall plate to the patch panel. As you can see, the actual conductor length is longer due to the twists in the wire.

Pair	Distance
1-2	145′
3-6	143′
4-5	141′
7-8	142′

WARP FACTOR ONE, PLEASE

Light travels almost 300,000,000 meters per second in a perfect vacuum, faster than nonphysicists can imagine. In a fiber-optic cable one kilometer long, data can travel from start to finish in about 3.3 microseconds (0.0000033 seconds).

Data does not travel through copper cabling quite as fast. One of the ways that the speed of data through a copper cable is measured is by how fast electricity can travel through the cable. This value is called the nominal velocity of propagation (NVP) and is expressed as a percentage of the speed of light. The value for most cables is between 60 and 90 percent. The cable manufacturer specifies NVP as part of the cable's design.

Take, for example, a cable that was recently measured using a handheld cable tester. The NVP for this cable was 67 percent, and the cable was 90 meters long. Electricity will travel through this cable at a speed of about 200,000,000 meters per second; it travels from one end of the cable to another in 450 nanoseconds (0.00000045 seconds).

Data Communications 101

Before we discuss more of the limitations involved with data communications and network cabling, some basic terms must be defined. Unfortunately, vendors, engineers, and network managers serve up high-tech and communications terms like balls in a tennis match. Worse, they often misuse the terms or don't even fully understand what they mean.

One common term is *bandwidth*. Does it mean maximum frequency or maximum data rate? Other terms are thrown at you as if you have a PhD in electrical engineering, including *impedance*, *resistance*, and *capacitance*.

Our favorite misunderstood term is *decibels*. We always thought decibels were used to measure sound, but that's not necessarily true when it comes to data communications. Over the next few pages, we will take you through a crash course in Data Communications 101 and get you up to speed on certain terms pertaining to cabling.

Bandwidth, Frequency, and Data Rate

One initially confusing aspect about cabling is that cables are rated in hertz rather than bits per second. Network engineers (and you, presumably) are more concerned with how much data can be pushed through the cable than with the frequency at which that data is traveling.

Frequency is the number of cycles completed per unit of time and is generally expressed in *hertz* (cycles per second). Figure 1.10 shows a cycle that took one second to complete; this is one hertz. Data cabling is typically rated in kilohertz (kHz) or megahertz (MHz). For a cable rated at 100MHz, the cycle would have to complete 100,000,000 times in a single second! The more cycles per second, the more noise the cable generates and the more susceptible the cable is to signal-level loss.

FIGURE 1.10
One cycle every
second or one hertz

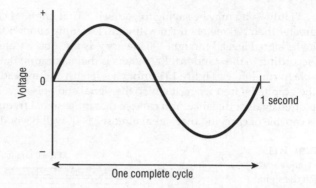

The bandwidth of a cable is the maximum frequency at which data can be effectively transmitted and received. The bit rate is dependent upon the network electronics, not the cable, provided the operating frequency of the network is within the cable's usable bandwidth. Put another way, the cable is just a pipe. Think of the bandwidth as the pipe's diameter. Network electronics provide the water pressure. Either a trickle comes through or a gusher, but the pipe diameter doesn't change.

Cable bandwidth is a difficult animal to corral. It is a function of three interrelated, major elements: distance, frequency, and signal-level-to-noise-level ratio (SNR). Changing any one element alters the maximum bandwidth available. As you increase the frequency, SNR gets worse, and the maximum bandwidth is decreased. As you increase distance, SNR worsens, thereby decreasing the maximum bandwidth. Conversely, reducing frequency or distance increases the maximum bandwidth because SNR improves.

To keep the same maximum bandwidth, increasing the frequency means you must either decrease distance or improve the signal level at the receiver. If you increase the distance, either the frequency must decrease, or, again, the signal level at the receiver must improve. If you improve signal level at the receiving end, you can either increase frequency or leave the frequency alone and increase distance. It's a tough bronc to ride.

STANDARDS PROVIDE A STRUCTURED APPROACH

With all this variability, how do you get anywhere with cable and network design? It helps to lasso one or more of the variables.

This is done for you via the IEEE network specifications and implemented through ANSI/TIA-568-C. A maximum horizontal run length of 100 meters (308´), including workstation and communication closet patch cords, is specified. This figure arises from some timing limitations of some Ethernet implementations. So distance is fixed.

The standards also define the maximum operating frequency. In the case of Category 3 cables, it is 16MHz. In the case of 5e, it is 100MHz; for Category 6, 250 MHz; and for Category 6A, 500 MHz.

Now that two of the three elements are firmly tied to the fence, you can rope in the last. Cable design focuses on improving the signal level and reducing the noise in the cable to achieve optimum transmission performance for given frequencies at a fixed length.

"Huh?" you may be saying to yourself. "That implies I could have horizontal run lengths greater than 100 meters if I'm willing to lower my bandwidth expectations or put up with a lower signal level. I thought 100 meters was the most a Category 5e (or better) cable could run." According to the standard, 100 meters is the maximum. But technically, the cabling might be able to run longer. Figure 1.11 unhitches length and instead ties down frequency and SNR. In the graph, the frequency at which the signal and noise level coincide (the "ACR=0" point) is plotted against distance. You can see that if the signal frequency is 10MHz, a Category 5e cable is capable of carrying that signal almost 2,500', well beyond the 100 meter (308') length specified.

FIGURE 1.11

The frequency at which the signal and noise level coincide (ACR=0 point) plotted against distance

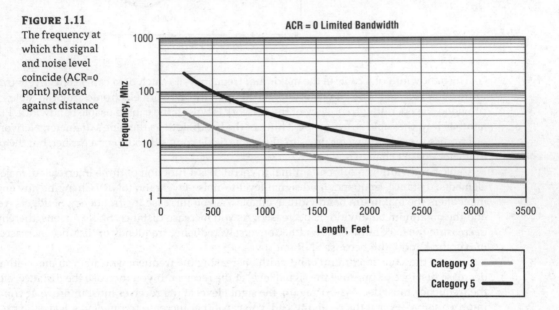

So why not do so? Because you'd be undermining the principle of structured wiring, which requires parameters that will work with many LAN technologies, not just the one you've got in mind for today. Some network architectures wouldn't tolerate it, and future upgrades might be impossible. Stick to the 100 meter maximum length specified.

The *data rate* (throughput or information capacity) is defined as the number of bits per second that move through a transmission medium. With some older LAN technologies, the data rate has a one-to-one relationship with the transmission frequency. For example, 4Mbps Token Ring operates at 4MHz.

It's tough to keep pushing the bandwidth of copper cables higher and higher. There are the laws of physics to consider, after all. So techniques have been developed to allow more than 1 bit per hertz to move through the cable. Table 1.6 compares the operating frequency of transmission with the throughput rate of various LAN technologies available today.

All the systems listed in the table except the last (which requires Category 6A cable) will work with Category 5 or higher cable. So how do techniques manage to deliver data at 1Gbps across a Category 5e cable whose maximum bandwidth is 100MHz? The next section gives you the answer.

TABLE 1.6   LAN Throughput vs. Operating Frequency

LAN System	Data Rate	Operating Frequency
Token Ring	4Mbps	4MHz
10Base-T Ethernet	10Mbps	10MHz
Token Ring	16Mbps	16MHz
100BaseT Ethernet	100Mbps	31.25MHz
ATM 155	155Mbps	38.75MHz
1000Base-T (Gigabit) Ethernet	1,000Mbps	Approximately 65MHz
10GBase-T (10Gb) Ethernet	10,000 Mbps	Approximately 400MHz

The Secret Ingredient: Encoding and Multipair Simultaneous Send and Receive

Consider the example illustrated in Figure 1.12. A street permits one car to pass a certain stretch of road each second. The cars are spaced a certain distance apart, and their speeds are limited so that only one is on the stretch of road at a time.

FIGURE 1.12
A street that allows one car to pass each second

Only a single car can pass through each second.

But suppose as in Figure 1.13 that the desired capacity for this particular part of the street is three cars per second. The cars can drive faster, and they can be spaced so that three at a time fit on the stretch of road. This is bit encoding. It is a technology for packing multiple data bits in each hertz to increase throughput.

FIGURE 1.13
A street that allows multiple cars through during each cycle

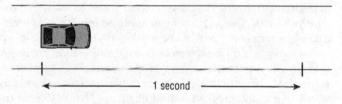

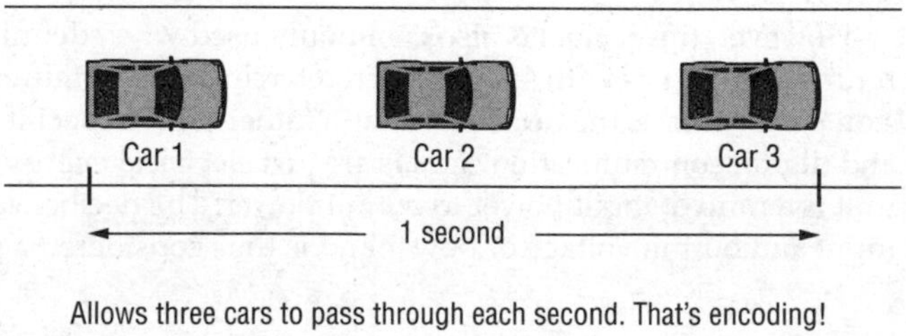

Car 1   Car 2   Car 3

Allows three cars to pass through each second. That's encoding!

Add a lane in each direction, and you can see how most LAN technologies work today. They use two of the four pairs of cable, one to transmit and one to receive—effectively, a two-lane highway.

At some point, though, a limit will be reached as to how fast the cars can travel. Plus, eventually the cars will be packed end-to-end in a lane and we just won't be able to fit any more cars (data bits) through that stretch in the available time.

What to do? How about building multiple lanes? Instead of using two lanes, one in each direction, four lanes (four pairs of cable) would ease the congestion.

Four lanes still might not be enough capacity to get all the cars needed down the highway. So all four lanes will be used, but instead of two being dedicated to send and two to receive, the cars will drive both directions in every lane. It takes accurate timing and nerves of steel, but it can be done. This is, in fact, how Gigabit Ethernet is implemented on Category 5e and higher cabling. Transmitting at an operating frequency of about 65MHz, data is simultaneously sent and received on all four pairs at a rate of 250Mbps each. Voilà! That's 1,000Mbps in less than 100MHz of bandwidth!

TIP For Gigabit Ethernet to work over Category 5e, 6, and 6A cabling, all four pairs must be used. The same is true for 10 Gigabit Ethernet: all four pairs must be used in a Category 6A cable.

What a Difference a dB Makes!

Suppose you are comparing cable performance. A manufacturer states that the attenuation (power loss) for a cable with a length of 90 meters, operating at 100MHz, is 20dB. What does the measurement mean? Would you be surprised to learn that the signal strength has dropped by a factor of 100? That's right; if you apply an input power level of 5 watts, the output level will be 0.05 watts! For every 3dB of attenuation, it's a 50 percent loss of power!

To summarize: Low decibel values of attenuation are desirable because then less of the signal is lost on its way to the receiver. Higher decibel values of crosstalk (NEXT, ELFEXT, etc.) and return loss are actually desirable because that means less signal has been measured on adjacent wires. (For more on NEXT and ELFEXT, see "Noise (Signal Interference)" later in this chapter.)

This section may be all you ever wanted to know about decibels. If you want to know more and get the technical details, read on!

DIGGING A LITTLE DEEPER INTO DECIBELS

You may think of a decibel in terms of audible noise. When referring to the domain of sound, a decibel is not actually a specific unit of measurement but rather is used to express a ratio of sound pressure.

However, the decibel is also commonly used when defining attenuation, crosstalk, and return loss. Just as with sound, when referring to communications and electrical transmission performance, the decibel is a ratio rather than a specific measurement. Because analog and digital communication signals are just electrical energy instead of sound pressure, the dB unit is a ratio of input power to output power. The decibel value is independent of the actual input and output voltage or power and is thus considered a generic performance specification.

Understanding what the decibel numbers mean is important when comparing one cabling media or performance measurement with another.

Decibels 101

The *bel* part of decibel was named after Alexander Graham Bell, the inventor of the telephone. A *decibel* is a tenfold logarithmic ratio of power (or voltage) output to power (or voltage) input. Keep in mind that the decibel is indicating a power ratio, not a specific measurement. The decibel is a convenient way to reflect the power loss or gain, regardless of the actual values.

NOTE   For measurements such as attenuation, NEXT, ELFEXT, ACR, and return loss, the decibel value is always negative because it represents a loss, but often the negative sign is ignored when the measurement is written. The fact that the number represents a loss is assumed.

Cable testers as well as performance specifications describe attenuation in decibels. Let's say, for example, that you measure two cables of identical length and determine that the attenuation is 15dB for one cable and 21dB for the other. Naturally, you know that because lower attenuation is better, the cable with an attenuation of 15dB is better than the one with a 21dB value. But how much better? Would you be surprised to learn that even though the difference between the two cables is only 6dB, there is 50 percent more attenuation of voltage or amperage (power is calculated differently) on the cable whose attenuation was measured at 21dB?

Knowing how a decibel is calculated is vital to appreciating the performance specifications that the decibel measures.

Decibels and Power

When referring to power (watts), decibels are calculated in this fashion:

$$dB = 10 \times \log10(P1/P2)$$

P1 indicates the measured power, and P2 is the reference power (or input power).

To expand on this formula, consider this example. The reference power level (P2) is 1.0 watts. The measured power level (P1) on the opposite side of the cable is 0.5 watts. Therefore, through this cable, 50 percent of the signal was lost due to attenuation. Now, plug these values into the power formula for decibels. Doing so yields a value of 3dB. What does the calculation mean? It means that:

◆ Every 3dB of attenuation translates into 50 percent of the signal power being lost through the cable. Lower attenuation values are desirable, as a higher power level will then arrive at the destination.

◆ Every 3dB of return loss translates into 50 percent less signal power being reflected back to the source. Higher decibel values for return loss are desirable, as *less* power will then be returned to the sender.

◆ Every 3dB of NEXT translates into 50 percent less signal power being allowed to couple to adjacent pairs. Higher decibel values for NEXT (and other crosstalk values) are desirable, as higher values indicate that less power will then couple with adjacent pairs.

An increase of 10dB means a tenfold increase in the actual measured parameter. Table 1.7 shows the logarithmic progression of decibels with respect to power measurements.

TABLE 1.7   Logarithmic Progression of Decibels

DECIBEL VALUE	ACTUAL INCREASE IN MEASURED PARAMETER
3dB	2
10dB	10
20dB	100
30dB	1,000
40dB	10,000
50dB	100,000
60dB	1,000,000

Decibels and Voltage

Most performance specifications and cable testers typically reference voltage ratios, not power ratios. When referring to voltage (or amperage), decibels are calculated slightly differently than for power. The formula is as follows:

dB = 20 × log10(P1/P2)

P1 indicates the measured voltage or amperage, and P2 is the reference (or output) voltage (amperage). Substituting a reference value of 1.0 volt for P2 and 0.5 volts for P1 (the measured output), you get a value of –6dB. What does the calculation mean? It means that:

◆   Every 6dB of attenuation translates into 50 percent of the voltage being lost to attenuation. Lower decibel attenuation values are desirable, as a higher voltage level will then arrive at the destination.

◆   Every 6dB of return loss translates into 50 percent less voltage being reflected back to the source. Higher decibel values for return loss are desirable, as *less* voltage will then be returned to the sender.

◆   Every 6dB of NEXT translates into 50 percent less voltage coupling to adjacent wire pairs. Higher decibel values for NEXT (and other crosstalk values) are desirable, as higher values indicate that less power will then couple with adjacent pairs.

Table 1.8 shows various decibel levels and the corresponding voltage and power ratios. Notice that (for the power ratio) if a cable's attenuation is measured at 10dB, only one-tenth of the signal transmitted will be received on the other side.

TABLE 1.8   Decibel Levels and Corresponding Power and Voltage Ratios

DB	VOLTAGE RATIO	POWER RATIO
1	1	1
−1	0.891	0.794
−2	0.794	0.631
−3	0.707	0.500
−4	0.631	0.398
−5	0.562	0.316
−6	0.500	0.250
−7	0.447	0.224
−8	0.398	0.158
−9	0.355	0.125
−10	0.316	0.100
−12	0.250	0.063
−15	0.178	0.031
−20	0.100	0.010
−25	0.056	0.003
−30	0.032	0.001
−40	0.010	0.000
−50	0.003	0.000

APPLYING KNOWLEDGE OF DECIBELS

Now that you have a background on decibels, look at the specified channel performance for Category 5e versus the channel performance for Category 6 cable at 100MHz.

Media Type	Attenuation	NEXT	Return Loss
Category 5e	24	30.1	10.0
Category 6	21.3	39.9	12.0

For the values to be meaningful, you need to look at them with respect to the actual percentage of loss. For this example, use voltage. If you take each decibel value and solve for the P1/P2 ratio using this formula, you would arrive at the following values:

Ratio = 1 / (Inverse log10(dB/20))

Media	Remaining Signal Due to Attenuation	Allowed to Couple (NEXT)	Signal Returned (NEXT)
Category 5e	6.3%	3.1%	39.8%
Category 6	8.6%	1%	31.6%

Existing standards allow a transmission to lose 99 percent of its signal to attenuation and still be received properly. For an Ethernet application operating at 2.5 volts of output voltage, the measured voltage at the receiver must be greater than 0.025 volts. In the Category 5e cable example, only 6.3 percent of the voltage is received at the destination, which calculates to about 0.16. For Category 6 cable it calculates to 0.22 volts, almost 10 times the minimum required voltage for the signal to be received.

Using such techniques for reversing the decibel calculation, you can better compare the performance of any media.

Speed Bumps: What Slows Down Your Data

The amount of data that even simple unshielded twisted-pair cabling can transfer has come a long way over the past dozen or so years. In the late 1980s, many experts felt that UTP cabling would never support data rates greater than 10Mbps. Today, data rates of 10Gbps are supported over cable lengths of 100 meters! And UTP may be able to support even greater data rates in the future.

Think back to the MIS director who mistakenly assumed "it is just wire." Could he be right? What is the big deal? Shouldn't data cabling be able to support even higher data rates?

Have you tried to purchase data-grade cable recently? Have you ever tested a cable run with an even mildly sophisticated cable tester? A typical cabling catalog can have over 2,000 different types of cables! You may have come away from the experience wondering if you needed a degree in electrical engineering in order to understand all the terms and acronyms. The world of modern cabling has become a mind-boggling array of communications buzzwords and engineering terms.

As the requirements for faster data rates emerge, the complexity of the cable design increases. As the data rates increase, the magic that happens inside a cable becomes increasingly mysterious, and the likelihood that data signals will become corrupted while traveling at those speeds also increases.

Ah! So it is not that simple after all! As data rates increase, electrical properties of the cable change, signals become more distorted, and the distance that a signal can travel decreases. Designers of both 1000Base-T (Gigabit Ethernet) and the cables that can support frequencies greater than 100MHz found electrical problems that they did not have to contend with at lower frequencies and data rates. These additional electrical problems are different types of crosstalk and arrival delay of electrons on different pairs of wires.

Hindrances to High-Speed Data Transfer

Electricity flowing through a cable is nothing more than electrons moving inside the cable and bumping into each other—sort of like dominoes falling. For a signal to be received properly by the receiver, enough electrons must make contact all the way through the cable from the sender to the receiver. As the frequency on a cable (and consequently the potential data rate) increases, a number of phenomena hinder the signal's travel through the cable (and consequently the transfer of data). These phenomena are important not only to the person who has to authorize cable purchase but also to the person who tests and certifies the cable.

The current specifications for Category 5e, 6, and 6A cabling outline a number of these phenomena and the maximum (or minimum) acceptable values that a cable can meet and still be certified as compliant.

Due to the complex modulation technology used by 1000Base-T Ethernet, and even more so with 10GBase-T, the TIA has specified cabling performance specifications beyond what was included in the original testing specification. These performance characteristics include power-sum and pair-to-pair crosstalk measurements, delay skew, return loss, and ELFEXT. Some of these newer performance characteristics are important as they relate to crosstalk—for example, AXT (alien crosstalk) to express the interaction between cables in a cable bundle. Although crosstalk is important in all technologies, faster technologies such as 1000Base-T and 10GBase-T are more sensitive to it because they use all four pairs in parallel for transmission.

All these requirements are built into the current version of the standard, ANSI/TIA-568-C.

Many transmission requirements are expressed as mathematical formulas. For the convenience of humans who can't do complex log functions in their heads (virtually everyone!), values are pre-computed and listed in the specification according to selected frequencies. But the actual requirement is that the characteristic must pass the "sweep test" across the full bandwidth specified for the cable category. So performance must be consistent and in accordance with the formula, at any given frequency level, from the lowest to the highest frequency specified.

The major test parameters for communication cables, and the general groupings they fall into, are as follows:

- Attenuation (signal-loss) related

  - Conductor resistance

  - Mutual capacitance

  - Return loss

  - Impedance

- Noise-related

  - Resistance unbalance

  - Capacitance unbalance

  - Near-end crosstalk (NEXT)

  - Far-end crosstalk (FEXT)

  - Power-sum NEXT

  - Power-sum FEXT

  - Alien crosstalk (AXT)

◆ Other

    ◆ Attenuation-to-crosstalk ratio (ACR)

    ◆ Power-sum ACR

    ◆ Propagation delay

    ◆ Delay skew

Attenuation (Loss of Signal)

As noted earlier, attenuation is loss of signal. That loss happens because as a signal travels through a cable, some of it doesn't make it all the way to the end of the cable. The longer the cable, the more signal loss there will be. In fact, past a certain point, the data will no longer be transmitted properly because the signal loss will be too great.

Attenuation is measured in decibels (dB), and the measurement is taken on the receiver end of the conductor. So if 10dB of signal were inserted on the transmitter end and 3dB of signal were measured at the receiver end, the attenuation would be calculated as 3 − 10 = −7dB. The negative sign is usually ignored, so the attenuation is stated as 7dB *of signal loss*. If 10dB were inserted at the transmitter and 6dB measured at the receiver, then the attenuation would be only 4dB of signal loss. So, the lower the attenuation value, the more of the original signal is received (in other words, the lower the better).

Figure 1.14 illustrates the problem that attenuation causes in LAN cabling.

FIGURE 1.14
The signal deterio-rates as it travels between a node on a LAN and the hub.

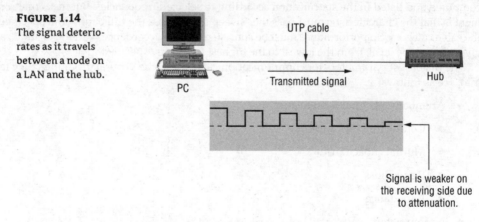

Attenuation on a cable will increase as the frequency used increases. A 100-meter cable may have a measured attenuation of less than 2dB at 1MHz but greater than 20dB at 100MHz!

Higher temperatures increase the effect of attenuation. For each higher degree Celsius, attenuation is typically increased 1.5 percent for Category 3 cables and 0.4 percent for Category 5e cables. Attenuation values can also increase by 2 to 3 percent if the cable is installed in metal conduit.

When the signal arrives at the receiver, it must still be recognizable to the receiver. Attenuation values for cables are very important.

Attenuation values are different for the categories of cables and the frequencies employed. As the bandwidth of the cable increases, the allowed attenuation values get lower (less loss), although the differences between Category 5e and 6 are negligible at the common frequency of 100MHz.

Characteristics that contribute to attenuation are detailed as follows:

Conductor resistance   Conductor resistance acts as a hindrance to the signal because it restricts the flow of electricity through the cable conductors. This causes some of the signal energy to be dissipated as heat, but the amount of heat generated by LAN cabling is negligible due to the low current and voltage levels. The longer the cable or the smaller the conductor diameters (actually, the cross-sectional area), the more resistance. After allowing for dimensional factors, resistance is more or less a fixed property of the conductor material. Copper, gold, and silver offer low resistance and are used as conductors.

Mutual capacitance   This characteristic is an electrical occurrence experienced when a cable has more than one wire and the wires are placed close together. The insulation material will steal and store some of the signal energy, acting as a capacitor between two conductors in the cable. A property of the insulating material called *dielectric constant* has a great influence over the mutual capacitance. Different materials have different dielectric constants. The lower the dielectric constant, the less signal loss. FEP and HDPE have low dielectric constants, along with other properties, that make them well suited for use in high-frequency cables.

Impedance   Impedance is a combination of resistance, capacitance, and inductance and is expressed in ohms; a typical UTP cable is rated at between 85 and 115 ohms. All UTP Category 3, 5e, 6, and 6A cables used in the United States are rated at 100 + 15 ohms. Impedance values are useful when testing the cable for problems, shorts, and mismatches. A cable tester could show three possible impedance readings that indicate a problem:

◆ An impedance value not between 85 and 115 ohms indicates a mismatch in the type of cables or components. This might mean that an incorrect connector type has been installed or an incorrect cable type has been cross-connected into the circuit.

◆ An impedance value of infinity indicates that the cable is open or cut.

◆ An impedance value of 0 indicates that the cable has been short-circuited.

Some electrons sent through a cable may hit an impedance mismatch or imperfection in the wire and be reflected back to the sender. Such an occurrence is known as *return loss*. If the electrons travel a great distance through the wire before being bounced back to the sender, the return loss may not be noticeable because the returning signal may have dissipated (due to attenuation) before reaching the sender. If the signal echo from the bounced signal is strong enough, it can interfere with ultra-high-speed technologies such as 1000Base-T.

Noise (Signal Interference)

Everything electrical in the cable that isn't the signal itself is noise and constitutes a threat to the integrity of the signal. Many sources of noise exist, from within and outside the cable. Controlling noise is of major importance to cable and connector designers because uncontrolled noise will overwhelm the data signal and bring a network to its knees.

Twisted-pair cables utilize balanced signal transmission. The signal traveling on one conductor of a pair should have essentially the same path as the signal traveling the opposite direction on the other conductor. (That's in contrast to coaxial cable, in which the center conductor provides an easy path for the signal but the braid and foil shield that make up the other conductor are less efficient and therefore a more difficult pathway for the signal.)

As signals travel along a pair, an electrical field is created. When the two conductors are perfectly symmetrical, everything flows smoothly. However, minute changes in the diameter of the copper, the thickness of the insulating layer, or the centering of conductors within that insulation cause disturbances in the electrical field called *unbalances*. Electrical unbalance means noise.

Resistance unbalance occurs when the dimensions of the two conductors of the pair are not identical. Mismatched conductors, poorly manufactured conductors, or one conductor that got stretched during installation will result in resistance unbalance.

Capacitance unbalance is also related to dimensions, but to the insulation surrounding the conductor. If the insulation is thicker on one conductor than on the other, then capacitance unbalance occurs. Or, if the manufacturing process is not well controlled and the conductor is not perfectly centered (like a bull's-eye) in the insulation, then capacitance unbalance will exist.

Both these noise sources are usually kept well under control by the manufacturer and are relatively minor compared to *crosstalk*.

You've likely experienced crosstalk on a telephone. When you hear another's conversation through the telephone, that is crosstalk. Crosstalk occurs when some of the signal being transmitted on one pair leaks over to another pair.

When a pair is in use, an electrical field is created. This electrical field induces voltage in adjacent pairs, with an accompanying transfer of signal. The more parallel the conductors, the worse this phenomenon is, and the higher the frequency, the more likely crosstalk will happen. Twisting the two conductors of a pair around each other couples the energy out of phase (that's electrical-engineer talk) and cancels the electrical field. The result is reduced transfer of signal. *But the twists must be symmetrical*; i.e., both conductors must twist around each other, not one wrapping around another that's straight, *and two adjacent pairs shouldn't have the same interval of twists*. Why? Because those twist points become convenient signal-transfer points, sort of like stepping-stones in a stream. In general, the shorter the twist intervals, the better the cancellation and the less crosstalk. That's why Category 5e and higher cables are characterized by their very short twist intervals.

Crosstalk is measured in decibels; the higher the crosstalk value, the less crosstalk noise in the cabling. See Figure 1.15.

FIGURE 1.15
Crosstalk

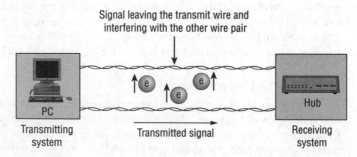

Signal leaving the transmit wire and interfering with the other wire pair

PC

Transmitting system

Transmitted signal

Hub

Receiving system

WAIT A MINUTE! HIGHER CROSSTALK VALUES ARE BETTER?

Yep, illogical as it seems at first, higher crosstalk values are better. Unlike attenuation, where you measure output signal at the receiving end of a single pair, crosstalk coupling is measured between two separate pairs. The way the testing is done, you measure how much signal energy *did not* transfer to the other pair. A pair (or pairs, in the case of power-sum measurements) is energized with a signal. This is the *disturber*. You "listen" on another pair called the *disturbed pair*. Subtracting what you inserted on the disturber from what you measured on the disturbed pair tells you how much signal stayed with the disturber. For example, a 10dB signal is placed on the disturber, but 6dB is detected on the disturbed pair. So −4dB of signal did not transfer (6 − 10). The minus sign is ignored, so the crosstalk is recorded as 4dB. If 2dB were measured on the disturbed pair, then 2 − 10 = −8dB of signal did not transfer, and the crosstalk value is recorded as 8dB. Higher crosstalk numbers represent less loss to adjacent pairs.

Types of Crosstalk

Crosstalk can occur from different elements of the cabling systems and in different locations. The industry has developed a comprehensive set of crosstalk measurements to ensure that cabling systems meet their intended applications. These measurements are covered in this section.

Near-End Crosstalk (NEXT)

When the crosstalk is detected on the same end of the cable that generated the signal, then near-end crosstalk has occurred. NEXT is most common within 20 to 30 meters (60 to 90 feet) of the transmitter. Figure 1.16 illustrates near-end crosstalk.

FIGURE 1.16
Near-end crosstalk
(NEXT)

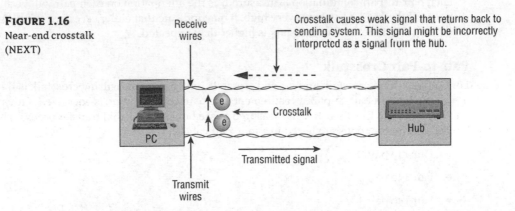

Crosstalk on poorly designed or poorly installed cables is a major problem with technologies such as 10Base-T and 100Base-TX. However, as long as the cable is installed correctly, NEXT is less of an issue when using 1000Base-T because the designers implemented technologies to facilitate NEXT cancellation. NEXT-cancellation techniques with 1000Base-T are necessary because all four pairs are employed for both transmitting and receiving data.

NOTE Cables that have had their twists undone (untwisted) can be problematic because the twists help cancel crosstalk. Twists are normally untwisted at the ends near the patch panels or connectors when the cable is connected. On the receiving pair of wires in a cable, the signal received at the end of the cable will be the weakest, so the signal there can be more easily interfered with. If the wires on adjacent transmit pairs are untwisted, this will cause a greater amount of crosstalk than normal. A cable should never have the wire pairs untwisted more than 0.5″ for Category 5e, and 0.375″ maximum for Category 6 cables.

Far-End Crosstalk (FEXT)

Far-end crosstalk (FEXT) is similar to NEXT except that it is detected at the opposite end of the cable from where the signal was sent. Due to attenuation, the signals at the far end of the transmitting wire pair are much weaker than the signals at the near end.

The measure of FEXT is used to calculate equal-level far-end crosstalk (ELFEXT) (discussed in the next section). More FEXT will be seen on a shorter cable than a longer one because the signal at the receiving side will have less distance over which to attenuate.

Equal-Level Far-End Crosstalk (ELFEXT)

Equal-level far-end crosstalk (ELFEXT) is the crosstalk coupling between cabling pairs measured at the end of the cable opposite to the end of the signal source, taking into account signal loss. ELFEXT is calculated, not measured, by subtracting the attenuation of the disturber pair from the far-end crosstalk (FEXT) on the disturbed pair. The calculation describes the ratio of disturbance to the level of the desired signal; it is another indication of signal-to-noise ratio. Another way of looking at it is that the value represents the ratio between the strength of the noise due to crosstalk from end signals compared to the strength of the received data signal. You could also think of ELFEXT as far-end ACR (attenuation-to-crosstalk ratio, described later in this chapter).

Each pair-to-pair combination is measured, as the attenuation on each pair will be slightly different. If the ELFEXT value is very high, it may indicate that either excessive attenuation has occurred or that the far-end crosstalk is higher than expected.

Pair-to-Pair Crosstalk

For both near-end crosstalk and far-end crosstalk, one way of measuring crosstalk is the *pair-to-pair* method. In pair-to-pair measurement, one pair, the disturber, is energized with a signal, and another pair, the disturbed, is measured to see how much signal transfer occurs. The following six combinations are tested in a four-pair cable:

◆ Pair 1 to pair 2

◆ Pair 1 to pair 3

◆ Pair 1 to pair 4

◆ Pair 2 to pair 3

◆ Pair 2 to pair 4

◆ Pair 3 to pair 4

The test is repeated from the opposite end of the cable, resulting in 12 pair-to-pair combinations tested. The worst combination is what is recorded as the cable's crosstalk value. See Figure 1.17.

Power-Sum Crosstalk

Power-sum crosstalk also applies to both NEXT and FEXT and must be taken into consideration for cables that will support technologies using more than one wire pair at the same time. When testing power-sum crosstalk, all pairs except one are energized as disturbing pairs, and the remaining pair, the disturbed pair, is measured for transferred signal energy. Figure 1.18 shows a cutaway of a four-pair cable. Notice that the energy from pairs 2, 3, and 4 can all affect pair 1. The sum of this crosstalk must be within specified limits. Because each pair affects all the other pairs, this measurement will have to be made four separate times, once for each wire pair against the others. Again, testing is done from both ends, raising the number of tested combinations to eight. The worst combination is recorded as the cable's power-sum crosstalk.

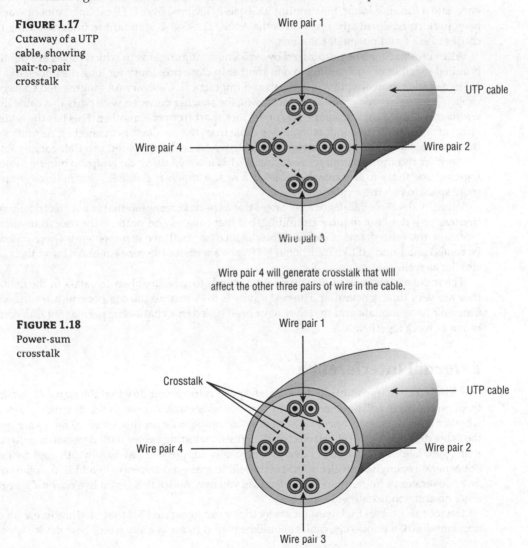

FIGURE 1.17
Cutaway of a UTP cable, showing pair-to-pair crosstalk

Wire pair 1

UTP cable

Wire pair 4

Wire pair 2

Wire pair 3

Wire pair 4 will generate crosstalk that will affect the other three pairs of wire in the cable.

FIGURE 1.18
Power-sum crosstalk

Crosstalk

Wire pair 1

UTP cable

Wire pair 4

Wire pair 2

Wire pair 3

Crosstalk from pairs 2, 3, and 4 will affect pair 1.

Alien Crosstalk (AXT)

Alien crosstalk (AXT) occurs when the signal being carried in one cable interferes with the signal being carried in another cable. This can occur in a cable that runs alongside one or more signal-carrying cables. The term *alien* arises from the fact that this form of crosstalk occurs between different cables in a bundle, rather than between individual wire pairs within a cable.

Alien crosstalk can be a problem because, unlike the simpler forms of crosstalk that take place within a single cable, it cannot be eliminated by traditional phase cancellation. Because AXT resembles noise rather than signals, alien crosstalk degrades the performance of the cabling system by reducing the signal-to-noise ratio of the link. As the signal rate increases in a cable, this form of crosstalk becomes more important. In fact, this is a major source of interference, and a limiting factor, for running 10GBase-T (10Gbps) over UTP cabling. A lot of work has been performed during the creation of the ANSI/TIA-568-C standard in understanding the causes of AXT and potential solutions.

Alien crosstalk can be minimized by avoiding configurations in which cables are tightly bundled together or run parallel to one another in close proximity for long distances. In a typical installation, however, this is difficult and impractical. Category 6A (augmented Category 6) cable tries to solve this problem by increasing the spacing between wire pairs in a cable using *separators* within a cable to space the conductors apart from one another. This has the added effect of separating the conductors in one cable from the conductors in another. As you can imagine, this increases the diameter of a Category 6A cable compared to a Category 6 cable.

Another recommendation for reducing AXT is to avoid using tie-wraps to bundle cable together and to try to separate the cables in a rack as much as possible. This in turn requires more space to run these cables.

Recently developed Category 6A cables use a special core wrap that is not electrically continuous, so it does not require grounding, but that isolates and protects the core from alien crosstalk and other forms of external interference (as you'll see in a moment). These cables can be routed and bundled like traditional UTP cables without the concern of AXT and their size is smaller as well.

The industry has created measurement methods to measure alien crosstalk in the field, but they are very time consuming. The best advice is to ensure all the components are verified to be Category 6A compliant and that they have been tested in a channel or permanent link configuration to work together.

External Interference

One hindrance to transmitting data at high speed is the possibility that the signals traveling through the cable will be acted upon by some outside force. Although the designer of any cable, whether it's twisted-pair or coaxial, attempts to compensate for this, external forces are beyond the cable designer's control. All electrical devices, including cables with data flowing through them, generate EMI. Low-power devices and cables supporting low-bandwidth applications do not generate enough of an electromagnetic field to make a difference. In addition, some equipment generates radio-frequency interference; you may notice this if you live near a TV or radio antenna and you use a cordless phone.

Devices and cables that use a lot of electricity can generate EMI that can interfere with data transmission. Consequently, cables should be placed in areas away from these devices.

Some common sources of EMI in a typical office environment include the following:

◆ Motors

◆ Heating and air-conditioning equipment

◆ Fluorescent lights

◆ Laser printers

◆ Elevators

◆ Electrical wiring

◆ Televisions

◆ Some medical equipment

NOTE   Talk about electromagnetic interference! An MRI (magnetic resonance imaging) machine, which is used to look inside the body without surgery or x-rays, can erase the magnetic strip on a credit card from 10´ away.

When running cabling in a building, do so a few feet away from these devices. Never install data cabling in the same conduit as electrical wiring.

In some cases, even certain types of businesses and environments have high levels of interference, including airports, hospitals, military installations, and power plants. If you install cabling in such an environment, consider using cables that are properly shielded, or use fiber-optic cable.

CABLING AND STANDARDS

Maximum acceptable values of attenuation, minimum acceptable values of crosstalk, and even cabling design issues—who is responsible for making sure standards are published? The group varies from country to country; in the United States, the predominant standards organization supervising data cabling standards is the Telecommunications Industry Association (TIA). The standard that covers Category 3, 5e, 6, and 6A cabling, for example, is ANSI/TIA-568-C, which is part of the guideline for building structured cabling systems. These standards are not rigid like an Internet RFC but are refined as needed via addendums. The ANSI/TIA-568-C document dictates the performance specifications for cables and connecting hardware. Chapter 2 discusses common cabling standards in more detail.

Attenuation-to-Crosstalk Ratio (ACR)

Attenuation-to-crosstalk ratio (ACR) is an indication of how much larger the received signal is when compared to the NEXT (crosstalk or noise) on the same pair. ACR is also sometimes referred to as the *signal-to-noise ratio* (SNR). It is a calculated value; you can't directly measure

ACR. Also, despite the name, it's not really a ratio. It is the mathematical difference you get when you subtract the crosstalk value from the attenuation value at a given frequency. Note that ACR, and its power-sum equivalent, PSACR, is not a required parameter in TIA/EIN-568-B. Technically, SNR also incorporates not only noise generated by the data transmission but also outside interference. For practical purposes, the ACR and true SNR are functionally identical, except in environments with high levels of EMI.

KEY TERM *headroom* Because ACR represents the minimum gap between attenuation and crosstalk, the *headroom* represents the difference between the minimum ACR and the actual ACR performance values. Greater headroom is desirable because it provides additional performance margin that can compensate for the sins of cheap connectors or sloppy termination practices. It also results in a slight increase in the maximum bandwidth of the cable.

The differential between the crosstalk (noise) and the attenuation (loss of signal) is important because it assures that the signal being sent down a wire is stronger at the receiving end than any interference that may be imposed by crosstalk or other noise.

Figure 1.19 shows the relationship between attenuation and NEXT and graphically illustrates ACR for Category 5. (Category 5e, Category 6 and 6A would produce similar graphs.) Notice that as the frequency increases, the NEXT values get lower while the attenuation values get higher. The difference between the attenuation and NEXT lines is the ACR. Note that for all cables, a theoretical maximum bandwidth exists greater than the specified maximum in the standards. This is appropriate conservative engineering.

FIGURE 1.19
Attenuation-to-crosstalk ratio for a Category 5e channel link

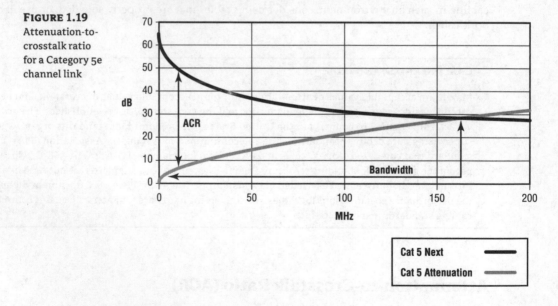

Solving problems relating to ACR usually means troubleshooting NEXT because, short of replacing the cable, the only way to reduce attenuation is to use shorter cables. PSACR is similar to ACR but is the difference between insertion loss and NEXT.

Propagation Delay

Electricity travels through a cable at a constant speed, expressed as a percentage of light speed called NVP (nominal velocity of propagation). For UTP cables, NVP is usually between 60 and 90 percent. The manufacturer of the cable controls the NVP value because it is largely a function of the dielectric constant of the insulation material. The difference between the time at which a signal starts down a pair and the time at which it arrives on the other end is the *propagation delay*.

Delay Skew

Delay skew is a phenomenon that occurs as a result of each set of wires being different lengths (as shown in Figure 1.20). Twisting the conductors of a pair around each other to aid in canceling crosstalk increases the actual length of the conductors relative to the cable length. Because the pairs each have a unique twist interval, the conductor lengths from pair to pair are unique as well. Signals transmitted on two or more separate pairs of wire will arrive at slightly different times, as the wire pairs are slightly different lengths. Cables that are part of a Category 5e, 6, or 6A installation cannot have more than a 45ns delay skew.

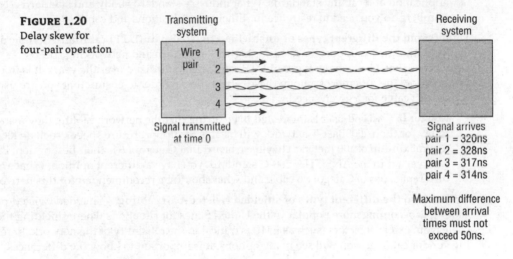

FIGURE 1.20
Delay skew for
four-pair operation

Excessive delay or delay skew may cause timing problems with network transceivers. These timing issues can either slow a link dramatically because the electronics are constantly requesting that the data be resent, or choke it off completely.

The Future of Cabling Performance

Category 6A, which was ratified in January 2008, supports 10Gbps Ethernet over 100 meters of UTP. It is conceivable that 10Gbps Ethernet will run to the desktop over twisted-pair cable, but it's unlikely to happen soon. While the current transceiver costs make 10Gbps over UTP more cost effective for data center applications, the costs of these transceivers will eventually make

10Gbs Ethernet to the desk over UTP a reasonable proposition. As materials and manufacturing techniques improve, who knows what types of performance future twisted-pair cabling may offer?

The Bottom Line

Identify the key industry standards necessary to specify, install, and test network cabling. Early cabling systems were unstructured and proprietary, and often worked only with a specific vendor's equipment. Frequently, vendor-specific cabling caused problems due to lack of flexibility. More important, with so many options, it was difficult to utilize a standard approach to the design, installation, and testing of network cabling systems. This often led to poorly planned and poorly implemented cabling systems that did not support the intended application with the appropriate cable type.

Master It In your new position as a product specification specialist, it is your responsibility to review the end users' requirements and specify products that will support them. Since you must ensure that the product is specified per recognized U.S. industry standards, you will be careful to identify that the customer request references the appropriate application and cabling standards. What industry standards body and standards series numbers do you need to reference for Ethernet applications and cabling?

Understand the different types of unshielded twisted-pair (UTP) cabling. Standards evolve with time to support the need for higher bandwidth and networking speeds. As a result, there have been many types of UTP cabling standardized over the years. It is important to know the differences among these cable types in order to ensure that you are using the correct cable for a given speed.

Master It An end user is interested in ensuring that the network cabling they install today for their 1000Base-T network will be able to support future speeds such as 10Gbps to a maximum of 100 meters. They have heard that Category 6 is their best option. Being well versed in the ANSI/TIA-568-C standard, you have a different opinion. What are the different types of Category 6 cable and what should be recommended for this network?

Understand the different types of shielded twisted-pair cabling. Shielded twisted-pair cabling is becoming more popular in the United States for situations where shielding the cable from external factors (such as EMI) is critical to the reliability of the network. In reviewing vendor catalogs, you will see many options. It is important to know the differences.

Master It Your customer is installing communications cabling in a factory full of stray EMI. UTP is not an option and a shielded cable is necessary. The customer wants to ensure capability to operate at 10GBase-T. What cable would you recommend to offer the best shielding performance?

Determine the uses of plenum- and riser-rated cabling. There are two main types of UTP cable designs: plenum and riser. The cost difference between them is substantial. Therefore, it's critical to understand the differences between the two.

Master It Your customer is building a traditional star network. They plan to route cable for horizontal links through the same space that is used for air circulation and HVAC systems. They plan to run cable vertically from their main equipment room to their telecommunications rooms on each floor of the building. What type of cable would you use for:

a) The horizontal spaces

b) The vertical links

Identify the key test parameters for communications cables. As you begin to work with UTP cable installation, you will need to perform a battery of testing to ensure that the cabling system was installed properly and meets the channel requirements for the intended applications and cable grades. If you find faults, you will need to identify the likely culprits and fix it.

Master It Crosstalk is one of the key electrical phenomena that can interfere with the signal. There are various types of crosstalk: NEXT, FEXT, AXT, among others. This amount of crosstalk can be caused in various ways. What would you look for in trying to find fault if you had the following failures:

a) NEXT and FEXT problems in 1Gbps links

b) Difficulty meeting 10Gbps performance requirements

Chapter 2

Cabling Specifications and Standards

In the past, companies often had several cabling infrastructures because no single cabling system would support all of a company's applications. Nowadays, a standardized cabling system is important not only for consumers but also for vendors and cabling installers. Vendors must clearly understand how to design and build products that will operate on a universal cabling system. Cable installers need to understand what products can be used, proper installation techniques and practices, and how to test installed systems.

This chapter covers some of the important topics related to cabling standards.

In this chapter, you will learn to:

◆ Identify the key elements of the ANSI/TIA-568-C Commercial Building Telecommunications Cabling Standard

◆ Identify other ANSI/TIA standards required to properly design the pathways and spaces and grounding of a cabling system

◆ Identify key elements of the ISO/IEC 11801 Generic Cabling for Customer Premises Standard

Structured Cabling and Standardization

Typical business environments and requirements change quickly. Companies restructure and reorganize at alarming rates. In some companies, the average employee changes work locations once every two years. During a two-year tenure, a friend changed offices at a particular company five times. Each time, his telephone, both networked computers, a VAX VT-100 terminal, and a networked printer had to be moved. The data and voice cabling system had to support these reconfigurations quickly and easily. Earlier cabling designs would not have easily supported this business environment.

Until the early 1990s, cabling systems were proprietary, vendor-specific, and lacking in flexibility. Some of the downsides of pre-1990 cabling systems included the following:

◆ Vendor-specific cabling locked the customer into a proprietary system.

◆ Upgrades or new systems often required a completely new cabling infrastructure.

◆ Moves and changes often necessitated major cabling plant reconfigurations. Some coaxial and twinax cabling systems required that entire areas (or the entire system) be brought down in order to make changes.

◆ Companies often had several cabling infrastructures that had to be maintained for their various applications.

◆ Troubleshooting proprietary systems was time consuming and difficult unless you were intimately familiar with a system.

Cabling has changed quite a bit over the years. Cabling installations have evolved from proprietary systems to flexible, open solutions that can be used by many vendors and applications. This change is the result of the adaptation of standards-based, structured cabling systems. The driving force behind this acceptance is due not only to customers but also to the cooperation between many telecommunications vendors and international standards organizations.

A properly designed *structured cabling system* is based around components or wiring units. An example of a wiring unit is a story of an office building, as shown in Figure 2.1. All the work locations on that floor are connected to a single wiring closet. All of the wiring units (stories of the office building) can be combined together using backbone cables as part of a larger system.

TIP　This point bears repeating: a structured cabling system is not designed around any specific application but rather is designed to be generic. This permits many applications to take advantage of the cabling system.

FIGURE 2.1
A typical small office with horizontal cabling running to a single wiring room (closet)

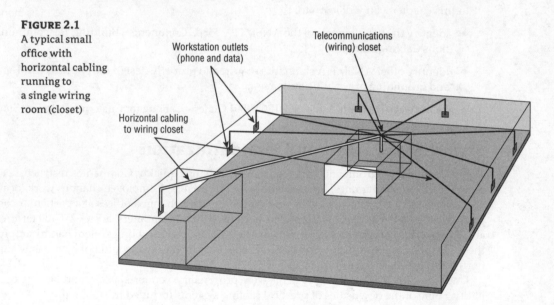

The components used to design a structured cabling system should be based on a widely accepted specification and should allow many applications (analog voice, digital voice, 10Base-T, 100Base-TX, 16Mbps Token Ring, RS-232, etc.) to use the cabling system. The components should also adhere to certain performance specifications so that the installer or customer will know exactly what types of applications will be supported.

A number of documents are related to data cabling. In the United States, the standard is ANSI/TIA-568-C, also known as the Commercial Building Telecommunications Cabling

Standard. The ANSI/TIA-568-C standard is a specification adopted by ANSI (American National Standards Institute), but the ANSI portion of the document name is commonly left out. In Europe, the predominant standard is the ISO/IEC 11801 Ed. 2 standard, also known as the International Standard on Information Technology Generic Cabling for Customer Premises.

WHEN IS A STANDARD NOT A STANDARD?

In the United States, a document is not officially a national standard until it is sanctioned by ANSI. In Canada, the CSA is the sanctioning body, and in Europe, it is the ISO. Until sanctioned by these organizations, a requirements document is merely a specification. However, many people use the words *specification* and *standard* interchangeably. (In Europe, the word *norm* also comes into play.) Just be aware that a "specification" can be created by anyone with a word processor, whereas a national standard carries the weight of governmental recognition as a comprehensive, fair, and objective document.

These two documents are quite similar, although their terminology is different, and the ISO/IEC 11801 Ed. 2 standard permits an additional type of UTP cabling. Throughout much of the rest of the world, countries and specification organizations have adopted one of these standards as their own. Both of these documents are discussed in more detail later in this chapter.

CABLING STANDARDS: A MOVING TARGET

This chapter briefly introduces the ANSI/TIA-568-C and the ISO/IEC 11801 Ed. 2 standards, but it is not intended to be a comprehensive guide to either. Even as you read this book, networking vendors and specifications committees are figuring out ways to transmit larger quantities of data, voice, and video over copper and fiber-optic cable. Therefore, the requirements and performance specifications for the standards are continually being updated. If you are responsible for large cabling-systems design and implementation, you should own a copy of the relevant documents.

Most of the TIA/EIA documents mentioned in this chapter are available for purchase through Global Engineering Documents at (877) 413-5184 or on the Web at http://global.ihs.com. Global Engineering Documents sells printed versions of the ISO, TIA, EIA, and ETSI specifications, as well as others. The ITU recommendations are available for purchase from the ITU's website at www.itu.int/.

CSA International Standards documents are available from the CSA at www.csa.ca.

Standards and Specification Organizations

If you pick up any document or catalog on data cabling, you will see acronyms and abbreviations for the names of specification organizations. If you want to know more about a particular specification, you should be familiar with the organization that publishes that particular document. These U.S.-based and international organizations publish hardware, software, and physical-infrastructure specifications to ensure interoperability between electrical, communications, and other technology systems. Your customers and coworkers may laugh at the elation you express when you get even simple networked devices to work, but you are not alone. In fact, the simple act of getting two stations communicating with one another on a 10Base-T network, for example, is a monumental achievement considering the number of components and vendors involved. Just think: Computers from two different vendors may use Ethernet adapters that also may be from different manufacturers. These Ethernet adapters may also be connected by cable and connectors provided by another manufacturer, which in turn may be connected to a hub built by still another manufacturer. Even the software that the two computers are running may come from different companies. Dozens of other components must work together.

That anything is interoperable at all is amazing. Thankfully, a number of organizations around the world are devoted to the development of specifications that encourage interoperability. These are often nonprofit organizations, and the people who devote much of their time to the development of these specifications are usually volunteers. These specifications include not only cabling specifications and performance and installation practices but also the development of networking equipment like Ethernet cards. As long as the manufacturer follows the appropriate specifications, their devices should be interoperable with other networking devices.

The number of organizations that provide specifications is still more amazing. It might be simpler if a single international organization were responsible for all standards. However, if that were the case, probably nothing would ever get accomplished—hence the number of specifications organizations. The following sections describe a number of these organizations, but the list is by no means exhaustive.

American National Standards Institute (ANSI)

Five engineering societies and three U.S. government agencies founded the American National Standards Institute (ANSI) in 1918 as a private, nonprofit membership organization sustained by its membership. ANSI's mission is to encourage voluntary compliance with standards and methods. ANSI's membership includes almost 1,400 private companies and government organizations in the United States as well as international members.

ANSI does not develop the American National Standards (ANS) documents, but it facilitates their development by establishing a consensus between the members interested in developing a particular standard.

To gain ANSI approval, a document must be developed by a representative cross section of interested industry participants. The cross section must include both manufacturers and end users. In addition, a rigorous balloting and revision process must be adhered to so that a single powerful member does not drive proprietary requirements through and establish a particular market advantage.

Through membership in various international organizations such as the International Organization for Standardization (ISO) and the International Electrotechnical Commission (IEC), ANSI promotes standards developed in the United States. ANSI was a founding member of the ISO and is one of the five permanent members of the ISO governing council and one of four permanent members on the ISO's Technical Management Board.

ANSI standards include a wide range of information-technology specifications, such as SCSI interface specifications, programming language specifications, and specifications for character sets. ANSI helped to coordinate the efforts of the Electronic Industries Alliance (EIA) and the Telecommunications Industry Association (TIA) to develop ANSI/TIA/EIA-568, *the* cabling specification in the United States. ANSI/TIA-568-C is discussed in more detail later in this chapter. You can find information on it and links to purchase the documents on ANSI's website at www.ansi.org.

Electronic Industries Alliance (EIA)

The Electronic Industries Alliance (EIA) was established in 1924 and was originally known as the Radio Manufacturers Association. Since then, the EIA has evolved into an organization that represents a wide variety of electronics manufacturers in the United States and abroad; these manufacturers make products for a wide range of markets. The EIA is organized along specific product and market lines that allow each EIA sector to be responsive to its specific needs. These sectors include components, consumer electronics, electronic information, industrial electronics, government, and telecommunications.

The EIA (along with the TIA) was the driving force behind the original ANSI/TIA/EIA-568 Commercial Building Telecommunications Cabling Standard. More information is available on the Web at www.eia.org.

Telecommunications Industry Association (TIA)

The Telecommunications Industry Association (TIA) is a trade organization that consists of a membership of over 1,100 telecommunications and electronics companies that provide services, materials, and products throughout the world. The TIA membership manufactures and distributes virtually all the telecommunication products used in the world today. TIA's mission is to represent its membership on issues relating to standards, public policy, and market development. The 1988 merger of the United States Telecommunications Suppliers Association (USTSA) and the EIA's Information and Telecommunications Technologies Group formed the TIA.

The TIA (along with the EIA) was instrumental in the development of the ANSI/TIA/EIA-568 Commercial Building Telecommunications Cabling Standard. TIA can be found on the Web at www.tiaonline.org.

TIA COMMITTEES

In the United States (and much of the world), the TIA is ultimately responsible for the standards related to structured cabling as well as many other technological devices used every day. If you visit the TIA website (www.tiaonline.org), you will find that committees develop the specifications. Often, a number of committees will contribute to a single specification. You may find a number of abbreviations that you are not familiar with. These include the following:

SFG Standards Formulation Group is a committee responsible for developing specifications.

FO Fiber Optics was a committee dedicated to fiber-optic technology. This has now been merged into the TR-42 engineering committee

TR Technical Review is an engineering committee.

WG Working Group is a general title for a subcommittee.

UPED The User Premises Equipment Division centers its activities on FCC (Federal Communications Commission) regulatory changes.

Some of the TIA committees and their responsibilities are as follows:

TR-30 Develops standards related to the functional, electrical, and mechanical characteristics of interfaces between data circuit terminating equipment (DCE), data terminal equipment (DTE) and multimedia gateways, the telephone and voice-over-Internet protocol (VoIP) networks, and other DCE and facsimile systems.

TR-41 User Premises Telecommunications Requirements is responsible for the specifications relating to technologies such as IP (Internet Protocol), telephony (VoIP), wireless telephones, caller ID, multimedia building distribution, and wireless user premises equipment.

TR-42 User Premises Telecommunications Infrastructure is responsible for specifications such as the Commercial Building Telecommunications Cabling (ANSI/TIA-568-C.1 or subcommittee TR-42.1), Residential Telecommunications Infrastructure (ANSI/TIA-570-B or subcommittee TR-42.2), Commercial Building Telecommunications Pathways and Spaces (TIA-569-B or subcommittee TR-42.3), Telecommunications Copper Cabling Systems (ANSI/TIA-568-C.2 or subcommittee TR-42.7), Telecommunications Optical Fiber Cabling Systems (ANSI/TIA-568-C.3 or subcommittee TR-42.8). The subcommittees of TR-42 formed the TIA/EIA-568-B specification ratified in 2001 and are presently updating this into the ANSI/TIA-568-C specification version. FO-4 was merged into TR-42 in 2008:

- Optical Systems (subcommittee TR-42.11)
- Optical Fibers and Cables (subcommittee TR-42.12)
- Passive Optical Devices and Components (subcommittee TR-42.13)
- Fiber Optic Metrology (subcommittee TR-42.15)
- Premises Telecommunications Bonding & Grounding (subcommittee TR-42.16).

Insulated Cable Engineers Association (ICEA)

The ICEA is a nonprofit professional organization sponsored by leading cable manufacturers in the United States. It was established in 1925 with the goal of producing cable specifications for telecommunication, electrical power, and control cables. The organization draws from the technical expertise of the representative engineer members to create documents that reflect the most current cable-design, material-content, and performance criteria. The group is organized in four sections: Power Cable, Control & Instrumentation Cable, Portable Cable, and Communications Cable.

The ICEA has an important role in relation to the ANSI/TIA/EIA standards for network cabling infrastructure. ICEA cable specifications for both indoor and outdoor cables, copper and fiber optic, are referenced by the TIA documents to specify the design, construction, and physical performance requirements for cables.

ICEA specifications are issued as national standards. In the Communications section, ANSI requirements for participation by an appropriate cross section of industry representatives in a document's development is accomplished through TWCSTAC (pronounced *twix-tak*), the Telecommunications Wire and Cable Standards Technical Advisory Committee. The TWCSTAC consists of ICEA members, along with other manufacturers, material suppliers, and end users. The ICEA maintains a website at www.icea.net.

National Fire Protection Association (NFPA)

The National Fire Protection Association (NFPA) was founded in 1896 as a nonprofit organization to help protect people, property, and the environment from fire damage. NFPA is now an international organization with more than 65,000 members representing over 100 countries. The organization is a world leader on fire prevention and safety. The NFPA's mission is to help reduce the risk of fire through codes, safety requirements, research, and fire-related education. The Internet home for NFPA is at www.nfpa.org.

Though not directly related to data cabling, the NFPA is responsible for the development and publication of the National Electrical Code (NEC). The NEC is published every three years (the next NEC will be published in 2011) and covers issues related to electrical safety requirements; it is not used as a design specification or an instruction manual.

Two sections of the NEC are relevant to data cabling, Articles 725 and 800. Many municipalities have adopted the NEC as part of their building codes, and consequently, electrical construction and wiring must meet the specifications in the NEC. Although the NEC is not a legal document, portions of the NEC become laws if municipalities adopt them as part of their local building codes. In Chapter 4, we will discuss the use of the NEC when considering the restrictions that may be placed on cabling design.

National Electrical Manufacturers Association (NEMA)

The National Electrical Manufacturers Association (NEMA) is a U.S.-based industry association that helps promote standardization of electrical components, power wires, and cables. The specifications put out by NEMA help to encourage interoperability between products built by different manufacturers. The specifications often form the basis for ANSI standards. NEMA can be found on the Internet at www.nema.org.

Federal Communications Commission (FCC)

The Federal Communications Commission (FCC) was founded in 1934 as part of the U.S. government. The FCC consists of a board of seven commissioners appointed by the President; this board has the power to regulate electrical-communications systems originating in the United States. These communications systems include television, radio, telegraph, telephone, and cable TV systems. Regulations relating to premises cabling and equipment are covered in FCC Part 68 rules. The FCC website is at www.fcc.gov.

Underwriters Laboratories (UL)

Founded in 1894, Underwriters Laboratories, Inc. (UL) is a nonprofit, independent organization dedicated to product safety testing and certification. Although not involved directly with cabling specifications, UL works with cabling and other manufacturers to ensure that electrical devices are safe. UL tests products for paying customers; if the product passes the specification for which the product is submitted, the UL listing or verification is granted. The UL mark of approval is applied to cabling and electrical devices worldwide. UL can be found on the Web at www.ul.com.

International Organization for Standardization (ISO)

The International Organization for Standardization (ISO) is an international organization of national specifications bodies and is based in Geneva, Switzerland. The specifications bodies that are members of the ISO represent over 130 countries from around the world; the U.S. representative to the ISO is the American National Standards Institute (ANSI). The ISO was established in 1947 as a nongovernmental organization to promote the development of standardization in intellectual, scientific, technological, and economic activities. You can find the ISO website at www.iso.org.

NOTE If the name is the International Organization for Standardization, shouldn't the acronym be IOS instead of ISO? It should be, if ISO were an acronym—but ISO is taken from the Greek word *isos*, meaning equal.

ISO standards include specifications for film-speed codes, telephone and banking-card formats, standardized freight containers, the universal system of measurements known as SI, paper sizes, and metric screw threads, just to name a few. One of the common standards that you may hear about is the ISO 9000 standard, which provides a framework for quality management and quality assurance.

ISO frequently collaborates with the IEC (International Electrotechnical Commission) and the ITU (International Telecommunications Union). One result of such collaboration is the ISO/IEC 11801 Ed. 2:2002 standard titled Generic Cabling for Customer Premises. ISO/IEC 11801 is the ISO/IEC equivalent of the ANSI/TIA-568-C standard.

International Electrotechnical Commission (IEC)

The International Electrotechnical Commission (IEC) is an international specifications and conformity-assessment body founded in 1906 to publish international specifications relating to electrical, electronic, and related technologies. Membership in the IEC includes more than 50 countries.

A full member has voting rights in the international standards process. The second type of member, an associate member, has observer status and can attend all IEC meetings.

The mission of the IEC is to promote international standards and cooperation on all matters relating to electricity, electronics, and related technologies. The IEC and the ISO cooperate on the creation of standards such as the Generic Cabling for Customer Premises (ISO/IEC 11801:2002). The IEC website is www.iec.ch.

Institute of Electrical and Electronic Engineers (IEEE)

The Institute of Electrical and Electronic Engineers (IEEE, pronounced *I triple-E*) is an international, nonprofit association consisting of more than 330,000 members in 150 countries. The IEEE was formed in 1963 when the American Institute of Electrical Engineers (AIEE, founded in 1884) merged with the Institute of Radio Engineers (IRE, founded in 1912). The IEEE is responsible for 30 percent of the electrical-engineering, computer, and control-technology literature published in the world today. They are also responsible for the development of over 800 active specifications and have many more under development. These specifications include the 10Base-*x* specifications (such as 10Base-T, 100Base-TX, etc.) and the 802.*x* specifications (such as 802.2, 802.3, etc.). You can get more information about the IEEE on the Web at www.ieee.org.

National Institute of Standards and Technology (NIST)

The U.S. Congress established the National Institute of Standards and Technology (NIST) with several major goals in mind, including assisting in the improvement and development of manufacturing technology, improving product quality and reliability, and encouraging scientific discovery. NIST is an agency of the U.S. Department of Commerce and works with major industries to achieve its goals.

NIST has four major programs through which it carries out its mission:

◆ Measurement and Standards Laboratories

◆ Advanced Technology Program

◆ Manufacturing Extension Partnership

◆ A quality outreach program associated with the Malcolm Baldrige National Quality Award called the Baldrige National Quality Program

Though not directly related to most cabling and data specifications, NIST's efforts contribute to the specifications and the development of the technology based on them. You can locate NIST on the Internet at www.nist.gov.

International Telecommunications Union (ITU)

The International Telecommunications Union (ITU), based in Geneva, Switzerland, is the specifications organization formerly known as the International Telephone and Telegraph Consultative Committee (CCITT). The origins of the CCITT can be traced back over 100 years; the ITU was formed to replace it in 1993. The ITU does not publish specifications per se, but it does publish recommendations. These recommendations are nonbinding specifications agreed to by consensus by one of the 14 technical study groups. The mission of the ITU is to study the technical and operations issues relating to telecommunications and to make recommendations on implementing standardized approaches to telecommunications.

The ITU currently publishes more than 2,500 recommendations, including specifications relating to telecommunications, electronic messaging, television transmission, and data communications. The ITU's web address is www.itu.int.

CSA International (CSA)

CSA International originated as the Canadian Standards Association but changed its name to reflect its growing work and influence on international standards. Founded in 1919, CSA International is a nonprofit, independent organization with more than 8,000 members worldwide; it is the functional equivalent of the UL. CSA International's mission is to develop standards, represent Canada on various ISO committees, and work with the IEC when developing the standards. Some of the common standards published by CSA International include:

◆ CAN/CSA-T524 Residential Wiring

◆ CAN/CSA-T527 Bonding and Grounding for Telecommunications

◆ CAN/CSA-T528 Telecommunications Administration Standard for Commercial Buildings

◆ CAN/CSA-T529 Design Guidelines for Telecommunications Wiring Systems in Commercial Buildings

◆ CAN/CSA-T530 Building Facilities Design Guidelines for Telecommunications

Many cabling and data products certified by the United States National Electrical Code (NEC) and Underwriters Laboratories (UL) are also certified by the CSA. Cables manufactured for use in the United States are often marked with the CSA electrical and flame-test ratings as well as the U.S. ratings, if they can be used in Canada. CSA International is on the Internet at www.csa.ca.

IP/MPLS Forum (ATM Forum)

Started in 1991, the ATM Forum is an international, nonprofit organization whose mission is to promote the use of ATM (Asynchronous Transfer Mode) products and services. In 2004 it merged with the IP/MPLS forum.

Specifications developed and published by the ATM Forum include LAN Emulation (LANE) over ATM (af-lane-0021.000) and ATM Physical Medium Dependent Interface Specification for 155Mbps over Twisted-Pair Cable (af-phy-0015.000). These documents are available free of charge on the ATM Forum's website at www.ipmplsforum.org.

European Telecommunications Standards Institute (ETSI)

The European Telecommunications Standards Institute (ETSI) is a nonprofit organization based in Sophia Antipolis, France. The ETSI currently consists of some 696 members from 50 countries and represents manufacturers, service providers, and consumers. The ETSI's mission is to determine and produce telecommunications specifications and to encourage worldwide standardization. The ETSI coordinates its activities with international standards bodies such as the ITU. You can find the organization at www.etsi.org.

Building Industry Consulting Services International (BICSI)

Though not explicitly a specifications organization, the Building Industry Consulting Services International (BICSI) deserves a special mention. BICSI is a nonprofit, professional organization founded in 1974 to support telephone company building-industry consultants (BICs) who are responsible for design and implementation of communications distribution systems in commercial and multifamily buildings. Currently, the BICSI serves over 25,000 members.

BICSI supports a professional certification program called the RCDD (Registered Communications Distribution Designer). Over 6,400 people with the RCDD certification have demonstrated competence and expertise in the design, implementation, and integration of telecommunications systems and infrastructure. For more information on the RCDD program or to learn how to become a member of the BICSI, check out its website at www.bicsi .org. Information on becoming a BICSI-accredited RCDD is detailed in Appendix B.

Occupational Safety and Health Administration (OSHA)

A division of the U.S. Department of Labor, the Occupational Safety and Health Administration (OSHA) was formed in 1970 with the goal of making workplaces in the United States the safest in the world. To this end, it passes laws designed to protect employees from many types of job hazards. OSHA adopted many parts of the National Electrical Code (NEC), which was not a law unto itself, giving those adopted portions of the NEC legal status. For more information on OSHA, look on the Web at www.osha.gov.

ANSI/TIA-568-C Cabling Standard

In the mid-1980s, consumers, contractors, vendors, and manufacturers became concerned about the lack of specifications relating to telecommunications cabling. Before then, all communications cabling was proprietary and often suited only to a single-purpose use. The Computer Communications Industry Association (CCIA) asked the EIA to develop a specification that would encourage structured, standardized cabling.

Under the guidance of the TIA TR-41 committee and associated subcommittees, the TIA and EIA in 1991 published the first version of the Commercial Building Telecommunications Cabling Standard, better known as ANSI/TIA/EIA-568 or sometimes simply as TIA/EIA-568.

A Little History Lesson

Sometimes you will see the Commercial Building Telecommunications Cabling Standard referred to as ANSI/TIA/EIA-568 and sometimes just as TIA/EIA-568 and now just TIA-568. You will also sometimes see the EIA and TIA transposed. The original name of the specification was ANSI/EIA/TIA-568-1991.

Over a period of several years, the EIA released a number of Telecommunications Systems Bulletins (TSBs) covering specifications for higher grades of cabling (TSB-36), connecting hardware (TSB-40), patch cables (TSB-40A), testing requirements for modular jacks (TSB-40A), and additional specifications for shielded twisted-pair cabling (TSB-53). The contents of these TSBs, along with other improvements, were used to revise ANSI/TIA/EIA-568; this revision was released in 1995 and was called ANSI/TIA/EIA-568-A.

Progress marched on, and communications technologies advanced faster than the entire specification could be revised, balloted, and published as a standard. But it is relatively easy to create ad hoc addendums to a standard as the need arises. Consequently, five official additions to the ANSI/TIA/EIA-568-A base standard were written after its publication in 1995:

ANSI/TIA/EIA-568-A-1, the Propagation Delay and Delay Skew Specifications for 100-Ohm Four-Pair Cable   Approved in August and published in September 1997, this addendum was created to include additional requirements with those in the base standard in support of high-performance networking, such as 100Base-T (100Mbps Ethernet).

ANSI/TIA/EIA-568-A-2, Corrections and Addition to ANSI/TIA/EIA-568-A   Approved in July and published in August 1998, this document contains corrections to the base document.

ANSI/TIA/EIA-568-A-3, Addendum 3 to ANSI/TIA/EIA-568-A   Approved and published in December 1998, the third addendum defines bundled, hybrid, and composite cables and clarifies their requirements.

ANSI/TIA/EIA-568-A-4, Production Modular Cord NEXT Loss Test Method for Unshielded Twisted-Pair Cabling   Approved in November and published in December 1999, this addendum provides a nondestructive methodology for NEXT loss testing of modular-plug (patch) cords.

ANSI/TIA/EIA-568-A-5, Transmission Performance Specifications for Four-Pair 100-Ohm Category 5e Cabling   Approved in January and published in February 2000, the latest addendum specifies additional performance requirements for the *cabling* (not just the *cable*) for Enhanced Category 5 installations. Additional requirements include minimum-return-loss, propagation-delay, delay-skew, NEXT, PSNEXT, FEXT, ELFEXT, and PSELFEXT parameters. Also included are laboratory measurement methods, component and field-test methods, and computation algorithms over the specified frequency range. In ANSI/TIA/EIA-568-A-5, performance requirements for Category 5e cabling do not exceed 100MHz, even though some testing is done beyond this frequency limit.

The ANSI/TIA/EIA-568-B revision was published in 2001 and incorporates all five of the addendums to the 568-A version. Among other changes, Category 4 and Category 5 cable are no longer recognized. In fact, Category 4 ceased to exist altogether, and Category 5 requirements were moved to a "for reference only" appendix. Category 5e and Category 6 replace Categories 4 and 5 as recognized categories of cable. The standard is currently being published in ANSI/TIA-568-C revision in 2009. This standard now includes a Category 6A cabling for 10Gbps applications.

Should I Use ANSI/TIA-568-C or ISO/IEC 11801 Ed. 2?

This chapter describes both the ANSI/TIA-568-C and ISO/IEC 11801 Ed. 2 cabling standards. You may wonder which standard you should follow. Though these two standards are quite similar (ISO/IEC 11801 was based on ANSI/TIA/EIA-568), the ISO/IEC 11801 standard was developed with cable commonly used in Europe and consequently contains some references more specific to European applications. Also, some terminology in the two documents is different and some of the internal channel requirements for CAT-6A are different.

If you are designing a cabling system to be used in the United States or Canada, you should follow the ANSI/TIA-568-C standard. You should know, however, that the ISO is taking the lead (with assistance from TIA, EIA, CSA, and others) in developing new international cabling specifications, so maybe in the future you will see only a single standard implemented worldwide that will be a combination of both specifications. Also, note that whereas 568-C contains both components, permanent link and channel specifications, 11801 contains only permanent link and channel requirements. The component requirements are contained in separate documents that are referenced in 11801.

ANSI/TIA-568-C Purpose and Scope

The ANSI/TIA/EIA-568 standard was developed and has evolved into its current form for several reasons:

◆ To establish a cabling specification that would support more than a single vendor application

◆ To provide direction of the design of telecommunications equipment and cabling products that are intended to serve commercial organizations

♦ To specify a cabling system generic enough to support both voice and data

♦ To establish technical and performance guidelines and provide guidelines for the planning and installation of structured cabling systems

The standard addresses the following:

♦ Subsystems of structured cabling

♦ Minimum requirements for telecommunications cabling

♦ Installation methods and practices

♦ Connector and pin assignments

♦ The life span of a telecommunications cabling system (which should exceed 10 years)

♦ Media types and performance specifications for horizontal and backbone cabling

♦ Connecting hardware performance specifications

♦ Recommended topology and distances

♦ The definitions of cabling elements (horizontal cable, cross-connects, telecommunication outlets, etc.)

The current configuration of ANSI/TIA-568-C subdivides the standard as follows:

♦ ANSI/TIA-568-C.0: Generic Telecommunications Cabling for Customer Premises

♦ ANSI/TIA-568-C.1: Commercial Building Telecommunications Cabling Standard

♦ ANSI/TIA-568-C.2: Balanced Twisted-Pair Telecommunications Cabling and Components Standard

♦ ANSI/TIA-568-C.3: Optical Fiber Cabling Components Standard

In this chapter, we'll discuss the standard as a whole, without focusing too much on specific sections.

TIP   This chapter provides an overview of the ANSI/TIA-568-C standard and is not meant as a substitute for the official document. Cabling professionals should purchase a full copy; you can do so at the Global Engineering Documents website (http://global.ihs.com).

WARNING   Welcome to the Nomenclature Twilight Zone. The ANSI/TIA-568-C standard contains two wiring patterns for use with UTP jacks and plugs. They indicate the order in which the wire conductors should be connected to the pins in modular jacks and plugs and are known as T568A and T568B. Do not confuse these with the documents ANSI/TIA/EIA-568-B and the previous version, ANSI/TIA/EIA-568-A. The wiring schemes are both covered in ANSI/TIA/EIA-568. To learn more about the wiring patterns, see Chapter 10, "Connectors."

Subsystems of a Structured Cabling System

The ANSI/TIA-568-C.1 standard breaks structured cabling into six areas:

◆ Horizontal cabling

◆ Backbone cabling

◆ Work area

◆ Telecommunications rooms and enclosures

◆ Equipment rooms

◆ Entrance facility (building entrance)

INTERPRETING STANDARDS AND SPECIFICATIONS

Standards and specification documents are worded with precise language designed to spell out exactly what is expected of an implementation using that specification. If you read carefully, you may notice that slightly different words are used when stating requirements.

If you see the word *shall* or *must* used when stating a requirement, it signifies a *mandatory* requirement. Words such as *should*, *may*, and *desirable* are *advisory* in nature and indicate *recommended* requirements.

In ANSI/TIA-568-C, some sections, specifically some of the annexes, are noted as being *normative* or *informative*. *Normative* means the content is a requirement of the standard. *Informative* means the content is for reference purposes only. For example, Category 5 cable is no longer a recognized media and Category 5 requirements have been placed in *informative* Annex M of 568-C.2 in support of "legacy" installations.

TIP This chapter provides an overview of the ANSI/TIA-568-C standard and is not meant as a substitute for the official document. Cabling professionals should purchase a full copy; you can do so at the Global Engineering Documents website (http://global.ihs.com).

HORIZONTAL CABLING

Horizontal cabling, as specified by ANSI/TIA-568-C.1, is the cabling that extends from horizontal cross-connect, intermediate cross-connect, or main cross-connect to the work area and terminates in telecommunications outlets (information outlets or wall plates). Horizontal cabling includes the following:

◆ Cable from the patch panel to the work area

◆ Telecommunications outlets

- Cable terminations

- Cross-connections (where permitted)

- A maximum of one transition point

- Cross-connects in telecommunications rooms or enclosures

Figure 2.2 shows a typical horizontal-cabling infrastructure spanning out in a star topology from a telecommunications room. The horizontal cabling is typically connected into patch panels and switches/hubs in telecommunications rooms or enclosures. A telecommunications room is sometimes referred to as a *telecommunications closet* or *wiring closet*. A telecommunications enclosure is essentially a small assembly in the work area that contains the features found in a telecommunications room.

FIGURE 2.2
Horizontal cabling
in a star topology
from the telecom-
munications room

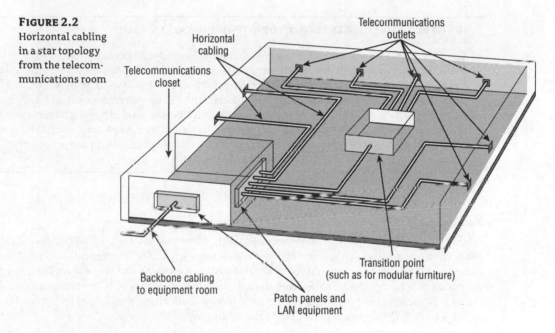

Transition point ANSI/TIA-568-C allows for one *transition point* in horizontal cabling. The transition point is where one type of cable connects to another, such as where round cable connects to under-carpet cable. A transition point can also be a point where cabling is distributed out to modular furniture. Two types of transition points are recognized:

MUTOA This acronym stands for multiuser telecommunications outlet assembly, which is an outlet that consolidates telecommunications jacks for many users into one area. Think of it as a patch panel located out in the office area instead of in a telecommunications room.

CP CP stands for consolidation point, which is an intermediate interconnection scheme that allows horizontal cables that are part of the building pathways to extend to telecommunication outlets in open-office pathways such as those in modular furniture. The ISO/IEC 11801 refers to the CP as a transition point (TP).

If you plan to use modular furniture or movable partitions, check with the vendor of the furniture or partitions to see if it provides data-cabling pathways within its furniture. Then ask what type of interface it may provide or require for your existing cabling system. You will have to plan for connectivity to the furniture in your wiring scheme.

Application-specific components (baluns, repeaters) should not be installed as part of the horizontal-cabling system (inside the walls). These should be installed in the telecommunication rooms or work areas.

Is There a Minimum Distance for UTP Horizontal Cable?

The ANSI/TIA/EIA-568-B does not specify a minimum length for UTP cabling, except when using a multiuser telecommunications outlet assembly (MUTOA). A *short-link phenomenon* occurs in cabling links usually less than 20 meters (60´) long that usually support 100Base-TX applications. The first 20 to 30 meters of a cable is where near-end crosstalk (NEXT) has the most effect. In higher-speed networks such as 100Base-TX, short cables may cause the signal generated by crosstalk or return loss reflections to be returned back to the transmitter. The transmitter may interpret these returns as collisions and cause the network not to function correctly at high speeds. To correct this problem, try extending problematic cable runs with extra-long patch cords.

Recognized Media

ANSI/TIA-568-C recognizes two types of media (cables) that can be used as horizontal cabling. More than one media type may be run to a single work-area telecommunications outlet; for example, a UTP cable can be used for voice, and a fiber-optic cable can be used for data. The maximum distance for horizontal cable from the telecommunications room to the telecommunications outlet is 90 meters (295´) regardless of the cable media used. Horizontal cables recognized by the ANSI/TIA-568-C standard are limited to the following:

◆ Four-pair, 100 ohm, unshielded or shielded twisted-pair cabling: Category 5e, Category 6 or Category 6A (ANSI/TIA-568-C.2)

◆ Two-fiber 62.5/125-micron or 50/125-micron optical fiber (or higher fiber count) multi-mode cabling (ANSI/TIA-568-C.3)

◆ Two-fiber (or higher fiber count) optical fiber single-mode cabling (ANSI/TIA-568-C.3)

Cabling @ Work: Maximum Horizontal Cabling Distance

If you ask someone what the maximum distance of cable is between a network hub (such as 10Base-T) and the computer, you are likely to hear "100 meters." But many people ignore the fact that patch cords are required and assume the distance is from the patch panel to the telecommunication outlet (wall plate). That is not the case.

The ANSI/TIA-568-C standard states that the maximum distance between the telecommunications outlet and the patch panel is 90 meters. The standard further allows for a patch cord up to 5 meters in length in the workstation area and a patch cord up to 5 meters in length in the telecommunications room. (If you did the math, you figured out that the actual maximum length is 99 meters, but what's one meter between friends?) The total distance is the maximum distance for a structured cabling system, based on ANSI/TIA-568-C, regardless of the media type (twisted-pair copper or optical fiber).

The *100 meter* maximum distance is not a random number; it was chosen for a number of reasons, including the following:

◆ The number defines transmissions distances for communications-equipment designers. This distance limitation assures them that they can base their equipment designs on the maximum distance of 100 meters between the terminal and the hub in the closet.

◆ It provides building architects a specification that states they should place telecommunications rooms so that no telecommunications outlet will be farther than 90 meters from the nearest wall outlet (that's in cable distance, which is not necessarily a straight line).

◆ The maximum ensures that common technologies (such as 10Base-T Ethernet) will be able to achieve reasonable signal quality and maintain data integrity. Much of the reasoning for the maximum was based on the timing required for a 10Base-T Ethernet workstation to transmit a minimum packet (64 bytes) to the farthest station on an Ethernet segment. The propagation of that signal through the cable had to be taken into account.

Can a structured cabling system exceed the 100 meter distance? Sure. Good-quality Category 5e or 6 cable will allow 10Base-T Ethernet to be transmitted farther than Category 3 cable will. When using 10Base-FL (10Mbps Ethernet over fiber-optic cable), multimode optical-fiber cable has a maximum distance of 2,000 meters; so a structured cabling system that will support exclusively 10Base-FL applications could have much longer horizontal cabling runs.

But (you knew there was a *but*, didn't you?) such a cabling infrastructure is no longer based on a standard. It will support the application it was designed to support, but it may not support others.

Further, for unshielded twisted-pair cabling, the combined effects of attenuation, crosstalk, and other noise elements increase as the length of the cable increases. Although attenuation and crosstalk do not drastically worsen immediately above the 100 meter mark, the signal-to-noise ratio (SNR) begins to approach zero. When the SNR equals zero, the signal is indistinguishable from the noise in the cabling. (That's analogous to a screen full of snow on a TV.) Then your cabling system will exceed the limits that your application hardware was designed to expect. Your results will be inconsistent, if the system works at all.

The moral of this story is not to exceed the specifications for a structured cabling system and still expect the system to meet the needs of specifications-based applications.

Telecommunications Outlets

ANSI/TIA-568-C.1 specifies that each work area shall have a minimum of two information-outlet ports. Typically, one is used for voice and another for data. Figure 2.3 shows a possible telecommunications outlet configuration. The outlets go by a number of names, including *equipment outlets*, *information outlets, wall jacks*, and *wall plates*. However, an information outlet is officially considered to be one jack on a telecommunications outlet; the telecommunications outlet is considered to be part of the horizontal-cabling system. Chapter 8, "Fiber-Optic Media," and Chapter 10, "Connectors," have additional information on telecommunications outlets.

FIGURE 2.3
A telecommunications outlet with a UTP for voice and a UTP/ScTP/fiber for data

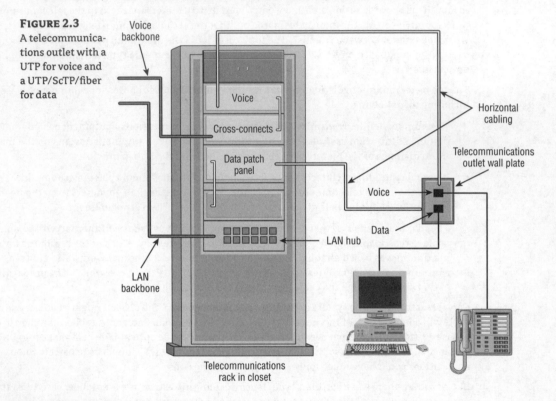

The information outlets wired for UTP should follow one of two conventions for wire-pair assignments or wiring patterns: T568A or T568B. They are nearly identical, except that pairs 2 and 3 are interchanged. Neither of the two is *the* correct choice, as long as the same convention is used at each end of a permanent link. It is best, of course, to always use the same convention throughout the cabling system. T568B used to be much more common in commercial installations, but T568A is now the recommended configuration. (T568A is the *required* configuration for residential installations, in accordance with ANSI/TIA-570-B.) The T568A configuration is partially compatible with an older wiring scheme called USOC, which was commonly used for voice systems.

Be consistent at both ends of the horizontal cable. When you purchase patch panels and jacks, you may be required to specify which pattern you are using, as the equipment may be color-coded to make installation of the wire pairs easier. However, most manufacturers now include options that allow either configuration to be punched down on the patch panel or jack.

Figure 2.4 shows the T568A and T568B pin-out assignments. For more information on wiring patterns, modular plugs, and modular jacks, see Chapter 10.

The wire/pin assignments in Figure 2.4 are designated by wire color. The standard wire colors are shown in Table 2.1.

FIGURE 2.4
Modular jack wire pattern assignments for T568A and T568B

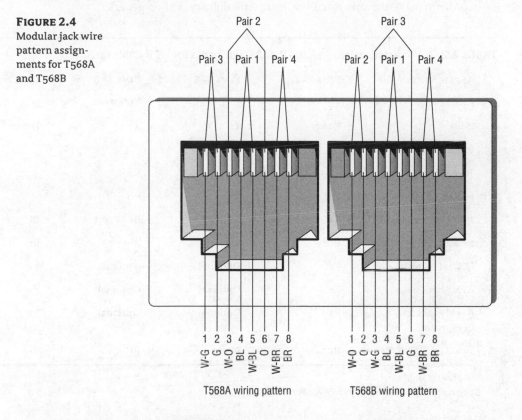

T568A wiring pattern

T568B wiring pattern

TABLE 2.1 Wire-Color Abbreviations

WIRE ABBREVIATION	WIRE COLOR
W/G	White/Green
G	Green
W/O	White/Orange
O	Orange
W/Bl	White/Blue
Bl	Blue
W/Br	White/Brown
Br	Brown

Although your application may not require all the pins in the information outlet, you should make sure that all wires are terminated to the appropriate pins if for no other reason than to ensure interoperability with future applications on the same media. Table 2.2 shows some common applications and the pins that they use, and clearly illustrates why all pairs should be terminated in order to make the structured-wiring installation application generic.

TABLE 2.2   Application-Specific Pair Assignments for UTP Cabling

APPLICATION	PINS 1–2	PINS 3–6	PINS 4–5	PINS 7–8
Analog voice	-	-	Tx/Rx	-
ISDN	Power	Tx	Rx	Power
10Base-T (802.3)	Tx	Rx	-	-
Token Ring (802.5)	-	Tx	Rx	-
100Base-TX (802.3u)	Tx	Rx	-	-
100Base-T4 (802.3u)	Tx	Rx	Bi	Bi
100Base-VG (802.12)	Bi	Bi	Bi	Bi
FDDI (TP-PMD)	Tx	Optional	Optional	Rx
ATM User Device	Tx	Optional	Optional	Rx
ATM Network Equipment	Rx	Optional	Optional	Tx
1000Base-T (802.3ab)	Bi	Bi	Bi	Bi

Bi = bidirectional
Optional = may be required by some vendors

TIP   A good structured-wiring system will include documentation printed and placed on each of the telecommunications outlets.

Pair Numbers and Color Coding

The conductors in a UTP cable are twisted in pairs and color coded so that each pair of wires can be easily identified and quickly terminated to the appropriate pin on the connecting hardware (patch panels or telecommunication outlets). With four-pair UTP cables, each pair of wire is coded with two colors, the tip color and the ring color (see also "Insulation Colors" in Chapter 1). In a four-pair cable, the tip color of every pair is white. To keep the tip conductors associated with the correct ring conductors, often the tip conductor has bands in the color of the ring conductor. Such positive identification (PI) color coding is not necessary in some cases, such as with Category 5 and higher cables, because the intervals between twists in the pair are very close together, making separation unlikely.

You identify the conductors by their color codes, such as white-blue and blue. With premises (indoor) cables, it is common to read the tip color first (including its PI color), then the ring color. Table 2.3 lists the pair numbers, color codes, and pin assignments for T568A and T568B.

TABLE 2.3   Four-Pair UTP Color Codes, Pair Numbers, and Pin Assignments for T568A and T568B

PAIR NUMBER	COLOR CODE	T568A PINS	T568B PINS
1	White-Blue (W-Bl)/Blue (Bl)	W-Bl=5/Bl=4	W-Bl=5/ Bl=4
2	White-Orange (W-O)/Orange (O)	W-O=3/O=6	W-O=1/O=2
3	White-Green (W-G)/Green (G)	W-G=1/G=2	W-G=3/G=6
4	White-Brown (W-Br)/Brown (Br)	W-Br=7/Br=8	W-Br=7/Br=8

BACKBONE CABLING

The next subsystem of structured cabling is called backbone cabling. (Backbone cabling is also sometimes called vertical cabling, cross-connect cabling, riser cabling, or intercloset cabling.) Backbone cabling is necessary to connect entrance facilities, equipment rooms, and telecommunications rooms and enclosures. Refer to Figure 2.7 later in the chapter to see backbone cabling that connects an equipment room with telecommunications rooms. Backbone cabling consists of not only the cables that connect the telecommunications rooms, equipment rooms, and building entrances but also the cross-connect cables, mechanical terminations, or patch cords used for backbone-to-backbone cross-connection.

PERMANENT LINK VS. CHANNEL LINK

ANSI/TIA-568-C defines two basic link types commonly used in the cabling industry with respect to testing: the permanent link and the channel link.

The *permanent link* contains only the cabling found in the walls (horizontal cabling), one transition point, the telecommunications outlet, and one cross-connect or patch panel. It is assumed to be the permanent portion of the cabling infrastructure. The permanent link is illustrated here:

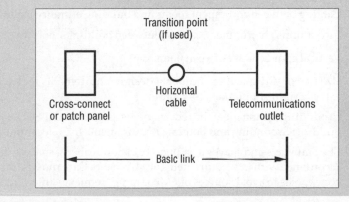

The *channel link* includes the basic link, as well as installed equipment, patch cords, and the cross-connect jumper cable; however, the channel does *not* include phones, PBX equipment, hubs, or network interface cards. Two possible channel link configurations are shown here; one is the channel link for a 10Base-T Ethernet workstation, and one is for a telephone.

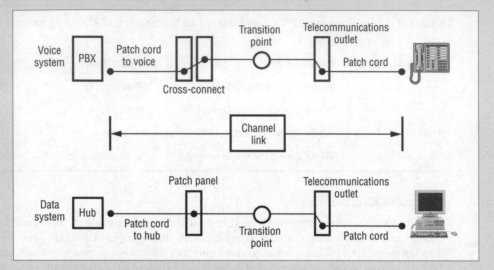

Permanent and channel link performance requirements are provided in Chapter 15, "Cable System Testing and Troubleshooting."

KEY TERM *cross-connect* A *cross-connect* is a facility or location within the cabling system that permits the termination of cable elements and their interconnection or cross-connection by jumpers, termination blocks, and/or cables to another cabling element (another cable or patch panel).

Basic Requirements for Backbone Cabling

Backbone cabling includes:

◆ Cabling between equipment rooms and building entrance facilities

◆ In a campus environment, cabling between buildings' entrance facilities

◆ Vertical connections between floors

ANSI/TIA-568-C.1 specifies additional design requirements for backbone cabling, some of which carry certain stipulations, as follows:

◆ Grounding should meet the requirements as defined in J-STD-607-A, the Commercial Building Grounding and Bonding Requirements for Telecommunications.

◆ The pathways and spaces to support backbone cabling shall be designed and installed in accordance with the requirements of TIA-569-B. Care must be taken when running backbone cables to avoid sources of EMI or radio frequency interference.

◆ No more than two hierarchical levels of cross-connects are allowed, and the topology of back-bone cable will be a hierarchical star topology. (A hierarchical star topology is one in which all cables lead from their termination points back to a central location. Star topology is explained in more detail in Chapter 3, "Choosing the Correct Cabling.") Each horizontal cross-connect should be connected directly to a main cross-connect or to an intermediate cross-connect that then connects to a main cross-connect. No more than one cross-connect can exist between a main cross-connect and a horizontal cross-connect. Figure 2.5 shows multiple levels of equipment rooms and telecommunications rooms.

◆ Centralized optical fiber cabling is designed as an alternative to the optical cross-connection located in the telecommunications room or telecommunications enclosure when deploying recognized optical fiber to the work area from a centralized cross-connect. This is explained in more detail in Chapter 3.

◆ The length of the cord used to connect telecommunications equipment directly to the main or intermediate cross-connect should not exceed 30 meters (98´).

◆ Unlike horizontal cabling, backbone cabling lengths are dependent on the application and on the specific media chosen. (See ANSI/TIA-568-C.0 Annex D.) For optical fiber, this can be as high as 10,000 meters depending on the application! However, distances of ≤ 550 meters are more likely inside a building. This distance is for uninterrupted lengths of cable between the main cross-connect and intermediate or horizontal cross-connect.

◆ Bridge taps or splices are not allowed.

◆ Cables with more than four pairs may be used as long as they meet additional performance requirements such as for power-sum crosstalk. These requirements are specified in the standard. Currently, only Category 5e cables are allowed to have more than four pairs.

FIGURE 2.5
Star topology of equipment room and telecommunication rooms connected via backbone cabling

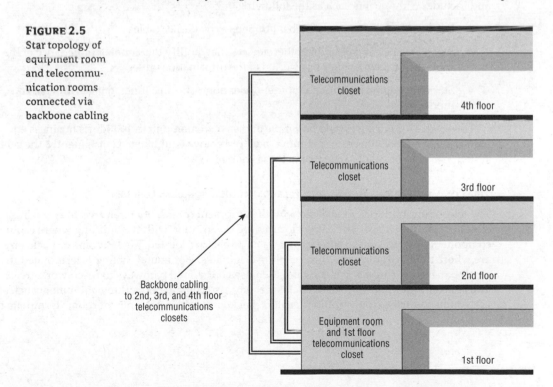

Telecommunications closet

4th floor

Telecommunications closet

3rd floor

Telecommunications closet

2nd floor

Backbone cabling to 2nd, 3rd, and 4th floor telecommunications closets

Equipment room and 1st floor telecommunications closet

1st floor

Recognized Backbone Media

ANSI/TIA-568-C recognizes several types of media (cable) for backbone cabling. These media types can be used in combination as required by the installation. The application and the area being served will determine the quantity and number of pairs required. The maximum distances permitted depend on the application standard and are available in ANSI/TIA-568-C.0 Annex D. In general, the higher the speed, the shorter the distance. Also, optical fiber maximums can range from 220 to 10,000 meters depending on the media and application, whereas UTP is limited to 100 meters.

KEY TERM *media* The term *media* is used in the cabling business to denote the type of cabling used. Media can include fiber-optic cable, twisted-pair cable, or coaxial cable. The definition of media can also be broadened to include wireless networking.

The distances for recognized media are dependent on the application and are shown in ANSI/TIA-568-C.0 Annex D. (Note: distances are the total cable length allowed between the main cross-connect and the horizontal cross-connect, allowing for one intermediate cross-connect.)

WARNING Coaxial cabling is not recognized by the ANSI/TIA-568-C version of the standard.

WORK AREA

The work area is where the horizontal cable terminates at the wall outlet, also called the telecommunications outlet. In the work area, the users and telecommunications equipment connect to the structured-cabling infrastructure. The work area begins at the telecommunications area and includes components such as the following:

- ◆ Patch cables, modular cords, fiber jumpers, and adapter cables

- ◆ Adapters such as baluns and other devices that modify the signal or impedance of the cable (these devices must be external to the information outlet)

- ◆ Station equipment such as computers, telephones, fax machines, data terminals, and modems

The work area wiring should be simple and easy to manipulate. In today's business environments, it is frequently necessary to move, add, or remove equipment. Consequently, the cabling system needs to be easily adaptable to these changes.

TELECOMMUNICATIONS ROOMS AND TELECOMMUNICATIONS ENCLOSURES

The telecommunications rooms (along with equipment rooms, often referred to as wiring closets) and telecommunications enclosures are the location within a building where cabling components such as cross-connects and patch panels are located. These rooms or enclosures are where the horizontal structured cabling originates. Horizontal cabling is terminated in patch panels or termination blocks and then uses horizontal pathways to reach work areas. The telecommunications room or enclosure may also contain networking equipment such as LAN hubs, switches, routers, and repeaters. Backbone-cabling equipment rooms terminate in the telecommunications room or enclosure. Figures 2.5 and 2.7 illustrate the relationship of a telecommunications room to the backbone cabling and equipment rooms.

NOTE A telecommunications enclosure is intended to serve a smaller floor area than a telecommunications room.

TIA's Fiber Optics LAN Section (www.fols.org) has compared the cost differences between network cabling systems using either telecommunications rooms or telecommunications enclosures on each floor of a commercial building and has found as much as 30 percent savings when using multiple telecommunications enclosures.

CABLING @ WORK: PLANNING FOR SUFFICIENT OUTLETS AND HORIZONTAL CABLE

Do you have enough horizontal cabling? Company XYZ recently moved to a new location. In its old location, the company continually suffered from a lack of data and voice outlets. Users wanted phones, modems, and fax machines located in areas that no one ever imagined would have that equipment. The explosion of users with multiple computers in their offices and networked printers only compounded the problem.

XYZ's director of information services vowed that the situation would never happen to her again. Each work area was wired with a four-port telecommunications outlet. Each of these outlets could be used for either voice or data. In the larger offices, she had telecommunications outlets located on opposite walls. Even the lunchrooms and photocopier rooms had telecommunications outlets. This foresight gave Company XYZ the ability to add many more workstations, printers, phones, and other devices that require cabling without the additional cost of running new cables. The per-cable cost to install additional cables later is far higher than installing additional cables during the initial installation.

TIA-569-B discusses telecommunications room design and specifications, and a further discussion of this subsystem can be found in Chapter 5, "Cabling System Components." TIA 569-B recommends that telecommunications rooms be stacked vertically between one floor and another. ANSI/TIA-568-C further dictates the following specifications relating to telecommunications rooms:

◆ Care must be taken to avoid cable stress, tight bends, staples, cable wrapped too tightly, and excessive tension. You can avoid these pitfalls with good cable-management techniques.

◆ Use only connecting hardware that is in compliance with the specifications you want to achieve.

◆ Horizontal cabling should terminate directly not to an application-specific device but rather to a telecommunications outlet. Patch cables or equipment cords should be used to connect the device to the cabling. For example, horizontal cabling should never come directly out of the wall and plug in to a phone or network adapter.

ENTRANCE FACILITY

The entrance facility (building entrance) as defined by ANSI/TIA-568-C.1 specifies the point in the building where cabling interfaces with the outside world. All external cabling (campus backbone, interbuilding, antennae pathways, and telecommunications provider) should enter the building and terminate in a single point. Telecommunications carriers are usually required to terminate within 50′ of a building entrance. The physical requirements of the interface equipment are defined in TIA-569-B, the Commercial Building Standard for Telecommunications Pathways and Spaces. The specification covers telecommunications room design and cable pathways.

TIA-569-B recommends a dedicated entrance facility for buildings with more than 20,000 usable square feet. If the building has more than 70,000 usable square feet, TIA-569-B requires a dedicated, locked room with plywood termination fields on two walls. The TIA-569-B standard also specifies recommendations for the amount of plywood termination fields, based on the building's square footage.

KEY TERM *demarcation point* The *demarcation point* (also called the *demarc*, pronounced *dee-mark*) is the point within a facility, property, or campus where a circuit provided by an outside vendor, such as the phone company, terminates. Past this point, the customer provides the equipment and cabling. Maintenance and operation of equipment past the demarc is the customer's responsibility.

The entrance facility may share space with the equipment room, if necessary or possible. Telephone companies often refer to the entrance facility as the demarcation point. Some entrance facilities also house telephone or PBX (private branch exchange) equipment. Figure 2.6 shows an example of an entrance facility.

FIGURE 2.6
Entrance facility for campus and telecommunications wiring

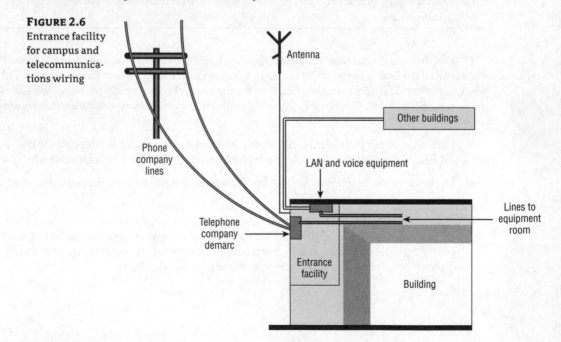

Antenna

Other buildings

Phone company lines

LAN and voice equipment

Telephone company demarc

Lines to equipment room

Entrance facility

Building

TIP To improve data and voice security, the entrance facility should be located in an area that can be physically secured, such as a locked room.

EQUIPMENT ROOM

The next subsystem of structured cabling defined by ANSI/TIA-568-C.1 is the equipment room, which is a centralized space specified to house more sophisticated equipment than the entrance facility or the telecommunications rooms. Often, telephone equipment or data networking equipment such as routers, switches, and hubs are located there. Computer equipment may possibly be stored there. Backbone cabling is specified to terminate in the equipment room.

In smaller organizations, it is desirable to have the equipment room located in the same area as the computer room, which houses network servers and possibly phone equipment. Figure 2.7 shows the equipment room.

For information on the proper design of an equipment room, refer to TIA-569-B.

FIGURE 2.7
Equipment room, backbone cabling, and telecommunications rooms

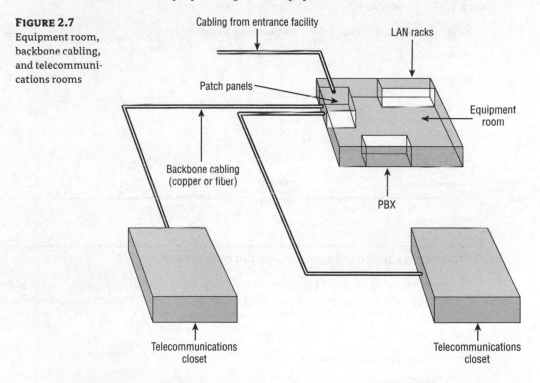

Cabling from entrance facility

LAN racks

Patch panels

Equipment room

Backbone cabling (copper or fiber)

PBX

Telecommunications closet

Telecommunications closet

TIP Any room that houses telecommunications equipment, whether it's a telecommunications room or equipment room, should be physically secured. Many data and voice systems have had security breaches because anyone could walk in off the street and gain physical access to the voice/data network cabling and equipment. Some companies go so far as to put alarm and electronic access systems on their telecommunication rooms and equipment rooms.

Media and Connecting Hardware Performance

ANSI/TIA-568-C specifies performance requirements for twisted-pair cabling and fiber-optic cabling. Further, specifications are laid out for length of cable and conductor types for horizontal, backbone, and patch cables.

100 OHM UNSHIELDED TWISTED-PAIR CABLING

ANSI/TIA-568-C.2 recognizes four categories of UTP cable to be used with structured cabling systems. These UTP cables are specified to have a characteristic impedance of 100 ohms, plus or minus 15 percent, from 1MHz up to the maximum bandwidth supported by the cable. They are commonly referred to by their category number and are rated based on the maximum frequency bandwidth. The categories are found in Table 2.4.

TABLE 2.4   ANSI/TIA-568-C.2 Category Classifications

ANSI/TIA-568-C.2 CATEGORY	MAXIMUM BANDWIDTH
Not defined	100KHz
Not defined	4MHz
Category 3	16MHz
Category 5e	100MHz
Category 6	250MHz
Category 6A	600MHz

ENSURING A SPECIFIC LEVEL OF CABLING PERFORMANCE

UTP cabling systems cannot be considered Category 3–, 5e–, 6–, or 6A–compliant (and consequently certified) unless all components of the cabling system satisfy the specific performance requirements of the particular category. The components include the following:

◆ All backbone and horizontal cabling

◆ Telecommunications outlets

◆ Patch panels

◆ Cross-connect wires and cross-connect blocks

All patch panel terminations, wall-plate terminations, crimping, and cross-connect punch-downs also must follow the specific recommendations for the respective category.

In other words, a network link will perform only as well as the lowest category-compliant component in the link.

Connecting Hardware: Performance Loss

Part of the ANSI/TIA-568-C.2 standard is intended to ensure that connecting hardware (cross-connects, patch panels, patch cables, telecommunications outlets, and connectors) does not have an adverse effect on attenuation and NEXT. To this end, the standard specifies requirements for connecting hardware to ensure compatibility with cables.

Patch Cables and Cross-Connect Jumpers

ANSI/TIA-568-C.1 also specifies requirements that apply to cables used for patch cables and cross-connect jumpers. The requirements include recommendations for maximum-distance limitations for patch cables and cross-connects, as shown here:

Cable Type	Maximum Distance
Main cross-connect, Intermediate cross-connect (used with voice and other low-bandwidth applications)	20 meters (66′)
Telecommunications room	6 meters (20′)
Work area	3 meters (10′)

The total maximum distance of the channel should not exceed the maximum distance recommended for the application being used. For example, the channel distance for 100Base-TX Ethernet should not exceed 100 meters.

TIP   Patch cables should use stranded conductors rather than solid conductors so that the cable is more flexible. Solid conductor cables are easily damaged if they are bent too tightly or too often.

Patch cables usually have a slightly higher attenuation than horizontal cables because they are stranded rather than solid conductors. Though stranded conductors increase patch cable flexibility, they also increase attenuation.

TIP   Detailed requirements for copper cabling and connectivity components are found in ANSI/TIA 568-C.2. Fiber-optic cabling and connectivity components are contained in ANSI/TIA 568-C.3. We highly recommend that you familiarize yourself with cabling requirements if you need to specify performance to a cabling contractor. You should only have to reference the standard for purposes of the Request for Quotation, but your knowledge will help in your discussions with the contractor.

FIBER-OPTIC CABLING

The ANSI/TIA-568-C standard permits both single-mode and multimode fiber-optic cables.

Two connectors were formerly widely used with fiber-optic cabling systems: the ST and SC connectors. Many installations have employed the ST connector type, but the standard now recognizes only the 568SC-type connector. This was changed so that the fiber-optic specifications in ANSI/TIA-568-C.3 could agree with the ISO 11801 standard used in Europe.

The ANSI/TIA-568-C.3 standard also recognizes small-form-factor connectors such as the MT-RJ and LC connectors as well as array connectors such as MPO connectors.

KEY TERM *fiber modes* Fiber-optic cable is referred to as either single-mode or multimode fiber. The term *mode* refers to the number of independent subcomponents of light that propagate through distinct areas of the fiber-optic cable core. Single-mode fiber-optic cable uses only a single mode of light to propagate through the fiber cable, whereas multimode fiber allows multiple modes of light to propagate.

DECIPHERING OPTICAL FIBER CORE AND CLADDING DIAMETERS

What do those numbers mean: 62.5/125, 8.7/125, 50/125? Is this math class?

An optical fiber consists of two primary layers. In the center is the core, where the light is actually transmitted. Surrounding the core is a layer known as the cladding. The cladding material has a lower optical index than the core, acting as a reflective barrier so the light stays in the center. The numbers are the diameters of the core and the cladding, respectively, measured in microns, or one-thousandth of a millimeter. So, a 62.5/125 optical fiber has a core diameter of 62.5 microns with a cladding layer 125 microns in diameter.

Why are all the cladding diameters the same when the core diameters are different? That's so stripping and termination devices such as connector ferrules can be used with all types of fiber strands.

Multimode Optical-Fiber Cable

Multimode optical fiber is most often used as backbone cable inside a building and for horizontal cable. Multimode cable permits multiple modes of light to propagate through the cable and thus lowers cable distances and has a lower available bandwidth. Devices that use multimode fiber-optic cable typically use light-emitting diodes (LEDs) to generate the light that travels through the cable; however, higher-bandwidth network devices such as Gigabit Ethernet are now using lasers with multimode fiber-optic cable. ANSI/TIA-568-C.3 recognizes two types of multimode optical fiber cable:

♦ Two-fiber (duplex) 62.5/125-micron (aka OM1 per ISO 11801)

♦ 50/125-micron multimode fiber-optic cable

Within the 50/125-micron multimode fiber-optic classification, there are two options:

♦ A standard 50-micron fiber (aka OM2 per ISO 11801)

♦ A higher bandwidth option known as 850nm laser-optimized 50/125-micron (aka OM3)

ANSI/TIA-568-C.3 recommends the use of 850nm laser-optimized 50/125-micron (OM3) since it has much higher bandwidth and supports all Gigabit Ethernet applications to the longest distances.

The same connectors and transmission electronics are used on both 62.5/125-micron and 50/125-micron multimode fiber-optic cable. Since multimode fiber has a large core diameter, the connectors and transmitters do not need the same level of precision required with single-mode connectors and transmitters. As a result, they are less expensive than single-mode parts.

Single-Mode Optical-Fiber Cable

Single-mode optical-fiber cable is commonly used as backbone cabling outside the building and is also usually the cable type for long-distance phone systems. Light travels through single-mode fiber-optic cable using only a single mode, meaning it travels straight down the fiber and does not "bounce" off the cable walls. Because only a single mode of light travels through the cable, single-mode fiber-optic cable supports higher bandwidth and longer distances than multimode fiber-optic cable. Devices that use single-mode fiber-optic cable typically use lasers to generate the light that travels through the cable. Since the core size of single-mode cable is much smaller than multimode fiber, the connecting hardware and especially the lasers are much more expensive than those used for multimode fiber. As a result, single-mode based systems (cable plus electronics) are more costly than multimode systems.

ANSI/TIA-568-C.3 recognizes OSI and OS2 single-mode optical fiber cables.

Optical Fiber and Telecommunications Rooms

The ANSI/TIA-568-C standard specifies that certain features of telecommunications must be adhered to in order for the installation to be specifications-compliant:

◆ The telecommunications outlet(s) must have the ability to terminate a minimum of two fibers into 568SC couplings.

◆ To prevent damage to the fiber, the telecommunications outlet(s) must provide a means of securing fiber and maintaining a minimum bend radius of 30 millimeters.

◆ The telecommunications outlet(s) must be able to store at least one meter of two-fiber (duplex) cable.

◆ The telecommunications outlet(s) supporting fiber cable must be a surface-mount box that attaches on top of a standard 4″ × 4″ electrical box.

TIA-569-B

Although the ANSI/TIA-568-C standard describes the subsystems of a structured cabling system, the TIA has published a more thorough document called TIA-569-B Commercial Building Standard for Telecommunications Pathways and Spaces. The purpose of the TIA-569-B standard is to provide a flexible and standardized support system for a structured cabling system, along with the detail necessary to design and build these facilities. The detail pertains to both single and multitenant buildings.

NOTE   This TIA-569-B document is especially important because network managers, architects, and even cable installers often don't give enough forethought to the spaces and infrastructure that will support structured cabling systems or data communications equipment.

Although repetitive to large degree with respect to ANSI/TIA-568-C, TIA-569-B does define and detail pathways and spaces used by a commercial cabling system. The elements defined include:

◆ Entrance facility

◆ Equipment room

◆ Main terminal space

◆ Telecommunications rooms and enclosures

◆ Horizontal pathways

◆ Backbone pathways

◆ Work areas

◆ Data centers

WARNING When planning telecommunications pathways and spaces, make sure you allow for future growth.

TIA-569-B provides some common design considerations for the entrance facility, equipment room, and telecommunications rooms with respect to construction, environmental considerations, and environmental controls:

◆ One of the key features of TIA-569-B is the necessity for pathway diversity and redundancy to assure continuous operation in the case of a catastrophic event.

◆ The door (without sill) should open outward, slide sideways, or be removable. It should be fitted with a lock and be a minimum of 36 inches (.91 meters) wide by 80 inches (2 meters) high.

◆ Electrical power should be supplied by a minimum of two dedicated 120V-20A nominal, nonswitched, AC-duplex electrical outlets. Each outlet should be on separate branch circuits. The equipment room may have additional electrical requirements based on the telecommunications equipment that will be supported there (such as LAN servers, hubs, PBXs, or UPS systems).

◆ Sufficient lighting should be provided (500 lx or 50′ candles). The light switches should be located near the entrance door.

◆ Grounding should be provided and used per ANSI/TIA/EIA-607 (the Commercial Building Grounding and Bonding Requirements for Telecommunications Standard) and either the NEC or local code, whichever takes precedence.

◆ These areas should not have false (drop) ceilings.

◆ Slots and sleeves that penetrate firewalls or that are used for riser cables should be firestopped per the applicable codes.

◆ Separation of horizontal and backbone pathways from sources of electromagnetic interference (EMI) must be maintained per NEC Article 800.154.

◆ Metallic raceways and conduits should be grounded.

Based on our own experiences, we recommend the following:

◆ Equip all telecommunications rooms, the entrance facility, and the equipment room with electrical surge suppression and a UPS (uninterruptible power supply) that will supply that area with at least 15 minutes of standby AC power in the event of a commercial power failure.

◆ Equip these areas with standby lighting that will last for at least an hour if the commercial power fails.

◆ Make sure that these areas are sufficiently separated from sources of EMI such as antennas, medical equipment, elevators, motors, and generators.

◆ Keep a flashlight or chargeable light in an easy-to-find place in each of these areas in case the commercial power fails and the battery-operated lights run down.

NOTE For full information, consult the TIA-569-B standard, which may be purchased through Global Engineering Documents on the Web at http://global.ihs.com.

NOTE The Canadian equivalent of ANSI/TIA/EIA-568-B is CSA T529.

ENTRANCE FACILITY

The entrance facility is usually located either on the first floor or in the basement of a building. This facility must take into consideration the requirements of the telecommunications services and other utilities (such as CATV, water, and electrical power).

TIA-569-B specifies the following design considerations for an entrance facility:

◆ When security, continuity, or other needs dictate, an alternate entrance facility may need to be provided.

◆ One wall at a minimum should have ¾″ (20 mm) A-C plywood.

◆ It should be a dry area not subject to flooding.

◆ It should be as close to the actual entrance pathways (where the cables enter the building) as possible.

◆ Equipment not relating to the support of the entrance facility should not be installed there.

WARNING The entrance facility should not double as a storage room or janitor's closet.

CABLING @ WORK: BAD EQUIPMENT ROOM DESIGN

One company we are familiar with spent nearly a million dollars designing and building a high-tech equipment room, complete with raised floors, cabling facilities, power conditioning, backup power, and HVAC. The room was designed to be a showcase for its voice and computer systems. On the delivery day, much of the HVAC equipment could not be moved into the room because of lack of clearance in the outside hallway. Several walls had to be torn out (including the wall of an adjacent tenant) to move the equipment into the room.

Another company located its equipment room in a space that used to be part of a telecommunications room. The space had core holes drilled to the floor above, but the holes had not been filled in after the previous tenant vacated. The company installed its computer equipment but did not have the core holes filled. A few months later, a new tenant on the second floor had a contractor fill the holes. The contractor's workers poured nearly a ton of concrete down the core and on top of the computer equipment in the room below before someone realized the hole was not filling up.

Many organizations have experienced the pain of flooding from above. One company's computer room was directly below bathrooms. An overflowing toilet caused hundreds of gallons of water to spill down into the computer room. Don't let this kind of disaster occur in your equipment rooms!

MAIN TERMINAL SPACE

The main terminal space is a facility that is commonly a shared space in a multitenant building. The main cross-connects are in this room. This room is generally a combination of an equipment room and a telecommunications room, though the TIA specifies that the design for a main terminal space should follow the design considerations laid out for an equipment room. Customer equipment may or may not be located here. However, our opinion is that it is not desirable to locate your own equipment in a room shared with other tenants. One reason is that you may have to get permission from the building manager to gain access to this facility.

EQUIPMENT ROOM

Considerations to think about when designing an equipment room include the following:

◆ Environmental controls must be present to provide HVAC at all times. A temperature range of 64–75 degrees Fahrenheit (or 18–24 degrees Celsius) should be maintained, along with 30–55 percent relative humidity. An air-filtering system should be installed to protect against pollution and contaminants such as dust.

◆ Seismic and vibration precautions should be taken.

◆ The ceiling should be at least 8′ (2.4 meters) high.

- ◆ A double door is recommended. (See also door design considerations at the beginning of the section "TIA-569-B," earlier in this chapter.)

- ◆ The entrance area to the equipment room should be large enough to allow delivery of large equipment.

- ◆ The room should be above water level to minimize danger of flooding.

- ◆ The backbone pathways should terminate in the equipment room.

- ◆ In a smaller building, the entrance facility and equipment room may be combined into a single room.

TELECOMMUNICATIONS ROOMS

Here are some design considerations for telecommunications rooms, suggested by TIA-569-B:

- ◆ Each floor of a building should have at least one telecommunications room, depending on the distance to the work areas. The rooms should be close enough to the areas being served so that the horizontal cable does not exceed a maximum of 90 meters (as specified by the ANSI/TIA-568-C standard).

- ◆ Environmental controls are required to maintain a temperature that is the same as adjacent office areas. Positive pressure should be maintained in the telecommunications rooms, with a minimum of one air change per hour (or per local code).

- ◆ Ideally, closets should "stack" on top of one another in a multifloor building. Then, backbone cabling (sometimes called vertical or riser cable) between the closets merely goes straight up or down.

- ◆ Two walls of the telecommunications room must have ¾" (20 mm) A-C plywood mounted on the walls, and the plywood should be 8′ (2.4 meters) high.

- ◆ Vibration and seismic requirements should be taken into consideration for the room and equipment installed there.

- ◆ Two closets on the same floor must be interconnected with a minimum of one 78(3) trade-size conduit or equivalent pathway. The 78(3) trade-size conduit has a sleeve size of 3″, or 78 mm.

HORIZONTAL PATHWAYS

The horizontal pathways are the paths that horizontal cable takes between the wiring closet and the work area. The most common place in which horizontal cable is routed is in the space between the structural ceiling and the false (or drop) ceiling. Hanging devices such as J hooks should be secured to the structural ceiling to hold the cable. The cable should be supported at intervals not greater than 60 inches. For long runs, this interval should be varied slightly so that structural harmonics (regular physical anomalies that may coincide with transmission frequency intervals) are not created in the cable, which could affect transmission performance.

SHAKE, RATTLE, AND ROLL

A company that a friend worked for was using metal racks and shelving in the equipment rooms and telecommunications rooms. The metal racks were not bolted to the floors or supported from the ceiling. During the 1989 San Francisco earthquake, these racks all collapsed forward, taking with them hubs, LAN servers, tape units, UPSs, and disk subsystems. Had the racks been secured to the wall and ceilings, some or all of the equipment would have been saved. If you live in an area prone to earthquakes, be sure to take seismic precautions.

NOTE   Cable installers often install cable directly on the upper portion of a false ceiling. This is a poor installation practice because cable could then also be draped across fluorescent lights, power conduits, and air-conditioning ducts. In addition, the weight of cables could collapse the false ceiling. Some local codes may not permit communications cable to be installed without conduit, hangers, trays, or some other type of pathway.

WARNING   In buildings where the ceiling space is also used as part of the environmental air-handling system (i.e., as an air return), plenum-rated cable must be installed in accordance with Article 800 of the NEC.

Other common types of horizontal pathways include conduit and trays, or wireways. *Trays* are metal or plastic structures that the cable is laid into when it is installed. The trays can be rigid or flexible. *Conduit* can be metal or plastic tubing and is usually rigid but can also be flexible (in the case of fiber-optic cable, the flexible tubing is sometimes called *inner duct*). Both conduit and trays are designed to keep the cable from resting on top of the false ceiling or being exposed if the ceiling is open.

Other types of horizontal pathways include the following:

♦   Access floor, which is found in raised-floor computer rooms. The floor tile rests on pedestals, and each tile can be removed with a special tool. Some manufacturers make cable management systems that can be used in conjunction with access floors.

♦   Under floor or trenches, which are in concrete floors. They are usually covered with metal and can be accessed by pulling the metal covers off.

♦   Perimeter pathways, which are usually made of plastic or metal and are designed to mount on walls, floors, or ceilings. A pathway contains one or more cables. Many vendors make pathway equipment (see Chapter 5 for more information).

When designing or installing horizontal pathways, keep the following considerations in mind:

♦   Horizontal pathways are not allowed in elevator shafts.

♦   Make sure that the pathways will support the weight of the cable you plan to run and that they meet seismic requirements.

- Horizontal pathways should be grounded.

- Horizontal pathways should not be routed through areas that may collect moisture.

KEY TERM   *Drawstring* A *drawstring* is a small nylon cord inserted into a conduit when the conduit is installed; it assists with pulling cable through. Larger conduits will have multiple drawstrings.

BACKBONE PATHWAYS

Backbone pathways provide paths for backbone cabling between the equipment room, telecommunications rooms, main-terminal space, and entrance facility. The TIA suggests in TIA-569-B that the telecommunications rooms be stacked on top of one another from floor to floor so that cables can be routed straight up through a riser. They also recommend pathway diversity and redundancy. ANSI/TIA-568-C defines a few types of backbone pathways:

Ceiling pathways   These pathways allow the cable to be run loosely through the ceiling space.

Conduit pathways   Conduit pathways have the cable installed in a metallic or plastic conduit.

Tray pathways   These are the same types of trays used for horizontal cabling.

KEY TERM   *sleeves, slots, and cores* *Sleeves* are circular openings that are cut in walls, ceilings, and floors; a *slot* is the same but rectangular in shape. A *core* is a circular hole that is cut in a floor or ceiling and is used to access the floor above or below. Cores, slots, and sleeves cut through a floor, ceiling, or wall designed as a firestopping wall must have firestopping material inserted in the hole after the cable is installed through it.

Some points to consider when designing backbone pathways include the following:

- Intercloset conduit must be 78(3) trade size (3″ or 78 mm sleeve).

- Backbone conduit must be 103(4) trade size (4″ or 103 mm sleeve).

- Firestopping material must be installed where a backbone cable penetrates a firewall (a wall designed to stop or hinder fire).

- Trays, conduits, sleeves, and slots need to penetrate at least 1″ (25 mm) into telecommunication rooms and equipment rooms.

- Backbone cables should be grounded per local code, the NEC, and ANSI/TIA-607-B.

- Backbone pathways should be dry and not susceptible to water penetration.

WARNING   Devices such as cable trays, conduit, and hangers must meet requirements of the NEC with regard to their placement. For example, flexible-metal conduit is not allowed in plenum spaces except under restricted circumstances.

WORK AREAS

TIA-569-B recommendations for work areas include the following:

♦ A power outlet should be nearby but should maintain minimum power/telecommunications separation requirements (see NEC Article 800-154 for specific information).

♦ Each work area should have at least one telecommunications outlet box. ANSI/TIA-568-C recommends that each telecommunications outlet box have a minimum of two outlets (one for voice and one for data).

♦ For voice applications, the PBX control center, attendant, and reception areas should have independent pathways to the appropriate telecommunications rooms.

♦ The minimum bend radius of cable should not be compromised at the opening in the wall.

TIA-569-B also makes recommendations for wall openings for furniture pathways.

ANSI/TIA-607-B

The ANSI/TIA-607-B Commercial Building Grounding and Bonding Requirements for Telecommunications Standard covers grounding and bonding to support a telecommunications system. This document should be used in concert with Article 250 and Article 800 of the NEC. ANSI/TIA-607-B does *not* cover building grounding; it only covers the grounding of telecommunications systems.

ANSI/TIA-607-B specifies that the telecommunications ground must tie in with the building ground. Each telecommunications room must have a telecommunications grounding system, which commonly consists of a *telecommunications bus bar* tied back to the building grounding system. All shielded cables, racks, and other metallic components should be tied into this bus bar.

ANSI/TIA-607-B specifies that the minimum ground-wire size must 6 AWG, but, depending on the distance that the ground wire must cover, it may be up to 3/0 AWG (a pretty large copper wire!). Ground-wire sizing is based on the distance that the ground wire must travel; the farther the distance, the larger the wire must be. ANSI/TIA-607-B supplements (and is supplemented by) the NEC. For example, Article 800-100 specifies that telecommunications cables *entering* a building must be grounded as near as possible to the point at which they enter the building.

WARNING   When protecting a system with building ground, don't overlook the need for lightning protection. Network and telephone components are often destroyed by a lightning strike. Make sure your grounding system is compliant with the NEC.

Grounding is one of the most commonly overlooked components during the installation of a structured cabling system. An improperly grounded communications system, although supporting low-voltage applications, can result in, well, a shocking experience. Time after time we have heard stories of improperly grounded (or ungrounded) telecommunications cabling systems that have generated mild electrical or throw-you-off-your-feet shocks; they have even resulted in some deaths.

Grounding is not to be undertaken by the do-it-yourselfer or an occasional cable installer. A professional electrician *must* be involved. He or she will know the best practices to follow, where to ground components, which components to ground, and the correct equipment to be used.

Further, electricians must be involved when a telecommunications bus bar is tied into the main building ground system.

WARNING   Grounding to a water pipe may not provide you with sufficient grounding, as many water systems now tie in to PVC-based (plastic) pipes. It may also violate NEC and local-code requirements.

ANSI/TIA-570-B

ANSI and TIA published ANSI/TIA-570-B-2004, or the Residential and Light Commercial Telecommunications Cabling Standard, to address the growing need for "data-ready" homes and multidwelling residential buildings. Just a few years ago, only the most serious geeks would have admitted to having a network in their homes. Today, more and more homes have small networks consisting of two or more home computers, a cable modem, and a shared printer. Even apartment buildings and condominiums are being built or remodeled to include data outlets; some apartment buildings and condos even provide direct Internet access.

CABLING @ WORK: AN EXAMPLE OF POOR GROUNDING

One of the best examples we can think of that illustrates poor grounding practices was a very large building that accidentally had two main grounds installed, one on each side. (A building should only have one main ground.) A telecommunications backbone cable was then grounded to each main ground.

Under some circumstances, a ground loop formed that caused this cable to emit EMI at specific frequencies. This frequency just so happened to be used by air traffic control beacons. When the building cable emitted signals on this frequency, it caused pilots to think they were closer to the airport than they really were. One plane almost crashed as a result of this poorly grounded building. The FAA (Federal Aviation Administration) and the FCC (Federal Communications Commission) closed the building and shut down all electrical systems for weeks until the problem was eventually found.

The ANSI/TIA-570-B standard provides requirements for residential telecommunications cabling for two grades of information outlets: basic and multimedia cabling. This cabling is intended to support applications such as voice, data, video, home automation, alarm systems, environmental controls, and intercoms. The two grades are as follows:

Grade 1   This grade supports basic telephone and video services. The standard specifies twisted-pair cable and coaxial cable placed in a star topology. Grade 1 cabling requirements consist of a minimum of one four-pair UTP cable that meets or exceeds the requirements for Category 5e, a minimum of one 75-ohm coaxial cable, and their respective connectors at each telecommunications outlet and the DD. Installation of Category 6 cable in place of Category 5e cable is recommended.

Grade 2   Grade 2 provides a generic cabling system that meets the minimum requirements for basic and advanced telecommunications services such as high-speed Internet and in-home generated video. Grade 2 specifies twisted-pair cable, coaxial cable, and optionally optical fiber cable, all placed in a star topology. Grade 2 cabling minimum requirements consist of two four-pair UTP cables and associated connectors that meet or exceed the requirements for Category 5e cabling; two 75-ohm coaxial and associated connectors at each telecommunications outlet and the DD; optionally, two-fiber optical fiber cabling. Installation of Category 6 cabling in place of Category 5e cabling is recommended.

The standard further dictates that a central location within a home or multitenant building be chosen at which to install a central cabinet or wall-mounted rack to support the wiring. This location should be close to the telephone company demarcation point and near the entry point of cable TV connections. Once the cabling system is installed, you can use it to connect phones, televisions, computers, cable modems, and EIA 6000–compliant home automation devices.

ANSI/TIA-942

The ANSI/TIA-942 standard provides requirements and guidelines for the design and installation of a data center or computer room. ANSI/TIA-942 enables the data center design to be taken into account early in the building development process, contributing to the architectural considerations, by providing information that cuts across the multidisciplinary design efforts and thus promoting cooperation in the design and construction.

Other TIA/EIA Standards and Bulletins

The TIA/EIA alliance published additional specifications and bulletins relating to data and voice cabling as well as performance testing.

If you want to keep up on the latest TIA/EIA specifications and the work of the various committees, visit the TIA website at www.tiaonline.org and go to the TR-42 page.

ISO/IEC 11801

The International Organization for Standardization (ISO) and the International Electrotechnical Commission (IEC) publish the ISO/IEC 11801 standard predominantly used in Europe. This standard was released in 1995 and is similar in many ways to the ANSI/TIA-568-C standard upon which it is based. The second edition was released in 2002 and is largely in harmony with ANSI/TIA-568-C. However, the ISO/IEC 11801 standard has a number of differences in terminology. Table 2.5 shows the common codes and elements of an ISO/IEC 11801 structured cabling system.

Differences Between ANSI/TIA-568-C and ISO/IEC 11801

Differences between ANSI/TIA-568-C and ISO/IEC 11801 Ed. 2 include the following:

◆ ISO/IEC 11801 allows for an additional media type for use with backbone and horizontal cabling and 120 ohm UTP.

◆ The term *consolidation point* is much broader in ISO/IEC 11801; it includes not only transition points for under-carpet cable to round cable (as defined by ANSI/TIA-568-C), but also consolidation point connections.

TABLE 2.5   Common Codes and Elements Defined by ISO/IEC 11801

ELEMENT	CODE	DESCRIPTION
Building distributor	BD	A distributor in which building-to-building backbone cabling terminates and where connections to interbuilding or campus backbone cables are made.
Building entrance facilities	BEF	Location provided for the electrical and mechanical services necessary to support telecommunications cabling entering a building.
Campus distributor	CD	Distributor location from which campus backbone cabling emanates.
Equipment room	ER	Location within a building dedicated to housing distributors and application-specific equipment.
Floor distributor	FD	A distributor used to connect horizontal cable to other cabling subsystems or equipment.
Horizontal cable	HC	Cable from the floor distributor to the telecommunications outlet.
Telecommunications room	TC	Cross-connection point between backbone cabling and horizontal cabling. May house telecommunications equipment, cable terminations, cross-connect cabling, and data networking equipment.
Telecommunications outlet	TO	The point where the horizontal cabling terminates on a wall plate or other permanent fixture. The point is an interface to the work area cabling.
Consolidation point	CP	The location in horizontal cabling where a cable may end, which is not subject to moves and changes, and another cable starts leading to a telecommunications outlet, which easily adapts to changes.
Work-area cable	None	Connects equipment in the work area (phones, computers, etc.) to the telecommunications outlet.

ISO/IEC 11801 Ed. 2 specifies a maximum permanent link length of 90 meters and a maximum channel link of 100 meters. Patch and equipment cord maximum lengths may be adjusted by formulas depending on the actual link lengths. Terminology differences between ANSI/TIA-568-C and ISO/IEC 11801 Ed. 2 include the following:

◆ The ISO/IEC 11801 Ed. 2 definition of the campus distributor (CD) is similar to the ANSI/TIA-568-C definition of a main cross-connect (MC).

◆ The ISO/IEC 11801 Ed. 2 definition of a building distributor (BD) is equal to the ANSI/TIA-568-C definition of an intermediate cross-connect (IC).

◆ The ISO/IEC 11801 Ed. 2 definition of a floor distributor (FD) is defined by ANSI/TIA-568-C as the horizontal cross-connect (HC).

Classification of Applications and Links

ISO/IEC 11801 Ed. 2 defines classes of applications and links based on the type of media used and the frequency requirements. ISO/IEC 11801 Ed. 2 specifies the following classes or channels of applications and links:

Class A   For voice and low-frequency applications up to 100kHz.

Class B   For low-speed data applications operating at frequencies up to 1MHz.

Class C   For medium-speed data applications operating at frequencies up to 16MHz.

Class D   Concerns high-speed applications operating at frequencies up to 100MHz.

Class E   Concerns high-speed applications operating at frequencies up to 250MHz.

Class E_A   Concerns high-speed applications operating at frequencies up to 500MHz.

Class F   Concerns high-speed applications operating at frequencies up to 600MHz.

Class F_A   Concerns high-speed applications operating at frequencies up to 1000MHz.

Optical Class   An optional class for applications where bandwidth is not a limiting factor.

The Bottom Line

Identify the key elements of the ANSI/TIA-568-C Commercial Building Telecommunications Cabling Standard.   The ANSI/TIA-568-C Commercial Building Telecommunications Cabling Standard is the cornerstone design requirement for designing, installing and testing a commercial cabling system. It is important to obtain a copy of this standard and to keep up with revisions. The 2009 standard is divided into several sections, each catering to various aspects.

Master It

1. Which subsection of the ANSI/TIA-568-C standard would you reference for UTP cabling performance parameters?

2. Which subsection of the ANSI/TIA-568-C standard would you reference for optical fiber cabling and component performance parameters?

3. What is the recommended multimode fiber type per the ANSI/TIA-568-C.1 standard for backbone cabling?

4. Which is typically a more expensive total optical fiber system solution: single-mode or multimode?

5. The ANSI/TIA-568-C.1 standard breaks structured cabling into six areas. What are these areas?

Identify other ANSI/TIA/EIA standards required to properly design the pathways and spaces and grounding of a cabling system. The ANSI/TIA-568-C Commercial Building Telecommunications Cabling Standard is necessary for designing, installing, and testing a cabling system, but there are specific attributes of the pathways and spaces of the cabling systems in ANSI/TIA-568-C that must be considered. In addition, these systems need to follow specific grounding regulations. It is just as important to obtain copies of these standards.

Master It Which other TIA standards need to be followed for:

1. Pathways and spaces?

2. Grounding and bonding?

3. Data centers?

Identify key elements of the ISO/IEC 11801 Generic Cabling for Customer Premises Standard. The International Organization for Standardization (ISO) and the International Electrotechnical Commission (IEC) publish the ISO/IEC 11801 standard predominantly used in Europe. It is very similar to ANSI/TIA-568-C, but they use different terminology for the copper cabling. It is important to know the differences if you will be doing any work internationally.

Master It What are the ISO/IEC 11801 copper media classifications and their bandwidths (frequency)?

Chapter 3

Choosing the Correct Cabling

Technically, when you begin the planning stages of a new cabling installation, you should not have to worry about the types of applications used. The whole point of structured cabling standards such as ANSI/TIA-568-C and ISO/IEC 11801 Ed. 2 is that they will support almost any networking or voice application in use today.

Still, it is a good idea to have an understanding of the networking application you are cabling for and how that can affect the use of the cabling system.

In this chapter, you will learn to:

◆ Identify important network topologies for commercial buildings

◆ Understand the basic differences of UTP and optical fiber cabling and their place in future-proofing networks

◆ Identify key network applications and the preferred cabling media for each

Topologies

The network's *topology* refers to the physical layout of the elements of the telecommunications cabling system structure (e.g., main equipment room, cross-connections, telecommunications rooms and enclosures, switches, and hubs) that make up the network. Choosing the right topology is important because the topology affects the type of networking equipment, cabling, growth path, and network management.

Today's networking architectures fall into one of three categories:

◆ Hierarchical star

◆ Bus

◆ Ring

Topologies are tricky because some networking architectures appear to be one type of technology but are in reality another. Token Ring is a good example of this because Token Ring uses hubs (multistation access units or MAUs). All stations are connected to a central hub, so physically it is a hierarchical star topology; logically, though, it is a ring topology. Often two topology types will be used together to expand a network.

> ## PHYSICAL VS. VIRTUAL TOPOLOGY
>
> *Topology* refers to the physical layout of the wiring and key connection points of a network, and it also refers to the network's method of transmitting data and to the *logical*, or *virtual*, layout of its connection points. Before the advent of structured wiring, physical and logical topology were often the same. For example, a network that had a ring topology actually had the wiring running from point to point in a ring. This can be confusing these days. The implementation of structured wiring standardized a hierarchical star configuration as the physical topology for modern networks, and network electronics take care of the logical topologies.

NOTE   *Topology* and *architecture* are often used interchangeably. They are not exactly synonymous but are close enough for purposes of this book.

Hierarchical Star Topology

When implementing a hierarchical *star topology*, all computers are connected to a single, centrally located point. This central point is usually a hub of servers and switches located in the main equipment room and interconnected through the main cross-connection. All cabling used in a hierarchical star topology is run from the equipment outlets back to this central location. Typically in commercial buildings, there is a horizontal cross-connection with a workgroup switch located in a telecommunications room (TR) that allows backbone cabling to interconnect with horizontal cabling. Usually, the utilization of switch ports is low. In other words, only 70–80 percent of the ports in a switch are connected. Essentially, you are paying for more ports than you are using.

A lower-cost method involves placing horizontal cross-connections and workgroup switches in telecommunications enclosures. This is standardized in ANSI/TIA-568-C and is commonly referred to as fiber-to-the-telecommunications enclosure (FTTE). These house mini-patch panels and switches and are located in enclosures, installed overhead or on wall space, very close to clusters of equipment outlets. One benefit of this implementation of the hierarchical star is that the utilization of switch ports is typically 90 percent or greater. Another benefit is that TRs can be smaller, reducing power and HVAC requirements since TRs do not house the equipment and patch panels. Studies performed by TIA's Fiber Optic LAN Section have shown that the reduction of TR space and utility requirements, combined with lower costs in switches and ports, can lead to savings of 25 percent or more compared to a traditional implementation of hierarchical star where switches are located in TRs. One practical disadvantage of FTTE that has been raised by some users is the need to service equipment out in the work space environment (for example, above office cubes) as opposed to in a TR.

Another alternative implementation of a hierarchical star topology per ANSI/TIA-568-C is called *centralized cabling*. Centralized cabling is a hierarchical star topology that extends from the main cross-connection in an equipment room directly to an equipment outlet by allowing a cable to be pulled through a telecommunications room (or enclosure) without passing through a switch. The cable can be a continuous sheath of cable from the equipment room, or two separate cables may be spliced or interconnected in the TR. In either case, there is no need to use a workgroup switch in a TR to interconnect a backbone cable to a horizontal cable since all electronics are centralized in the main equipment room. This subset of the hierarchical star topology is commonly referred to as fiber-to-the-desk (FTTD) since it employs fiber to support the greater than 100 meter distances from the main equipment room/cross-connection to the equipment/ telecommunications outlet. However, media converters or optical NICs are required to convert from optical to electrical near the equipment outlets. The centralized cabling topology can produce savings by reducing the size of TRs and need for HVAC since no active equipment is used in the TR. TIA's Fiber Optic LAN Section (www.fols.org) has studied this topology in detail and provides good information on its benefits. Figure 3.1 and Figure 2.5 (in Chapter 2) show a simple hierarchical star topology using a central hub to support centralized cabling/FTTD.

FIGURE 3.1
Hierarchical star topology with a central hub to support FTTD

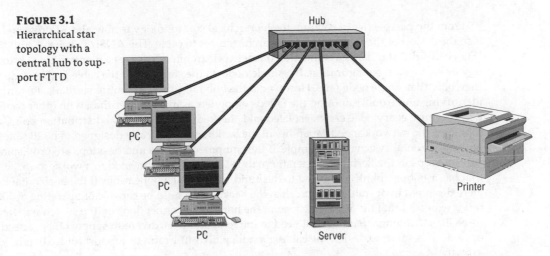

CABLING @ WORK: HIGH SCHOOL DECIDES TO USE OPTICAL FIBER CABLING

A new school recently designed a new communications network. They decided to use an optical fiber network. Why? Let's take a look.

WHY ABC HIGH USES OPTICAL FIBER CABLING

ABC High chose to use optical fiber cabling for two primary reasons:

Link lengths   By taking advantage of the long data runs supported by multimode fiber, network planners were able to consolidate electronics into a single, centrally located wiring closet. This approach drastically reduced the amount of electronics needed for the installation and lowered the cost of (and simplified) network maintenance.

Network longevity   School administrators realized that additional "headroom" would become essential in future programs and/or applications. This cabling system provides a quick and easy way to upgrade the system in years to come.

INSTALLATION DETAILS

The new high school has 32 classrooms and accommodates approximately 650 students in grades 9–12. School administrators chose a centralized cabling network topology with a switch serving as the backbone.

Within the building, a single, centrally located wiring closet forms the heart of the network cabling plant, with OM3 50/125 micron (per ISO 11801 Ed. 2) multimode fiber links radiating to each of the classrooms and other areas throughout the building. There are at least seven network drops in each classroom.

The use of small form-factor connectors is valuable as it allows large numbers of fibers to be terminated in a small panel space.

From the perspective of cabling, the hierarchical star topology is now almost universal. It is also the easiest of the three networking architectures to cable. The ANSI/TIA-568-C and ISO/IEC 11801 Ed. 2 standards assume that the network architecture uses a hierarchical star topology as its physical configuration. If a single node on the star fails or the cable to that node fails, then only that single node fails. However, if the hub fails, then the entire star fails. Regardless, identifying and troubleshooting the failed component is much easier than with other configurations because every node can be isolated and checked from the central distribution point. In this topology, the network speed capability of the backbone is typically designed to be 10 times that of the horizontal system. For example, if the equipment outlets and desktops are provisioned to run 100Mbps in the horizontal, then these links are usually connected to a workgroup switch that has a 1Gbps uplink connected to the building riser backbone cable. If the horizontal links were designed to operate at 1Gbps, then the backbone should be capable of operating at 10Gbps if it is expected that the aggregate load of the horizontal connections on the switch will be close to 10Gbps. Although you may not need to implement the hardware to support these speeds on day one, it is smart to design the cabling media with a 10:1 ratio to support future bandwidth growth.

From this point on in the chapter, we will assume you understand that the physical layout of a modern network is a hierarchical star topology and that when we discuss bus and ring topologies we're referring to the logical layout of the network.

KILLING AN ENTIRE HIERARCHICAL STAR TOPOLOGY

Although a single node failure cannot usually take down an entire star topology, sometimes it can. In some circumstances, a node fails and causes interference for the entire star. In other cases, shorts in a single cable can send disruptive electrical signals back to the hub and cause the entire star to cease functioning. Of course, failure of the hub will also affect all nodes in a star topology. (Just as you need diversity and redundancy in cabling/network paths, you need diversity and redundancy in your switches as well.)

Bus Topology

The *bus topology* is the simplest network topology. Also known as a *linear bus*, in this topology all computers are connected to a contiguous cable or a cable joined together to make it contiguous. Figure 3.2 illustrates a bus topology.

FIGURE 3.2
Bus topology

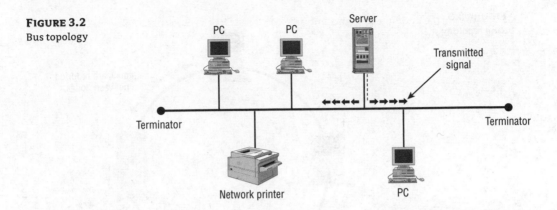

Ethernet is a common example of a bus topology. Each computer determines when the network is not busy and transmits data as needed. Computers in a bus topology listen only for transmissions from other computers; they do not repeat or forward the transmission on to other computers.

The signal in a bus topology travels to both ends of the cable. To keep the signal from bouncing back and forth along the cable, both ends of the cable in a bus topology must be terminated. A component called a *terminator*, essentially nothing more than a resistor, is placed on both ends of the copper (coax) cable. The terminator absorbs the signal and keeps it from ringing, which is also known as overshoot or resonance; this is referred to as *maximum impedance*. If either terminator is removed or if the cable is cut anywhere along its length, all computers on the bus will fail to communicate.

Coaxial cabling was most commonly used in true bus-topology networks such as thin/thick Ethernet. However, UTP-based cabling Ethernet applications like 10Base-T, 100Base-T, and above still function as if they were a bus topology even though they are wired as a hierarchical star topology.

Ring Topology

A *ring topology* requires that all computers be connected in a contiguous circle, as shown in Figure 3.3. The ring has no ends or hub. Each computer in the ring receives signals (data) from its neighbor, repeats the signal, and passes it along to the next node in the ring. Because the signal has to pass through each computer on the ring, a single node or cable failure can take the entire ring down.

A true ring topology is a pain in the neck to install cable for because the circular nature of the ring makes it difficult to expand a ring over a large physical area. Token Ring is a ring topology. Even though Token Ring stations may be connected to a central MAU (and thus appear to be a star topology), the data on the Token Ring travels from one node to another. It passes through the MAU each time.

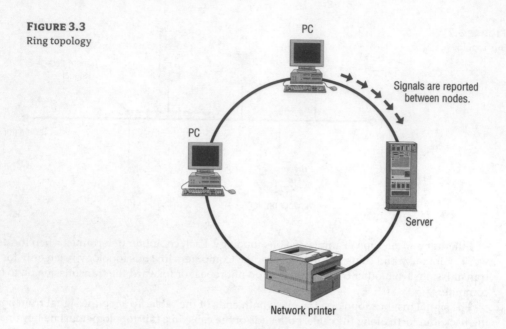

FIGURE 3.3
Ring topology

UTP, Optical Fiber, and Future-Proofing

The common networking technologies today (Ethernet, Token Ring, FDDI, and ATM) can all use either UTP or optical fiber cabling, and IT professionals are faced with the choice. MIS managers and network administrators hear much about "future-proofing" their cabling infrastructures. The claim is that installing particular grades of cable and components will guarantee that you won't have to ever update your cabling system again. However, you should keep in mind that in the early 1990s network managers thought they were future-proofing their cabling system when they installed Category 4 rather than Category 3 cabling.

Today, decision makers who must choose between Category 6 and 6A cabling components are thinking about future-proofing. Each category is an improvement in potential data throughput and therefore a measure of future-proofing. Deciding whether to use optical fiber adds to the decision. Here are some of the advantages of using optical fiber:

◆ It has much higher potential bandwidth, which means that the data throughput is much greater than with copper cable. Optical fiber cable has the *potential* for higher bandwidth, because it requires a transceiver to deliver the bandwidth. If the highest bandwidth optical transceiver currently available is 10GB, then its actual bandwidth is no better 10GBase-T, notwithstanding the distance capability.

◆ It's not susceptible to electromagnetic interference.

◆ It can transmit over longer distances, which is useful for centralized cabling topologies and backbone cabling, although distance is set at 100 meters for the run of cable from the TR to equipment outlets regardless of media, according to ANSI/TIA-568-C.

◆ Optical fiber also allows the use of telecommunications enclosures, since it can support longer backbone distances than UTP. This essentially places the switches closer to the equipment outlets and can provide savings of approximately 25 percent over the use of switches in TRs. (For more information, see the TIA Fiber Optics LAN Section at www.fols.org.)

◆ Improved termination techniques and equipment make it easier to install and test.

◆ Cable, connectors, and patch panels are now cheaper than before.

◆ It's valuable in situations where EMI is especially high.

◆ It offers better security (because the cable cannot be easily tapped or monitored).

UTP cabling is still popular in a traditional hierarchical topology where an intermediate switch is used in a TR, and you may want to consider remaining with UTP cabling for the following reasons:

◆ The TIA estimates that the combined installation and hardware costs result in a finished centralized cabling fiber optic network that is 30 percent more expensive than a Category 5e or 6 copper cable network using a traditional hierarchical star topology. However, these cost differences are expected to decrease with time.

◆ If higher bandwidth (more than a gigabit per second) requirements are not an issue for you, you may not need optical fiber.

◆ Fiber optics is the medium of choice for security only if security concerns are unusually critical.

◆ EMI interference is only an issue if it is extreme.

Fiber-optic cabling and transmission media are likely to outpace copper for 100 meter links as speeds increase; however, when considering optical fiber cable, remember that you are trying to guarantee that the cabling system will not have to be replaced for a very long time, regardless of future networking technologies. Some questions you should ask yourself when deciding if fiber optic is right for you include the following:

◆ Do you rent or own your current location?

◆ If you rent, how long is your lease, and will you be renewing your lease when it is up?

◆ Are there major renovations planned that would cause walls to be torn out and rebuilt?

As network applications are evolving, better UTP and optical fiber cabling media are required to keep up with bandwidth demand. As you will see from standards, the end user has many options in a media category. There are many types of UTP and optical fiber cabling. Standards will continue to evolve, but it's always a good idea to install the best grade of cabling since the cost of the structured cabling systems (excluding installation cost) is usually only 5–10 percent of the total project cost. Therefore, making the right decisions today can greatly future-proof the network.

Network Applications

The ANSI/TIA-568-C cabling standard covers most combinations of cable necessary to take advantage of the current network applications found in the business environment. These network applications include Ethernet, Token Ring, Fiber Distributed Data Interface (FDDI), and Asynchronous Transfer Mode (ATM). Although the predominant cabling infrastructure is UTP, many of these applications are capable of operating on optical fiber media as well. Understanding the different types of cable that these architectures utilize is important.

Ethernet

Ethernet is the most mature and common of the network applications. According to technology analysts IDC (International Data Corporation, 2007), Ethernet is used in over 80 percent of all network installations.

In some form, Ethernet has been around for over 30 years. A predecessor to Ethernet was developed by the University of Hawaii (and was called, appropriately, the Alohanet) to connect geographically dispersed computers. This radio-based network operated at 9600Kbps and used an access method called CSMA/CD (Carrier Sense Multiple Access/Collision Detection), in which computers "listened" to the cable and transmitted data if there was no traffic. If two computers transmitted data at exactly the same time, the nodes needed to detect a collision and retransmit the data. Extremely busy CSMA/CD-based networks became very slow because collisions were excessive.

In the early 1970s, Robert Metcalfe and David Boggs, scientists at Xerox's Palo Alto Research Center (PARC), developed a cabling and signaling scheme that used CSMA/CD and was loosely based on the Alohanet. This early version of Ethernet used coaxial cable and operated at 2.94Mbps. Even early on, Ethernet was so successful that Xerox (along with Digital Equipment Corporation and Intel) updated it to support 10Mbps. Ethernet was the basis for the IEEE 802.3 specification for CSMA/CD networks.

NOTE Ever seen the term *DIX*? Or *DIX connector*? DIX is an abbreviation for Digital, Intel, and Xerox. The DIX connector is also known as the AUI (attachment unit interface), which is the 15-pin connector that you see on older Ethernet cards and transceivers.

Over the past 25 years, despite stiff competition from more modern network architectures, Ethernet has flourished. In the past 10 years alone, Ethernet has been updated to support speeds of 100Mbps, 1Gbps (about 1000Mbps) and 10Gbps; currently 40 and 100 Gigabit Ethernet are being standardized in the IEEE 802.3ba committee. Forty and 100 Gigabit Ethernet will be deployed over optical fiber for 100 meters or greater, and research is progressing to make it available over UTP for distances up to 10 meters.

Ethernet has evolved to the point that it can be used on a number of different cabling systems. Table 3.1 lists some of the Ethernet technologies. The first number in an Ethernet designator indicates the speed of the network, the second portion (the *base* portion) indicates baseband, and the third indicates the maximum distance or the media and transmitter type.

TABLE 3.1   Cracking the Ethernet Designation Codes

DESIGNATION	DESCRIPTION
10Base-2	10Mbps Ethernet over thinnet (50 ohm) coaxial cable (RG-58) with a maximum segment distance of 185 meters (it was rounded up to 10Base-2 instead of 10Base185).
10Base-5	10Mbps Ethernet over thick (50 ohm) coaxial cable with a maximum segment distance of 500 meters.
10Broad-36	A 10Mbps broadband implementation of Ethernet with a maximum segment length of 3,600 meters.
10Base-T	10Mbps Ethernet over unshielded twisted-pair cable. Maximum cable length (network device to network card) is 100 meters.

TABLE 3.1   Cracking the Ethernet Designation Codes   *(CONTINUED)*

DESIGNATION	DESCRIPTION
10Base-FL	10Mbps Ethernet over multimode optical fiber cable. Designed for connectivity between network interface cards on the desktop and a fiber-optic Ethernet hub. Maximum cable length (hub to network card) is 2,000 meters.
10Base-FB	10Mbps Ethernet over multimode optical fiber cable. Designed to use a signaling technique that allows a 10Base-FB backbone to exceed the maximum number of repeaters permitted by Ethernet. Maximum cable length is 2,000 meters.
10Base-FP	10Mbps Ethernet over multimode optical fiber cable designed to allow linking multiple computers without a repeater. Not commonly used. Maximum of 33 computers per segment, and the maximum cable length is 500 meters.
100Base-TX	100Mbps Ethernet over Category 5 or better UTP cabling using two wire pairs. Maximum cable distance is 100 meters.
100Base-T2, 100Base-T4	100Mbps Ethernet over Category 3 or better UTP. T2 uses two cable pairs; T4 uses four cable pairs. Maximum distance using Category 3 cable is 100 meters.
100Base-FX	100Mbps Ethernet over multimode optical fiber cable. Maximum cable distance is 400 meters.
100Base-VG	More of a first cousin of Ethernet. This is actually 100VG-AnyLAN, which is described later in this chapter.
1000Base-SX	Gigabit Ethernet over multimode optical fiber cable, designed for workstation-to-hub implementations using 850nm short-wavelength light sources.
1000Base-LX	Gigabit Ethernet over single-mode optical fiber cable, designed for backbone implementations using 1310nm long-wavelength light sources.
1000Base-CX	Gigabit Ethernet over STP Type 1 cabling designed for equipment interconnection such as clusters. Maximum distance is 25 meters.
1000Base-T	Gigabit Ethernet over Category 5 or better UTP cable where the installation has passed performance tests specified by ANSI/TIA/EIA-568-B. Maximum distance is 100 meters from equipment outlet to switch.
1000Base-TX	Gigabit Ethernet over Category 6 cable. Maximum distance is 100 meters from equipment outlet to switch.
10Gbase-	10 Gigabit Ethernet over optical fiber cable. Several implementations exist, designated as -SR, -LR, -ER, -SW, -LW, or -EW, depending on the light wavelength and transmission technology employed. See Table 6 of Annex D in ANSI/TIA-568-C.0 for distances.
10Gbase-T	10 Gigabit Ethernet over Category 6A copper cable. Maximum distance is 100 meters from equipment outlet to switch. Shorter distances are allowed over Category 6.

KEY TERM *baseband and broadband* *Baseband* network equipment transmits digital information (bits) using a single analog signal frequency. *Broadband* networks transmit the bits over multiple signal frequencies. Think of a baseband network as a single-channel TV set. The complete picture is presented to you on one channel. Think of a broadband network as one of those big matrix TV displays, where parts of the picture are displayed on different sets within a rectangular grid. The picture is being split into pieces and, in effect, transmitted over different channels where it is reassembled for you to see. The advantage of a broadband network is that much more data throughput can be achieved, just as the advantage of the matrix TV display is that a much larger total picture can be presented.

10Mbps Ethernet Systems

Why is Ethernet so popular? The reason is that on a properly designed and cabled network, Ethernet is fast, easy to install, reliable, and inexpensive. Ethernet can be installed on almost any type of structured cabling system, including unshielded twisted-pair and fiber-optic cable. Please be aware that 10Mbps systems are rarely installed today, but we will provide some historical information.

10Base-5: "Standard Ethernet Cable"

The earliest version of Ethernet ran on a rigid coaxial cable that was called Standard Ethernet cable, but was more commonly referred to as *thicknet*. To connect a node to the thicknet cable, a specially designed connector was attached to the cable (called a *vampire tap* or *piercing tap*).

When the connector was tightened down onto the cable, the tap pierced the jacket, shielding, and insulation to make contact with the inner core of the cable. This connector had a transceiver attached, to which a transceiver cable (or drop cable) was linked. The transceiver cable connected to the network node.

While thicknet was difficult to work with (because it was not very flexible and was hard to install and connect nodes to), it was reliable and had a usable cable length of 500 meters (about 1,640´). That is where the *10* and *5* in 10Base-5 come from: *10Mbps, baseband, 500 meters*.

Although you never see new installations of 10Base-5 systems anymore, it can still be found in older installations, typically used as backbone cable. The 10Base-T hubs and coaxial (thinnet) cabling are attached at various places along the length of the cable. Given the wide availability of fiber-optic equipment and inexpensive hubs and UTP cabling, virtually no reason exists for you to install a new 10Base-5 system today.

10Base-T Ethernet

For over 10 years, 10Base-T (the *T* stands for *twisted pair*) Ethernet reigned as king of the network applications; however, it is less common today and has been overtaken by 100Base-T.

TIP Even though 10Base-T uses only two pairs of a four-pair cable, all eight pins should be connected properly in anticipation of future upgrades or other network architectures.

TIP If you are cabling a facility for 100Base-T, plan to use, at a minimum, Category 5e cable and components. The incremental price is only slightly higher than Category 3, and you will provide a growth path to faster network technologies. If you've got the cabling in place to handle it, it's hard to say no to 10 times your current bandwidth when the only obstacles in the way are inexpensive hubs and NICs.

Here are some important facts about 10Base-T:

◆ The maximum cable length of a 10Base-T segment is 100 meters (328´) when using Category 3 cabling. Somewhat longer distances may be achieved with higher grades of equipment, but remember that you are no longer following the standard if you attempt to stretch the distance.

◆ The minimum length of a 10Base-T cable (node to hub) is 2.5 meters (about 8´).

◆ A 10Base-T network can have a maximum of 1,024 computers on it; however, performance may be extremely poor on large networks.

◆ For older network devices that have only AUI-type connectors, transceivers can be purchased to convert to 10Base-T.

◆ Although a 10Base-T network appears to operate like a star topology, internally it is a bus architecture. Unless a technology like switching or bridging is employed, a signal on a single network segment will be repeated to all nodes on the network.

◆ 10Base-T requires only two wire pairs of an eight-pin modular jack. Figure 3.4 shows the pin layout and usage.

FIGURE 3.4
An eight-pin
modular jack used
with 10Base-T

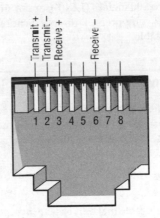

10Base-F Ethernet

Specifications for using Ethernet over fiber-optic cable existed back in the early 1980s. Originally, fiber-optic cable was simply used to connect repeaters whose separation exceeded the distance limitations of thicknet cable. The original specification was called Fiber-Optic Inter-Repeater Link (FOIRL), which described linking two repeaters together with fiber-optic cable up to 1,000 meters (3,280´) in length.

NOTE   Unless stated otherwise, all fiber-optic devices are assumed here to use multimode optical fiber cable.

The cost of fiber-optic repeaters and fiber-optic cabling dropped greatly during the 1980s, and connecting individual computers directly to the hub via fiber-optic cable became more common. Originally, the FOIRL specification was not designed with individual computers in mind, so the IEEE developed a series of fiber-optic media specifications. These specifications are collectively known as 10Base-F. It is uncommon to use optical fiber at these slow speeds today. For historical purposes, the individual specifications for (and methods for implementing) 10Base-F Ethernet include the following:

10Base-FL   This specification is an updated version of the FOIRL specification and is designed to interoperate with existing FOIRL equipment. Maximum distance used between 10Base-FL and an FOIRL device is 1,000 meters, but it is 2,000 meters (6,561′) between two 10Base-FL devices. The 10Base-FL is most commonly used to connect network nodes to hubs and to interconnect hubs. Most modern Ethernet equipment supports 10Base-FL; it is the most common of the 10Base-F specifications.

10Base-FB   The 10Base-FB specification describes a synchronous signaling backbone segment. This specification allows the development of a backbone segment that exceeds the maximum number of repeaters that may be used in a 10Mbps Ethernet system. The 10Base-FB is available only from a limited number of manufacturers and supports distances of up to 2,000 meters.

10Base-FP   This specification provides the capability for a fiber-optic mixing segment that links multiple computers on a fiber-optic system without repeaters. The 10Base-FP segments may be up to 500 meters (1,640′), and a single 10Base-FP segment (passive star coupler) can link up to 33 computers. This specification has not been adopted by many vendors and is not widely available.

WHY USE 10BASE-FL?

In the past, fiber-optic cable was considered expensive, but it is becoming more affordable. In fact, fiber-optic installations are becoming nearly as inexpensive as UTP copper installations. The major point that causes some network managers to cringe is that the network equipment can be more expensive.

However, fiber-optic cable, regardless of the network architecture, has key benefits for many businesses:

◆ Fiber-optic cable makes it easy to incorporate newer and faster technologies in the future.

◆ Fiber-optic cable is not subject to electromagnetic interference, nor does it generate interference.

◆ Fiber-optic cable is difficult to tap or monitor for signal leakage, so it is more secure.

◆ Potential data throughput of fiber-optic cable is greater than any current or forecast copper technologies.

So fiber-optic cable is more desirable for customers who are concerned about security, growth, or electromagnetic interference. Fiber is commonly used throughout the networks of hospital, education, and military environments.

Getting the Fiber-Optic Cable Right

A number of manufacturers make equipment that supports Ethernet over fiber-optic cabling. One of the most important elements of the planning of a 10Base-F installation is to pick the right cable and connecting hardware. Here are some pointers:

◆ Use 62.5/125-micron or 50/125-micron multimode fiber-optic cable.

◆ Each horizontal run should have at least two strands of multimode fiber.

◆ Make sure that the connector type for your patch panels and patch cables matches the hardware you choose. Some older equipment uses the ST connector exclusively, whereas newer equipment uses the more common SC connector. Connections between equipment with different types of connectors can be made using a patch cable with an ST connector at one end and an SC connector at the other. Follow the current standard requirements when selecting a connector type for new installations.

10Base-2 Ethernet

Although not as common as it once was, 10Base-2 is still an excellent way to connect a small number of computers together in a small physical area such as a home office, classroom, or lab. The 10Base-2 Ethernet uses thin coaxial (RG-58/U or RG-58 A/U) to connect computers together. This thin coaxial cable is also called *thinnet*.

Coaxial cable and network interface cards (NICs) use a special connector called a *BNC connector*. On this type of connector, the male is inserted into the female, and then the male connector is twisted 90 degrees to lock it into place. A BNC T-connector allows two cables to be connected on each side of it, and the middle of the T-connector plugs into the network-interface card. The thinnet cable never connects directly to the NIC. This arrangement is shown in Figure 3.5.

FIGURE 3.5
The 10Base-2
network

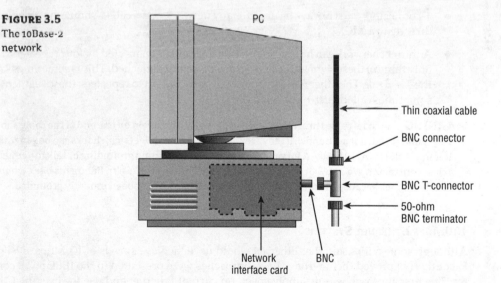

PC

Thin coaxial cable

BNC connector

BNC T-connector

50-ohm
BNC terminator

Network
interface card

BNC

NOTE *BNC* is an abbreviation for Bayonet Neill-Concelman. The *B* indicates that the connector is a bayonet-type connection, and Neill and Concelman are the inventors of the connector. You may also hear this connector called a British Naval Connector.

The ANSI/TIA-568-C standard does not recognize the use of coaxial cabling. From our own experience, here are some reasons *not* to use coax-based 10Base-2:

◆ The 10Base-2 network isn't suited for connecting more than 10 computers on a single segment.

◆ Ethernet cards with thinnet (BNC) connections are not as common as they once were. Usually you have to pay extra for network interface cards with thinnet connectors.

◆ The network may not be the best choice if you want to use Ethernet switching technologies.

◆ If your network spans more than one or two rooms or building floors, 10Base-2 isn't for you.

◆ If you are building a home network and plan to connect to the Internet using a cable modem or DSL, investing in a simple UTP or wireless Ethernet router is a better choice.

◆ UTP cabling, 10Base-T routers, and 10Base-T network interface cards are plentiful and inexpensive.

Though 10Base-2 is simple to install, you should keep a number of points in mind if you choose to implement it:

◆ Both ends of the cable must be terminated.

◆ A cable break anywhere along the length of the cable will cause the entire segment to fail.

◆ The maximum cable length is 185 meters and the minimum is 0.5 meters.

◆ T-connectors must always be used for any network node; cables should never be connected directly to a NIC.

◆ A thinnet network can have as many as five segments connected by four repeaters. However, only three of these segments can have network nodes attached. This is sometimes known as the *5-4-3 rule*. The other two segments will only connect to repeaters; these segments are sometimes called *inter-repeater links*.

WARNING Coaxial cables must be grounded properly (the shield on one end of the cable should be grounded, but not both ends). If they aren't, possibly lethal electrical shocks can be generated. Refer to ANSI/TIA/EIA-607-A for more information on building grounding or talk to your electrical contractor. We know of one network manager who was thrown flat on his back when he touched a rack because the cable and its associated racks had not been properly grounded.

100Mbps Ethernet Systems

Although some critics said that Ethernet would never achieve speeds of 100Mbps, designers of Fast Ethernet proved them wrong. Two approaches were presented to the IEEE 802.3 committee. The first approach was to simply speed up current Ethernet and use the existing CSMA/

CD access-control mechanism. The second was to implement an entirely new access-control mechanism called *demand priority*. In the end, the IEEE decided to create specifications for both approaches. The 100Mbps version of 802.3 Ethernet specifies a number of different methods of cabling a Fast Ethernet system, including 100Base-TX, 100Base-T4, and 100Base-FX. Fast Ethernet and the demand-priority approach is called 100VG-AnyLAN.

100Base-TX Ethernet

The 100Base-TX specification uses physical-media specifications developed by ANSI that were originally defined for FDDI (ANSI specification X3T9.5) and adapted for twisted-pair cabling. The 100Base-TX requires Category 5e or better cabling but uses only two of the four pairs. The eight-position modular jack (RJ-45) uses the same pin numbers as 10Base-T Ethernet.

Although a typical installation requires hubs or switches, two 100Base-TX nodes can be connected together "back to back" with a crossover cable made exactly the same way as a 10Base-T crossover cable. (See Chapter 10, "Connectors," for more information on making a 10Base-T or 100Base-TX crossover cable.) Understand the following when planning a 100Base-TX Fast Ethernet network:

◆ All components must be Category 5e or better certified, including cables, patch panels, and connectors. Proper installation practices must be followed.

◆ If you have a Category 5 "legacy" installation, the cabling system must be able to pass tests specified in ANSI/TIA-568-C.2.

◆ The maximum segment cable length is 100 meters. With higher-grade cables, longer lengths of cable may work, but proper signal timing cannot be guaranteed.

◆ The network uses the same pins as 10Base-T, as shown previously in Figure 3.4.

100Base-T4 Ethernet

The 100Base-T4 specification was developed as part of the 100Base-T specification so that existing Category 3–compliant systems could also support Fast Ethernet. The designers accomplish 100Mbps throughput on Category 3 cabling by using all four pairs of wire; 100Base-T4 requires a minimum of Category 3 cable. The requirement can ease the migration path to 100Mbps technology.

The 100Base-T4 is not used as frequently as 100Base-TX, partially due to the cost of the NICs and network equipment. The 100Base-T4 NICs are generally 50 to 70 percent more expensive than 100Base-TX cards. Also, 100Base-T4 cards do not automatically negotiate and connect to 10Base-T hubs, as most 100Base-TX cards do. Therefore, 100Base-TX cards are more popular. However, 100Base-TX does require Category 5 or better cabling.

If you plan to use 100Base-T4, understand the following:

◆ Maximum cable distance is 100 meters using Category 3, although distances of up to 150 meters can be achieved if Category 5e or better cable is used. Distances greater than 100 meters are not recommended, however, because round-trip signal timing cannot be ensured even on Category 5e cables.

◆ All eight pins of an eight-pin modular jack must be wired. Older Category 3 systems often wired only the exact number of pairs (two) necessary for 10Base-T Ethernet. Figure 3.6 shows the pins used.

◆ The 100Base-T4 specification recommends using Category 5e or better patch cables, panels, and connecting hardware wherever possible.

FIGURE 3.6
The eight-pin modular-jack wiring pattern for 100Base-T4

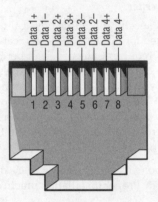

Table 3.2 shows the usage of each of the pins in a 100Base-T4 connector. Either the T568A or T568B pinout configurations can be used, but you must be consistent.

TABLE 3.2 Pin Usage in an Eight-Pin Modular Jack Used by 100Base-T4

PIN	NAME	USAGE	ABBREVIATION
1	Data 1 +	Transmit +	Tx_D1+
2	Data 1 −	Transmit −	Tx_D1−
3	Data 2 +	Receive +	Rx_D2+
4	Data 3 +	Bidirectional Data 3 +	Bi_D3+
5	Data 3 −	Bidirectional Data 3 −	Bi_D3−
6	Data 2 −	Receive −	Rx_D2−
7	Data 4 +	Bidirectional Data 4 +	Bi_D4+
8	Data 4 −	Bidirectional Data 4 −	Bi_D4−

100Base-FX Ethernet

Like its 100Base-TX copper cousin, 100Base-FX uses a physical-media specification developed by ANSI for FDDI. The 100Base-FX specification was developed to allow 100Mbps Ethernet to be used over fiber-optic cable. Although the cabling plant is wired in a star topology, 100Base-FX is a bus architecture.

If you choose to use 100Base-FX Ethernet, consider the following:

◆ Cabling-plant topology should be a star topology and should follow ANSI/TIA-568-C or ISO 11801 Ed. 2 recommendations.

◆ Each network node location should have a minimum of two strands of multimode fiber (MMF).

◆ Maximum link distance is 2,000 meters.

◆ The most common fiber connector type used for 100Base-FX is the SC connector, but the ST connector and the FDDI MIC connector may also be used. Make sure you know which type of connector(s) your hardware vendor will require.

GIGABIT ETHERNET (1000MBPS)

The IEEE approved the first Gigabit Ethernet specification in June 1998—IEEE 802.3z. The purpose of IEEE 802.3z was to enhance the existing 802.3 specification to include 1000Mbps operation (802.3 supported 10Mbps and 100Mbps). The new specification covers media access control, topology rules, and the gigabit media-independent interface. IEEE 802.3z specifies three physical layer interfaces: 1000Base-SX, 1000Base-LX, and 1000Base-CX.

In July 1999, the IEEE approved an additional specification known as IEEE 802.3ab, which adds an additional Gigabit Ethernet physical layer for 1000Mbps over UTP cabling, 1000Base-T. When the standard was written, the UTP cabling, all components, and installation practices had to be Category 5 or greater. New installations must now meet Category 5e or greater performance requirements outlined in ANSI/TIA-568-C.

Gigabit Ethernet deployment is widely used as a backbone application and, in some cases, directly to the desktop. The very low cost of 1000Base-T Gigabit Ethernet NICs (less than $50), hubs, and switches (less than $80 for an eight-port switch) has made this possible.

Initially, the most common uses for Gigabit Ethernet were for intra-building or campus backbones; however, Gigabit Ethernet is the de facto application for building intra-backbones. Figure 3.7 shows a before-and-after illustration of a simple network with Gigabit Ethernet deployed. Prior to deployment, the network had a single 100Mbps switch as an intra-building backbone for several 10Mbps and 100Mbps segments. All servers were connected to the 100Mbps backbone switch, which was sometimes a bottleneck.

During deployment of Gigabit Ethernet, the 100Mbps backbone switch is replaced with a Gigabit Ethernet switch. The NICs in the servers are replaced with Gigabit NICs. The 10Mbps and 100Mbps hubs connect to ports on the Gigabit switch that will accommodate 10- or 100Mbps segments. In this simple example, the bottleneck on the backbone has been relieved. The hubs and the computers did not have to be disturbed.

TIP To take full advantage of Gigabit Ethernet, computers that have Gigabit Ethernet cards installed should have a 64-bit PCI bus. The 32-bit PCI bus will work with Gigabit Ethernet, but it is not nearly as fast as the 64-bit bus.

FIGURE 3.7

Moving to a
Gigabit Ethernet
backbone

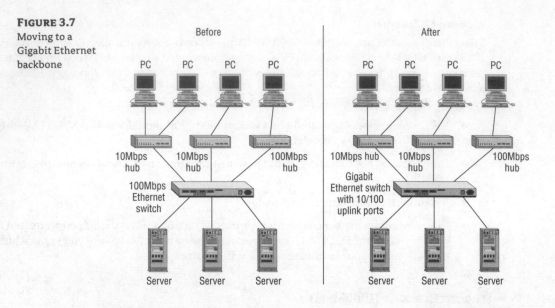

Gigabit Ethernet and Fiber-Optic Cables

Initially, 1000Mbps Ethernet was supported only on fiber-optic cable. The IEEE 802.3z specification included support for three physical-media options (PHYs), each designed to support different distances and types of communications:

1000Base-SX Targeted to intra-building backbones and horizontal cabling applications such as to workstations and other network nodes, 1000Base-SX is designed to work with multimode fiber-optic cable at the 850nm wavelength. Traditionally, 850nm electronics are less expensive than 1300nm versions for several reasons. First, the 850nm lasers, vertical cavity surface emitting lasers (VCSELs), are less expensive than their 1300nm counterparts. Second, because of the large diameter of the multimode fiber core (62.5 or 50 microns), the 850nm VCSEL transceivers require much lower precision and tolerance to focus their output onto the fiber. While it is possible to use multimode fiber with 1300nm VCSEL, the costs outweigh the benefits. Single-mode fiber is not used at 850nm due to specific optical differences, and large insertion loss arises since the spot size of 850nm emitters is much larger than the size of the single-mode core.

1000Base-LX Designed to support backbone-type cabling such as inter-building campus backbones, 1000Base-LX is for single-mode fiber-optic cable at 1310nm, though multimode fiber can be used for short inter-building backbones and intra-building cabling applications. As far as cost, 1310nm electronics are more expensive than the 850nm versions because they require very high precision and tolerances to focus on the very small core diameters of single-mode fiber (8.3 microns). Since the electronics are typically the most expensive portion of the network, the use of 1000Base-LX with either multimode or single-mode cabling will be more expensive than the use of 1000Base-SX with multimode cabling.

1000Base-CX Designed to support interconnection of equipment clusters, this specification uses 150 ohm STP cabling similar to IBM Type 1 cabling over distances no greater than 25 meters.

When cabling for Gigabit Ethernet using fiber, you should follow the ANSI/TIA-568-C standards for 62.5/125 micron or 50/125 micron multimode fiber for horizontal cabling and 8.3/125 micron single-mode fiber for backbone cabling. See Table 6 of Annex D in ANSI/TIA-568-C.0.

1000Base-T Ethernet

The IEEE designed 1000Base-T with the intention of supporting Gigabit Ethernet to the desktop. One of the primary design goals was to support the existing base of Category 5 cabling.

In July 1999, the IEEE 802.3ab task force approved IEEE specification 802.3ab, which defines using 1000Mbps Ethernet over Category 5 unshielded twisted-pair cable. Unlike 10Base-T and 100Base-TX, all four pairs must be used with 1000Base-T. Network electronics simultaneously send and receive 250Mbps over each pair using a transmission frequency of about 65MHz. These special modulation techniques are employed to "stuff" 1000Mbps through a cable that is only rated to 100MHz.

In 1999, the TIA issued TSB-95 to define additional performance parameters (above and beyond those specified in TSB-67) that should be performed in order to certify an existing Category 5 cabling installation for use with 1000Base-T. Structural return loss was dropped and the additional criteria of far-end crosstalk, delay skew, and return loss have been incorporated into ANSI/TIA-568-C as Category 5e.

When deploying 1000Base-T, make sure that you use Category 5e or better cable, that solid installation practices are used, and that all links are tested and certified using ANSI/TIA-568-C performance criteria.

10 GIGABIT ETHERNET (10,000MBPS)

The IEEE approved the first Gigabit Ethernet specification in June, 2002: IEEE 802.3ae. It defines a version of Ethernet with a nominal data rate of 10 Gbit/s. Over the years the following 802.3 standards relating to 10GbE have been published: 802.3ae-2002 (fiber -SR, -LR, -ER, and -LX4 physical-media-dependent devices[PMDs]), 802.3ak-2004 (-CX4 copper twin-ax *InfiniBand* type cable), 802.3an-2006 (10GBASE-T copper twisted pair), 802.3ap-2007 (copper backplane -KR and -KX4 PMDs), and 802.3aq-2006 (-LRM over legacy multimode fiber -LRM PMD with electronic dispersion compensation [EDC]). The 802.3ae-2002 and 802.3ak-2004 amendments were consolidated into the *IEEE 802.3-2005* standard. IEEE 802.3-2005 and the other amendments have been consolidated into IEEE Standard 802.3-2008.

In the premises environment, 10 Gigabit Ethernet is mostly used in data center storage servers, high-performance servers, and in some cases for intra-building backbones. It can be used for connection directly to the desktop. The applications most relevant for network cabling systems in commercial buildings are explained next.

10GBASE-SR (Short Range)

10GBASE-SR (short range) uses 850nm VCSEL lasers over multimode fibers. Low-bandwidth 62.5/125 micron (OM1) and 50/125 micron (OM2) multimode fiber support limited distances of 33–82 meters. To support 300 meters, the fiber-optics industry developed a higher bandwidth version of 50/125 micron fiber optimized for use at 850nm. This fiber, standardized in TIA-492-AAAC-A, is called *850nm laser-optimized 50/125 micron multimode fiber*. It is commonly referred to as OM3 fiber per the channel requirements identified in ISO 11801 Ed. 2. Most new structured cabling installations use OM3 multimode fiber since it is optimized for use with low-cost

850nm-based 10GBASE-SR optics. 10GBASE-SR delivers the lowest cost, lowest power, and smallest form-factor optical modules.

10GBASE-LR (Long Range)

10GBASE-LR (long range) uses 1310nm lasers to transmit over single-mode fiber up to 10 kilometers. Fabry-Pérot lasers are commonly used in 10GBASE-LR optical modules. Fabry-Pérot lasers are more expensive than 850nm VCSELs because they require the precision and tolerances to focus on very small single-mode core diameters (8.3 microns). 10GBASE-LR ports are typically used for long-distance communications.

10GBASE-LX4

10GBASE-LX4 uses coarse wavelength division multiplexing (WDM) to support 300 meters over standard, low-bandwidth 62.5/125 micron (OM1) and 50/125 micron (OM2) multimode fiber cabling. This is achieved through the use of four separate laser sources operating at 3.125Gbps in the range of 1300nm on unique wavelengths. This standard also supports 10 kilometers over single-mode fiber.

10GBASE-LX4 is used to support both standard multimode and single-mode fiber with a single optical module. When used with standard multimode fiber, an expensive mode conditioning patch cord is needed. The mode conditioning patch cord is a short length of single-mode fiber that connects to the multimode in such a way as to move the beam away from the central defect in legacy multimode fiber. Because 10GBASE-LX4 uses four lasers, it is more expensive and larger in size than 10GBASE-LR. To decrease the footprint of 10GBASE-LX4, a new module, 10GBASE-LRM, was standardized in 2006.

10GBASE-LRM

10GBASE-LRM (long reach multimode) supports distances up to 220 meters on standard, low-bandwidth 62.5/125 micron (OM1) and 50/125 micron (OM2) using a 1310nm laser. Expensive mode conditioning patch cord may also be needed over standard fibers. 10GBASE-LRM does not reach quite as far as the older 10GBASE-LX4 standard. However, it is hoped that 10GBASE-LRM modules will be lower cost and lower power consumption than 10GBASE-LX4 modules. (It will still be more expensive than 10GBASE-SR.)

10GBASE-T

10GBASE-T supports 10Gbps over Category 6A UTP or Category 7 shielded (per ISO/IEC 11801 Ed. 2) twisted-pair cables over distances of 100 meters. Category 5e is supported to much lower distances due to its limited bandwidth. Special care needs to be taken in installing Category 6A cables in order to minimize alien cross-talk on signal performance.

What's Next?

At the time of this writing, the IEEE 802.3ba committee is standardizing 40 and 100 Gigabit Ethernet. This will be deployed over OM3 50/125 micron multimode optical fiber for 100–200 meters, and research is progressing to make it available over UTP for distances up to 10 meters. This could be the speed point at which there is mass conversion of copper to fiber-based systems.

Token Ring

Developed by IBM, *Token Ring* uses a ring architecture to pass data from one computer to another.

Token Ring employs a sophisticated scheme to control the flow of data. If no network node needs to transmit data, a small packet, called the *free token*, continually circles the ring. If a node needs to transmit data, it must have possession of the free token before it can create a new Token Ring data frame. The token, along with the data frame, is sent along as a *busy token*. Once the data arrives at its destination, it is modified to acknowledge receipt and sent along again until it arrives back at the original sending node. If there are no problems with the correct receipt of the packet, the original sending node releases the free token to circle the network again. Then another node on the ring can transmit data if necessary.

WHAT HAPPENED TO TOKEN RING?

Token Ring is perhaps a superior technology compared to Ethernet, but Token Ring has not enjoyed widespread success since the early 1990s. IBM was slow to embrace structured wiring using UTP and eight-position (RJ-45 type) plugs and jacks, so cabling and components were relatively expensive and difficult to implement. When IBM finally acknowledged UTP as a valid media, 4Mbps Token Ring ran on Category 3 UTP, but 16Mbps Token Ring required a minimum of Category 4. In the meantime, a pretty quick and robust 10Mbps Ethernet network could be put in place over Category 3 cables that many offices already had installed. So, while Token Ring was lumbering, Ethernet zoomed by, capturing market share with the ease and economy of its deployment.

This scheme, called *token passing*, guarantees equal access to the ring and ensures that no two computers will transmit at the same time. Token passing is the basis for IEEE specification 802.5. This scheme might seem pretty slow since the free token must circle the ring continually, but keep in mind that the free token is circling at speeds approaching 70 percent of the speed of light. A smaller Token Ring network may see a free token circle the ring up to 10,000 times per second!

Because a ring topology is difficult to cable, IBM employs a hybrid star/ring topology. All nodes in the network are connected centrally to a hub (media attachment unit [MAU] or multistation access unit [MSAU]), as shown in Figure 3.8. The transmitted data still behaves like a ring topology, traveling down each cable (called a *lobe*) to the node and then returning to the hub, where it starts down the next cable on the MAU.

Even a single-node failure or lobe cable can take down a Token Ring. The designers of Token Ring realized this and designed the MAU with a simple electromechanical switch (a relay switch) that adds a new node to the ring when it is powered on. If the node is powered off or if the lobe cable fails, the electromechanical switch disengages, and the node is removed from the ring. The ring continues to operate as if the node were not there.

Token Ring operates at either 4Mbps or 16Mbps; however, a ring only operates at a single speed. (That's unlike Ethernet, where 10Mbps and 100Mbps nodes can coexist on the same network.) Care must be taken on older Token Ring hardware that a network adapter operating at the wrong speed is not inserted into a ring because doing so can shut down the entire network.

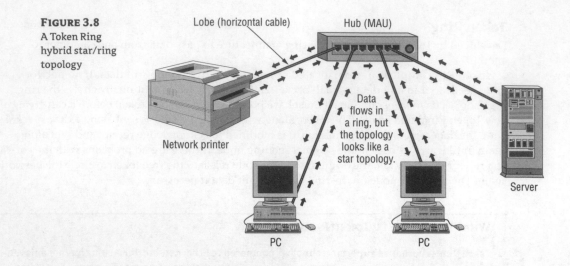

FIGURE 3.8
A Token Ring hybrid star/ring topology

Lobe (horizontal cable)

Hub (MAU)

Data flows in a ring, but the topology looks like a star topology.

Network printer

PC

PC

Server

TOKEN RING AND SHIELDED TWISTED PAIR (STP)

Token Ring originally operated on shielded twisted-pair (STP) cabling. IBM designed a cabling system that included a couple of types of shielded twisted-pair cables; the most common of these was IBM Type 1 cabling (later called IBM Type 1A). STP cabling is a recognized cable type in the ANSI/TIA-568-C specification.

The IBM cabling system used a unique, hermaphroditic connector that is commonly called an *IBM data connector*. The IBM data connector has no male and female components, so two IBM patch cables can be connected together to form one long patch cable.

Unless your cabling needs specifically require an STP cabling solution for Token Ring, we recommend you don't use STP cabling. Excellent throughput is available today over UTP cabling; the only reason to implement STP is if electromagnetic interference is too great to use UTP, in which case fiber-optic cable might be your best bet anyway.

TOKEN RING AND UNSHIELDED TWISTED PAIR (UTP)

Around 1990, vendors started releasing unshielded twisted-pair solutions for Token Ring. The first of these solutions was simply to use media filters or baluns on the Token Ring NICs, which connected to the card's nine-pin interface and allowed a UTP cable to connect to the media filter. The balun matches the impedance between the 100 ohm UTP and the network device, which is expecting 150 ohms.

KEY TERM *baluns and media filters* Baluns and *media filters* are designed to match impedance between two differing types of cabling, usually unbalanced coaxial cable and balanced two-wire twisted pair. Although baluns can come in handy, they can also be problematic and should be avoided if possible.

The second UTP solution for Token Ring was NICs equipped with eight-pin modular jacks (RJ-45) that supported 100 ohm cables, rather than a DB9 connector.

Any cabling plant certified Category 3 or better should support 4Mbps Token Ring.

NOTE   A number of vendors make Token Ring NICs that support fiber-optic cable. Although using Token Ring over fiber-optic cables is uncommon, it is possible.

Fiber Distributed Data Interface (FDDI)

Fiber Distributed Data Interface (FDDI) is a networking specification that was produced by the ANSI X3T9.5 committee in 1986. It defines a high-speed (100Mbps), token-passing network using fiber-optic cable. In 1994, the specification was updated to include copper cable. The copper cable implementation was designated TP-PMD, which stands for Twisted Pair-Physical Media Dependent. FDDI was slow to be widely adopted, but for a while it found a niche as a reliable, high-speed technology for backbones and applications that demanded reliable connectivity.

Although at first glance FDDI appears to be similar to Token Ring, it is different from both Token Ring and Ethernet. A Token Ring node can transmit only a single frame when it gets the free token; it must wait for the token to transmit again. An FDDI node, once it possesses the free token, can transmit as many frames as it can generate within a predetermined time before it has to give up the free token.

FDDI can operate as a true ring topology, or it can be physically wired like a star topology. Figure 3.9 shows an FDDI ring that consists of *dual-attached stations (DASs)*; this is a true ring topology. A DAS has two FDDI interfaces, designated as an A port and a B port. The A port is used as a receiver for the primary ring and as a transmitter for the secondary ring. The B port does the opposite: it is a transmitter for the primary ring and a receiver for the secondary ring. Each node on the network in Figure 3.9 has an FDDI NIC that has two FDDI attachments. The card creates both the primary and secondary rings. Cabling for such a network is a royal pain because the cables have to form a complete circle.

FIGURE 3.9
An FDDI ring

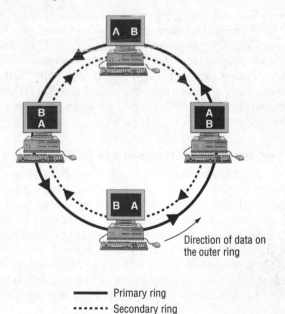

Direction of data on
the outer ring

——— Primary ring
······· Secondary ring

FDDI networks can also be cabled as a hierarchical star topology, though they would still behave like a ring topology. FDDI NICs may be purchased with either a single FDDI interface (*single-attached station* [SAS]) or with two FDDI interfaces (DAS). Single-attached stations must connect to an FDDI concentrator or hub. A network can also be mixed and matched, with network nodes such as workstations using only a single-attached station connection and servers or other critical devices having dual-attached station connections. That configuration would allow the critical devices to have a primary and secondary ring.

FDDI has specific terminology and acronyms, including the following:

MAC The media access control is responsible for addressing, scheduling, and routing data.

PHY The physical protocol layer is responsible for coding and timing of signals, such as clock synchronization of the ring. The actual data speed on an FDDI ring is 125Mbps; an additional control bit is added for every four bits.

PMD The physical media dependent layer medium is responsible for the transmission between nodes. FDDI includes two PMDs: Fiber-PMD for fiber-optic networks and TP-PMD for twisted-pair networks.

SMT The station management is responsible for handling FDDI management, including ring management (RMT), configuration management (CFM), connection management (CMT), physical-connection management (PCM), and entity-coordination management (ECM). SMT coordinates neighbor identification, insertion to and removal from the ring, traffic monitoring, and fault detection.

CABLING AND FDDI

When planning cabling for an FDDI network, practices recommended in ANSI/TIA-568-C or ISO 11801 should be followed. FDDI using fiber-optic cable for the horizontal links uses FDDI medium interface connectors (MICs). Care must be taken to ensure that the connectors are keyed properly for the device they will connect to.

CDDI using copper cabling (Copper-Distributed Data Interface, or CDDI) requires Category 5 or better cable and associated devices. Horizontal links should at a minimum pass performance tests specified in ANSI/TIA-568-C. Of course, a Category 5e or better installation is a better way to go.

Asynchronous Transfer Mode (ATM)

ATM (*asynchronous transfer mode*, not to be confused with automated teller machines) first emerged in the early 1990s. If networking has an equivalent to rocket science, then ATM is it. ATM was designed to be a high-speed communications protocol that does not depend on any specific LAN topology. It uses a high-speed cell-switching technology that can handle data as well as real-time voice and video. The ATM protocol breaks up transmitted data into 48-byte cells that are combined with a 5-byte header. A cell is analogous to a data packet or frame.

ATM is designed to "switch" these small, fixed-size cells through an ATM network very quickly. It does this by setting up a virtual connection between the source and destination nodes; the cells may go through multiple switching points before ultimately arriving at their final destination. If the cells arrive out of order, and if the implementation of the receiving system is set up to do so, the receiving system may have to correctly order the arriving cells. ATM is a connection-oriented service, in contrast to many network applications, which are broadcast based. Connection orientation simply means that the existence of the opposite end is established through manual setup or automated control information before user data is transmitted.

Data rates are scalable and start as low as 1.5Mbps, with other speeds of 25, 51, 100, and 155Mbps and higher. The most common speeds of ATM networks today are 51.84Mbps and 155.52Mbps. Both of these speeds can be used over either copper or fiber-optic cabling. A 622.08Mbps ATM is also becoming common but is currently used exclusively over fiber-optic cable, mostly as a network backbone architecture.

ATM supports very high speeds because it is designed to be implemented by hardware rather than software and is in use at speeds as high as 10Gbps.

In the United States, the specification for synchronous data transmission on optical media is SONET (Synchronous Optical Network); the international equivalent of SONET is SDH (Synchronous Digital Hierarchy). SONET defines a base data rate of 51.84Mbps; multiples of this rate are known as optical carrier (OC) levels, such as OC-3, OC-12, and so on. Table 3.3 shows common OC levels and their associated data rates.

TABLE 3.3   Common Optical Carrier Levels (OC-*X*)

LEVEL	DATA RATE
OC-1	51.84Mbps
OC-3	155.52Mbps
OC-12	622.08Mbps
OC-48	2.488Gbps
OC-96	4.976Gbps
OC-192	9.953Gbps
OC-768	39.813Gbps

ATM was designed as a WAN protocol. However, due to the high speeds it can support, many organizations are using it to attach servers (and often workstations) directly to the ATM network. To do this, a set of services, functional groups, and protocols was developed to provide LAN emulation via MPoA (MultiProtocol over ATM). MPoA also provides communication between network nodes attached to a LAN (such as Ethernet) and ATM-attached nodes. Figure 3.10 shows an ATM network connecting to LANs using MPoA. Note that the ATM network does not have to be in a single physical location and can span geographic areas.

FIGURE 3.10
An ATM network

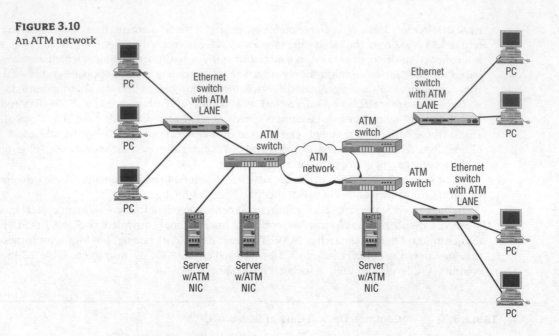

NOTE For more information on ATM, check out the IP/MPLS Forum's website at www.ipmplsforum.org.

CABLING AND ATM

What sort of cabling should you consider for ATM networks? Fiber-optic cabling is still the medium of choice for most ATM installations. ATM to the desktop is still not terribly common.

For fiber-optic cable, as long as you follow the ANSI/TIA-568-C or the ISO 11801 Ed. 2 standard, you should not have problems. ATM equipment and ATM NICs use 62.5/125 micron multimode optical fiber.

If you plan on using 155Mbps ATM over copper, plan to use Category 5e cabling at minimum.

The Bottom Line

Identify important network topologies for commercial buildings. Over the years, various network topologies have been created: bus, ring, and star. From the perspective of cabling, the hierarchical star topology is now almost universal. It is also the easiest to cable. The ANSI/TIA-568-C and ISO/IEC 11801 Ed. 2 standards assume that the network architecture uses a hierarchical star topology as its physical configuration. This is the configuration you will most likely be involved in as you begin cabling commercial buildings. The ANSI/TIA-568-C standard specifies the use of the hierarchical star network; however, there are several ways of implementing this.

Master It

1. What is the most typical implementation of the hierarchical star? Specifically, where are the horizontal cross-connection and workgroup switches typically placed?

2. To reduce the cost of a hierarchical star, the network designer has the option to locate network elements closer to equipment outlets. What is this called and what are the benefits?

3. In certain situations it makes sense to install all network equipment in a central location. What is this implementation of the hierarchical star topology called and what are the pros and cons?

Understand the basic differences between UTP and optical fiber cabling and their place in future-proofing networks.   As network applications are evolving, better UTP and optical fiber cabling media are required to keep up with bandwidth demand. As you will see from standards, the end user has many options within a media category. There are many types of UTP and optical fiber cabling. Standards will continue to evolve, but it's always a good idea to install the best grade of cabling since the cost of the structured cabling systems (excluding installation cost) is usually only 5–10 percent of the total project cost. Therefore, making the right decisions today can greatly future-proof the network.

Master It

1. What is the preferred UTP cabling media to support 10Gbps network speeds?

2. What is the preferred optical fiber cabling media to support low-cost transmission at 10Gbps network speeds?

Identify key network applications and the preferred cabling media for each.   In your network design planning and installation you will most likely be running Ethernet-based network applications. There is an "alphabet soup" full of jargon associated with naming the application for a given speed, the reach range, and media type. This chapter has provided a good starting point.

Master It   You are asked to design a commercial building network. The owner wants to ensure that the desktops are able to operate using UTP with a maximum of 1Gbps capability. The distance between the workgroup switches located in the telecommunications rooms and the equipment room is 250 meters. What would you recommend for the following?

1. What Ethernet network application ports/modules should you use for the horizontal links?

2. What are your horizontal cabling options?

3. What backbone speed and network application should you use to obtain the lowest cost assuming a 10:1 rule?

4. What backbone cabling should you use?

Chapter 4

Cable System and Infrastructure Constraints

What constrains you when building a structured cabling system? Can you install cable anywhere you please? You probably already realize some of the restrictions of your cabling activities, including installing cable too close to electrical lines and over fluorescent lights. However, many people don't realize that documents and codes help dictate how cabling systems (electrical as well as communications) must be designed and installed to conform to your local laws.

In the United States, governing bodies issue codes for minimum safety requirements to protect life, health, and property. Once adopted by the local regulating authority, codes have the force of law. Standards, which are guidelines to ensure system functionality after installation, are issued to ensure construction quality.

Codes for a specific locality will be issued by the governing body with local jurisdiction. The codes for an area are written or adopted by and under control of the jurisdiction having authority (JHA). Sometimes these codes are called *building codes* or simply *codes*. This chapter discusses codes and how they affect the installation of communications cabling.

In this chapter, you will learn to:

◆ Identify the key industry codes necessary to install a safe network cabling system

◆ Understand the organization of the National Electrical Code

◆ Identify useful resources to make knowing and following codes easier

Where Do Codes Come From?

Building, construction, and communications codes originate from a number of sources. Usually, these codes originate nationally rather than at the local, city, or county level. Local municipalities usually adopt these national codes as local laws. Other national codes are issued that affect the construction of electrical and communications equipment.

Two of the predominant national code governing bodies in the United States are the Federal Communications Commission (FCC) and the National Fire Protection Association (NFPA). The Americans with Disabilities Act (ADA) also affects the construction of cabling and communications facilities because it requires that facilities be constructed to provide universal access.

The Federal Communications Commission

The United States Federal Communications Commission (FCC) issues guidelines that govern the installation of telecommunications cabling and the design of communications devices built or used in the United States. The guidelines help prevent problems relating to communications equipment, including interference with the operation of other communications equipment. Part 68 of the FCC rules provides regulations that specifically address connecting premises cabling and customer-provided equipment to the regulated networks.

The FCC also publishes numerous reports and orders that deal with specific issues regarding communications cabling, electromagnetic emissions, and frequency bandwidths. The following is a list of some of the important documents issued by the FCC:

Part 68 of the FCC rules   Governs the connection of premises equipment and wiring to the national network.

Telecommunications Act of 1996   Establishes new rules for provisioning and additional competition in telecommunications services.

CC Docket No. 81-216   Establishes rules for providing customer-owned premises wiring.

CC Docket No. 85-229   Includes the Computer Inquiry III review of the regulatory framework for competition in telecommunications.

Part 15 of the FCC rules   Addresses electromagnetic radiation of equipment and cables.

CC Docket No. 87-124   Addresses implementing the ADA.

CC Docket No. 88-57   Defines the location of the demarcation point on a customer premise.

Fact Sheet ICB-FC-011   Deals with connection of one- and two-line terminal equipment to the telephone network and the installation of premises wiring.

Memorandum Opinion and Order FCC 85-343   Covers the rights of users to access embedded complex wire on customer premises.

TIP   Most of the FCC rules, orders, and reports can be viewed on the FCC website at www.fcc.gov. Since rules can change over time, it's wise to monitor updates.

The National Fire Protection Association

In 1897, a group of industry professionals (insurance, electrical, architectural, and other allied interests) formed the National Association of Fire Engineers with the purpose of writing and publishing the first guidelines for the safe installation of electrical systems and providing guidance to protect people, property, and the environment from fire. The guidelines are called the *National Electrical Code (NEC)*. Until 1911, the group continued to meet and update the NEC. The National Fire Protection Association (NFPA), an international, nonprofit, membership organization representing over 65,000 members and 100 countries, now sponsors the NEC. The NFPA continues to publish the NEC as well as other recommendations for a variety of safety concerns.

The NEC is updated by various committees and code-making panels, each responsible for specific articles in the code.

TIP   You can find information about NFPA and many of its codes and standards at www.nfpa.org. You can purchase NFPA codes through Global Engineering Documents (http://global.ihs .com); major codes, such as the National Electrical Code, can be purchased at most bookstores.

The NEC is called NFPA 70 by the National Fire Protection Association, which also sponsors more than 600 other fire codes and standards that are used in the United States and throughout the world. The following are some examples of these documents:

NFPA 1 (Fire Prevention Code)   Addresses basic fire-prevention requirements to protect buildings from hazards created by fire and explosion.

NFPA 13 (Installation of Sprinkler Systems)   Addresses proper design and installation of sprinkler systems for all types of fires.

NFPA 54 (National Fuel Gas Code)   Provides safety requirements for fuel-gas equipment installations, piping, and venting.

NFPA 70 (National Electrical Code)   Deals with proper installation of electrical systems and equipment.

NFPA 70B (Recommended Practice for Electrical Equipment Maintenance)   Provides guidelines for maintenance and inspection of electrical equipment such as batteries.

NFPA 70E (Standard for Electrical Safety in the Workplace)   A basis for evaluating and providing electrical safety–related installation requirements, maintenance requirements, requirements for special equipment, and work practices. This document is compatible with OSHA (Occupational Safety and Health Administration) requirements.

NFPA 72 (National Fire Alarm Code)   Provides a guide to the design, installation, testing, use, and maintenance of fire-alarm systems.

NFPA 75 (Standard for the Protection of Information Technology Equipment) Establishes requirements for computer room installations that require fire protection.

NFPA 101 (Life Safety Code)   Deals with minimum building design, construction, operation, and maintenance requirements needed to protect building occupants from fire.

NFPA 262 (Standard Method of Test for Flame Travel and Smoke of Wires and Cables for Use in Air-Handling Spaces)   Describes techniques for testing visible smoke and fire-spreading characteristics of wires and cables.

NFPA 780 (Standard for the Installation of Lightning Protection Systems)   Establishes guidelines for protection of buildings, people, and special structures from lightning strikes.

NFPA 1221 (Standard for the Installation, Maintenance, and Use of Emergency Services Communications System)   Provides guidance for fire service communications systems used for emergency notification. This guide incorporates NFPA 297 (Guide on Principles and Practices for Communications Systems).

These codes are updated every few years; the NEC, for example, is updated every three years. It was updated in 2008 and will be updated again in 2011.

You can purchase guides to the NEC that make the code easier for the layperson to understand. Like the NEC, these guides may be purchased at almost any technical or large bookstore. You can also purchase the NEC online from the NFPA's excellent website at www.nfpa.org.

If you are responsible for the design of a telecommunications infrastructure, a solid understanding of the NEC is essential. Otherwise, your installation may run into all sorts of red tape from your local municipality.

TIP A good reference on the Internet for the NEC is the National Electrical Code Internet Connection maintained by Mike Holt at www.mikeholt.com. You'll find useful information there for both the beginner and expert. Mike Holt is also the author of the book *Understanding the NEC, Volume 1* (Delmar Learning, 2002), which is an excellent reference for anyone trying to make heads or tails of the NEC.

Underwriters Laboratories

Underwriters Laboratories, Inc. (UL) is a nonprofit product safety testing and certification organization. Once an electrical product has been tested, UL allows the manufacturer to place the UL listing mark on the product or product packaging.

KEY TERM *UL listed and UL recognized* The UL mark identifies whether a product is *UL listed* or *UL recognized*. If a product carries the *UL Listing Mark* (UL in a circle) followed by the word *LISTED*, an alphanumeric control number, and the product name, it means that the complete (all components) product has been tested against the UL's nationally recognized safety standards and found to be reasonably free of electrical shock risk, fire risk, and other related hazards. If a product carries the *UL Recognized Component Mark* (the symbol looks like a backward *R* and *J*), it means that individual components may have been tested but not the complete product. This mark may also indicate that testing or evaluation of all the components is incomplete.

You may find a number of different UL marks on a product listed by the UL (all UL listing marks contain UL inside a circle). Some of these include:

UL The most common of the UL marks, this mark indicates that samples of the complete product have met UL's safety requirements.

C-UL This UL mark is applied to products that have been tested (by Underwriters Laboratories) according to Canadian safety requirements and can be sold in the Canadian market.

C-UL-US This is a relatively new listing mark that indicates compliance with both Canadian and U.S. requirements.

UL-Classified This mark indicates that the product has been evaluated for a limited range of hazards or is suitable for use under limited or special conditions. Specialized equipment such as firefighting gear, industrial trucks, and other industrial equipment carry this mark.

C-UL-Classified This is the classification marking for products that the UL has evaluated for specific hazards or properties, according to Canadian standards.

C-UL-Classified-US Products with this classification marking meet the classified compliance standards for both the United States and Canada.

Recognized Component Mark (backward *R* and *J*) Products with the backward *R* and *J* have been evaluated by the UL but are designed to be part of a larger system. Examples are the power supply, circuit board, disk drives, CD-ROM drive, and other components of a computer. The Canadian designator (a *C* preceding the Recognized Component Mark) is the Canadian equivalent.

C-Recognized Component-US The marking indicates a component certified by the UL according to both the U.S. and Canadian requirements.

International EMC mark   The electromagnetic compatibility mark indicates that the product meets the electromagnetic requirements for Europe, the United States, Japan, and Australia (or any combination of the four). In the United States, this mark is required for some products, including radios, microwaves, medical equipment, and radio-controlled equipment.

Other marks on equipment include the Food Service Product Certification mark, the Field Evaluated Product mark, the Facility Registration mark, and the Marine UL mark.

TIP   To see examples of the UL marks we've described, visit www.ul.com.

The NEC requires that Nationally Recognized Test Laboratories (NRTL) rate communications cables used in commercial and residential products as "listed for the purpose." Usually UL is used to provide listing services, but the NEC only requires that the listing be done by an NRTL; other laboratories, therefore, can provide the same services. One such alternate testing laboratory is Intertek ETL SEMKO (www.usa.intertek-etlsemko.com).

More than 750 UL standards and standard safety tests exist; some of the ones used for evaluating cabling-related products are:

UL 444   Applies to testing multiple conductors, jacketed cables, single or multiple coaxial cables, and optical fiber cables. This test applies to communications cables intended to be used in accordance with the NEC Article 800 or the Canadian Electrical Code (Part I) Section 60.

NFPA 262 (formerly UL 910)   Applies to testing the flame spread and smoke density (visible smoke) for electrical and optical fiber cables used in spaces that handle environmental air (that's a fancy way to say the *plenum*). This test does not investigate the level of toxic or corrosive elements in the smoke produced, nor does it cover cable construction or electrical performance. NEC Article 800 specifies that cables that have passed this test can carry the NEC flame rating designation CMP (communications multipurpose plenum).

UL 1581   Applies to testing flame-spread properties of a cable designed for general-purpose or limited use. This standard contains details of the conductors, insulation, jackets, and coverings, as well as the methods for testing preparation. The measurement and calculation specifications given in UL 1581 are used in UL 44 (Standards for the Thermoset-Insulated Wires and Cables), UL 83 (Thermoplastic-Insulated Wires and Cables), UL 62 (the Standard for Safety of Flexible Cord), and UL 854 (Service-Entrance Cables). NEC Article 800 specifies that cables that have passed these tests can carry the NEC flame-rating designation CMG, CM, or CMX (all of which mean communications general-purpose cable).

UL 1666   Applies to testing flame-propagation height for electrical and optical fiber cables installed in vertical shafts (the riser). This test only makes sure that flames will not spread from one floor to another. It does *not* test for visible smoke, toxicity, or corrosiveness of the products' combustion. It does not evaluate the construction for any cable or the cable's electrical performance. NEC Article 800 specifies that cables that have passed this test may carry a designation of CMR (communications riser).

UL has an excellent website that has summaries of all the UL standards and provides access to its newsletters. The main UL website is www.ul.com; a separate website for the UL Standards Department is located at http://ulstandardsinfonet.ul.com. UL standards may be purchased through IHS/Global at http://global.ihs.com.

Codes and the Law

At the state level in the United States, many public utility/service commissions issue their own rules governing the installation of cabling and equipment in public buildings. States also monitor tariffs on the state's service providers.

At the local level, the state, county, city, or other authoritative jurisdiction issues codes. Most local governments issue their own codes that must be adhered to when installing communications cabling or devices in the jurisdictions under their authority. Usually, the NEC is the basis for electrical codes, but often the local code will be stricter.

Over whom the jurisdiction has authority must be determined prior to any work being initiated. Most localities have a code office, a fire marshal, or a permitting office that must be consulted.

The strictness of the local codes will vary from location to location and often reflects a particular geographic region's potential for or experience with a disaster. For example:

◆ Some localities in California have strict earthquake codes regarding how equipment and racks must be attached to buildings.

◆ In Chicago, some localities require that all cables be installed in metal conduits so that cables will not catch fire easily. This is also to help prevent flame spread that some cables may cause.

◆ Las Vegas has strict fire-containment codes that require firestopping of openings between floors and firewalls. These openings may be used for running horizontal or backbone cabling.

WARNING　Local codes take precedence over all other installation guidelines. Ignorance of local codes could result in fines, having to reinstall all components, or the inability to obtain a Certificate of Occupancy.

Localities may adopt any version of the NEC or write their own codes. Don't assume that a specific city, county, or state has adopted the NEC word for word. Contact the local building codes, construction, or building permits department to be sure that what you are doing is legal.

Historically, telecommunications cable installations were not subject to local codes or inspections. However, during several commercial building fires, the communications cables burned and produced toxic smoke and fumes, and the smoke obscured the building's exit points. This contributed to deaths. When the smoke mixed with the water vapor, hydrochloric acid was produced, resulting in significant property damage. Because of these fires, most JHAs now issue permits and perform inspections of the communications cabling.

It is impossible to completely eliminate toxic elements in smoke. Corrosive elements, although certainly harmful to people, are more a hazard to electronic equipment and other building facilities. The NEC flame ratings for communications cables are designed to limit the spread of the fire and, in the case of plenum cables, the production of visible smoke that could obscure exits. The strategy is to allow sufficient time for people to exit the building and to minimize potential property damage. By specifying acceptable limits of toxic or corrosive elements in the smoke and fumes, NFPA is not trying to make the burning cables "safe." Note, however, that there are exceptions to the previous statement, notably cables used in transportation tunnels, where egress points are limited.

TIP If a municipal building inspector inspects your cabling installation and denies you a permit (such as a Certificate of Occupancy), he or she must tell you exactly which codes you are not in compliance with.

The National Electrical Code

This section summarizes the information in the National Electrical Code (NFPA 70). All information contained in this chapter is based on the 2008 edition of the NEC; the code is reissued every three years. Prior to installing any communications cable or devices, consult your local JHAs to determine which codes apply to your project.

Do not assume that local jurisdictions automatically update local codes to the most current version of the NEC. You may find that local codes reference older versions with requirements that conflict with the latest NEC. Become familiar with the local codes. Verify all interpretations with local code enforcement officials, as enforcement of the codes is their responsibility. If you are responsible for the design of a telecommunications infrastructure or if you supervise the installation of such an infrastructure, you should own the official code documents and be intimately familiar with them.

The following list of NEC articles is not meant to be all-inclusive; it is a representation of some of the articles that may impact telecommunications installations.

The NEC is divided into chapters, articles, and sections. Technical material often refers to a specific article or section. Section 90-3 explains the arrangement of the NEC chapters. The NEC currently contains nine chapters; most chapters concern the installation of electrical cabling, equipment, and protection devices. The pertinent chapter for communications systems is Chapter 8. The rules governing the installation of communications cable differ from those that govern the installation of electrical cables; thus, the rules for electrical cables as stated in the NEC do not generally apply to communications cables. Section 90-3 states this by saying that Chapter 8 is independent of all other chapters in the NEC except where they are specifically referenced in Chapter 8.

This section only summarizes information from the 2008 National Electrical Code relevant to communications systems.

NOTE If you would like more information about the NEC, you should purchase the NEC in its entirety or a guidebook.

NEC Chapter 1 General Requirements

NEC Chapter 1 includes definitions, usage information, and descriptions of spaces about electrical equipment. Its articles are described in the following sections.

ARTICLE 100—DEFINITIONS

Article 100 contains definitions for NEC terms that relate to the proper application of the NEC.

ARTICLE 110.3 (B)—INSTALLATION AND USE

Chapter 8 of the NEC references this article, among others. It states that any equipment included on a list acceptable to the local jurisdiction having authority and/or any equipment labeled as having been tested and found suitable for a specific purpose shall be installed and used in accordance with any instructions included in the listing or labeling.

ARTICLE 110.26—SPACES ABOUT ELECTRICAL EQUIPMENT

This article calls for a minimum of 3′ to 4′ of clear working space around all electrical equipment, to permit safe operation and maintenance of the equipment. Article 110.26 is not referenced in Chapter 8 of the NEC, but many standards-making bodies address the need for 3′ of clear working space around communications equipment.

NEC Chapter 2 Wiring and Protection

NEC Chapter 2 includes information about conductors on poles, installation requirements for bonding, and grounding.

Grounding is important to all electrical systems because it prevents possibly fatal electrical shock. Further information about grounding can be found in TIA's J-STD-607-A, which is the Commercial Building Grounding and Bonding Requirements for Telecommunications standard. The grounding information in NEC Chapter 2 that affects communications infrastructures includes the following articles.

ARTICLE 225.14 (D)—CONDUCTORS ON POLES

This article is referenced in Chapter 8 of the NEC and states that conductors on poles shall have a minimum separation of 1′ when not placed on racks or brackets. If a power cable is on the same pole as communications cables, the power cable (over 300 volts) shall be separated from the communications cables by not less than 30″. Historically, power cables have always been placed above communications cables on poles because when done so communications cables cannot inflict bodily harm to personnel working around them. Power cables, though, *can* inflict bodily harm, so they are put at the top of the pole out of the communications workers' way.

ARTICLE 250—GROUNDING

Article 250 covers the general requirements for the bonding and grounding of electrical-service installations. Communications cables and equipment are bonded to ground using the building electrical-entrance service ground. Several subsections in Article 250 are referenced in Chapter 8 of the NEC; other subsections not referenced in Chapter 8 will be of interest to communications personnel both from a safety standpoint and for effective data transmission. Buildings not properly bonded to ground are a safety hazard to all personnel. Communications systems not properly bonded to ground will not function properly.

ARTICLE 250.4 (A)(4)—BONDING OF ELECTRICALLY CONDUCTIVE
MATERIALS AND OTHER EQUIPMENT

Electrically conductive materials (such as communications conduits, racks, cable trays, and cable shields) likely to become energized in a transient high-voltage situation (such as a lightning strike) shall be bonded to ground in such a manner as to establish an effective path to ground for any fault current that may be imposed.

ARTICLE 250.32—TWO OR MORE BUILDINGS OR STRUCTURES SUPPLIED BY A FEEDER OR BRANCH CIRCUIT

This article is referenced in Chapter 8 of the NEC. In multibuilding campus situations, the proper bonding of communications equipment and cables is governed by several different circumstances as follows:

Section 250.32 (A)—Grounding Electrode Each building shall be bonded to ground with a grounding electrode (such as a ground rod), and all grounding electrodes shall be bonded together to form the grounding-electrode system.

Section 250.32 (B)—Grounded Systems In remote buildings, the grounding system shall comply with either (1) or (2):

(1) Equipment-Grounding Conductor Rules here apply where the equipment-grounding conductor is run with the electrical-supply conductors and connected to the building or structure disconnecting means and to the grounding-electrode conductors.

(2) Grounded Conductor Rules here apply where the equipment-grounding conductor is not run with the electrical-supply conductors.

Section 250.32 (C)—Ungrounded Systems The electrical ground shall be connected to the building disconnecting means.

Section 250.32 (D)—Disconnecting Means Located in Separate Building or Structure on the Same Premises The guidelines here apply to installing grounded circuit conductors and equipment-grounding conductors and bonding the equipment-grounding conductors to the grounding-electrode conductor in separate buildings when one main electrical service feed is to one building with the service disconnecting means and branch circuits to remote buildings. The remote buildings do not have a service disconnecting means.

Section 250.32 (E)—Grounding Conductor The size of the grounding conductors per NEC Table 250.66 is discussed here.

ARTICLE 250.50—GROUNDING-ELECTRODE SYSTEM

This article is referenced in Chapter 8 of the NEC. On premises with multiple buildings, each electrode at each building shall be bonded together to form the grounding-electrode system. The bonding conductor shall be installed in accordance with the following:

Section 250.64 (A) Aluminum or copper-clad aluminum conductors shall not be used.

Section 250.64 (B) This section deals with grounding-conductor installation guidelines.

Section 250.64 (E) Metallic enclosures for the grounding-electrode conductor shall be electrically continuous.

The bonding conductor shall be sized per Section 250.66; minimum sizing is listed in NEC Table 250.66. The grounding-electrode system shall be connected per Section 250.70. An unspliced (or spliced using an exothermic welding process or an irreversible compression connection) grounding-electrode conductor shall be run to any convenient grounding electrode. The grounding electrode shall be sized for the largest grounding-electrode conductor attached to it.

WARNING Note that interior metallic above-ground water pipes shall not be used as part of the grounding-electrode system. This is a change from how communications workers historically bonded systems to ground.

ARTICLE 250.52—GROUNDING ELECTRODES

This article defines the following structures that can be used as grounding electrodes:

Section 250.52 (1)—Metal Underground Water Pipe An electrically continuous metallic water pipe, running a minimum of 10′ in direct contact with the earth, may be used in conjunction with a grounding electrode. The grounding electrode must be bonded to the water pipe.

Section 250.52 (2)—Metal Frame of the Building or Structure The metal frame of a building may be used as the grounding electrode, where effectively grounded.

Section 250.52 (3)—Concrete-Encased Electrode Very specific rules govern the use of steel reinforcing rods, embedded in concrete at the base of the building, as the grounding-electrode conductor.

Section 250.52 (4)—Ground Ring A ground ring that encircles the building may be used as the grounding-electrode conductor if the minimum rules of this section are applied.

Section 250-52 (5)—Rod and Pipe Electrodes Rods and pipes of not less than 8′ in length shall be used. Rods or pipes shall be installed in the following manner (the letters correspond to NEC subsections):

(a) Electrodes of pipe or conduit shall not be smaller than ¾″ trade size and shall have an outer surface coated for corrosion protection.

(b) Electrodes of rods of iron or steel shall be at least ⅝ inch in diameter.

Section 250-52 (6)—Other Listed Electrodes Other listed grounding electrodes shall be permitted.

Section 250-52 (7)—Plate Electrodes Each plate shall be at least ¼″ in thickness installed not less than 2 ½″ below the surface of the earth.

Section 250.52 (8)—Other Local Metal Underground Systems or Structures Underground pipes, tanks, or other metallic systems may be used as the grounding electrode. In certain situations, vehicles have been buried and used for the grounding electrode.

WARNING Metal underground gas piping systems or aluminum electrodes shall not be used for grounding purposes.

ARTICLE 250.60—USE OF AIR TERMINALS

This section is referenced in Article 800. *Air terminals* are commonly known as lightning rods. They must be bonded directly to ground in a specific manner. The grounding electrodes used for the air terminals shall not replace a building grounding electrode. Article 250.60 does not prohibit the bonding of all systems together. FPN (fine print note) number 2: Bonding together of all separate grounding systems will limit potential differences between them and their associated wiring systems.

ARTICLE 250.70—METHODS OF GROUNDING CONDUCTOR CONNECTION TO ELECTRODES

This section is referenced in Article 800 of Chapter 8. All conductors must be bonded to the grounding-electrode system. Connections made to the grounding-electrode conductor shall be made by exothermic welding, listed lugs, listed pressure connectors, listed clamps, or other listed means. Not more than one conductor shall be connected to the electrode by a single clamp.

For indoor telecommunications purposes only, a listed sheet-metal strap-type ground clamp, which has a rigid metal base and is not likely to stretch, may be used.

ARTICLE 250.94—BONDING TO OTHER SERVICES

An accessible means for connecting intersystem bonding and grounding shall be provided at the service entrance. This section is also referenced in Article 800, as telecommunications services must have an accessible means for connecting to the building bonding and grounding system where the telecommunications cables enter the building. The three acceptable means are as follows:

(1) A set of terminals securely mounted to the meter enclosure and electrically connected to the meter enclosure

(2) A bonding bar near the service equipment enclosure, meter enclosure, or raceway for service conductors

(3) A bonding bar near the grounding electrode conductor

ARTICLE 250.104—BONDING OF PIPING SYSTEMS AND EXPOSED STRUCTURAL STEEL

Article 250.104 concerns the use of metal piping and structural steel. The following section is relevant here:

Section 250.104 (A)—Metal Water Piping The section is referenced in Article 800. Interior metal water-piping systems may be used as bonding conductors as long as the interior metal water piping is bonded to the service-entrance enclosure, the grounded conductor at the service, or the grounding-electrode conductor or conductors.

ARTICLE 250.119—IDENTIFICATION OF EQUIPMENT-GROUNDING CONDUCTORS

Equipment-grounding conductors may be bare, covered, or insulated. If covered or insulated, the outer finish shall be green or green with yellow stripes. The following section is relevant:

Section 250.119 (A)—Conductors Larger than 6 AWG A conductor larger than 6 AWG (American Wire Gauge) shall be permitted. The conductor shall be permanently identified at each end and at each point where the conductor is accessible. The conductor shall have one of the following:

(1) Stripping on the insulation or covering for the entire exposed length

(2) A green coloring or covering

(3) Marking with green tape or adhesive labels

NOTE The bonding and grounding minimum specifications listed here are for safety. Further specifications for the bonding of telecommunication systems to the building grounding electrode are in TIA's J-STD-607-A standard, which is discussed in detail in Chapter 2, "Cabling Specifications and Standards."

NEC Chapter 3 Wiring Methods and Materials

NEC Chapter 3 covers wiring methods for all wiring installations. Certain articles are of special interest to telecommunication installation personnel and are described as follows.

ARTICLE 300.11—SECURING AND SUPPORTING

This article covers securing and supporting electrical and communications wiring. The following section is of interest:

Section 300.11 (A)—Secured in Place Cables and raceways shall not be supported by ceiling grids or by the ceiling support-wire assemblies. All cables and raceways shall use an independent means of secure support and shall be securely fastened in place. This section was a new addition in the 1999 code. Currently, any wires supported by the ceiling assembly are "grandfathered" and do not have to be rearranged. So if noncompliant ceiling assemblies existed before NEC 1999 was published, they can remain in place. A ceiling or the ceiling support wires cannot support new installations of cable; the cables must have their own independent means of secure support.

Ceiling support wires may be used to support cables; however, those support wires shall not be used to support the ceiling. The cable support wires must be distinguished from the ceiling support wires by color, tags, or other means. Cable support wires shall be secured to the ceiling assembly.

ARTICLE 300.21—SPREAD OF FIRE OR PRODUCTS OF COMBUSTION

Installations of cable in hollow spaces such as partition walls, vertical shafts, and ventilation spaces such as ceiling areas shall be made so that the spread of fire is not increased. Communications cables burn rapidly and produce poisonous smoke and gasses. If openings are created or used through walls, floors, ceilings, or fire-rated partitions, they shall be firestopped. If a cable is not properly firestopped, a fire can follow the cable. A basic rule of thumb is this: if a hole exists, firestop it. Firestop manufacturers have tested and approved design guidelines that must be followed when firestopping any opening.

WARNING Consult with your local jurisdiction having authority prior to installing any firestop.

ARTICLE 300.22—WIRING IN DUCTS, PLENUMS, AND OTHER AIR-HANDLING SPACES

This article applies to using communications and electrical cables in air ducts and the plenum. The following sections go into detail:

Section 300.22 (A)—Ducts for Dust, Loose Stock, or Vapor Removal No wiring of any type shall be installed in ducts used to transport dust, loose stock, or flammable vapors or for ventilation of commercial cooking equipment.

Section 300.22 (B)—Ducts or Plenums Used for Environmental Air If cables will be installed in a duct used to transport environmental air, the cable must be enclosed in a metal conduit or metallic tubing. Flexible metal conduit is allowed for a maximum length of 4′.

Section 300.22 (C)—Other Space Used for Environmental Air The space over a hung ceiling, which is used for the transport of environmental air, is an example of the type of space to which this section applies. Cables and conductors installed in environmental air-handling spaces must be listed for the use; for example, a plenum-rated cable must be installed in a plenum-rated space. Other cables or conductors that are not listed for use in environmental air-handling spaces shall be installed in electrical metallic tubing, metal conduit, or solid-bottom metal cable tray with solid metal covers.

Section 300.22 (D)—Information Technology Equipment Electric wiring in air-handling spaces beneath raised floors for information technology equipment shall be permitted in accordance with Article 645.

NEC Chapter 5 Special Occupancy

NEC Chapters 1 through 3 apply to residential and commercial facilities. NEC Chapter 5 deals with areas that may need special consideration, including those that may be subject to flammable or hazardous gas and liquids and that have electrical or communications cabling.

NEC Chapter 7 Special Conditions

NEC Chapter 7 deals with low-power emergency systems such as signaling and fire control systems.

ARTICLE 725.1—SCOPE

This article covers remote control, signaling, and power-limited circuits that are not an integral part of a device or appliance (for example, safety control equipment and building management systems). The article covers the types of conductors to be used, their insulation, and conductor support.

ARTICLE 760—FIRE-ALARM SYSTEMS

Fire-alarm systems are not normally considered part of the communications infrastructure, but the systems and wiring used for fire-alarm systems are becoming increasingly integrated into the rooms and spaces designated for communications. As such, all applicable codes must be followed. Codes of particular interest to communications personnel are as follows:

Section 760.154 (D)—Cable Uses and Permitted Substitutions Multiconductor communications cables CMP, CMR, CMG, and CM are permitted substitutions for Class 2 and 3 general- and limited-use communication cable. Class 2 or 3 riser cable can be substituted for CMP or CMR cable. Class 2 or 3 plenum cable can only be substituted for CMP plenum cable. Coaxial, single-conductor cable CMP (multipurpose plenum), CMR (multipurpose riser), CMG (multipurpose general), and CM (multipurpose) are permitted substitutions for FPLP (fire-protective signal cable plenum), FPLR (fire-protective signal cable riser), and FPL (fire-protective signal cable general use) cable, respectively.

Section 760.179 (B)—Conductor Size The size of conductors in a multiconductor cable shall not be smaller than AWG 26. Single conductors shall not be smaller than AWG 18.

Standard multiconductor communications cables are AWG 24 or larger. Standard coaxial cables are AWG 16 or larger.

ARTICLE 770—OPTICAL FIBER CABLES AND RACEWAYS

The provisions of this article apply to the installation of optical fiber cables, which transmit light for control, signaling, and communications. This article also applies to the raceways that contain and support the optical fiber cables. The provisions of this article are for the safety of the installation personnel and users coming in contact with the optical fiber cables; as such, installation personnel should follow the manufacturers' guidelines and recommendations for the installation specifics on the particular fiber being installed. Three types of optical fiber are defined in the NEC:

Nonconductive Optical fiber cables that contain no metallic members or other conductive materials are nonconductive. It is important for personnel to know whether a cable contains metallic members. Cables containing metallic members may become energized by transient voltages or currents, which may cause harm to the personnel touching the cables.

Conductive Cables that contain a metallic strength member or other metallic armor or sheath are conductive. The conductive metallic members in the cable are for the support and protection of the optical fiber—not for conducting electricity or signals—but they may become energized and should be tested for foreign voltages and currents prior to handling.

Composite Cables that contain optical fibers and current-carrying electrical conductors, such as signaling copper pairs, are composite. Composite cables are classified as electrical cables and should be tested for voltages and currents prior to handling. All codes applying to copper conductors apply to composite optical fiber cables.

WARNING The 2008 version of the NEC defines abandoned optical fiber cable as "installed optical fiber cable that is not terminated at equipment other than a connector and not identified for future use with a tag." This is important to you because 800.25 requires that abandoned cable be removed during the installation of any additional cabling. Make sure you know what abandoned cable you have and who will pay for removal when you have cabling work performed.

ARTICLE 770.12—RACEWAYS FOR OPTICAL FIBER CABLES

Plastic raceways for optical fiber cables, otherwise known as *innerducts*, shall be listed for the space they occupy; for example: a general listing for a general space, a riser listing for a riser space, or a plenum listing for a plenum space. The optical fiber occupying the innerduct must also be listed for the space. Unlisted-underground or outside-plant innerduct shall be terminated at the point of entrance.

ARTICLE 770.24—MECHANICAL EXECUTION OF WORK

Optical fiber cables shall be installed in a neat and workmanlike manner. Cables and raceways shall be supported by the building structure. The support structure for the optical fibers and raceways must be attached to the structure of the building, not attached to a ceiling, lashed to a pipe or conduit, or laid in on ductwork.

ARTICLE 770.154 (E)—CABLE SUBSTITUTIONS

In general, a cable with a higher (better) flame rating can always be substituted for a cable with a lower rating (see Table 4.1).

TABLE 4.1   Optical Fiber Cable Substitutions from NEC Table 770.154 (E)

CABLE TYPE	PERMITTED SUBSTITUTIONS
OFNP	None
OFCP	OFNP
OFNR	OFNP
OFCR	OFNP, OFCP, OFNR
OFNG, OFN	OFNP, OFNR
OFCG, OFC	OFNP, OFCP, OFNR, OFCR, OFNG, OFN

ARTICLE 770.179—LISTINGS, MARKING, AND INSTALLATION OF OPTICAL FIBER CABLES

Optical fiber cables shall be listed as suitable for the purpose; cables shall be marked in accordance with NEC Table 770.179. Most manufacturers put the marking on the optical fiber cable jacket every 2′ to 4′. The code does not tell you what type of cable to use (such as single-mode or multimode), just that the cable should be resistant to the spread of fire. For fire resistance and cable markings for optical cable, see Table 4.2.

TABLE 4.2   Optical Cable Markings from NEC Table 770.179

MARKING	DESCRIPTION
OFNP	Nonconductive optical fiber plenum cable
OFCP	Conductive optical fiber plenum cable
OFNR	Nonconductive optical fiber riser cable
OFCR	Conductive optical fiber riser cable
OFNG	Nonconductive optical fiber general-purpose cable
OFCG	Conductive optical fiber general-purpose cable
OFN	Nonconductive optical fiber general-purpose cable
OFC	Conductive optical fiber general-purpose cable

ARTICLE 770.179—LISTING REQUIREMENTS FOR OPTICAL FIBER CABLES AND RACEWAYS

Below are the listing requirements for optical fiber cables and raceways per article 770.179.

Section 770.179 (A)—Types OFNP and OFCP   Cables with these markings are for use in plenums, ducts, and other spaces used for handling environmental air. These cables have adequate fire resistance and low smoke-producing characteristics.

Section 770.179 (B)—Types OFNR and OFCR   These markings indicate cable for use in a vertical shaft or from floor to floor. These cables have fire-resistant characteristics capable of preventing the spread of fire from floor to floor.

Section 770.179 (C)—Types OFNG and OFCG   Cables with these designations are for use in spaces not classified as a plenum and are for general use on one floor. These cables are fire resistant.

Section 770.179 (D)—Types OFN and OFC   These cables are for the same use as OFNG and OFCG cables. OFN and OFC have the same characteristics as OFNG and OFCG, though they must meet different flame tests.

ARTICLE 770.182—OPTICAL FIBER RACEWAYS

Below are the listing requirements for optical fiber cables and raceways per article 770.182.

Section 770.182 (A)—Plenum raceways   These raceways have adequate fire-resistant and low smoke-producing characteristics. Plenum raceways must be used in plenum-rated areas. Plenum-rated cable is the only type of cable that may occupy the plenum-rated raceway.

Section 770.182 (B)—Riser raceways   Riser raceways have fire-resistant characteristics to prevent the spread of fire from floor to floor and must be used in the riser.

Section 770.182 (C)—General-Purpose Raceways   General-purpose raceways are fire resistant and used in general nonplenum areas, or they travel from floor to floor (if they travel from floor to floor, they have to be at least riser).

NEC Chapter 8 Communications Systems

NEC Chapter 8 is the section of the NEC that directly relates to the design and installation of a telecommunications infrastructure.

ARTICLE 800.1—SCOPE

This article covers telephone systems, telegraph systems, outside wiring for alarms, paging systems, building management systems, and other central station systems.

For the purposes of this chapter, we define *cable* as a factory assembly of two or more conductors having an overall covering.

WARNING   The 2008 version of the NEC defines abandoned communication cable as "installed communications cable that is not terminated at both ends at a connector or other equipment and not identified for future use with a tag." This is important to you because 800.25 requires that abandoned cable be removed during the installation of any additional cabling. Make sure you know what abandoned cable you have and who will pay for removal when you have cabling work performed.

ARTICLE 800.3 (A)—HAZARDOUS LOCATIONS

Cables and equipment installed in hazardous locations shall be installed in accordance with Article 500.

ARTICLE 800.24—MECHANICAL EXECUTION OF WORK

Communications circuits and equipment shall be installed in a neat and workmanlike manner. Cables installed exposed on the outer surface of ceiling and sidewalls shall be supported by the structural components of the building structure in such a manner that the cable is not damaged by normal building use. Such cables shall be attached to structural components by straps, staples, hangers, or similar fittings designed and installed so as not to damage the cable.

ARTICLE 800.44—OVERHEAD WIRES AND CABLES

Cables entering buildings from overhead poles shall be located on the pole on different crossarms from power conductors; the crossarms for communications cables shall be located below the crossarms for power. Sufficient climbing space must be between the communications cables in order for someone to reach the power cables. A minimum distance separation of 12″ must be maintained from power cables.

ARTICLE 800.47—UNDERGROUND CIRCUITS ENTERING BUILDINGS

In a raceway system underground, such as one composed of conduits, communications raceways shall be separated from electric-cable raceways with brick, concrete, or tile partitions.

ARTICLE 800.90—PROTECTIVE DEVICES

A listed primary protector shall be provided on each circuit run partly or entirely in aerial wire and on each circuit that may be exposed to accidental contact with electric light or power. Primary protection shall also be installed on circuits in a multibuilding environment on premises in which the circuits run from building to building and in which a lightning exposure exists. A circuit is considered to have lightning exposure unless one of the following conditions exists:

(1) Circuits in large metropolitan areas where buildings are close together and sufficiently high to intercept lightning.

(2) Direct burial or underground cable runs 140′ or less with a continuous metallic shield or in a continuous metallic conduit where the metallic shield or conduit is bonded to the building grounding-electrode system. An underground cable with a metallic shield or in metallic conduit that has been bonded to ground will carry the lightning to ground prior to its entering the building. If the conduit or metallic shields have not been bonded to ground, the lightning will be carried into the building on the cable, which could result in personnel hazards and equipment damage.

(3) The area has an average of five or fewer thunderstorm days per year with an earth resistance of less than 100 ohm meters.

Very few areas in the United States meet any one of these criteria. It is required that customers in areas that do meet any one of these criteria install primary protection. Primary protection is inexpensive compared to the people and equipment it protects. When in doubt, install primary protection on all circuits entering buildings no matter where the cables originate or how they travel.

Types of Primary Protectors

Several types of primary protectors are permitted by the National Electrical Code:

Fuseless primary protectors   Fuseless primary protectors are permitted under any of the following conditions:

♦ Noninsulated conductors enter the building through a cable with a grounded metallic sheath, and the conductors in the cable safely fuse on all currents greater than the current-carrying capacity of the primary protector. This protects all circuits in an overcurrent situation.

♦ Insulated conductors are spliced onto a noninsulated cable with a grounded metallic sheath. The insulated conductors are used to extend circuits into a building. All conductors or connections between the insulated conductors and the exposed plant must safely fuse in an overcurrent situation.

♦ Insulated conductors are spliced onto noninsulated conductors without a grounded metallic sheath. A fuseless primary protector is allowed in this case only if the primary protector is listed for this purpose or the connections of the insulated cable to the exposed cable or the conductors of the exposed cable safely fuse in an overcurrent situation.

♦ Insulated conductors are spliced onto unexposed cable.

♦ Insulated conductors are spliced onto noninsulated cable with a grounded metallic sheath, and the combination of the primary protector and the insulated conductors safely fuse in an overcurrent situation.

Fused primary protectors   If the requirements for fuseless primary protectors are not met, a fused type primary protector shall be used. The fused-type protector shall consist of an arrester connected between each line conductor and ground.

The primary protector shall be located in, on, or immediately adjacent to the structure or building served and as close as practical to the point at which the exposed conductors enter or attach to the building. In a residential situation, primary protectors are located on an outside wall where the drop arrives at the house. In a commercial building, the primary protector is located in the space where the outside cable enters the building. The primary-protector location should also be the one that offers the shortest practicable grounding conductor to the primary protector to limit potential differences between communications circuits and other metallic systems. The primary protector shall not be located in any hazardous location nor in the vicinity of easily ignitable material.

Requirements for Secondary Protectors

Secondary protection shunts to ground any currents or voltages that are passed through the primary protector. Secondary protectors shall be listed for this purpose and shall be installed behind the primary protector. Secondary protectors provide a means to safely limit currents to less than the current-carrying capacity of the communications wire and cable, listed telephone-line cords, and listed communications equipment that has ports for external communications circuits.

ARTICLE 800.100—CABLE GROUNDING

The metallic sheath of a communications cable entering a building shall be grounded as close to the point of entrance into the building as practicably possible. The sheath shall be opened to expose the metallic sheath, which shall then be grounded. In some situations, it may be necessary to remove a section of the metallic sheath to form a gap. Each section of the metallic sheath shall then be bonded to ground.

ARTICLE 800.100—PRIMARY-PROTECTOR GROUNDING

Primary protectors shall be grounded in one of the following ways:

Section 800.100 (A)—Grounding Conductor　The grounding conductor shall be insulated and listed as suitable for the purpose. The following criteria apply:

- Material: The grounding conductor shall be copper or other corrosion-resistant conductive material and either stranded or solid.

- Size: The grounding conductor shall not be smaller than 14 AWG.

- Run in a straight line: The grounding conductor shall be run in as straight a line as possible.

- Physical protection: The grounding conductor shall be guarded from physical damage. If the grounding conductor is run in a metal raceway (such as conduit), both ends of the metal raceway shall be bonded to the grounding conductor.

Section 800.100 (B)—Electrode　The grounding conductor shall be attached to the grounding electrode as follows:

- It should be attached to the nearest accessible location on the building or structure grounding-electrode system, the grounded interior metal water-pipe system, the power-service external enclosures, the metallic power raceway, or the power-service equipment enclosure.

- If a building has no grounding means from the electrical service, install the grounding conductor to an effectively grounded metal structure or a ground rod or pipe of not less than 5′ in length and ½″ in diameter, driven into permanently damp earth and separated at least 6′ from lightning conductors or electrodes from other systems.

Section 800.100 (D)—Bonding of Electrodes　If a separate grounding electrode is installed for communications, it must be bonded to the electrical electrode system with a conductor not smaller than 6 AWG. Bonding together of all electrodes will limit potential differences between them and their associated wiring systems.

ARTICLE 800.154—INSTALLATION OF COMMUNICATIONS WIRES, CABLES, AND EQUIPMENT

This article defines the installation of communications wires, cables, and equipment with respect to electrical-power wiring. The following summarizes important sections within Article 800.154:

Communications wires and cables are permitted in the same raceways and enclosures with the following power-limited types: remote-control circuits, signaling circuits, fire-alarm systems, nonconductive and conductive optical fiber cables, community antenna and radio distribution systems, and low-power, network-powered, broadband-communications circuits.

Communications cables or wires shall not be placed in any raceway, compartment, outlet box, junction box, or similar fitting with any conductors of electrical power.

Communications cables and wires shall be separated from electrical conductors by at least 2″, but the more separation the better. The NEC and the ANSI standards no longer give minimum power separations from high-voltage power and equipment because it has been found that separation is generally not enough to shield communications wires and cables from the induced noise of high power. Concrete, tiles, grounded metal conduits, or some other form of insulating barrier may be necessary to shield communications from power.

Installations in hollow spaces, vertical shafts, and ventilation or air-handling ducts shall be made so that the possible spread of fire or products of combustion is not substantially increased. Openings around penetrations through fire resistance-rated walls, partitions, floors, or ceilings shall be firestopped using approved methods to maintain the fire resistance rating.

The accessible portion of abandoned communications cables shall not be permitted to remain.

WARNING That previous sentence, simple as it sounds, is actually an earth shaker. Especially since the infrastructure expansion boom of the 1990s, commercial buildings are full of abandoned cable. Much of this cable is old Category 1 and Category 3 type cable, some of it with inadequate flame ratings. The cost of removing these cables as part of the process of installing new cabling could be substantial. Make sure your RFQ clearly states this requirement and who is responsible for the removal and disposal costs.

Section 800.154 (E)—Cable Substitutions In general, a cable with a higher (better) flame rating can always be substituted for a cable with a lower rating. Table 4.3 shows permitted substitutions.

TABLE 4.3 Cable Uses and Permitted Substitutions from NEC Article 800.154 (E), Table 800.154 (E)

CABLE TYPE	USE	REFERENCES	PERMITTED SUBSTITUTIONS
CMR	Communications riser cable	800.154 (B)	CMP
CMG, CM	Communications general-purpose cable	800.154 (C)(1)	CMP, CMR
CMX	Communications cable, limited use	800.53 (C)	CMP, CMR, CMG, CM

Article 179—Listings, Markings, and Installation of Communications Wires and Cables

Communications wires and cables installed in buildings shall be listed as suitable for the purpose and marked in accordance with NEC Table 800.179. Listings and markings shall not be required on a cable that enters from the outside and where the length of the cable within the building, measured from its point of entrance, is less than 50′. It is possible to install an unlisted cable more than 50′ into a building from the outside, but it must be totally enclosed in rigid metal conduit. Outside cables may not be extended 50′ into a building if it is feasible to place the primary protector closer than 50′ to the entrance point. Table 4.4 refers to the contents of NEC Table 800.179.

TABLE 4.4	Copper Communications-Cable Markings from NEC Table 800.179
MARKING	**DESCRIPTION**
CMP	Communications plenum cable
CMR	Communications riser cable
CMG	Communications general-purpose cable
CM	Communications general-purpose cable
CMX	Communications cable, limited use
CMUC	Under-carpet communications wire and cable

Conductors in communications cables, other than coaxial, shall be copper. The listings are described as follows:

Section 800.179 (A)—Type CMP   Type CMP cable is suitable for use in ducts, plenums, and other spaces used for environmental air. CMP cable shall have adequate fire-resistant and low smoke-producing characteristics.

Section 800.179 (B)—Type CMR   Type CMR cable is suitable for use in a vertical run from floor to floor and shall have fire-resistant characteristics capable of preventing the spreading of fire from floor to floor.

Section 800.179 (C)—Type CMG   Type CMG is for general use, not for use in plenums or risers. Type CMG is resistant to the spread of fire.

Section 800.179 (D)—Type CM   Type CM is suitable for general use, not for use in plenums or risers; it is also resistant to the spread of fire.

Section 800.179 (E)—Type CMX   Type CMX cable is used in residential dwellings. It is resistant to the spread of fire.

Section 800.179 (F)—Type CMUC   Type CMUC is a cable made specifically for under-carpet use; it may not be used in any other place, nor can any other cable be installed under carpets. It is resistant to flame spread.

Section 800.179 (G)—Communications Circuit Integrity Cables (CI) Type CI is a cable suitable for use in communications systems to ensure survivability of critical circuits.

Section 800.179 (H)—Communications Wires Wires and cables used as cross-connects or patch cables in communications rooms or spaces shall be listed as being resistant to the spread of fire.

Section 800.179 (I)—Hybrid Power and Communications Cable Hybrid power and communications cables are permitted if they are listed and rated for 600 volts minimum and are resistant to the spread of fire. These cables are allowed only in general-purpose spaces, not in risers or plenums.

ARTICLE 800.182—COMMUNICATIONS RACEWAYS

Requirements for communications raceways per Article 800.182 are described below.

Section 800.182 (A)—Plenum Communications Raceway Plenum-listed raceways are allowed in plenum areas; they shall have low smoke-producing characteristics and be resistant to the spread of fire.

Section 800.182 (B)—Riser Communications Raceway Riser-listed raceways have adequate fire-resistant characteristics capable of preventing the spread of fire from floor to floor.

Section 800.182 (C)—General-Purpose Communications Raceway General-purpose communications raceways shall be listed as being resistant to the spread of fire.

CABLING @ WORK: KNOWING AND FOLLOWING THE CODES

A perfect example of the importance of knowing and following the NEC code was recently displayed at a local university network cabling installation. The local installation company arrived to the job site with drawings and cabling specifications in hand. The specifications required the use of CMG type cable in some noncritical overhanging space in an auditorium. Unfortunately, the crew did not have this type of cable with them. They did, however, have plenty of CMR-rated cable that was being used to wire components through the university's riser system.

One of the junior members highlighted this problem and went on to estimate the total hours it would take to order the correct wiring and the extra time to disrupt the activities in the university. The cost seemed high. He asked the foreman, who was recently promoted, if they could just use the CMR cable. He was told no.

Luckily, the junior crew member had just finished a course on the NEC and had his trusty handbook available. He opened his 2008 edition code to Article 800.154 (E) and section 800.179 and pointed out to the foreman that CMR cable can be substituted for CMG cable. The foreman noted that the cost of CMR cable is much greater than CMG, but when the junior crew member showed him the financial analysis he had done, the foreman quickly agreed that the extra time, labor, and customer disruption was easily minimized by using the more expensive cable.

This is a great example that shows that things don't always go as planned, but good knowledge of codes can enable you to make good and efficient decisions!

Knowing and Following the Codes

Knowing and following electrical and building codes is of utmost importance. If you don't, at the very least you may have problems with building inspectors. But more importantly, an installation that does not meet building codes may endanger the lives of the building's occupants.

Furthermore, even if you are an information technology director or network manager, being familiar with the codes that affect the installation of your cabling infrastructure can help you when working with cabling and electrical contractors. Knowing your local codes can also help you when working with your local, city, county, or state officials.

The Bottom Line

Identify the key industry codes necessary to install a safe network cabling system.　In the United States, governing bodies issue codes for minimum safety requirements to protect life, health, and property. Building, construction, and communications codes originate from a number of different sources. Usually, these codes originate nationally rather than at the local city or county level.

Master It　What industry code is essential in ensuring that electrical and communications systems are safely installed in which an industry listing body is typically used to test and certify the performance of key system elements?

Understand the organization of the National Electrical Code.　The NEC is divided into chapters, articles, and sections. It is updated every three years. As of the time of this writing, the NEC is in its 2008 edition. This is a comprehensive code book, and it is critical to know where to look for information.

Master It

1.　What section describes the organization of the NEC?

2.　What is the pertinent chapter for communications systems?

Identify useful resources to make knowing and following codes easier.　It can take years to properly understand the details of the NEC. Luckily there are some useful tools and Internet websites that can make this a lot easier.

Master It

1.　Where can you view the NEC online?

2.　What is one of the best references on the Internet for the NEC?

3.　Where should you go for UL information?

4.　Where can you go to buy standards?

Chapter 5

Cabling System Components

Pick up any cabling catalog, and you will find a plethora of components and associated buzzwords that you never dreamed existed. Terms such as *patch panel, wall plate, plenum, 110-block, 66-block, modular jacks, raceways,* and *patch cords* are just a few. What do they all mean, and how are these components used to create a structured cabling system?

In this chapter, we'll provide an overview and descriptions of the inner workings of a structured cabling system so that you won't feel so confused next time you pick up a cabling catalog or work with professional cabling installers.

In this chapter, you will learn to:

◆ Differentiate between the various types of network cable and their application in the network

◆ Estimate the appropriate size of a telecommunications room based on the number of workstations per office building floor

◆ Identify the industry standard required to properly design the pathways and spaces of your cabling system

The Cable

In Chapter 2, "Cabling Specifications and Standards," we discussed the various cable media recommended by the ANSI/TIA-568-C.1 Commercial Building Telecommunications Cabling Standard and some of the cables' performance characteristics. Rather than repeating the characteristics of available cable media, we'll describe the components involved in transmitting data from the work area to the telecommunications room or enclosure. These major cable components are horizontal cable, backbone cable, and patch cords used in cross-connections and for connecting to network devices.

Horizontal and Backbone Cables

The terms *horizontal cable* and *backbone* (sometimes called *vertical* or *riser*) *cable* have nothing to do with the cable's physical orientation toward the horizon. Horizontal cables run between a cross-connect panel in a telecommunications room and a telecommunications outlet located near the work area. Backbone cables run between telecommunications rooms, and enclosures, and the main cross-connect point of a building (usually located in the *equipment room*). Figure 5.1 illustrates the typical components found in a structured cabling environment, including the horizontal cable, backbone cable, telecommunication outlets, and patch cords.

FIGURE 5.1
Typical components found in a structured cabling system

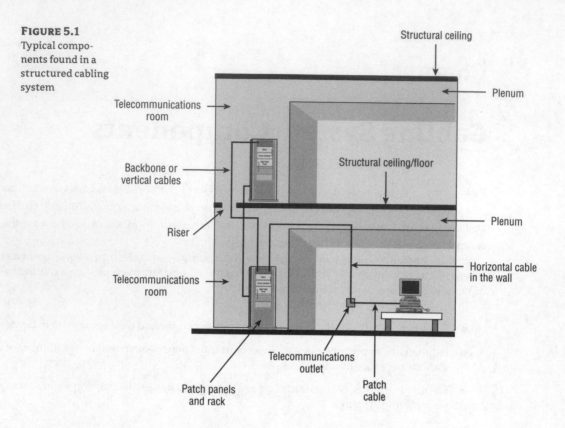

More information on horizontal and backbone cabling can be found in Chapter 2. Installing copper cabling for use with horizontal or backbone cabling is discussed in Chapter 7, "Copper Cable Media." Installing optical fiber cabling for use with horizontal or backbone cabling is discussed in Chapter 8, "Fiber-Optic Media."

HORIZONTAL CABLES

Horizontal runs are most often implemented with 100 ohm, four-pair, unshielded twisted-pair (UTP), solid-conductor copper cables, as specified in the ANSI/TIA-568-C.2 standard for commercial buildings. The standard also provides for horizontal cabling to be implemented using 62.5/125 micron or 50/125 micron multimode optical fiber. Optical fiber is typically used when electromagnetic interference (EMI) or radio-frequency interference (RFI) is a problem and when security is critical. Coaxial cable is not a recognized horizontal cable type for voice or data installations.

BACKBONE CABLES

Backbone cables can be implemented using 100 ohm UTP, ScTP or STP; 62.5/125 micron or 50/125 micron multimode optical fiber; or 8.3/125 micron single-mode optical cable. Neither 150 ohm STP nor coaxial cable is allowed. Optical fiber is the preferred cabling medium because of distance limitations associated with copper wiring (90 meters is the maximum distance). Optical fiber cable can transmit over distances up to 40,000 meters depending on fiber type and transmission speed! Another plus for running a fiber backbone is that glass does not conduct electricity, nor is it subject to EMI and RFI like copper is.

Patch Cords

Patch cords are used in patch panels to provide the connection between field-terminated horizontal cables and network connectivity devices (such as switches and hubs) and connections between the telecommunications outlets and network devices (such as computers, printers, and other Ethernet-based devices). They are the part of the network wiring you can actually see. As the saying goes, a chain is only as strong as its weakest link. Because of their exposed position in structured cable infrastructures, patch cords are almost always the weakest link.

Whereas horizontal UTP cables contain solid conductors, patch cords are made with stranded conductors because they are more flexible. The flexibility allows them to withstand the abuse of frequent flexing and reconnecting. Although you could build your own field-terminated patch cords, we strongly recommend against it.

The manufacture of patch cords is very exacting, and even under controlled factory conditions it is difficult to achieve and guarantee consistent transmission performance. The first challenge lies within the modular plugs themselves. The parallel alignment of the contact blades forms a capacitive plate, which becomes a source of signal coupling or crosstalk. Further, the untwisting and splitting of the pairs as a result of the termination process increases the cable's suscep tibility to crosstalk interference. If that weren't enough, the mechanical crimping process that secures the plug to the cable could potentially disturb the cable's normal geometry by crushing the conductor pairs. This is yet another source of crosstalk interference and a source of attenuation.

TIP   Patch cords that have been factory terminated and tested are required to achieve consistent transmission performance.

At first glance, patch cords may seem like a no-brainer, but they may actually be the most crucial component to accurately specify. When specifying patch cords, you may also require that your patch cords be tested to ensure that they meet the proper transmission-performance standards for their category.

Picking the Right Cable for the Job

Professional cable installers and cable-plant designers are called upon to interpret and/or draft cable specifications to fulfill businesses' structured-cabling requirements. Anyone purchasing cable for business or home use may also have to make a decision regarding what type of cable to use. Installing inappropriate cable could be unfortunate in the event of a disaster such as a fire.

What do we mean by unfortunate? It is conceivable that the cable-plant designer or installer could be held accountable in court and held responsible for damages incurred as a result of substandard cable installation. Cables come in a variety of ratings, and many of these ratings have to do with how well the cable will fare in a fire.

Using the general overview information provided in Chapter 1 and the more specific information in Chapters 2 through 4, you should now have adequate information to specify the proper cable for your installation.

First, you must know the installation environment and what the applicable NEC and local fire-code requirements will allow regarding the cables' flame ratings. In a commercial building, this usually comes down to where plenum-rated cables must be installed and where a lower rating (usually CMR) is acceptable.

Your second decision on cabling must be on media type. The large majority of new installations use fiber-optic cable in the backbone and UTP cable for the horizontal.

For fiber cable, you will need to specify single-mode or multimode, and if it is multimode, you will need to specify core diameter—that is, 62.5/125 or 50/125. The large majority of new installations utilize an 850nm, laser-optimized 50/125 multimode fiber (TIA-492AAAC-A); better known to the industry as OM3 fiber (per ISO/IEC 11801 Ed. 2). For UTP cables, you need to specify the appropriate transmission-performance category. Most new installations today use Category 6, and there is a growing migration to Category 6A. Make sure that you specify that patch cords be rated in the same category as, or higher than, the horizontal cable.

Wall Plates and Connectors

Telecommunications outlets can be located on a wall or surface and/or floor-mounted boxes in your work area. Telecommunications outlets located on a wall are commonly referred to as *wall plates*. Wall plates, or surface and/or floor-mounted boxes, and connectors serve as the work-area endpoints for horizontal cable runs. Using these telecommunications outlets helps you organize your cables and aids in protecting horizontal wiring from end users. Without the modularity provided by telecommunications outlets, you would wind up wasting a significant amount of cable trying to accommodate all the possible computer locations within a client's work area—the excess cable would most likely wind up as an unsightly coil in a corner. Modular wall plates can be configured with outlets for UTP, optical fiber, coaxial, and audio/visual cables.

NOTE   Refer to Chapter 9 for more information on wall plates.

Wall plates and surface- and floor-mounted boxes come in a variety of colors to match your office's decor. Companies such as Leviton, Ortronics, Panduit, and The Siemon Company also offer products that can be used with modular office furniture. The Siemon Company even went one step further and integrated its telecommunications cabling system into its own line of office furniture called MACsys. Figure 5.2 shows a sample faceplate from the Ortronics TracJack line of faceplates.

To help ensure that a cable's proper bend radius is maintained, Leviton, Panduit and The Siemon Company all offer angled modules to snap into their faceplates. Figure 5.3 shows The Siemon Company's CT faceplates and MAX series angled modules. Faceplates with angled modules for patch cords keep the cord from sticking straight out and becoming damaged.

FIGURE 5.2
An Ortronics
TracJack faceplate
configured with
UTP and audio/
visual modular
outlets

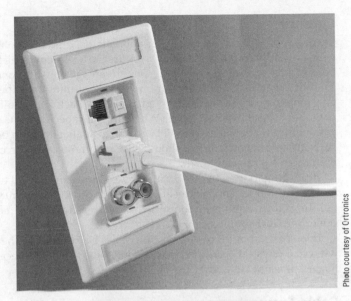

Photo courtesy of Ortronics

FIGURE 5.3
A Siemon Com-
pany's CT faceplate
configured with
UTP and optical
fiber MAX series
angled modules

Photo courtesy of The Siemon Company/

Cabling Pathways

In this section, we'll look at the cabling system components outlined by the TIA-569-B
Commercial Building Telecommunications Pathways and Spaces Standard for concealing,
protecting, and routing your cable plant. In particular, we'll describe the components used
in work areas and telecommunications rooms and for horizontal and backbone cable runs.
As you read these descriptions, you'll notice all components must be electrically grounded
per the ANSI/TIA-607-B Commercial Building Grounding and Bonding Requirements for
Telecommunications.

Conduit

Conduit is pipe. It can be metallic or nonmetallic, rigid or flexible (as permitted by the applicable electrical code), and it runs from a work area to a telecommunications room and a telecommunications room to an equipment room. One advantage of using conduit to hold your cables is that conduit may already exist in your building. Assuming the pipe has enough space, it shouldn't take long to pull your cables through it. A drawback to conduit is that it provides a finite amount of space to house cables. When drafting specifications for conduit, we recommend that you require that enough conduit be installed so that it would be only 40 percent full by your current cable needs. Conduit should be a maximum of 60 percent full. This margin leaves you with room for future growth.

According to the TIA-569-B standard, conduit can be used to route horizontal and backbone cables. Firestopped conduit can also be used to connect telecommunications rooms in multistoried buildings to an equipment room. Some local building codes require the use of conduit for *all* cable, both telecommunication and electrical.

In no cases should communication cables be installed in the same conduit as electrical cables without a physical barrier between them. Aside from (and because of) the obvious potential fire hazard, it is not allowed by the NEC.

Cable Trays

As an alternative to conduit, *cable trays* can be installed to route your cable. Cable trays are typically wire racks specially designed to support the weight of a cable infrastructure. They provide an ideal way to manage a large number of horizontal runs. Cables simply lie within the tray, so they are very accessible when it comes to maintenance and troubleshooting. The TIA-569-B standard provides for cable trays to be used for both horizontal and backbone cables.

Figure 5.4 shows a cable runway system. This type of runway looks like a ladder that is mounted horizontally inside the ceiling space or over the top of equipment racks in a telecommunications or equipment room. In the ceiling space, this type of runway keeps cables from being draped over the top of fluorescent lights, HVAC equipment, or ceiling tiles; the runway is also helpful in keeping cable from crossing electrical conduit. Separating the cable is especially useful near telecommunications and equipment rooms where there may be much horizontal cable coming together. When used in a telecommunications or equipment room, this runway can keep cables off the floor or can run from a rack of patch panels to an equipment rack.

Another type of cable-suspension device is the CADDY CatTrax from ERICO. These cable trays are flexible and easy to install, and they can be installed in the ceiling space, telecommunications room, or equipment room. The CatTrax (shown in Figure 5.5) also keeps cables from being laid directly onto the ceiling tile of a false ceiling or across lights and electrical conduit because it provides continuous support for cables.

TIP Numerous alternatives to cable-tray supports are available. One of the most common is a J hook. J hooks are metal supports in the shape of an *L* or *J* that attach to beams, columns, walls, or the structural ceiling. Cables are simply draped from hook to hook. Spacing of hooks should be from 4´ to 5´ maximum, and the intervals should vary slightly to avoid the creation of harmonic intervals that may affect transmission performance.

FIGURE 5.4
A runway system
used to suspend
cables overhead

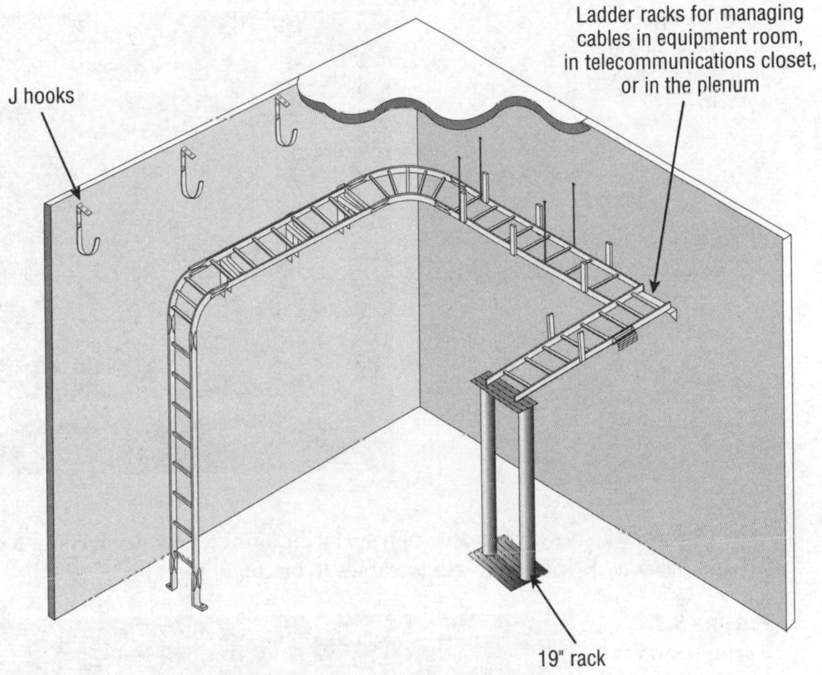

J hooks

Ladder racks for managing
cables in equipment room,
in telecommunications closet,
or in the plenum

19" rack

FIGURE 5.5
The CADDY
CatTrax flexible
cable tray from
ERICO

Photo courtesy of ERICO

Raceways

Raceways are special types of conduits used for surface-mounting horizontal cables. Raceways
are usually pieced together in a modular fashion with vendors providing connectors that do
not exceed the minimum bend radius. Raceways are mounted on the outside of a wall in places
where cable is not easily installed inside the wall; they are commonly used on walls made of brick
or concrete where no telecommunications conduit has been installed. To provide for accessibility
and modularity, raceways are manufactured in components (see Figure 5.6).

FIGURE 5.6
A surface-mounted
modular raceway
system

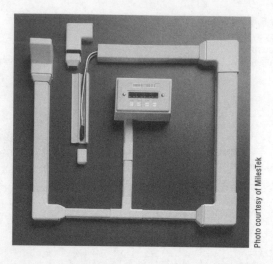

Photo courtesy of MilesTek

Figure 5.7 shows a sample of a surface-mount raceway carrying a couple of different cables; this raceway is hinged to allow cables to be easily installed.

FIGURE 5.7
A sample surface-
mount raceway
with cables

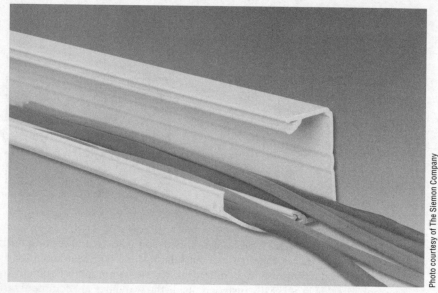

Photo courtesy of The Siemon Company

One-piece systems usually provide a flexible joint for opening the raceway to access cables; after opening, the raceway can be snapped shut. To meet information-output needs, raceway vendors often produce modular connectors to integrate with their raceway systems.

Fiber-Protection Systems

As with raceways, *fiber-protection systems* (see Figure 5.8) are special types of conduits and cable-management systems designed specifically to address the special protection needs of optical fiber cable. Although maintaining proper bend radius is important for all cable media, severe bends in optical fiber cable will result in attenuation and eventual signal loss, which translates to lost data, troubleshooting, downed network connections, and lost productivity. Severe bends can also lead to cracking and physical failure of the fiber. By employing rounded surfaces and corners, fiber-protection systems essentially limit the degree of bending put on an optical fiber cable. To protect your fiber investment, we recommend that you consider investing in a fiber-protection system.

FIGURE 5.8
The Siemon Company's LightWays fiber-protection system

Photo courtesy of The Siemon Company

KEY TERM *inner duct* Inner duct is a flexible plastic conduit system often used inside a larger conduit; fiber-optic cable is run through it for an additional layer of protection.

When evaluating a prospective fiber-protection system, you should account for the total cost of the installation rather than just the cost of materials. Also ensure that it will support the weight of your cable without sagging. In addition, because your network will grow with time, you should consider how flexible the solution will be for future modifications. Will you be able to add new segments or vertical drops without having to move existing cable? The most expensive part of your system will be the labor costs associated with the installation. Does the system require special tools to install, or does it snap together in a modular fashion?

Telecommunications Rooms, Enclosures, and Equipment Rooms

The *telecommunications room* or enclosure is where your network devices are aggregated into switches. These switches route the signals from these devices to the equipment room where servers and storage equipment are located. In this section, we'll cover the function of these rooms, along with suggested design elements. From there, we'll discuss the pieces of equipment found within a typical telecommunications and equipment room. We'll conclude with a brief discussion on network devices.

CABLING @ WORK: A TELECOMMUNICATIONS ROOM BY ANY OTHER NAME

Telecommunications rooms or enclosures are known by a number of names and acronyms. Although some cabling professionals use the term *wiring closets*, others call them *intermediate cross-connects* (ICCs) or *intermediate distribution frames* (IDFs). The ANSI/TIA-568-C.1 standard refers to them as *telecommunications rooms*. They are usually remote locations in a large or multistory building.

The telecommunications rooms or enclosures are all connected to a central wiring center known by the ANSI/TIA-568-C.1 standard as an *equipment room*. Other cabling professionals call this the *main distribution frame* (MDF) or the *main cross-connect* (MCC).

The terms *intermediate cross-connect, main distribution frame,* and *main cross-connect* are incomplete descriptions of the rooms' purposes because modern systems need to house electronic gear in addition to the cross-connect frames, main or intermediate.

Horizontal cabling is run from telecommunications rooms to the workstation areas. Backbone cabling runs from the telecommunications rooms to the equipment rooms and between telecommunications rooms.

Typically a telecommunications room is a closed space; however, the ANSI/TIA-568-C.1 standard allows for workspace equipment to be connected to telecommunications enclosures. Telecommunications enclosures are typically located in open areas of an office building closer to the workspace than a telecommunications room. These enclosures contain mini-switches and patch panels and are small enough in size to be placed above the ceiling of the workspace area or on walls. These telecommunications enclosures have the same functionality as the telecommunications room, but since they are smaller in size and more efficient, using them allows for less expensive network designs. (For more information, visit TIA's Fiber Optics LAN Section at www.fols.org.)

Three types of wiring locations exist: telecommunications rooms, telecommunications enclosures, and equipment rooms.

Depending on the size of your organization and the size of your building, you may have one or more telecommunications rooms connecting into an equipment room. Telecommunications rooms are strategically placed throughout a building to provide a single point for termination from your work areas. In a multistory building, you should have at least one telecommunications room per floor. As the distances between your end devices and telecommunications room approach their recommended maximum limits (90 meters), you should consider implementing additional telecommunications rooms. Ideally, these are included during the planning stage prior to construction or remodeling.

Telecommunications rooms are connected to the equipment room in a star configuration by either fiber or copper backbone cables. As we mentioned in our discussion of backbone cabling, fiber is preferred because fiber allows for distances from the equipment room to the last telecommunications room of up to 2,000 meters for multimode and 3,000 meters for single mode. When connecting with UTP copper, the backbone run lengths must total no more than 800 meters for telephone systems and no more than 90 meters for data systems.

A telecommunications enclosure is essentially a mini-telecommunications room. These enclosures contain active switches and patch panels and have the same functionality as the equipment located in a telecommunications room. The advantage of using telecommunications enclosures instead of telecommunications rooms is in the higher switch port utilization and cost savings obtained from eliminating the construction of dedicated rooms and associated HVAC loading. The Fiber Optics LAN Section of TIA (www.fols.org) has conducted extensive cost modeling showing the advantages of using telecommunications enclosures. However, telecommunications enclosures do not come without disadvantages, such as the need to service certain types of telecommunications enclosures in an open workspace.

TIA/EIA Recommendations for Telecommunications Rooms

The TIA/EIA does not distinguish between the roles of telecommunications rooms for its published standards. The following is a summary of the minimum standards for a telecommunications wiring room per the ANSI/TIA-569-B Commercial Building Telecommunications Pathways and Spaces Standard:

- The telecommunications room must be dedicated to telecommunications functions.

- Equipment not related to telecommunications shall not be installed in or enter the telecommunications room.

- Multiple rooms on the same floor shall be interconnected by a minimum of one 78(3) (3″ or 78mm opening) trade-size conduit or equivalent pathway.

- The telecommunications room must support a minimum floor loading of 2.4 kilopascals (50 lbf/ft^2).

The equipment room is used to contain the main distribution frame (the main location for backbone cabling), phone systems, power protection, uninterruptible power supplies, LAN equipment (such as bridges, routers, switches, and hubs), and any file servers and data processing equipment. TIA-569-B recommends a minimum of 0.75 square feet of floor space in the equipment room for every 100 square feet of user workstation area. You can also estimate the requirements for square footage using Table 5.1, which shows estimated equipment-room square footage based on the number of workstations.

TABLE 5.1   Estimated Square Foot Requirements Based on the Number of Workstations

NUMBER OF WORKSTATIONS	ESTIMATED EQUIPMENT ROOM FLOOR SPACE
1 to 100	150 square feet
101 to 400	400 square feet
401 to 800	800 square feet
801 to 1,200	1,200 square feet

TIP Further information about the TIA-569-B standard can be found in Chapter 2.

NOTE The floor space required in any equipment room will be dictated by the amount of equipment that must be housed there. Use Table 5.1 for a base calculation, but don't forget to take into account equipment that may be in this room, such as LAN racks, phone switches, and power supplies.

Here are some additional requirements:

◆ There shall be a minimum of two dedicated 120V 20A nominal, nonswitched, AC duplex electrical-outlet receptacles, each on separate branch circuits.

◆ Additional convenience duplex outlets shall be placed at 1.8 meter (6′) intervals around the perimeter, 150mm (6″) above the floor.

◆ There shall be access to the telecommunications grounding system, as specified by ANSI/TIA-607-B.

◆ HVAC requirements to maintain a temperature the same as the adjacent office area shall be met. A positive pressure shall be maintained with a minimum of one air change per hour or per code.

◆ There shall be a minimum of one room per floor to house telecommunications equipment/cable terminations and associated cross-connect cable and wire.

◆ The telecommunications room shall be located near the center of the area being served.

◆ Horizontal pathways shall terminate in the telecommunications room on the same floor as the area served.

◆ The telecommunications room shall accommodate seismic requirements.

◆ Two walls should have 20mm (¾″) A-C plywood 2.44m (8′) high.

◆ Lighting shall be a minimum of 500 lx (50 footcandles) and mounted 2.6m (8.5′) above the floor.

◆ False ceilings shall not be provided.

◆ There shall be a minimum door size of 910mm (36″) wide and 2,000mm (80″) high without sill, hinged to open outward or slide side-to-side or be removable, and it shall be fitted with a lock.

Although these items are suggestions, we recommend that you strive to fulfill as many of these requirements as possible. If your budget only allows for a few of these suggestions, grounding, separate power, and the ventilation and cooling requirements should be at the top of your list.

NOTE As noted in Chapter 2, telecommunications rooms and equipment rooms should be locked. If your organization's data is especially sensitive, consider putting an alarm system on the rooms.

Cabling Racks and Enclosures

Racks are the pieces of hardware that help you organize cabling infrastructure. They range in height from 39″ to 84″ and come in two widths: 19″ and 23″. Nineteen-inch widths are much more commonplace and have been in use for nearly 60 years. These racks are commonly called just 19″ racks or, sometimes, EIA racks. Mounting holes are spaced between ⅝″ and 2″ apart, so you can be assured that no matter what your preferred equipment vendor is, its equipment will fit in your rack. In general, three types of racks are available for purchase: wall-mounted brackets, skeletal frames, and full equipment cabinets.

TIP   Not all racks use exactly the same type of mounting screws or mounting equipment. Make sure that you have sufficient screws or mounting gear for the types of racks you purchase.

WALL-MOUNTED BRACKETS

For small installations and areas where economy of space is a key consideration, *wall-mounted brackets* may provide the best solution. Wall-mounted racks such as MilesTek's Swing Gate wall rack in Figure 5.9 have a frame that swings out 90 degrees to provide access to the rear panels and includes wire guides to help with cable management.

FIGURE 5.9
MilesTek's Swing
Gate wall rack

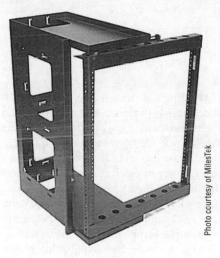

Photo courtesy of MilesTek

Racks such as the one in Figure 5.9 are ideal for small organizations that may only have a few dozen workstations or phone outlets but are still concerned about building an organized cabling infrastructure.

TIP   Prior to installing wall-mounted racks with swinging doors, be sure to allow enough room to open the front panel.

SKELETAL FRAMES (19″ RACKS)

Skeletal frames, often called *19″ racks* or *EIA racks*, are probably the most common type of rack. These racks, like the one shown in Figure 5.10, are designed and built based on the EIA/ECA-310-E standard, issued in 2005. These skeletal frames come in sizes ranging from 39″ to 84″ in height with a 22″ base plate to provide stability. Their open design makes it easy to work on both the front and back of the mounted equipment.

FIGURE 5.10
A skeletal frame
(19″ rack)

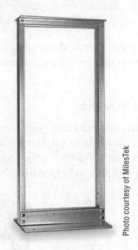

Photo courtesy of MilesTek

When installing a skeletal frame, you should leave enough space between the rack and the wall to accommodate the installed equipment (most equipment is 6″ to 18″ deep). You should also leave enough space behind the rack for an individual to work (at least 12″ to 18″). You will also need to secure the rack to the floor so that it does not topple over.

These racks can also include cable management. If you have ever worked with a rack that has more than a few dozen patch cords connected to it with no cable management devices, then you understand just how messy skeletal racks can be. Figure 5.11 shows an Ortronics Mighty Mo II wall-mount rack that includes cable management.

Racks are not limited to just patch panels and network connectivity devices. Server computers, for example, can be installed into a rack-mountable chassis. Many accessories can be mounted into rack spaces, including utility shelves, monitor shelves, and keyboard shelves. Figure 5.12 shows some of the more common types of shelves available for 19″ racks. If you have a need for some sort of shelf not commercially available, most machine shops are equipped to manufacture it.

FULL EQUIPMENT CABINETS

The most expensive of your rack options, *full equipment cabinets*, offer the security benefits of locking cabinet doors. Full cabinets can be as simple as the ones shown in Figure 5.13, but they can also become quite elaborate, with Plexiglas doors and self-contained cooling systems. Racks such as the one in Figure 5.13 provide better physical security, cooling, and protection against electromagnetic interference than standard 19″ rack frames. In some high-security environments, this type of rack is required for LAN equipment and servers.

FIGURE 5.11
The Ortronics
Mighty Mo II wall-
mount rack with
cable management

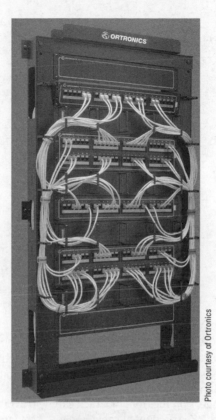

Photo courtesy of Ortronics

FIGURE 5.12
Shelves available
for 19″ racks

Photo courtesy of MilesTek

FIGURE 5.13
A full equipment
cabinet

Photo courtesy of MilesTek

CABLE MANAGEMENT ACCESSORIES

If your rack equipment does not include wire management, numerous cable management accessories, as shown in Figure 5.14, can suit your organizational requirements. Large telecommunications rooms can quickly make a rat's nest out of your horizontal cable runs and patch cables. Cable hangers on the front of a rack can help arrange bundles of patch cables to keep them neat and orderly. Rear-mounted cable hangers provide strain-relief anchors and can help to organize horizontal cables that terminate at the back of patch panels.

FIGURE 5.14
Cable management
accessories from
MilesTek

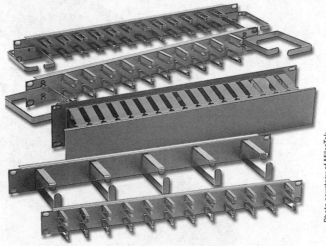

Photo courtesy of MilesTek

ELECTRICAL GROUNDING

In our discussion on conduit, we stated that regardless of your conduit solution, you will have to make sure that it complies with the ANSI/TIA-607-B Commercial Building Grounding and Bonding Requirements for Telecommunications Standard for electrical grounding. The same holds true for your cable-rack implementations. Why is this so important? Well, to put it bluntly, your network can kill you, and in this case, we're not referring to the massive coronary brought on by users' printing challenges!

For both alternating- and direct-current systems, electrons flow from a negative to a positive source, with two conductors required to complete a circuit. If a difference in resistance exists between a copper wire path and a grounding path, a voltage potential will develop between your hardware and its earth ground. In the best-case scenario, this voltage potential will form a Galvanic cell, which will simply corrode your equipment. This phenomenon is usually demonstrated in freshman chemistry classes by using a potassium-chloride salt bridge to complete the circuit between a zinc anode and a copper cathode. If the voltage potential were to become great enough, however, simply touching your wiring rack could complete the circuit and discharge enough electricity to kill you or one of your colleagues.

WARNING   Grounding is serious business and should not be undertaken by the layperson. Low voltage does not mean large shocks cannot be generated.

We recommend working with your electrical contractor and power company to get the best and shortest ground you can afford. One way to achieve this is to deploy separate breaker boxes for each office area. Doing so will shorten the grounding length for each office or group.

Cross-Connect Devices

Fortunately for us, organizations seem to like hiring consultants; however, most people are usually less than thrilled to see some types of consultants—in particular, space-utilization and efficiency experts. Why? Because they make everyone move! *Cross-connect devices* are cabling components you can implement to make changes to your network less painful.

THE 66 PUNCH-DOWN BLOCKS

For more than 25 years, *66 punch-down blocks*, shown in Figure 5.15, have been used as telephone system cross-connect devices. They support 50 pairs of wire. Wires are connected to the terminals of the block using a punch-down tool. When a wire is "punched down" into a terminal, the wire's insulation is pierced and the connection is established to the block. Separate jumpers then connect blocks. When the need arises, jumpers can be reconfigured to establish the appropriate connections.

The use of 66 punch-down blocks has dwindled significantly in favor of 110-blocks.

FIGURE 5.15
A 66 punch-down
block

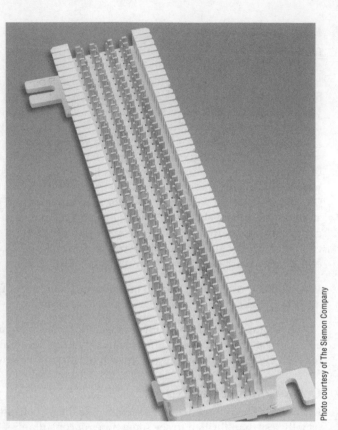

Photo courtesy of The Siemon Company

THE 110 AND S-210 PUNCH-DOWN BLOCKS

Figure 5.16 shows *110-blocks*, another flavor of punch-down media; they are better suited for use with data networks. The 110-blocks come in sizes that support anywhere from 25 to 500 wire pairs. Unlike 66-blocks, which use small metal jumpers to bridge connections, 110-blocks are not interconnected via jumpers but instead use 24 AWG cross-connect wire. Both Leviton and The Siemon Company produce 110-blocks capable of delivering Category 6 performance.

Some installations of data and voice systems require the use of 25-pair connectors. Some network hubs and phone systems use these 25-pair connectors, rather than modular-type plugs like the RJ-45, to interface with their hardware. You can purchase 110-style connector blocks prewired with 25-pair connector cables, such as the one shown in Figure 5.17.

TIP If you purchase a 110- or 66-style block wired to 25-pair connectors, make sure the equipment is rated to the appropriate category of cable performance that you intend to use it with. The 66-blocks are rarely used for data.

FIGURE 5.16
Another type
of punch-down
media, 110 punch-
down blocks

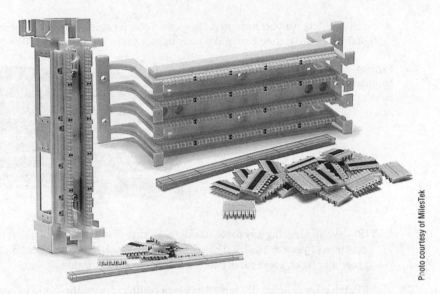

Photo courtesy of MilesTek

FIGURE 5.17
The Siemon Com-
pany's prewired
110-block with
25-pair connectors

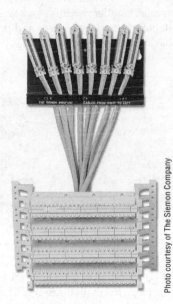

Photo courtesy of The Siemon Company

MODULAR PATCH PANELS

As an alternative to punch-down blocks, you can terminate your horizontal cabling directly into RJ-45 patch panels (see Figure 5.18). This approach is becoming increasingly popular because it lends itself to exceptionally easy reconfigurations. To reassign a network client to a new port on the switch, all you have to do is move a patch cable. Another benefit is that when they're installed cleanly, they can make your telecommunications room look great!

FIGURE 5.18
Modular patch
panels

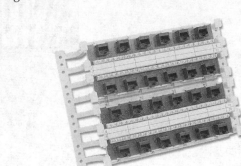

Photo courtesy of MilesTek

TIP When ordering any patch panel, make sure that you order one that has the correct wiring pattern (T568A or T568B). The wiring pattern is usually color-coded on the 110-block. As with modular jacks, some patch panels allow either configuration.

Patch panels normally have 110-block connectors on the back. In some environments, only a few connections are required, and a large patch panel is not needed. In other environments, it may not be possible to mount a patch panel with a 110-block on the back because of space constraints. In this case, smaller modular-jack wall-mount blocks (see Figure 5.19) may be useful. These are available in a variety of sizes and port configurations. You can also get them in either horizontal or backbone configurations.

FIGURE 5.19
The Siemon
Company's S-110
modular-jack wall-
mount block

Photo courtesy of The Siemon Company

CONSOLIDATION POINTS

Both the ANSI/TIA-568-C and ISO/IEC 11801 Ed. 2 standards allow for a single transition point or *consolidation point* in horizontal cabling. The consolidation point is usually used to transition between a 25-pair UTP cabling (or separate four-pair UTP cables) that originated in the telecommunications room and cable that spreads out to a point where many networked or voice devices may be, such as with modular furniture. An example of a typical consolidation point (inside a protective cabinet) is shown in Figure 5.20.

FIGURE 5.20
A consolidation
point

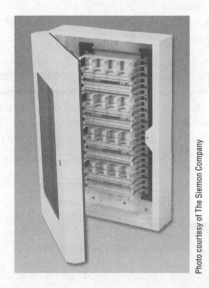

Photo courtesy of The Siemon Company

NOTE One type of consolidation point is a multiuser telecommunications outlet assembly (MUTOA). Basically, this is a patch-panel device located in an open office space. Long patch cords are used to connect workstations to the MUTOA. When using a MUTOA, the 90 meter horizontal cabling limit must be shortened to compensate for the longer patch cords.

FIBER-OPTIC CONNECTOR PANELS

If your organization is using optical fiber cabling (either for horizontal or backbone cabling), then you may see *fiber-optic connector panels*. These will sometimes look similar to the UTP RJ-45 panels seen earlier in this chapter, but they are commonly separate boxes that contain space for cable slack. A typical 24-port fiber-optic panel is pictured in Figure 5.21.

FIGURE 5.21
A fiber-optic
connector panel

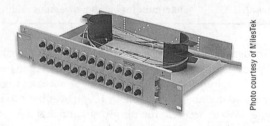

Photo courtesy of NilesTek

Administration Standards

After troubleshooting a network issue and figuring out that it's a problem with the physical layer, have you ever found complete spaghetti in a telecommunications room? In our consulting practices, we see this all too often. Our clients then pay two to three times the regular consulting fees because it takes so much time to sort through the mess.

NOTE Network administrators should be judged by the neatness of their telecommunications rooms.

To provide a standard methodology for the labeling of cables, pathways, and spaces, the TIA published the ANSI/TIA/EIA-606-A Administration Standard for the Telecommunications Infrastructure of Commercial Buildings. In addition to guidelines for labeling, the standard recommends the color-coding scheme shown in Table 5.2. This scheme applies not only to labeling of cables and connections but also to the color of the cross-connect backboards in the telecommunication rooms. It does not necessarily apply to the colors of cable jackets, although some installations may attempt to apply it.

TABLE 5.2   Color-Coding Schemes

COLOR CODE	USAGE
Black	No termination type assigned
White	First-level backbone (MC/IC or MC/TC terminations)
Red	Reserved for future use
Gray	Second-level backbone (IC/TC terminations)
Yellow	Miscellaneous (auxiliary, security alarms, etc.)
Blue	Horizontal-cable terminations
Green	Network connections
Purple	Common equipment (PBXs, host LANs, muxes)
Orange	Demarcation point (central office terminations)
Brown	Interbuilding backbone (campus cable terminations)

Besides labeling and color coding, you should consider bundling groups of related cables with plastic cable ties (tie-wraps). Plastic cable ties come in a variety of sizes for all kinds of applications. When bundling cables, however, be sure not to cinch them too tightly, as you could disturb the natural geometry of the cable. If you ever have to perform maintenance on a group of cables, all you have to do is cut the plastic ties and add new ones when you're finished. Many companies make hook-and-loop (Velcro) type tie-wraps, and these are recommended over tie-wraps for both copper and optical fiber cables as they typically prevent over-cinching.

TIP   While hook-and-loop cable wraps are more expensive than traditional thin plastic tie wraps, they more than pay for themselves by assuring that cables are not over-cinched; be sure to have plenty on hand.

Whether you implement the ANSI/TIA/EIA-606-A standard or come up with your own methodology, the most import aspect of cable administration is to have accurate documentation of your cable infrastructure.

STOCKING YOUR TELECOMMUNICATIONS ROOMS

The wiring equipment discussed in this chapter is commonly found in many cabling installations; larger, more complex installations may have additional components that we did not mention here. The components mentioned in this chapter can be purchased from just about any cabling or telecommunications supplier. Some of the companies that were very helpful in the production of this chapter have much more information online. You can find more information about these companies and their products by visiting them on the Web:

ERICO	www.erico.com
Leviton	www.leviton.com
MilesTek	www.milestek.com
Ortronics	www.ortronics.com
The Siemon Company	www.siemon.com

The Bottom Line

Differentiate between the various types of network cable and their application in the network. ANSI/TIA-568-C.1 defines the various types of network cabling typically used in an office building environment. Much like major arteries and capillaries in the human body, backbone and horizontal cabling serve different functions within a network. Each of these cable types can use either fiber-optic or copper media. In specifying the network cable it is important to understand which type of cable to use and the proper media to use within it.

Master It You are asked to design a network for a company fully occupying a three-story office building using a traditional star network topology. The equipment room, containing the company's server, is located on the first floor, and telecommunications rooms are distributed among each of the floors of the building. The shortest run between the equipment room and the telecommunications rooms is 200 meters. What type of network cable would you use to connect the telecommunications rooms to the equipment room and what type of media would you specify inside the cable?

Estimate the appropriate size of a telecommunications room based on the number of workstations per office building floor. The equipment room is like the "heart" (and brain) of the office building LAN. This is where servers, storage equipment, and the main cross-connection are located, and it also typically serves as the entrance of an outside line to the wide area network. It is important that the size of the equipment room be large enough to contain the key network devices.

Master It   In the previous example, you are asked to design a network for a company fully occupying a three-story office building using a traditional star network topology. The company presently has 300 workstations, but they expect to grow to 500 workstations in 1–2 years. Which standard should you reference for the answer and what is the estimate for the minimum size (in square feet) of the equipment room?

Identify the industry standard required to properly design the pathways and spaces of your cabling system.   Providing the proper pathways and spaces for your cabling system is critical to ensuring that you obtain the level of performance the media is rated to. Too many bends, sharp bends, or twists and turns can significantly impair the bandwidth of the media and could prevent your network from operating

Master It   Which of these industry standards provides you with the recommendations you must follow to ensure a properly working network?

A. J-STD-607-A

B. ANSI/TIA/EIA-606-A

C. TIA-569-B

D. ANSI/TIA-568-C.1

Chapter 6

Tools of the Trade

This chapter discusses tools that are essential to proper installation of data and video cabling. It also describes tools, many of which you should already have, that make the job of installing cables easier.

If you're reading this book, it is likely that you're a do-it-yourselfer or you're managing people who are hands-on. So be advised: don't start any cabling job without the proper tools. You might be able to install a data cabling system with nothing but a knife and screwdriver, but doing so may cost you many hours of frustration and diminished quality.

In addition to saving time, using the appropriate tools will save money. Knowing what the right tools are and where to use them is an important part of the job.

In this chapter, you will learn to:

◆ Select common cabling tools required for cabling system installation

◆ Identify useful tools for basic cable testing

◆ Identify cabling supplies to make cable installations easy

Building a Cabling Toolkit

Throughout this chapter, a number of tools are discussed, and photos illustrate them. Don't believe for a minute that we've covered all the models and permutations available! This chapter is an introduction to the types of tools you may require, and will help you recognize a particular tool so you can get the one that best suits you. It is impossible for us to determine your exact tool needs. Keeping your own needs in mind, read through the descriptions that follow, and choose those tools that you anticipate using.

THE RIGHT TOOL AND THE RIGHT PRICE

Just as the right tools are important for doing a job well, so is making sure that you have high-quality equipment. Suppose you see two punch-down tools advertised in a catalog and one of them is $20 and the other is $60. Ask why one is more expensive than the other is. Compare the features of the two; if they seem to be the same, you can usually assume that more expensive tool is designed for professionals.

With all tools, there are levels of quality and a range of prices you can choose from. It's trite but true: you get what you pay for, generally speaking, so our advice is to stay away from the really cheap stuff. On the other hand, if you only anticipate light to moderate use, you needn't buy top-of-the-line equipment.

Myriad online catalog houses and e-commerce sites sell the tools and parts you need to complete your cabling tool kit. A few of these include:

◆   IDEAL DataComm at www.idealindustries.com

◆   Jensen Tools at www.stanleysupplyservices.com

◆   Labor Saving Devices Inc. at www.lsdinc.com

◆   MilesTek at www.milestek.com

◆   The Siemon Company at www.siemon.com

If you have to scratch and sniff before buying, visit a local distributor in your area. Check your local phone book for vendors such as Anicom, Anixter, CSC, Graybar, and many other distributors that specialize in servicing the voice/data market; many of these vendors have counter sale areas where you can see and handle the merchandise before purchasing.

We can't describe in precise detail how each tool works or all the ways you can apply it to different projects. We'll supply a basic description of each tool's use, but because of the wide variety of manufacturers and models available, you'll have to rely on the manufacturer's instructions to learn how to use a particular device.

Common Cabling Tools

A number of tools are common to most cabling tool kits: wire strippers, wire cutters, cable crimpers, punch-down tools, fish tape, and toning tools. Most of these tools are essential for installing even the most basic of cabling systems.

TOOLS CAN BE EXPENSIVE

Most people who are not directly involved in the installation of telecommunications cabling systems don't realize how many tools you might need to carry or their value. A do-it-yourselfer can get by with a few hundred dollars' worth of tools, but a professional may need to carry many thousands of dollars' worth, depending on the job that is expected.

A typical cabling team of three or four installers may carry as much as $12,000 in installation gear and tools. If this team carries sophisticated testing equipment such as a fiber-optic OTDR (optical time-domain reflectometer), the value of their tools may jump to over $50,000. A fully equipped fiber-optic team carrying an OTDR and optical fiber fusion splicer could be responsible for over $100,000 worth of tools. And some people wonder why cabling teams insist on taking their tools home with them each night!

Wire Strippers

The variety of cable strippers represented in this section is a function of the many types of cable you can work with, various costs of the cable strippers, and versatility of the tools.

TWISTED-PAIR STRIPPERS

Strippers for UTP, ScTP, and STP cables are used to remove the outer jacket and have to accommodate the wide variation in the geometry of UTP cables. Unlike coax, which is usually consistently smooth and round, twisted-pair cables can have irregular surfaces due to the jacket shrinking down around the pairs. Additionally, the jacket thickness can differ greatly depending on brand and flame rating. The trick is to aid removal of the jacket without nicking or otherwise damaging the insulation on the conductors underneath.

The wire stripper in Figure 6.1 uses an adjustable blade so that you can fix the depth, matching it to the brand of cable you are working with. Some types use spring tension to help keep the blade at the proper cutting depth.

FIGURE 6.1
A wire stripper

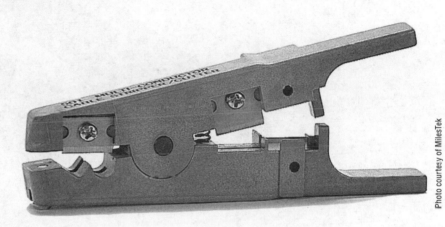

Photo courtesy of MilesTek

In both cases, the goal is to score (lightly cut) the jacket without penetrating it completely. Then, you flex the cable to break the jacket along the scored line. This ensures that the wire insulation is nick-free. In some models, the tool can also be used to score or slit the jacket lengthwise in the event you need to expose a significant length of conductors.

NOTE When working with UTP, ScTP, or STP cables, you will rarely need to strip the insulation from the conductors. Termination of these cable types on patch panels, cross-connections, and most wall plates employs the use of *insulation displacement connectors* (IDCs) that make contact with the conductor by slicing through the insulation. In case you need to strip the insulation from a twisted-pair cable, keep a pair of common electrician's strippers handy. Just make sure it can handle the finer-gauge wires such as 22, 24, and 26 AWG that are commonly used with LAN wiring.

COAXIAL WIRE STRIPPERS

Coaxial cable strippers are designed with two or three depth settings. These settings correspond to the different layers of material in the cable. Coaxial cables are pretty standardized in terms of central-conductor diameter, thickness of the insulating and shielding layers, and thickness of the outer jacket, making this an effective approach.

In the inexpensive (but effective for the do-it-yourself folks) model shown in Figure 6.2, the depth settings are fixed. The wire stripper in Figure 6.2 can be used to strip coaxial cables (RG-59 and RG-6) to prepare them for F-type connectors.

FIGURE 6.2
Inexpensive coaxial wire strippers

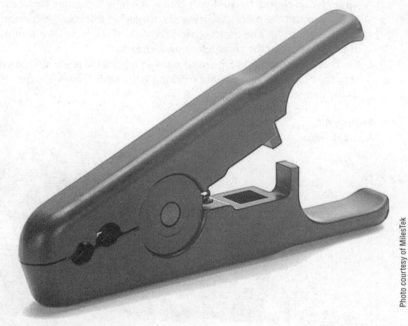

Photo courtesy of MilesTek

To strip the cable, you insert it in a series of openings that allows the blade to penetrate to different layers of the cable. At every step, you rotate the tool around the cable and then pull the tool toward the end of the cable, removing material down to where the blade has penetrated. To avoid nicking the conductor, the blade is notched at the position used to remove material.

One problem with the model shown in Figure 6.2 is that you end up working pretty hard to accomplish the task. For its low price, the extra work may be a good trade-off if stripping coax isn't a day-in, day-out necessity. However, if you are going to be working with coaxial cables on a routine basis, you should consider some heftier equipment. Figure 6.3 shows a model that accomplishes the task in a more mechanically advantageous way (that means it's easier on your hands). In addition, it offers the advantage of adjustable blades so that you can optimize the cutting thickness for the exact brand of cable you're working with.

Coaxial strippers are commonly marked with settings that assist you in removing the right amount of material at each layer from the end of the cable so it will fit correctly in an F- or BNC-type connector.

FIBER-OPTIC CABLE STRIPPERS

Fiber-optic cables require very specialized tools. Fortunately, the dimensions of fiber coatings, claddings, and buffers are standardized and manufactured to precise tolerances. This allows tool manufacturers to provide tools such as the one shown in Figure 6.4 that will remove material to the exact thickness of a particular layer without damage to the underlying layer. Typically, these look like a conventional multigauge wire stripper with a series of notches to provide the proper depth of penetration.

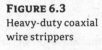

FIGURE 6.3
Heavy-duty coaxial
wire strippers

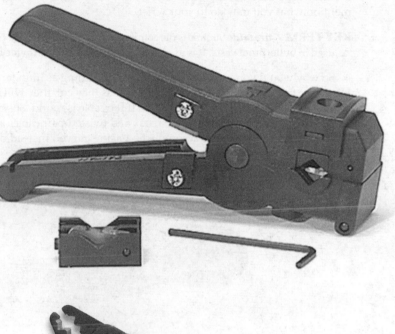

Photo courtesy of M esTek

FIGURE 6.4
A fiber-optic
cable stripper

Photo courtesy of IDEAL DataComm

Wire Cutters

You can, without feeling very guilty, use a regular set of lineman's pliers to snip through coaxial and twisted-pair cables. You can even use them for fiber-optic cables, but cutting through the aramid yarns used as strength members can be difficult; you will dull your pliers quickly, not to mention what you may do to your wrist.

KEY TERM *aramid* *Aramid* is the common name for the material trademarked as Kevlar that's used in bulletproof vests. It is used in optical fiber cable to provide additional strength.

So why would you want a special tool for something as mundane as cutting through the cable? Here's the catch regarding all-purpose pliers: as they cut, they will mash the cable flat. All the strippers described previously work best if the cable is round. Specialized cutters such as the one shown in Figure 6.5 are designed for coax and twisted-pair cables and preserve the geometry of the cable as they cut. This is accomplished using curved instead of flat blades.

FIGURE 6.5
Typical wire
cutters

Photo courtesy of MilesTek

For fiber-optic cables, special scissors are available that cut through aramid with relative ease. Figure 6.6 shows scissors designed for cutting and trimming the Kevlar strengthening members found in fiber-optic cables.

FIGURE 6.6
IDEAL
DataComm's
Kevlar scissors

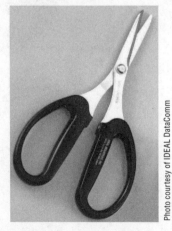

Photo courtesy of IDEAL DataComm

Cable Crimpers

Modular plugs and coaxial connectors are attached to cable ends using crimpers, which are essentially very specialized pliers. So why can't you just use a pair of pliers? Crimpers are designed to apply force evenly and properly for the plug or connector being used. Some crimpers use a ratchet mechanism to ensure that a complete crimp cycle has been made. Without this special design, your crimp job will be inconsistent at best, and it may not work at all. In addition, you'll damage connectors and cable ends, resulting in wasted time and materials. Remember that the right tool, even if it's expensive, can save you money!

TWISTED-PAIR CRIMPERS

Crimpers for twisted-pair cable must accommodate various-sized plugs. The process of crimping involves removing the cable jacket to expose the insulated conductors, inserting the conductors in the modular plug (in the proper order!), and applying pressure to this assembly using the crimper. The contacts for the modular plug (such as the ones shown in Figure 6.7) are actually blades that cut through the insulation and make contact with the conductor. The act of crimping not only establishes this contact but also pushes the contact blades down into proper position for insertion into a jack. Finally, the crimping die compresses the plug strain-relief indentations to hold the connector on the cable.

FIGURE 6.7
An eight-position
modular plug
(a.k.a. RJ-45
connector)

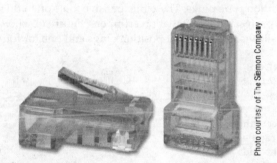

Photo courtesy of The Siemon Company

NOTE Modular plugs for cables with solid conductors (horizontal wiring) are sometimes different from plugs for cables with stranded conductors (patch cords). The crimper fits either, and some companies market a universal plug that works with either. Make sure you select the proper type when you buy plugs and make your connections.

The crimper shown in Figure 6.8 is designed so that a specific die is inserted, depending on the modular plug being crimped. If you buy a flexible model like this, you will need dies that fit an eight-conductor position (data, a.k.a. RJ-45) and a six-position type (voice, a.k.a. RJ-11 or RJ-12) plug at a minimum. If you intend to do any work with telephone handset cords, you should also get a die for four-position plugs.

FIGURE 6.8
A crimper with multiple dies for RJ-11, RJ-45, and MMJ modular connectors

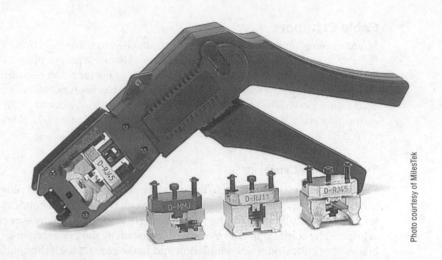

Photo courtesy of MilesTek

Other twisted-pair crimpers are configured for specific plug sizes and don't offer the flexibility of changeable dies. Inexpensive models available at the local home improvement center for less than $15 usually have two positions; these are configured to crimp eight-, six-, or four-position type plugs. These inexpensive tools often do not have the ratchet mechanism found on professional installation crimpers. Figure 6.9 shows a higher-quality crimper that has two positions: one for eight-position plugs and one for four-position plugs.

FIGURE 6.9
An IDEAL Ratchet Telemaster crimper with crimp cavities for eight- and four-position modular plug

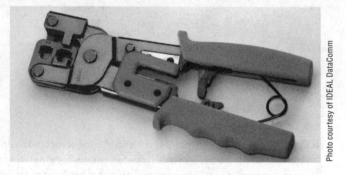

Photo courtesy of IDEAL DataComm

Less expensive crimpers are targeted at the do-it-yourself market—those who are doing a little phone-extension work around the house on a weekend or who only crimp a few cables at a time. Better-quality units targeted for the intermediate user will usually have one opening for eight-position and one opening for six-position plugs. If you work with data connectors such as the eight-position modular jack (RJ-45), your crimping tool must have a crimp cavity for eight-position plugs.

COAXIAL-CABLE CRIMPERS

Coaxial-cable crimpers also are available either with changeable dies or with fixed-size crimp openings. Models aimed strictly at the residential installer will feature dies or openings suitable for applying F-type connectors to RG-58, RG-59, and RG-6 series coax. For the commercial installer, a unit that will handle dies such as RG-11 and thinnet with BNC-type connectors is also necessary. Figure 6.10 shows IDEAL DataComm's Crimpmaster crimp tool, which can be configured with a variety of die sets such as RG-6, RG-9, RG-58, RG-59, RG-62, cable-TV F-type connectors, and others.

FIGURE 6.10
An IDEAL Crimp-master crimp tool

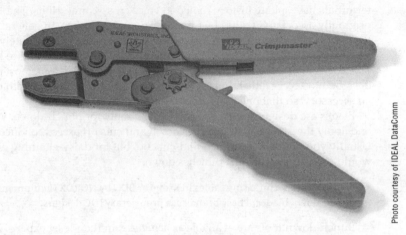

Photo courtesy of IDEAL DataComm

There's a very functional item that is used in conjunction with your crimper to install F-type RG-59 and RG-6 connectors. Figure 6.11 shows an F-type connector installation tool. One end is used to ream space between the outer jacket and the dielectric layer of the coax. On the other end, you thread the connector and use the tool to push the connector down on the cable. This accessory speeds installation of F-type connectors and reduces wear and tear on your hands.

FIGURE 6.11
A MilesTek F-type connector installation tool

Photo courtesy of MilesTek

Punch-Down Tools

Twisted-pair cables are terminated in jacks, cross-connect blocks (66-blocks), or patch panels (110-blocks) that use insulation displacement connectors (IDCs). Essentially, IDCs are little knife blades with a V-shaped gap or slit between them. You force the conductor down into the V and the knife blades cut through the insulation and make contact with the conductor. Although you could accomplish this using a small flat-blade screwdriver, doing so is not recommended. It would be sort of like hammering nails with a crescent wrench. The correct device for inserting a conductor in the IDC termination slot is a punch-down tool.

NOTE You can find more information on 66-blocks and 110-blocks in Chapter 5, "Cabling System Components," and Chapter 7, "Copper Cable Media." Additional information about wall plates can be found in Chapter 9, "Wall Plates."

A punch-down tool is really just a handle with a special "blade" that fits a particular IDC. There are two main types of IDC terminations: the 66-block and the 110-block. The 66-block terminals have a long history rooted in voice cross-connections. The 110-block is a newer design, originally associated with AT&T but now generic in usage. In general, 110-type IDCs are used for data, and 66-type IDCs are used for voice, but neither is absolutely one or the other.

Different blades are used depending on whether you are going to be terminating on 110-blocks or 66-blocks. Although the blades are very different, most punch-down tools are designed to accept either. In fact, most people purchase the tool with one and buy the other as an accessory, so that one tool serves two terminals.

Blades are designed with one end being simply for punch-down. When you turn the blade and apply the other end, it punches down and cuts off excess conductor in one operation. Usually you will use the punch-and-cut end, but for daisy-chaining on a cross-connection, you would use the end that just punches down.

TIP If you are terminating cables in Krone or BIX (by NORDX) equipment, you will need special punch-down blades. These brands use proprietary IDC designs.

Punch-down tools are available as *nonimpact* in their least expensive form. Nonimpact tools generally require more effort to make a good termination, but they are well suited for people who only occasionally perform punch-down termination work. Figure 6.12 shows a typical nonimpact punch-down tool.

FIGURE 6.12
IDEAL DataComm's nonimpact punch-down tool

Photo courtesy of IDEAL DataComm

The better-quality punch-down tools are spring-loaded *impact* tools. When you press down and reach a certain point of resistance, the spring gives way, providing positive feedback that the termination is made. Typically, the tool will adjust to high- and low-impact settings. Figure 6.13 shows an impact punch-down tool. Notice the dial near the center of the tool—it allows the user to adjust the impact setting. The manufacturer of the termination equipment you are using will recommend the proper impact setting.

With experience, you can develop a technique and rhythm that lets you punch down patch panels and cross-connections very quickly. However, nothing is so frustrating as interrupting your sequence rhythm because the blade stayed on the terminal instead of in the handle of the tool. The better punch-down tools have a feature that locks the blade in place, rather than just holding it in with friction. For the occasional user, a friction-held blade is okay, but for the professional, a lock-in feature is a must that will save time and, consequently, money.

FIGURE 6.13
IDEAL DataComm's impact tool with adjustable impact settings

TIP You should always carry at least one extra blade for each type of termination that you are doing. Once you get the hang of punch-downs, you'll find that the blades don't break often, but they do break occasionally. The cutting edge will also become dull and stop cutting cleanly. Extra blades are inexpensive and can be easily ordered from the company you purchased your punch-down tool from.

Some brands of 110-block terminations support the use of special blades that will punch down multiple conductors at once, instead of one at a time.

If you are punching down IDC connectors on modular jacks from The Siemon Company that fit into modular wall plates, a tool from that company may be of use to you. Rather than trying to find a surface to hold the modular jack against, you can use the Palm Guard (see Figure 6.14) to hold the modular jack in place while you punch down the wires.

FIGURE 6.14
The Palm Guard

TIP A 4˝ square of carpet padding or mouse pad makes a good palm protector when punching down cable on modular jacks.

Fish Tapes

A good fish tape is the best friend of the installer who does MACs (moves, adds, changes) or retrofit installations on existing buildings. Essentially, it is a long wire, steel tape, or fiberglass rod that is flexible enough to go around bends and corners but retains enough stiffness so that it can be pushed and worked along a pathway without kinking or buckling.

Like a plumber's snake, a fish tape is used to work blindly through an otherwise inaccessible area. For example, say you needed to run a cable from a ceiling space down inside a joist cavity in a wall to a new wall outlet. From within the ceiling space, you would thread the fish tape down into the joist cavity through a hole in the top plate of the wall. From this point, you would maneuver it in front of any insulation and around any other obstacles such as electrical cables that might also be running in the joist cavity. When the tape becomes visible through the retrofit outlet opening, you would draw the tape out. Then you would attach either a pull string or the cable itself and withdraw the fish tape.

Fish tapes (see Figure 6.15) are available in various lengths, with 50′ and 100′ lengths common. They come in spools that allow them to be reeled in and out as necessary and are available virtually anywhere electrical supplies are sold, in addition to those sources mentioned earlier.

FIGURE 6.15
IDEAL
DataComm's
fish tape

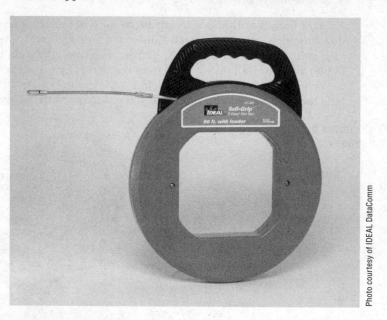

Photo courtesy of IDEAL DataComm

An alternative to fish tapes that is often helpful when placing cable in existing wall or ceiling spaces is the fiberglass pushrod, as shown in Figure 6.16. These devices are more rigid than fish tapes but are still able to flex when necessary. Their advantage is that they will always return to a straight orientation, making it easier to probe for "hidden" holes and passageways. The rigidity also lets you push a cable or pull string across a space. Some types are fluorescent or reflective so that you can easily see their position in a dark cavity. They typically come in 48″ sections that connect together as you extend them into the space.

FIGURE 6.16
Fiberglass
pushrods

Photo courtesy of Labor Saving Devices, Inc.

A number of accessories (see Figure 6.17) are available to place on the tip of a pushrod that make it easier to push the rod over obstructions, to aid in retrieval through a hole at the other end, or to attach a pull string or cable for pulling back through the space.

FIGURE 6.17
Pushrod
accessories

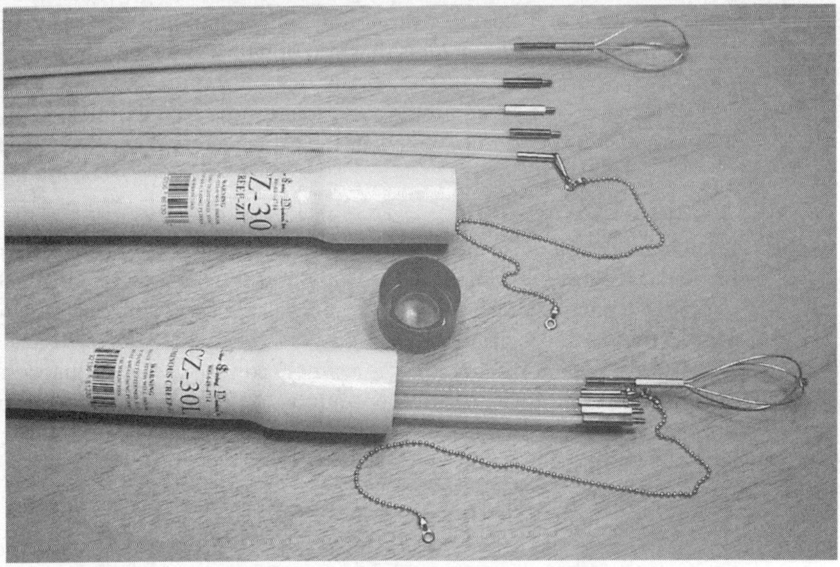

Photo courtesy of Labor Saving Devices, Inc.

Voltage Meter

There is a right way and a wrong way to determine if an electrical circuit has a live voltage on it. Touching it is the wrong way. A simple voltage meter such as the one pictured in Figure 6.18 is a much better solution, and it won't put your health plan to work. Though not absolutely necessary in the average data-cabling tool kit, a voltage meter is rather handy.

FIGURE 6.18
A voltage/
continuity tester

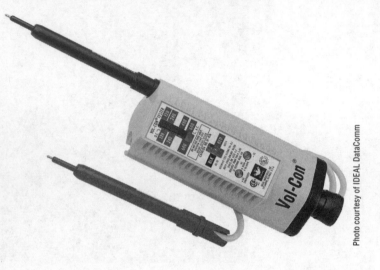

Photo courtesy of IDEAL DataComm

Cable Testing

Dozens of cable testers are available on the market. Some of them sell for less than $100; full-featured ones sell for over $5,000. High-end fiber-optic testers can cost more than $30,000! Chapter 15, "Cable System Testing and Troubleshooting," discusses cable testing and certification, so we won't steal any thunder from that chapter here. However, in your toolkit you should include some basic tools that you don't need to get a second mortgage on your house to purchase.

Cable testers can be as simple as a cable-toning tool that helps you to identify a specific cable; they can also be continuity testers or the cable testers that cost thousands of dollars.

A Cable-Toning Tool

A cable toner is a device for determining if the fundamental cable installation has been done properly. It should be noted that we are not discussing the sophisticated type of test set required to certify a particular level of performance, such as a Category 5e link or channel. These are discussed in detail in Chapter 15.

In its simplest form, the toner is a simple continuity tester that confirms that what is connected at one end is electrically continuous to the other end. An electrical signal, or tone, is injected on the circuit being tested and is either received and verified on the other end or looped back for verification on the sending end. Some tools provide visual feedback (with a meter), whereas others utilize audio feedback. Testing may require that you have a partner at the far end of the cable to administer the inductive probe or loop-back device (this can also be accomplished with a lot of scurrying back and forth on your part). Figure 6.19 shows a tone generator, and Figure 6.20 shows the corresponding amplifier probe.

More sophisticated testers will report not only continuity but also length of run, and will check for shorts and crosses (accidental contact of one conductor with another), reversed pairs, transposed pairs, and split pairs.

FIGURE 6.19
A tone generator

Photo courtesy of ICEAL DataComm

FIGURE 6.20
An amplifier probe

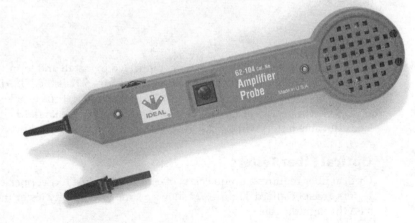

Photo courtesy of IDEAL DataComm

Twisted-Pair Continuity Tester

Many of the common problems of getting cables to work are simple ones. The $5,000 cable testers are nice, but for simple installations they are overkill. If the cable installer is not careful during installation, the cable's wire pairs may be reversed, split, or otherwise incorrectly wired. A simple continuity tester can help you solve many of the common problems of data and voice twisted-pair cabling by testing for open circuits and shorts.

Figure 6.21 shows a simple continuity tester from IDEAL DataComm; this tester (the LinkMaster Tester) consists of the main testing unit and a remote tester. The remote unit is patched into one side of the cable, and the main unit is patched into the other side. It can quickly and accurately detect common cabling problems such as opens, shorts, reversed pairs, or split pairs. Cable testers such as the one shown in Figure 6.19 earlier are available from many vendors and sell for under $100. Testers such as these can save you many hours of frustration as well as the hundreds or even thousands of dollars that you might spend on a more sophisticated tester.

FIGURE 6.21
IDEAL's
LinkMaster Tester

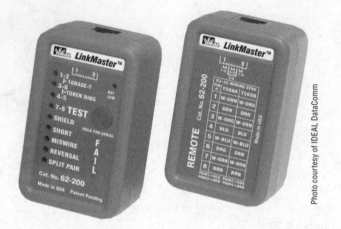

Photo courtesy of IDEAL DataComm

Coaxial Tester

Although coaxial cable is a little less complicated to install and terminate, problems can still arise during installation. The tester shown in Figure 6.22 is the IDEAL DataComm Mini Coax Tester. This inexpensive, compact tester is designed to test coax-cable runs terminated with BNC-style connectors. It can test two modes of operation: standard and Hi-Z for long runs. Coaxial-cable testers will quickly help you identify opens and shorts.

Optical Fiber Testers

Optical fiber requires a unique class of cable testers. Just like copper-cable testers, optical fiber testers are specialized. Figure 6.23 shows a simple continuity tester that verifies light transmission through the cable.

FIGURE 6.22
IDEAL's Mini Coax
Tester

Photo courtesy of IDEAL DataComm

FIGURE 6.23
An optical fiber
continuity tester

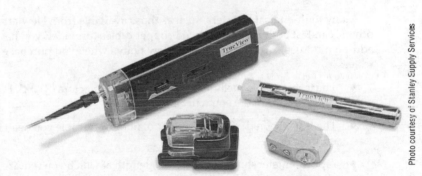

Photo courtesy of Stanley Supply Services

Another type of optical fiber test device is the power meter (also known as an attenuation tester), such as the one shown in Figure 6.24. Like the continuity tester, the power meter tests whether light is making its way through the cable, but it also tests how much of the light signal is being lost. Anyone installing fiber-optic cable should have a power meter. Most problems with optical fiber cables can be detected with this tool. Good optical fiber power meters can be purchased for less than $1,000.

NOTE   An attenuation tester checks for how much signal is lost on the cable, whereas a continuity tester only measures whether light is passing through the cable.

FIGURE 6.24
An optical fiber
power meter

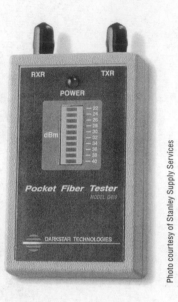

Many high-end cable testers, such as those available from Hewlett-Packard, Microtest, and others, can test both optical fiber and copper cables (provided you have purchased the correct add-on modules). You need to know a few points when you purchase any type of optical fiber tester:

◆ The tester should include the correct fiber connectors (ST, SC, FC, LC, MT-RJ, etc.) for the types of connectors you will be using.

◆ The tester should support the type of fiber-optic cable you need to test (single-mode or multimode).

◆ A power meter should test the wavelength at which you require the cable to be used (usually 850 or 1,300nm).

Professional fiber-optic cable installers usually carry tools such as an OTDR that perform more advanced tests on optical fiber cable. OTDRs are not for everyone, as they can easily cost in excess of $30,000. However, they are an excellent tool for locating faults.

Cabling Supplies and Tools

When you think of cabling supplies, you probably envision boxes of cables, wall plates, modular connectors, and patch panels. True, those are all necessary parts of a cabling installation, but you should have other key consumable items in your cabling tool kit that will make your life a little easier.

Some of the consumable items you may carry are fairly generic. A well-equipped cabling technician carries a number of miscellaneous items essential to a cabling install, including the following:

- Electrician's tape—multiple colors are often desirable

- Duct tape

- Plastic cable ties (tie-wraps) for permanent bundling and tie-offs

- Hook and loop cable ties for temporarily segregating and bundling cables

- Adhesive labels or a specialized cable-labeling system

- Sharpies or other type of permanent markers

- Wire nuts or crimp-type wire connectors

An item that most cable installers use all the time is the tie-wrap. Tie-wraps help to make the cable installation neater and more organized. However, most tie-wraps are permanent; you have to cut them to release them. Hook-and-loop (Velcro-type) cable wraps (shown in Figure 6.25) give you the ability to quickly wrap a bundle of cable together (or attach it to something else) and then to remove it just as easily. The hook-and-loop variety also has the advantage of not over-cinching or pinching the cable, which could cause failure in both optical fiber and copper category cables. Hook-and-loop cable wraps come in a variety of colors and sizes and can be ordered from most cable equipment and wire management suppliers.

FIGURE 6.25
Reusable cable
wraps

Photo courtesy of MilesTek

Cable-Pulling Tools

One of the most tedious tasks that a person pulling cables will face is the process of getting the cables through the area between the false or drop-ceiling tiles and the structural ceiling. This is where most horizontal cabling is installed. One method is to pull out every ceiling tile, pull the cable a few feet, move your stepladder to the next open ceiling tile, and pull the cable a few more feet. Some products that are helpful in the cabling-pulling process are telescoping pull tools and pulleys that cable can be threaded through so that more cable can be pulled without exceeding the maximum pull tension.

Figure 6.26 shows the Gopher Pole, which is a telescoping pole that compresses to a minimum length of less than 5′ and extends to a maximum length of 22′. This tool can help when you are pulling or pushing cable through hard-to-reach places.

FIGURE 6.26
The Gopher Pole

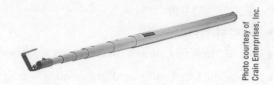

Photo courtesy of
Crain Enterprises, Inc.

Another useful set of items to carry are cable pulleys (shown in Figure 6.27); these pulleys help a single person to do the work of two people when pulling cable. We recommend carrying a set of four pulleys if you are pulling a lot of cable.

FIGURE 6.27
Cable pulleys

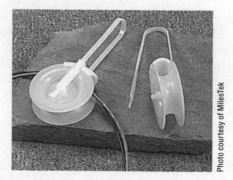

Photo courtesy of MilesTek

Though not specifically in the cable-pulling category, equipment to measure distance is especially important. A simple tape measure will suffice for most of you, but devices that can record long distances quickly may also be useful if you measure often. Sophisticated laser-based tools measure distances at the click of a button; however, a less expensive tool would be something like one of the rolling measure tools pictured in Figure 6.28. This tool has a measuring wheel that records the distance as you walk.

FIGURE 6.28
Professional
rolling measure
tools

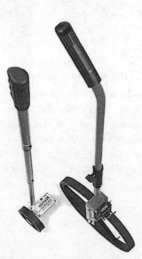

Photo courtesy of MilesTek

If you do much work fishing cable through tight, enclosed spaces, such as stud or joist cavities in walls and ceilings, the Labor Saving Devices Wall-Eye, shown in Figure 6.29, can be an indispensable tool. This device is a periscope with a flashlight attachment that fits through small openings (like a single-gang outlet cutout) and lets you view the inside of the cavity. You can look for obstructions or electrical cables, locate fish-tapes, or spot errant cables that have gotten away during the pulling process.

FIGURE 6.29
The Wall-Eye

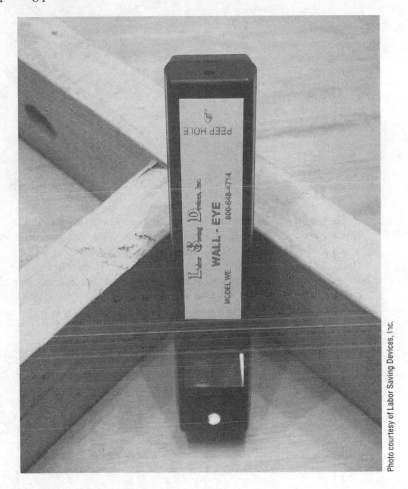

Another great tool for working in cavities and enclosed spaces is a length of bead chain and a magnet. When you drop the chain into a cavity from above (holding on to one end, of course), the extremely flexible links will "pour" over any obstructions and eventually end up in the bottom of the cavity. You insert the magnet into an opening near the bottom of the cavity and snag and extract the chain. Attach a pull-string, retract the chain, and you're ready to make the cable pull. One model of such a device is the Wet Noodle, marketed by Labor Saving Devices Inc., and shown in Figure 6.30.

FIGURE 6.30
The Wet Noodle

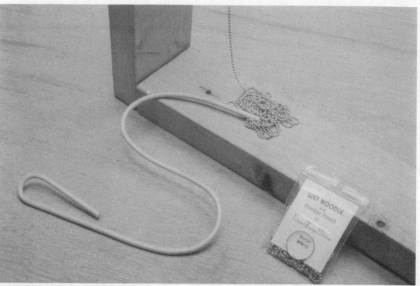

Photo courtesy of Labor Saving Devices, Inc.

Retrofit installations in residences require specialized drill bits. These bits come on long, flexible shafts that let you feed them through restricted openings in order to drill holes in studs, joists, and sole and top plates. Attachable extensions let you reach otherwise inaccessible locations. Examples of these bits and extensions are shown in Figure 6.31. Because of their flexibility, they can bend or curve, even during the drilling process. You can almost literally drill around corners! Most models have holes in the ends of the bit for attaching a pull string so that when you retract the bit, you can pull cable back through. The bits can be purchased in lengths up to 72″, with extensions typically 48″ each.

FIGURE 6.31
Specialized drill
bits and extensions

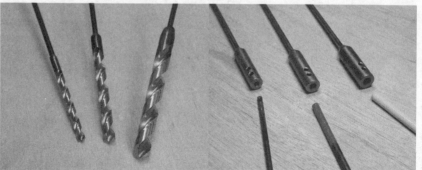

Photos courtesy of Labor Saving Devices, Inc.

Controlling flexible drill bits requires an additional specialized tool: the bit directional tool shown in Figure 6.32. It has loops that hook around the shaft of the drill bit and a handle you use to flex the bit to its proper path—simple in design, essential in function.

FIGURE 6.32
A bit
directional tool

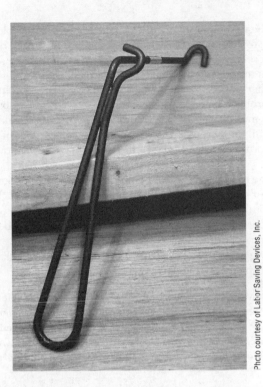

Photo courtesy of Labor Saving Devices, Inc.

Wire-Pulling Lubricant

Wire- or cable-pulling lubricant is a slippery, viscous liquid goop that you apply to the cable jacket to allow it to slide more easily over surfaces encountered during the cable pull. Wire lubricant (see Figure 6.33) is available in a variety of quantities, from less than a gallon to five-gallon buckets.

FIGURE 6.33
Wire-pulling
lubricant

Photo courtesy of IDEAL DataComm

The vast majority of cable jackets for premises cables in the United States are made out of some form of PVC. One characteristic of PVC is that, depending on the specific compound, it has a relatively high coefficient of friction. This means that at the microscopic level, the material is rough, and the rough surface results in drag resistance when the cable jacket passes over another surface. Where two PVC-jacketed cables are in contact, or where PVC conduit is used, the problem is made worse. Imagine two sandpaper blocks rubbing against each other.

In many cases, the use of pulling lubricant is not necessary. However, for long runs through conduit or in crowded cable trays or raceways, you may find either that you cannot complete the pull or that you will exceed the cable's maximum allowable pulling tension unless a lubricant is used.

The lubricant is applied either by continuously pouring it over the jacket near the start of the run, or by wiping it on by hand as the cable is pulled. Where conduit is used, the lubricant can be poured in the conduit as the cable is pulled.

Lubricant has some drawbacks. Obviously, it can be messy; some types also congeal or harden over time, which makes adjustment or removal of cables difficult because they are effectively glued in place. Lubricant can also create a blockage in conduit and raceways that prevents new cables from being installed in the future.

TIP Make sure the lubricant you are using is compatible with the insulation and jacket material of which the cables are made (hint: don't use 10W30 motor oil). The last thing you need is a call back because the pulling lubricant you used dissolved or otherwise degraded the plastics in the cable, leaving a bunch of bare conductors or fibers. Also check with the manufacturer of the cable, as the use of certain lubricants may void the warranty of the cable.

CABLING @ WORK: ONE PERSON'S TRASH IS ANOTHER'S TREASURE!

A couple of years ago the owner of an office building in Dallas leased a floor to a high-profile financial company. The IT director quickly realized that the UTP copper cabling would not support the bandwidth required for its workstations. A local data communications firm was brought in to specify and install new Category 6 cabling and equipment infrastructure.

Luckily there was plenty of space left in the ducting, which contained Category 3 cable. The installers had a bright idea: instead of leaving the Category 3 cable in place and adding new Category 6 on top of it, they decided to use the old cabling as pulling cords for the new cables. They tied the new Category 6 to the Category 3 and pulled the Category 3 out of the ducting, thereby installing the Category 6 cable in its place. Then, instead of throwing away the Category 3, they brought the Category 3 cable to a recycler.

Cable-Marking Supplies

One of the worst legacies of installed cabling systems (and those yet to be installed) is a profound lack of documentation. If you observe a professional data cable installer in action, you will notice that the cabling system is well documented. Though some professionals will even use color coding on cables, the best start for cable documentation is assigning each cable a number. ANSI/TIA-606-B provides specifications about how to document and label cables.

CABLE-MARKING LABELS

The easiest way to number cables is to use a simple numbering system consisting of strips of numbers. These strips are numbered 0 through 9 and come in a variety of colors. Colors include black, white, gray, brown, red, orange, yellow, green, blue, and violet. You can use these strips to create your own numbering system. The cable is labeled at each end (the patch panel and the wall plate), and the cable number is recorded in whatever type of documentation is being used.

The numbered strips are often made of Tyvek, a material invented by DuPont that is well suited for making strong, durable products of high-density polyethylene fibers. Tyvek is nontoxic and chemically inert, so it will not adversely affect cables that it is applied to.

These wire-marking labels are available in two flavors: rolls and sheets. The rolls can be used without dispensers. Figure 6.34 shows a 3M dispenser that holds rolls of wire markers; the dispenser also provides a tear-off cutting blade.

FIGURE 6.34
A 3M dispenser for rolls of wire-marking strips

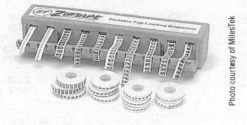

Photo courtesy of MilesTek

Figure 6.35 shows a booklet of wire-marker sheets that allow you to pull off individual numbers.

FIGURE 6.35
A booklet of wire-marker sheets

Photo courtesy of IDEAL DataComm

WALL-PLATE MARKING SUPPLIES

Some wall-plate and patch-panel systems provide their own documentation tools, but others don't. A well-documented system includes identifying labels on the wall plates. Figure 6.36 shows self-adhesive letters, numbers, and icons that can be used with wall plates and patch panels. Check with the manufacturer of your wall plates and patch panels to see if these are part of the system you are using; if they are not, you should use some such labeling system.

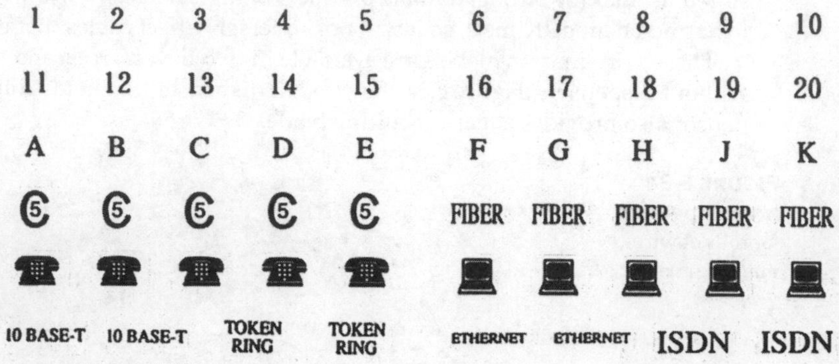

FIGURE 6.36
Letters, numbers, and icons on self-adhesive strips

Tools That a Smart Data Cable Technician Carries

Up to this point, all the tools we've described are specific to the wire-and-cable installation industry. But you'll also need everyday tools in the course of the average install. Even if you don't carry all of these (you'd clank like a knight in armor and your tool belt would hang around your knees if you did), you should at least have them handy in your arsenal of tools:

- A flat blade screwdriver and number 1 and number 2 Phillips screwdrivers. Power screwdrivers are great time-and-effort savers, but you'll still occasionally need the hand types.

- A hammer.

- Nut drivers.

- Wrenches.

- A flashlight (a no-hands or headband model is especially handy).

- A drill and bits up to 1.5".

- A saw that can be used to cut rectangular holes in drywall for electrical boxes.

- A good pocket knife, electrician's knife, or utility knife.

- Electrician's scissors.

- A tape measure.

◆ Face masks to keep your lungs from getting filled with dust when working in dusty areas.

◆ A stud finder to locate wooden or steel studs in the walls.

◆ A simple continuity tester or multitester.

◆ A comfortable pair of work gloves.

◆ A sturdy stepladder, preferably one made of nonconductive materials.

◆ A tool belt with appropriate loops and pouches for the tools you use most.

◆ Two-way radios or walkie-talkies. They are indispensable for pulling or testing over even moderate distances or between floors. Invest in the hands-free models that have a headset, and you'll be glad you did.

◆ Extra batteries (or recharging stands) for your flashlights, radios, and cable testers.

TIP Here's an installation tip: Wall-outlet boxes are often placed one hammer length from the floor, especially in residences (this is based on a standard hammer, not the heavier and longer framing hammers). It's a real time-saver, but check the boxes installed by the electricians before you use this quick measuring technique for installing the data communications boxes so that they'll all be the same height.

A multipurpose tool is also very handy. One popular choice is a Leatherman model with a coax crimper opening in the jaws of the pliers. It's just the thing for those times when you're on the ladder looking down at the exact tool you need lying on the floor where you just dropped it.

One of the neatest ideas for carrying tools is something that IDEAL DataComm calls the Bucket Bag (pictured in Figure 6.37). This bag sits over a five-gallon bucket and allows you to easily organize your tools.

FIGURE 6.37
IDEAL
DataComm's
Bucket Bag

Photo courtesy of IDEAL DataComm

A Preassembled Kit Could Be It

Finally, don't ignore the possibility that a preassembled kit might be just right for you. It may be more economical and less troublesome than buying the individual components. IDEAL DataComm, Jensen Tools from Stanley Supply Services, and MilesTek all offer a range of tool kits for the voice and data installer. These are targeted for the professional installer, and they come in a variety of configurations customized for the type of installation you'll do most often. They are especially suitable for the intermediate to expert user. Figure 6.38 shows a tool kit from MilesTek, and Figure 6.39 shows a Jensen toolkit from Stanley Supply Services.

FIGURE 6.38
MilesTek toolkit

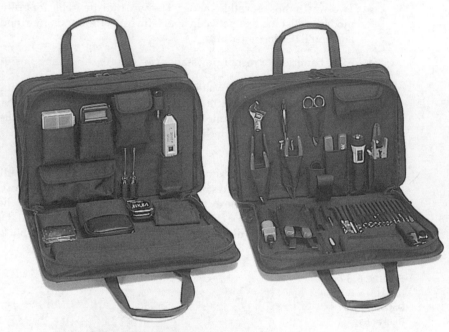

Photo courtesy of MilesTek

FIGURE 6.39
Jensen Tools
Master Cable
Installer's Kit

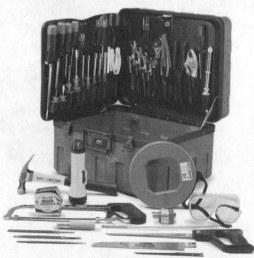

Photo courtesy of Stanley Supply Services

The Bottom Line

Select common cabling tools required for cabling system installation.   Don't start any cabling job without the proper tools. Having the proper tools not only makes the installation quicker and easier, but can also improve the quality of the finished product, making the testing portion go with minimal faults. Therefore, picking tools specific for the applications will make life much easier.

Master It   What are three basic tools required to install a plug for UTP wire?

Identify useful tools for basic cable testing.   You will be installing fiber-optic cable in your next infrastructure project; however, a separate group will be performing the final testing. At a minimum, you want to make sure the fiber-optic cable was installed without any fault and that enough light is being transmitted through the fiber.

Master It   What type of fiber-optic tester should you purchase to make sure the fiber-optic cable was installed without any fault and that enough light is being transmitted through the fiber, and what are the three key attributes to know before you purchase a tester?

Identify cabling supplies to make cable installations easy.   Copper and fiber-optic cables are sensitive to the tension applied during the installation of these cables over long lengths through conduits and cabling trays. In fact, industry standards have maximum pulling loads that can be applied.

Master It   Which of the following supplies enables the reduction of friction on the cables while being pulled?

1. Cable pulling tool

2. Cable pulley

3. Wire-pulling lubricant

4. Gopher Pole

Chapter 7

Copper Cable Media

Although optical fiber cabling continues to make inroads toward becoming the cabling medium of choice for horizontal cable (cable to the desktop), copper-based cabling, specifically unshielded twisted-pair (UTP), is still more popular. This is mainly due to the fact that networking devices required to support copper cabling are inexpensive compared with their fiber-optic counterparts. Cost is almost always the determining factor when deciding whether to install copper or optical fiber cable—unless you have a high-security or really high-bandwidth requirement, in which case optical fiber becomes more desirable.

This chapter will focus on the use of Category 5e and Category 6 UTP cable for data and telecommunications infrastructures. When installing a copper-based cabling infrastructure, one of your principal concerns should be adhering to whichever standard you have decided to use, either the ANSI/TIA-568-C Commercial Building Telecommunications Cabling Standard or the ISO/IEC 11801 Ed. 2 Generic Cabling for Customer Premises Standard. In North America, the ANSI/TIA-568-C standard is preferred. Both these documents are discussed in Chapter 2, "Cabling Specifications and Standards."

In this chapter, you will learn to:

◆ Recognize types of copper cabling

◆ Understand best practices for copper installation

◆ Perform key aspects of testing

Types of Copper Cabling

Pick up any larger cabling catalog, and you will find myriad types of copper cables. However, many of these cables are unsuitable for data and voice communications. Often, cable is manufactured with specific purposes in mind, such as audio, doorbell, remote equipment control, or other low-speed, low-voltage applications. Cable used for data communications must support high-bandwidth applications over a wide frequency range. Even for digital telephones, the cable must be chosen correctly.

Many types of cable are used for data and telecommunications. The application you are using must be taken into consideration when choosing the type of cable you will install. Table 7.1 lists some of the historic and current copper cables and common applications run on them. With the

UTP cabling types found in Table 7.1, applications that run on lower-grade cable will also run on higher grades of cable (for example, digital telephones can be used with Category 3, 4, 5, 5e, 6, or 6A cabling). Category 1, 2, 4, and 5 are no longer recognized by ANSI/TIA-568-C and should be avoided, but they are described below for historical purposes.

TABLE 7.1   Common Types of Copper Cabling and the Applications That Run on Them

CABLE TYPE	COMMON APPLICATIONS
UTP Category 1 (not supported by ANSI/TIA-568-C)	Signaling, doorbells, alarm systems
UTP Category 2 (not supported by ANSI/TIA-568-C)	Digital phone systems, Apple LocalTalk
UTP Category 3	10Base-T, 4Mbps Token Ring
UTP Category 4 (not supported by ANSI/TIA-568-C)	16Mbps Token Ring
UTP Category 5 (not supported by ANSI/TIA-568-C)	100Base-TX, 1000Base-T
UTP Category 5e	100Base-TX, 1000Base-T
UTP Category 6	100Base-TX, 1000Base-T
UTP Category 6A	100Base-TX, 1000Base-T, 10 Gigabit Ethernet
Multi-pair UTP Category 3 cable	Analog and digital voice applications; 10Base-T
25 pair UTP Category 5e cable	10Base-T, 100Base-T, 1000Base-T
Shielded twisted-pair (STP or U/FTP)	4Mbps and 16Mbps Token Ring
Screened twisted-pair (ScTP or F/UTP)	100Base-TX, 1000Base-T, 10 Gigabit Ethernet
Coaxial RG-8	Thick Ethernet (10Base-5), video
Coaxial RG-58	Thin Ethernet (10Base-2)
Coaxial RG-59	CATV (community antenna television, or cable TV)
Coaxial RG-6/U	CATV, CCTV (Closed Circuit TV), satellite, HDTV, cable modem
Coaxial RG-6/U Quad Shield	Same as RG-6 with extra shielding
Coaxial RG-62	ARCnet, video, IBM 3270

Major Cable Types Found Today

When you plan to purchase cable for a new installation, the decisions you have to make are mind-boggling. What cable will support 100Base-TX or 10GBase-T? Will this cable support even faster applications in the future? Do you choose stranded-conductor cable or solid-conductor cable? Should you use different cable for voice and data? Do you buy a cable that only supports present standards or one that is designed to support future standards? The list of questions goes on and on.

NOTE Up to 400 pairs are supported by Categories 3, 4, and 6, and 25 pairs are supported by Category 5e. Only four-pair cables are supported by Categories 6 and 6A.

NOTE Solid-conductor cable is used for horizontal cabling. The entire conductor is one single piece of copper. Stranded-conductor cable is used for patch cords and shorter cabling runs; the conductor consists of strands of smaller wire. These smaller strands make the cable more flexible but also cause it to have higher attenuation. Any cable that will be used for horizontal cabling (in the walls) should be solid conductor.

We'll review the different types of cable listed in Table 7.1 and expand on their performance characteristics and some of their possible uses.

UTP cables are 100 ohm plus or minus 15 percent, 23 or 24 AWG (American Wire Gauge), twisted-pair cables. Horizontal cabling typically uses unshielded, four-pair cables (as shown in Figure 7.1), but voice applications can use Category 3 cables with 25, 50, 100, or more pairs bundled together. UTP cables may contain a slitting cord or rip cord that makes it easier to strip back the jacket. Each of the wires is color coded to make it easier for the cable installer to identify and correctly terminate the wire.

FIGURE 7.1
Common
UTP cable

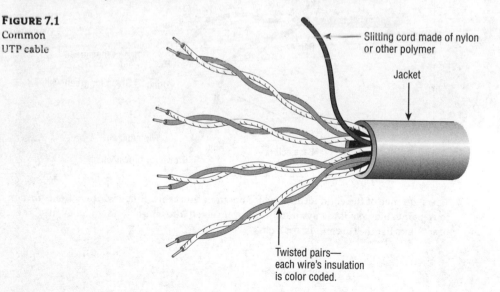

Slitting cord made of nylon or other polymer

Jacket

Twisted pairs—
each wire's insulation
is color coded.

COLOR CODES FOR UTP CABLES

The individual wires in a UTP cable are color coded for ease of identification and termination. A four-pair cable has eight conductors. Four of these conductors are each colored blue, orange, green, or brown, and are called "ring" conductors. Four of the conductors are colored white. These are the "tip" conductors. Each tip conductor is mated with a ring conductor and twisted together to form a pair. So, with all those white conductors, how do you tell which tip conductor goes with which ring conductor when they are untwisted prior to termination? Each tip conductor is marked with either a band of its ring-mate's color at regular intervals, or has a stripe of its ring-mate's color running its length. One company, Superior Essex, uses a pastel shade of the ring color on the tip so that the mate can be seen at any angle and almost any light level. This becomes even more important when working with 25-pair or larger cables, and in larger cables, the ring conductors may also have PI (pair identification) markings. A four-pair UTP cable with band marks on the tip conductors is shown here:

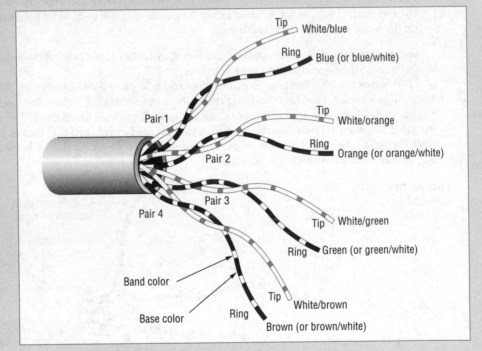

The sequence of the conductor pairs is shown in this table. Since white is the common color in four-pair cables and is always numbered or inserted into a punch-down block first, it is common practice to list the tip conductors first.

Pair	Tip Conductor	Ring Conductor
Pair 1	White/blue (white with blue PI)	Blue (or blue/white)
Pair 2	White/orange (white with orange PI)	Orange (or orange/white)
Pair 3	White/green (white with green PI)	Green (or green/white)
Pair 4	White/brown (white with brown PI)	Brown (or brown/white)

For example, in pair 1 (the blue pair), the two wires are white/blue and blue. Depending on whom you ask, you may get different answers as to which wire is considered primary and which is considered secondary. In the United States and much of the world, premises-cabling people consider the tip wire to be primary because that wire is connected to a connecting block first. Others consider the ring wire to be the primary. However, as long as they are wired correctly, it does not matter what you call the tip and ring wires.

CATEGORY 1 UTP CABLE

Category 1 UTP cable only supports applications operating at 100kHz or less. Applications operating at less than 100kHz are very low-speed applications, such as analog voice, doorbells, alarm systems, RS-232, and RS-422. Category 1 cable is not used very often due to its limited use with data and voice applications, and although it is cheap to install, it cannot be used for anything other than low-speed applications. Category 1 was never recognized by any version of the ANSI/TIA/EIA-568 standard.

CATEGORY 2 UTP CABLE

Category 2 UTP cable was designed to support applications that operate at a frequency rate of less than 4MHz. If you could find any these days, this cable could be used for low-speed applications such as digital voice, Apple LocalTalk, serial applications, ARCnet, ISDN, some DSL applications, and T-1. Most telecommunications designers choose a minimum of Category 3 cable for digital voice. Because of its very limited capabilities, Category 2 was never recognized in ANSI/TIA/EIA-568 and is now extinct.

CATEGORY 3 UTP CABLE

In the early 1990s, Category 3 UTP cable was the workhorse of the networking industry for a few years after it was approved as a standard. It is designed to support applications requiring bandwidth up to 16MHz, including digital and analog voice, 10Base-T Ethernet, 4Mbps Token Ring, 100Base-T4 Fast Ethernet, 100VG-AnyLAN, ISDN, and DSL applications. Most digital-voice applications use a minimum of Category 3 cabling.

Category 3 cable is usually four-pair twisted-pair cable, but some multi-pair (bundled) cables (25-pair, 50-pair, etc.) are certified for use with Category 3 applications. Those multi-pair cables are sometimes used with 10Base-T Ethernet applications; however, they are not recommended.

The FCC published 99-405, effective July 2000, for residential cabling. It says in part: "For new installations and modifications to existing installations, copper conductors shall be, at a minimum, solid, 24 gauge or larger, twisted pairs that comply with the electrical specifications for Category 3, as defined in the ANSI EIA/TIA Building Wiring Standards." This is not a recommendation or a suggested practice—it is the law!

NOTE The industry trend is toward installing Category 5e or Category 6 cabling instead of a combination of Category 3 cabling for voice and Category 5e or 6 for data.

Category 5e Cable

Category 5e UTP cable is one of the most popular in existing installations of UTP cabling for data applications. It is also available in a shielded version, Category 5e ScTP (F/UTP), for applications where there is a high amount of electromagnetic or radio frequency, like industrial applications.

Category 5 cable was invented to support applications requiring bandwidth up to 100MHz. In addition to applications supported by Category 4 and earlier cables, Category 5 supported 100Base-TX, TP-PMD (FDDI over copper), ATM (155Mbps), and under certain conditions, 1000Base-T (Gigabit Ethernet). However, ANSI/TIA-568-C no longer recognizes Category 5.

In the fall 1999, the ANSI/TIA/EIA ratified an addendum to the ANSI/TIA/EIA-568-A standard to approve additional performance requirements for Category 5e cabling to support applications requiring bandwidth up to 100MHz. Category 5e has superseded Category 5 as the recognized cable for new UTP data installations, and this is reflected in the current version of the standard.

Category 6 Cable

With the publication of ANSI/TIA/EIA-568-B.2-1, Category 6 UTP became a recognized cable type. With bandwidth up to 250MHz, this cable category will support any application that Category 5e and lower cables will support. Further, it is designed to support 1000Base-T (Gigabit Ethernet). Category 6 designs typically incorporate an inner structure that separates each pair from the others in order to improve crosstalk performance.

Like Category 5e, Category 6 cables are available in shielded versions. Two types are typically available: Category 6 ScTP (F/UTP) and Category 6 STP (U/FTP). The ScTP version has a PET-backed aluminum foil that covers the entire four-pair core, hence F/UTP, and the STP version has each individual pair shielded with the same type of tape, hence U/FTP.

Category 6A Cable

Category 6A cabling was officially recognized with the publication of ANSI/TIA/EIA-568-B.2-10 in February 2008. In addition to more stringent performance requirements as compared to Category 6, it extends the usable bandwidth to 500MHz. Its intended use is for 10 Gigabit Ethernet. Like Category 6, successful application of Category 6A cabling requires closely matched components in all parts of the transmission channel, such as patch cords, connectors, and cable.

In addition to having specific internal performance requirements (near-end crosstalk(NEXT), return loss, insertion loss, etc.) up to 500MHz, Category 6A cables are required to meet alien crosstalk (AXT) requirements. Alien crosstalk is the noise that is seen by one cable (the disturbed cable) and produced by adjacent cables (the disturber cables). Unlike internal crosstalk noise, AXT cannot be compensated for by the electronic circuitry of a NIC. As a result, Category 6A cables must pass AXT in a six-around-one configuration, where the disturbed cable is surrounded and in direct contact with six disturber cables. This ensures that in the field, AXT will not be an issue. The connectivity must pass similar AXT tests with configurations that are typically used in the field.

Like Categories 5e and 6 cables, shielded Category 6A cables are allowed by ANSI/TIA-568-C. And like Category 6 cables, they are available in either an F/UTP or a U/FTP configuration. These shielded Category 6A cables have the advantage over their UTP versions of being completely immune to AXT.

150 Ohm Shielded Twisted-Pair Cable (IBM Type 1A)

Originally developed by IBM to support applications such as Token Ring and the IBM Systems Network Architecture (SNA), shielded twisted-pair (STP) cable can currently support applications requiring bandwidth up to 600MHz. Though many types of shielded cable are on the market, the Type 1A cable is the most shielded. An IBM Type 1A (STP-A) cable, shown in Figure 7.2, has an outer shield that consists of braided copper; this shield surrounds the 150 ohm, 22 AWG, two-pair conductors. Each conductor is insulated and then each twisted pair is individually shielded.

Figure 7.2
An STP-A cable and an ScTP cable

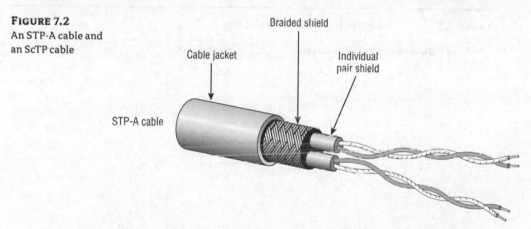

All the shielding in an STP-A cable provides better protection against external sources of EMI than UTP cable does, but the shielding makes the cable thicker and bulkier. Typical applications are 4Mbps and 16Mbps Token Ring and IBM terminal applications (3270 and 5250 terminals). STP cabling is expensive to install, and many people think that it provides only marginally better EMI protection than a well-made Category 5e or higher UTP. If you are considering STP cabling solely because it provides better EMI protection and higher potential bandwidth, consider using fiber-optic cable instead.

Multi-pair UTP Cable

Multi-pair UTP cable is cable that has more than four pairs. Often called *backbone, bundled,* or *feeder cable,* multi-pair cable usually comes in 25-, 50-, or 100-pair increments, though higher pair counts are available. While it is sometimes called *backbone cabling,* this term can be misleading if you are looking at cabling from a data-cabling perspective. High-pair-count multi-pair cabling is typically used with voice applications only.

Some vendors sell 25- and 50-pair cable that is intended for use with Category 5e applications, but all those individual wire pairs supporting data in the same sheath will generate crosstalk that affects all the other pairs. The ANSI/TIA-568-C standard does not recognize such cables for horizontal applications, but it does recognize both multi-pair Category 3 and multi-pair Category 5e for backbone cables and has requirements for meeting the internal crosstalk performance necessary for the appropriate Ethernet application.

Applications with voice and 10Base-T Ethernet data in the same multi-pair cable also exist. Sharing the same sheath with two different applications is not recommended either.

Color Codes and Multi-pair Cables

Color codes for 25-pair cables are a bit more sophisticated than for four-pair cables due to the many additional wire pairs. In the case of 25-pair cables, there is one additional ring color (slate) and four additional tip colors (red, black, yellow, and violet). Table 7.2 lists the color coding for 25-pair cables.

TABLE 7.2 Color Coding for 25-Pair Cables

PAIR NUMBER	TIP COLOR	RING COLOR
1	White	Blue
2	White	Orange
3	White	Green
4	White	Brown
5	White	Slate
6	Red	Blue
7	Red	Orange
8	Red	Green
9	Red	Brown
10	Red	Slate
11	Black	Blue
12	Black	Orange
13	Black	Green

TABLE 7.2   Color Coding for 25-Pair Cables   *(CONTINUED)*

Pair Number	Tip Color	Ring Color
14	Black	Brown
15	Black	Slate
16	Yellow	Blue
17	Yellow	Orange
18	Yellow	Green
19	Yellow	Brown
20	Yellow	Slate
21	Violet	Blue
22	Violet	Orange
23	Violet	Green
24	Violet	Brown
25	Violet	Slate

NOTE   Often, with high-pair-count UTP cable, both the tip and the ring conductor bear PI markings. For example, in pair 1, the white tip conductor would have PI markings of blue, and the blue ring conductor would have PI markings of white.

As with four-pair UTP cable, the tip color is always connected first. For example, when terminating 25-pair cable to a 66-block, white/blue would be connected to pin 1, then blue/white would be connected to pin 2, and so forth.

Binder Groups

When cable pair counts exceed 25 pairs, the cable is broken up into *binder groups* consisting of 25 pairs of wire. Within each binder group, the color code for the first 25 pairs is repeated. So how do you tell pair 1 in one binder group from pair 1 in another? Each binder group within the larger bundle of pairs that make up the total cable is marked with uniquely colored plastic binders wrapped around the 25-pair bundle. The binder colors follow the same color-code sequence as the pairs, so installers don't have to learn two color systems; for example, the first 25-pair binder group has white/blue binders, the second has white/orange binders, and so on.

COAXIAL CABLE

Coaxial cable has been around since local area networking was in its infancy. The original designers of Ethernet picked coaxial cable as their "ether" because coaxial cable is well shielded, has high bandwidth capabilities and low attenuation, and is easy to install. Coaxial cables are identified by their *RG* designation. Coaxial cable can have a solid or stranded core and impedance of 50, 75, or 92 ohms. Coaxial such as the one shown in Figure 7.3 is called coaxial cable because it has one wire that carries the signal surrounded by a layer of insulation and another concentric shield; both the shield and the inner conductor run along the same axis. The outer shield also serves as a ground and should be grounded to be effective.

FIGURE 7.3
Coaxial cable

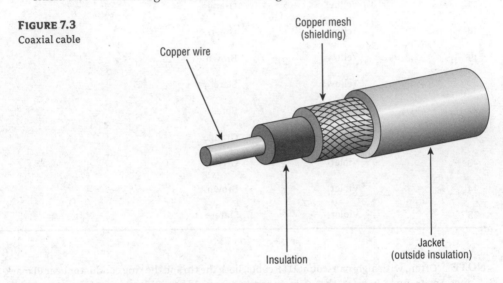

Copper wire

Copper mesh (shielding)

Insulation

Jacket (outside insulation)

NOTE Coaxial cable is still widely used for satellite TV and CCTV video applications. However, it is not recommended for data installations and is not recognized by the standard for such.

A number of different types of coaxial cable were formerly used for data; these are shown in Table 7.3.

TABLE 7.3   Common Coaxial-Cable Types

RG NUMBER	CENTER WIRE GAUGE	IMPEDANCE	CONDUCTOR
RG-6/U	18 AWG	75 ohms	Solid
RG-6/U QS	18 AWG	75 ohms	Solid
RG-8/U	10 AWG	50 ohms	Solid
RG-58/U	20 AWG	53.5 ohms	Solid

TABLE 7.3   Common Coaxial-Cable Types   *(CONTINUED)*

RG NUMBER	CENTER WIRE GAUGE	IMPEDANCE	CONDUCTOR
RG-58C/U	20 AWG	50 ohms	Solid
RG-58A/U	20 AWG	50 ohms	Stranded
RG-59/U	20 AWG	75 ohms	Solid
RG-62/U	22 AWG	93 ohms	Solid

NOTE   Sometimes you will see coaxial cable labeled as 802.3 Ethernet thinnet or 802.3 Ethernet thicknet. Thin Ethernet cable is RG-58 and is used for 10Base-2 Ethernet; thick Ethernet cable is RG-8 and is used for 10Base-5 Ethernet.

HYBRID OR COMPOSITE CABLE

You may hear the term *hybrid* or *composite cable* used. This cable is not really a special type of cable but is one that contains multiple smaller cables within a common cable jacket or spiral wrap. The smaller cables can either be the same type or a mixture of cable types. For example, a common cable that is manufactured now contains four-pair Category 5e UTP cable and two strands of multimode fiber-optic cable. What is nice about these cable types is that you get two different types of media to a single location by pulling only one cable. A number of manufacturers, including Berk-Tek (www.berktek.com), CommScope (www.commscope.com), and Superior Essex (www.superioressex.com), build hybrid cables. For more information, check out their websites. Requirements for these cables are called out in several sections of ANSI/TIA-568-C.

Composite cables are also offered by a number of distributors. Keep in mind that bundling Category 5e, 6 or 6A UTP cables adversely affects their performance, so make sure that all the cables have been tested for standards compliance after they have been bundled whether the bundling was done by the manufacturer or the distributor.

Picking the Right Patch Cables

Although not really part of a discussion on picking cable types for horizontal cable, the subject of patch cords should be addressed. *Patch cables* (or *patch cords*) are the cables that are used to connect 110-type connecting blocks, patch-panel ports, or telecommunication outlets (wall-plate outlets) to network equipment or telephones.

We've mentioned this before but it deserves repeating: you should purchase factory-made patch cables. Patch cables are a critical part of the link between a network device (such as a PC) and the network equipment (such as a hub). Determining appropriate transmission requirements and testing methodology for patch cords was one of the holdups in completing the ANSI/TIA/EIA-B.2-1 Category 6 specification. Low-quality, poorly made, and damaged patch cables very frequently contribute to network problems. Often the patch cable is considered the weakest link in the structured cabling system. Poorly made patch cables will contribute to attenuation loss and increased crosstalk.

Factory-made patch cables are constructed using exacting and controlled circumstances to assure reliable and consistent transmission-performance parameters. These patch cables are tested and guaranteed to perform correctly.

Patch cables are made of stranded-conductor cable to give them additional flexibility. However, stranded cable has up to 20 percent higher attenuation values than solid-conductor cable, so lengths should be kept to a minimum. The ANSI/TIA-568-C standard allows for a 5-meter (16′) maximum-length patch cable in the wiring closet and a 5-meter (16′) maximum-length patch cable at the workstation area.

Here are some suggestions to consider when purchasing patch cables:

◆ Don't make them yourself. Many problems result from bad patch cables.

◆ Choose the correct category for the performance level you want to achieve.

◆ Make sure the patch cables you purchase use stranded conductors for increased flexibility.

◆ Purchase a variety of lengths and make sure you have a few extra of each length.

◆ Consider purchasing patch cords from the same manufacturer that makes the cable and connecting hardware, or from manufacturers who have teamed up to provide compatible cable, patch cords, and connecting hardware. Many manufacturers are a part of such alliances.

◆ Consider color-coding your patch cords in the telecommunication closet. Here's an example:

  ◆ Blue cords for workstations

  ◆ Gray cords for voice

  ◆ Red cords for servers

  ◆ Green cords for hub-to-hub connections

  ◆ Yellow for other types of connections

NOTE The suggested color coding for patch cords loosely follows the documentation guidelines described in Chapter 5, "Cabling System Components."

Why Pick Copper Cabling?

Copper cabling has been around and in use since electricity was invented. And the quality of copper wire has continued to improve. Over the past 100 years, copper manufacturers have developed the refining and drawing processes so that copper is even more high quality than when it was first used for communication cabling.

High-speed technologies (such as 155Mbps ATM and 10 Gigabit Ethernet) that experts said would never run over copper wire are running over copper wiring today.

Network managers pick copper cabling for a variety of reasons: Copper cable (especially UTP cable) is as inexpensive as optical fiber and easy to install, the installation methods are well understood, and the components (patch panels, wall-plate outlets, connecting blocks, etc.) are inexpensive. Further, UTP-based equipment (PBX systems, Ethernet routers, etc.) that uses the copper cabling is much more affordable than comparable fiber equipment.

NOTE The main downsides to using copper cable are that copper cable can be susceptible to outside interference (EMI), copper cable provides less bandwidth than optical fiber, and the data on copper wire is not as secure as data traveling through an optical fiber. This is not an issue for the typical installation.

Table 7.4 lists some of the common technologies that currently use unshielded twisted-pair Ethernet. With the advances in networking technology and twisted-pair cable, it makes you wonder what applications you will see on UTP cables in the future.

TABLE 7.4   Applications That Use Unshielded Twisted-Pair Cables

APPLICATION	DATA RATE	ENCODING SCHEME*	PAIRS REQUIRED
10Base-T Ethernet	10Mbps	Manchester	2
100Base-TX Ethernet	100Mbps	4B5B/NRZI/MLT-3	2
100Base-T4 Ethernet	100Mbps	8B6T	4
1000Base-T Gigabit Ethernet	1000Mbps	PAM5	4
10GBase-T Gigabit Ethernet	10,000Mbps	PAM16/DSQ128	4
100Base-VG AnyLAN	100Mbps	5B6B/NRZ	4
4Mbps Token Ring	4Mbps	Manchester	2
16Mbps Token Ring	16Mbps	Manchester	2
ATM-25	25Mbps	NRZ	2
ATM-155	155Mbps	NRZ	2
TP-PMD (FDDI over copper)	100Mbps	MLT-3	2

Encoding is a technology that allows more than one bit to be passed through a wire during a single cycle (hertz).

Best Practices for Copper Installation

Based on our experience installing copper cabling, we created guidelines for you to follow to ensure that your UTP cabling system will support all the applications you intend it to. These guidelines include the following:

- ◆ Following standards
- ◆ Making sure you do not exceed distance limits
- ◆ Good installation techniques

Following Standards

One of the most important elements to planning and deploying a new telecommunications infrastructure is to make sure you are following a standard. In the United States, this standard is the ANSI/TIA-568-C Commercial Building Telecommunications Cabling Standard. It may be purchased from Global Engineering Documents on the Internet at www.global.ihs.com. We highly recommend that anyone designing a cabling infrastructure own this document.

TIP Have you purchased or do you plan to purchase the ANSI/TIA-568-C standard? We recommend that you buy the entire TIA/EIA Telecommunications Building Wiring Standards collection on CD from Global Engineering (www.global.ihs.com). This is a terrific resource (especially from which to cut and paste sections into an RFP) and can be purchased with a subscription that lets you receive updates as they are published.

Following the ANSI/TIA-568-C standard will ensure that your cabling system is interoperable with any networking or voice applications that have been designed to work with that standard.

Standards development usually lags behind what is available on the market, as manufacturers try to advance their technology to gain market share. Getting the latest innovations incorporated into a standard is difficult because these technologies are often not tested and deployed widely enough for the standards committees to feel comfortable approving them.

TIP If a vendor proposes a solution to you that has a vendor-specific performance spin on it, make sure it is backward-compatible with the current standards. Also ask the vendor to explain how their product will be compatible with what is still being developed by the standards workgroups.

CABLE DISTANCES

One of the most important things that the ANSI/TIA-568-C standard defines is the maximum distance that a horizontal cable should traverse. The maximum distance between the patch panel (or cross-connection, in the case of voice) and the wall plate (the horizontal portion of the cable) *must* not exceed 90 meters (285´). Further, the patch cord used in the telecommunications closet (patch panel to hub or cross-connection) cannot exceed 5 meters (16´), and the patch cord used on the workstation side must not exceed 5 meters (16´).

You may find that higher-quality cables will allow you to exceed this distance limit for older technologies such as 10Base-T Ethernet or 100VG-AnyLAN. However, you are not guaranteed that those horizontal cable runs that exceed 90 meters will work with future technologies designed to work with TIA/EIA standards, so it is strongly recommended that you follow the standard and not "customize" your installation.

Some tips relating to distance and the installation of copper cabling include:

◆ Never exceed the 90-meter maximum distance for horizontal cables.

◆ Horizontal cable rarely goes in a straight line from the patch panel to the wall plate. Don't forget to account for the fact that horizontal cable may be routed up through walls, around corners, and through conduit. If your horizontal cable run is 90 meters (295´) as the crow flies, it's too long.

♦ Account for any additional cable distance that may be required as a result of trays, hooks, and cable management.

♦ Leave some slack in the ceiling above the wiring rack in case re-termination is required or the patch panel must be moved; cabling professionals call this a *service loop*. Some professional cable installers leave as much as an extra 10′ in the ceiling bundled together or looped around a hook (as seen in Figure 7.4).

FIGURE 7.4
Leaving some cable
slack in the ceiling

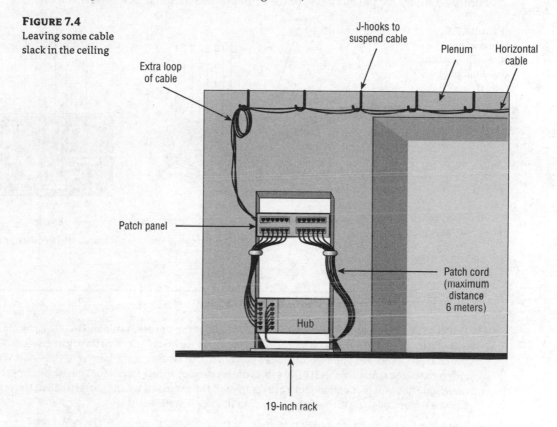

Labels: J-hooks to suspend cable · Plenum · Horizontal cable · Extra loop of cable · Patch panel · Patch cord (maximum distance 6 meters) · Hub · 19-inch rack

WIRING PATTERNS

The ANSI/TIA-568-C standard recommends one of two wiring patterns for modular jacks and plugs: T568-A and T568-B. The only difference between these wiring patterns is that pin assignments for pairs 2 and 3 are reversed. However, these two wiring patterns are constantly causing problems for end users and weekend cable installers. What is the problem? Older patch panels and modular wall-plate outlets came in either the T568-A or T568-B wiring patterns. The actual construction of these devices is exactly the same, but they are color coded according to either the T568-A wiring standard or the T568-B wiring standard. Newer connecting hardware is usually color coded so that either configuration can be used. The confusion comes from people wondering which one to use. It doesn't matter—they both work the same way. But you have to be consistent at each end of the cable. If you use T568-A at one end, you must use it at the other; likewise with T568-B.

The cable pairs are assigned to specific pin numbers. The pins are numbered from left to right if you are looking into the modular jack outlet or down on the top of the modular plug. Figure 7.5 shows the pin numbers for the eight-position modular jack (RJ-45) and plug.

FIGURE 7.5

Pin positions for the eight-position modular plug and jack

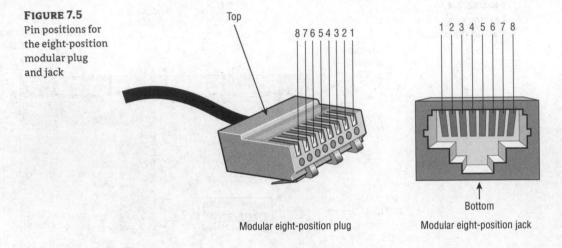

Modular eight-position plug   Modular eight-position jack

WHICH WIRING PATTERN SHOULD YOU CHOOSE?

The T568-A wiring pattern is most prevalent outside of the United States and in U.S. government installations. T568-B used to be more prevalent in private installations in the United States. This has changed, however. The recommended pattern to use for new installations is T568-A. It is the only pattern recognized by ANSI/TIA-570-B, the residential wiring standard. The reason for recommending T568-A is that pairs 1 and 2 are configured the same as a wiring pattern called Universal Service Order Code (USOC), which is prevalent in voice installations.

The wiring pattern chosen makes no difference to the applications used. The signal does not care what color wire it is running through.

The most important factor is to choose one wiring configuration and stick with it. This means when you're purchasing patch panels, 110-blocks, and wall plates, they should all be capable of using the same wiring pattern.

NOTE More information about the T568-A and T568-B wiring configurations can be found in Chapter 10, "Connectors."

Planning

Planning plays an essential role in any successful implementation of a technology; structured cabling systems are no exception. If you are planning to install a larger structured cabling system (more than a few hundred cable runs), consider hiring a professional consultant to assist you with the planning phases.

NOTE Chapter 16, "Creating a Request for Proposal," has information on planning and preparing a request for proposal (RFC) for a structured cabling system. Chapter 13, "Cabling System Design and Installation," covers the essential design issues you must consider when building a structured cabling system.

The following are some questions you should ask when planning a cabling infrastructure that includes copper cabling:

◆ How many cables should be run to each location?

◆ Should you use cable trays, J hooks, or conduit where the cable is in the ceiling space?

◆ Will the voice system use patch panels, or will the voice cable be cross-connected via 66-blocks directly to the phone system blocks?

◆ Is there a danger of cable damage from water, rodents, or chemicals?

◆ Has proper grounding been taken care of for equipment racks and cable terminations requiring grounding?

◆ Will you use the same category of cable for voice and data?

◆ Will new holes be required between floors for backbone cable or through firewalls for horizontal or backbone cable?

◆ Will any of the cables be exposed to the elements or outdoors?

CABLING @ WORK: CRITTER DAMAGE

Cabling folklore is full of stories of cabling being damaged by termites, rats, and other vermin. This might have been hard for us to believe if we had not seen such damage ourselves. One such instance of this type of damage occurred because rats were using a metal conduit to run on the cable between walls. Additional cable was installed, which blocked the rats' pathway, so they chewed holes in the cable.

We have heard numerous stories of cable damaged by creatures with sharp teeth. Outside plant (OSP) cables typically have metal tape surrounding the jacket. Some are thick and strong enough to resist chew-through by rodents.

Consider any area that cable may be run through and take into account what you may need to do to protect the cable.

NOTE The NEC requires that any cable that is in contact with earth or in a concrete slab in contact with earth—whether in a conduit or not—must be rated for a wet environment; that is, the cable must pass ICEA S-99-689, which specifies the requirements for water-blocked cable designs.

Good cable management starts with the design of the cabling infrastructure. When installing horizontal cable, consider using cable trays or J hooks in the ceiling to run the cable. They will prevent the cable from resting on ceiling tiles, power conduits, or air-conditioning ducts, all of which are not allowed according to ANSI/TIA/EIA-568-B.

Further, make sure that you plan to purchase and install cable management guides and equipment near patch panels and on racks so that when patch cables are installed, cable management will be available.

Installing Copper Cable

When you start installing copper cabling, much can go wrong. Even if you have adequately planned your installation, situations can still arise that will cause you problems either immediately or in the long term. Here are some tips to keep in mind for installing copper cabling:

♦ Do not untwist the twisted pairs at the cable connector or anywhere along the cable length any more than necessary (less than 0.5″ for Category 5e, and less than 0.375″ for Category 6).

♦ Taps (bridged taps) are not allowed.

♦ Use connectors, patch panels, and wall plates that are compatible with the cable.

♦ When tie-wrapping cables, do not overtighten cable bundles. Instead of tie-wraps, use Velcro® type wraps. While they are more expensive, they are easily reused if the cables require rearrangement.

♦ Staples are not recommended for fastening cables to supports.

♦ Never splice a data cable if it has a problem at some point through its length; run a new cable instead.

♦ When terminating, remove as little of the cable's jacket as possible, preferably less than three inches. When finally terminated, the jacket should be as close as possible to where the conductors are punched down.

♦ Don't lay data cables directly across ceiling tiles or grids. Use a cable tray, J hook, horizontal ladder, or other method to support the cables. Avoid any sort of cable-suspension device that appears as if it will crush the cables.

♦ Follow proper grounding procedures for all equipment to reduce the likelihood of electrical shock and reduce the effects of EMI.

♦ All voice runs should be home-run, not daisy-chained. When wiring jacks for home or small office telephone use, the great temptation is to daisy-chain cables together from one jack to the next. Don't do it. For one thing, it won't work with modern PBX (private branch exchange) systems. For another, each connection along the way causes attenuation and crosstalk, which can degrade the signal even at voice frequencies.

♦ If you have a cable with damaged pairs, replace it. You will be glad you did. Don't use another unused pair from the same cable because other pairs may be damaged to the point where they only cause intermittent problems, which are difficult to solve. Substituting pairs also prevents any future upgrades that require the use of all four pairs in the cable.

PULLING CABLE

If you are just starting out in the cabling business or if you have never been around cable when it is installed, the term *pulling cable* is probably not significant. However, any veteran installer will tell you that *pulling* is exactly what you do. Cable is pulled from boxes or spools, passed up into the ceiling, and then, every few feet, the installers climb into the ceiling and pull the cable along a few more feet. In the case of cable in conduit, the cable is attached to a drawstring and pulled through.

While the cable is pulled, a number of circumstances can happen that will cause irreparable harm to the cable. But you can take steps to make sure that damage is avoided. Here is a list of copper-cabling installation tips:

◆ Do not exceed the cable's minimum bend radius by making sharp bends. The bend radius for four-pair UTP cables should not be less than four times the cable diameter and not less than 10 times the cable diameter for multi-pair (25-pair and greater cable). Avoid making too many 90-degree bends.

◆ Do not exceed maximum cable pulling tension (110N or 25 pounds of force for four-pair UTP cable).

◆ When pulling a bundle of cables, do not pull cables unevenly. It is important that all the cables share the pulling tension equally.

◆ When building a system that supports both voice and data, run the intended voice lines to a patch panel separate from the data lines.

◆ Be careful not to twist the cable too tightly; doing so can damage the conductors and the conductor insulation.

◆ Avoid pulling the cable past sources of heat such as hot-water pipes, steam pipes, or warm-air ducts.

◆ Be aware that damage can be caused by all sorts of other evil entities such as drywall screws, wiring-box edges, and other sharp objects found in ceilings and walls.

New cable is shipped in reels or coils. Often the reels are in boxes and the cable easily unspools from the boxes as you pull on it. Other times, the cable reels are not in a box, and you must use some type of device to allow the reel to turn freely while you pull the cable. In these cases, a device similar to the one pictured in Figure 7.6 may be just the ticket. These are often called *wire-spool trees*. For emergency or temporary use, a broomstick or piece of conduit through a stepladder will work.

FIGURE 7.6
A reel for holding spools of cable to make cable pulling easier

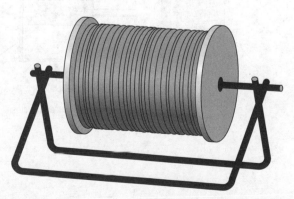

When the coils are inside a box, you dispense the cable directly from the box by pulling on it. You should never take these coils from the box and try to use them. The package is a special design and without the box the cable will tangle hopelessly.

TIP When troubleshooting any wiring system, disconnect the data or voice application from *both* sides (the phone, PC, hub, and PBX). This goes for home telephone wiring, too!

SEPARATING VOICE AND DATA PATCH PANELS

Some installations of voice and data cabling will terminate the cabling on the same patch panel. Although this is not entirely frowned upon by cabling professionals, many will tell you that it is more desirable to have a separate patch panel dedicated to voice applications. This is essential if you use a different category of cable for voice than for data (such as if you use Category 5e cable for data but Category 3 cable for voice).

In the example in Figure 7.7, the wall plate has two eight-position modular outlets (one for voice and one for data). The outlets are labeled V1 for voice and D1 for data. In the telecommunications closet, these two cables terminate on different patch panels, but each cable goes to position 1 on the patch panel. This makes the cabling installation much easier to document and to understand. The assumption in Figure 7.7 is that the voice system is terminating to a patch panel rather than a 66-block. The voice system is then patched to another patch panel that has the extensions from the company's PBX, and the data port is patched to a network hub.

FIGURE 7.7
Using separate
patch panels for
voice and data

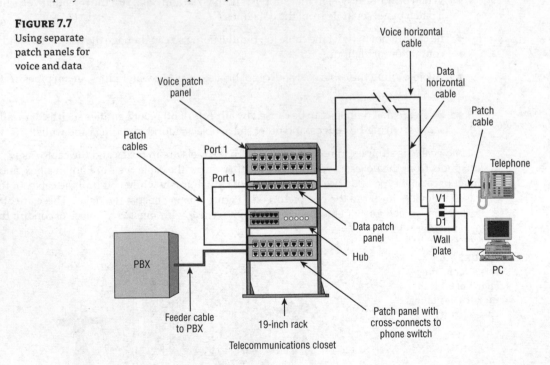

SHEATH SHARING

The ANSI/TIA-568-C standard does not specifically prohibit sheath sharing—that is, when two applications share the same sheath—but its acknowledgment of this practice is reserved for cables with more than four pairs. Occasionally, though, someone may decide that he or she cannot afford to run two separate four-pair cables to a single location and may use different pairs of the cable for different applications. Table 7.5 shows the pin arrangement that might be used if a splitter (such as the one described in Chapter 10) were employed. Some installations may split the cable at the wall outlet and patch panel rather than using a splitter.

TABLE 7.5   Shared-Sheath Pin Assignments

PIN NUMBER	USAGE	T568-A WIRE COLOR	T568-B WIRE COLOR
Pin 1	Ethernet transmit +	White/green	White/orange
Pin 2	Ethernet transmit –	Green	Orange
Pin 3	Ethernet receive +	White/orange	White/green
Pin 4	Phone inner wire 1	Blue	Blue
Pin 5	Phone inner wire 2	White/blue	White/blue
Pin 6	Ethernet receive –	Orange	Green
Pin 7	Phone inner wire 3	White/brown	White/brown
Pin 8	Phone inner wire 4	Brown	Brown

When two applications share the same cable sheath, performance problems can occur. Two applications (voice and data or data and data) running inside the same sheath may interfere with one another. Applications operating at lower frequencies such as 10Base-T may work perfectly well, but higher-frequency applications such as 100Base-TX will operate with unpredictable results. Also, as previously noted, two applications sharing the same four-pair cable sheath will prevent future upgrades to faster LAN technologies such as Gigabit Ethernet.

Because results can be unpredictable, and because you probably want to future-proof your installation, we strongly recommend that you never use a single four-pair cable for multiple applications. Even for home applications where you may want to share voice and data applications (such as Ethernet and your phone service), we recommend separate cables. The ringer voltage on a home telephone can disrupt data transmission on adjacent pairs of wire, and induced voltage could damage your network electronics.

AVOIDING ELECTROMAGNETIC INTERFERENCE

All electrical devices generate electromagnetic fields in the radio frequency (RF) spectrum. These electromagnetic fields produce EMI and interfere with the operation of other electric devices and the transmission of voice and data. You will notice EMI if you have a cordless or cell phone and you walk near a microwave oven or other source of high EMI.

Data transmission is especially susceptible to disruption from EMI, so it is essential that cabling installed with the intent of supporting data (or voice) transmissions be separated from EMI sources. Here are some tips that may be helpful when planning pathways for data and voice cabling:

◆ Data cabling must never be installed in the same conduit with power cables. Aside from the EMI issue, it is not allowed by the NEC.

◆ If data cables must cross power cables, they should do so at right angles.

◆ Power and data cables should never share holes bored through concrete, wood, or steel. Again, it is an NEC violation as well as an EMI concern.

◆ Telecommunication outlets should be placed at the same height from the floor as power outlets, but they should not share stud space.

◆ Maintain at least 2″ of separation from open electrical cables up to 300 volts. Six inches is a preferred minimum separation.

◆ Maintain at least 6″ of separation from lighting sources or fluorescent-light power supplies.

◆ Maintain at least 4″ of separation from antenna leads and ground wires.

◆ Maintain at least 6″ of separation from neon signs and transformers.

◆ Maintain at least 6′ of separation from lightning rods and wires.

◆ Other sources of EMI include photocopiers, microwave ovens, laser printers, electrical motors, elevator shafts, generators, fans, air conditioners, and heaters.

Copper Cable for Data Applications

In this section, we'll discuss using the cable you have run for data applications, and you'll see some samples of ways that these applications can be wired. An important part of any telecommunications cabling system that supports data is the 110-block, which is a great place to start.

110-Blocks

The telecommunications industry used the 66-style block for many years, and it was considered the mainstay of the industry. The 66-blocks were traditionally used only for voice applications; though we have seen them used to cross-connect data circuits, this is not recommended. The 110-blocks are newer than 66-blocks and have been designed to overcome some of the problems associated with 66-blocks. The 110-blocks were designed to support higher-frequency applications, accommodate higher-density wiring arrangements, and better separate the input and output wires.

The standard 66-block enabled you to connect 25 pairs of wires to it, but the 110-blocks are available in many different configurations supporting not only 25 pairs of wire but 50, 100, 200, and 300 pairs of wires as well. The 110-block has two primary components: the 110 wiring block on which the wires are placed, and the 110-connecting block (shown in Figure 7.8), which is used to terminate the wires. A 110-wiring block will consist of multiple 110-connector blocks; there will be one 110-connector block for each four-pair cable that must be terminated.

FIGURE 7.8
The 110-connector
block

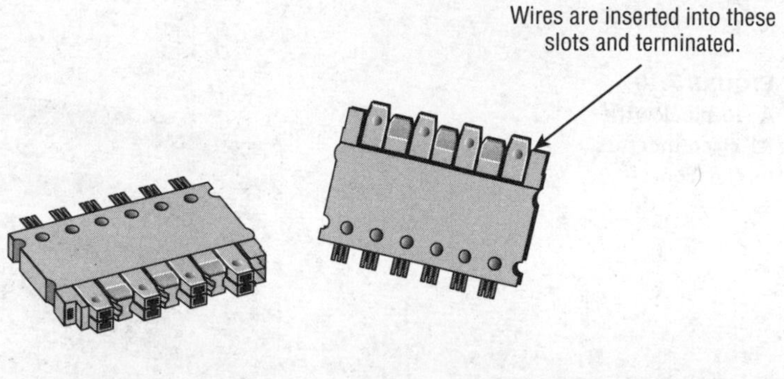

Wires are inserted into these
slots and terminated.

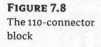

The 110-wiring block will consist of a few or many rows of 110-connector blocks. The wires are inserted into the connecting block and terminated by a punch-down tool or vendor-specific tool. These blocks are a type of IDC (insulation displacement connector); as the wires make contact with the metal on the blocks, the insulation is sliced, and the metal makes contact with the conductor. Remember, to prevent excessive crosstalk, don't untwist the pairs more than 0.5 inches for Category 5e, and 0.375 inches for Category 6 cable, when terminating onto a 110-connecting block.

The 110-blocks come in a wide variety of configurations. Some simply allow the connection of 110-block jumper cables. Figure 7.9 shows a 110-block jumper cable; one side of the cable is connected to the 110-block, and the other side is a modular eight-pin plug (RJ-45).

FIGURE 7.9
A 110-block to
RJ-45 patch cable

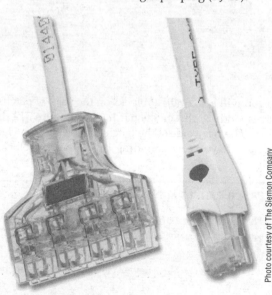

Photo courtesy of The Siemon Company

Other 110-blocks have RJ-45 connectors adjacent to the 110-blocks, such as the one shown in Figure 7.10. If the application uses the 50-pin Telco connectors such as some Ethernet equipment and many voice applications do, 110-blocks such as the one shown in Figure 7.11 can be purchased that terminate cables to the 110-connecting blocks but then connect to 50-pin Telco connectors.

FIGURE 7.10
A 110-block with RJ-45 connectors on the front

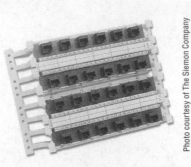

Photo courtesy of The Siemon Company

FIGURE 7.11
A 110-block with 50-pin Telco connectors

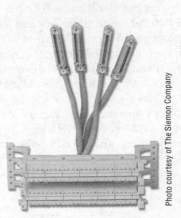

Photo courtesy of The Siemon Company

You will also find 110-blocks on the back of patch panels; each 110-connecting block has a corresponding port on the patch panel. Figure 7.12 shows the 110-block on the back of a patch panel. The front side of the patch panel shown in Figure 7.13 shows a 96-port patch panel; each port will have a corresponding 110-connecting block.

PATCH PANELS AND 110-BLOCKS

The patch panel with the 110-block on the back is the most common configuration in modern data telecommunication infrastructures.

When purchasing patch panels and 110-blocks, make sure you purchase one that has the correct wiring pattern. Most newer 110-blocks are color coded for either the T568-A or T568-B wiring pattern.

The 110-connecting blocks are almost always designed for solid-conductor wire. Make sure that you use solid-conductor wire for your horizontal cabling.

FIGURE 7.12
A 110-block on
the back side of
a patch panel

Photo courtesy of Computer Training Academy

FIGURE 7.13
A 96-port
patch panel

Photo courtesy of MileeTek

Sample Data Installations

As long as you follow the ANSI/TIA-568-C standard, most of your communications infrastructure will be pretty similar and will not vary based on whether it is supporting voice or a specific data application. The horizontal cables will all follow the same structure and rules. However, when you start using the cabling for data applications, you'll notice some differences. We will now take a look at a couple of possible scenarios for using a structured cabling system.

The first scenario, shown in Figure 7.14, shows the typical horizontal cabling terminated to a patch panel. The horizontal cable terminates to the 110-block on the back of the patch panel.

When a workstation is connected to the network, it is connected to the network hub by means of a RJ-45 patch cable that connects the appropriate port on the patch panel to a port on the hub.

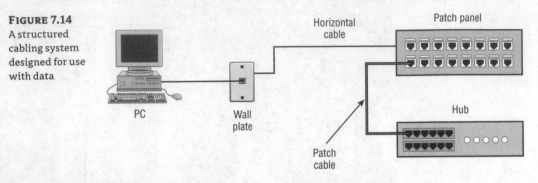

FIGURE 7.14
A structured cabling system designed for use with data

The use of a generic patch panel in Figure 7.14 allows this cabling system to be the most versatile and expandable. Further, the system can also be used for voice applications if the voice system is also terminated to patch panels.

Another scenario involves the use of 110-blocks with 50-pin Telco connectors. These 50-pin Telco connectors are used to connect to phone systems or to hubs that are equipped with the appropriate 50-pin Telco interface. These are less versatile than patch panels because each connection must be terminated directly to a connection that connects to a hub.

In past years, we have worked with these types of connections, and network administrators have reported to us that these are more difficult to work with. Further, these 50-pin Telco connectors may not be interchangeable with equipment you purchase in the future. Figure 7.15 shows the use of a 110-block connecting to network equipment using a 50-pin Telco connector.

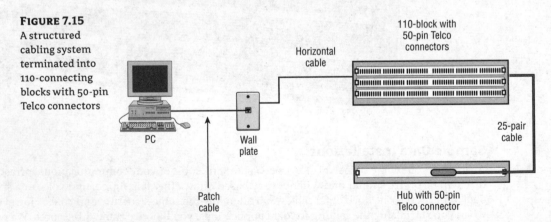

FIGURE 7.15
A structured cabling system terminated into 110-connecting blocks with 50-pin Telco connectors

A final scenario that is a combination of the patch-panel approach and the 110-block approach is the use of a 110-block (such as the one shown previously in Figure 7.9) and 110-block patch cables. This is almost identical to the patch-panel approach, except that the patch cables used in the telecommunications closet have a 110-block connector on one side and an RJ-45 on the other. This configuration is shown in Figure 7.16.

FIGURE 7.16
Structured cabling using 110-blocks and 110-block patch cords

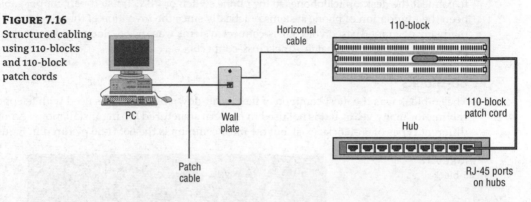

The previous examples are fairly simple and involve only one wiring closet. Any installation that requires more than one telecommunications closet and also one equipment room will require the service of a data backbone. Figure 7.17 shows an example where data backbone cabling is required. Due to distance limitations on horizontal cable when it is handling data applications, all horizontal cable is terminated to network equipment (hubs) in the telecommunications closet. The hub is then linked to other hubs via the data backbone cable.

FIGURE 7.17
Structured cabling that includes data backbone cabling

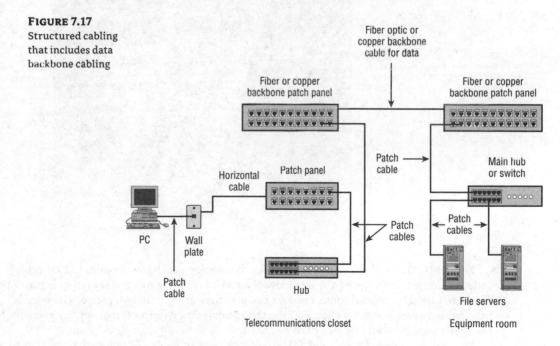

Copper Cable for Voice Applications

Unless you have an extraordinarily expensive phone system, it probably uses copper cabling to connect the desktop telephones to the phone switch or PBX. Twisted-pair copper cables have been the foundation of phone systems practically since the invention of the telephone. The mainstay of copper-based voice cross-connect systems was the 66-block, but it is now being surpassed by 110-block and patch-panel cross-connects.

66-Blocks

The 66-block was the most common of the punch-down blocks. It was used with telephone cabling for many years, but is not used in modern structured wiring installations. A number of different types of 66-blocks exist, but the most common is the 66M1-50 pictured in Figure 7.18.

FIGURE 7.18
A 66-block

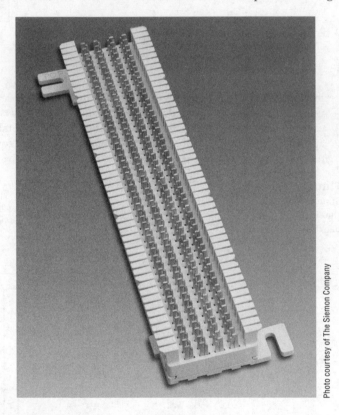

Photo courtesy of The Siemon Company

The 66M1-50 has 50 horizontal rows of IDC connectors; each row consists of four prongs called *bifurcated contact prongs*. A side view of a row of contact prongs is shown in Figure 7.19. They are called bifurcated contact prongs because they are split in two pieces. The wire is inserted between one of the clips, and then the punch-down (impact) tool applies pressure to insert the wire between the two parts of the clip.

The clips are labeled *1, 2, 3,* and *4.* The 66-block clips in Figure 7.19 show that the two clips on the left (clips 1 and 2) are electrically connected together, as are the two clips (clips 3 and 4) on the right. However, the two sets of clips are not electrically connected to one another. Wires can be terminated on both sides of the 66-block, and a metal "bridging" clip is inserted between clips 1 and 2 and clips 3 and 4. This bridging clip mechanically and electrically joins the two sides together. The advantage to this is that the sides can be disconnected easily if you need to troubleshoot a problem.

FIGURE 7.19
The 66-block con-
tact prongs

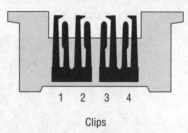

Clips

NOTE   Some 66-blocks have a 50-pin Telco connector on one side of the 66-block.

Figure 7.20 shows a common use of the 66-block; in this diagram, the phone lines from the phone company are connected to one side of the block. The lines into the PBX are connected on the other side. When the company is ready to turn the phone service on, the bridge clips are inserted, which makes the connection.

FIGURE 7.20
A 66-block separat-
ing phone com-
pany lines from the
phone system

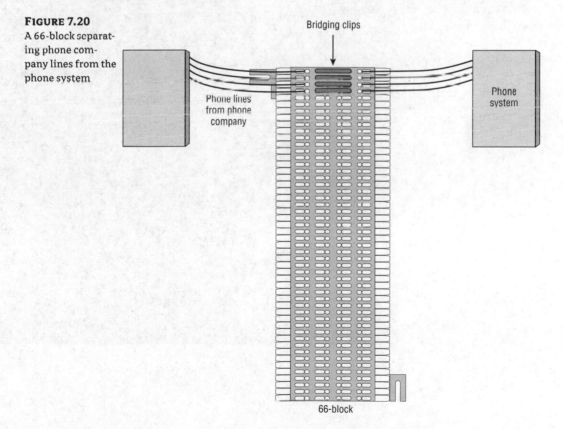

NOTE The 66-blocks are typically designed for solid-conductor cable. Stranded-conductor cables will easily come loose from the IDC-style connectors. Stranded-conductor 66-blocks are available, however.

Figure 7.21 shows a 66-block in use for a voice system. In this picture, you can see that parts of the 66-block connectors have bridging clips connecting them. This block also has a door that can be closed to protect the front of the block and prevent the bridging clips from being knocked off.

FIGURE 7.21
A 66-block used for voice applications

The most typical type of cable connected to a 66-block is the 25-pair cable. The wiring pattern used with the 66-block is shown in Figure 7.22. If you look at a 66-block, you will notice notches in the plastic clips on the left and right side. These notches indicate the beginning of the next binder group.

FIGURE 7.22

The 66-block wire color/pin assignments for 25-pair cables

Row #	Wire Color
1	White/blue
2	Blue/white
3	White/orange
4	Orange/white
5	White/green
6	Green/white
7	White/brown
8	Brown/white
9	White/slate
10	Blue/white
11	Red/blue
12	Blue/red
13	Red/orange
14	Orange/red
15	Red/green
16	Green/red
17	Red/brown
18	Brown/red
19	Red/slate
20	Slate/red
21	Black/blue
22	Blue/black
23	Black/orange
24	Orange/black
25	Black/green
26	Green/black
27	Black/brown
28	Brown/black
29	Black/slate
30	Slate/black
31	Yellow/blue
32	Blue/yellow
33	Yellow/orange
34	Orange/yellow
35	Yellow/green
36	Green/yellow
37	Yellow/brown
38	Brown/yellow
39	Yellow/slate
40	Slate/yellow
41	Violet/blue
42	Blue/violet
43	Violet/orange
44	Orange/violet
45	Violet/green
46	Green/violet
47	Violet/brown
48	Brown/violet
49	Violet/slate
50	Slate/Violet

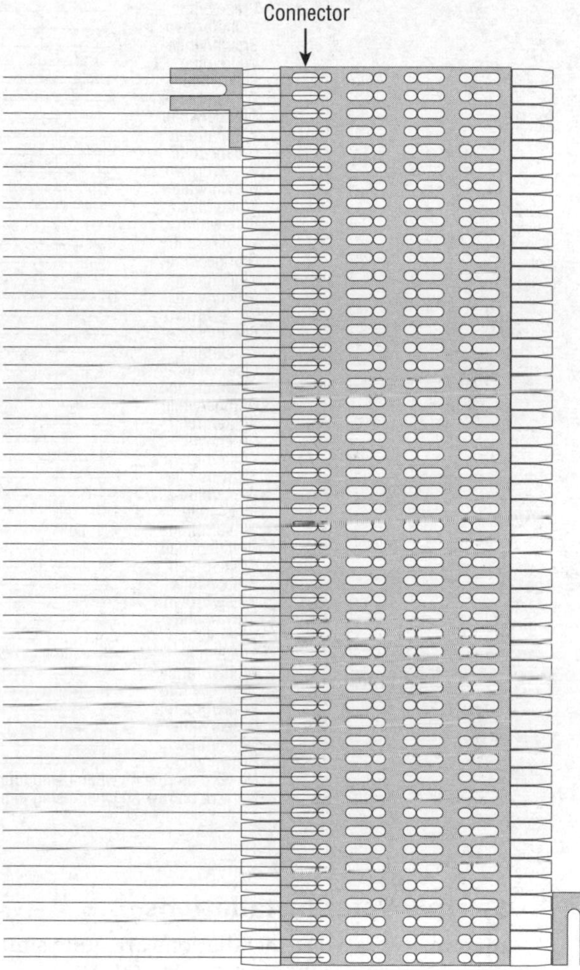

Connector

NOTE The T568-A and T568-B wiring patterns do *not* apply to 66-blocks.

If you were to use 66-blocks and four-pair UTP cables instead of 25-pair cables, then the wire color/pin assignments would be as shown in Figure 7.23.

FIGURE 7.23

The 66-block wire color/pin assignments for four-pair cables

Row #	Wire Color
1	White/blue
2	Blue/white
3	White/orange
4	Orange/white
5	White/green
6	Green/white
7	White/brown
8	Brown/white
9	White/blue
10	Blue/white
11	White/orange
12	Orange/white
13	White/green
14	Green/white
15	White/brown
16	Brown/white
17	White/blue
18	Blue/white
19	White/orange
20	Orange/white
21	White/green
22	Green/white
23	White/brown
24	Brown/white
25	White/blue
26	Blue/white
27	White/orange
28	Orange/white
29	White/green
30	Green/white
31	White/brown
32	Brown/white
33	White/blue
34	Blue/white
35	White/orange
36	Orange/white
37	White/green
38	Green/white
39	White/brown
40	Brown/white
41	White/blue
42	Blue/white
43	White/orange
44	Orange/white
45	White/green
46	Green/white
47	White/brown
48	Brown/white
49	No wire on row 49 when using four-pair wire
50	No wire on row 50 when using four-pair wire

Connector

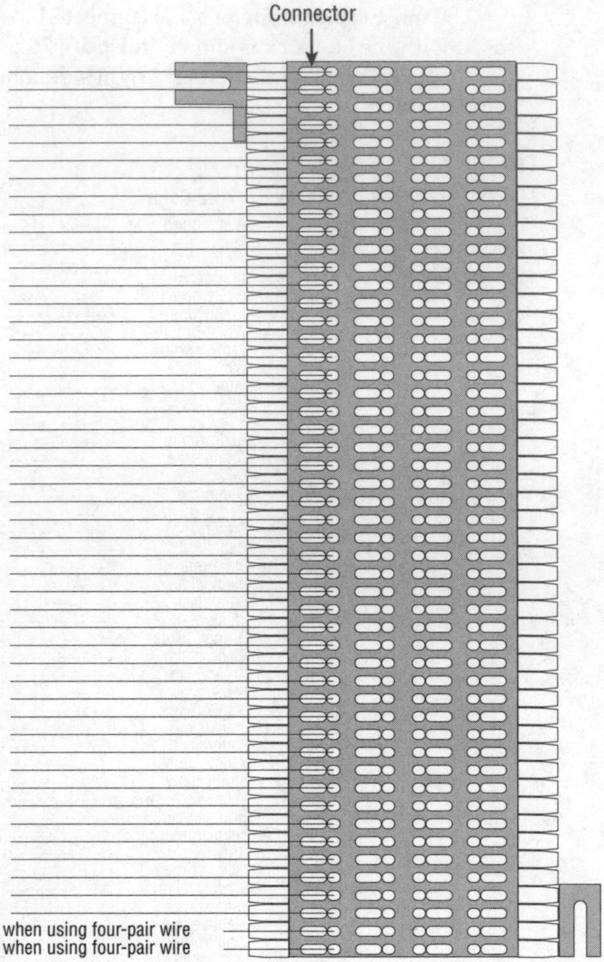

Sample Voice Installations

In many ways, voice installations are quite similar to data installations. The differences are the type of equipment that each end of the link is plugged into and, sometimes, the type of patch cables used. The ANSI/TIA-568-C standard requires at least one four-pair, unshielded twisted-pair cable to be run to each workstation outlet installed. This cable is to be used for voice applications. We recommend using a minimum of Category 3 cable for voice applications; however, if you will purchase Category 5e or higher cable for data, we advise using the same category of cable for voice. This potentially doubles the number of outlets that can be used for data.

Some sample cabling installations follow; we have seen them installed to support voice and data. Because so many possible combinations exist, we will only be able to show you a few. The first one (shown in Figure 7.24) is common in small- to medium-sized installations. In this example, each horizontal cable designated for voice terminates to an RJ-45 patch panel. A second patch panel has RJ-45 blocks terminated to the extensions on the phone switch or PBX. This makes moving a phone extension from one location to another as simple as moving the patch cable. If this type of flexibility is required, this configuration is an excellent choice.

FIGURE 7.24

A voice application using RJ-45 patch panels

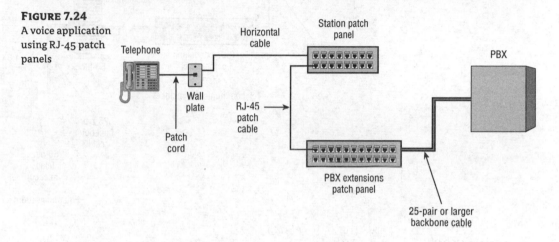

TIP Any wiring system that terminates horizontal wiring into an RJ-45-type patch panel will be more versatile than traditional cross-connect blocks because any given wall-plate port/patch-panel port combination can be used for either voice or data. However, cabling professionals generally recommend separate patch panels for voice and data. Separate panels prevent interference that might occur as a result of incompatible systems and different frequencies used on the same patch panels.

The next example illustrates a more complex wiring environment, which includes backbone cabling for the voice applications. This example could employ patch panels in the telecommunications closet or 66-blocks, depending on the flexibility desired. The telecommunications closet is connected to the equipment room via twisted-pair backbone cabling. Figure 7.25 illustrates the use of patch panels, 66-blocks, and backbone cabling.

The final example is the most common for voice installations; it uses 66-blocks exclusively. You will find many legacy installations that have not been modernized to use 110-block connections. Note that in Figure 7.26 two 66-blocks are connected by cross-connected cable. Cross-connect cable is simple single-pair, twisted-pair wire that has no jacket. You can purchase cross-connect wire, so don't worry about stripping a bunch of existing cables to get it. The example shown in Figure 7.26 is not as versatile as it would be if you used patch panels because 66-blocks require either reconnecting the cross-connect or reprogramming the PBX.

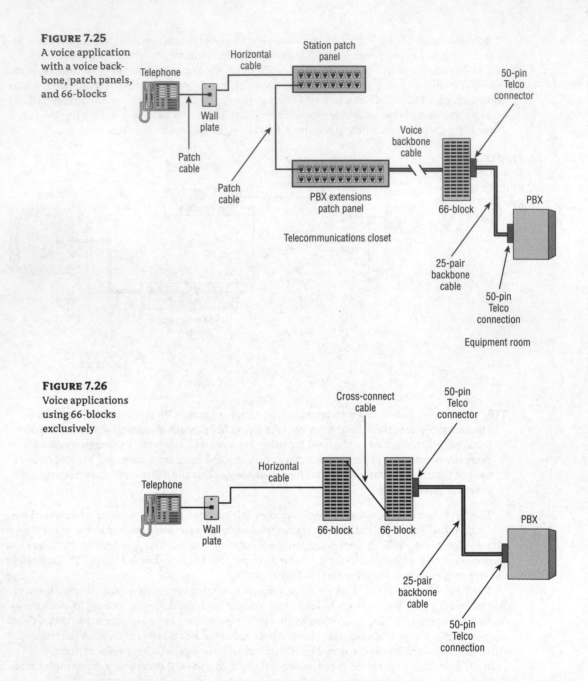

FIGURE 7.25
A voice application with a voice backbone, patch panels, and 66-blocks

FIGURE 7.26
Voice applications using 66-blocks exclusively

Figure 7.27 shows a 66-block with cross-connect wires connected to it. Though you cannot tell it from the figure, cross-connect wires are often red and white.

FIGURE 7.27
A 66-block with
cross-connect
wires

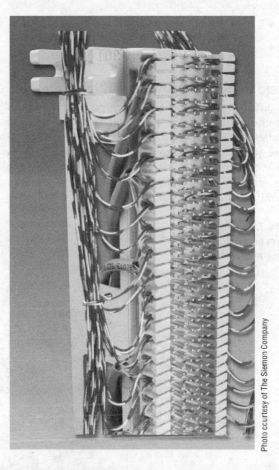

The examples of 66-blocks and 110-blocks in this chapter are fairly common, but we could not possibly cover every possible permutation and usage of these types of blocks. We hope we have given you a representative view of some possible configurations.

Testing

Every cable run must receive a minimum level of testing. You can purchase $5,000 cable testers that will provide you with many statistics on performance, but the most important test is simply determining that the pairs are connected properly.

The $5,000 testers provide you with much more performance data than the simple cable testers and will also certify that each cable run will operate at a specific performance level. Some customers will insist on viewing results on the $5,000 cable tester, but the minimum tests you should run will determine continuity and ascertain that the wire map is correct. You can perform a couple of levels of testing. The cable testers that you can use include the following:

◆ Tone generators and amplifier probes

◆ Continuity testers

- Wire-map testers
- Cable-certification testers

Tone Generators and Amplifier Probes

If you have a bundle of cable and you need to locate a single cable within the bundle, using a tone generator and amplifier is the answer. Often, cable installers will pull more than one cable (sometimes dozens) to a single location, but they will not document the ends of the cables. The tone generator is used to send an electrical signal through the cable. On the other side of the cable, the amplifier (a.k.a. the inductive amplifier) is placed near each cable until a sound from the amplifier is heard, indicating that the cable is found. Figure 7.28 shows a tone generator and amplifier probe from IDEAL DataComm.

FIGURE 7.28
A tone generator and amplifier probe

Continuity Testing

The simplest test you can perform on a cable is the continuity test. It ensures that electrical signals are traveling from the point of origin to the receiving side. Simple continuity testers only guarantee that a signal is being received; they do not test attenuation or crosstalk.

Wire-Map Testers

A wire-map tester is capable of examining the pairs of wires and indicating whether or not they are connected correctly through the link. These testers will also indicate if the continuity of each wire is good. As long as good installation techniques are used and the correct category of cables, connectors, and patch panels are used, many of the problems with cabling can be solved by a simple wire-map tester. Figure 7.29 shows a simple tester from IDEAL DataComm that performs both wire-map testing and continuity testing.

FIGURE 7.29
A simple cable-testing tool

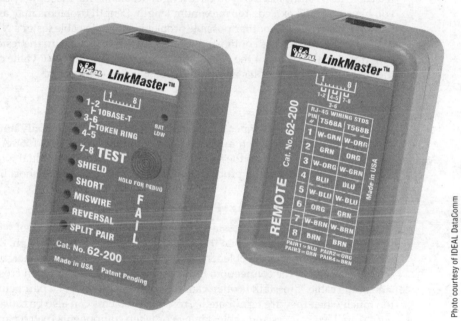

Photo courtesy of IDEAL DataComm

Cable Certification

If you are a professional cable installer, you may be required to certify that the cabling system you have installed will perform at the required levels. Testing tools more sophisticated than a simple continuity tester or wire-map tester perform these tests. The tools have two components, one for each side of the cable link. Tools such as the DTX CableAnalyzer Series from Fluke perform, analyze, and document many sophisticated tests that the less expensive scanners cannot. Cable testing and certification is covered in more detail in Chapter 15, "Cable System Testing and Troubleshooting."

Common Problems with Copper Cabling

Sophisticated testers may provide a reason for a failed test. Some of the problems you may encounter include:

- Length problems
- Wire-map problems
- NEXT and FEXT (crosstalk) problems
- Attenuation problems

LENGTH PROBLEMS

If a cable tester indicates that you have length problems, the most likely cause is that the cable you have installed exceeds the maximum length. Length problems may also occur if the cable has an open or short. Another possible problem is that the cable tester's NVP (Nominal Velocity of Propagation) setting is configured incorrectly. To correct it, run the tester's NVP diagnostics or setup to make sure that the NVP value is set properly. The NVP value can be obtained from the cable manufacturer if it's not properly installed in your tester.

WIRE-MAP PROBLEMS

When the cable tester indicates a wire-map problem, pairs are usually transposed in the wire. This is often a problem when mixing equipment that supports the T568-A and T568-B wiring patterns; it can also occur if the installer has split the pairs (individual wires are terminated on incorrect pins). A wire-map problem may also indicate an open or short in the cable.

NEXT AND FEXT (CROSSTALK) PROBLEMS

If the cable tester indicates crosstalk problems, the signal in one pair of wires is "bleeding" over into another pair of wires; when the crosstalk values are strong enough, this can interfere with data transmission. NEXT problems indicate that the cable tester has measured too much crosstalk on the near end of the connection. FEXT problems indicate too much crosstalk on the opposite side of the cable. Crosstalk is often caused by the conductors of a pair being separated, or "split," too much when they are terminated. Crosstalk problems can also be caused by external interference from EMI sources and cable damage or when components (patch panels and connectors) that are only supported for lower categories of cabling are used.

NEXT failures reported on very short cable runs, 15 meters (50′) and less, require special consideration. Such failures are a function of signal harmonics, resulting from imbalance in either the cable or the connecting hardware or induced by poor-quality installation techniques. The hardware or installation (punch-down) technique is usually the culprit, and you can fix the problem by either re-terminating (taking care not to untwist the pairs) or by replacing the connecting hardware with a product that is better electrically balanced. It should be noted that most quality NICs are constructed to ignore the "short-link" phenomenon and may function just fine under these conditions.

ATTENUATION PROBLEMS

When the cable tester reports attenuation problems, the cable is losing too much signal across its length. This can be a result of the cabling being too long. Also check to make sure the cable is terminated properly. When running horizontal cable, make sure that you use solid-conductor cable; stranded cable has higher attenuation than solid cable and can contribute to attenuation problems over longer lengths. Other causes of attenuation problems include high temperatures, cable damage (stretching the conductors), and the wrong category of components (patch panels and connectors).

The Bottom Line

Recognize types of copper cabling.   Pick up any larger cabling catalog, and you will find a myriad of copper cable types. However, many of these are unsuitable for data or voice communications. It's important to understand which copper cables are recognized by key standards.

Master It   What are the recognized copper cables in ANSI/TIA-568-C? Which cable is necessary to support 100 meters (328′) for 10GBase-T?

Understand best practices for copper installation.   This discipline is loaded with standards that define key requirements for copper cable installation, but it is important to know and follow some key "best practices."

Master It   What are the three primary classes of best practices for copper installation?

Perform key aspects of testing.   Every cable run must go through extensive testing to certify the installation to support the intended applications. The ANSI/TIA-568-C standard specifies they key elements. It's important to understand the basic types of testers required.

Master It   What are the cable testers that you can use to perform testing?

Chapter 8

Fiber-Optic Media

Fiber-optic media (or optical fiber, or fiber, for short) are any network transmission media that use glass, or in some special cases, plastic, fiber to transmit network data in the form of light pulses.

Within the last decade, fiber optics has become an increasingly popular type of network transmission media as the need for higher bandwidth and longer spans continues. We'll begin this chapter with a brief look at how fiber-optic transmissions work.

In this chapter, you will learn to:

◆ Understand the basic aspects of fiber-optic transmission

◆ Identify the advantages and disadvantages of fiber-optic cabling

◆ Understand the types and composition of fiber-optic cables

◆ Identify the key performance factors of fiber optics

Introducing Fiber-Optic Transmission

Fiber-optic technology is different in its operation than standard copper media because the transmissions are "digital" light pulses instead of electrical voltage transitions. Very simply, fiber-optic transmissions encode the ones and zeroes of a digital network transmission by turning on and off the light pulses of a laser light source, of a given wavelength, at very high frequencies. The light source is usually either a laser or some kind of light-emitting diode (LED). The light from the light source is flashed on and off in the pattern of the data being encoded. The light travels inside the fiber until the light signal gets to its intended destination and is read by an optical detector, as shown in Figure 8.1.

FIGURE 8.1
Reflection of a light signal within a fiber-optic cable

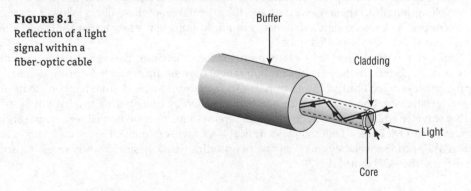

Buffer

Cladding

Light

Core

Fiber-optic cables are optimized for one or more *wavelengths* of light. The wavelength of a particular light source is the length, measured in nanometers (billionths of a meter, abbreviated nm), between wave peaks in a typical light wave from that light source (as shown in Figure 8.2). You can think of a wavelength as the color of the light, and it is equal to the speed of light divided by the frequency. In the case of single-mode fiber, many different wavelengths of light can be transmitted over the same optical fiber at any one time. This is useful for increasing the transmission capacity of the fiber-optic cable since each wavelength of light is a distinct signal. Therefore, many signals can be carried over the same strand of optical fiber. This requires multiple lasers and detectors and is referred to as *wavelength-division multiplexing (WDM)*.

FIGURE 8.2
A typical light
wave

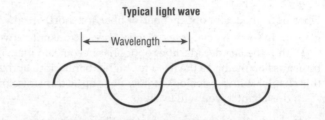

Typically, optical fibers use wavelengths between 850 and 1550nm, depending on the light source. Specifically, multimode fiber is used at 850 or 1300nm and single-mode fiber is typically used at 1310, 1490, and 1550nm (and, in WDM systems, in wavelengths around these primary wavelengths). The latest technology is extending this to 1625nm for single-mode fiber that is being used for next-generation passive optical networks (PON) for FTTH (fiber-to-the-home) applications. Silica-based glass is most transparent at these wavelengths, and therefore the transmission is more efficient (there is less attenuation of the signal) in this range. For a reference, visible light (the light that you can see) has wavelengths in the range between 400 and 700nm. Most fiber-optic light sources operate within the near infrared range (between 750 and 2500nm). You can't see infrared light, but it is a very effective fiber-optic light source.

NOTE Most traditional light sources can only operate within the visible wavelength spectrum and over a range of wavelengths, not at one specific wavelength. Lasers (light amplification by stimulated emission of radiation) and LEDs produce light in a more limited, even single-wavelength, spectrum.

Figure 8.3 shows the typical attenuation of single-mode and multimode fibers as a function of wavelength in this range. As you can see, the attenuation of these fibers is lower at longer wavelengths. As a result, longer distance communications tends to occur at 1310 and 1550nm wavelengths over single-mode fibers.

Notice that typical fibers have a larger attenuation at 1385nm. This *water peak* is a result of very small amounts (in the part-per-million range) of water incorporated during the manufacturing process. Specifically it is a terminal –OH (hydroxyl) molecule that happens to have its characteristic vibration at the 1385nm wavelength; thereby contributing to a high attenuation at this wavelength. Historically, communications systems operated on either side of this peak. However, in 1999 Lucent Technologies's optical fiber division (now OFS) created a *zero water peak (ZWP)* process whereby this water peak was eliminated by significantly reducing and then modifying the –OH molecule.

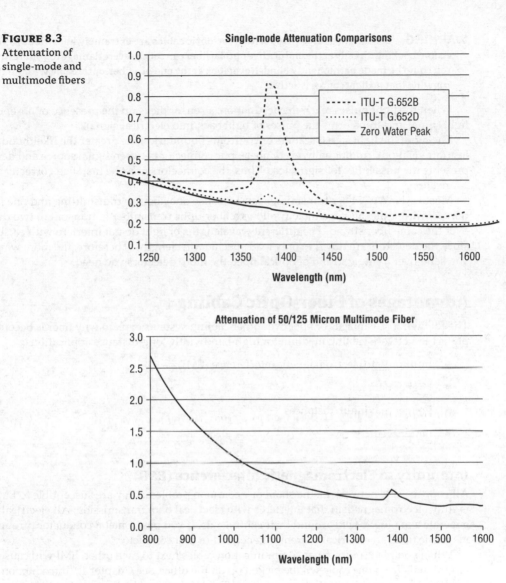

FIGURE 8.3
Attenuation of
single-mode and
multimode fibers

To help you understand, let's use a very simple spring and weight analogy: If you replace the hydrogen with deuterium (an isotope of hydrogen that weighs twice as much) the molecule would now have a characteristic vibration that is not at a frequency of 1385nm and therefore does not cause high attenuation—still there, but out of the way. This invention opened up this wavelength range to transmission systems and allowed the International Telecommunication Union (ITU) to create a new operating band referred to as the E-band. This type of fiber is commonly referred to as *low water peak (LWP)* and has been standardized in the industry as ITU-T G.652D fiber. Earlier fibers had much larger attenuations at 1385nm and are referred to as ITU-T G.652B fiber.

WARNING Laser light sources used with fiber-optic cables are extremely hazardous to your vision. Looking directly at the end of a live optical fiber can cause severe damage to your retinas. You could be made permanently blind. Never look at the end of a fiber-optic cable without first knowing that no light source is active.

When the light pulses reach the destination, a sensor picks up the presence or absence of the light signal and transforms the pulses of light back into electrical signals.

The more the light signal scatters or confronts boundaries, the greater the likelihood of signal loss (attenuation). Additionally, every fiber-optic connector between signal source and destination presents the possibility for signal loss. Thus, the connectors must be installed correctly at each connection.

Most LAN/WAN fiber transmission systems use one fiber for transmitting and one for reception. However, the latest technology allows a fiber-optic transmitter to transmit in two directions over the same fiber strand. The different wavelengths of light do not interfere with each other since the detectors are tuned to only read specific wavelengths. Therefore, the more wavelengths you send over a single strand of optical fiber, the more detectors you need.

Advantages of Fiber-Optic Cabling

The following advantages of fiber over other cabling systems explain why fiber is becoming the preferred network cabling medium for high-bandwidth, long-distance applications:

◆ Immunity to electromagnetic interference (EMI)

◆ Higher data rates

◆ Longer maximum distances

◆ Better security

Immunity to Electromagnetic Interference (EMI)

All copper-cable network media share one common problem: they are susceptible to EMI. EMI is stray electromagnetism that interferes with electrical data transmission. All electrical cables generate a magnetic field around their central axis. If you pass a metal conductor through a magnetic field, an electrical current is generated in that conductor.

When you place two copper communication cables next to each other, EMI will cause crosstalk; signals from one cable will be picked up on the other. See Chapter 1, "Introduction to Data Cabling," for more information on crosstalk. The longer a particular copper cable is, the more chance for crosstalk.

Fiber-optic cabling is immune to crosstalk because optical fiber does not conduct electricity and uses light signals in a glass fiber, rather than electrical signals along a metallic conductor, to transmit data. So it cannot produce a magnetic field and thus is immune to EMI. Fiber-optic cables can therefore be run in areas considered to be "hostile" to regular copper cabling (such as elevator shafts, electrical transformers, in tight bundles with other electrical cables, and industrial machinery).

Higher Possible Data Rates

Because light is immune to interference, can be modulated at very high frequencies, and travels almost instantaneously to its destination, much higher data rates are possible with fiber-optic cabling technologies than with traditional copper systems. Data rates far exceeding the gigabit per second (Gbps) range and higher are possible, and the latest IEEE standards body is working on 100Gbps fiber-based applications over much longer distances than copper cabling. Multimode is the preferred fiber-optic type for 100–550 meters seen in LAN networks, and since single-mode fiber-optic cables are capable of transmitting at these multi-gigabit data rates over very long distances, they are the preferred media for transcontinental and oceanic applications.

You will often encounter the word "bandwidth" when describing fiber-optic data rates. In Chapter 3, "Choosing the Correct Cabling," we described copper bandwidth as a function of analog frequency range. With optical fiber, bandwidth does not refer to channels or frequency, but rather just the bit-throughput rate.

Longer Maximum Distances

Typical copper data-transmission media are subject to distance limitations of maximum segment lengths no longer than 100 meters. Because they don't suffer from the EMI problems of traditional copper cabling and because they don't use electrical signals that can degrade substantially over long distances, single-mode fiber-optic cables can span distances up to 75 kilometers (about 46.6 miles) without using signal-boosting repeaters.

Better Security

Copper-cable transmission media are susceptible to eavesdropping through taps. A *tap* (short for wiretap) is a device that punctures through the outer jacket of a copper cable and touches the inner conductor. The tap intercepts signals sent on a LAN and sends them to another (unwanted) location. Electromagnetic (EM) taps are similar devices, but rather than puncturing the cable, they use the cable's magnetic fields, which are similar to the pattern of electrical signals. If you'll remember, simply placing a conductor next to a copper conductor with an electrical signal in it will produce a duplicate (albeit lower-power) version of the same signal. The EM tap then simply amplifies that signal and sends it on to the person who initiated the tap.

Because fiber optic cabling uses light instead of electrical signals, it is immune to most types of eavesdropping. Traditional taps won't work because any intrusion on the cable will cause the light to be blocked and the connection simply won't function. EM taps won't work because no magnetic field is generated. Because of its immunity to traditional eavesdropping tactics, fiber-optic cabling is used in networks that must remain secure, such as government and research networks.

Disadvantages of Fiber-Optic Cabling

With all of its advantages, many people use fiber-optic cabling. However, fiber-optic cabling does have a couple of disadvantages, including higher cost and a potentially more difficult installation in some cases.

Cost

It's ironic, but the higher cost of fiber-optic cabling has little to do with the cable these days. Increases in available fiber-optic cable–manufacturing capacity have lowered cable prices to levels comparable to high-end UTP on a per-foot basis, and the cables are no harder to pull. Modern fiber-optic connector systems have greatly reduced the time and labor required to terminate fiber. At the same time, the cost of connectors and the time it takes to terminate UTP have increased because Category 5e and Category 6 require greater diligence and can be harder to work with than Category 5. This is even more of a concern for Category 6A and STP cabling. So the installed cost of the basic link, patch panel to wall outlet, is roughly the same for fiber and UTP.

Here's where the costs diverge. Ethernet hubs, switches, routers, NICs, and patch cords for UTP are relatively (no, not relatively, *very*) inexpensive. A good-quality UTP-based 10/100/1000 autosensing Ethernet NIC for a PC can be purchased for less than $25. A fiber-optic NIC for a PC costs at least four times as much. Similar price differences exist for hubs, routers, and switches. For an IT manager who has several hundred workstations to deploy and support, that translates to megabucks and keeps UTP a viable solution. The cost of network electronics keeps the total system cost of fiber-based networks higher than UTP, and ultimately, it is preventing a mass stampede to fiber-to-the-desk. This is why hierarchical star, the typical topology in a commercial building, involves running fiber backbone cabling between equipment and telecommunications rooms or enclosures, and copper UTP horizontal cabling between telecommunications rooms or enclosures and telecommunications outlets near workstations. However, as mentioned in Chapter 3, optical fiber offers some options in network topologies that can make the overall network cost lower than a traditional hierarchical star network wired with more copper cabling (also see TIA's Fiber Optics LAN Section: www.fols.org).

Installation

Depending on the connector system you select, the other main disadvantage of fiber-optic cabling is that it can be more difficult to install. Copper-cable ends simply need a mechanical connection, and those connections don't have to be perfect. Most often, the plug connectors for copper cables are crimped on (as discussed in Chapter 9, "Wall Plates") and are punched down in an insulation displacement connector (IDC) connection on the jack and patch-panel ends.

Fiber-optic cables can be much trickier to make connections for, mainly because of the nature of the glass or plastic core of the fiber cable. When you cut or cleave (in fiber-optic terms) the fiber, the unpolished end consists of an irregular finish of glass that diffuses the light signal and prevents it from guiding into the receiver correctly. The end of the fiber must be polished with a special polishing tool to make it perfectly flat so that the light will shine through correctly. Figure 8.4 illustrates the difference between a polished and an unpolished fiber-optic cable end. The polishing step adds extra time to the installation of cable ends and amounts to a longer, and thus more expensive, cabling-plant installation.

FIGURE 8.4
The difference between a freshly cut and a polished end

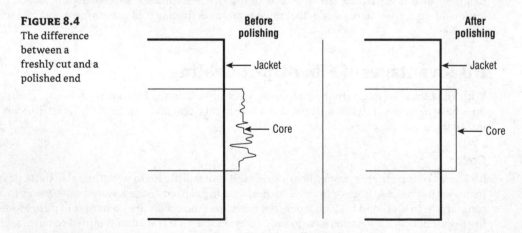

Connector systems are available for multimode fiber-optic cables that don't require the polishing step. Using specially designed guillotine cleavers, you can make a sufficiently clean cleave in the fiber to allow a good end-to-end mate when the connector is plugged in. And, instead of using epoxy or some other method to hold the fiber in place, you can position the fibers in the connector so that dynamic tension holds them in the proper position. Using an index-matching gel in such connectors further improves the quality of the connection. Such systems greatly reduce the installation time and labor required to terminate fiber cables.

Types of Fiber-Optic Cables

Fiber-optic cables come in many configurations. The fiber strands can be either single-mode or multimode, step index or graded index, and the cable jacketing can be either tight buffered or loose-tube buffered. The fiber strands have a variety of core diameters. Most often, the fiber strands are glass, but plastic optical fiber (POF) exists as well. Finally, the cables can be strictly for outdoor use, strictly for indoor use, or a "universal" type that works both indoors and out. These cables also have various fiber ratings.

Composition of a Fiber-Optic Cable

A typical fiber-optic cable consists of several components:

◆ Optical-fiber strand

◆ Buffer

◆ Strength members

◆ Optional shield materials for mechanical protection

◆ Outer jacket

Each of these components has a specific function within the cable to help ensure that the data gets transmitted reliably.

OPTICAL FIBER

An *optical-fiber strand* (also called an *optical waveguide*) is the basic element of a fiber-optic cable. All fiber strands have at least three components to their cross sections: the core, the cladding, and the coating. Figure 8.5 depicts the three layers of the strand.

FIGURE 8.5
Elemental layers in
a fiber-optic strand

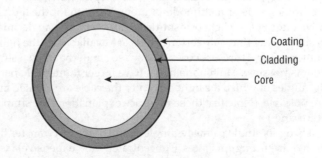

Coating

Cladding

Core

OPTICAL FIBER IS TINY—AND HAZARDOUS

Fiber strands have elements so small that it is hard to imagine the scale. You're just not used to dealing with such tiny elements in everyday life. The components of a fiber strand are measured in microns. A micron is one thousandth of a millimeter, or about 0.00004″. A typical single-mode fiber strand has a core only 8.3 microns, or 0.0003″, in diameter. A human hair is huge by comparison. The core of multimode fiber is typically either 50 or 62.5 microns, or 0.002″ in diameter. For both single-mode and multimode, the cladding usually has a diameter of 125 microns, or 0.005″. And finally, commonly used single-mode and multimode fiber strands have a coating layer that is 250 microns, or 0.01″, in diameter. Now we're getting somewhere, huh? That's all the way to one hundredth of an inch.

The tiny diameter of fiber strands makes them extremely dangerous. When stripped of their coating layer, as must be done for some splicing and connectorizing techniques, the strands can easily penetrate the skin. Shards, or broken pieces of strand, can even be carried by blood vessels to other parts of the body (or the brain), where they could wreak serious havoc. They are especially dangerous to the eye because small pieces can pierce the eyeball, doing damage to the eye's surface and possibly getting trapped inside. Safety glasses and special shard-disposal containers are a must when connecting or splicing fibers.

The fiber core and cladding is usually made of some type of plastic or glass. Several types of materials make up the glass or plastic composition of the optical-fiber core and cladding. Each material differs in its chemical makeup and cost as well as its *index of refraction*, which is a number that indicates how much light will bend when passing through a particular material. The number also indicates how fast light will travel through a particular material. The refractive index of the core is higher than the cladding.

A fiber-optic strand's *cladding* is a layer around the central core that has a lower refractive index. The index difference between the core and cladding is what allows the light inside the core to stay in the core and not escape into the cladding. The cladding thus permits the signal to travel in angles from source to destination—it's like shining a flashlight onto one mirror and having it reflect into another, then another, and so on.

The protective *coating* around the cladding protects the fiber core and cladding from mechanical damage. It does not participate in the transmission of light but is simply a protective material against fracture. It protects the cladding from abrasion damage, adds additional strength to the core, and builds up the diameter of the strand.

The most basic differentiation of fiber-optic cables is whether the fiber strands they contain are single mode or multimode. A mode is a path for the light to take through the cable. The wavelength of the light transmitted, the acceptance angle, and the numerical aperture interact in such a way that only certain paths are available for the light. Single-mode fibers have a lower numerical aperture and cores that are so small that only a single pathway, or mode, for the light is possible. Multimode fibers have larger numerical apertures and cores; the options for the angles at which the light can enter the cable are greater, and so multiple pathways, modes, are possible. (Note that these ray-trace explanations are simplifications of what is actually occurring.)

Using its single pathway, single-mode fibers can transfer light over great distances with high data-throughput rates. Concentrated (and expensive) laser light sources are required to send data down single-mode fibers, and the small core diameters make connections expensive. This is because the mechanical tolerances required to focus the lasers into the core and to hold the fiber in connectors without moving the core away from the laser are extremely precise and require expensive manufacturing methods.

Multimode fibers can accept light from less intense and less expensive sources, usually LEDs or 850nm vertical cavity surface-emitting lasers (VCSELs). In addition, connections are easier to align properly due to larger core diameters. Since the core diameters are larger than single-mode fiber, the tolerances required to manufacture these parts are less precise and less expensive as a result. This is why lasers that are used to operate over only multimode fiber—that is, 850nm sources—are less expensive than single-mode sources. However, distance and bandwidth are more limited than with single-mode fibers. Since multimode cabling and electronics are generally a less expensive solution, multimode is the preferred cabling for short distances found in buildings and on campuses.

Single-mode fibers are usually used in long-distance transmissions or in backbone cables, so you find them mostly in outdoor cables. These applications take advantage of the extended distance and high-bandwidth properties of single-mode fiber.

Multimode fibers are usually used in an indoor LAN environment in the building backbone and horizontal cables. They are also often used in the backbone cabling where great distances are not a problem.

Single-mode and multimode fibers come in a variety of flavors. Some of the types of optical fibers, listed from highest bandwidth and distance potential to least, include the following:

- Single-mode step-index glass
- Multimode graded-index glass
- Multimode plastic

Single-Mode Step-Index Glass

A *single-mode glass fiber core* is very narrow (usually around 8.3 microns) and made of a core of silica glass (SiO_2) with a small amount of Germania glass (GeO_2) to increase the index of refraction relative to the all-silica cladding. To keep the cable size manageable, the cladding for a single-mode glass core is usually about 15 times the size of the core (around 125 microns). Single-mode fibers systems are expensive, but because of the lack of attenuation (less than 0.35dB per kilometer), very high speeds are possible over very long distances. Figure 8.6 shows a single-mode glass-fiber core. The latest class of single-mode fibers have very low loss at the 1385nm (water peak) region and are insensitive to bending. These are known as ITU-T G.652D and G.657, respectively.

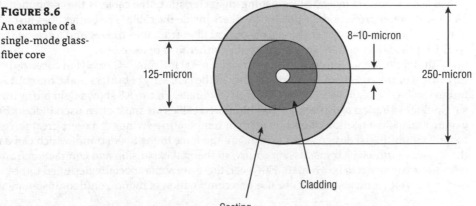

FIGURE 8.6
An example of a single-mode glass-fiber core

125-micron

8–10-micron

250-micron

Cladding

Coating

Multimode Graded-Index Glass

A *graded-index glass-fiber core* is made of core of silica glass (SiO_2) and a small amount of Germania glass (GeO_2) in order to increase the index of refraction relative to the all-silica cladding. Graded-index multimode fibers have an index of refraction that changes gradually from the center outward to the cladding. The center of the core has the highest index of refraction. The most commonly used multimode graded-index glass fibers have a core that is either 50 microns or 62.5 microns in diameter. Figure 8.7 shows a graded-index glass core. Notice that the core is bigger than the single-mode core. The ANSI/TIA-568-C standard recommends the use of an 850nm laser-optimized 50/125 micron multimode fiber. This is commonly referred to OM3, as per the ISO 11801 cabling standard. This class of 50 micron fiber has higher bandwidth than 62.5 micron fiber and other versions of 50 micron fiber. It is recommended for the laser-driven 1–100Gbps applications.

FIGURE 8.7
A graded-index
glass-fiber core

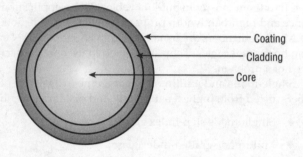

Coating

Cladding

Core

Multimode Plastic

Plastic optical fibers (POF) consist of a plastic core of anywhere from 50 microns on up, surrounded by a plastic cladding of a different index of refraction. Generally speaking, these are the lowest-quality optical fibers and are seldom sufficient to transmit light over long distances. Plastic optical cables are used for very short-distance data transmissions or for transmission of visible light in decorations or other specialty lighting purposes not related to data transmission. Recently, POF has been promoted as a horizontal cable in LAN applications for residential systems. However, the difficulty in manufacturing a graded-index POF, combined with a low bandwidth-for-dollar value, has kept POF from being accepted as a horizontal medium in commercial applications.

BUFFER

The *buffer*, the second-most distinguishing characteristic of the cable, is the component that provides additional protection for the optical fibers inside the cable. The buffer does just what its name implies: it buffers, or cushions, the optical fiber from the stresses and forces of the outside world. Optical fiber buffers are categorized as either *tight* or *loose tube*.

With tight buffers, a protective layer (usually a 900 micron PVC or Nylon covering) is applied directly over the coating of each optical fiber in the cable. Tight buffers make the entire cable more durable, easier to handle, and easier to terminate. Figure 8.8 shows tight buffering in a single-fiber (simplex) construction. Tight-buffered cables are most often used indoors because expansion and contraction caused by outdoor temperature swings can exert great force on a cable. Tight-buffered designs tend to transmit the force to the fiber strand, which can damage the strand or inhibit its transmission ability, so thermal expansion and contraction from temperature extremes is to be avoided. However, there are some specially designed tight-buffered designs for either exclusive outdoor use or a combination of indoor/outdoor installation.

FIGURE 8.8
A simplex fiber-
optic cable using
tight buffering

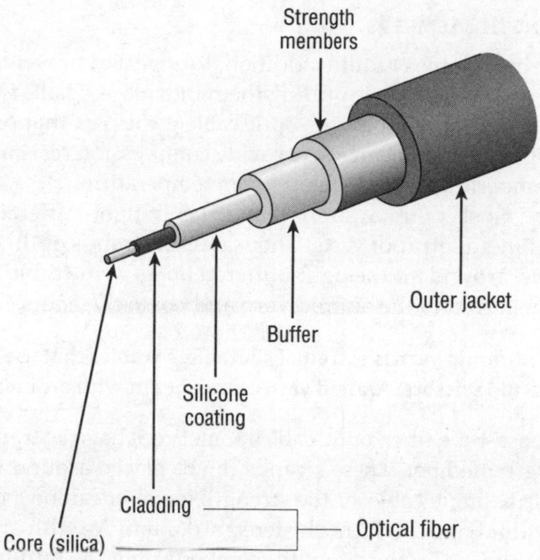

Strength
members

Outer jacket

Buffer

Silicone
coating

Cladding ——————— Optical fiber

Core (silica) ———————

A loose-tube buffer, on the other hand, is essentially a tough plastic pipe about 0.125″ in diameter. One or several coated fibers can be placed inside the tube, depending on the cable design. The tube can then be filled with a protective substance, usually a water-blocking gel, to provide cushioning, strength, and protection from the elements if the cabling is used outdoors. More commonly, water-blocking powders and tapes are used to waterproof the cable. A loose-tube design is very effective at absorbing forces exerted on the cable so that the fiber strands are isolated from the damaging stress. For this reason, loose-tube designs are almost always seen in outdoor installations.

Multiple tubes can be placed in a cable to accommodate a large fiber count for high-density communication areas such as large cities. They can also be used as trunk lines for long-distance telecommunications.

Figure 8.9 shows a loose-buffered fiber-optic cable. Notice that the cable shown uses water-blocking materials.

FIGURE 8.9
A fiber-optic cable
using loose buffer-
ing with water-
blocking materials

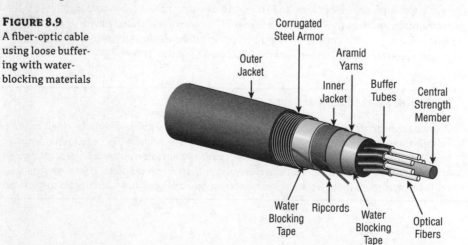

Corrugated
Steel Armor

Outer
Jacket

Aramid
Yarns

Inner
Jacket

Buffer
Tubes

Central
Strength
Member

Water
Blocking
Tape

Ripcords

Water
Blocking
Tape

Optical
Fibers

STRENGTH MEMBERS

Fiber-optic cables require additional support to prevent breakage of the delicate optical fibers within the cable while pulling them into place. That's where the *strength members* come in. The strength member of a fiber-optic cable is the part that provides additional tensile (pull) strength. Strength elements can also provide compression resistance. Compression is encountered when the temperature drops below room temperature.

The most common strength member in tight-buffered cables is aramid yarn, the same material found in bulletproof vests. Thousands of strands of this material are placed in a layer, called a *serving*, around all the tight-buffered fibers in the cable. When pulling on the cable, tensile force is transferred to the aramid yarn and not to the fibers.

TIP Aramid yarn is extremely durable, so cables that use it require a special cutting tool, called aramid scissors. Aramid yarn cannot be cut with ordinary cutting tools.

Loose-tube fiber-optic cables sometimes have a strand of either fiberglass or steel wire as a strength member. These strands can be placed around the perimeter of a bundle of optical fibers within a single cable, or the strength member can be located in the center of the cable with the individual optical fibers clustered around it. As with aramid yarn in tight-buffered cable, tensile force is borne by the strength member(s), not the buffer tubes or fiber strands. Unlike aramid yarns, glass or steel strength members also have the ability to prevent compression-induced microbending caused by temperatures as low as −40°C.

SHIELD MATERIALS

In fiber-optic cables designed for outdoor use, or for indoor environments with the potential for mechanical damage, metallic shields are often applied over the inner components but under the jacket. The shield is often referred to as *armor*. A common armoring material is 0.006˝ steel with a special coating that adheres to the cable jacket. This shield should not be confused with shielding to protect against EMI. However, when present, the shield must be properly grounded at both ends of the cable in order to avoid an electrical-shock hazard should it inadvertently come into contact with a voltage source such as a lightning strike or a power cable.

CABLE JACKET

The *cable jacket* of a fiber-optic cable is the outer coating of the cable that protects all the inner components from the environment. It is usually made of a durable plastic material and comes in various colors. As with copper cables, fiber-optic cables designed for indoor applications must meet fire-resistance requirements of the NEC (see Chapter 1).

EXTERIOR PROTECTION OF FIBER-OPTIC CABLES

If you ever need to install fiber-optic cabling outdoors, the cable should be rated for an exterior installation. An exterior rating means that the cable is specifically designed for outdoor use. It will have features such as UV protection, superior crush and abrasion protection, protection against the extremes of temperature, and an extremely durable strength member. If you use standard indoor cable in an outdoor installation, the cables could get damaged and stop functioning properly.

Additional Designations of Fiber-Optic Cables

Once you've determined whether you need single-mode or multimode fiber strands, loose tube or tight-buffered cable types, and indoor or outdoor cable capability, you still have a variety of fiber-optic cable options from which to choose. When buying fiber-optic cables, you will have to decide which fiber ratings you want for each type of cable you need. Some of these ratings are:

◆ Core/cladding sizes

◆ Number of optical fibers

◆ LAN/WAN application

CORE/CLADDING SIZE

The individual fiber-optic strands within a cable are most often designated by a ratio of *core size/ cladding size*. This ratio is expressed in two numbers. The first is the diameter of the optical-fiber core, given in microns (μm). The second number is the outer diameter of the cladding for that optical fiber, also given in microns.

Three major core/cladding sizes are in use today:

◆ 8.3/125

◆ 50/125

◆ 62.5/125

We'll examine what each one looks like as well as its major use(s).

NOTE Sometimes, you will see a third number in the ratio (e.g., 8.3/125/250). The third number is the outside diameter of the protective coating around the individual optical fibers.

8.3/125

An 8.3/125 optical fiber is shown in Figure 8.10. It is almost always designated as single-mode fiber because the core size is only about 10 times larger than the wavelength of the light it's carrying. Thus, the light doesn't have much room to bounce around. Essentially, the light is traveling in a straight line through the fiber.

FIGURE 8.10
An 8.3/125
optical fiber

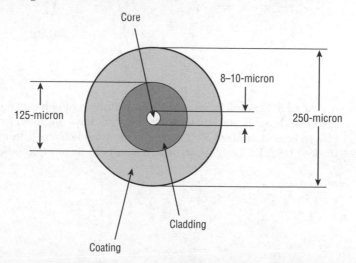

As discussed earlier, 8.3/125 optical fibers are used for high-speed long-distance applications, like backbone fiber architectures for metro, fiber-to-the-home, transcontinental, and transoceanic applications. Single-mode fibers are standardized in ITU, IEC, and TIA.

50/125

In recent years, several fiber manufacturers have been promoting 50/125 multimode fibers instead of the 62.5/125 for use in structured wiring installations. This type of fiber has advantages in bandwidth and distance over 62.5/125 fiber, with about the same expense for equipment and connectors. ANSI/TIA-568-C.3, the fiber-optic-specific segment of the standard, recommends the use of 850nm laser-optimized 50/125 fiber instead of the alternate 62.5/125 type.

62.5/125

Until the introduction of 50/125, the most common multimode-fiber cable designation was 62.5/125 because it was specified in earlier versions of ANSI/TIA/EIA-568 as the multimode media of choice for fiber installations. It had widespread acceptance in the field. A standard multimode fiber with a 62.5 micron core with 125 micron cladding is shown in Figure 8.11.

FIGURE 8.11
A sample 62.5/125
optical fiber

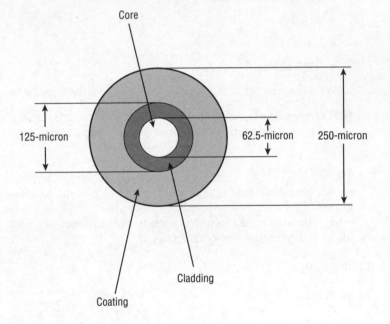

The 62.5/125 optical fibers are used mainly in LED-based, lower-transmission-rate, FDDI LAN/WAN applications. However, as speeds have migrated to Gigabit Ethernet and above, 850nm laser-optimized 50/125 micron multimode fibers (commonly referred to as OM3 per the ISO 11801 Ed 2 cabling standard) have become more common.

NUMBER OF OPTICAL FIBERS

Yet another difference between fiber-optic cables is the number of individual optical fibers within them. The number depends on the intended use of the cable and can increase the cable's size, cost, and capacity.

Because the focus of this book is network cabling and the majority of fiber-optic cables you will encounter for networking are tight buffered, we will limit our discussions here to tight-buffered cables. These cables can be divided into three categories based on the number of optical fibers:

- ◆ Simplex cables

- ◆ Duplex cables

- ◆ Multifiber cables

A *simplex* fiber-optic cable has only one tight-buffered optical fiber inside the cable jacket. An example of a simplex cable was shown earlier in this chapter in Figure 8.8. Because simplex cables only have one fiber inside them, only aramid yarn is used for strength and flexibility; the aramid yarns along with the protective jacket allow the simplex cable to be connectorized and crimped directly to a mechanical connector. Simplex fiber-optic cables are typically categorized as interconnect cables and are used to make interconnections in front of the patch panel (also known as "in front of shelf" connections).

Duplex cables, in contrast, have two tight-buffered optical fibers inside a single jacket (as shown in Figure 8.12). The most popular use for duplex fiber-optic cables is as a fiber-optic LAN backbone cable, because all LAN connections need a transmission fiber and a reception fiber. Duplex cables have both inside a single cable, and running a single cable is of course easier than running two.

FIGURE 8.12
A sample duplex fiber-optic cable

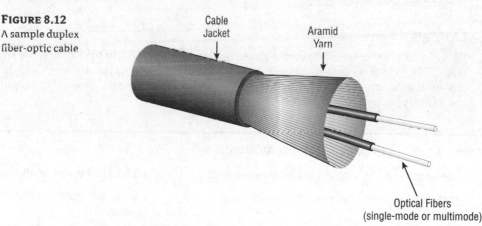

Cable Jacket

Aramid Yarn

Optical Fibers
(single-mode or multimode)

One type of fiber-optic cable is called a duplex cable but technically is not one. This cable is known as zip cord. Zip cord is really two simplex cables bonded together into a single flat optical-fiber cable. It's called a duplex because there are two optical fibers, but it's not really duplex because the fibers aren't covered by a common jacket. Zip cord is used primarily as a duplex patch cable. It is used instead of true duplex cable because it is cheaper to make and to use. Most importantly, however, it allows each simplex cable to be connectorized and crimped directly to a mechanical connector for both strength and durability. Figure 8.13 shows a zip-cord fiber-optic cable.

FIGURE 8.13
A zip-cord cable

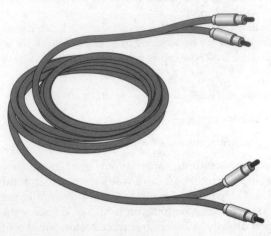

Finally, *multifiber* cables contain more than two optical fibers in one jacket. Multifiber cables have anywhere from three to several hundred optical fibers in them. More often than not, however, the number of fibers in a multifiber cable will be a multiple of two because, as discussed earlier, LAN applications need a send and a receive optical fiber for each connection. Six, twelve, and twenty-four fiber cables are the most commonly used for backbone applications. These cables are typically used for making connections behind the patch-panel (also known as "behind the shelf" connections).

LAN/WAN APPLICATION

Different fiber cable types are used for different applications within the LAN/WAN environment. Table 8.1 shows the relationship between the fiber network type, the wavelength, and fiber size for both single-mode and multimode fiber-optic cables. Table 8.2 shows the recognized fiber and cable types in ANSI/TIA-568-C.3.

TABLE 8.1   Network-Type Fiber Applications

Network Type	Single-Mode Wavelength/Size	Multimode Wavelength/Size
10 Gigabit Ethernet	1300nm–8.3/125 micron	850nm–50/125 micron-OM3 (preferred)
	1550nm–8.3/125 micron	1300nm–62.5/125 or 50/125 micron to 220m (using –LRM)

TABLE 8.1 Network-Type Fiber Applications *(CONTINUED)*

Network Type	Single-Mode Wavelength/Size	Multimode Wavelength/Size
Gigabit Ethernet	1300nm–8/125 micron 1550nm–8.3/125 micron	850nm–62.5/125 or 50/125 micron 1300nm–62.5/125 or 50/125 micron
Fast Ethernet	1300nm–8.3/125 micron	1300nm–62.5/125 or 50/125 micron
Ethernet	1300nm–8.3/125 micron	850nm–62.5/125 or 50/125 micron
10Gbase	1300nm–8.3/125 micron 1550nm–8.3/125 micron	850nm–62.5/125 or 50/125 micron 1300nm–62.5/125 or 50/125 micron
Token Ring	Proprietary–8.3/125 micron	Proprietary–62.5/125 or 50/125 micron
ATM 155Mbps	1300nm–8.3/125 micron	1300nm–62.5/125 or 50/125 micron
FDDI	1300nm–8.3/125 micron	1300nm–62.5/125 or 50/125 micron

TABLE 8.2 ANSI/EIA-568-C.3 Recognized Fiber and Cable Types

Optical Fiber and Relevant Standard	Wavelengths (nm)	Maximum Cable Attenuation (dB/km)	Minimum Overfilled Modal Bandwidth (MHz · km)	Minimum Effective Modal Bandwidth (MHz · km)
62.5/125μ micron Multimode TIA 492AAAA (OM1)	850 1300	3.5 1.5	200 500	Not specified Not specified
50/125μ micron Multimode TIA 492AAAB (OM2)	850 1300	3.5 1.5	500 500	Not specified Not specified
850nm Laser-Optimized 50/125μ micron multimode TIA 492AAAC (OM3)	850 1300	3.5 1.5	1500 500	2000 Not specified
Single-mode Indoor-Outdoor TIA 492CAAA (OS1) TIA 492CAAB (OS2)3	1310 1550	0.5 0.5	N/A N/A	N/A N/A

TABLE 8.2   ANSI/EIA-568-C.3 Recognized Fiber and Cable Types   *(CONTINUED)*

OPTICAL FIBER AND RELEVANT STANDARD	WAVELENGTHS (NM)	MAXIMUM CABLE ATTENUATION (DB/KM)	MINIMUM OVERFILLED MODAL BANDWIDTH (MHZ · KM)	MINIMUM EFFECTIVE MODAL BANDWIDTH (MHZ · KM)
Single-mode Inside plant				
TIA 492CAAA (OS1)	1310	1.0	N/A	N/A
TIA 492CAAB (OS2)3	1550	1.0	N/A	N/A
Single-mode Outside plant				
TIA 492CAAA (OS1)	1310	0.5	N/A	N/A
TIA 492CAAB (OS2)3	1550	0.5	N/A	N/A

NOTE   The philosophy of a generic cable installation that will function with virtually any application led the industry standard, ANSI/TIA-568-C, to cover all the applications by specifying 50/125 multimode or 62.5/125 multimode as a medium of choice (in addition to single-mode). The revised standard, ANSI/TIA-568-C.3, continues to recognize single-mode as well because it also effectively covers all the applications.

CABLING @ WORK: SO MANY FLAVORS OF OPTICAL FIBER, HOW DO I CHOOSE?

Our customers often become overwhelmed by the many types of optical fiber. Although there are various types of UTP cabling, the designation of categories makes it easier to know which is better and there are fewer options customers typically choose (for example, Category 5e or 6 are the popular choices). For the most part all UTP cables support 100 meters, with the application speed being the main difference.

For fiber-based applications, we typically narrow down the choices by asking two key questions:

◆ What is the span distance or link length between active equipment?

◆ What applications (transmission speeds) will you operate on day one and in the future?

Based on these answers, it's very easy to recommend a fiber type. Let's look at some rules of thumb:

◆ If the link length is greater than 1,000 meters, and transmissions speeds are Gigabit Ethernet, 10 Gigabit Ethernet, or greater, we recommend choosing a single-mode fiber that is ITU-T G.652D or OS2 compliant.

◆ If the link length is less than or equal to 300 meters, and transmissions speeds are Gigabit Ethernet, 10 Gigabit Ethernet, or greater, we recommend choosing an 850nm laser-optimized 50/125 micron multimode fiber (also known as OM3).

◆ If the link length is less than 1,000 meters, and transmissions speeds will only go as high as Gigabit Ethernet, we also recommend choosing an 850nm laser-optimized 50/125 micron multimode fiber (also known as OM3).

So this basically narrows down the choice to single-mode or OM3 multimode fiber, depending on distance. As a result of the move to Gigabit Ethernet and higher, 62.5 micron fiber is declining in usage globally. If you can boil it down to these questions, you'll have an easier time recommending a fiber cable type and you will take the complexity out of the decision making.

Fiber Installation Issues

Now that we've discussed details about the fiber-optic cable, we must cover the components of a typical fiber installation and fiber-optic performance factors.

We should also mention here that choosing the right fiber-optic cable for your installation is critical. If you don't, your fiber installation is doomed from the start. Remember the following:

Match the rating of the fiber you are installing to the equipment.   It may seem a bit obvious, but if you are installing fiber for a hub and workstations with 850nm-based optical ports, you cannot use single-mode fiber.

Use fiber-optic cable appropriate for the locale.   Don't use interior cable outside. The interior cable doesn't have the protection features that the exterior cable has.

Un-terminated fiber is dangerous.   Fiber can be dangerous in two ways: you can get glass slivers in your hands from touching the end of a glass fiber, and laser light is harmful to unprotected eyes. Many fiber-optic transmitters use laser light that can damage the cornea of the eyeball when looked at. Bottom line: Protect the end of an un-terminated fiber cable.

Components of a Typical Installation

Just like copper-based cabling systems, fiber-optic cabling systems have a few specialized components, including enclosures and connectors.

FIBER-OPTIC ENCLOSURES

Because laser light is dangerous, the ends of every fiber-optic cable must be encased in some kind of enclosure. The enclosure not only protects humans from laser light but also protects the fiber from damage. Wall plates and patch panels are the two main types of fiber enclosures. You will learn about wall plates in Chapter 9, so we'll discuss patch panels here.

When most people think about a fiber enclosure, a *fiber patch panel* comes to mind. It allows connections between different devices to be made and broken at the will of the network administrator. Basically, a bunch of fiber-optic cables will terminate in a patch panel. Then, short fiber-optic patch or interconnect cables are used to make connections between the various cables. Figure 8.14 shows an example of a fiber-optic patch panel. Note that dust caps are on all the fiber-optic ports; they prevent dust from getting into the connector and interfering with a proper connection.

FIGURE 8.14
An example of a fiber-optic patch panel

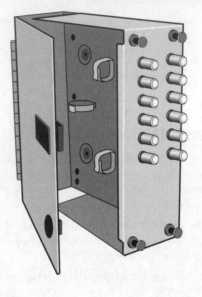

In addition to the standard fiber patch panels, a fiber-optic installation may have one or more *fiber distribution panels*, which are very similar to patch panels, in that many cables interconnect there. However, in a distribution panel (see Figure 8.15), the connections are more permanent. Distribution panels usually have a lock and key to prevent end users from making unauthorized changes. Generally speaking, a patch panel is found wherever fiber-optic equipment (i.e., hubs, switches, and routers) is found. Distribution panels are found wherever multifiber cables are split out into individual cables.

FIBER-OPTIC CONNECTORS

Fiber-optic connectors are unique in that they must make both an optical and a mechanical connection. Connectors for copper cables, like the RJ-45 type connector used on UTP, make an electrical connection between the two cables involved. However, the pins inside the connector only need to be touching to make a sufficient electrical connection. Fiber-optic connectors, on the other hand, must have the fiber internally aligned almost perfectly in order to make a connection. The common fiber-optic connectors use various methods to accomplish this, and they are described in Chapter 10, "Connectors."

FIGURE 8.15
A sample fiber-optic distribution panel

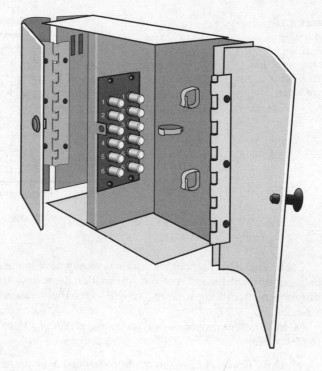

Fiber-Optic Performance Factors

During the course of a normal fiber installation, you must be aware of a few factors that can negatively affect performance. They are as follows:

◆ Attenuation

◆ Acceptance angle

◆ Numerical aperture (NA)

◆ Modal dispersion

◆ Chromatic dispersion

ATTENUATION

The biggest negative factor in any fiber-optic cabling installation is attenuation, or the loss or decrease in power of a data-carrying signal (in the case of fiber, the light signal). It is measured in decibels (dB or dB/km as shown in Figure 8.3). In real-world terms, a 3dB attenuation loss in a fiber connection is equal to about a 50 percent loss of signal. Figure 8.16 graphs attenuation in decibels versus percent signal loss. Notice that the relationship is exponential.

FIGURE 8.16

Relationship of
attenuation to
percent signal loss
of a fiber-optic
transmission

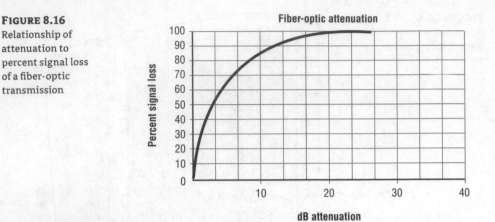

Cables that have higher attenuation typically have lower maximum supportable distances between transmitter and receiver. Attenuation negatively affects transmission speeds and distances of all cabling systems, but fiber-optic transmissions are particularly sensitive to attenuation.

Many different problems can cause attenuation of a light signal in an optical fiber, including the following:

◆ Dirty fiber end-faces (accounts for 85 percent of attenuation loss issues)

◆ Excessive gap between fibers in a connection

◆ Improperly installed connectors leading to offsets of the fiber cores when mated

◆ Impurities in the fiber

◆ Excessive bending of the cable

◆ Excessive stretching of the cable (note, however, that only bending causes light loss, not stretching; stretching the cable can cause the fiber to bend, causing light loss or attenuation)

These problems will be covered in Chapter 15, "Cable System Testing and Troubleshooting." For now, just realize that these problems cause attenuation, an undesirable effect.

ACCEPTANCE ANGLE

Another factor that affects the performance of a fiber-optic cabling system is the acceptance angle of the optical-fiber core. The acceptance angle (as shown in Figure 8.17) is the angle through which a particular (multimode) fiber can accept light as input.

The greater the acceptance angle difference between two or more signals in a multimode fiber, the greater the effect of modal dispersion (see the section "Modal Dispersion and Bandwidth," later in this chapter). The modal-dispersion effect also has a negative effect on the total performance of a particular cable segment.

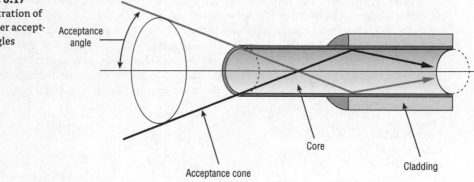

FIGURE 8.17
An illustration of multifiber acceptance angles

Acceptance angle

Acceptance cone

Core

Cladding

NUMERICAL APERTURE (NA)

A characteristic of fiber-optic cable that is related to the acceptance angle is the *numerical aperture* (NA). The NA is calculated from the refractive indexes of the core and cladding. The result of the calculation is a decimal number between 0 and 1 that reflects the ability of a particular optical fiber to accept light.

A value for NA of 0 indicates that the fiber accepts, or gathers, no light. A value of 1 for NA indicates that the fiber will accept all light it's exposed to. A higher NA value means that light can enter and exit the fiber from a wide range of angles, including severe angles that will not reflect inside the core, but be lost to refraction. A lower NA value means that light can enter and exit the fiber only at shallow angles, which helps assure the light will be properly reflected within the core. Multimode fibers typically have higher NA values than single-mode fibers. This is a reason why the less focused light from LEDs can be used to transmit over multimode fibers as opposed to the focused light of a laser that is required for single-mode fibers.

MODAL DISPERSION AND BANDWIDTH

Multimode cables suffer from a unique problem known as *modal dispersion*, which is similar in effect to delay skew, described in Chapter 1 relative to twisted-pair cabling. Modal dispersion, also known as *differential mode delay* (DMD), causes transmission delays in multimode fibers. Modal dispersion is the main property affecting the multimode fiber's bandwidth.

Here's how it occurs. The modes (signals) enter the multimode fiber at varying angles, so the signals will enter different portions of the optical core. In simple terms, the light travels over different paths inside the fiber and can arrive at different times as a result of different refractive indexes over the cross-sectional area of the core (as shown in Figure 8.18). The more severe the difference between arrival times of the modes, the more delay between the modes, and the lower the bandwidth of the fiber system.

In Figure 8.18, mode A will exit the fiber first because it has a shorter distance to travel inside the core than mode B. Mode A has a shorter distance to travel because its entrance angle is less severe (i.e., it's of a *lower order*) than that of mode B. The difference between the time that mode

A exits and the time that mode B exits is the modal dispersion or DMD. Modal dispersion gets larger, or worse, as the fiber transports more modes or if there is an imperfection in the graded index refractive index profile. Multimode fiber with a 62.5 micron core carriers more modes than 50 micron fiber. As a result, 62.5 micron fiber has higher modal dispersion and lower bandwidth than 50 micron fiber. Fiber manufacturers pay careful attention to manufacturing the refractive index profiles of graded-index multimode fibers in order to reduce modal dispersion and thereby increase bandwidth. Differential mode delay is measured by fiber manufacturers and is directly related to bandwidth.

FIGURE 8.18
Illustration of modal dispersion

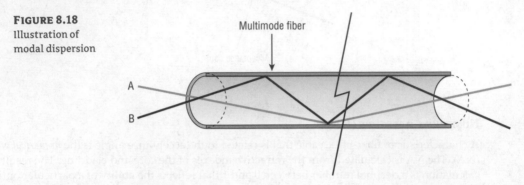

CHROMATIC DISPERSION

The last fiber-optic performance factor is *chromatic dispersion*, which limits the bandwidth of certain single-mode optical fibers. Contrary to popular belief, lasers cannot emit one pure wavelength of light. Lasers emit a distribution of wavelengths with the characteristic laser wavelength in the center. (Granted, the distribution is very tight and narrow; it is a distribution, nonetheless.) These various wavelengths of light transmitted by the laser spread out in time as they travel through an optical fiber. This happens because different wavelengths of light travel at different speeds through the same media. As they travel through the fiber, the various wavelengths will spread apart (as shown in Figure 8.19). The wavelengths will spread farther and farther apart until they arrive at the destination at completely different times. This will cause the resulting pulse of light to be wider, less sharp, and lower in overall power. As consecutive pulses begin to overlap, the signal detection will be compromised. Multimode fiber can also have this problem, and this combined with modal dispersion lowers the bandwidth of multimode fibers compared to single-mode, which is mainly impaired only by chromatic dispersion.

FIGURE 8.19
Single-mode optical fiber chromatic dispersion

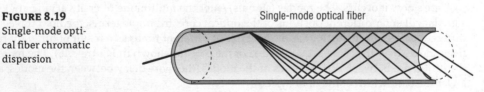

The Bottom Line

Understand the basic aspects of fiber-optic transmission. Optical-fiber based systems work differently than copper-based systems. Optical-fiber systems are based on transmitting a digital signal by modulating light pulses on and off. Since information is transmitted optically, you must understand certain factors.

Master It

A. What are the common wavelengths of operation for multimode fiber?

B. What are the common operating wavelengths of single-mode fiber?

Identify the advantages and disadvantages of fiber-optic cabling. There are pros and cons to fiber-optic cabling. However, with fiber-based systems, the list of pros continues to increase and the list of cons decreases with time.

Master It

A. What are the key advantages of fiber-based systems?

B. Based on these advantages, in which applications would you use fiber instead of UTP copper?

Understand the types and composition of fiber-optic cables. Optical fiber comes in many flavors. There are key features of optical fiber that dictate a specific choice of ancillary components, such as optical transmitters and connectors. When choosing an optical cabling system, look at the full system cost.

Master It In most short-distance applications of 100–550 meters using optical fiber, multimode fiber is the preferred choice over single-mode fiber. This is because single-mode–based systems are more expensive than multimode. Although single-mode cable is less expensive than multimode cable, what factors lead to single-mode systems being more expensive?

Identify the key performance factors of fiber optics. You have learned several performance factors that affect the ability of optical fiber to transport signals. Since multimode fibers are the preferred optical media for commercial buildings, you must understand key performance factors affecting these types of optical fiber.

Master It What is the primary performance factor affecting multimode fiber transmission capability and how can you minimize this?

Chapter 9

Wall Plates

In Chapter 5, "Cabling System Components," you learned about the basic components of a structured cabling system. One of the most visible of these components is the wall plate (also called a *workstation outlet* or *station outlet* because it is usually placed near a workstation). As its name suggests, a wall plate is a flat plastic or metal plate that usually mounts in or on a wall (although some "wall" plates actually are mounted in floors, office cubicles, and ceilings). Wall plates include one or more *jacks*. A jack is the connector outlet in the wall plate that allows a workstation to make a physical and electrical connection to the network cabling system. *Jack* and *outlet* are often used interchangeably.

Wall plates come in many different styles, types, brands, and yes, even colors (in case you want to color-coordinate your wiring system). In this chapter, you will learn about the different types of wall plates available and their associated installation issues.

In this chapter, you will learn to:

♦ Identify key wall-plate design and installation aspects

♦ Identify the industry standards required to ensure proper design and installation of wall-plates

♦ Understand the different types of wall plates and their benefits and suggested uses

Wall Plate Design and Installation Issues

When you plan your cabling system installation, you must be aware of a few wall plate installation issues to make the most efficient installation. The majority of these installation issues come from compliance with the ANSI/TIA-570-B (for residential) and ANSI/TIA-568-C.1 (for commercial installations) telecommunications standards. You'll have to make certain choices about how best to conform to these standards based on the type of installation you are doing. These choices will dictate the different steps you'll need to take during the installation of the different kinds of wall plates.

WARNING   The National Electrical Code dictates how various types of wiring (including power and telecommunications wiring) must be installed, but be aware that NEC compliance varies from state to state. The NEC code requirements described in this chapter should be verified against your local code requirements before you do any structured cable system design or installation.

The main design and installation issues you must deal with for wall plates are as follows:

◆ Manufacturer system

◆ Wall plate location

◆ Wall plate mounting system

◆ Fixed-design or modular plate

In this section, you will learn what each of these installation issues is and how each will affect your cabling system installation.

Manufacturer System

There is no "universal" wall plate. Hundreds of different wall plates are available, each with its design merits and drawbacks. It would be next to impossible to detail every type of manufacturer and wall plate, so in this chapter we'll just give a few examples of the most popular types. The most important thing to remember about using a particular manufacturer's wall plate system in your structured-cabling system is that it is a *system*. Each component in a wall plate system is designed to work with the other components and, generally speaking, can't be used with components from other systems. A *wall plate system* consists of a wall plate and its associated jacks. When designing your cabling system, therefore, you must choose the manufacturer and wall plate system that best suits your present and future needs. It is best to stay with one common wall plate system design for all of the workstation's outlets.

Wall Plate Location

When installing wall plates, you must decide the best location on the wall. Obviously, the wall plate should be fairly near the workstation, and in fact, the ANSI/TIA-568-C.1 standard says that the maximum length from the workstation to the wall plate patch cable can be no longer than 5 meters (16′). This short distance will affect exactly where you place your wall plates in your design. If you already have your office laid out, you will have to locate the wall plates as close as possible to the workstations so that your wiring system will conform to the standard.

Additionally, you want to keep wall plates away from any source of direct heat that could damage the connector or reduce its efficiency. In other words, don't place a wall plate directly above a floor heating register or baseboard heater.

A few guidelines exist for where to put your wall plates on a wall for code compliance and the most trouble-free installation. You must account for the vertical and horizontal positions of the wall plate. Both positions have implications, and you must understand them before you design your cabling system. We'll examine the vertical placement first.

Vertical Position

When deciding the vertical position of your wall plates, you must take into account either the residential or commercial National Electrical Code (NEC) sections. Obviously, which section you go by depends on whether you are performing a residential or commercial installation.

In residential installations, you have some flexibility. You can place a wall plate in almost any vertical position on a wall, but the NEC suggests that you place it so that the top of the plate is no more than 18″ from the subfloor (the same distance as electrical outlets). If the wall plate is intended to service a countertop or a wall phone, the top of the plate should be no more than 48″ from the subfloor. These vertical location requirements are illustrated in Figure 9.1.

FIGURE 9.1
Wall plate vertical location

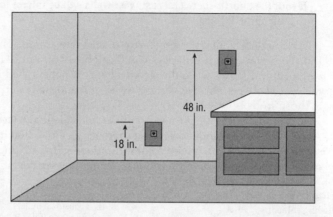

NOTE   The vertical heights may be adjusted, if necessary, for elderly or disabled occupants, according to the Americans with Disabilities Act (ADA) guidelines.

TIP   Remember that the vertical heights may vary from city to city and from residential to commercial electrical codes.

HORIZONTAL POSITION

Wall plates should be placed horizontally so that they are as close as possible to work-area equipment (computers, phones, etc.). As previously mentioned, the ANSI/TIA-568-C.1 standard requires that work-area cables should not exceed 5 meters (16′). Wall plates should therefore be spaced so that they are within 5 meters of any possible workstation location. This means you will have to know where the furniture is in a room before you can decide where to put the wall plates for the network and phone. Figure 9.2 illustrates this horizontal-position requirement.

FIGURE 9.2
Horizontal wall plate placement

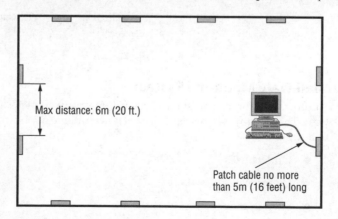

TIP When placing telecommunications outlets, consider adding more than one per room to accommodate rearrangement of the furniture. It usually helps to "mirror" the opposing wall-outlet layout (i.e., north-south and east-west walls will be mirror images of each other with respect to their outlet layout).

Another horizontal-position factor to take into account is the proximity to electrical fixtures. Data communications wall plates and wall boxes cannot be located in the same *stud cavity* as electrical wall boxes when the electrical wire is not encased in metal conduit. (A stud cavity is the space between the two vertical wood or metal studs and the drywall or wallboard attached to those studs.)

The stud-cavity rule primarily applies to residential telecommunications wiring as per the ANSI/TIA-570-B standard. The requirement, as illustrated in Figure 9.3, keeps stray electrical signals from interfering with communications signals. Notice that even though the electrical outlets are near the communications outlets, they are never in the same stud cavity.

FIGURE 9.3

Placing telecommunications outlets and electrical wall boxes in different stud cavities

Notice that telecom outlet and electrical outlets are located in separate stud cavities.

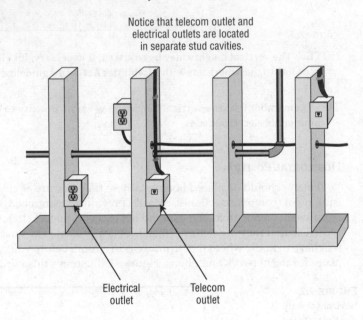

Electrical outlet

Telecom outlet

Wall Plate Mounting System

Another decision you must make regarding your wall plates is how you will mount them to the wall. Three main systems, each with its own unique applications, are used to attach wall plates to a wall:

- ◆ Outlet boxes
- ◆ Cut-in plates
- ◆ Surface-mount outlet boxes

The following sections describe each of these mounting systems and their various applications.

OUTLET BOXES

The most common wall plate mounting in commercial applications is the *outlet box*, which is simply a plastic or metal box attached to a stud in a wall cavity. Outlet boxes have screw holes in them that allow a wall plate to be attached. Additionally, they usually have some provision (either nails or screws) that allows them to be attached to a stud. These outlet boxes, as their name suggests, are primarily used for electrical outlets, but they can also be used for telecommunications wiring because the wall plates share the same dimensions and mountings.

Plastic boxes are cheaper than metal ones and are usually found in residential or light commercial installations. Metal boxes are typically found in commercial applications and usually use a conduit of some kind to carry electrical or data cabling. Which you choose depends on the type of installation you are doing. Plastic boxes are fine for simple, residential Category 3 copper installations. However, if you want to install Category 5e or higher, you must be extremely careful with the wire so that you don't kink it or make any sharp bends in it. This is especially true for Category 6A UTP cables, which can be as large as 0.35˝. Make sure you use boxes that are designed for the cable that you will be using. Also, if you run your network cable before the drywall is installed (and in residential wiring with plastic boxes, you almost always have to), it is likely that the wires will be punctured or stripped during the drywall installation. Open-backed boxes are often installed to avoid bend-radius problems and to allow cable to be pushed back into the cavity and out of reach of the dry-wall installers' tools. If you can't find open-backed boxes, buy plastic boxes and cut the backs off with a saw.

Metal boxes can have the same problems, but these problems are minimized if the metal boxes are used with conduit—that is, a plastic or metal pipe that attaches to the box. In commercial installations, a metal box to be used for telecommunications wiring is attached to a stud. Conduit is run from the box to a 45-degree elbow that terminates in the airspace above a dropped ceiling. This installation technique is the most common wiring method in new commercial construction and is illustrated in Figure 9.4. This method allows you to run the telecommunications wire after the wallboard and ceiling have been installed, thus minimizing the chance of damage to the cable.

FIGURE 9.4
A common metal box with conduit, in a commercial installation

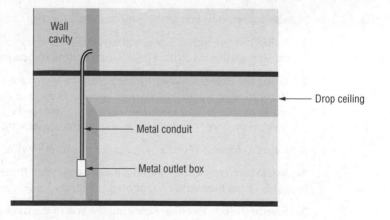

CUT-IN MOUNTING

Outlet boxes work well as wall plate supports when you are able to access the studs during the construction of a building. But what type of wall plate mounting system do you use once the drywall is in place and you need to put a wall plate on that wall? In this case, you should use some kind of *cut-in* mounting hardware (also called *remodeling* or *retrofit* hardware), so named because you cut a hole in the drywall and place into it some kind of mounting box or plate that will support the wall plate. This type of mounting is used when you need to run a cable into a particular stud cavity of a finished wall.

Cut-in mountings fall into two different types: *remodel boxes* and *cover-plate mounting brackets*.

Remodel Boxes

Remodel boxes are simply plastic or metal boxes that mount to the hole in the drywall using screws or special friction fasteners. The main difference between remodel boxes and regular outlet boxes is that remodel boxes are slightly smaller and can only be mounted in existing walls. Some examples of remodel boxes are shown in Figure 9.5.

FIGURE 9.5
Examples of
common remodel
boxes

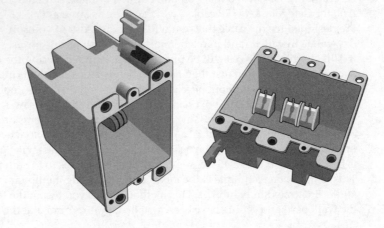

Installing a remodel box so that you can use it for data cabling is simple. Just follow these steps:

1. Using the guidelines discussed earlier in this chapter, determine the location of the new cabling wall plate. With a pencil, mark a line indicating the location for the top of the box.

2. Using the hole template provided with the box, trace the outline of the hole to be cut onto the wall with a pencil or marker, keeping the top of the hole aligned with the mark you made in step 1. If no template is provided, use the box as a template by flipping the box over so the face is against the wall and tracing around the box.

3. Using a drywall keyhole saw, cut out a hole, following the lines drawn using the template.

4. Insert the remodel box into the hole you just cut. If the box won't go in easily, trim the sides of the hole with a razor blade or utility knife.

5. Secure the box by either screwing the box to the drywall or by using the friction tabs. To use the friction tabs (if your box has them), just turn the screw attached to the tabs until the tabs are secured against the drywall.

Cover-Plate Mounting Brackets

The other type of cut-in mounting device for data cabling is the cover-plate mounting bracket. Also called a *cheater bracket*, this mounting bracket allows you to mount a wall plate directly to the wallboard without installing an outlet box. Figure 9.6 shows some examples of preinstalled cover-plate mounting brackets.

FIGURE 9.6
Cover-plate
mounting bracket
examples

These brackets are usually made of steel or aluminum and contain flexible tabs that you push into a precut hole in the drywall. The tabs fold over into the hole and hold the bracket securely to the drywall. Additionally, some brackets allow you to put a screw through both the front and the tabs on the back, thus increasing the bracket's hold on the drywall. Plastic models are becoming popular as well; these use tabs or ears that you turn to grip the drywall. Some also have ratchet-type gripping devices.

Figure 9.7 shows a cover-plate mounting bracket installed in a wall ready to accept a wall plate. Once the mounting bracket is installed, the data cable(s) can be pulled through the wall and terminated at the jacks for the wall plate, and the wall plate can be mounted to the bracket.

FIGURE 9.7
An installed cover-
plate mounting
bracket

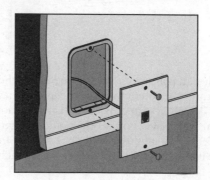

SURFACE-MOUNT OUTLET BOXES

The final type of wall plate mounting system is the surface-mount outlet box, which is used where it is not easy or possible to run the cable inside the wall (in concrete, mortar, or brick walls, for example). Cable is run in a surface-mount raceway (a round or flat conduit) to an outlet box mounted (either by adhesive or screws) on the surface of the wall. This arrangement is shown in Figure 9.8.

FIGURE 9.8

A surface-mount outlet box and conduit

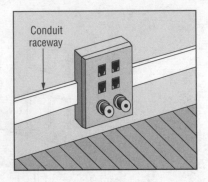

The positive side to surface-mount outlet boxes is their flexibility—they can be placed just about anywhere. The downside is their appearance. Surface-mount installations, even when performed with the utmost care and workmanship, still look cheap and inelegant. But sometimes they are the only choice.

Fixed-Design or Modular Plate

Another design and installation decision you have to make is whether to use *fixed-design* or *modular* wall plates. Fixed-design wall plates (as shown in Figure 9.9) are available with multiple kinds of jacks, but the jacks are molded as part of the wall plate. You cannot remove the jack and replace it with a different type of connector.

FIGURE 9.9

A fixed-design wall plate

Fixed-design plates are usually used in telephone applications rather than LAN wiring applications because, although they are cheap, they have limited flexibility. Fixed-design plates have a couple of advantages and disadvantages (as shown in Table 9.1).

Modular wall plates, on the other hand, are generic and have multiple jack locations (as shown in Figure 9.10). In a modular wall plate system, this plate is known as a *faceplate* (it's not a wall plate until it has its jacks installed). Jacks for each faceplate are purchased separately from the wall plates.

TABLE 9.1   Advantages and Disadvantages of Fixed-Design Wall Plates

ADVANTAGES	DISADVANTAGES
Inexpensive	Configuration cannot be changed
Simple to install	Usually not compatible with high-speed networking systems

FIGURE 9.10
Modular wall plates with multiple jack locations

Modular wall plates

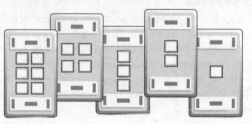

TIP   When using modular wall plates, make sure to use the jacks designed for that wall plate system. Generally speaking, jacks from different wall plate systems are not interchangeable.

You will learn more about these types of wall plates in the next sections.

Fixed-Design Wall Plates

A fixed-design wall plate cannot have its jack configuration changed. In this type of wall plate, the jack configuration is determined at the factory, and the jacks are molded as part of the plate assembly.

You must understand a few issues before choosing a particular fixed-design wall plate for your cabling installation, including the following:

◆   Number of jacks

◆   Types of jacks

◆   Labeling

Number of Jacks

Because fixed-design wall plates have their jacks molded into the faceplate assembly, the number of jacks that can fit into the faceplate is limited. It is very unusual to find a fixed-design faceplate with more than two jacks (they are usually in an over-under configuration, with one jack above the other). Most fixed-design wall plates are for UTP or coaxial copper cable only; but fiber-optic fixed-design wall plates are available for fiber-to-the-desk applications. Figure 9.11 shows some examples of fixed-design wall plates with various numbers and types of sockets.

FIGURE 9.11
Fixed-design wall plates

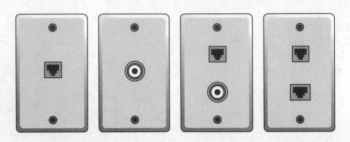

Types of Jacks

Fixed-design wall plates can accommodate many different types of jacks for different types of data communications media. However, you cannot change a wall plate's configuration once it is in place; instead, you must install a completely new wall plate with a different configuration.

The most common configuration of a fixed-design wall plate is the single six-position (RJ-11) or eight-position (RJ-45) jack (as shown in Figure 9.12), which is most often used for home or office telephone connections. This type of wall plate can be found in your local hardware store or home center.

FIGURE 9.12
Fixed-design plates with a single RJ-11 or RJ-45 jack

WARNING   Fixed-design wall plates that have eight-position jacks must be carefully checked to see if they are data-capable. We know of retail outlets that claim their eight-position, fixed-design wall plates are "CAT 5e" compliant. They're not. They use screw terminals instead of 110-type IDC connections. If it's got screws, folks, it ain't Category 5e.

Other types of fixed-design wall plates can include any combination of socket connectors, based on market demand and the whims of the manufacturer. Some of the connector combinations commonly found are as follows:

◆ Single RJ-11 type

◆ Single RJ-45 type

- Single coax (TV cable)

- Single BNC

- Dual RJ-11 type

- Dual RJ-45 type

- Single RJ-11 type, single RJ-45 type

- Single RJ-11 type, single coax (TV cable)

- Single RJ-45 type, single BNC

Labeling

Not all wall plate connectors are labeled. Most fixed-design wall plates don't have special preparations for labeling (unlike modular plates). However, that doesn't mean it isn't important to label each connection; on the contrary, it is extremely important so that you can tell which connection is which (vital when troubleshooting). Additionally, some jacks, though they look the same, may serve a completely different purpose. For example, RJ-45 jacks can be used for both PBX telephone and Ethernet networking, so it's helpful to label which is which if a fixed-design plate has two RJ-45 jacks.

For these reasons, structured-cabling manufacturers have come up with different methods of labeling fixed-design wall plates. The most popular method is using adhesive-backed stickers or labels of some kind. There are alphanumeric labels (e.g., *LAN* and *Phone*) as well as icon labels with pictures of computers for LAN ports and pictures of telephones for telephone ports. Instead of printed labels, sometimes the manufacturer will mold the labels or icons directly into the wall plate.

Modular Wall Plates

Modular wall plates have individual components that can be installed in varying configurations depending on your cabling needs. The wall plates come with openings into which you install the type of jack you want. For example, when you have a cabling-design need for a wall plate that can have three RJ-45 jacks in one configuration and one RJ-45 jack and two fiber-optic jacks in another configuration, the modular wall plate fills that design need very nicely.

Just like fixed-design wall plates, modular wall plates have their own design and installation issues, including:

- Number of jacks

- Wall plate jack considerations

- Labeling

Number of Jacks

The first decision you must make when using modular wall plates is how many jacks you want in each wall plate. Each opening in the wall plate can hold a different type of jack for a different type of cable media, if necessary. Additionally, each jack must be served by its own cable, and at least one of those should be a four-pair, 100 ohm, UTP cable.

The number of jacks a plate can have is based on the size of the plate. Fixed-design wall plates mainly come in one size. Modular plates come in a couple of different sizes. The smallest size is single-gang, which measures 4.5″ high and 2.75″ wide. The next largest size is called double-gang, which measures 4.5″ by 4.5″ (the same height as single-gang plates but almost twice as wide). There are also triple- and quad-gang plates, but they are not used as often as single- and double-gang plates. Figure 9.13 shows the difference between a single- and double-gang wall plate.

FIGURE 9.13
Single- and double-gang wall plates

Single gang   Double gang

Each manufacturer has different guidelines about how many openings for jacks fit into each type of wall plate. Most manufacturers, however, agree that six jacks are the most you can fit into a single-gang wall plate.

With the advent of technology and applications such as videoconferencing and fiber-to-the-desktop, users need more jacks and different types of cabling brought to the desktop. You can bring Category 3, Category 5e, or Category 6 UTP cable, fiber-optic and coaxial cable all to the desktop for voice, data, and video with 6-, 12- and 16-jack wall plates.

Wall Plate Jack Considerations

Modular wall plates are the most common type of wall plate in use for data cabling because they meet the various ANSI/TIA and NEC standards and codes for quality data communications cabling. So modular wall plates have the widest variety of jack types available. All the jacks available today differ based on a few parameters, including the following:

- Wall plate system type
- Cable connection
- Jack orientation
- ANSI/TIA-568-C.2 and -C.3 wiring pattern

WALL PLATE SYSTEM TYPE

Remember how the type of wall plate you use dictates the type of jacks for that wall plate? Well, logically, the reverse is also true. The interlocking system that holds the jack in place in the wall plate differs from brand to brand. So, when you pick a certain brand and manufacturer for a jack, you must use the same brand and manufacturer of wall plate.

CABLE CONNECTION

Jacks for modern communication applications use insulation displacement connectors (IDCs), which have small metal teeth or pins in the connector that press into the individual wires of a UTP cable (or the wires are pressed into the teeth). The teeth puncture the outer insulation of the individual wires and make contact with the conductor inside, thus making a connection. This process (known as *crimping* or *punching down*, depending on the method or tool used) is illustrated in Figure 9.14.

FIGURE 9.14
Using insulation displacement connectors (IDCs)

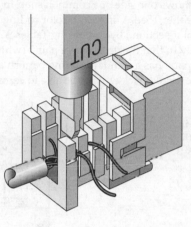

Though they may differ in methods, any connector that uses some piece of metal to puncture through the insulation of a strand of copper cable is an IDC connector.

JACK ORIENTATION

Yes, jack orientation. The individual wall plate systems use many different types of jacks, and some of those systems use jacks with positions other than straight ahead (which is the "standard" configuration). These days, a popular configuration is a jack that's angled approximately 45 degrees down. There are many reasons that this jack became popular. Because it's angled, the cable-connect takes up less room (which is nice when a desk is pushed up tight against the wall plate). The angled connector works well in installations with high dust content because it's harder for dust to rest inside the connector. It is especially beneficial in fiber-to-the-desktop applications because it avoids damage to the fiber-optic patch cord by greatly reducing the bend radius of the cable when the cable is plugged in. Figure 9.15 shows an example of an angled connector.

FIGURE 9.15
A faceplate with angled RJ-45 and coaxial connectors

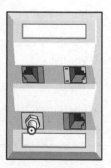

NOTE Angled connectors are found in many different types of cabling installations, including ScTP, UTP, and fiber optic.

Wiring Pattern

When connecting copper RJ-45 jacks for universal applications (according to the standard, of course), you must wire all jacks and patch points according to either the T568-A or T568-B pattern. Figure 9.16 shows one side of a common snap-in jack to illustrate that the same jack can be terminated with either T568-A or T568-B color coding. (You may want to see the color version of this figure in the color section.) By comparing Tables 9.2 and 9.3, you can see that the wiring schemes are different only in that the positions of pair 2 (white/orange) and pair 3 (white/green) are switched. If your company has a standard wiring pattern and you wire a single jack with the opposing standard, that particular jack will not be able to communicate with the rest of the network.

FIGURE 9.16
A common snap-in jack showing both T568-A and T568-B wiring schemes

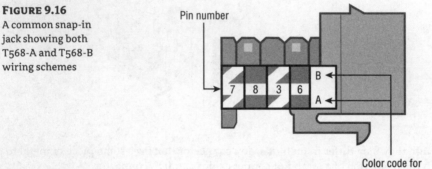

Table 9.2 shows the wiring color scheme for the T568-A pattern. Notice how the wires are paired and which color goes to which pin. Table 9.3 shows the same for T568-B.

TABLE 9.2   Wiring Scheme for T568-A

Pin Number	Wire Color
1	White/green
2	Green
3	White/orange
4	Blue
5	White/blue
6	Orange
7	White/brown
8	Brown

TABLE 9.3 Wiring Scheme for T568-B

Pin Number	Wire Color
1	White/orange
2	Orange
3	White/green
4	Blue
5	White/blue
6	Green
7	White/brown
8	Brown

Labeling

Just like fixed-design wall plates, modular wall plates use labels to differentiate the different jacks by their purpose. In fact, modular wall plates have the widest variety of labels—every modular wall plate manufacturer seems to pride itself on its varied colors and styles of labeling. However, as with fixed-design plates, the labels are either text (e.g., *LAN*, *Phone*) or pictures of their intended use, perhaps permanently molded in the plate or on the jack.

CABLING @ WORK: WE DON'T KNOW WHAT TOMORROW WILL BRING, BUT ADDING SOME FLEXIBILITY TODAY MAY MAKE YOU MORE PREPARED

A few years ago during a hotel's remodeling project, the hotel wired its rooms with only the minimum network connection points using fixed wall plates for its new local area network. Later that year the hotel had to rewire each room to accommodate an IP phone system upgrade, and to offer in-room fax machines and additional telephones on additional lines. Each room had to have additional cabling installed as well as new wall plates.

Though no precise figures have been calculated to see exactly how much would have been saved by installing modular wall plates initially instead of fixed wall plates, it was determined that the fixed wall plate design would not enable the hotel to keep up with the ever-growing number of additional network points and media types that its guests require.

Biscuit Jacks

No discussion of wall plates would be complete without a discussion of *biscuit jacks*, or surface-mount jacks that look like small biscuits (see Figure 9.17). They were originally used in residential and light commercial installations for telephone applications. In fact, you may have some in your home if it was built before 1975. Biscuit jacks are still used when adding phone lines in residences, especially when people can't put a hole in the wall where they want the phone jack to go.

FIGURE 9.17
An example of a
biscuit jack

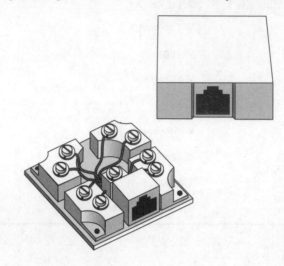

Types of Biscuit Jacks

The many different types of biscuit jacks differ primarily by size and number of jacks they can support. The smaller type measures 2.25 inches wide by 2.5˝ high and is mainly used for residential-telephone applications. The smaller size can generally support up to a maximum of two jacks.

The larger-sized biscuit jacks are sometimes referred to simply as *surface-mount boxes* because they don't have the shape of the smaller biscuit jacks. These surface-mount boxes are primarily used for data communications applications and come in a variety of sizes. They also can have any number or type of jacks and are generally modular. Figure 9.18 shows an example of a larger biscuit jack that is commonly used in surface-mount applications.

FIGURE 9.18
Example of a larger
biscuit jack

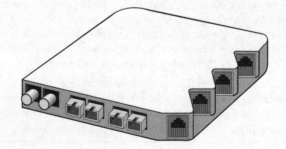

NOTE Generally speaking, the smaller biscuit jacks are not rated for Category 5e (or any higher categories). They must be specifically designed for a Category 5e application. Some companies offer a modular-design biscuit jack that lets you snap in high-performance, RJ-45-type jacks for Category 5e and better compliance.

Advantages of Biscuit Jacks

Biscuit jacks offer a few advantages in your structured-cabling design. First of all, they are very inexpensive compared to other types of surface-mount wiring systems, which is why many houses that had the old four-pin telephone systems now have biscuit jacks—you could buy 20 of them for around $25. Even the biscuits that support multiple jacks are still fairly inexpensive.

Another advantage of biscuit jacks is their ability to work in situations where standard modular or fixed-design wall plates won't work and other types of surface-mount wiring are too bulky. The best example of this is office cubicles (i.e., modular furniture). A biscuit jack has an adhesive tab on the back that allows it to be mounted anywhere, so you can run a telephone or data cable to a biscuit jack and mount it under the desk where it will be out of the way.

Finally, biscuit jacks are easy to install. The cover is removed with one screw. Inside many of the biscuit jacks are screw terminals (one per pin in each jack), as shown in Figure 9.19. To install the jack, you just strip the insulation from each conductor and wrap it clockwise around the terminal and between the washers and tighten the screw. Repeat this process for each conductor in the cable. These jacks are not high-speed data compatible and are capable of Category 3 performance at best.

FIGURE 9.19
Screw terminals
inside a
biscuit jack

Screw terminals

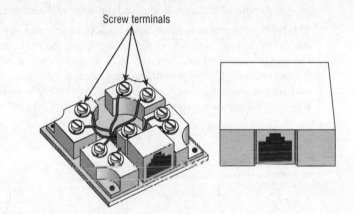

NOTE Not all biscuit jacks use screw terminals. The more modern data communications jacks use IDC connectors to attach the wire to the jack.

Disadvantages of Biscuit Jacks

The main disadvantage to biscuit jacks is that the older biscuit jacks are not rated for high-speed data communications. Notice the bunch of screw terminals in the biscuit jack shown in Figure 9.19. When a conductor is wrapped around these terminals, it is exposed to stray electro-magnetic interference (EMI) and other interference, which reduces the effective ability of this type of jack to carry data. At most, the older biscuit jacks with the screw terminals can be rated as Category 3 and are not suitable for the 100Mbps and faster communications today's wiring systems must be able to carry.

The Bottom Line

Identify key wall plate design and installation aspects.   The wall plate is the "gateway" for the workstation (or network device) to make a physical and electrical/optical connection to the network cabling system. There are important aspects to consider when designing and installing wall plates.

Master It   What are the main aspects that must be addressed when designing and installing wall plates?

Identify the industry standards required to ensure proper design and installation of wall plates.   You'll have to make certain choices about how best to conform to the relevant industry standards when designing and installing wall plates.

Master It   Which TIA standards should you refer to for ensuring proper design and installation of wall plates?

Understand the different types of wall plates and their benefits and suggested uses. There are two basic types of wall plate designs: fixed-design and modular wall plates. Each of these has some advantages and disadvantages and recommended uses.

Master It   Your commercial office building environment is expected to change over the next few years as some of the network devices that are presently installed with coax connections need to be converted to RJ-45 and some network devices that are currently connected with RJ-45 connections may need to be converted to fiber-optic connections. In addition, the number of connections for each wall plate may increase. Which type of wall plate would provide you with the most flexibility and why?

Chapter 10

Connectors

Have you ever wired a cable directly into a piece of hardware? Some equipment in years past provided terminals or termination blocks so that cable could be wired directly into a direct component. In modern times, this approach is considered bad; it is fundamentally against the precepts of a structured cabling system to attach directly to active electronic components, either at the workstation or in the equipment closet. On the ends of the cable you install, something must provide access and transition for attachment to system electronics. Thus, you have connectors.

Connectors generally have a male component and a female component, except in the case of hermaphroditic connectors such as the IBM data connector. Usually jacks and plugs are symmetrically shaped, but sometimes they are *keyed*. This means that they have a unique, asymmetric shape or some system of pins, tabs, and slots that ensure that the plug can be inserted only one way in the jack. This chapter covers many of the connector types you will encounter when working with structured cabling systems.

In this chapter, you will learn to:

- ◆ Identify key twisted-pair connectors and associated wiring patterns
- ◆ Identify coaxial cable connectors
- ◆ Identify types of fiber-optic connectors and basic installation techniques

Twisted-Pair Cable Connectors

Many people in the cabling business use twisted-pair connectors more than any other type of connector. The connectors include the modular RJ types of jacks and plugs and the hermaphroditic connector employed by IBM that is used with shielded twisted-pair cabling.

Almost as important as the cable connector is the connector used with patch panels, punch-down blocks, and wall plates; this connector is called an IDC, or *insulation displacement connector*.

Patch-Panel Terminations

Most unshielded twisted-pair (UTP) and screened twisted-pair (ScTP) cable installations use patch panels and, consequently, 110-style termination blocks. The 110-block (shown in Figure 10.1) contains rows of specially designed slots in which the cables are terminated using a punch-down tool. Patch panels and 110-blocks are described in more detail in Chapter 5, "Cabling System Components," and Chapter 7, "Copper Cable Media."

When terminating 66-blocks, 110-blocks, and often, wall plates, both UTP and ScTP connectors use IDC technology to establish contact with the copper conductors. You don't strip the wire insulation off the conductor as you would with a screw-down connection. Instead, you force the conductor either between facing blades or onto points that pierce the plastic insulation and make contact with the conductor.

FIGURE 10.1
An S-110-block
with wire
management

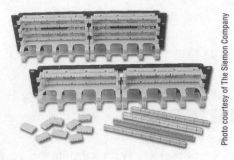

Photo courtesy of The Siemon Company

Solid- vs. Stranded-Conductor Cables

UTP and ScTP cables have either solid copper conductors or conductors made of several tiny strands of copper. Solid conductors are very stable geometrically and, therefore, electrically superior, but they will break if flexed very often. Stranded conductors are very flexible and resistant to bend-fatigue breaks, but their cross-sectional geometry changes as they are moved, and this can contribute to electrical anomalies. Stranded cables also have a higher attenuation (signal loss) than solid-conductor cables.

NOTE Solid-conductor cables are usually used in backbone and horizontal cabling where, once installed, there won't be much movement. Stranded-conductor cables are used in patch cords, where their flexibility is desirable and their typically short lengths mitigate transmission problems.

The differences in conductors mean a difference in IDC types. You have to be careful when you purchase plugs, wall plates, and patch panels because they won't work interchangeably with solid- and stranded-core cables—the blade designs are different.

WARNING Using the wrong type of cable/connector combination can be a major source of intermittent connection errors after your system is running.

With a solid-conductor IDC, you are usually forcing the conductor between two blades that form a V-shaped notch. The blades slice through the plastic and into the copper conductor, gripping it and holding it in place. This makes a very reliable electrical contact. If you force a stranded conductor into this same opening, contact may still be made. But, because one of the features of a stranded design is that the individual copper filaments can move (this provides the flexibility), they will sort of mush into an elongated shape in the *V*. Electrical contact may still be made, but the grip on the conductor is not secure and often becomes loose over time.

The blade design of IDC connectors intended for stranded-core conductors is such that forcing a solid-core conductor onto the IDC connector can break the conductor or fail to make contact entirely. Broken conductors can be especially problematic because the two halves of the break can be close enough together that contact is made when the temperature is warm, but the conductor may contract enough to cause an open condition when cold.

Some manufacturers of plugs advertise that their IDC connectors are universal and may be used with either solid or stranded conductors. Try them if you like, but if you have problems, switch to a plug specifically for the type of cable you are using.

Jacks and termination blocks are almost exclusively solid-conductor devices. You should never punch down on a 66, 110, or modular jack with stranded conductors.

Modular Jacks and Plugs

Twisted-pair cables are most commonly available as UTP, but occasionally, a customer or environmental circumstances may require that ScTP cable be installed. In an ScTP cable, the individual twisted pairs are not shielded, but all the pairs collectively have a thin shield around the shield of foil around them. Both UTP and ScTP cables use modular jacks and plugs. For decades, modular jacks have been commonplace in the home for telephone wiring.

Modular connectors come in four-, six-, and eight-position configurations. The number of positions defines the width of the connector. However, often only some of the positions have metal contacts installed. Make sure that the connectors you purchase are properly populated with contacts for your application.

Commercial-grade jacks are made to snap into modular cutouts in faceplates. (More information is available on modular wall plates in Chapter 9, "Wall Plates.") This gives you the flexibility of using the faceplate for voice, data, coax, and fiber connections, or combinations of them. Figure 10.2 shows a modular plug (male end of the connection), and Figure 10.3 shows the modular jack (female end of the connection) used for UTP. Figure 10.4 shows a modular jack for ScTP cables. Note the metal shield around the jack; it is designed to help reduce both EMI emissions and interference from outside sources, but it must be connected properly to the cable shield to be effective.

FIGURE 10.2
An eight-position
modular plug for
UTP cable

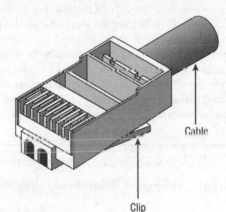

Cable

Clip

FIGURE 10.3
An eight-position
modular jack for
UTP cable

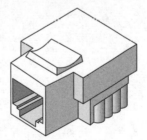

FIGURE 10.4
An eight-position modular jack for ScTP cable

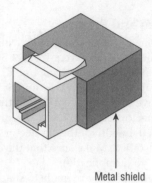

Metal shield

NOTE The quality of plugs and jacks varies widely. Make sure that you use plugs and jacks that are rated to the category of cabling you purchase.

Though the common name is *modular jack*, these components are specifically referred to as *RJ-type connectors* (e.g., RJ-45). The RJ (registered jack) prefix is one of the most widely (and incorrectly) used prefixes in the computer industry; nearly everyone, including people working for cabling companies, is guilty of referring to an eight-position modular jack (sometimes called an 8P8C) as an RJ-45. Bell Telephone originated the RJ prefix and the Universal Service Order Code (USOC) to indicate to telephone technicians the type of service to be installed and the wiring pattern of the jack. Since the breakup of AT&T and the divestiture of the Regional Bell Operating Companies, the term *registered* has lost most of its meaning with respect to these connectors. However, the FCC has codified a number of RJ-type connectors and detailed the designations and pinout configurations in FCC Part 68, Subpart F, Section 68.502. Table 10.1 shows some of the common modular-jack configurations.

TABLE 10.1   Common Modular-Jack Designations and Their Configuration

DESIGNATION	POSITIONS	CONTACTS	USED FOR	WIRING PATTERN
RJ-11	6	2	Single-line telephones	USOC
RJ-14	6	4	Single- or dual-line telephones	USOC
RJ-22	4	4	Phone-cord handsets	USOC
RJ-25	6	6	Single-, dual-, or triple-line telephones	USOC
RJ-31	8	4	Security and fire alarms	See note
RJ-45	8	8	Data (10Base-T, 100Base-TX, 1000Base-T, etc.)	T568A or T568B
RJ-48	8	4	1.544Mbps (T1) connections	System dependent
RJ-61	8	8	Single- through quad-line telephones	USOC

NOTE The RJ-31 connection is not specifically a LAN or phone-service jack. It's used for remote monitoring of a secured installation via the phone lines. The monitoring company needs first access to the incoming phone line in case of a security breach. (An intruder then couldn't just pick up a phone extension and interrupt the security-alert call.) USOC, T568A, or T568B wiring configuration schemes will all work with an RJ-31, but additional shorting circuitry is needed, which is built into the modules that use RJ-31 jacks.

The standard six- and eight-position modular jacks are not the only ones that you may find in use. Digital Equipment Corporation (DEC) designed its own six-position modular jack called the MMJ (*modified modular jack*). The MMJ moved the clip portion of the jack to the right to reduce the likelihood that phone equipment would accidentally be connected to a data jack. The MMJ and DEC's wiring scheme for it are shown in Figure 10.5. Although the MMJ is not as common as standard six-position modular connectors (a.k.a. RJ-11) are, the displaced clip connector on the MMJ, when combined with the use of plugs called the MMP (*modified modular plug*), certainly helps reduce accidental connections by phone or non-DEC equipment.

FIGURE 10.5
The DEC MMJ jack
and wiring scheme

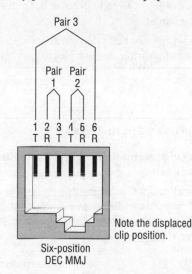

Another connector type that may occasionally be lumped in the category of eight-position modular-jack architecture is called the eight-position *keyed* modular jack (see Figure 10.6). This jack has a key slot on the right side of the connector. The keyed slot serves the same purpose as the DEC MMJ when used with keyed plugs; it prevents the accidental connection of equipment that should be not be connected to a particular jack.

FIGURE 10.6
The eight-position
keyed modular jack

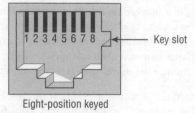

Eight-position keyed

CAN A SIX-POSITION PLUG BE USED WITH AN EIGHT-POSITION MODULAR JACK?

The answer is maybe. First, consider how many of the pairs of wires the application requires. If the application requires all eight pairs, or if it requires the use of pins 1 and 8 on the modular jack, then it will not work.

In addition, keep in mind that repeated inserting and extracting of a six-position modular plug into and from an eight-position modular jack may eventually damage pins 1 and 8 in the jack.

DETERMINING THE PIN NUMBERS

Which one is pin or position number 1? When you start terminating wall plates or modular jacks, you will need to know.

Wall-plate jacks usually have a printed circuit board that identifies exactly which IDC connector you should place each wire into. However, to identify the pins on a jack, hold the jack so that you are facing the side that the modular plug connects to. Make sure that the clip position is facing down. Pin 1 will be on the leftmost side, and pin 8 will be on the rightmost side.

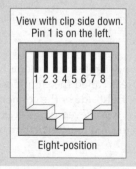

For modular plugs, hold the plug so that the portion that connects to a wall plate or network equipment is facing away from you. The clip should be facing down and you should be looking down at the connector. Pin 1 is the leftmost pin, and thus pin 8 will be the rightmost pin.

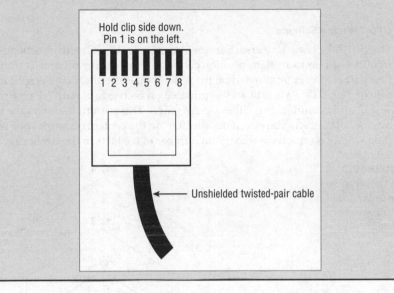

Hold clip side down.
Pin 1 is on the left.

1 2 3 4 5 6 7 8

← Unshielded twisted-pair cable

WIRING SCHEMES

The *wiring scheme* (also called the *pinout scheme*, *pattern*, or *configuration*) that you choose indicates in what order the color-coded wires will be connected to the jacks. These schemes are an important part of standardization of a cabling system. Almost all UTP cabling uses the same color-coded wiring schemes for cables; the color-coding scheme uses a solid-color conductor, and it has a mate that is white with a stripe or band the same color as its solid-colored mate. The orange pair, for example, is often called "orange and white/orange." Table 10.2 shows the color coding and wire-pair numbers for each color code.

TABLE 10.2   Wire Color Codes and Pair Numbers

PAIR NUMBER	COLOR CODE
Pair 1	White/blue and blue
Pair 2	White/orange and orange
Pair 3	White/green and green
Pair 4	White/brown and brown

NOTE When working with a standardized, structured cabling system, the only wiring patterns you will need to worry about are the T568A and T568B patterns recognized in the ANSI/TIA-568-C standard.

USOC Wiring Scheme

The Bell Telephone Universal Service Order Code (USOC) wiring scheme is simple and easy to terminate in up to an eight-position connector; this wiring scheme is shown in Figure 10.7. The first pair is always terminated on the center two positions. Pair 2 is split and terminated on each side of pair 1. Pair 3 is split and terminated on each side of pair 2. Pair 4 continues the pattern; it is split and terminated on either side of pair 3. This pattern is always the same regardless of the number of contacts you populate. You start in the center and work your way to the outside, stopping when you reach the maximum number of contacts in the connector.

FIGURE 10.7
The Universal Service Order Code (USOC) wiring scheme

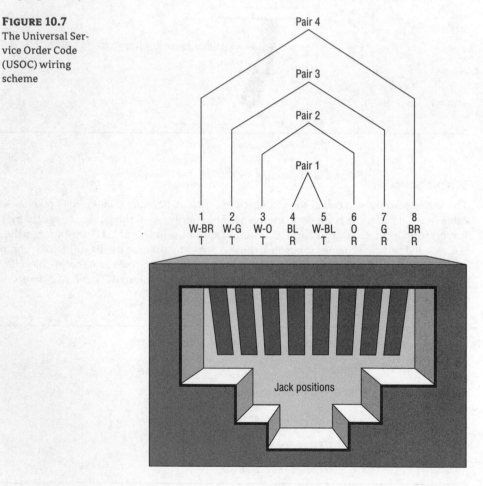

TIP AND RING COLORS

When looking at wiring schemes for modular plugs and jacks, you may see the letters *T* and *R* used, as in Figure 10.7. The *T* identifies the tip color, and the *R* identifies the ring color. In a four-pair cable, the cable pairs are coded in a standard color coding, which is on the insulation of the individual wires. In a four-pair cable, the tip is the wire that is predominantly white, and the ring identifies the wire that is a predominantly solid color.

The wire colors and associated pin assignments for USOC look like this:

Pin	Wire Color
1	White/brown
2	White/green
3	White/orange
4	Blue
5	White/blue
6	Orange
7	Green
8	Brown

WARNING   Do not use the USOC wiring scheme for systems that will support data transmission.

USOC is used for analog and digital voice systems but should *never* be used for data installations. Splitting the pairs can cause a number of transmission problems when used at frequencies greater than those employed by voice systems. These problems include excessive crosstalk, impedance mismatches, and unacceptable signal-delay differential.

ANSI/TIA -568-C Wiring Schemes T568A and T568B

ANSI/TIA-568-C does not sanction the use of the USOC scheme. Instead, two wiring schemes are specified, both of which are suitable for either voice or high-speed LAN operation. These are designated as T568A and T568B wiring schemes.

Both T568A and T568B are universal in that all LAN systems and most voice systems can utilize either wiring sequence without system errors. After all, the electrical signal really doesn't care if it is running on pair 2 or pair 3, as long as a wire is connected to the pin it needs to use. The TIA/EIA standard specifies eight-position, eight-contact jacks and plugs and four-pair cables, fully terminated, to facilitate this universality.

The T568B wiring configuration was at one time the most commonly used scheme, especially for commercial installations; it is shown in Figure 10.8. The TIA/EIA adopted the T568B wiring scheme from the AT&T 258A wiring scheme.

FIGURE 10.8
The T568B wiring
pattern

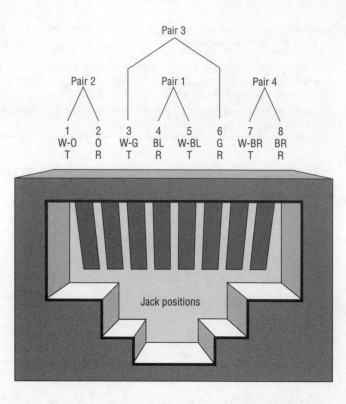

The T568A scheme (shown in Figure 10.9) is well suited to upgrades and new installations in residences because the wire-termination pattern for pairs 1 and 2 is the same as for USOC. Unless a waiver is granted, the U.S. government requires all government cabling installations to use the T568A wiring pattern. The current recommendation according to the standard is for all new installations to be wired with the T568A scheme.

The wire colors and the associated pin assignments for the T568B wiring scheme look like this:

Pin	Wire Color
1	White/orange
2	Orange
3	White/green
4	Blue
5	White/blue
6	Green
7	White/brown
8	Brown

FIGURE 10.9
The T568A wiring
pattern

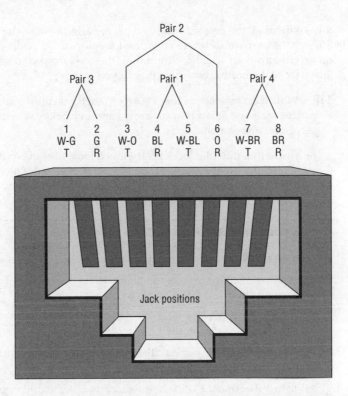

Jack positions

The pin assignments for the T568A wiring schemes are identical to the assignments for the T568B pattern except that wire pairs 2 and 3 are reversed. The T568A pattern looks like this:

Pin	Wire Color
1	White/green
2	Green
3	White/orange
4	Blue
5	White/blue
6	Orange
7	White/brown
8	Brown

Note that when you buy eight-position modular jacks, you may need to specify whether you want a T568A or T568B scheme because the jacks often have IDC connections on the back where you punch the pairs down in sequence from 1 to 4. The jacks have an internal PC board that takes care of all the pair splitting and proper alignment of the cable conductors with the pins in the jack. Most manufacturers now provide color-coded panels on the jacks that let you punch down either pinout scheme, eliminating the need for you to specify (and for them to stock) different jacks depending on which pinout you use.

TIP Whichever scheme you use, T568A or T568B, you must also use that same scheme for your patch panels and follow it in any cross-connect blocks you install. Consistency is the key to a successful installation.

Be aware that modular jacks pretty much look alike even though their performance may differ dramatically. Be sure you also specify the performance level (e.g., Category 3, Category 5e, Category 6, Category 6A, etc.) when you purchase your jacks.

TIPS FOR TERMINATING UTP CONNECTORS

Keep the following points in mind when terminating UTP connectors:

◆ When connecting to jacks and plugs, do not untwist UTP more than 0.5″ for Category 5e and not more than 0.375″ for Category 6.

◆ Always use connectors, wall plates, and patch panels that are compatible (same rating or higher) with the grade of cable used.

◆ To "future-proof" your installation, terminate all four pairs, even if the application requires only two of the pairs.

◆ Remember that the T568A wiring scheme is compatible with USOC wiring schemes that use pairs 1 and 2.

◆ When terminating ScTP cables, always terminate the drain wire on both ends of the connection.

When working with ScTP wiring, the drain wire makes contact with the cable shield along its entire length; this provides a ground path for EMI energy that is collected by the foil shield. When terminating ScTP, the drain wire within the cable is connected to a metal shield on the jack. This must be done at both ends of the cable. If left floating or if connected only on one end, instead of providing a barrier to EMI the cable shield becomes a very effective antenna for both emitting and receiving stray signals.

In a cable installation that utilizes ScTP, the plugs, patch cords, and patch panels must be shielded as well.

Other Wiring Schemes

You may come across other wiring schemes, depending on the demands of the networking or voice application to be used. UTP Token Ring requires that pairs 1 and 2 be wired to the inside four pins, as shown in Figure 10.10. The T568A, T568B, and USOC wiring schemes can be used. It is possible to use a six-position modular jack rather than an eight-position modular jack, but we recommend against that because your cabling system would not follow the ANSI/TIA-568-C standard.

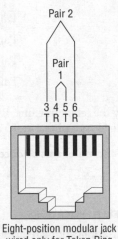

FIGURE 10.10
The Token Ring wiring scheme

Eight-position modular jack
wired only for Token Ring
(pairs 1 and 2)

The ANSI X3T9.5 TP-PMD standard uses the two outer pairs of the eight-position modular jack; this wiring scheme (shown in Figure 10.11) is used with FDDI over copper and is compatible with both the T568A and T568B wiring patterns.

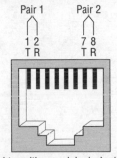

FIGURE 10.11
The ANSI X3T9.5
TP-PMD wiring
scheme

Eight-position modular jack wired
for TP-PMD (ANSI X3T9.5)

If you are wiring a six-position modular jack (RJ-11) for home use, be aware that a few points are not covered by the ANSI/TIA-568-C standard. First, the typical older-design home-telephone cable uses a separate color-coding scheme. The wiring pattern used is the USOC wiring pattern, but the colors are different. The wiring pattern and colors you might find in a home telephone cable and RJ-11 are as follows:

Pin Number	Pair Number	Wire Color
1	Pair 3	White
2	Pair 2	Yellow

3	Pair 1	Green
4	Pair 1	Red
5	Pair 2	Black
6	Pair 3	Blue

Pins 3 and 4 carry the telephone line. Pair 3 is rarely used in home wiring for RJ-11 jacks. Splitters are available to split pins 2 and 5 into a separate jack for use with a separate phone line.

WARNING If you encounter this color code in your home wiring, its performance is likely Category 3 at best.

PINS USED BY SPECIFIC APPLICATIONS

Common networking applications require the use of specific pins in the modular connectors. The most common of these are 10Base-T and 100Base-TX. Table 10.3 shows the pin assignments and what each pin is used for.

TABLE 10.3   10Base-T and 100Base-TX Pin Assignments

PIN	USAGE
1	Transmit +
2	Transmit −
3	Receive +
4	Not used
5	Not used
6	Receive −
7	Not used
8	Not used

USING A SINGLE HORIZONTAL CABLE RUN FOR TWO 10BASE-T CONNECTIONS

Let's face it, you will sometimes fail to run enough cable to a certain room. You will need an extra workstation in an area, and you won't have enough connections. Knowing that you have a perfectly good four-pair UTP cable in the wall, and that only two of those pairs are in use, makes your mood even worse. Modular Y-adapters can come to your rescue.

Several companies make Y-adapters that function as splitters. They take the four pairs of wire that are wired to the jack and split them off into two separate connections. The Siemon

Company makes a variety of modular Y-adapters (see Figure 10.12) for splitting 10Base-T, Token Ring, and voice applications. This splitter will split the four-pair cable so that it will support two separate applications, provided that each application requires only two of the pairs. You must specify the type of splitter you need (voice, 10Base-T, Token Ring, etc.). Don't forget, for each horizontal cable run you will be splitting, you will need two of these adapters: one for the patch-panel side and one for the wall plate.

FIGURE 10.12

A modular Y-adapter for splitting a single four-pair cable into a cable that will support two separate applications

Photo courtesy of The Siemon Company

WARNING   Many cabling professionals are reluctant to use Y-adapters because the high-speed applications such as 10Base-T Ethernet and Token Ring may interfere with one another if they are operating inside the same sheath. Certainly you should not use Y-adapters for applications such as 100Base-TX. Furthermore, Y-adapters eliminate any chance of migrating to a faster LAN system that may utilize all four pairs.

CROSSOVER CABLES

One of the most frequently asked questions on wiring newsgroups and forums is "How do I make a crossover cable?" Computers that are equipped with 10Base-T or 100Base-TX network adapters can be connected "back-to-back"; this means they do not require a hub to be networked together. Back-to-back connections via crossover cables are really handy in a small or home office. Crossover cables are also used to link together two pieces of network equipment (e.g., hubs, switches, and routers) if the equipment does not have an uplink or crossover port built-in.

A crossover cable is just a patch cord that is wired to a T568A pinout scheme on one end and a T568B pinout scheme on the other end. To make a crossover cable, you will need a crimping tool, a couple of eight-position modular plugs (a.k.a. RJ-45 plugs), and the desired length of cable. Cut and crimp one side of the cable as you would normally, following whichever wiring pattern you desire, T568A or T568B. When you crimp the other end, just use the other wiring pattern.

WARNING   As mentioned several times elsewhere in this book, we recommend that you buy your patch cords, either straight through or crossover, instead of making them yourself. Field-terminated patch cords can be time-consuming (i.e., expensive) to make and may result in poor system performance.

Table 10.4 shows the pairs that cross over. The other two pairs wire straight through.

TABLE 10.4   Crossover Pairs

SIDE-ONE PINS	WIRE COLORS	SIDE-TWO PINS
1 (Transmit +)	White/green	3 (Receive +)
2 (Transmit –)	Green	6 (Receive –)
3 (Receive +)	White/orange	1 (Transmit +)
6 (Receive –)	Orange	2 (Receive –)

Shielded Twisted-Pair Connectors

In the United States, the most common connectors for cables that have individually shielded pairs in addition to an overall shield are based on a pre-1990 proprietary cabling system specified by IBM. Designed originally to support Token Ring applications using a two-pair cable (shielded twisted-pair, or STP), the connector is hermaphroditic. In other words, the plug looks just like the jack, but in mirror image. Each side of the connection has a connector and a receptacle to accommodate it. Two hermaphroditic connectors are shown in Figure 10.13. This connector is known by a number of other names, including the STP connector, the IBM data connector, and the universal data connector.

FIGURE 10.13
Hermaphroditic
data connectors

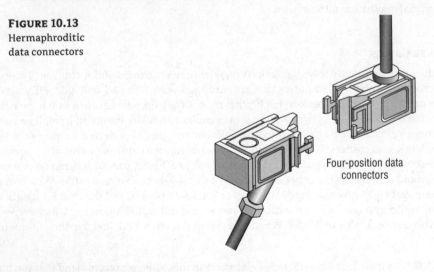

Four-position data
connectors

The original Token Ring had a maximum throughput of 4Mbps (and later 16Mbps) and was designed to run over STP cabling. The 16Mbps Token Ring used a 16MHz spectrum to achieve its throughput. Cables and connectors rated to 20MHz were required to allow the system to operate reliably, and the original STP hermaphroditic connectors were limited to a 20MHz bandwidth. Enhancements to these connectors increased the bandwidth limit to 300MHz. These higher-rated connectors (and cable) are designated as STP-A.

STP connectors are the Jeeps of the connector world. They are large, rugged, and versatile. Both the cable and connector are enormous compared to four-pair UTP and RJ-type modular plugs. They also have to be assembled and have more pieces than an Erector set. Cabling contractors used to love the STP connectors because of the premium they could charge based on the labor required to assemble and terminate them.

Darwinian theory prevailed, however, and now the STP and STP-A connectors are all but extinct—they've been crowded out by the smaller, less expensive, and easier-to-use modular jack and plug.

Coaxial Cable Connectors

Unless you have operated a 10Base-2 or 10Base-5 Ethernet network, you are probably familiar only with the coaxial connectors you have in your home for use with televisions and video equipment. Actually, a number of different types of coaxial connectors exist.

F-Series Coaxial Connectors

The coax connectors used with video equipment are referred to as *F-series connectors* (shown in Figure 10.14). The F-connector consists of a ferrule that fits over the outer jacket of the cable and is crimped in place. The center conductor is allowed to project from the connector and forms the business end of the plug. A threaded collar on the plug screws down on the jack, forming a solid connection. F-connectors are used primarily in residential installations for RG-58, RG-59, and RG-6 coaxial cables to provide CATV, security-camera, and other video services.

FIGURE 10.14
The F-type coaxial-cable connector

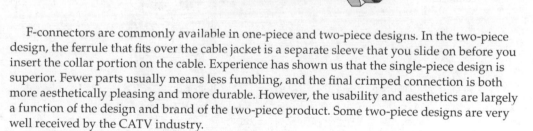

F-connectors are commonly available in one-piece and two-piece designs. In the two-piece design, the ferrule that fits over the cable jacket is a separate sleeve that you slide on before you insert the collar portion on the cable. Experience has shown us that the single-piece design is superior. Fewer parts usually means less fumbling, and the final crimped connection is both more aesthetically pleasing and more durable. However, the usability and aesthetics are largely a function of the design and brand of the two-piece product. Some two-piece designs are very well received by the CATV industry.

A cheaper F-type connector available at some retail outlets attaches to the cable by screwing the outer ferrule onto the jacket instead of crimping it in place. These are very unreliable and pull off easily. Their use in residences is not recommended, and they should never be used in commercial installations.

N-Series Coaxial Connectors

The *N-connector* is very similar to the F-connector but has the addition of a pin that fits over the center conductor; the N-connector is shown in Figure 10.15. The pin is suitable for insertion in the jack and must be used if the center conductor is stranded instead of solid. The assembly is attached to the cable by crimping it in place. A screw-on collar ensures a reliable connection with the jack. The N-type connector is used with RG-8, RJ-11U, and thicknet cables for data and video backbone applications.

The BNC Connector

When coaxial cable distributes data in commercial environments, the *BNC connector* is often used. BNC stands for Bayonet Neill-Concelman, which describes both the method of securing the connection and its inventors. Many other expansions of this acronym exist, including British Naval Connector, Bayonet Nut Coupling, Bayonet Navy Connector, and so forth. Used with RG-6, RG-58A/U thinnet, RG-59, and RG-62 coax, the BNC utilizes a center pin, as in the N-connector, to accommodate the stranded center conductors usually found in data coax.

The BNC connector (shown in Figure 10.16) comes as a crimp-on or a design that screws onto the coax jacket. As with the F-connector, the screw-on type is not considered reliable and should not be used. The rigid pin that goes over the center conductor may require crimping or soldering in place. The rest of the connector assembly is applied much like an F-connector, using a crimping die made specifically for a BNC connector.

Connector

To secure a connection to the jack, the BNC has a rotating collar with slots cut into it. These slots fit over combination guide and locking pins on the jack. Lining up the slots with the pins, you push as you turn the collar in the direction of the slots. The slots are shaped so that the plug is drawn into the jack, and locking notches at the end of the slot ensure positive contact with the jack. This method allows quick connection and disconnection while providing a secure match of plug and jack.

Be aware that you must buy BNC connectors that match the impedance of the coaxial cable to which they are applied. Most commonly, they are available in 75 ohm and 50 ohm types, with 93 ohm as a less-used option.

TIP With all coaxial connectors, be sure to consider the dimensions of the cable you will be using. Coaxial cables come in a variety of diameters that are a function of their transmission properties, series rating, and number of shields and jackets. Buy connectors that fit your cable.

Fiber-Optic Cable Connectors

Whereas the RJ-type connector is the most commonly used connector for twisted-pair copper data communications and voice connections, a variety of choices exist for fiber-optic connections you need to use.

This section of the chapter focuses on the different types of fiber connectors and discusses how they are installed onto fiber-optic cable.

SC, ST, and FC Fiber-Optic Connector Types

Fiber-optic connections use different terminology than copper based connectors. The male end of the connection in a fiber-optic system is termed the *connector*, in contrast to the *plug* in a copper-based system. The female end of the connection is termed the *receptacle* or *adapter*, in contrast to the *jack* in a copper-based system.

To transmit data up to 10Gbps, two fibers are typically required: one to send and the other to receive. For 40Gbps and 100Gbps over multimode, as many as 24 fibers will be required. Fiber-optic connectors fall into one of three categories based on how the fiber is terminated:

◆ Simplex connectors terminate only a single fiber in the connector assembly.

◆ Duplex connectors terminate two fibers in the connector assembly.

◆ Array connectors terminate more than two fibers (typically 12 or 24 fibers) in the connector assembly.

The disadvantage of simplex connectors is that you have to keep careful track of polarity. In other words, you must always make sure that the connector on the "send" fiber is always connected to the "send" receptacle (or adapter) and that the "receive" connector is always connected to the "receive" receptacle (or adapter). The real issue is when normal working folk need to move furniture around and disconnect from the receptacle in their work area and then get their connectors mixed up. Experience has shown us that the connectors are not always color coded or labeled properly. Getting them reversed means, at the least, that link of the network won't work.

Array and duplex connectors and adapters take care of this issue. Once terminated, color coding and keying ensures that the connector can be inserted only one way in the adapter and will always achieve correct polarity.

Table 10.5 lists some common fiber-optic connectors, along with their corresponding figure numbers. These connectors can be used for either single-mode or multimode fibers, but make sure you order the correct model connector depending on the type of cable you are using.

TABLE 10.5   Fiber-Optic Connectors

Designation	Connection Method	Configuration	Figure
MPO	Snap-in	Array	Figure 10.17
LC	Snap-in	Simplex	Figure 10.18
Duplex LC	Snap-in	Duplex	Figure 10.19
SC	Snap-in	Simplex	Figure 10.20
Duplex SC	Snap-in	Duplex	Figure 10.21
ST	Bayonet	Simplex	Figure 10.22
Duplex ST	Snap-in	Duplex	Figure 10.23
FDDI (MIC)	Snap-in	Duplex	Figure 10.24
FC	Screw-on	Simplex	Figure 10.25

FIGURE 10.17
An MPO array con-
nector, courtesy
of OFS

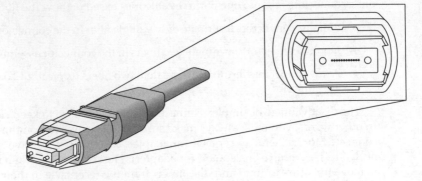

FIGURE 10.18
An LC connector
(compared to the
larger SC connec-
tor), courtesy
of OFS

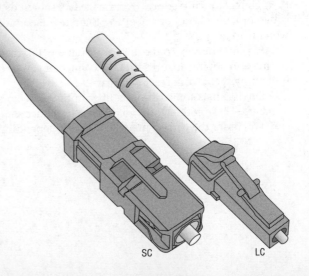

SC   LC

FIGURE 10.19
A duplex LC fiber-optic connector, courtesy of OFS

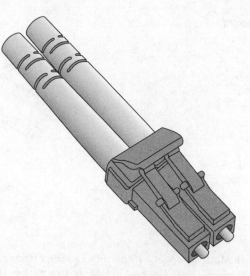

FIGURE 10.20
An SC fiber-optic connector

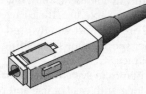

FIGURE 10.21
A duplex SC fiber-optic connector

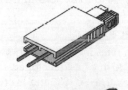

FIGURE 10.22
An ST connector

FIGURE 10.23
A duplex ST fiber-optic connector

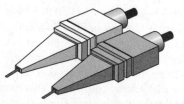

FIGURE 10.24
An FDDI fiber-
optic connector

FIGURE 10.25
An FC fiber-optic
connector

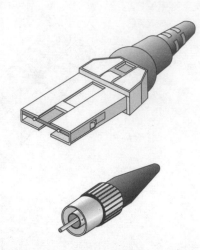

Of the four layers of a tight-buffered fiber (the core, cladding, coating, and buffer), only the core where the light is actually transmitted differs in diameter. The cladding, coating, and buffer diameters are mostly identical, allowing universal use of stripping tools and connectors.

Of the connectors in Table 10.5, the ST used to be the most widely deployed, but now the duplex SC and LC are the most widely used connectors. Other connector styles are allowed, but not specified. Other specifications, including those for ATM, FDDI, and broadband ISDN, now also specify the duplex SC.

This wide acceptance in system specifications and standards (acceptance in one begets acceptance in others), along with ease of use and positive assurance that polarity will be maintained, are all reasons why the duplex SC and LC are the current connectors of choice.

The SFF Connector

As transmission rates increase and networks require the cramming in of a greater number of connections, the industry has developed *small-form-factor* (SFF) connectors and adapter systems for fiber-optic cables. The SC, ST, and FC connectors shown in Table 10.5 all take up more physical space than their RJ-45 counterparts on the copper side. This makes multimedia receptacle faceplates a little crowded and means that you get fewer terminations (lower density) in closets and equipment rooms than you can get with copper in the same space. The goal for the designers of the SFF connector was to create an optical-fiber connector with the same or lower cross-sectional footprint as an RJ-45-style connector in order to increase the number of connections per area (higher density).

The LC, the VF-45, and the MT-RJ SFF fiber-optic connectors were initially developed to support the increase in density of fiber connections. The LC connector (the connector on the lower part of Figure 10.26) is gaining greater use and is regarded by many optical-fiber professionals as the superior simplex or duplex SFF connector.

The SFF connector was taken several steps forward in increasing connection density with the creation of the MPO array connector. This connector typically has 12 fibers lined up side by side (hence, use of the term *array*) in a single connector housing. These connectors may have multiple arrays stacked up on top of each other to produce connectors with as many as 48 or more fibers in a single connector housing. These types of connectors are typically used in data center and super-computing applications and to support the migration to parallel optical transmission systems for 40Gbps and 100Gbps.

FIGURE 10.26
Duplex SC (top), simplex ST (middle), and LC (bottom) connectors

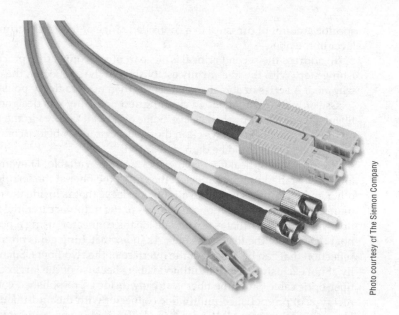

Photo courtesy of The Siemon Company

Installing Fiber-Optic Connectors

With twisted-pair and coax cables, connectors are joined to the cable and conductors using some form of crimping or punch down, forcing the components into place. With fiber-optic cables, a variety of methods can join the fiber with its connector. Each manufacturer of connectors, regardless of type, specifies the method to be used, the materials that are acceptable, and sometimes, the specialized tools required to complete the connection. Connectors can be installed onto cable in the field or they can be factory-terminated.

When the fiber connector is inserted into the receptacle, the fiber-optic core in the connector is placed in end-to-end contact with the core of a mating fiber in the adapter of a wall plate or transceiver. Two issues are of vital importance:

♦ The fiber-optic cores must be properly aligned. The end-to-end contact must be perfectly flush with no change in the longitudinal axis. In other words, they can't meet at an angle.

♦ The surfaces must be free of defects such as scratches, pits, protrusions, and cracks.

To address the first critical issue, fiber connector systems must incorporate a method that both aligns and fixes the fiber in its proper position. The alignment is usually accomplished by inserting the fiber in a built-in sleeve or ferrule. Some technique—either gluing or crimping—is then applied to hold it in place.

CRIMPING

Crimp-style connector systems for fiber-optic cable are always manufacturer-specific regarding the tools and materials required. Follow the manufacturer's instructions carefully. With crimp connectors, the fiber is inserted into the connector, and the assembly is then placed in a crimping tool that holds the fiber and connector in proper position. The tool is then used to apply a specific amount of pressure in a controlled range of motion to crimp the connector to the buffer layer of the fiber.

To address the second critical issue, part of the connecting process usually involves a polishing step. With the fiber firmly established in the connector, the end of the fiber is rough-trimmed. A series of abrasive materials is then used to finely polish the end of the fiber.

Use only the recommended cleaning and polishing kits designed to be used with optical fibers: do not use your shirt sleeve. Some 85% of all failures in the field are due to dirt or debris on the face of a fiber. Always clean the end face properly before connecting; otherwise, the fiber on the receiving side could be damaged as well.

Pre-polished optical fiber connectors are also available. They are connectors that have a fiber already installed in the connector: the end of the fiber at the ferrule end is polished and the other end terminates inside an alignment sleeve that is inside the connector. Following the recommended fiber preparation procedures of the manufacturer, the fiber from the cable is butted-up against the fiber that is already inside the connector, using a positive mechanical force to hold the ends of the fibers together. An index matching gel is often used to reduce the amount of reflection that can be caused at the interface of the two fibers. Such connectors are used primarily, if not exclusively, with multimode fibers because of the larger core diameter of multimode fiber-optic cable. Since the fiber is already inside a pre-polished connector, you must use the correct type of pre-polished multimode connector with the multimode fiber that you are installing. Use only 62.5 micron MMF with 62.5 MMF pre-polished connectors and 50 micron MMF with 50 micron MMF pre-polished connectors. Mixing MMF types will cause severe insertion loss where they are mated together.

GLUING

Three types of adhesives can glue the fiber into position:

Heat-cured adhesives   After the material is injected and the fiber is inserted into the connector assembly, it is placed in a small oven to react with the adhesive and harden it. This is time-consuming—heat-cured adhesives require as much as 20 minutes of hardening. Multiple connectors can be done at one time, but the time required to cure the adhesive still increases labor time, and the oven is, of course, extra baggage to pack to the job site.

UV-cured adhesives   Rather than hardening the material in an oven, an ultraviolet light source is used. You may have had something similar done at your dentist the last time you had a tooth filled. Only about a minute of exposure to the UV light is required to cure the adhesive, making this a more time-effective process.

Anaerobic-cured adhesives   This method relies on the chemical reaction of two elements of an epoxy to set up and harden. A resin material is injected in the ferrule. Then a hardener catalyst is applied to the fiber. When the fiber is inserted in the ferrule, the hardener reacts with the resin to cure the material. No extra equipment is required beyond the basic materials and tools. Hardening can take place in as little as 15 seconds.

Note that while you can use a single-mode optical fiber connector with a multimode fiber, you cannot use a multimode connector with a single-mode fiber. The reason is that the single-mode connecter is manufactured to tighter tolerances to permit the accurate matching of the cores of two single-mode fibers. Single-mode fiber connectors are much more expensive for this reason.

> ### Cabling @ Work: Save Space and Installation Time
>
> Data centers, generically referred to as main equipment rooms, are the heart and brains of a commercial enterprise network. More and more of the large Fortune 100 companies are consolidating their data centers in an effort to minimize operating costs. Since network uptime is critical, any interruption of ongoing systems can cause significant costs. Installing thousands of connectors in the field using adhesives can take a lot of labor time. To address this, a growing trend in data center cabling involves pre-connectorized cables. These are typically high-fiber-count cables that are pre-connected with multiple MPO connectors. The IT planning group would survey the data center and order specific lengths of these pre-terminated cables. On installation day, the cables can be simply plugged into factory-assembled patch panel systems. These MPO pre-terminated high-fiber-count cables are commonly called "plug-'n'-play" solutions. In designing and costing your networks, it would be wise to compare the material and labor cost of these cables to the cost of bare cables, which require connector installation in the field.

The Bottom Line

Identify key twisted-pair connectors and associated wiring patterns.   Table 10.1 showed the common modular jack designations for copper cabling. As business systems, phones, and cameras continue to converge to IP/Ethernet-based applications, the types of jacks and plugs will narrow down quickly.

Master It   What is the most common modular jack designation for IP/Ethernet-based systems (for example, 100Base-TX)? What are your wiring pattern options? Which of these wiring patterns is necessary for government installations?

Identify coaxial cable connectors.   Although coaxial connectors are not commonly used in Ethernet-based operations of 100Base-T or above, there is still a need to use coaxial connectors for coaxial cables connecting security cameras to the building security systems.

Master It   What is the common coaxial connector type to support security-camera service?

Identify types of fiber-optic connectors and basic installation techniques.   You will be using optical-fiber connectors more and more in years to come. It's important to understand the differences and the basic installation options to mount the connectors. An understanding of both of these aspects will enable savings in time and money as well as provide a more space-efficient installation.

Master It

A.   What class of optical-fiber connector would you use to increase the density of fiber connections?

B.   Which one of these types would you use for single fiber connections?

C.   Which one of these types would you use for 12 or more fiber connections in one footprint?

D.   What are the three types of adhesives that can be used to glue fiber into position of a connector housing, and which one provides the quickest cure?

Chapter 11

Network Equipment

Thus far, this book has dealt with common elements of structured cabling systems (cabling, connectors and pathways), network topologies (hierarchical star, bus, and ring) and applications (Ethernet, ATM) that you may encounter. We've looked at the products you can use to bring your communication endpoints to a central location. But is there any communication taking place over your infrastructure? What you need now is a way to tie everything together. In this chapter, the common types of network equipment used in LAN networks are covered.

In this chapter, you will learn to:

♦ Identify the basic active components of a hierarchical star network for commercial buildings and networks

♦ Identify differences between various types of transceiver modules

♦ Determine if your workgroup switching system is blocking or nonblocking

Network Connectivity Devices

Active network equipment is connected by the structured cabling system to support the topology and applications required for the network. More simply, the active equipment is what sends, aggregates, directs, and receives actual data. This equipment is called *active* because it involves devices that transmit and receive electrical and optical signals. As a result, they are powered devices, as opposed to cabling, which is considered *passive*. Active devices include ports on your workstation, hubs, bridges, workgroup switches, core switches, servers, and storage media. These devices are an integral part of the commercial building network. Figure 11.1 shows how these devices are integrated to the cabling systems of a network in a traditional hierarchical stat topology.

Workstation Ports

Workstation ports are the active interfaces located in desktop computers, laptops, printers, and other network-connected devices that are located near the telecommunications outlets. Essentially, workstation ports allow these devices to be connected to the network. They are the ports that *speak* to the network, since they send and process received data. Other network connectivity devices like switches, repeaters, and hubs simply pass the data along the network.

FIGURE 11.1
Active equipment used in a hierarchical star network.

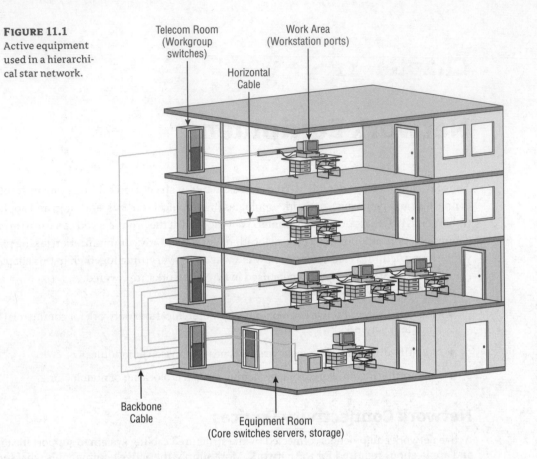

Patch cords are typically used to connect the workstation ports in desktop computers, laptops, printers, and other network devices, to the telecommunications outlets. Telecommunications outlets are connected to the structured cabling system using horizontal cabling. Since most horizontal cabling systems in commercial buildings use UTP wiring, the workstation ports support copper cabling–based connections using RJ-45 connectors. The two most popular connections on workstation equipment are 10Base-T and 100Base-TX, with 1000Base-T close behind. We will see the 10Base-T connections declining in use in the not-too-distant future. Workstation ports are the *visible* portion of network interface cards that allows connection to the patchcords.

Network Interface Cards

A network interface card (NIC), also called *network card* for short, is the full component installed in a computer, or a network device such as a printer or label maker, that allows the device to communicate over the network. Most of this card is situated inside the computer or other device. What is visible to the user is the workstation port.

NICs are both an Open Systems Interconnection (OSI) Layer 1 (Physical layer) and Layer 2 (Data Link layer) component since they provide access to the network and have a unique address.

Since most networks operate using Ethernet applications, most network interface cards are based on Ethernet. Every Ethernet NIC has a unique 48-bit serial number called a *MAC address*. Every computer or other network device on an Ethernet network must have a card with a unique MAC address to communicate on the network.

Network cards can either be in the form of an expansion card that plugs into a computer PCI bus, or be built in directly on the motherboard of the computer or other network device. Most new computers have a network interface card built directly into the motherboard. Copper-based 10Base-T and 100Base-TX network interface cards, or varieties that auto-negotiate between 10, 100, and 1000Mbps, are the most common NICs installed in computers today because of their very low prices. Prices of 10, 100, and 1000Mbps NICs range from $15 to $30.

Network interface cards are also available with fiber-optic connections to support optical-based applications. The visible portion of a NIC has optical ports that would be connected using a fiber-optic cable with ST, SC, or LC connectors. The optical signal is converted to electrical on this board and then passed through to the computer bus. Optical NICs are less common in computers in commercial building networks because of their high cost relative to copper-based NICs. Prices of 100Base-FX (100Mbps) NICs range from $100 to $140.

Media Converters

Media converters allow cost-effective conversion of signals from one cabling media type to another. The most common type converts signals running over copper UTP cabling to fiber-optic cabling. Media converters also exist to support conversion of coax cable to UTP cable or fiber-optic cabling.

Media converters are application specific. For example, you would need an Ethernet converter to convert 1000Base-T to 1000Base-SX. Media converters do not convert one application protocol to another, such as Token Ring to Ethernet. However, within any one of these applications and more, media converters exist to convert from one cabling media to another.

Media converters typically come in a stand-alone device. However, they can also be ordered in a multiport model, similar to a workgroup switch, or in a modular chassis. A media converter has two connections: one for one cabling media type and the other for the other media type. One of the more common Ethernet media converters converts an Ethernet signal over copper UTP cabling to an optical signal. The front plate has an RJ-45 connection for the UTP cabling and an optical connection. Optical connections can come in a variety of types. Media converters are also available with a pluggable optical GBIC (Gigabit Ethernet interface converter).

As LAN networks grow and expand beyond the original plans of a commercial building or campus, workstations are often placed in locations beyond the reach of UTP copper cabling. Since the copper cabling cannot support a distance long enough to connect these devices to the network, IT managers are faced with a choice. They can add an additional telecommunications room near this new set of users and run a fiber-optic cable back to the main equipment room; alternatively, they can run fiber-optic cable directly to the desk. An IT manager can buy a new workgroup switch with a fiber-optic uplink or use media converters. Media converters become useful when there are only a few connections, which means that a stand-alone workgroup switch and closet would be inefficient and cost prohibitive. (Keep in mind that these installations would not be covered by the standards and should be considered temporary.) A media converter connecting a 10, 100, or 1000Base-T copper connection to a 10, 100, or 1000Base-SX multimode connection can cost approximately $200.

Another area where media converters are becoming useful is in fiber-to-the-desk applications where the end-user workstations only come fitted with RJ-45 workstation ports. A media converter can translate between the copper-based connection and a fiber-optic connection. Media converters are also useful in large factory settings with equipment that produces large amounts of EMI. The EMI can have adverse effects on long UTP copper cabling runs. Since fiber-optic cables are resistant to EMI, fiber-optic cabling can be used to interconnect equipment with copper-based ports and NICs. In this case, the media converters are used at the ends of these links.

Repeaters and Hubs

Nowadays, the terms *repeater* and *hub* are used synonymously, but they are actually not the same. Prior to the days of twisted-pair networking, network backbones carried data across coaxial cable, similar to what is used for cable television.

Computers would connect into these by means of either BNC connectors, in the case of thinnet, or vampire taps, in the case of thicknet. Everyone would be connected to the same coaxial backbone. Unfortunately, when it comes to electrical current flowing through a solid medium, you have to contend with the laws of physics. A finite distance exists in which electrical signals can travel across a wire before they become too distorted to carry information. Repeaters were used with coaxial cable to overcome this challenge.

Repeaters work at the physical layer of the OSI reference model. Digital signals decay due to attenuation and noise. A repeater's job is to regenerate the digital signal and send it along in its original state so that it can travel farther across a wire. Theoretically, repeaters could be used to extend cables infinitely, but due to the underlying limitations of communication architectures like Ethernet's collision domains, repeaters were originally used to tie together a maximum of five coaxial-cable segments.

Because repetition of signals is a function of repeating hubs, *hub* and *repeater* are used interchangeably when referring to twisted-pair networking. The semantic distinction between the two terms is that a repeater joins two backbone coaxial cables, whereas a hub joins two or more twisted-pair cables.

In twisted-pair networking, each network device is connected to an individual network cable. In coaxial networking, all network devices are connected to the same coaxial backbone. A hub eliminates the need for BNC connectors and vampire taps. Figure 11.2 illustrates how network devices connect to a hub compared to coaxial backbones.

Hubs work the same way as repeaters in that they regenerate incoming signals before they are retransmitted across its ports. Like repeaters, hubs operate at the OSI Physical layer, which means they do not alter or look at the contents of a frame traveling across the wire. When a hub receives an incoming signal, it regenerates it and sends it out over all its ports. Figure 11.3 shows a hub at work.

FIGURE 11.2
Twisted-pair networking versus coaxial networking

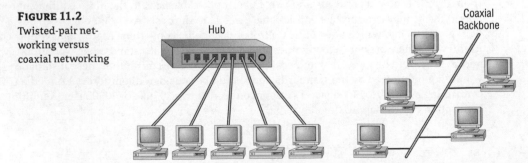

FIGURE 11.3
Hubs at work

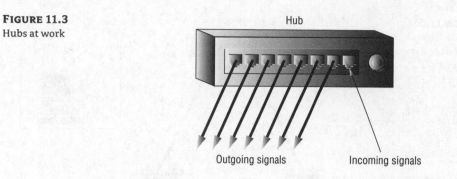

Hub

Outgoing signals   Incoming signals

Hubs typically provide from 8 to 24 twisted-pair connections, depending on the manufacturer and model of the hub (although some hubs support several dozen ports). Hubs can also be connected to each other (cascaded) by means of BNC, AUI ports, or crossover cables to provide flexibility as networks grow. The cost of this flexibility is paid for in performance.

As a media-access architecture, Ethernet is built on carrier-sensing and collision-detection mechanisms (CSMA/CD). Prior to transmitting a signal, an Ethernet host listens to the wire to determine if any other hosts are transmitting. If the wire is clear, the host transmits. On occasion, two or more hosts will sense that the wire is free and try to transmit simultaneously or nearly simultaneously. Only one signal is free to fly across the wire at a time, and when multiple signals meet on the wire, they become corrupted by the collision. When a collision is detected, the transmitting hosts wait a random amount of time before retransmitting, in the hopes of avoiding another data collision. Figure 11.4 shows a situation where a data collision is produced, and Figure 11.5 shows how Ethernet handles these situations.

FIGURE 11.4
An Ethernet data collision

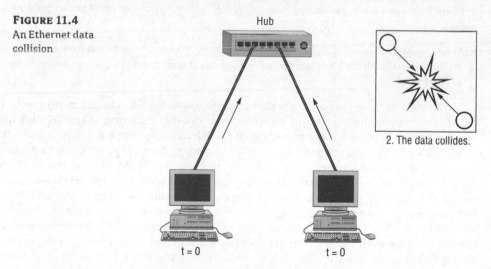

Hub

2. The data collides.

t = 0   t = 0

Two stations transmit
at exactly the same time.

FIGURE 11.5
How Ethernet
responds to data
collisions

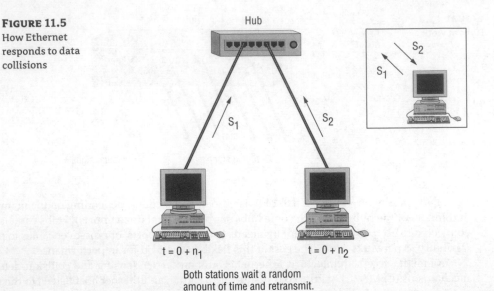

Both stations wait a random
amount of time and retransmit.

So what are the implications of collision handling on performance? If you recall from our earlier explanation of how a hub works, a hub, after it receives an incoming signal, simply passes it across all its ports. For example, with an eight-port hub, if a host attached to port 1 transmits, the hosts connected to ports 2 through 8 will all receive the signal. Consider the following: if a host attached to port 8 wants to communicate with a host attached to port 7, the hosts attached to ports 1 through 6 will be barred from transmitting when they sense signals traveling across the wire.

NOTE   Hubs pass incoming signals across all their ports, preventing two hosts from transmitting simultaneously. All the hosts connected to a hub are therefore said to share the same amount of bandwidth.

On a small scale, such as our eight-port example, the shared-bandwidth performance implications may not be that significant. However, consider the cascading of four 24-port hubs, where 90 nodes (6 ports are lost to cascade and backbone links) share the same bandwidth. The bandwidth that the network provides is finite (limited by the cable plant and network devices). Therefore, in shared-bandwidth configurations, the amount of bandwidth available to a connected node is inversely proportional to the number of actively transmitting nodes sharing that bandwidth. For example, if 90 nodes are connected to the same set of Fast Ethernet (100Mbps) hubs and are all actively transmitting at the same time, they potentially have only 1.1Mbps available each. (The situation is actually worse than that because of the overhead bandwidth that each node requires.) For Ethernet (10Mbps), the situation is even worse, with potentially only 0.1Mbps available each. These 100 percent utilization examples are worst-case scenarios, of course. Your network would have given up and collapsed before it reached full saturation, probably at around 80 percent utilization, and your users would have been loudly complaining long before that.

All hope is not lost, however. We'll look at ways of overcoming these performance barriers through the use of switches and routers. As a selling point, hubs are relatively inexpensive to implement. Hubs are still widely used in many low-traffic situations such as a home or small office when fewer than 10 computer systems are connected to a network.

Bridges

When we use the terms *bridge* and *bridging*, we are generally describing functionality provided by modern switches. Just like a repeater, a bridge is a network device used to connect two network segments. The main difference between a repeater and a bridge is that bridges operate at the Link layer of the OSI reference model and can therefore provide translation services required to connect dissimilar media access architectures such as Ethernet and Token Ring. Therefore, bridging is an important internetworking technology.

In general, there are four types of bridging:

Transparent bridging Typically found in Ethernet environments, the transparent bridge analyzes the incoming frames and forwards them to the appropriate segments one hop at a time (see Figure 11.6).

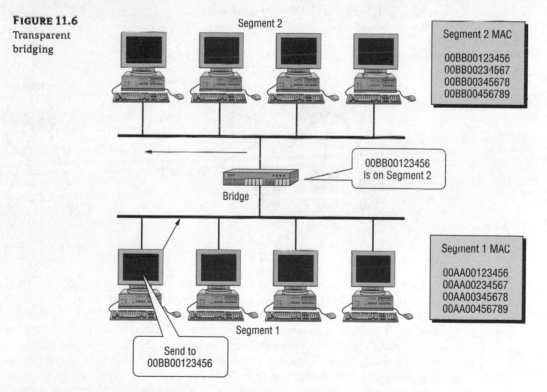

FIGURE 11.6
Transparent bridging

Source-route bridging Typically found in Token Ring environments, source-route bridging provides an alternative to transparent bridging for NetBIOS and SNA protocols. In source-route bridging, each ring is assigned a unique number on a source-route bridge port. Token Ring frames contain address information, including a ring and bridge numbers, which each bridge analyzes to forward the frame to the appropriate ring (see Figure 11.7).

FIGURE 11.7
Source-route
bridging

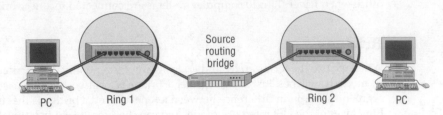

Source-route transparent bridging   Source-route transparent bridging is an extension of source-route bridging, whereby nonroutable protocols such as NetBIOS and SNA receive the routing benefits of source-route bridging and a performance increase associated with transparent bridging.

Source-route translation bridging   Source route translation bridging is used to connect network segments with different underlying media-access technologies such as Ethernet to Token Ring or Ethernet to FDDI (see Figure 11.8).

FIGURE 11.8
Translation
bridging

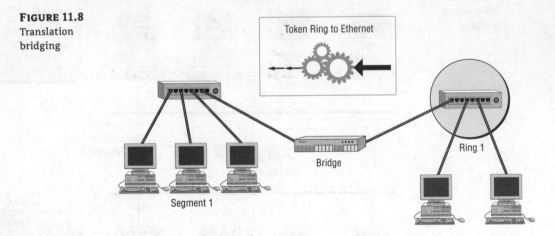

Compared to modern routers, bridges are not complicated devices; they consist simply of network interface cards and the software required to forward packets from one interface to another. As previously mentioned, bridges operate at the Link layer of the OSI reference model, so to understand how bridges work, a brief discussion of link-layer communication is in order.

How are network nodes uniquely identified? In general, OSI Network layer protocols, such as the Internet Protocol (IP), are assumed. When you assign an IP address to a network node, one of the requirements is that the address must be unique on the network. At first, you might think every computer in the world must have a unique IP address in order to communicate, but such is not the case. This is because of the Internet Assigned Numbers Authority's (IANA) specification for the allocation of private address spaces, in RFC 1918. For example, Company XYZ and Company WXY could both use IP network 192.168.0.0/24 to identify network devices on their private networks. However, networks that use a private IP address specified in RFC 1918 cannot communicate over the Internet without network address translation or proxy server software and hardware.

IP as a protocol merely provides for the logical grouping of computers as networks. Because IP addresses are logical representations of groups of computers, how does communication between two endpoints occur? IP as a protocol provides the rules governing addressing and routing. IP requires the services of the Data Link layer of the OSI reference model to communicate.

Every network interface card has a unique 48-bit address, known as its *MAC address*, assigned to the adapter. For two nodes to converse, one computer must first resolve the MAC address of its destination. In IP, this is handled by a protocol known as the *Address Resolution Protocol* (ARP). Once a MAC address is resolved, the frame gets built and is transmitted on the wire as a unicast frame. (Both a source and a destination MAC address exist.) Each network adapter on that segment hears the frame and examines the destination MAC address to determine if the frame is destined for them. If the frame's destination MAC address matches the receiving system's MAC address, the frame gets passed up to the Network layer; otherwise, the frame is simply discarded.

So how does the communication relate to bridging, you may ask? In transparent bridging, the bridge listens to all traffic coming across the lines and analyzes the source MAC addresses to build tables that associate a MAC address with a particular network segment. When a bridge receives a frame destined for a remote segment, it then forwards that frame to the appropriate segment so that the clients can communicate seamlessly.

Bridging is one technique that can solve the shared-bandwidth problem that exists with hubs. Consider the hub example where we cascaded four 24-port hubs. Through the use of bridges, we can physically isolate each segment so that only 24 hosts compete for bandwidth; throughput is therefore increased. Similarly, with the implementation of bridges, you can also increase the number of nodes that can transmit simultaneously from one (in the case of cascading hubs) to four. Another benefit is that collision domains can be extended; that is, the physical distance between two nodes can exceed the physical limits imposed if the two nodes exist on the same segment. Logically, all of these nodes will appear to be on the same network segment.

Bridging does much for meeting the challenges of internetworking, but its implementation is limited. For instance, source-route bridges will accommodate a maximum of seven physical segments. And although you will have made more efficient use of available bandwidth through segmentation, you can still do better with switching technologies.

Switches

A *switch* is the next rung up the evolutionary ladder from bridges. In modern star-topology networking, when you need bridging functionality you often buy a switch. But bridging is not the only benefit of switch implementation. Switches also provide the benefit of micro-LAN segmentation, which means that every node connected to a switched port receives its own dedicated bandwidth. And with switching, you can further segment the network into virtual LANs.

Like bridges, switches also operate at the Link layer of the OSI reference model and, in the case of Layer 3 switches, extend into the Network layer. The same mechanisms are used to build dynamic tables that associate MAC addresses with switched ports. However, whereas bridges implement store-and-forward bridging via software, switches implement either store-and-forward or cut-through switching via hardware, with a marked improvement of speed.

Switches can identify which network devices are connected to it since the Ethernet protocol assigns a unique network address to each network card or workstation port associated with it. As data is transmitted, an Ethernet switch acknowledges and *remembers* the IP address of each packet of data sent by the transmitting workstation port. When a data packet is sent over the network for that workstation port, the switch only sends the packet to that specific workstation port. Switches work very well in large commercial building networks with many computers and network-connected devices because much more aggregate traffic can be moved through a switch.

LAN segmentation is the key benefit of switches, and most organizations have either completely phased out hubs or are in the process of doing so to accommodate the throughput requirements for multimedia applications. Switches are becoming more affordable, ranging in price from $10 to slightly over $20 per port, and are allowing most networks to migrate to completely switched infrastructures.

Workgroup Switches

Workgroup switches are responsible for interconnecting a cluster of workstations to the main network. As shown previously in Figure 11.1, workstations including desktops computers or laptops, printers, and other network devices are connected to workgroup switches using horizontal cabling. These switches aggregate data signals from a cluster of workstations and pass them over the backbone cabling to the servers, located in the main equipment room of a commercial building.

Workgroup switches are typically located in a telecommunications room (TR) located on each floor of a commercial building in the traditional implementation of a hierarchical star network. These switches are installed in a rack above or below the patch panel cross-connection. They are usually plugged into UPS power backups also located in the TR.

When implementing the hierarchical star network using fiber-to-the-telecom enclosure (FTTE), these switches are located in enclosures closer to clusters of workstations. In some cases, you can find an FTTE enclosure housing a switch located in the ceiling above a cluster or workstations.

Horizontal links in a traditional hierarchical star network are typically cabled using Category 5e or 6. The front panel of the workgroup switch is fitted with a number of copper ports. Twelve- and 24-port 10 and 100Base-TX switches are very common. The workgroup switch is typically connected to the servers in the main equipment room using a fiber-optic cable uplink since backbone distances can exceed those that can be supported over UTP cabling. Therefore, a workgroup switch also accommodates a fiber-optic port for the connection to the backbone cabling. Nowadays, two 1000Base-SX fiber-optic uplink ports are common in workgroup switches.

Workgroup switches come in a variety of performance levels. There are four main aspects that need to be considered in specifying workgroup switches:

◆ Number of ports for horizontal connections (ports)

◆ Speed/application of horizontal ports

◆ Number of uplink connections for connection to backbone cabling

◆ Speed/application of uplink ports

In the planning phase of the network design, it's important to determine the number of users on each floor of a commercial building upon implementation of the network, and also the possibility of additions, moves, and changes. Ideally, the network should be designed with some room for growth. Racks in a TR should be ordered, with the ability to include additional switches and patch panels to support network growth. Backbone cabling to the telecommunications should have enough fiber-optic strands to support the addition of switches. Twelve fiber-optic strands of OM3 multimode fiber is common for backbone cabling to each TR.

Blocking vs. Nonblocking

Blocking and nonblocking has to do with the *effective* bandwidth performance of the switch in carrying all of the information to and from the workstation ports. Depending on the number and application speed of the horizontal ports in a workgroup switch and the number and speed of the uplink ports, the switch may be considered *blocking* or *nonblocking*. To illustrate this, let's look at some examples.

A blocking situation is created when the total bandwidth capability of the horizontal ports exceeds the bandwidth of the uplink to the switch. For example, a 24-port 100Base-TX (100Mbps) switch would have the potential to transmit a total of 2,400Mbps (or 2.4Gbps) if all connections were sending and receiving 100Mbps at one time. However, if the uplink to the switch contains only one 1000Base-SX fiber-optic port, the switch only has the ability to send and receive 1,000Mbps (or 1Gbps) from the 24 ports. The maximum workstation throughput in Mbps per port would then be only 42Mbps per port (1,000Mbps uplink capability divided by 24 ports). If each of the 24 ports were trying to transmit 100Mbps, the uplink would be blocking the bandwidth by almost 58 percent.

A nonblocking situation is created when the total bandwidth capability of the horizontal ports is equal to or less than the bandwidth of the uplink to the switch. For example, an 8-port 100Base-TX (100Mbps) switch would have the potential to transmit a total of 800Mbps (or 0.8Gbps) if all connections were sending and receiving 100Mbps at one time. If the uplink to the switch contains one 1000Base-SX fiber-optic port, the switch now has the ability to send and receive a total of 1,000Mbps (or 1Gbps) from the 8 ports. The maximum workstation throughput would then be a full 125Mbps for each of the ports (1,000Mbps uplink capability divided by 8 ports). In reality, the throughput is only 100Mbps since the port is 100Base-TX (100Mbps). In this case, the uplink is not blocking the full potential of its horizontal connections.

For standard desktop computing, IT managers typically opt for high-density, high-port-count switches of the blocking type because they are the most cost effective in terms of dollar per port. The blocking potential is not typically an issue in this application because the computers are rarely sending and receiving the full potential of the application speed they are connected with. However, in situations where a lot of data is being transmitted back and forth over the network, a low-density, low-port-count switch of the nonblocking type is required. This is more common in situations where large CAD files or video streaming and transfer of video files is performed as a normal course of work. Although this increases the cost per port, the bandwidth performance is the driving force.

CABLING @ WORK: FLEXIBILITY IS THE SPICE OF LIFE

We recently came across a situation where an additional wing of a factory was added to an existing factory building. This wing required connection of three very large CNC machines to the network over 1000Base-T NIC cards. Of course, this was not initially planned. Even worse, when the planning for this new wing was done no provision was made to build a dedicated telecommunications room to service this new area. This isn't the first time this type of oversight has been made in our experience!

The building owner had to hire a consultant to provide a solution to their problem. The area where a new TR could be built in the new wing was approximately 250 meters from the main equipment room of the factory and 150 meters from the closest TR in the adjoining building. A new TR could be added in the new wing with a fiber-optic connection backhauled to the main equipment room. This would take approximately two weeks to accomplish and would lead to a mostly inefficient use of the TR and its equipment.

Luckily, this situation provided some flexibility in the way the problem could be solved because of the provisions made in the cabling system in the old building and some devices that are well suited to this type of problem.

Since this was an unplanned situation and the building owner needed to get online ASAP, the consultant recommended that a 150 meter fiber-optic cable be spliced between the closest TR in the adjoining building and a set of wall plates in the new wing. This was made possible because enough spare strands of fiber-optic cable were available in the closest TR. Therefore, the total run of fiber between the new wing and main equipment room was 250 meters. To support a nonblocking situation and avoid a new and costly switch in an enclosure, the building owner decided to use media converters to connect the 1000Base-T NICs to the fiber-optic cable. This is essentially a centralized cabling architecture. Fiber to the CNC machine!

Here is a perfect situation where there was a lot of flexibility in choices because of the planning for future growth in cabling media that was done and the option to use different active equipment.

Core Switches

Core or core-class switches, also known as backbone switches, are OSI Layer 2 and are responsible for interconnecting the various workgroup switches in a commercial building to the servers in the main network. Some core switches also have OSI Layer 3 capability and serve as routers as well. As shown previously in Figure 11.1, workgroup switches are connected to core switches using backbone cabling. These switches aggregate all of the data signals from the workgroup switches and pass them over main cross-connections to the servers, located in the main equipment room of a commercial building. A core switch is simply a switch that is designed to be placed at the *core* of a network setup. It handles a large portion of the network load and will need to be extremely fast and have a large bandwidth capacity to handle the traffic demand placed on it. Core switches are typically located in a main equipment room (ER) located in the data center or main computer room of a commercial building in the traditional implementation of a hierarchical star network. Usually there are at least two core switches to provide for redundancy.

Core switches typically have 1000Base-SX backbone ports since the uplink ports of workgroup switches are also 1000Base-SX. The server ports can be based on either copper or fiber-optic cabling since the distance to the servers is typically less than 100 meters. However, in very large data centers and equipment rooms, the distance between core-class switches and servers may be so large that fiber-optic connections are required. These switches are usually fully integrated with power and fan-cooling capability.

Larger core switches typically come in a set of components. A main switch chassis is installed and fitted with *blades* to be installed into the chassis to provide the switching capability. Although this option is very costly, its advantage is that it allows the end user to add blades as the network traffic grows, through the addition of many other workgroup switches.

Pluggable Transceivers and Form Factors

Pluggable transceivers are available in a variety of sizes and support a variety of cabling media and application speeds. You will hear phrases such as GBIC, mini-GBIC, SFP, XFP and others. These support different speeds and have different form factors.

A Gigabit Ethernet interface converter (GBIC) is a standard used for the type of optical transceivers that can be plugged into switches and media converters. This method of plugging a transceiver into a GBIC module slot of a switch, as opposed to having a fixed interface, is also called *hot-swappable* because the transceiver is loaded into the switch while the switch is powered on. The GBIC standard is defined in the SFF Committee in document number SFF-8053. GBIC transceivers are available in the –SX, –LX, and –ZX wavelengths and media ranges.

GBICs were developed to provide both a small form factor and the flexibility to plug a variety of transceiver types into a common electrical interface for Gigabit Ethernet speeds. When many different optical port types are required, an IT administrator can purchase GBICs as needed. This lowers the cost of the overall switch system and gives the IT administrator far more flexibility, since they have the choice of different transceiver types. However, if the switch is intended to mostly have one port type, then it will probably be cheaper and take up less space per port to use a switch with that port type built into it. The back of the device has an electrical adapter that plugs into a switch. The front of the transceiver has an SC connector interface for the two 850nm optical ports intended for use over multimode fiber. One is used for transmitting and the other for receiving signals. Figure 11.9 shows how a GBIC is plugged into a GBIC module slot.

FIGURE 11.9
Plugging in a GBIC

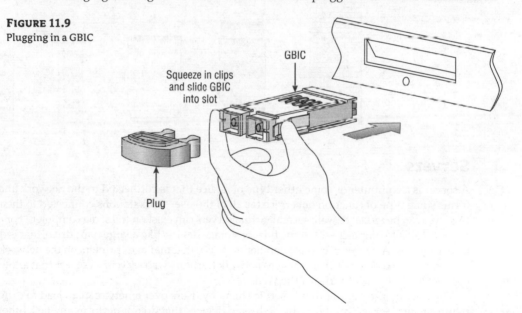

A smaller version of the GBIC is called a mini-GBIC or SFP (small form factor pluggable) module. It serves the same purpose as the GBIC but has a smaller footprint. To support the smaller footprint, SFPs use LC connector interfaces. SFPs are also available for a 1000-Base-T connection with an RJ-45 interface.

The XENPAK transceiver was standardized in 2001 to support 10 Gigabit Ethernet over fiber-optic media and copper media. These come in a variety of optical media types and wavelength ranges and also support copper media using a –CX4 interface. They are much larger in dimension than GBICs and SFPs. After XENPAK was standardized, XPAK and X2 were created to provide a smaller footprint for 10 Gigabit Ethernet. These devices share the same electrical interface as XENPAK but are lighter.

The XFP small form factor pluggable transceiver was initially developed in 2003 to provide a smaller form factor transceiver for 10Gbps applications than XENPAX, XPAK, and X2 transceivers. XFP was developed by the XFP Multi-Source Agreement Group prior to the development of the SFP+. XFPs can support 10 Gigabit Ethernet, SONET/SDH, and Fibre Channel applications. In keeping with their small form factor, similar to SFPs, XFPs use an LC connector interface. XFPs are installed in a similar fashion to GBICs and SFPs, and are available in a wide range of media type and wavelengths.

The SFP+ module was developed later to support 10 Gigabit Ethernet speeds. It is about 30 percent smaller than an XFP, consumes less power, and is less expensive. Figure 11.10 shows a comparison of these various transceiver types and provides a relative size comparison.

FIGURE 11.10
Relative sizes
of pluggable
transceivers

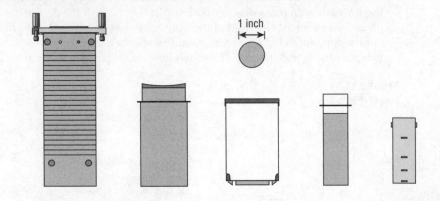

Servers

A server is a computer or some other type of device that is connected to the network and performs some type of function or service for all of the user workstations connected to the network. (A server can be *virtual* as well; several virtual servers can exist on the same computer.) For example, a file server is a computer with some type of storage device, like a large hard drive, that is dedicated to storing files. A print server is a computer or device that manages printers on the network. A network server is a computer that manages network traffic. A web server is a server that allows you to connect to the Internet and surf the Web.

The duty of a server to provide service to many users over a network can lead to different requirements. Servers are typically dedicated devices that do not perform any task other than their primary role of serving a single application, as mentioned earlier. The technical requirements of the server vary depending on the application it is designed for. The most important attributes are fast network connections and high throughput. How many times have you waited to access your file server? This can be a frustrating experience.

Servers need to be on and available to the network community all of the time. As a result they run for long periods without interruption. The reliability and durability of the server is critical to providing this level of service. Servers are typically built using special, very robust parts with low failure rates. For example, special servers typically have faster and higher capacity hard drives, larger fans and cooling systems to remove heat, and most importantly, some type of backup battery power supply to ensure that the server continues working in case of a short-term power failure. Some companies even have dedicated power to their severs. Redundancy is another measure of ensuring 24×7 operations. Installing more than one power supply, hard drive, and cooling system is not uncommon.

In a typical installation of racks of servers, core switches are connected to servers typically with a high speed connection of 1 or 10Gbps. Depending on the distance, the connection is made with either a UTP copper cable or a fiber-optic cable. In many high-tech firms, the cabling system is installed underneath a raised floor. Servers are typically rack-mounted and situated in a main equipment room along with all of the other equipment or in a dedicated server room to ensure a better environment for cooling and security.

Routers

Routers are packet-forwarding devices just like switches and bridges; however, routers allow transmission of data between network segments to the outside world. Unlike switches, which forward packets based on physical node addresses, routers operate at the Network layer of the OSI reference model, forwarding packets based on a network ID. Routers are necessary in the main equipment room to provide access to the WAN (wide area network) through the cabling brought in by a service provider like Verizon, AT&T, or any other carrier operating in your area.

So how do routers work? In the case of the IP protocol, an IP address is 32 bits long. Those 32 bits contain both the network ID and the host ID of a network device. IP distinguishes between network and host bits by using a subnet mask. The *subnet mask* is a set of contiguous bits with values of one from left to right, which IP considers to be the address of a network. Bits used to describe a host are masked out by a value of 0, through a binary calculation process called *ANDing*. Figure 11.11 shows two examples of network IDs calculated from an ANDing process.

We use IP as the basis of our examples because it is the industry standard for enterprise networking; however, TCP/IP is not the only routable protocol suite. Novell's IPX/SPX and Apple Computer's AppleTalk protocols are also routable.

FIGURE 11.11
Calculation of IP network IDs

192.168.145.27 / 24	192.168.136.147 / 29
Address: 11000000 10101000 10010001 00011011	Address: 11000000 10101000 10001000 10010011
Mask: 11111111 11111111 11111111 00000000	Mask: 11111111 11111111 11111111 11111000
Network ID: 11000000 10101000 10010001 00000000	Network ID: 11000000 10101000 10001000 10010000
192.168.145.0	192.168.136.144

We use IP as the basis of our examples because it is the industry standard for enterprise networking; however, TCP/IP is not the only routable protocol suite. Novell's IPX/SPX and Apple Computer's AppleTalk protocols are also routable.

Routers are simply specialized computers concerned with getting packets from point A to point B. When a router receives a packet destined for its network interface, it examines the destination address to determine the best way to get it there. It makes the decision based on information contained within its own routing tables. Routing tables are associations of network IDs and interfaces that know how to get to that network. If a router can resolve a means to get the packet from point A to point B, it forwards it to either the intended recipient or to the next router in the chain. Otherwise, the router informs the sender that it doesn't know how to reach the destination network. Figure 11.12 illustrates communication between two hosts on different networks.

FIGURE 11.12
Host communication between internetworked segments

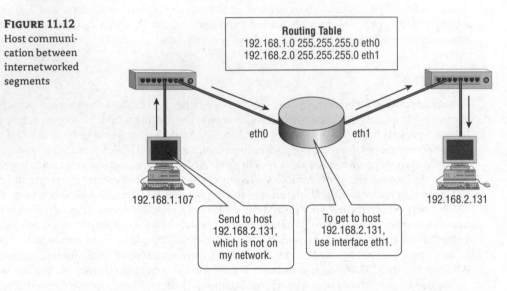

Routing Table
192.168.1.0 255.255.255.0 eth0
192.168.2.0 255.255.255.0 eth1

eth0   eth1

192.168.1.107   192.168.2.131

Send to host
192.168.2.131,
which is not on
my network.

To get to host
192.168.2.131,
use interface eth1.

Routers enabled with the TCP/IP protocol and all networking devices configured to use TCP/IP make some sort of routing decision. All decisions occur within the IP protocol framework. IP has other responsibilities that are beyond the scope of this book, but ultimately IP is responsible for forwarding or delivering packets. Once a destination IP address has been resolved, IP will perform an AND calculation on the IP address and subnet mask, as well as on the destination IP address to the subnet mask. IP then compares the results. If they are the same, then both devices exist on the same network segment, and no routing has to take place. If the results are different, then IP checks the devices routing table for explicit instructions on how to get to the destination network and forwards the frame to that address or sends the packet along to a default gateway (router).

A detailed discussion on the inner workings of routers is well beyond the scope of this book. Internetworking product vendors such as Cisco Systems offer certifications in the configuration and deployment of their products. If you are interested in becoming certified in Cisco products, Sybex also publishes excellent study guides for the CCNA and CCNP certification exams. For a more intimate look at the inner workings of the TCP/IP protocol suite, check out *TCP/IP Protocol Suite* by Behrouz A. Forouzan (McGraw-Hill 2009).

The Bottom Line

Identify the basic active components of a hierarchical star network for commercial buildings and networks. Active network equipment is connected by the structured cabling system to support the topology and applications required for the network. More simply, the active equipment is what sends, aggregates, directs and receives actual data. If you are responsible for specifying and procuring the active equipment for a commercial building network, it's important t understand the differences between equipment and their primary functions.

Master It

1. What is the active component that a patch cord is plugged into when it is attached to a network connected device, such as a desktop computer?

2. What type of switch, and over what type of cabling, are workstation connections funneled into in a hierarchical star network?

3. What type of switch, and over what type of cabling, are workgroup switches connected?

Identify differences between various types of transceiver modules. The emergence of 1 and 10Gbps Ethernet speeds brought a host of different transceiver modules. At times, there appeared to be a competition as to who could develop the smallest and fastest module. These modules involve an alphabet soup of terminology, and it's important to be able to navigate your way through this.

Master It

1. You are asked to specify a pluggable module for 850nm transmission over multimode fiber at 1Gbps with an LC connector interface. What module do you need to order?

2. You are asked to specify a pluggable module for 1310nm transmission over single-mode fiber at 10Gbps from one building to another. The IT manager wants you to order the smallest possible form factor. What module do you need to order?

Determine if your workgroup switching system is blocking or nonblocking. Certain situations may require you to design your workgroup switching system to support the maximum potential bandwidth to and from your end users' workstation ports. This is called a nonblocking configuration.

Master It You have nine users who have 100Base-T NIC cards on their computers. The 12-port workgroup switch they are connected to has a 1000Base-SX uplink connection. What is the maximum throughput per user that this switch is capable of? Is this presently a blocking or nonblocking configuration? How many additional users could be added before this situation changes?

Chapter 12

Wireless Networks

Wireless networks have network signals that are not directly transmitted by any type of fiber or cable. However, wireless switches and hubs are usually connected to the core network with some type of copper or fiber-optic backbone cabling media. Wireless LAN media are becoming extremely popular in modular office spaces for transmitting data over the "horizontal" portion of a traditional hierarchical star network.

You may ask, "Why talk about wireless technologies in a book about cabling?" The answer is that today's networks aren't composed of a single technology or wiring scheme—they are heterogeneous networks. Wireless technologies are just one way of solving a particular networking need in a heterogeneous cabling system. Although cabled networks are generally less expensive, more robust transmission-wise, and faster (especially in the horizontal environment), in certain situations wireless networks can carry data where traditional cabled networks cannot. This is particularly the case in some horizontal and certain backbone or WAN implementations.

Some pretty high bandwidth numbers are detailed in the sections that follow. These are typically for interbuilding or WAN implementations. The average throughput speed of installed wireless LANs in the horizontal work environment is 11Mbps. (Although 54Mbps and higher throughput is available, that is a relatively recent phenomenon.) By comparison, any properly installed Category 5e horizontal network is capable of 1000Mbps and higher. Wired and wireless are two different beasts, but the comparison helps to put the horizontal speed issue into perspective.

In this chapter, you will get a brief introduction to some of the wireless technologies found in both LANs and WANs and how they are used.

This chapter is only meant to introduce you to the different types of wireless networks. For more information, go to your favorite Internet search engine and type in **wireless networking**.

In this chapter, you will learn to:

- ◆ Understand how infrared wireless systems work

- ◆ Know the types of RF wireless networks

- ◆ Understand how microwave communication works and its advantages and disadvantages

Infrared Transmissions

Everyone who has a television with a remote control has performed an *infrared transmission*. Infrared (IR) transmissions are signal transmissions that use infrared radiation as their transmission method. Infrared radiation is part of the electromagnetic spectrum. It has a wavelength longer than visible light (actually, it's longer than the red wavelength in the visible spectrum) with less energy. Infrared is a very popular method of wireless networking. The sections that follow examine some of the details of infrared transmissions.

How Infrared Transmissions Work

Infrared transmissions are very simple. All infrared connections work similarly to LAN transmissions, except that no cable contains the signal. The infrared transmissions travel through the air and consist of infrared radiation that is modulated in order to encode the LAN data.

A *laser diode*, a small electronic device that can produce single wavelengths or frequencies of light or radiation, usually produces the infrared radiation. A laser diode differs from a regular laser in that it is much simpler, smaller, and lower powered; thus, the signals can only travel over shorter distances (usually less than 500′).

Besides needing an infrared transmitter, all devices that communicate via infrared need an infrared receiver. The receiver is often a photodiode, or a device that is sensitive to a particular wavelength of light or radiation and converts the infrared signals back into the digital signals that a computer will understand.

In some cases, the infrared transmitter and receiver are built into a single device known as an *infrared transceiver*, which can both transmit and receive infrared signals. Infrared transceivers are used primarily in short-distance infrared communications. For communications that must travel over longer distances (e.g., infrared WAN communications that must travel over several kilometers), a separate infrared transmitter and receiver are contained in a single housing. The transmitter is usually a higher-powered infrared laser. To function correctly, the lasers in both devices (sender and receiver) must be aligned with the receivers on the opposite device (as shown in Figure 12.1).

Point-to-point and *broadcast* are the two types of infrared transmission. We'll take a brief look at each.

FIGURE 12.1
Alignment of long-distance infrared devices

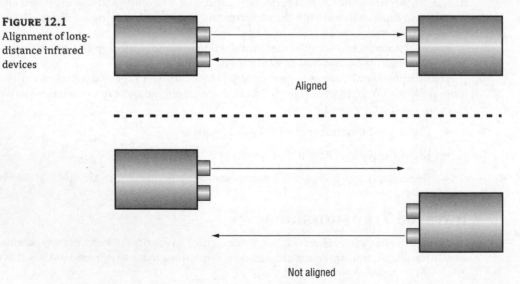

Aligned

Not aligned

POINT-TO-POINT INFRARED TRANSMISSION

The most common type of infrared transmission is the *point-to-point transmission*, also known as *line-of-sight transmission* or *free-space-optics transmission*. Point-to-point infrared transmissions are those infrared transmissions that use tightly focused beams of infrared radiation to send information or control information over a distance (i.e., from one "point" directly to another). The aforementioned infrared remote control for your television is one example of a point-to-point infrared transmission.

LANs and WANs can use point-to-point infrared transmissions to transmit information over short or long distances. Point-to-point infrared transmissions are used in LAN applications for communicating between separate buildings over short distances.

Using point-to-point infrared media reduces attenuation and makes eavesdropping difficult. Typical point-to-point infrared computer equipment is similar to that used for consumer products with remote controls, except they have much higher power. Careful alignment of the transmitter and receiver is required, as mentioned earlier. Figure 12.2 shows how a network might use point-to-point infrared transmission. Note that the two buildings are connected via a direct line-of-sight with infrared transmission and that the buildings are about 1,000′ apart.

FIGURE 12.2
Point-to-point
infrared usage

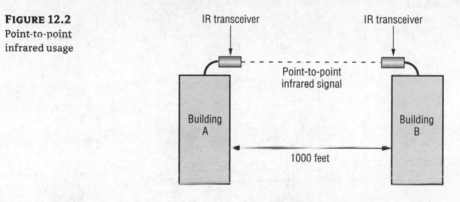

Point-to-point infrared systems have the following characteristics:

Frequency range   Infrared light usually uses the lowest range of light frequencies, between 100GHz and 1,000THz (terahertz).

NOTE   Frequency and wavelength are inversely proportional—that is, the longer (or larger) the wavelength, the slower (or smaller) the frequency.

Cost   The cost depends on the kind of equipment used. Long-distance systems, which typically use high-power lasers, can be very expensive. Equipment that is mass-produced for the consumer market and that can be adapted for network use is generally inexpensive.

Installation   Infrared point-to-point requires precise alignment. Take extra care if high-powered lasers are used, because they can damage or burn eyes.

Capacity   Data rates vary from 100Kbps to Gigabit Ethernet.

Attenuation The amount of attenuation depends on the quality of emitted light and its purity, as well as general atmospheric conditions and signal obstructions. Rain, fog, dust and heat can all affect the ability to transmit a signal.

EMI Infrared transmission can be affected by intense visible light. Tightly focused beams are fairly immune to eavesdropping because tampering usually becomes evident by the disruption in the signal. Furthermore, the area in which the signal may be picked up is very limited.

BROADCAST INFRARED TRANSMISSION

Broadcast infrared systems spread the signal to a wider area and allow reception of the signal by several receivers. One of the major advantages is mobility; the workstations or other devices can be moved more easily than with point-to-point infrared media. Figure 12.3 shows how a broadcast infrared system might be used.

FIGURE 12.3
An implementation of broadcast infrared media

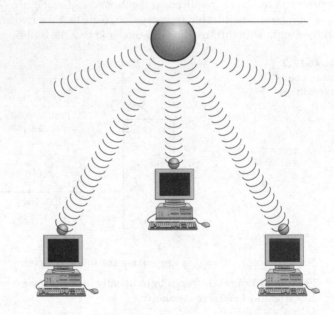

Because broadcast infrared signals (also known as *diffuse infrared*) are not as focused as point-to-point, this type of system cannot offer the high level of throughput that a point-to-point system can. Broadcast infrared is typically limited to less than 1Mbps, making it too slow for most network needs.

Broadcast infrared systems have the following characteristics:

Frequency range Infrared systems usually use the lowest range of light frequencies, from 100GHz to 1,000THz.

Cost The cost of infrared equipment depends on the quality of light required. Typical equipment used for infrared systems is quite inexpensive. High-power laser equipment is much more expensive.

Installation   Installation is fairly simple. When devices have clear paths and strong signals, they can be placed anywhere the signal can reach, making reconfiguration easy. One concern should be the control of strong light sources that might affect infrared transmission.

Capacity   Although data rates are most often less than 1Mbps, it is theoretically possible to reach much higher throughput.

Attenuation   Broadcast infrared, like point-to-point, is affected by the quality of the emitted light and its purity and by atmospheric conditions. Because devices can be moved easily, however, obstructions are generally not of great concern.

EMI   Intense light can dilute infrared transmissions. Because broadcast infrared transmissions cover a wide area, they are more easily intercepted for eavesdropping.

Advantages of Infrared

As a medium for LAN transmissions, infrared has many advantages that make it a logical choice for many LAN/WAN applications. These advantages include the following:

Relatively inexpensive   Infrared equipment (especially the short-distance broadcast equipment) is relatively inexpensive when compared to other wireless methods like microwave or radio frequency (RF). Because of its low cost, many laptop and portable-computing devices contain an infrared transceiver to connect to other devices and transfer files. Additionally, infrared is a cost-effective WAN transmission method because you only pay for the equipment once, and there are no recurring line charges.

High bandwidths   Point-to-point infrared transmissions support fairly high (up to 2.5Gbps) bandwidths. They are often used as WAN links because of their speed and efficiency.

No FCC license required   In the United States, if a wireless transmission is available for the general public to listen to, the Federal Communications Commission (FCC) probably governs it. The FCC licenses certain frequency bands for use for radio and satellite transmission. Because infrared transmissions are short range and their frequencies fall outside the FCC bands, you don't need to apply for an FCC-licensed frequency (a long and costly process) to use them.

NOTE   More information on the FCC can be found at its website: www.fcc.gov.

Ease of installation   Installation of most infrared devices is very simple. Connect the transceiver to the network (or host machine) and point it at the device you want to communicate with. Broadcast infrared devices don't even need to be pointed at their host devices. Long-distance infrared devices may need a bit more alignment, but the idea is the same.

High security on point-to-point connections   Because point-to-point infrared connections are line-of-sight and any attempt to intercept a point-to-point infrared connection will block the signal, point-to-point infrared connections are very secure. The signal can't be intercepted without the knowledge of the sending equipment.

Portability   Short-range infrared transceivers and equipment are usually small and have lower power requirements. Thus, these devices are great choices for portable, flexible networks. Broadcast infrared systems are often set up in offices where the cubicles are rearranged often. This does *not* mean that the computers can be in motion while connected, however. As discussed later in this section, infrared requires a constant line-of-sight. If you should walk behind an object and obstruct the line-of-sight between the two communicating devices, the connection will be interrupted.

Disadvantages of Infrared

Just as with any other network technology, infrared has its disadvantages. Some of these are the following:

Line-of-sight needed for focused transmissions Infrared transmissions require an unobstructed path between sender and receiver. Infrared transmissions are similar to regular light transmissions in that the signals don't "bend" around corners without help, nor can the transmissions go through walls. Some transmissions are able to bounce off surfaces, but each bounce takes away from the total signal strength (usually halving the effective strength for each bounce).

NOTE Some products achieve non-line-of-sight infrared transmissions by bouncing the signal off walls or ceilings. You should know that for every "bounce," the signal can degrade as much as 50 percent. For that reason, we have stated here that focused infrared is primarily a line-of-sight technology.

Weather attenuation Because infrared transmissions travel through the air, any change in the air can cause degradation of the signal over a distance. Humidity, temperature, and ambient light can all negatively affect signal strength in low-power infrared transmissions. In outdoor, higher-power infrared transmissions, fog, rain, and snow can all reduce transmission effectiveness.

Examples of Infrared Transmissions

Infrared transmissions are used for other applications in the PC world besides LAN and WAN communication. The other applications include the following:

◆ IrDA ports

◆ Infrared laser devices

We'll briefly examine these two examples of infrared technology.

IrDA Ports

More than likely, you've seen an IrDA port. IrDA ports are the small, dark windows on the backs of laptops and handheld PCs that allow two devices to communicate via infrared. IrDA is actually an abbreviation for the standards body that came up with the standard method of short-range infrared communications, the Infrared Data Association. Based out of Walnut Creek, California, and founded in 1993, it is a membership organization dedicated to developing standards for wireless, infrared transmission systems between computers. With IrDA ports, a laptop or PDA (personal digital assistant) can exchange data with a desktop computer or use a printer without a cable connection at rates up to 1.5Mbps.

NOTE For more information about the IrDA, its membership, and the IrDA port, see its website at www.irda.org.

Computing products with IrDA ports began to appear in 1995, and the LaserJet 5P was one of the first printers with a built-in IrDA port. You could print to the LaserJet 5P from any laptop or handheld device (as long as you had the correct driver installed) simply by pointing the IrDA port on the laptop or handheld device at the IrDA port on the 5P. This technology became known as *point and print*. Figure 12.4 shows an example of an IrDA port on a handheld PC. Notice how small it is compared to the size of the PC.

FIGURE 12.4
An example of an IrDA port

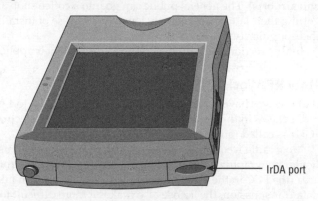

IrDA port

INFRARED-LASER DEVICES

Longer-distance communications via infrared transmissions are possible, but they require the use of a special class of devices, known as *infrared-laser devices*. These devices have a transmitting laser, which operates in the infrared range (a wavelength from 750nm to 2500nm and a frequency of around 1THz) and an infrared receiver to receive the signal. Infrared-laser devices usually connect multiple buildings within a campus or multiple sites within a city. One such example of this category of infrared devices is the TereScope free-space-optics system (as shown in Figure 12.5) from MRV. This system provides data rates from 1.5Mbps to 2.5Gbps.

NOTE You can find out more information about the TereScope system on MRV's website at www.mrv.com.

FIGURE 12.5
The TereScope infrared laser device

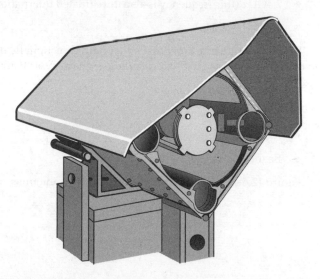

Radio Frequency (RF) Systems

Radio frequency (RF) transmission systems are those network transmission systems that use radio waves to transmit data. In late 1999, RF transmission systems saw a sharp increase in use. Many companies are installing RF access points in their networks to solve certain mobility issues. The year 2003 saw the explosion of *wireless hot spots* (especially in coffee shops, hotels, and airports). The general public can go into a coffee shop and check their e-mail while they're getting their latte. The relatively low cost and ease of installation of RF systems play a part in their popularity.

In this section, we will cover RF systems and their application to LAN and WAN uses.

How RF Works

Radio waves have frequencies from 10 kilohertz (kHz) to 1 gigahertz (GHz), and RF systems use radio waves in this frequency band. The range of the electromagnetic spectrum from 10kHz to 1GHz is called radio frequency (RF).

Most radio frequencies are regulated; some are not. To use a regulated frequency in a particular geographical location, you must receive a license from the regulatory body covering that area (the FCC in the United States). Getting a license can take a long time and can be costly; for data transmission, the license also makes it more difficult to move equipment. However, licensing guarantees that, within a set area, you will have clear radio transmission.

The advantage of unregulated frequencies is that few restrictions are placed on them. One regulation, however, does limit the usefulness of unregulated frequencies: unregulated-frequency equipment must operate at less than one watt. The point of this regulation is to limit the range of influence a device can have, thereby limiting interference with other signals. In terms of networks, this makes unregulated radio communication bandwidths of limited use.

WARNING Because unregulated frequencies are available for use by others in your area, you cannot be guaranteed a clear communications channel.

In the United States, the following frequencies are available for unregulated use:

◆ 902 to 928MHz

◆ 2.4GHz (this frequency is also unregulated internationally)

◆ 5.72 to 5.85GHz

Radio waves can be broadcast either omnidirectionally or directionally. Various kinds of antennas can be used to broadcast radio signals. Typical antennas include the following:

◆ Omnidirectional towers

◆ Half-wave dipole

◆ Random-length wire

◆ Beam (such as the Yagi)

Figure 12.6 shows these common types of RF antennas.

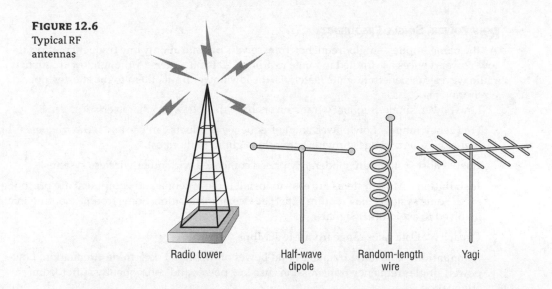

FIGURE 12.6
Typical RF antennas

Radio tower   Half-wave dipole   Random-length wire   Yagi

The antenna and transceiver determine the power of the RF signal. Each range has character-istics that affect its use in computer networks. For computer network applications, radio waves are classified in three categories:

♦ Low power, single frequency

♦ High power, single frequency

♦ Spread spectrum

Table 12.1 summarizes the characteristics of the three types of radio wave media that are described in the following sections.

TABLE 12.1   Radio Wave Media

FACTOR	LOW POWER	HIGH POWER	SPREAD SPECTRUM
Frequency range	All radio frequencies (typically low GHz range)	All radio frequencies (typically low GHz range)	All radio frequencies (typically 902 to 928MHz, 2.4 to 2.4835GHz, and 5.725 to 5.85GHz in the U.S., where 2.4 and 5.8GHz are the most popular today)
Cost	Moderate for wireless	Higher than low-power, single-frequency	Moderate
Installation	Simple	Difficult	Moderate
Capacity	From below 1 to 10Mbps	From below 1 to 10Mbps	3 to 11Mbps
Attenuation	High (25 meters)	Low	High
EMI	Poor	Poor	Fair

Low Power, Single Frequency

As the name implies, single-frequency transceivers operate at only one frequency. Typical low-power devices are limited in range to around 20 to 30 meters. Although low-frequency radio waves can penetrate some materials, the low power limits them to the shorter, open environments.

Low-power, single-frequency transceivers have the following characteristics:

Frequency range Low-power, single-frequency products can use any radio frequency, but higher gigahertz ranges provide better throughput (data rates).

Cost Most systems are moderately priced compared with other wireless systems.

Installation Most systems are easy to install if the antenna and equipment are preconfigured. Some systems may require expert advice or installation. Some troubleshooting may be involved to avoid other signals.

Capacity Data rates range from 1 to 10Mbps.

Attenuation The radio frequency and power of the signal determine attenuation. Low-power, single-frequency transmissions use low power and consequently suffer from attenuation.

EMI Resistance to EMI is low, especially in the lower bandwidths where electric motors and numerous devices produce noise. Susceptibility to eavesdropping is high, but with the limited transmission range, eavesdropping is generally limited to within the building where the LAN is located.

High Power, Single Frequency

High-power, single-frequency transmissions are similar to low-power, single-frequency transmissions but can cover larger distances. They can be used in long-distance outdoor environments. Transmissions can be line-of-sight or can extend beyond the horizon as a result of being bounced off the earth's atmosphere. High-power, single-frequency can be ideal for mobile networking, providing transmission for land-based or marine-based vehicles as well as aircraft. Transmission rates are similar to low-power rates but at much longer distances.

High-power, single-frequency transceivers have the following characteristics:

Frequency range As with low-power transmissions, high-power transmissions can use any radio frequency, but networks favor higher gigahertz ranges for better throughput (data rates).

Cost Radio transceivers are relatively inexpensive, but other equipment (antennas, repeaters, and so on) can make high-power, single-frequency radio moderately to very expensive.

Installation Installations are complex. Skilled technicians must be used to install and maintain high-power equipment. The radio operators must be licensed by the FCC, and their equipment must be maintained in accordance with FCC regulations. Equipment that is improperly installed or tuned can cause low data-transmission rates, signal loss, and even interference with local radio.

Capacity Bandwidth is typically from 1 to 10Mbps.

Attenuation   High-power rates improve the signal's resistance to attenuation, and repeaters can be used to extend signal range. Attenuation rates are fairly low.

EMI   Much like low-power, single-frequency transmission, vulnerability to EMI is high. Vulnerability to eavesdropping is also high. Because the signal is broadcast over a large area, it is more likely that signals can be intercepted.

SPREAD SPECTRUM

Spread-spectrum transmissions use the same frequencies as other radio-frequency transmissions, but they use several frequencies simultaneously rather than just one. Two modulation schemes can be used to accomplish this: *direct frequency modulation* and *frequency hopping*.

Direct frequency modulation is the most common modulation scheme. It works by breaking the original data into parts (called *chips*), which are then transmitted on separate frequencies. To confuse eavesdroppers, spurious signals can also be transmitted. The transmission is coordinated with the intended receiver, which is aware of which frequencies are valid. The receiver can then isolate the chips and reassemble the data while ignoring the decoy information. Figure 12.7 illustrates how direct frequency modulation works.

FIGURE 12.7
Direct frequency
modulation

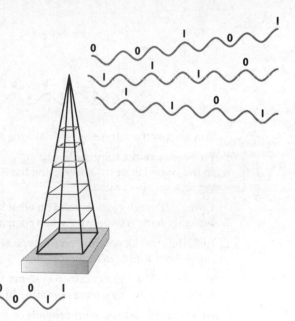

The signal can be intercepted, but it is difficult to watch the right frequencies, gather the chips, know which chips are valid data, and find the right message. This makes eavesdropping difficult.

Current 900MHz direct-sequence systems support data rates of 2 to 6Mbps. Higher frequencies offer the possibility of higher data rates.

Frequency hopping rapidly switches among several predetermined frequencies. In order for this system to work, the transmitter and receiver must be in nearly perfect synchronization. Bandwidth can be increased by simultaneously transmitting on several frequencies. Figure 12.8 shows how frequency hopping works.

FIGURE 12.8
Frequency hopping

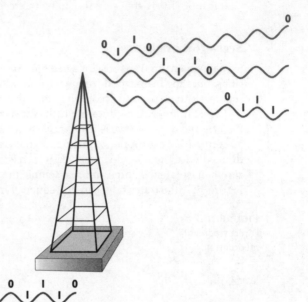

Spread-spectrum transceivers have the following characteristics:

Frequency range Spread spectrum generally operates in the unlicensed-frequency ranges. In the United States, devices using the 902 to 928MHz range are most common, but 2.4GHz devices are also available.

Cost Although costs depend on what kind of equipment you choose, spread spectrum is typically fairly inexpensive when compared with other wireless media.

Installation Depending on the type of equipment you have in your system, installation can range from simple to fairly complex.

Capacity The most common systems, the 900MHz systems, support data rates of 2 to 6Mbps, but newer systems operating in gigahertz produce higher data rates.

Attenuation Attenuation depends on the frequency and power of the signal. Because spread-spectrum transmission systems operate at low power, which produces a weaker signal, they usually have high attenuation.

EMI Immunity to EMI is low, but because spread spectrum uses different frequencies, interference would need to be across multiple frequencies to destroy the signal. Vulnerability to eavesdropping is low.

Advantages of RF

As mentioned earlier, RF systems are widely used in LANs today because of many factors:

No line-of-sight needed   Radio waves can penetrate walls and other solid obstacles, so a direct line-of-sight is not required between sender and receiver.

Low cost   Radio transmitters have been around since the early twentieth century. After 100 years, high-quality radio transmitters have become extremely cheap to manufacture.

Flexible   Some RF LAN systems allow laptop computers with wireless PC NICs to roam around the room while remaining connected to the host LAN.

Disadvantages of RF

As with the other types of wireless networks, RF networks have their disadvantages:

Susceptible to jamming and eavesdropping   Because RF signals are broadcast in all directions, it is very easy for someone to intercept and interpret a LAN transmission without the permission of the sender or receiver. Those RF systems that use spread-spectrum encoding are less susceptible to this problem.

Susceptible to RF interference   All mechanical devices with electric motors produce stray RF signals, known as *RF noise*. The larger the motor, the stronger the RF noise. These stray RF signals can interfere with the proper operation of an RF-transmission LAN.

Limited range   RF systems don't have the range of satellite networks (although they can travel longer distances than infrared networks). Because of their limited range, RF systems are normally used for short-range network applications (e.g., from a PC to a bridge, or short-distance building-to-building applications).

Examples of RF

RF systems are being used all over corporate America. The RF networking hardware available today makes it easy for people to connect wirelessly to their corporate network as well as to the Internet.

One popular type of RF network today is known as an *ad hoc RF network*, which is created when two or more entities with RF transceivers that support ad hoc networking are brought within range of each other. The entities send out radio waves to each other and recognize that they can communicate with another RF device close by. Ad hoc networks allow people with laptops or handheld devices to create their own networks on-the-fly and transfer data. Figure 12.9 shows an example of an ad hoc network between three notebooks. These three notebooks all have the same RF devices that support ad hoc and have been configured to talk to each other.

Another style of RF network is a *multipoint RF network*, which has many stations. Each station has an RF transmitter and receiver that communicate with a central device known as a *wireless bridge*. A wireless bridge (also known as an *RF access point* in RF systems) is a device that provides a transparent connection to the host LAN via an Ethernet or Token Ring connection wired with copper or fiber-optic cable and uses some wireless method (e.g., infrared, RF, or microwave) to connect to the individual nodes. This type of network is mainly used for two applications: office "cubicle farms" and metropolitan-area wireless Internet access. Both applications require that the wireless bridge be installed at some central point and that the stations accessing the network be within the operating range of the bridge device. Figure 12.10 shows an example of this type of network. Note that the workstations at the top of the figure can communicate wirelessly to the server and printer connected to the same network as the bridge device.

FIGURE 12.9
An example of an
ad hoc RF network

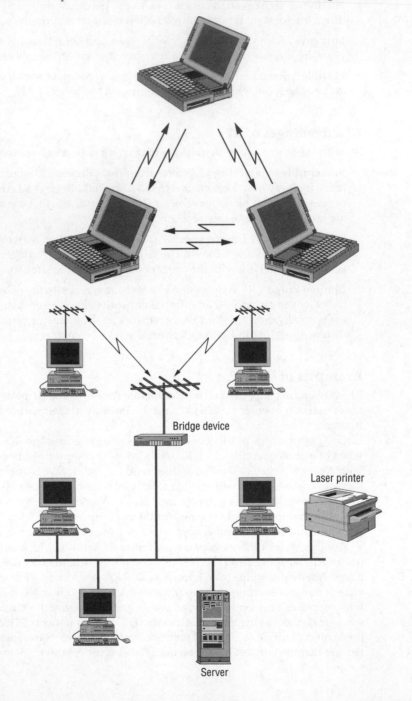

FIGURE 12.10
An example of an
RF multipoint
network

CABLING @ WORK: REDUNDANT NETWORKS WILL MINIMIZE DOWNTIME

Uptime of enterprise networks is critical to support business functions. Having the flexibility to use both wireless and cabled networks will ensure that downtime is minimized.

We recently worked with a customer who had a unique building structure and floor plan who demanded that all workstations be supported solely by a wireless network. The owners felt they would save considerable expenses by avoiding the installation of horizontal cable to their workstation outlets. This proved to be a bad idea.

Although things seemed to work well on day one, as time went on and more users were added, the capacity of the existing wireless bridges proved insufficient and the network access of individual users became intermittent. Adding more bridges was one solution, but as the office environment began to change and as more bandwidth-critical applications like videoconferencing began to be added, it was soon realized that a higher-bandwidth cabled network was necessary and should have been installed from the beginning.

The lesson learned by this user, and many others, is that you should consider a wireless network in an office environment as an overlay to a directly wired network to provide mobile flexibility. Fixed, cabled lines should always work and provide the bandwidth you need in case the wireless network proves to be insufficient. Therefore, always install a cabled network and consider a wireless network as an additional feature.

Many different brands, makes, and models of RF LAN equipment are available. The variety of equipment used to be a source of difficulty with LAN installers. In the infancy of wireless networking, every company used different frequencies, different encoding schemes, different antennas, and different wireless protocols. The marketplace was screaming for a standard. So the IEEE 802.11 standard was developed. Standard 802.11 specifies various protocols for wireless networking. It does, in fact, specify that either infrared or RF can be used for the wireless network, but for the most part, RF systems are the only ones advertising IEEE 802.11 compliance.

SO WHAT IS WI-FI?

Wireless fidelity (Wi-Fi) is a trade name given by the Wi-Fi Alliance to those devices that pass certification tests for strictest compliance to the IEEE 802.11 standards and for interoperability. Any equipment labeled with the Wi-Fi logo will work with any other Wi-Fi equipment, regardless of manufacturer. For more information, see www.wi-fi.org.

Table 12.2 shows some examples of RF wireless networking products available at the time of the writing of this book. This table shows which RF technology each product uses as well as its primary application.

TABLE 12.2   Examples of RF Wireless Networking Products

PRODUCT	RF TECHNOLOGY	APPLICATION	SPEED
Breezecom BreezeNET	Spread spectrum	Multipoint and ad hoc	Up to 11Mbps
ORiNOCO	Spread spectrum	Multipoint	1 to 11Mbps
Cisco Aironet	Spread spectrum	Multipoint	Up to 11Mbps
Apple AirPort	Spread spectrum	Multipoint	11Mbps

The 802.11 standards contain many subsets that define different wireless RF technologies that are used for different purposes. Table 12.3 details these subsets. IEEE is currently working on 802.11n. Likely to be released in 2009, 802.11n has the potential to deliver up to twice the range and five times the throughput of 802.11a, 802.11b, and 802.11g.

TABLE 12.3   802.11 Subsets

SUBSET	MAX RANGE (Indoors at Maximum Speed)	Frequency	Speeds	Compatible with Other 802.11 Subsets?
802.11a	18 meters	5GHz	6, 12, 24, 54Mbps	No
802.11b	50 meters	2.4GHz	1, 11Mbps	Yes (g)
802.11g	100 meters	2.4GHz	1, 11, 54Mbps	Yes (b)
802.11n	300 meters	2.4 and/or 5GHz	600Mbps	Yes: a, b, g

Microwave Communications

You've seen them: the satellite dishes on the tops of buildings in larger cities. These dishes are most often used for microwave communications. Microwave communications use powerful, focused beams of energy to send communications over long distances.

In this section, we will cover the details of microwave communications as they apply to LAN and WAN communications.

How Microwave Communication Works

Microwave communication makes use of the lower gigahertz frequencies of the electromagnetic spectrum. These frequencies, which are higher than radio frequencies, produce better throughput and performance than other types of wireless communications. Table 12.4 shows a brief comparison of the two types of microwave data-communications systems: terrestrial and satellite. A discussion of both follows.

TABLE 12.4   Terrestrial Microwave and Satellite Microwave

FACTOR	TERRESTRIAL MICROWAVE	SATELLITE MICROWAVE
Frequency range	Low gigahertz (typically from 4 to 6GHz or 21 to 23GHz)	Low gigahertz (typically 11 to 14GHz)
Cost	Moderate to high	High
Installation	Moderately difficult	Difficult
Capacity	1 to 100Mbps	1 to 100Mbps
Attenuation	Depends on conditions (affected by atmospheric conditions)	Depends on conditions (affected by atmospheric conditions)
EMI resistance	Poor	Poor

TERRESTRIAL MICROWAVE SYSTEMS

Terrestrial microwave systems typically use directional parabolic antennas to send and receive signals in the lower gigahertz frequency range. The signals are highly focused and must travel along a line-of-sight path. Relay towers extend signals. Terrestrial microwave systems are typically used in situations where cabling is cost-prohibitive.

TIP   Because they do not use cable, microwave links often connect separate buildings where cabling would be too expensive, difficult to install, or prohibited. For example, if a public road separates two buildings, you may not be able to get permission to install cable over or under the road. Microwave links would then be a good choice.

Because terrestrial microwave equipment often uses licensed frequencies, licensing commissions or government agencies such as the FCC may impose additional costs and time constraints.

Figure 12.11 shows a microwave system connecting separate buildings. Smaller terrestrial microwave systems can be used within a building as well. Microwave LANs operate at low power, using small transmitters that communicate with omnidirectional hubs. Hubs can then be connected to form an entire network.

FIGURE 12.11
A terrestrial microwave system connecting two buildings

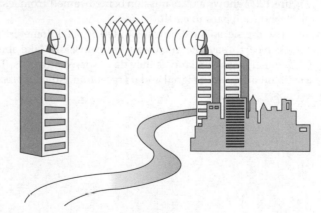

Terrestrial microwave systems have the following characteristics:

Frequency range   Most terrestrial microwave systems produce signals in the low gigahertz range, usually at 4 to 6GHz and 21 to 23GHz.

Cost   Short-distance systems can be relatively inexpensive, and they are effective in the range of hundreds of meters. Long-distance systems can be very expensive. Terrestrial systems may be leased from providers to reduce startup costs, although the cost of the lease over a long term may prove more expensive than purchasing a system.

Installation   Line-of-sight requirements for microwave systems can make installation difficult. Antennas must be carefully aligned. A licensed technician may be required. Suitable transceiver sites can be a problem. If your organization does not have a clear line-of-sight between two antennas, you must either purchase or lease a site.

Capacity   Capacity varies depending on the frequency used, but typical data rates are from 1 to 100Mbps.

Attenuation   Frequency, signal strength, antenna size, and atmospheric conditions affect attenuation. Normally, over short distances, attenuation is not significant, but rain and fog can negatively affect higher-frequency microwaves.

EMI   Microwave signals are vulnerable to EMI, jamming, and eavesdropping (although microwave transmissions are often encrypted to reduce eavesdropping). Microwave systems are also affected by atmospheric conditions.

SATELLITE MICROWAVE SYSTEMS

Satellite microwave systems transmit signals between directional parabolic antennas. Like terrestrial microwave systems, they use low gigahertz frequencies and must be in line-of-sight. The main difference with satellite systems is that one antenna is on a satellite in geosynchronous orbit about 50,000 kilometers (22,300 miles) above the earth. Therefore, satellite microwave systems can reach the most remote places on earth and communicate with mobile devices.

Here's how it usually works. A LAN sends a signal through cable media to an antenna (commonly known as a *satellite dish*), which beams the signal to the satellite in orbit above the earth. The orbiting antenna then transmits the signal to another location on the earth or, if the destination is on the opposite side of the earth, to another satellite, which then transmits to a location on earth.

Figure 12.12 shows a transmission being beamed from a satellite dish on earth to an orbiting satellite and then back to earth.

Because the signal must be transmitted 50,000 kilometers to the satellite and 50,000 kilometers back to earth, satellite microwave transmissions take about as long to reach a destination a few kilometers away on land as they do to span continents. The delay between the transmission of a satellite microwave signal and its reception, called a *propagation delay*, ranges from 0.5 to 5 seconds.

FIGURE 12.12
Satellite micro-
wave transmission

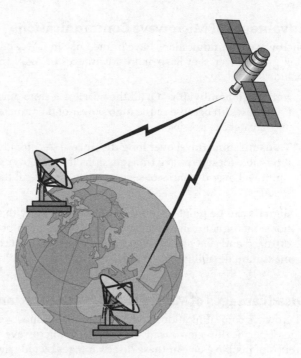

Satellite microwave systems have the following characteristics:

Frequency range Satellite links operate in the low gigahertz range, typically from 11 to 14GHz.

Cost The cost of building and launching a satellite is extremely expensive—as high as several hundred million dollars or more. Companies such as AT&T, Hughes Network Systems, and Scientific-Atlanta (now part of Cisco) lease services, making them affordable for a number of organizations. Although satellite communications are expensive, the cost of cable to cover the same distance may be even more expensive.

Installation Satellite microwave installation for orbiting satellites is extremely technical and difficult and certainly should be left to professionals in that field. The earth-based systems may require difficult, exact adjustments. Commercial providers can help with installation.

Capacity Capacity depends on the frequency used. Typical data rates are 1 to 10Mbps.

Attenuation Attenuation depends on frequency, power, antenna size, and atmospheric conditions. Higher-frequency microwaves are more affected by rain and fog.

EMI Microwave systems are vulnerable to EMI, jamming, and eavesdropping (although the transmissions are often encrypted to reduce eavesdropping). Microwave systems are also affected by atmospheric conditions.

Advantages of Microwave Communications

Microwave communications have limited use in LAN communications. However, because of their great power, they have many advantages in WAN applications. Some of these advantages include:

Very high bandwidth   Of all the wireless technologies, microwave systems have the highest bandwidth because of the high power of the transmission systems. Speeds of 100Mbps and greater are possible.

Transmissions travel over long distances   As already mentioned, their higher power makes it possible for microwave transmissions to travel over very long distances. Transmissions can travel over distances of several miles (or several thousand miles, in the case of satellite systems).

Signals can be point-to-point or broadcast   As with other types of wireless communications, the signals can be focused tightly for point-to-point communications, or they can be diffused and sent to multiple locations via broadcast communications. This allows for the maximum flexibility for the most applications.

Disadvantages of Microwave Communications

Microwave communications are not an option for most users because of their many disadvantages. Specifically, a few disadvantages make microwave communications viable for only a few groups of people. Some of these disadvantages include the following:

Equipment is expensive   Microwave transmission and reception equipment is the most expensive of all the types of wireless transmission equipment discussed in this chapter. A microwave transmitter/receiver combo can cost upward of $5,000—and two transmitters are required for communications to take place. Cheaper microwave systems are available, but their distance and features are more limited.

Line-of-sight required   Microwave communications require a line-of-sight between sender and receiver. Generally speaking, the signal can't be bounced off any objects.

Atmospheric attenuation   As with other wireless technologies (such as infrared laser), atmospheric conditions (e.g., fog, rain, snow) can negatively affect microwave transmissions. For example, a thunderstorm between sender and receiver can prevent reliable communication between the two. Additionally, the higher the microwave frequency, the more susceptible to attenuation the communication will be.

Propagation delay   This is primarily a disadvantage of satellite microwave. When sending between two terrestrial stations using a satellite as a relay station, it can take anywhere from 0.5 to 5 seconds to send from the first terrestrial station through the satellite to the second station.

Safety   Because the microwave beam is very high powered, it can pose a danger to any human or animal that comes between transmitter and receiver. Imagine putting your hand in a microwave on low power. It may not kill you, but it will certainly not be good for you.

Examples of Microwave Communications

Microwave equipment differs from infrared and RF equipment because it is more specialized and is usually only used for WAN connections. The high power and specialization makes it a poor choice for a LAN media (you wouldn't want to put a microwave dish on top of every PC in an office!). Because microwave systems are very specialized, instead of listing a few of the common microwave products, Table 12.5 lists a few microwave-product companies and their website addresses so you can examine their product offerings for yourself.

TABLE 12.5   Microwave-Product Companies and Websites

COMPANY	WEBSITE
Axxcelera Broadband Wireless	www.axxcelera.com
M/A-COM	www.macom.com
Southwest Microwave	www.southwestmicrowave.com

The Bottom Line

Understand how infrared wireless systems works.   The use of "free-space-optic" infrared wireless systems to communicate between buildings in large campuses is getting more attention. It's important to understand how these work and their key advantages and disadvantages.

Master It   Point-to-point systems use tightly focused beams of infrared radiation to send information between transmitters and receivers located some distance apart. What is the key parameter in how these individual devices are installed? What are some of the advantages and disadvantages of these systems?

Know the types of RF wireless networks.   The use of wireless networks in LAN applications has grown substantially since 1999. The IEEE 802.11 standard was created to support standardization and interoperability of wireless equipment used for this purpose. It's important to understand this standard.

Master It   What are the three existing 802.11 standard subsets and which one would provide the highest range and speed?

Understand how microwave communication works and its advantages and disadvantages. Microwave networks are not typically used for LAN systems; however, you might come across them for satellite TV or broadband systems.

Master It   What frequency range is used for microwave systems? What are the two main types of microwave systems? What are some of the general advantages and disadvantages of microwave systems?

Chapter 13

Cabling System Design and Installation

The previous chapters in this book were designed to teach you the basics of telecommunications cabling procedures. You learned about the various components of a typical telecommunications installation and their functions.

They're good to know, but it is more important to understand how to put the components together into a cohesive cabling system design. That is, after all, why you bought this book. Each of the components of a cabling system can fit together in many different ways. Additionally, you must design the cabling system so that each component of that system meets or exceeds the goals of the cabling project.

In this chapter, you will learn to apply the knowledge you learned in the previous chapters to designing and installing a structured cabling system.

In this chapter, you will learn to:

◆ Identify and understand the elements of a successful cabling installation

◆ Identify the pros and cons of network topologies

◆ Understand cable installation procedures

Elements of a Successful Cabling Installation

Before designing your system, you should understand how the following elements contribute to a successful installation:

◆ Using proper design

◆ Using quality materials

◆ Practicing good workmanship

Each of these aspects can drastically affect network performance.

Proper Design

A proper cabling system design is paramount to a well-functioning cabling infrastructure. As with any other major project, the key to a successful cabling installation is that four-letter word: P-L-A-N. A proper cabling system design is simply a plan for installing the cable runs and their associated devices.

So what is a *proper* design? A proper cabling system design takes into account five primary criteria:

- Desired standards and performance characteristics
- Flexibility
- Longevity
- Ease of administration
- Economy

Failure to take these criteria into account can cause usability problems and poor network performance. We'll take a brief look at each of these factors.

DESIRED STANDARDS AND PERFORMANCE CHARACTERISTICS

Of the proper cabling design criteria listed, standards and performance characteristics are the most critical. As discussed in Chapter 1, "Introduction to Data Cabling," standards ensure that products from many different vendors can communicate. When you design your cabling layout, you should decide on standards for all aspects of your cabling installation so that the various products used will interconnect. Additionally, you should choose products for your design that will meet desired performance characteristics. For example, if you will be deploying a broadcast video system over your LAN in addition to the everyday file and print traffic, it is important that the cabling system be designed with a higher-capacity network in mind (e.g., Fast Ethernet or fiber optic).

FLEXIBILITY

No network is a stagnant entity. As new technologies are introduced, companies will adopt them at different rates. When designing a cabling system, you should plan for MACs (moves, additions, and changes) so that if your network changes your cabling design will accommodate those changes. In a properly designed cabling system, a new device or technology will be able to connect to any point within the cabling system.

One aspect of flexibility that many people overlook is the number of cabling outlets or *drops* in a particular room. Many companies take a minimalist approach; that is, they put only the number of drops in each room that is currently necessary. That design is fine for the time being, but what happens when an additional device or devices are needed? It is usually easier to have an extra drop or two (or five) installed while all of the others are being installed than it is to return later to install a single drop.

LONGEVITY

Let's face it: cabling is hard work. You must climb above ceilings and, on occasion, snake through crawlspaces to properly run the cables. Therefore, when designing a cabling system, you want to make sure that the design will stand the test of time and last for a number of years without having to be replaced. A great case in point: many companies removed their coaxial cable–based networks in favor of the newer, cheaper, more reliable UTP cabling. Others are removing their UTP cabling in favor of fiber-optic cable's higher bandwidth. Now, wouldn't it make more sense for those companies that already had coaxial cable to directly upgrade to fiber-optic cable (or at least to a newly released, high-end, high-quality copper UTP cabling system) rather than having to "rip and replace" again in a few years? Definitely. If you have to upgrade your cabling system or are currently designing your system, it is usually best to upgrade to the most current technology you can afford. But you should also keep in mind that budget is almost always the limiting factor.

EASE OF ADMINISTRATION

Another element of a proper cabling design is ease of administration. This means that a network administrator (or subcontractor) should be able to access the cabling system and make additions and changes, if necessary. Some of these changes might include the following:

- Removing a station from the network
- Replacing hubs, routers, and other telecommunications equipment
- Installing new cables
- Repairing existing wires

Many elements make cabling system administration easier, the most important of which is documentation (discussed later in this chapter). Another element is neatness. A rat's nest of cables is difficult to manage because it is difficult to tell which cable goes where.

ECONOMY

Finally, how much money you have to spend will play a part in your cabling system design. If you had an unlimited budget, you'd go fiber-to-the-desktop without question. All your future-proofing worries would be over (at least until the next fiber-optic innovation).

The reality is you probably don't have an unlimited budget, so the best cabling system for you involves compromise—taking into account the four elements listed previously and deciding how to get the most for your investment. You have to do some very basic value-proposition work, factoring in how long you expect to be tied to your new cabling system, what your bandwidth needs are now, and what your bandwidth needs might be in the future.

Quality Materials

Another element of a successful cabling installation is the use of quality materials. The quality of the materials used in a cabling installation will directly affect the transmission efficiency of a network. Many times, a vendor will sell many different cabling product lines, each with a different price point. The old adage that you get what you pay for really does apply to cabling supplies.

Other factors to consider:

- Does your cable or cabling system come with an extended warranty? A good warranty often means that the manufacturer is confident that when their product is properly installed and tested, it will provide the performance that you purchased for years to come.

- In the case of Category 5e, 6, and 6A cables, is it verified by an independent third-party testing laboratory such as UL or ETL?

◆ Consider fire safety; the materials used in both copper and fiber cables to pass the required fiber safety tests, such as UL 1666 (riser) or NFPA 262 (plenum), can be expensive. If the price of a cable is unusually low, it may not be just the electrical or optical performance that is being compromised. While cables produced in the United States are closely monitored by the listing agencies, like UL and ETL, for compliance to the safety standards, cables produced elsewhere are not necessarily certified.

All the components that make up a cabling plant can be purchased in both high- and low-quality product lines. For example, you can buy RJ-45 connectors from one vendor that are 3 cents apiece but rated at only Category 3 (i.e., they won't work for 100Mbps networks). Another vendor's RJ-45 connectors may cost twice as much but be rated for Category 6 (155Mbps and above, over copper).

That doesn't always mean that low price means low quality. Some vendors make low-price, high-quality cabling supplies. It doesn't hurt to shop around when buying your cabling supplies. Check the Internet sites of many different cabling vendors to compare prices.

In addition to price, you should check how the product is assembled. Quality materials are sturdy and well constructed. Low-quality materials will not be durable and may actually break while you are handling them.

Good Workmanship

There is a saying that any job worth doing is worth doing correctly. When installing cabling, this saying is especially true because shoddy workmanship can cause data-transmission problems and thus lower the network's effective throughput. If you try to rush a cabling job to meet a deadline, you will usually end up doing some or the entire job over again. For example, when punching down the individual wires in a UTP installation, excessive untwisting of the individual wires can cause excessive near-end crosstalk (NEXT), thus lowering the effective data-carrying capacity of that connection. The connection must be removed and re-terminated to correct the problem.

The same holds true for fiber-optic cable connections. If you rush any part of the connector installation, the effective optical transmission capacity of that connection will probably be reduced. A reduced capacity means that you may not be able to use that connection at all. Eighty-five percent of all optical fiber cabling system failures are associated with dirty or damaged optical fiber connector end faces.

Cabling Topologies

As discussed in Chapter 3, "Choosing the Correct Cabling," a topology is basically a map of a network. The physical topology of a network describes the layout of the cables and workstations and the location of all network components. Choosing the layout of how computers will be connected in a company's network is critical. It is one of the first choices you will make during the design of the cabling system, and it is an important one because it tells you how the cables are to be run during the installation. Making a wrong decision regarding physical topology and media is costly and disruptive because it means changing an entire installation once it is in place. The typical organization changes the physical layout and physical media of a network only once every three to seven years, so it is important to choose a configuration that you can live with and that allows for growth.

Chapter 3 described the basics of the hierarchical star, bus, and ring topologies. Here, we'll look at some of their advantages and disadvantages and introduce a fourth, seldom-used topology: the mesh topology.

Bus Topology

A bus topology has the following advantages:

♦ It is simple to install.

♦ It is relatively inexpensive.

♦ It uses less cable than other topologies.

On the other hand, a bus topology has the following disadvantages:

♦ It is difficult to move and change.

♦ The topology has little fault tolerance (a single fault can bring down the entire network).

♦ It is difficult to troubleshoot.

Hierarchical Star Topology

Just as with the bus topology, the hierarchical star topology has advantages and disadvantages. The increasing popularity of the star topology is mainly due to the large number of advantages, which include the following:

♦ It can be reconfigured quickly.

♦ A single cable failure won't bring down the entire network.

♦ It is relatively easy to troubleshoot.

♦ It allows the ability to centralize electronics and run fiber-to-the-desk using the centralized cabling option.

♦ It also allows an end user to run backbone cables to telecommunications enclosures, also called FTTE.

♦ It is the only recognized topology in the industry standard, ANSI/TIA-568-C.

The disadvantages of a star topology include the following:

♦ The total installation cost can be higher than that of bus and ring topologies because of the larger number of cables.

♦ It has a single point of failure: the main hub.

Ring Topology

The ring topology has a few pros but many more cons, which is why it is seldom used. On the pro side, the ring topology is relatively easy to troubleshoot. A station will know when a cable fault has occurred because it will stop receiving data from its upstream neighbor.

The cons are as follows:

◆ It is expensive because multiple cables are needed for each workstation.

◆ It is difficult to reconfigure.

◆ It is not fault tolerant. A single cable fault can bring down the entire network.

NOTE Keep in mind that these advantages and disadvantages are for a *physical* ring, of which there are few, if any, in use. Logical ring topologies exist in several network types, but they are usually laid out as a physical star.

Mesh Topology

In a mesh topology (as shown in Figure 13.1), a path exists from each station to every other station in the network. Although not usually seen in LANs, a variation on this type of topology, the *hybrid mesh*, is used in a limited fashion on the Internet and other WANs. Hybrid mesh topology networks can have multiple connections between some locations, but this is done for redundancy. Also, it is not a true mesh because there is not a connection between each and every node; there are just a few, for backup purposes.

FIGURE 13.1
A typical mesh topology

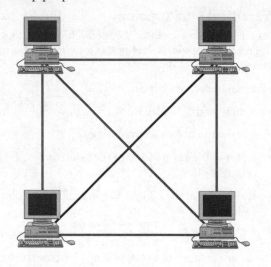

As you can see in Figure 13.1, a mesh topology can become quite complex because wiring and connections increase exponentially. For every *n* stations, you will have $n(n-1)/2$ connections. For example, in a network of four computers, you will have $4(4-1)/2$ connections, or six connections. If your network grows to only 10 devices, you will have 45 connections to manage! Given this impossible overhead, only small systems can be connected this way. The advantage to all the work this topology requires is a more fail-safe or fault-tolerant network, at least as far as cabling is concerned. On the con side, the mesh topology is expensive and, as you have seen, quickly becomes too complex. Today, the mesh topology is rarely used, and then only in a WAN environment because it is fault tolerant. Computers or network devices can switch between these multiple, redundant connections if the need arises.

NOTE Even though a mesh topology is not recommended, in critical network applications, redundancy is recommended.

Backbones and Segments

When discussing complex networks, you must be able to intelligently identify their parts. For this reason, networks are commonly broken into backbones and segments. Figure 13.2 shows a sample network with the backbone and segments identified.

FIGURE 13.2
The backbone and segments on a sample network

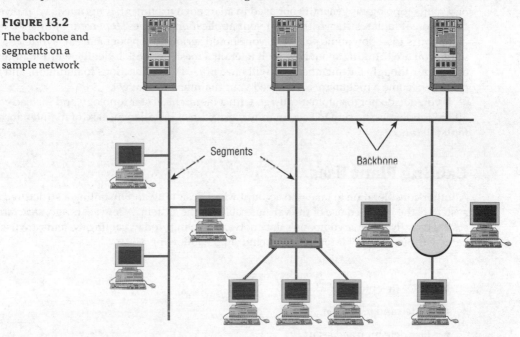

UNDERSTANDING THE BACKBONE

The backbone is the part of the network to which all segments and servers connect. A backbone provides the structure for a network and is considered the main part of any network. It usually uses a high-speed communications technology of some kind (such as FDDI, ATM, or Gigabit or 10Gb Ethernet). All servers and all network segments typically connect directly to the backbone so that any segment is only one segment away from any server on that backbone. Having all segments close to the servers makes the network efficient. Notice in Figure 13.2 that the three servers and three segments all connect to the backbone.

UNDERSTANDING SEGMENTS

Segment is a general term for any short section of the cabling infrastructure that is not part of the backbone, such as a horizontal link. Just as servers connect to the backbone, workstations connect to segments. Segments, or horizontal links, are connected to the backbone to allow the workstations on them access to the rest of the network. Figure 13.2 shows three segments. Segments are more commonly referred to as the horizontal cabling.

Selecting the Right Topology

From a practical standpoint, which topology to use has been decided for you. Because of its clear-cut advantages, the hierarchical star topology is the only recognized physical layout in ANSI/TIA-568-C. Unless you insist that your installation defy the standard, this will be the topology selected by your cabling system designer.

If you choose not to go with the hierarchical star topology, the bus topology is usually the most efficient choice if you're creating a simple network for a handful of computers in a single room because it is simple and easy to install. Because MACs are easier to manage in a star topology, a bus topology is generally not used in a larger environment. If uptime is your primary definition of fault resistance (that is, if your application requires 99 percent uptime, or less than eight hours total downtime per year), you should seriously consider a mesh layout. However, while you are thinking about how fault tolerant a mesh network is, let the word maintenance enter your thoughts. Remember, you will have $n(n - 1)/2$ connections to maintain, and this can quickly become a nightmare and exceed your maintenance budget.

If you decide not to automatically go with a hierarchical star topology and instead consider all the topologies, be sure to keep in mind cost, ease of installation, ease of maintenance, and fault tolerance.

Cabling Plant Uses

Another consideration to take into account when designing and installing a structured cabling system is the intended use of the various cables in the system. A few years ago, *structured cabling system* usually meant a company's data network cabling. Today, cabling systems are used to carry various kinds of information, including the following:

- Data
- Voice (telephone)
- Video and television
- Fire detection and security

When designing and installing your cabling system, you must keep in mind what kind of information is going to be traveling on the network and what kinds of cables are required to carry that information.

Because this book is mainly about data cabling, we'll assume you know that cables can be run for data. So, we'll start this discussion with a discussion of telephone wiring.

Telephone

The oldest (and probably most common) use for a cabling system is to carry telephone signals. In the old days, pairs of copper wires were strung throughout a building to carry the phone signal from a central telephone closet to the individual telephone handsets. In the telephone closet, the individual wires were brought together and mechanically and electrically connected to all the incoming telephone lines so that the entire building was connected to the outside world. Surprisingly, the basic layout for a telephone cabling system has changed very little. The major difference today is that telephone systems have become digital. So most require a private branch exchange (PBX), a special device that connects all the individual telephones together so the telephone calls can go out over one high-speed line (called a *trunk line*) rather than over multiple individual lines. Figure 13.3 shows how a current telephone network is arranged.

Generally speaking, today's telephone networks are run along the same cabling paths as the data cabling. Additionally, telephone systems use the same UTP cable that many networks use for carrying data. They will usually share the same wiring closets with the data and television cabling. The wires from telephone connections can be terminated almost identically to data cabling.

FIGURE 13.3
An example of a telephone network

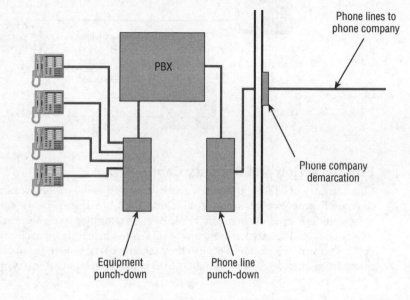

Television

With the increase in the use of on-demand video technology and videoconferencing, it is now commonplace to ensure your network can run televisions. In businesses where local cable access is possible, television cable will be run into the building and distributed to many areas to provide cable access. You may be wondering what cable TV has to do with business. The answer is plenty. News, stock updates, technology access, public-access programs, and, most important, Internet connections can all be delivered through television cable. Additionally, television cable is used for security cameras in buildings. Today, most video systems are IP/Ethernet based and operate over the UTP cable network. However, coaxial cable is still used for television sets and older cameras.

Like telephone cable, television cables can share the wiring pathways with their data counterparts. Television cable typically uses coaxial cable (usually RG-6/U cable) along with F-type, 75 ohm coaxial connectors. The cables to the various outlets are run back to a central point where they are connected to a distribution device. This device is usually an unpowered splitter, but it can also be a powered, complex device known as a *television distribution frame*. Figure 13.4 shows how a typical television cabling system might look. Notice the similarities between Figures 13.4 and 13.3. The topology is basically the same.

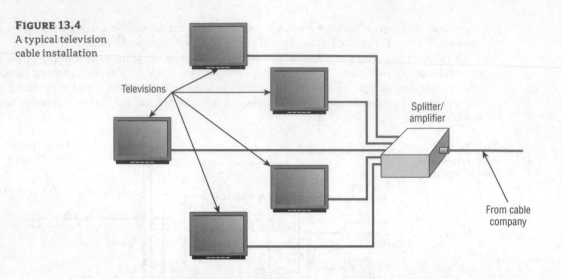

FIGURE 13.4
A typical television cable installation

Televisions

Splitter/amplifier

From cable company

Fire Detection and Security Cabling

One category of cabling that often gets overlooked is the cabling for fire detection and security devices. Examples of these devices include glass-breakage sensors, smoke alarms, motion sensors, and door-opening sensors. These devices usually run on DC current anywhere from +12 to +24 volts. Cables, which are usually UTP, must be run from each of these devices back to the central security controller. Because they usually carry power, these cables should be run separately from, or at least perpendicular to, copper cables that are carrying data.

Choice of Media

A very important consideration when designing a cabling system for your company is which media to use. Different media have different specifications that make them better suited for a particular type of installation. For example, for a simple, low-budget installation, some types of copper media might be a better choice because of their ease of installation and low cost. In previous chapters, you learned about the various types of cabling media available and their different communication aspects, so we won't describe them in detail here, but they are summarized in Table 13.1.

TABLE 13.1   Summary of Cabling Media Types

MEDIA	ADVANTAGES	DISADVANTAGES
UTP	Enables the use of relatively inexpensive switches.	May be susceptible to EMI and eavesdropping.
	Widely available.	Only covers short (up to 100 meters) distances.

TABLE 13.1   Summary of Cabling Media Types   *(CONTINUED)*

MEDIA	ADVANTAGES	DISADVANTAGES
Fiber	High data rates and long distances possible.	Moderately expensive electronics.
	Immune to EMI and largely immune to eavesdropping.	
Wireless	Few distance limitations.	More expensive than cabled media.
	Relatively easy to install.	Some wireless frequencies require an FCC license.
	Atmospheric attenuation.	Limited in bandwidth.

Telecommunications Rooms

Some components and considerations that pertain to telecommunications rooms must be taken into account during the design stage. In this section, we'll go over the LAN and telephone wiring found there, as well as the rooms' power and HVAC requirements. For more information on the functions and requirements of telecommunications rooms, see Chapter 2, "Cabling Specifications and Standards," and Chapter 5, "Cabling System Components." Figure 13.5 shows how telecommunications rooms are placed in a typical building.

FIGURE 13.5
Placement of telecommunications rooms

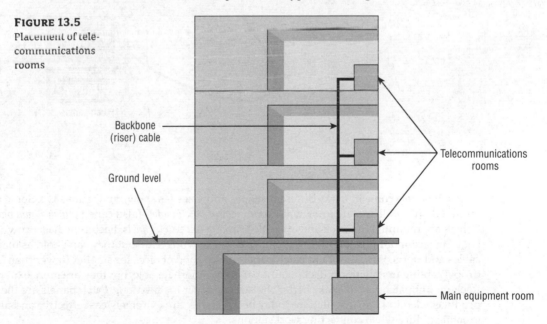

Backbone (riser) cable

Ground level

Telecommunications rooms

Main equipment room

LAN Wiring

The first item inside a telecommunications room that will draw your attention is the large bundle of cables coming into the room. The bundle contains the cables that run from the closet to the individual stations and may also contain cables that run from the room to other rooms in the building. The bundle of cables is usually bound together with straps and leads the LAN cables to a patch panel, which connects the individual wires within a particular cable to network ports on the front of the panel. These ports can then be connected to the network equipment (hubs, switches, routers, and so on), or two ports can be connected together with a patch cable. Figure 13.6 shows an example of the hardware typically found in a telecommunications room.

FIGURE 13.6
Hardware typically found in a telecommunications room and major networking items

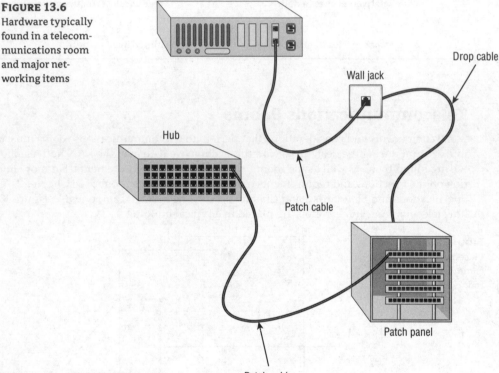

Drop cable

Wall jack

Hub

Patch cable

Patch panel

Patch cable

Patch panels come in many different shapes and sizes (as shown in Figure 13.7). Some are mounted on a wall and are known as *surface-mount patch panels* (also called *punch-down blocks*). Others are mounted in a rack and are called *rack-mount patch panels*. Each type has its own benefits. Surface-mount panels are cheaper and easier to work with, but they can't hold as many cables and ports. Surface-mount patch panels make good choices for smaller (fewer than 50 drops) cabling installations. Rack-mount panels are more flexible, but they are more expensive. Rack-mount patch panels make better choices for larger installations. Patch panels are the main products used in LAN installations today because they are extremely cost effective and allow great flexibility when connecting workstations.

FIGURE 13.7
Patch-panel
examples

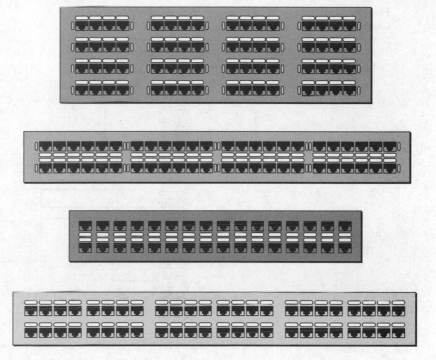

Telephone Wiring

In addition to the LAN wiring components found in the telecommunications room, you will usually find all of the wiring for the telephone system, because the two are interrelated. In most companies, a computer and a telephone are on every desk. Software programs are even available that can connect the two technologies and allow you to receive all of your voicemails as emails. These programs integrate with your current email system to provide integrated messaging services (a technology known as *unified messaging*).

The telephone cables from the individual telephones will come into the telecommunications room in approximately the same location as the data cables. They will then be terminated in some kind of patch panel (cross-connection). In many older installations, the individual wires will be punched down in *66-blocks*, a type of punch-down block that uses small "fingers" of metal to connect different UTP wires together. The wires on one side of the 66-block are from the individual wires in the cables for the telephone system. Newer installations use a type of cross-connection known as a 110-block. Although it looks different than a 66-block, it functions the same way. Instead of using punch-down blocks, it is also possible to use the same type of patch panel as is used for the UTP data cabling for the telephone cross-connection. As with the data cabling, that option enhances the flexibility of your cabling system.

The wires on the other side of the block usually come from the telephone PBX. The PBX controls all the incoming and outgoing calls as well as which pair of wires goes with which telephone extension. The PBX has connectors on the back that allow 25 telephone pairs to be connected to a single 66-block at a time using a single 50-pin connector (as shown in Figure 13.8).

FIGURE 13.8
Connecting a PBX
to a 66-block

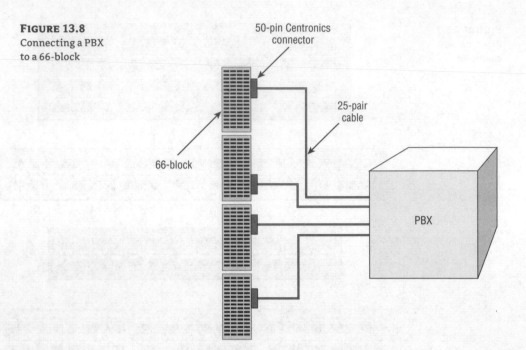

Typically, many of these 66-blocks are placed on a large piece of plywood fastened to the wall (as shown in Figure 13.9). The number of 66-blocks is as many as required to support the number of cables required for the number of telephones in the telephone system.

FIGURE 13.9
Multiple 66-blocks
in a wiring closet

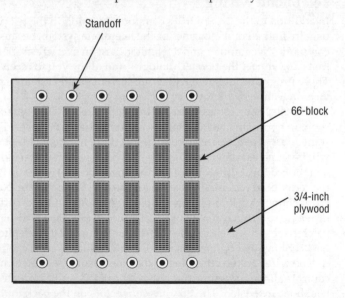

Power Requirements

With all of these devices in the wiring closet, it stands to reason that you are going to need some power receptacles there. Telecommunications rooms have some unique power requirements. First of all, each of the many small electronic devices will need power, and a single-duplex outlet will not have enough outlets. Additionally, these devices should all be on an electrical circuit dedicated to that wiring closet and separate from the rest of the building. And, in some cases, devices within the same room may require their own circuit, separate from other devices in that room. The circuit should have its own *isolated ground*. An isolated ground in commercial wiring is a ground wire for the particular isolated outlet that is run in the same conduit as the electrical-supply connectors. This ground is called *isolated* because it is not tied into the grounding of the conduit at all. The wire runs from the receptacle back to the point where the grounds and neutrals are tied together in the circuit panel. You can identify isolated-ground outlets in a commercial building because they are orange with a small green triangle on them.

NOTE Most, if not all, residential outlets have an isolated ground because conduit is not used, and these outlets must have a ground wire in the cable.

The wiring closet should be equipped with a minimum of two dedicated three-wire 120 volt AC duplex outlets, each on its own 20 amp circuit, for network and system equipment power. In addition, separate 120 volt AC duplex outlets should be provided as convenience outlets for tools, test equipment, and so forth. Convenience outlets should be a minimum of 6″ off the floor and placed at 6′ intervals around the perimeter of the room. None of the outlets should be switched—that is, controlled by a wall switch or other device that might accidentally interrupt power to the system.

HVAC Considerations

Computer and networking equipment generates much heat. If you place enough equipment in a telecommunications room without ventilation, the temperature will quickly rise to dangerous levels. Just as sunstroke affects the human brain, high temperatures are the downfall of electronic components. The room temperature should match the ambient temperature of office space occupied by humans, and should be kept at that temperature year-round.

For this reason, telecommunications rooms should be sufficiently ventilated. At the very least, some kind of fan should exchange the air in the closet. Some telecommunications rooms are pretty good-sized rooms with their own HVAC (heating, ventilation, and air-conditioning) controls.

Centralized cabling and fiber-to-the-enclosure are both standards-based implementations of the hierarchical star, per ANSI/TIA-568-C, which enable the ability to reduce the HVAC and electrical loads found in traditional implementations of the hierarchical star. This is accomplished by centralizing equipment into one room or miniaturizing switches and locating them closer to workstation outlets. For more information on these cost-saving approaches, visit TIA's Fiber Optics LAN Section at www.fols.org.

Cabling Management

Cabling management is guiding the cable to its intended destination without damaging it or its data-carrying capabilities. Many different cabling products protect cable, make it look good, and help you find the cables faster. They fall into three categories:

◆ Physical protection

◆ Electrical protection

◆ Fire protection

In this section, we will look at the various devices used to provide each level of protection and the concepts and procedures that go along with them.

Physical Protection

Cables can be fragile—easily cut, stretched, and broken. When performing a proper cabling installation, cables should be protected. Many items are currently used to protect cables from damage, including the following:

◆ Conduit

◆ Cable trays

◆ Standoffs

◆ D-rings

We'll take a brief look at each and the different ways they are used to protect cables from damage.

CONDUIT

The simplest form of cable protection is a metal or plastic conduit to protect the cable as it travels through walls and ceilings. Conduit is really nothing more than a thin-walled plastic or metal pipe. Conduit is used in many commercial installations to contain electrical wires. When electricians run conduit for electrical installation in a new building, they can also run additional conduit for network wiring. Conduit is put in place, and the individual cables are run inside it.

The main advantage to conduit is that it is the simplest and most robust protection for a network cable. Also, if you use plastic conduit, it can be a relatively cheap solution (metal conduit is more expensive).

WARNING   The flame rating of plastic conduit must match the installation environment. In other words, plastic conduit in a plenum space must have a plenum rating, just like the cable.

NOTE   Rigid metal conduit (steel pipe) exceeds all flame-test requirements. Any cable can be installed in any environment if it is enclosed in rigid metal conduit. Even cable that burns like crazy can be put in a plenum space if you put it in this type of conduit.

NOTE   There are many methods of cable support. Cable trays are popular in larger installations and as a method of supporting large numbers of cables or multiple trunks. However, there are also smaller support systems available (such as J hooks that mount to a wall or suspend from a ceiling).

CABLE TRAYS

When running cable, the cable must be supported every 48″ to 60″ when hanging horizontally. Supporting the cable prevents it from sagging and putting stress on the conductors inside. For this reason, special devices known as *cable trays* (also sometimes called *ludder racks*, because of their appearance) are installed in ceilings. The horizontal cables from the telecommunications rooms that run to the individual telecommunications outlets are usually placed into this tray to support them as they run horizontally. Figure 13.10 shows an example of a cable tray. This type of cable-support system hangs from the ceiling and can support hundreds of individual cables.

FIGURE 13.10
An example of a cable tray

STANDOFFS

When terminating UTP wires for telephone applications in a telecommunications room, you will often see telephone wires run from a multipair cable to the 66-punch-down block. To be neat, the individual conductors are run around the outside of the board that the punch-down blocks are mounted to (as shown in Figure 13.11). To prevent damage to the individual conductors, they are bent around devices known as standoffs. These objects look like miniature spools, are usually made of plastic, and are screwed to the mounting board every foot or so (also shown in Figure 13.11).

FIGURE 13.11
A telecommunications board using standoffs

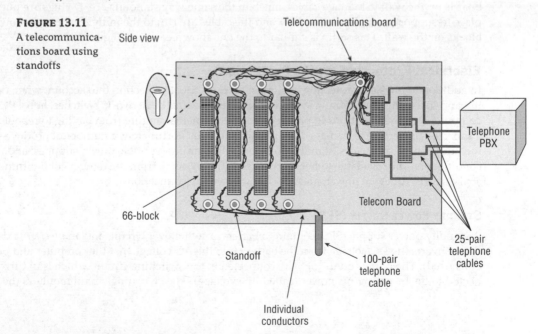

D-Rings

For LAN installations that use racks to hold patch panels, you need some method of keeping the cables together and organized as they come out of the cable trays and enter the telecommunications room to be terminated. On many racks, special metal rings called *D-rings* (named after their shape) are used to keep the individual cables in bundles and keep them close to the rack (as shown in Figure 13.12).

FIGURE 13.12
D-rings in a cabling closet for a cabling rack

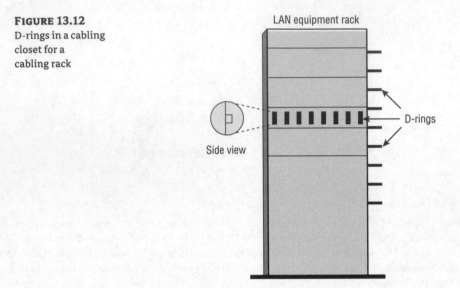

In addition to managing cable for a cabling rack, D-rings are also used on punch-down boards on the wall to manage cables, much in the same way standoffs are. D-rings are put in place to support the individual cables, and the cables are run to the individual punch-down blocks on the wall. This setup is similar to the one shown earlier in Figure 13.11.

Electrical Protection (Spike Protection)

In addition to physical protection, you must take electrical protection into account when designing and installing your cabling system. Electricity powers the network, switches, hubs, PCs, and computer servers. Variations in power can cause problems ranging from having to reboot after a short loss of service to damaged equipment and data. Fortunately, a number of products—including surge protectors, standby power supplies, uninterruptible power supplies, and line conditioners—are available to help protect sensitive systems from the dangers of lightning strikes, dirty (uneven) power, and accidental power disconnection.

Standby Power Supply (SPS)

A standby power supply (SPS) contains a battery, a switchover circuit, and an *inverter* (a device that converts the DC voltage from the battery into the AC voltage that the computer and peripherals need). The outlets on the SPS are connected to the switching circuit, which is in turn connected to the incoming AC power (called line voltage). The switching circuit monitors the line voltage. When it drops below a factory preset threshold, the switching circuit switches from line voltage to the battery and inverter. The battery and inverter power the outlets (and, thus, the computers or devices plugged into them) until the switching circuit detects line voltage at the correct level. The switching circuit then switches the outlets back to line voltage.

NOTE   Power output from battery-powered inverters isn't exactly perfect. Normal power output alternates polarity 60 times a second (60Hz). When graphed, this output looks like a sine wave. Output from inverters is stepped to approximate this sine-wave output, but it really never duplicates it. Today's inverter technology can come extremely close, but the differences between inverter and true AC power can cause damage to computer power supplies over the long run.

UNINTERRUPTIBLE POWER SUPPLY (UPS)

A UPS is another type of battery backup often found on computers and network devices today. It is similar to an SPS in that it has outlets, a battery, and an inverter. The similarities end there, however.

A UPS uses an entirely different method to provide continuous AC voltage to the equipment it supports. When a UPS is used, the equipment is always running off the inverter and battery. A UPS contains a charging/monitoring circuit that charges the battery constantly. It also monitors the AC line voltage. When a power failure occurs, the charger stops charging the battery, but the equipment never senses any change in power. The monitoring part of the circuit senses the change and emits a beep to tell the user the power has failed.

NOTE   The power output of some UPSs (usually lower quality ones) resembles more of a square wave than the true sine wave of AC. Over time, equipment can be damaged by this nonstandard power.

Fire Protection

All buildings and their contents are subject to destruction and damage if a fire occurs. The cabling in a building is no exception. You must keep in mind a few cabling-design concerns to prevent fire, smoke, or heat from damaging your cabling system, the premises on which they are installed, and any occupants.

As discussed in Chapter 1, you should make sure you specify the proper flame rating for the cable according to the location in which it will be installed.

Another concern is the puncturing of fire barriers. In most residential and commercial buildings, firewalls are built specifically to stop the spread of a fire within a building. Whenever there is an opening in a floor or ceiling that could possibly conduct fire, the opening is walled over with fire-rated drywall to make a firewall that will prevent the spread of fire (or at least slow it down). In commercial buildings, cinder-block walls are often erected as firewalls between rooms.

Because firewalls prevent the spread of fire, it is important not to compromise the protection they offer by punching holes in them for network cables. If you need to run a network cable through a firewall, first try to find another route that won't compromise the integrity of the firewall. If you can't, you must use an approved firewall penetration device (see Figure 13.13). These devices form a tight seal around each cable that passes through the firewall. One type of seal is made of material that is *intumescent*; that is, it expands several times its normal size when exposed to very high heat (fire temperatures), sealing the hole in the firewall. That way, the gases and heat from a fire won't pass through.

FIGURE 13.13
An example of a
firewall penetra-
tion device

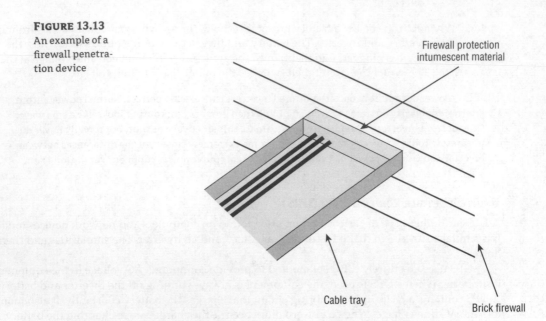

Firewall protection
intumescent material

Cable tray

Brick firewall

Data and Cabling Security

Your network cables carry all the data that crosses your network. If the data your cables carry is sensitive and should not be viewed by just anyone, you may need to take extra steps when designing and installing your cabling system to ensure that the data stays where it belongs: inside the cables. The level of protection you employ depends on how sensitive the data is and how serious a security breach could be. Cabling security measures can range from the simple to the absurdly complex.

Two ways to prevent data from being intercepted are EM (electromagnetic) transmission regulation and tapping prevention.

EM (Electromagnetic) Transmission Regulation

You should know that the pattern of the magnetic field produced by any current-carrying conductor matches the pattern of the signals being transmitted. Based on this concept, devices exist that can be placed around a cable to intercept these magnetic signals and turn them back into electrical signals that can be sent to another (unwanted) location. This process is known as *EM signal interception*. Because the devices pick up the magnetic signals surrounding the cable, they are said to be noninvasive.

Susceptibility to EM signal interception can be minimized by using shielded cables or by encasing all cabling runs from source to destination in a grounded metal conduit. These shielding methods reduce the amount of stray EM signals.

Tapping Prevention

Tapping is the interception of LAN EM signals through listening devices placed around the cable. Some tapping devices are invasive and will actually puncture the outer jacket of a cable, or the insulation of individual wires, and touch the metal inner conductor to intercept all signals sent along that conductor. Of course, taps can be applied at the cross-connections if security access to your equipment rooms and telecommunications rooms is lax.

To prevent taps, the best course of action is to install the cables in metal conduit or to use interlocked armored cable, where practical. Grounding of the metal conduit will provide protection from both EM and invasive taps but not from taps at the cross-connection. When these steps are not practical, otherwise securing the cables can make tapping much more difficult. If the person trying to tap your communications can't get to your cables, they can't tap them. So you must install cables in secure locations and restrict access to them by locking the cabling closets. Remember, if you don't have physical security, you don't have network security.

Cabling Installation Procedures

Now that we've covered some of the factors to take into account when designing a cabling system, it's time to discuss the process of installing an entire cabling system, from start to finish. A cabling installation involves five steps:

1. Design the cabling system.
2. Schedule the installation.
3. Install the cables.
4. Terminate the cables.
5. Test the installation.

Design the Cabling System

We've already covered this part of the installation in detail in this chapter. However, it's important enough to reiterate: following proper cabling design procedures will ensure the success of your cabling system installation. Before you pull a single cable, you should have a detailed plan of how the installation will proceed. You should also know the scope of the project (how many cable runs need to be made, what connections need to be made and where, how long the project will take, and so on). Finally, you should have the design plan available to all people involved with the installation of the cable. That list of people includes the cabling installer, the electrical inspector, the building inspector, and the customer (even if you are the customer). Be sure to include anyone who needs to refer to the way the cabling is being installed. At the very least, this information should contain a blueprint of how the cables will be installed.

Schedule the Installation

In addition to having a proper cabling design, you should know approximately how long the installation will take and pick the best time to do it. For example, the best time for a new cabling installation is while the building studs are still exposed and electrical boxes can be easily installed. From a planning standpoint, this is approximately the same time in new construction when the electrical cabling is installed. In fact, because of the obvious connection between electrical and telecommunications wiring, many electrical contractors are now doing low-voltage (data) wiring so they can contract the wiring for both the electrical system and the telecommunications system.

WARNING If you use an electrical contractor to install your communications cabling, make sure he or she is well trained in this type of installation. Many electricians are not aware of the subtleties required to properly handle network wiring. If they treat it like the electrical wire, or run it along with the electrical wire, you're going to have headaches in your network performance. We recommend that the communication wiring be installed after the electrical wiring is done so that they can be kept properly segregated.

For a postconstruction installation, you should schedule it so as to have the least impact on the building's occupants and on the existing network or existing building infrastructure. It also works to schedule it in phases or sections.

Install the Cabling

Once you have a design and a proper schedule, you can proceed with the installation. We'll start with a discussion of the tools you will need.

Cabling Tools

Just like any other industry, cable installation has its own tools, some not so obvious, including the following:

◆ Pen and paper

◆ Hand tools

◆ Cable spool racks

◆ Fish tape

◆ Pull string

◆ Cable-pulling lubricant

◆ Two-way radio

◆ Labeling materials

◆ Tennis ball

We'll briefly go over how each is used during installation.

NOTE Tools are covered in more detail in Chapter 6, "Tools of the Trade."

Pen and Paper

Not every cabling installer may think of pen and paper as tools, but they are. It is a good idea to have a pen and paper handy when installing the individual cables so that you can make notes about how particular cables are routed and installed. You should also note any problems that occur during installation.

These notes are invaluable when it's time to troubleshoot an installation, especially when you have to trace a particular cable. You'll know exactly where particular wires run and how they were installed.

Hand Tools

It's fairly obvious that a variety of hand tools are needed during the course of a cabling installation. You will need to remove and assemble screws, hit and cut things, and perform various types of construction and destruction tasks. Some of the hand tools you should make sure to include in your toolkit are (but are not limited to) the following:

- Screwdrivers (Phillips, slotted, and Torx drivers)
- Cordless drill (with drill bits and screwdriver bits)
- Hammer
- Cable cutters
- Wire strippers
- Punch-down tool
- Drywall saw (hand or power)

Cable Spool Racks

It is usually inefficient to pull one cable at a time during installation. Typically, more than one cable will be going from the cabling closet (usually the source of a cable run) to a workstation outlet. So a cable installer will tape several cables together and pull them as one bundle.

The tool used to assist in pulling multiple cables is the *cable spool rack* (see Figure 13.14). As you can see, the spools of cable are mounted on the rack. These racks can hold multiple spools to facilitate the pulling of multiple cables simultaneously. They allow the cable spools to rotate freely, thus reducing the amount of resistance to the pull.

FIGURE 13.14

A cable spool rack

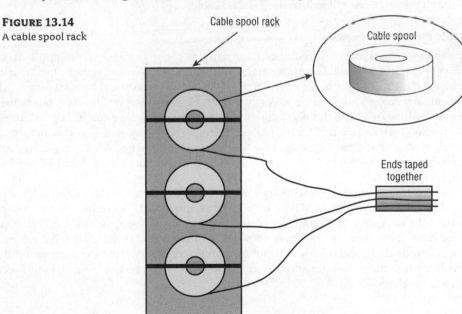

Cable spool rack

Cable spool

Ends taped together

Most manufacturers now offer their copper and optical fiber premises cables in a reel-in-a-box package where each box contains a reel with end plates that suspend the reel in the box allowing for easy and smooth payout. Some manufacturers are now offering these reel-in-a-box packages with adjustable resistance mechanisms to prevent over-spinning of the reel, which can cause tangling. One example of this is Superior Essex's BrakeBox, which has an adjustable brake on both sides of the reel.

Fish Tape

Many times, you will have to run cable into narrow conduits or narrow crawl spaces. Cables are flexible, much like rope. Just like rope, when you try to stuff a cable into a narrow space, it simply bunches up inside the conduit. You need a way of pulling the cable through that narrow space or providing some rigid backbone. A *fish tape* is one answer. It is really nothing more than a roll of spring steel or fiberglass with a hook on the end. A bunch of cables can be hooked and pulled through a small area, or the cables can be taped to the fish tape and pushed through the conduit or wall cavity.

Pull String

Another way to pull cables through small spaces is with a nylon *pull string* (also called a *fish cord*), a heavy-duty cord strong enough to pull several cables through a conduit or wall cavity. The pull string is either put in place before all the cables are pulled, or it is run at the same time as the cables. If it is put in place before the cables are pulled, such as when the conduit is assembled or in a wall cavity before the drywall is up, you can pull through your first cables with another string attached to the cables. The second string becomes the pull string for the next bundle, and so on. For future expansion, you leave one string in with the last bundle you pull. If the pull string is run at the same time as the cables, it can be used to pull additional cables through the same conduit as already-installed cables.

Cable-Pulling Lubricant

It is important not to put too much stress on network cables as they are being pulled. (The maximum is 25 pounds of pull for category cables; 50 pounds for plenum optical fiber distribution cables consisting of 12 or fewer fibers; and 100 lbs. for all riser optical fiber cables and plenum optical fiber cables with fiber counts of more than 12 fibers.) To prevent stress on the cable during the pulling of a cable through a conduit, a *cable-pulling lubricant* can be applied. It reduces the friction between the cable being pulled and its surroundings and is specially formulated so as not to plug up the conduit or dissolve the jackets of the other cables. It can be used any time cable needs to be pulled in tight quarters. See Chapter 6 for more details, including some drawbacks of lubricant.

A preferred method of reducing the amount of friction between cables and conduit is the use of fabric inner ducts, such as MaxCell. These fabric inner ducts not only provide a low-friction surface through which the cables are pulled, but they also help organize the cables. Installing more inner ducts than you need will allow you to go back and install additional cables in the unused inner ducts without the need for lubricants, keeping in mind the maximum fill-ratio of the duct or conduit. These inner ducts are available for both premises installations (riser and plenum) and outside plant (OSP) installations.

Labeling Materials

With the hundreds of cables that need to be pulled in large cabling installations, it makes a great deal of sense to label both ends of each cable while it's being pulled. That way, when it's time to terminate each individual cable, you will know which cable goes where, and you can document that fact on your cabling map.

So you will need some labeling materials. The most common are the sticky numbers sold by Panduit and other companies (check with your cabling supplier to see what it recommends). You should pick a numbering standard, stick with it, and record all the numbered cables and their uses in your cabling documentation. A good system is to number the first cable as *1*, with each subsequent cable the next higher number. You could also use combinations of letters and numbers. To label the cables, stick a number on each of the cables you are pulling and stick another of the same number on the corresponding box or spool containing the cable. When you are finished pulling the cable, you can cut the cable and stick the number from the cable spool onto the cut end of the cable. Voilà! Both ends are numbered. Don't forget to record on your notepad and in your network analyzer the number of each cable and where it's going.

The ANSI/TIA/EIA-606-A standard defines a labeling system to label each cable and workstation port with its exact destination in a wiring closet using three sets of letters and numbers separated by dashes. The label is in the following format:

BBBB–RR–PORT

BBBB is a four-digit building code (usually a number), *RR* is the telecommunications room number, and *PORT* is the patch panel and port number that the cable connects to. For example, 0001–01–W222 would mean building 1, closet 1, wall-mounted patch panel 2 (W2), and port 22.

Table 13.2 details the most commonly used labeling particulars.

TABLE 13.2   ANSI/TIA/EIA-606-A Labeling Particulars

SAMPLE: 0020-2B-B23		
Label	**Example**	**Notes**
Building	0020	Building 20 (comes from a standard campus or facilities map)
Closet	2B	Closet B, 2nd floor
Panel/Port	B23	Patch Panel B, port 23

TIP   This is just one example of the standard labeling system. For more information, you should read the ANSI/TIA/EIA-606-A standards document, which you can order from http://global .ihs.com.

Two-Way Radio

Two-way radios aren't used as often as some of the other tools listed here, but they come in handy when two people are pulling or testing cable as a team. Two-way radios allow two people who are cabling in a building to communicate with each other without having to shout down a hallway or use cell phones. The radios are especially useful if they have hands-free headset microphones. Many two-way radios have maximum operating ranges of greater than several kilometers (or, about one mile), which makes them effective for cabling even very large factories and buildings.

WARNING   If you need to use radios, be aware that you may need to obtain permission to use them in places like hospitals or other high-security environments.

Tennis Ball

You may be saying, "Okay. I know why these other tools are listed here, but a tennis ball?" Think of this situation. You've got to run several cables through the airspace above a suspended ceiling. Let's say the cable run is 75 meters (around 225′) long. The conventional way to run this cable is to remove the ceiling tiles that run underneath the cable path, climb a ladder, and run the cable as far as you can reach (or throw). Then you move the ladder, pull the cable a few feet farther, and repeat until you reach the end. An easier way is to tie a pull string to a tennis ball (using duct tape, nails, screws, or whatever) and throw the tennis ball from source to destination. The other end of the pull string can be tied to the bundle of cables so that it can be pulled from source to destination without you going up and down ladders too many times.

TIP   You may think using a tennis ball is a makeshift tool, but cabling installers have been making their own tools for as long as there have been installers. You may find that a tool you make yourself works better than any tool you can buy.

PULLING CABLE

Keep in mind the following points when pulling cable to ensure the proper operation of the network:

◆ Tensile strength

◆ Bend radius

◆ Protecting the cable while pulling

Tensile Strength

Contrary to popular opinion, network cables are fragile. They can be damaged in any number of ways, especially during the pulling process. The most important consideration to remember when pulling cable is the cable's *tensile strength*, a measure of how strong a cable is along its axis. The higher the tensile strength, the more resistant the cable is to stretching and, thus, breaking. Obviously, you can pull harder without causing damage on cables with higher tensile strength. A cable's tensile strength is normally given in either pounds, or in pounds per square inch (psi).

WARNING When pulling cable, don't exert a pull force on the cable greater than the tensile rating of the cable. If you do, you will cause damage to the cable and its internal conductors or optical fibers. If a conductor or optical fiber break occurs, you may not know it until you terminate and test the cable. If it breaks, you will have to replace the whole cable. Standards and the manufacturer's recommendations should be reviewed for tensile-strength information. It is also especially important to point out here that for optical fiber cables, the cable is pulled using the aramid or GRP (glass-reinforced plastic) rod strength elements, *not* the jacket. Pulling eyes are especially helpful for larger cables.

NOTE Four-pair UTP should not have more than 25 pounds of tension applied to it (note that this is 25 pounds, not 25 psi). This number is based on a calculation using the elongation properties of copper. When you are exerting pulling force on all four pairs of 24 AWG conductors in a UTP cable, 25 pounds is the maximum tensile load they can withstand before the copper starts to stretch. Once stretched, a point of high attenuation has been created that will also cause impedance and structural return-loss reflections.

Bend Radius

Most cables are designed to flex, and that makes them easy to use and install. Unfortunately, just because they can flex doesn't mean that they should be bent as far as possible around corners and other obstacles. Both copper and fiber-optic cables have a value known as the *minimum bend radius* of that cable. ANSI/TIA-568-C specifies that copper cables should be bent no tighter than the arc of a circle that has a radius four times the cables' diameter. For example, if a cable has a ¼″ diameter, it should be bent no tighter than the arc of a circle 2″ in diameter. Four times a ¼″ cable equals a 1″ radius. The continuous arc created using a 1″ radius creates a circle 2″ in diameter. Figure 13.15 illustrates how bend radius is measured.

Keep in mind as well that the bend radius of a cable can be compromised by an overzealous tightening of a cable tie wrap. Pinching will compromise the performance of both copper category cables and optical fiber cables. For this reason, hook-and-loop cable wraps are recommended. While they are more expensive, the amount of money they may save you in troubleshooting will far outweigh the cost of the wrap.

FIGURE 13.15
The bend radius for cable installation

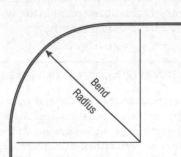

TIP   You can purchase some devices from cabling products vendors that aid in the pulling of cable so that the minimum bend radius is not exceeded. These devices are basically plastic or metal corners with large bend radii to help guide a cable around a corner.

Protection While Pulling

In addition to being careful not to exceed either the tensile strength or bend radius of a particular cable when pulling it, you should also be careful not to pull the cable over or near anything that could damage it. For example, never pull cables over sharp, metal corners, as these could cut into the outside jacket of the cable and, possibly, the interior conductors.

Many things could damage the cable during its installation. Just use common sense. If you would damage your finger (or any other body part) by running it across the surface you want to pull the cable across, chances are that it's not a good idea to run a cable over it either.

CABLING SYSTEM DOCUMENTATION

The most often overlooked item during cable installation is the documentation of the new cabling system. Cabling system documentation includes information about what components make up a cabling system, how it is put together, and where to find individual cables. This information is compiled in a set of documents that can be referred to by the network administrator or cabling installer any time moves, additions, or changes need to be made to the cabling system.

The most useful piece of cabling system documentation is the *cabling map*. Just as its name implies, a cabling map indicates where every cable starts and ends. It also indicates approximately where each cable runs. Additionally, a cabling map can indicate the location of workstations, segments, hubs, routers, closets, and other cabling devices.

NOTE   A map can be as simple as a listing of the run numbers and where they terminate at the workstation and patch-panel ends. Or it can be as complex as a street map, showing the exact cable routes from patch panel to workstation outlet.

To make an efficient cabling map, you need to have specific numbers for *all* parts of your cabling system. For example, a single cable run from a cabling closet to wall plate should have the same number on the patch panel port, patch cable, wall cable, and wall plate. This way, you can refer to a specific run of cable at any point in the system, and you can put numbers on the cabling map to refer to each individual cable run.

CABLING @ WORK: DEVELOP AND UTILIZE A PROCESS OF MONITORING CHANGES

In our experience, most performance or faulty issues with network cabling can be traced back to not following a standard process of installing a cabling system. As you learned in this chapter, this process starts with designing the cabling system. At the same time, it's rare to have the cabling installed the same way as dictated in the design. There are so many things that happen along the way that lead to more ports and cabling than initially anticipated, addition of locations and components, and so on. Often these don't fit into the logical, well-laid-out labeling plan that you envisioned.

Our advice is to make sure that you have a process for recording changes to the design and actual installation. This means having a process for documenting change orders, and a means of creating drawings that reflect the actual installation instead of the original or modified design. We also advise that you meet with the key project managers on a regular basis to review these changes. Good luck—nothing ever goes as planned!

Terminate the Cable

Now that you've learned about installing the cable, you need to know what to do with both ends of the cable. Terminating the cables involves installing some kind of connector on each end (either a connector or a termination block) so that the cabling system can be accessed by the devices that are going to use it. This is the part of cabling system installation that requires the most painstaking attention to detail, because the quality of the termination greatly affects the quality of the signal being transmitted. Sloppy termination will yield an installation that won't support higher-speed technologies.

Though many termination methods are used, they can be classified one of two ways: connectorizing or patch-panel termination. *Connectorizing* (putting some kind of connector directly on the end of the cable in the wall) is covered in detail in Chapter 14, "Cable Connector Installation." Here, we'll briefly discuss patch-panel termination.

There are many different types of patch panels, some for copper, some for fiber. Copper-cable patch panels for UTP all have a few similar characteristics, for the most part. First off, most UTP LAN patch panels (as shown in Figure 13.16) have UTP ports on the front and punch-down blades (see Figure 13.17) in the back. During termination, the individual conductors in the UTP cable are pressed between the metal blades to make both the mechanical and electrical connection between the cable and the connector on the front of the patch panel. This type of patch panel is a 110-punch-down block (or 110-block, for short).

FIGURE 13.16
A sample patch panel

FIGURE 13.17
A punch-down blade on a 110-block

The procedure for connecting an individual cable is as follows:

1. Route the cable to the back of the punch-down block.

2. Strip off about ¼–½ inch of the cabling jacket. (Be careful not to strip off too much, as that can cause interference problems.)

3. Untwist each pair of UTP conductors and push each conductor onto its slot between the color-coded "finger," as shown here:

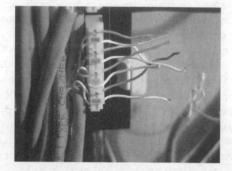

NOTE   Each category rating has standards for termination. For example, each category rating has a standard for how much length can be untwisted at the termination point. Make sure you follow these standards when terminating cable.

WARNING   Make sure that no more than ½" of each twisted-conductor pair is untwisted when terminated.

4. Using a 110-punch-down tool, push the conductor into the 110-block so that the metal fingers of the 110-block cut into the center of each conductor, thus making the connection, as shown here:

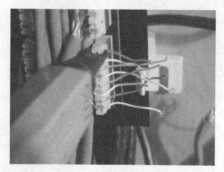

5. Repeat steps 3 and 4 for each conductor.

The process described here works only for UTP cables. Fiber-optic cables use different termination methods. For the most part, fiber-optic cables do use patch panels, but you can't punch down a fiber-optic cable because of the delicate nature of the optical fibers. Instead, the individual fiber-optic cables are simply connectorized and connected to a special "pass-through" patch panel (as shown in Figure 13.18).

FIGURE 13.18
A fiber-optic patch panel

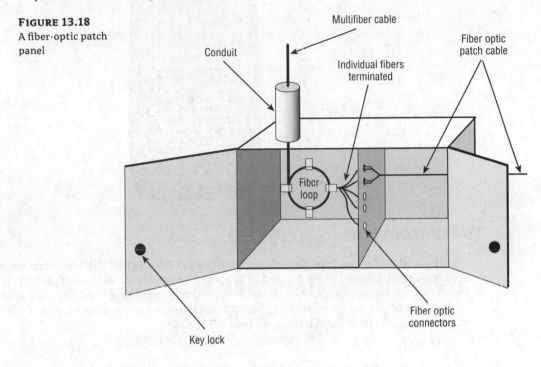

Multifiber cable

Conduit

Individual fibers terminated

Fiber optic patch cable

Fiber loop

Fiber optic connectors

Key lock

NOTE   Fiber-optic connectorization is covered in Chapter 14.

Test the Installation

Once you have a cable or cables installed and terminated, your last installation step is to test the connection. Each connection must be tested for proper operation, category rating, and possible connection problems. If the connection has problems, it must either be re-terminated or, in the worst-case scenario, the entire cable must be re-pulled.

Testing individual cables is done most effectively and quickly with a LAN cable tester (as shown in Figure 13.19). This cable tester usually consists of two parts: the tester itself and a signal injector. The tester is a very complex electronic device that measures not only the presence of a signal but also the quality and characteristics of the signal. Cable testers are available for both copper and fiber-optic cables.

You should test the entire cabling installation before installing any other hardware (hubs, PCs, etc.). That way, you avoid having to troubleshoot cabling-related problems later (or at least you minimize possible later problems).

NOTE   Testing tools and procedures are covered in more detail in Chapter 15, "Cable System Testing and Troubleshooting."

FIGURE 13.19
A LAN cable tester

The Bottom Line

Identify and understand the elements of a successful cabling installation. These may seem basic, but most networks succeed or fail based on how much care was taken in identifying the key elements of a successful installation and understanding how they are performed. If at the end of a network installation you find that the network does not operate as planned, chances are that one of these was not addressed carefully.

Master It

1. What are the key elements of a successful cabling installation?

2. What are the key aspects of a proper design?

Identify the pros and cons of network topologies. At this point, most commercial building network designs and installations will follow a hierarchical star network topology per the ANSI/TIA-568-C standard. There are many reasons why this is the only recognized cabling topology in this standard.

Master It What are the key reasons why the hierarchical star topology is the only recognized topology in the ANSI/TIA-568-C standard?

Understand cabling installation procedures. Many of the "microscopic" components necessary in a cabling network have been discussed in this chapter, from choice of media, telecommunications rooms, and cabling management devices to means of ensuring security. However, it is first important to understand the cable installation from a macro-level.

Master It

1. What are the five basic steps of the cabling installation?

2. What are some useful cabling tools?

3. Which TIA standard should you use to ensure that your labeling is done in a standardized fashion?

Chapter 14

Cable Connector Installation

So far, you have learned about the installation of cables and the termination process. In today's cabling installation, the cables you install into the walls and ceilings are usually terminated at either punch-down blocks or patch panels and wall outlets. In some cases (as with patch cables, for example), you may need to put a connector on the end of a piece of cable. Installing connectors, or *connectorizing*, is an important skill for the cabling installer.

This chapter will cover the basics of cable connector installation and teach you how to install the connectors for each type of cable.

In this chapter, you will learn to:

- ◆ Install twisted-pair cable connectors
- ◆ Install coaxial cable connectors
- ◆ Install fiber-optic cable connectors

Twisted-Pair Cable Connector Installation

For LAN and telephone installations, no cable type is currently more ubiquitous than twisted-pair copper cabling, particularly UTP cabling. The main method to put connectors on twisted-pair cables is *crimping*. You use a tool called a *crimper* to push the metal contacts inside the connector onto the individual conductors in the cable, thus making the connection.

NOTE The topic of this chapter is *not* cable termination (which we discussed in Chapter 13, "Cabling System Design and Installation"). Connectorization is normally done for patch and drop cables, whereas termination is done for the horizontal cables from the patch panel in the wiring closet to the wall plate at the workstation.

Types of Connectors

Two main types of connectors (often called *plugs*) are used for connectorizing twisted-pair cable in voice and data communications installations: the RJ-11 and RJ-45 connectors. As discussed in Chapter 10, "Connectors," these are more accurately referred to as six-position and eight-position modular plugs, but the industry is comfortable with the RJ labels. Figure 14.1 shows examples of RJ-11 and RJ-45 connectors for twisted-pair cables. Notice that these connectors are basically the same, except the RJ-45 accommodates more conductors and thus is slightly larger. Also note that the RJ-11 type connector shown in Figure 14.1, while having six positions, is only configured with two metal contacts instead of six. This is a common cost-saving practice on RJ-11 type plugs when only two conductor contacts will be needed for a telephone application. Conversely, you rarely see an RJ-45 connector with less than all eight of its positions configured with contacts.

FIGURE 14.1
RJ-11 and RJ-45
connectors

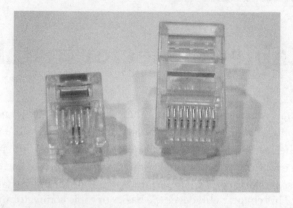

RJ-11 connectors, because of their small form factor and simplicity, were historically used in both business and residential telephone applications, and they remain in widespread use in homes. RJ-45 connectors, on the other hand, because of the number of conductors they support (eight total), are used primarily in LAN applications. Current recommendations are to install RJ-45 jacks for telephone applications because those jacks support both RJ-11 and RJ-45 connectors.

Both types of connectors are made of plastic with metal "fingers" inside them (as you can see in Figure 14.1). These fingers are pushed down into the individual conductors in a twisted-pair cable during the crimping process. Once these fingers are crimped and make contact with the conductors in the twisted-pair cable, they are the contact points between the conductors and the pins inside the RJ-11 or RJ-45 jack.

The different RJ connectors each come in two versions, for stranded and solid conductors. Stranded-conductor twisted-pair cables are made up of many tiny hairlike strands of copper twisted together into a larger conductor. These conductors have more surface area to make contact with but are more difficult to crimp because they change shape easily. Because of their difficulty to connectorize, they are usually used as patch cables.

Most UTP cable installed in the walls and ceilings between patch panels and wall plates is solid-conductor cable. Although they are not normally used as patch cables, solid-conductor cables are easiest to connectorize, so many people make their own patch cords out of solid-conductor UTP.

TIP   As discussed several times in this book, we do not recommend attaching your own UTP and STP plugs to make patch cords. Field-terminated modular connectors are notoriously time consuming to apply and are unreliable. Special circumstances may require that you make your own, but whenever possible, buy your UTP and STP patch cords.

Conductor Arrangement

When making solid-conductor UTP patch cords with crimped ends, you can make many different configurations, determined by the order in which their color-coded wires are arranged. Inside a normal UTP cable with RJ-45 ends are four pairs of conductors (for a total of eight conductors). Each pair is color coded blue, orange, green, or brown. Each wire will either be the solid color or a white wire with a mark of its pair's solid color (e.g., the orange and the white/orange pair). Table 14.1 illustrates some of the many ways the wires can be organized.

TABLE 14.1   Color-Coding Order for Various Configurations

Wiring Configuration	Pin #	Color Order
568A	1	White/green
	2	Green
	3	White/orange
	4	Blue
	5	White/blue
	6	Orange
	7	White/brown
	8	Brown
568B	1	White/orange
	2	Orange
	3	White/green
	4	Blue
	5	White/blue
	6	Green
	7	White/brown
	8	Brown

TABLE 14.1   Color-Coding Order for Various Configurations   *(CONTINUED)*

WIRING CONFIGURATION	PIN #	COLOR ORDER
10Base-T only	1	White/blue
	2	Blue
	3	White/orange
	6	Orange
Generic USOC	1	White/brown
	2	White/green
	3	White/orange
	4	Blue
	5	White/blue
	6	Orange
	7	Green
	8	Brown

TIP   A straight-through patch cord for data applications has both ends wired the same—that is, both ends T568-A or both ends T568-B. Straight-through patch cords connect PCs to wall outlets and patch panels to network equipment such as hubs, switches, and routers. A crossover patch cord is wired with one end T568-A and one end T568-B.

TIP   For Ethernet networking, crossover cords can connect two PCs directly together without any intermediate network equipment. To connect hubs, routers, or switches to each other, either a straight-through or crossover cable will be required, depending on device-type combination. Check the equipment documentation to determine what type of patch cord you require.

When connectorizing cables, make sure you understand which standard your cabling system uses and stick to it.

Connector Crimping Procedures

The installation procedure is pretty straightforward. Knowing what "hiccups" you might run into is the only difficult part.

PREREQUISITES

As with any project, you must first gather all the items you will need. These items include the following:

- Cable
- Connectors
- Stripping and crimping tools

By now, you know about the cable and connectors, so we'll discuss the tools you'll need for RJ-connector installation. The first tool you're going to need is a cable-jacket stripper, as shown in Figure 14.2. It will only cut through the outer jacket of the cable, not through the conductors inside. Many different kinds of cable strippers exist, but the most common are the small, plastic ones (as in Figure 14.2) that easily fit into a shirt pocket. They are cheap to produce and purchase.

FIGURE 14.2
A common twisted-pair cable stripper

NOTE   Common strippers don't work well (if at all) on flat cables, like silver satin. But then, technically, those cables aren't twisted-pair cables and should never be used for data applications.

Another tool you're going to need when installing connectors on UTP or STP cable is a cable-connector crimper. Many different styles of crimpers can crimp connectors on UTP or STP cables. Figure 14.3 shows an example of a crimper that can crimp both RJ-11 and RJ-45 connectors. Notice the two holes for the different connectors and the cutting bar.

FIGURE 14.3
A crimper for RJ-11 and RJ-45 connectors

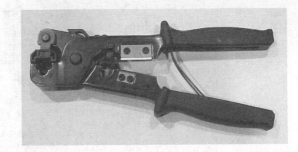

The last tool you're going to use is a cable tester. This device tests for a continuous signal from the source connector to the destination and also tests the quality of that connection. We won't devote much space to it in this chapter, as it will be covered in Chapter 15, "Cable System Testing and Troubleshooting."

INSTALLING THE CONNECTOR

Now we'll go over the steps for installing the connectors. Pay particular attention to the order of these steps and be sure to follow them exactly.

WARNING   Equipment from some manufacturers may require you to perform slightly different steps. Check the manufacturer's instructions before installing any connector.

1. Measure the cable you want to put ends on and trim it to the proper length using your cable cutters (as shown here). Cut the cable about 3″ longer than the final patch-cable length. For example, if you want a 10′ patch cable, cut the cable to 10′-3″.

2. Using your cable stripper, strip about 1.5″ of the jacket from the end of the cable. To do this, insert the cable into the stripper so that the cutter bar in the stripper is 1.5″ from the end of the cable (as shown in the graphic). Then, rotate the stripper around the cable twice. This will cut through the jacket.

3. Remove the stripper from the cable and pull the trimmed jacket from the cable, exposing the inner conductors (as shown). If a jacket slitting cord (usually a white thread) is present, separate it from the conductors and trim it back to the edge of the jacket.

TIP Most strippers only score the jacket to avoid cutting through and damaging the conductor insulation. The jacket is easily removed, as bending the cable at the score mark will cause the jacket to break evenly, and then it can be pulled off.

4. Untwist all the inner conductor pairs and spread them apart so that you can see each individual conductor, as shown here:

5. Line up the individual conductors so that the color code matches the color-coding standard you are using (see Table 14.1, earlier in this chapter). The alignment in the graphic shown here is for 568-B, with number 1 at the top.

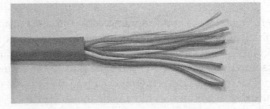

6. Trim the conductors so that the ends are even with each other, making sure that the jacket of the cable will be inside the connector (as shown here). The total length of exposed connectors after trimming should be no longer than ½″ to ⅝″ (as shown in the second graphic).

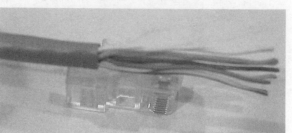

7. Insert the conductors in the connector, ensuring that all conductors line up properly with the pins as they were aligned in the last step. If they don't line up, pull them out and line them up. Do this carefully, as it's the last step before crimping on the connector.

8. Carefully insert the connector and cable into the crimping tool (as shown in the following graphic). Squeeze the handle firmly as far as it will go and hold it with pressure for three seconds. As you will see in the second graphic, the crimping tool has two dies that will press into the connector and push the pins in the connector into the conductors inside the connector. A die in the crimping tool will also push a plastic retainer into the cable jacket of the cable to hold it securely inside the connector.

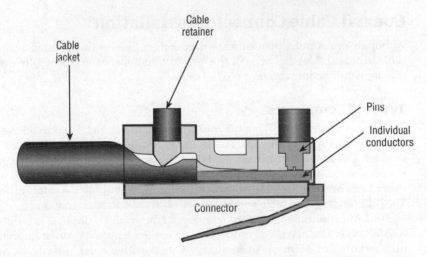

9. Now that you've crimped the connector, remove it from the crimping tool and examine it (as shown in the next graphic). Check to ensure all conductors are making contact and that all pins have been crimped into their respective conductors. If the connector didn't crimp properly, cut off the connector and redo it.

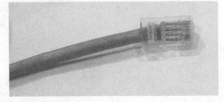

10. To finish the patch cable, put a connector on the other end of the cable and follow these steps again, starting with step 2.

TESTING

You should ensure that the connectorization was done correctly by testing the cable with a cable tester. Put the injector on one end of the cable and put the tester on the other end. Once you have the tester hooked up, you can test the cable for continuity (no breaks in the conductors), near-end crosstalk (NEXT), and category rating (all quality-of-transmission issues). The specific procedures for testing a cable vary depending on the cable tester. Usually you select the type of cable you are testing, hook up the cable, and then press a button labeled something like *Begin Test*. If the cable does not work or meet the testing requirements, re-connectorize the cable.

NOTE   Cable testers are covered in more detail in Chapter 15.

Coaxial Cable Connector Installation

Although less popular than either twisted-pair or fiber-optic cables, you'll encounter coaxial cable in older LANs and in modern video installations. After reading this section, you should be able to install a connector on a coaxial cable.

Types of Connectors

As discussed in Chapter 7, many types of coaxial cable exist, including RG-6, RG-58, and RG-62. LAN applications primarily use RG-62- and RG-58-designated coaxial cables. RG-6 is used chiefly in video and television cable installations. The preparation processes for connectorizing RG-6, RG-58, and RG-62 are basically the same; different connectors are used for different applications, either LAN or video. You can identify the cable by examining the printing on the outer jacket. The different types of cable will be labeled with their RG designation.

For LAN applications, the BNC connector (shown in Figure 14.4) is used with RG-58 or RG-62 coaxial cable. The male BNC connectors are easily identified by their knurled grip and quarter-turn locking slot. Many video applications, on the other hand, use what is commonly known as a *coax cable TV connector* or *F connector* (as shown in Figure 14.5) and RG-6 cable.

FIGURE 14.4
Male and female
BNC connectors

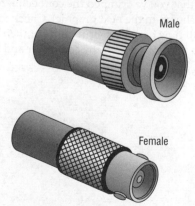

Male

Female

FIGURE 14.5
A coax cable TV F
connector

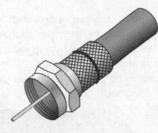

In addition to their physical appearance, coax connectors differ based on their installation method. Basically, two types of connectors exist: *crimp-on* and *screw-on* (also known as *threaded*). The crimp-on connectors require that you strip the cable, insert the cable into the connector, and then crimp the connector onto the jacket of the cable to secure it. Most BNC connectors used for LAN applications use this installation method. Screw-on connectors, on the other hand, have threads inside the connector that allow the connector to be screwed onto the jacket of the coaxial cable. These threads cut into the jacket and keep the connector from coming loose.

TIP   Screw-on connectors are generally unreliable because they can be pulled off with relative ease. Whenever possible, use crimp-on connectors.

Connector Crimping Procedures

Now that you understand the basic connector types, we can tell you how to install them. The basic procedural outline is similar to installing twisted-pair connectors.

PREREQUISITES

Make sure you have the right cable and connectors. For example, if you are making an Ethernet connection cable, you must have both RG-58 coaxial cable and BNC connectors available. You must also have the right tools—cable cutters, a cable stripper, a crimper for the type of connectors you are installing, and a cable tester. These tools were discussed in the last section and also in more detail in Chapter 6, "Tools of the Trade."

INSTALLING THE CONNECTOR

The connector you are going to learn how to install here is the most common crimp-on style that comes in three pieces: the center pin, the crimp sleeve, and the connector housing. Pay particular attention to the order of these steps and be sure to follow them exactly.

WARNING   Equipment from some manufacturers may require you to perform slightly different steps. Make sure you check the manufacturer's instructions before installing any connector.

1. Measure the cable you want to put ends on and trim it to the proper length using your cable cutters. Cut the cable to exactly the length you want the cable to be.

2. Put the crimp-on sleeve on the cable jacket on the end of the cable you are going to connectorize.

3. Using your cable stripper, strip about ⅝″ of the jacket from the end of the cable. To do this, insert the cable into the stripper so that the cutter bar in the stripper is 1″ from the end of the cable (as shown in the first graphic). Then, rotate the stripper around the cable twice (as shown in the second graphic). This will cut through the jacket. Remove the stripper from the cable and pull the trimmed jacket from the cable, exposing the braided shield and inner conductor.

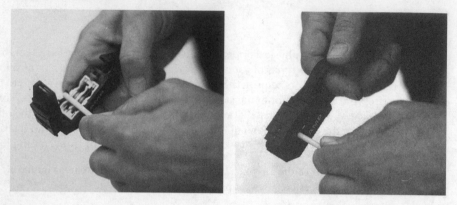

4. Trim the braided shielding so that ⁷⁄₃₂″ of braid is showing, as shown in the following graphic:

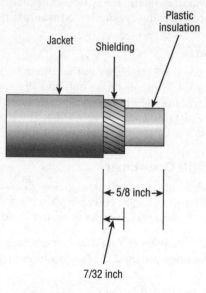

5. Strip the inner protective plastic insulation around the center conductor so that $\frac{7}{16}''$ of plastic is showing (thus, $\frac{3}{16}''$ of conductor is showing), as shown in the next graphic. Note that the shielding is folded back over the jacket.

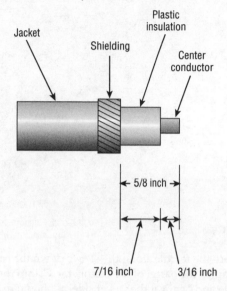

6. Insert the center conductor into the center pin of the connector, as shown in the next graphic. Crimp the pin twice with the ratcheting crimper. After crimping (shown in the second graphic), you shouldn't be able to twist the pin around the center conductor.

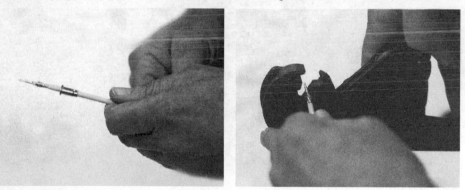

7. Push the connector onto the end of the cable. The barrel of the connector should slide under the shielding. Push the connector until the center pin clicks into the connector, as shown in the following graphic:

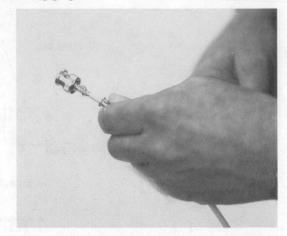

8. Slide the ferrule along the sleeve down the cable so that it pushes the braided shielding around the barrel of the connector. Crimp the ferule barrel twice, once at the connector side and again at the jacket side, as shown in the following two graphics:

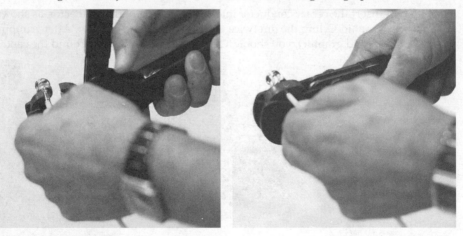

9. Now that you've crimped the connector, remove it from the crimping tool and examine it. Check to see that the connector is securely attached to the end of the cable—you should not be able to move it. If the connector didn't crimp properly, cut off the connector and redo it.

10. To finish the patch cable, put a connector on the other end of the cable and follow these steps again, starting with step 2.

TESTING

Once you have a tester hooked up, you can test the cable to ensure that the cable is connectorized properly and has no breaks. See the previous subsection "Testing" under the section on twisted-pair cable connector installation. (The procedure described there is the same as for coaxial cable.)

NOTE   Cable-testing procedures are covered in more detail in Chapter 15.

Fiber-Optic Cable Connector Installation

In the early days of fiber-optic connections, connectorizing a single fiber-optic cable could take up to a half hour. These days, due to improvements in connector design and materials, an experienced cable installer can put a connector on a fiber-optic cable in less than five minutes.

Connector Types

A number of different connector types exist for the different fiber-optic cables. Each connector type differs based on its form factor and the type(s) of fiber-optic cables it supports. Some of the most common fiber-optic connector types include the following:

- LC
- SC
- ST
- FDDI
- FC

Each of these types of connectors is discussed in detail in Chapter 10. MPO connectors are typically installed under factory conditions using specialized processes.

Connectorizing Methods

There are several methods of attaching a fiber-optic connector to the optical fiber in the field. Generally, connectors are attached with an adhesive (e.g., epoxy) or mechanically with a crimp-on-like connector.

For both methods, the particulars may vary by manufacturer. In general, adhesive-based connections typically are less expensive but take a longer amount of time to assemble. However, they provide a reliable connector that can withstand external pressure on the connector from disconnecting and reconnecting many times over the long term. Mechanical connectors are typically more expensive but can be mounted very quickly, saving a lot of labor and enabling rapid deployment. However, these connectors may be more susceptible to external pressure on the connector and may not perform well over time if connectors are disconnected and reconnected multiple times.

EPOXY CONNECTORS

The epoxy system uses, obviously, the two-part glue known as *epoxy*. The optical fibers are trimmed and the epoxy is applied. Then the fiber is inserted into the connector. Some epoxy systems don't include a tube of adhesive but have the adhesive preloaded into the connector. In this case, the adhesive is only activated by some outside element. For example, 3M's Hot Melt system uses a thermosetting adhesive, which means that high temperature must activate the adhesive and cause it to set. Other types of adhesive are activated by UV light.

Once the fiber is in the connector and the adhesive has been activated, the assembly is either placed aside to air dry, or the connector is inserted into a curing oven to speed the drying process.

The majority of fiber-optic connectors are installed by some type of epoxy method, mainly because of the method's long-term reliability.

MECHANICAL CONNECTORS

The main disadvantage to epoxy-based termination is the time and extra equipment needed to terminate a single connector—it may take up to 15 minutes. Because of this, many companies have developed connectors, called mechanical cam or *epoxy-less connectors*, that don't need any kind of adhesive to hold them together.

Instead of glue, some kind of friction method or cam is used to hold the fiber in place in the connector. The 3M CrimpLok, Panduit OptiCam, and Corning UniCam connectors are examples of epoxy-less systems. These connectors can typically be mounted in 3–5 minutes and are ideal for rapidly installing and deploying fiber-optic cables.

Connector Installation Procedures

In this section, you are going to learn how to connectorize a single multimode fiber-optic cable with an ST connector. Even though the SC connector is now the recommended fiber-optic connector, the installed base of ST connectors is significant and the procedures for installing different connectors differ only slightly. Where necessary, we'll point out where other connectorizing methods differ.

PREREQUISITES

As with the other types of connectorization, the first step is to gather all the tools and items you are going to need. You will need some specialized fiber-optic tools, including:

- Epoxy syringes
- Curing oven
- Cable-jacket stripper
- Fiber stripper
- Fiber-polishing tool (including a fiber-polishing puck and abrasive pad)
- Kevlar scissors
- Fiber-optic loss tester

You will also need a few consumable items, like:

◆ Cable

◆ Connectors

◆ Alcohol and wipes (for cleaning the fiber)

◆ Epoxy (self-curing or thermosetting, depending on the application)

◆ Polishing cloths

You can buy kits that contain all of these items.

If your fiber-termination system includes an oven or UV-curing device, plug it in ahead of time so that it will be ready. If possible, make sure you have adequate space to terminate the fiber, along with an adequate light source.

TIP If you can, work on a black surface. It makes the fiber easier to see while terminating it. It is hard to see optical fiber on a white space.

WARNING *Be extremely careful when dealing with bare fiber!* Most optical fibers are made of glass. The cutting or cleaving of optical fibers produces many sharp ends. Always wear safety glasses to protect your eyes from flying shards of glass. The very fine diameter of fiber-optic strands allows them to penetrate skin easily. Properly dispose of any cut fiber scraps in a specially designed fiber-optic trash bag.

Cabling @ Work: Safety Comes First!

The importance of safety cannot be overstated. When it comes to working with the sharp ends of glass fiber-optic cables, safety should not be overlooked and shortcuts should not be taken. A good example of this comes from firsthand experience in working in a fiber-optic manufacturing facility.

After glass is drawn into fiber, the fiber goes through a battery of mechanical, optical, and geometric tests. At each of these steps, fiber is cleaved prior to testing. This leaves many small segments of glass shards on the work surface. Improper handling often leads to the penetration of these shards into fingers and sometimes the eyes. As a result, manufacturing facilities are faced with employee work injuries, in some cases requiring medical care. To minimize this, managers spend a lot of time training the staff on the importance of handling glass fiber-optics and safety in general. We urge you to pay the same attention to safety and working with fiber-optics.

Finally, before you start, make sure you are familiar with the connector system you are using. If possible, have the directions from the fiber connector's manufacturer available while doing the termination.

INSTALLING THE CONNECTOR

Unlike the "quick and easy" copper termination procedures, terminating fiber requires more steps and greater care. You must take your time and perform the following steps correctly:

1. Cut and strip the cable.

2. Trim the aramid yarn.

3. Strip the optical-fiber buffer.

4. Prepare the epoxy.

5. Epoxy the connector.

6. Insert the fiber in the connector.

7. Dry the epoxy.

8. Scribe and remove extra fiber.

9. Polish the tip.

10. Perform a visual inspection.

11. Finish.

We have included several figures that show how to perform each operation.

Cut and Strip the Cable

Cutting the fiber is fairly simple: simply cut through the jacket and strength members using the fiber shears included in the fiber-optic termination kit. Optical fiber cannot be cut with regular cutters, mainly because many fiber-optic cables contain aramid-yarn strength members, which are next to impossible to cut with regular cutters. Trim the cable exactly to the length that you want.

TIP Before you get out the strippers, you should perform one operation to make the installation go smoothly. Open the fiber-optic-connector package and remove the strain-relief boot and crimp sleeve. Place them on the cable *before* you strip it. Slide them down, out of the way. That way, you don't have to try to push them over the optical fiber and aramid yarn.

Then strip one end of the cable in two steps. First, strip the outer jacket, exposing the buffered fiber and the aramid-yarn strength members. Set the jacket stripper to the size recommended by the manufacturer and squeeze the handle. Figure 14.6 shows a cable jacket stripper in action. The stripper will bite through the outer jacket only. Release the handle and remove the stripper. You should then be able to pull off the outer jacket, as shown in Figure 14.7.

FIGURE 14.6
Using a cable
jacket stripper

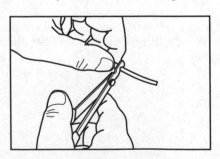

FIGURE 14.7
Pulling off the
outer jacket of a
fiber-optic cable

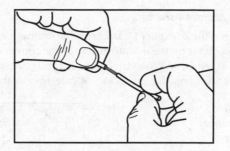

Second, carefully strip the buffer, exposing the optical fiber so you can insert it in the connector. A guide diagram on the back of the package containing the connector will show how much of the jacket and buffer to strip off, either with measurements or by a life-sized representation. Figure 14.8 shows such a diagram.

FIGURE 14.8
A strip guide for a
fiber-optic cable

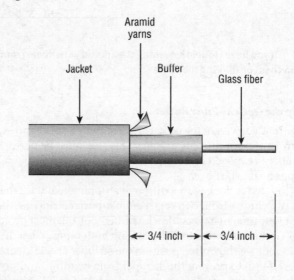

Aramid
yarns

Jacket   Buffer

Glass fiber

← 3/4 inch → ← 3/4 inch →

TIP   If you are stripping a relatively short cable (less than 25´) without connectors on either end, tie a knot in the end of the cable opposite of the end you are trying to strip. That way, you can't pull the strength members out of the cable while you strip it. Note that this will irreparably damage that portion of the cable, so make sure you can cut the knot out and still have enough cable.

WARNING   Never strip a fiber-optic cable as you would a copper cable (i.e., by twisting the stripper and pulling the stripper off with the end of the jacket). It can damage the cable.

Trim the Aramid Yarn

After removing the outer jacket, trim the aramid yarn, with the aramid-yarn scissors, to the length specified by the manufacturer of the connector system. To cut the yarn, grab the bundle of fibers together and loop them around your finger. Cut the fibers so that about ¼" (more or less, depending on the connector) of yarn fiber is showing. See Figure 14.9.

FIGURE 14.9
Cutting the aramid
yarn of a fiber-
optic cable

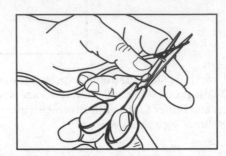

TIP If you have trouble loosening the aramid yarn fibers from the inside of the cable, try blowing on them or shaking the cable.

Strip the Optical-Fiber Buffer

Now that you've got the jacket and aramid yarn cut and stripped properly, you can strip the buffer from around the optical fiber. This step must be performed with *extreme* care. At this point, the fiber is exposed and fragile; if it is damaged or broken, you must cut off the ends you just stripped and start over.

This step is done with a different stripping tool than the stripper used to strip the cable jacket. You can choose from two types of fiber-buffer strippers: the Miller tool and the No-Nik stripper. Most first-time installers like the Miller tool, but most professionals prefer the No-Nik tool. Many fiber connectorization toolkits contain both types. We will show photos of the Miller tool.

To remove the buffer, position the stripper at a 45-degree angle to the fiber (as shown in Figure 14.10) to prevent the stripper from bending, and possibly breaking, the optical fiber. Position the stripper to only remove about ⅛" to ½" of buffer. Slowly but firmly squeeze the stripper to cut through the buffer. Make sure you have cut through the entire buffer. Then, using the stripper, pull the buffer from the fiber slowly and steadily, making sure to pull straight along the fiber without bending it. You will have to exert some pressure, as the buffer will not come off easily. Repeat this process to remove additional ⅛" to ¼" "bites" of buffer until sufficient buffer has been removed from the fiber and between ½" and 1" (depending on the type of connector being used) of fiber is exposed. See Figure 14.11.

FIGURE 14.10
Stripping buffer
from the optical
fiber with the
Miller tool

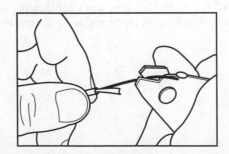

FIGURE 14.11
The optical fiber after stripping the buffer from the fiber

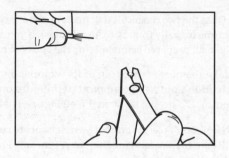

TIP It's better to have too much fiber exposed than not enough because you will trim off excess fiber in a later step.

Prepare the Epoxy

Now that the fiber-optic cable and optical fiber have been stripped and the cable is ready, set it aside and get the epoxy ready to use (assuming, of course, your connector system incorporates epoxy). Epoxy will not work unless both of its parts are mixed. The epoxy usually comes in packets with a syringe (see Figure 14.12) so that you can inject the epoxy into the connector.

FIGURE 14.12
An epoxy packet with a syringe

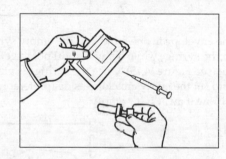

Open the bag that contains the plastic epoxy envelope and the syringe. Remove the divider from the envelope and mix the epoxy by kneading it with your fingers or running the flat side of a pencil over the envelope (as shown in Figure 14.13). The epoxy is mixed when it is a uniform color and consistency. It should take a couple of minutes to fully mix the epoxy, especially if you are using your fingers.

FIGURE 14.13
Mixing the epoxy

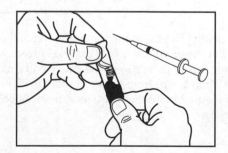

NOTE Once the two chemicals that make up the epoxy are mixed, it will remain workable for only a short time (usually from 15 to 30 minutes). If you are terminating several cables, you should have them all prepared before mixing the epoxy to make the best use of your time.

Then take the new syringe out of its wrapper and remove the plunger. Hold the epoxy envelope gently (don't put a large amount of pressure on the envelope) and cut one corner so a very small opening (⅟₁₆″ to ⅛″) is formed (see Figure 14.14).

WARNING Don't use the aramid-yarn scissors to cut the epoxy envelope! Epoxy is very sticky and will ruin the scissors (and they aren't cheap!). Find a pair of cheap scissors and put these in your fiber termination kit.

FIGURE 14.14
Opening the epoxy
envelope

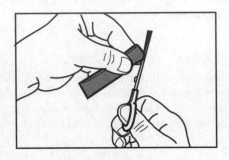

Hold the envelope in one hand and the empty syringe body in the other. Slowly pour the epoxy into the syringe while being careful not to get epoxy on your hands or on the outside of the syringe (see Figure 14.15). Once the syringe is almost full (leave a ⅛″ gap at the top), stop pouring and set the epoxy envelope aside (preferably on a wipe or towel, in case the epoxy spills) or throw it away if it's empty.

FIGURE 14.15
Pouring epoxy into
the syringe

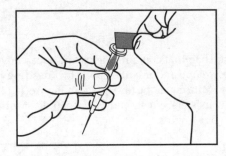

Next, gently place the plunger into the end of the syringe, but *don't* push it down. Just seat it in the end of the syringe to hold it in place. Invert the syringe so that the needle is at the top and then tap the side of the syringe. The epoxy will sink to the bottom, and the air bubbles will rise to the top. Grab a wipe from your termination kit and hold it above and around the needle (as shown in Figure 14.16). Slowly squeeze the air bubbles out of the syringe until only epoxy is left in the syringe.

FIGURE 14.16
Removing air from
the syringe

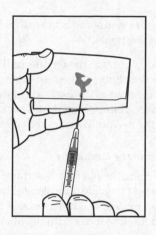

Once all the air is out of the syringe, stop pushing on the plunger. When no more epoxy comes out, pull very slightly on the plunger so a tiny bubble is at the tip of the needle. Put the cap on the needle (if there is one) and set the syringe aside, out of the way.

Epoxy the Connector

Now you have to put the connector on the fiber. Remove the rest of the components from the connector package (you already have the strain relief on the cable, remember?) and lay them out in front of you. Remove the dust cap from the end of the connector and the cap from the syringe. Push the plunger on the syringe lightly to expel the small air bubble in the needle. Insert the needle into the connector body on the cable side (the side that faces the cable, not the side that faces the equipment to be connected to).

Squeeze the plunger and expel epoxy into the inside of the connector. Continue to squeeze until a very small bead of epoxy appears at the ferrule inside the connector (as shown in Figure 14.17). The size of this bead is important, as too large of a bead means you will have to spend much time polishing off the extra epoxy. On the other hand, too small of a bead may not support the optical fiber inside the connector.

FIGURE 14.17
Putting epoxy
inside the
connector

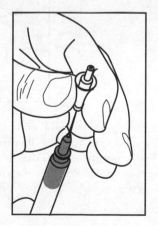

TIP   The proper size bead of epoxy to expel into the connector is approximately half the diameter of the inside of the ferrule.

Once the bead appears at the ferrule, pull the needle halfway out of the connector and continue to squeeze the plunger. Keep squeezing until the connector is filled with epoxy and the epoxy starts to come out of the backside of the connector (see Figure 14.18). Remove the needle completely from the connector and pull back slightly on the plunger to prevent the epoxy from dripping out of the needle. Then set the connector aside and clean the needle off with a wipe.

Insert the Fiber into the Connector

You will now prepare the fiber for insertion. The fiber must be free of all dirt, fingerprints, and oil to ensure the best possible adhesion to the epoxy. Most fiber termination kits come with special wipes soaked in alcohol, known as *Alco wipes*. Hold one of these wipes in one hand, between your thumb and forefinger, and run the fiber between them (see Figure 14.19).

FIGURE 14.18
Finishing epoxying
the connector

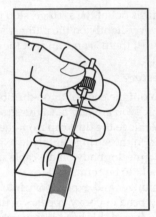

FIGURE 14.19
Cleaning the fiber

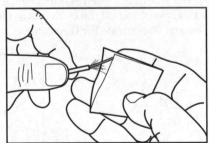

Pick up the connector in one hand and carefully slide the fiber into the epoxy-filled center (see Figure 14.20). While pushing the fiber in, rotate the connector back and forth. Doing so will spread the epoxy evenly around the outside of the optical fiber, and it will help to center the fiber in the connector. Don't worry if some epoxy leaks out onto the aramid yarn—that will actually help to secure the cable to the connector.

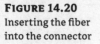

FIGURE 14.20
Inserting the fiber
into the connector

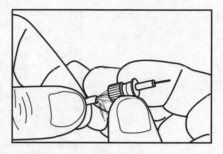

To secure the cable permanently to the connector, slide the crimp sleeve up from around the cable and over the aramid fibers so that it sits against the connector (see Figure 14.21). You must now use the crimper that comes with your fiber-optic termination kit and crimp the sleeve in two places, once at the connector and once at the fiber. The crimper has two holes for crimping, a larger and a smaller hole. Crimp the sleeve at the connector end using the larger hole and crimp the sleeve at the cable-jacket end using the smaller hole (as shown in Figure 14.22).

FIGURE 14.21
Putting on the
crimping sleeve

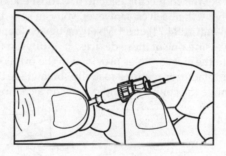

FIGURE 14.22
Crimping the
sleeve

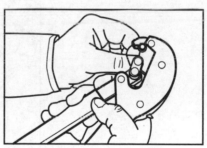

TIP While crimping, make sure to hold the connector against the jacket so that a tight connection is made.

After crimping the sleeve, slide the strain-relief boot up from the cable and over the crimp sleeve (see Figure 14.23). The connector is now secure to the cable. However, a short piece of fiber should protrude from the connector. Be careful not to break it off. It will be scribed and polished off in the next step.

FIGURE 14.23
Installing the
strain-relief boot

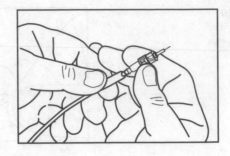

WARNING   If you do break the piece of protruding fiber, you will have to cut off the connector
and start over.

Dry the Epoxy

You must set the connector aside to dry. Most epoxies take anywhere from 12 to 24 hours to set
completely by themselves. However, you can speed up the process either by using a curing oven
(shown in Figure 14.24) or a UV setting device (depending on the type of epoxy used). To dry
the epoxy using one of these devices, carefully (so that you don't break the fiber) insert the connector into the slots or holes provided in the oven or setting device. Let the connector sit as long
as the manufacturer requires (usually somewhere between 5 and 15 minutes). Then, if using the
oven, remove the connector and place it on a cooling rack.

FIGURE 14.24
Drying the epoxy
with an oven

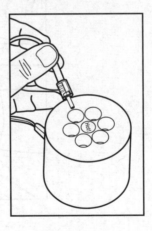

TIP   While the connectors are curing in the oven, you can connectorize more fibers. Remember,
you only have a short time before the epoxy is no longer usable.

Scribe and Remove Extra Fiber

After the connector has sufficiently cooled and the epoxy has dried in the connector, you are ready to remove the excess fiber. You do so with a special tool known as a *scribe*. It's impossible to get any kind of cutting tool close enough to the connector to cut off the remaining fiber and glass; instead, you remove the glass fiber by scratching one side of it and breaking off the fiber.

Hold the connector firmly in one hand and use the scribe to scratch the protruding fiber just above where it sticks out from the bead of epoxy on the connector ferrule (as shown in Figure 14.25). Use a *very* light touch. Remember, the glass is very small, and it doesn't take much to break it.

FIGURE 14.25
Scribing the pro-
truding fiber

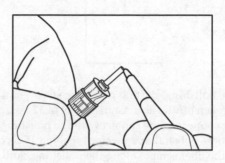

To remove the fiber, grab the protruding piece of fiber and sharply pull up and away from the connector (see Figure 14.26). The glass should break completely (it will still have a rough edge, although you may not be able to see it). Dispose of the fiber remnant properly in a specially designed fiber-optic trash bag.

FIGURE 14.26
Removing the fiber

Polish the Tip

After scribing the fiber, the end will look similar to the one shown at the left side of Figure 14.27. To make a proper connection, you must polish the end to a perfectly flat surface with varying grits of polishing films (basically the same idea as sandpaper, except that films are much, much finer). The idea is to use the polishing cloth to remove a little bit of the protruding fiber at a time until the fiber is perfectly flat and level, similar to the right side of Figure 14.27.

FIGURE 14.27
Fiber before and
after polishing

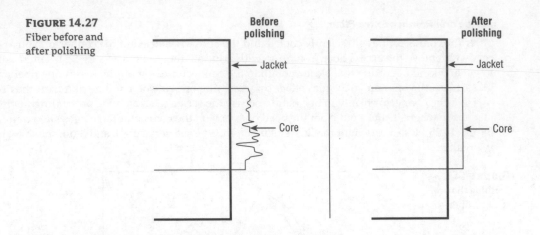

Coarse polishing, the first polishing step, removes the burrs and sharp ends present after you've broken off the fiber. Grab a sheet of 12 micron film and hold it as shown in Figure 14.28. Bring the connector into contact with the polishing film and move the connector in a figure-eight motion. Polish the connector for about 15 seconds or until you hear a change in the sound made as the fiber scrapes along the polishing cloth. This process is known as *air polishing* because you aren't using a backing for the polishing film.

FIGURE 14.28
Air polishing the
fiber after scribing

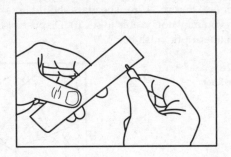

WARNING Air polishing will take some practice. Do not overpolish the fiber! If you do, the fiber will not transmit light correctly and will have to be cut off and re-terminated.

When you are done, a small amount of epoxy should be left, and the glass will not be completely smooth, as shown in Figure 14.29. Don't worry—this will be taken care of in the next part of the polishing procedure. Before proceeding, clean the end of the fiber with an Alco wipe to remove any loose glass shards or epoxy bits that might scratch the fiber during the next polishing step.

Next, with the polishing puck in one hand, insert the connector into the puck (as shown in Figure 14.30). Then, very gently place the puck with the connector in it on some 3 micron polishing film placed on the polishing pad. Move the puck in a figure-eight motion four or five times (see Figure 14.31). Stop polishing when the connector fiber doesn't scrape along the polishing cloth and feels somewhat slick.

FIGURE 14.29
Results of air
polishing

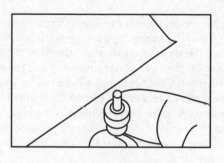

FIGURE 14.30
Insert the
connector into the
polishing puck

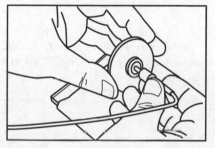

FIGURE 14.31
Polishing the tip of
the fiber

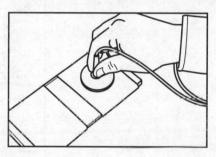

WARNING   Do not overpolish the conductor with the 3 micron polishing film. Overpolishing will cause the glass-fiber end to be undercut (concave) and cause light loss at the optic connection.

Then clean the connector with an Alco wipe to remove any debris before polishing again. Once clean, gently place the puck on some 0.3 micron film, which is 10 times finer than the 3 micron polishing film used in the initial polishing step above, and give it five or six quick figure-eights with little or no pressure to fine-polish the fiber. Remove the connector from the polishing puck and wipe it with an Alco wipe. You're done. It's time to test the connector to see how you did.

Inspect the Connector

At this stage, you should check the connector with a fiber-optic microscope for any flaws that might cause problems. A fiber-optic microscope allows you to look very closely at the end of the fiber you just terminated (usually magnifying the tip 100 times or more). Different microscopes work somewhat differently, but the basic procedure is the same.

Insert the connector into the fiber microscope (as shown in Figure 14.32). Look into the eyepiece and focus the microscope using the thumb wheel or slider so that you can see the tip of the fiber. Under 100 times magnification, the fiber should look like the image shown in Figure 14.33. What you see is the light center (the core) and the darker ring around it (the cladding). Holding the opposite end of the fiber near a light source will increase the contrast between the light center and the darker perimeter. Any cracks or imperfections will show up as very dark blotches. If you see any cracks or imperfections in the cladding, it's no problem because the cladding doesn't carry a signal. However, if you see cracks in the core, first try re-polishing the fiber on the 0.3 micron polishing film. If the crack still appears, you may have to cut the connector off and re-terminate it.

Finish

You can now terminate the other end of the cable. Then you can use a standard fiber-cable tester or optical time-domain reflectometer (OTDR) to test the cable. You will learn more about optical fiber testing in Chapter 15.

FIGURE 14.32
Insert the fiber into the fiber microscope.

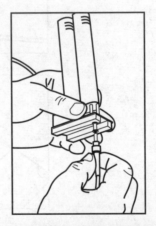

FIGURE 14.33
A sample fiber-tip image in a fiber-optic microscope

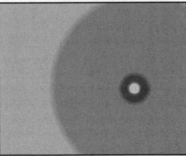

The Bottom Line

Install twisted-pair cable connectors.   This chapter covered detailed installation procedures for twisted-pair cable connectors. Although these are pretty straightforward, it is important to understand some key aspects that will make this easier.

> **Master It**
>
> **1.** When cutting a cable to meet a certain length, what is the recommended additional length of cable you should cut from the master reel? For example, if you are trying to connectorize a 5′ patch cable, how much should you cut?
>
> **2.** In step 5 of the process, what is the recommended total length of exposed conductor?

Install coaxial cable connectors.   Although less popular than either twisted-pair or fiber-optic cable connectors, you may still encounter the need for installing coaxial connectors for modern video systems. There are several types of connector mounting systems, and they have their advantages and disadvantages.

> **Master It**   Basically, two types of connectors exist: *crimp-on* and *screw-on* (also known as *threaded*). The crimp-on connectors require that you strip the cable, insert the cable into the connector, and then crimp the connector onto the jacket of the cable to secure it. Most BNC connectors used for LAN applications use this installation method. Screw-on connectors, on the other hand, have threads inside the connector that allow the connector to be screwed onto the jacket of the coaxial cable. These threads cut into the jacket and keep the connector from coming loose. Which one of these methods is considered less reliable and why? Which is the recommended connector type?

Install fiber-optic cable connectors.   This chapter covered detailed installation procedures for fiber-optic cable connectors. Similar to twisted-pair connectors, it is important to understand some key aspects that will make this easier.

> **Master It**
>
> **1.** What color surface should you use to perform the connector mounting and why?
>
> **2.** Fiber-optic cables typically have a specific yarn that will need to be cut back. What is this made of and what types of scissors are required?
>
> **3.** What is the proper size of the bead of epoxy to expel into the connector?
>
> **4.** Polishing is a key step when using epoxy connectors. What is the first step of the polishing process?

Chapter 15

Cable System Testing and Troubleshooting

Testing a cable installation is an essential part of both installing and maintaining a data network. This chapter will examine the cable testing procedures that you should integrate into the installation process and that you are likely to need afterward to troubleshoot network communication problems. We will also examine the standards with respect to cable testing.

In this chapter, you will learn to:

♦ Identify important cable plant certification guidelines

♦ Identify cable testing tools

♦ Troubleshoot cabling problems

Installation Testing

As you've learned in earlier chapters, installing the cable plant for a data network incorporates a large number of variables. Not only must you select the appropriate cable and other hardware for your applications and your environment, but you must also install the cable so that environmental factors have as little effect on the performance of the network as possible. Part of the installation process should include an individual test of each cable run to eliminate the cables as a possible cause of any problems that might occur later when you connect the computers to the network and try to get them to communicate. Even if you are not going to be installing or testing the cabling yourself, you should be familiar with the tests that the installers perform and the types of results that they receive.

Incorporating a cable test into the installation will help to answer several questions:

Connections   Have the connectors been attached to the cable properly? Have the wires been connected to the correct pins at both ends?

Cable performance   Is the cable free from defects that can affect performance?

Environment   Has the cable been properly routed around possible sources of interference, such as light fixtures and electrical equipment?

Certification   Does the entire end-to-end cable run, including connectors, wall plates, and other hardware, conform to the desired standard?

The following sections examine the tests that you can perform on copper and fiber-optic cables, the principles involved, and the tools needed. Realize, though, that you needn't perform every one of these tests on every cable installation. To determine which tests you need to perform and what results you should expect, see the section "Creating a Testing Regimen" later in this chapter.

Copper Cable Tests

Most of the copper cable installed today is twisted-pair of one form or another, and the number of individual wire connections involved makes the installation and testing process more complicated than for other cable, particularly in light of the various standards available for the connector pinouts. The following sections list the tests for copper cables and how they work.

NOTE For more information on the equipment used to perform these tests, see the section "Cable Testing Tools" later in this chapter. For more information on correcting the problems detected by these tests, see the section "Troubleshooting Cabling Problems," also later in this chapter.

WIRE MAPPING

Wire mapping is the most basic and obvious test for any twisted-pair cable installation. For twisted-pair cables, you must test each cable run to make sure that the individual wires within the cable are connected properly, as shown in Figure 15.1. As mentioned earlier in this book, you can select either the T568-A or T568-B pinout configurations for a twisted-pair installation. Because all of the pairs are wired straight through and the difference between the two configurations is minimal, there is no functional difference between them. However, you should select one pinout and stick to it throughout your entire installation. This way you can perform end-to-end tests as needed without being confused by mixed wire-pair colors.

FIGURE 15.1
A properly connected four-pair cable, using the T568-A pinout

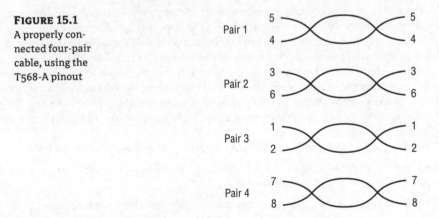

A perfunctory wire-mapping test can be performed visually by simply checking the pinouts at both ends of the cable. However, problems can occur that are not visible to the naked eye. A proper wire-mapping tester can detect any of the following faults:

Open pair An open pair occurs when one or more of the conductors in the pair are not connected to a pin at one or the other end. In other words, the electrical continuity of the conductor is interrupted. This can occur if the conductor has been physically broken, or because of incomplete or improper punch down on the IDC connector.

Shorted pair   A short occurs when the conductors of a wire pair are connected to each other at any location in the cable.

Short between pairs   A short between pairs occurs when the conductors of two wires in different pairs are connected at any location in the cable.

Reversed pair   A reversed pair (sometimes called a *tip/ring reversal*) occurs when the two wires in a single pair are connected to the opposite pins of the pair at the other end of the cable. For example, if the W-BL/BL pair is connected on one end with W-BL on pin 5 and BL on pin 4 of the connector, and at the other end of the cable, W-BL is connected to pin 4 and BL is punched down on pin 5, the W-BL/BL pair is reversed.

Crossed pairs   Crossed (or transposed) pairs occur when both wires of one color pair are connected to the pins of a different color pair at the opposite end.

Split pairs   Split pairs occur when the wire from one pair is split away from the other and crosses over a wire in an adjacent pair. Because this type of fault essentially requires that the same mistake be made at both ends of the connection, accidental occurrence of split pairs is relatively rare.

Figure 15.2 illustrates these faults.

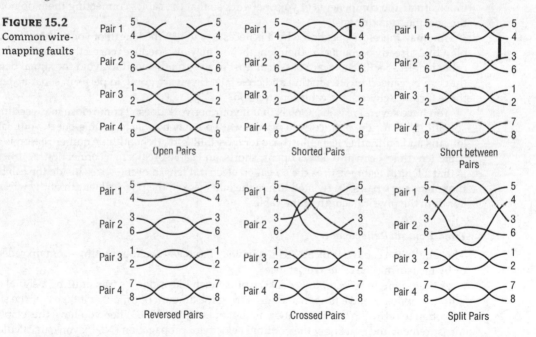

FIGURE 15.2
Common wire-mapping faults

NOTE   Figures 15.1 and 15.2 show the T568-A pinout configuration. If you are using the T568-B pinout, pairs 2 and 3 switch positions from the T568-A pinout.

Wire-mapping faults are usually caused by improper installation practices, although some problems like opens and shorts can result from faulty or damaged cable or connectors. The process of testing a connection's wire mapping is fairly straightforward and requires a remote unit that you attach at one end of the connection and a main unit for the other end. Wire-map testing is usually included in multifunction cable testers, but you can also purchase dedicated wire-map testers that are far less expensive.

The main unit simply transmits a signal over each wire and detects which pin at the remote unit receives the signal. The problem of split pairs (two wires in different pairs transposed at both ends of the connection) is the only one not immediately detectable using this method. Because each pin is connected to the correct pin at the other end of the connection, the wire map may appear to be correct and the connection may appear to function properly when it is first put into service. However, the transposition causes two different signals to run over the wires in a single twisted pair. This can result in an excess of near-end crosstalk that will cause the performance of the cable to degrade at high data rates. Although the occurrence of split pairs is relatively unlikely compared to the other possible wire-mapping faults, the ability to detect split pairs is a feature that you may want to check for when evaluating cable-testing products.

CABLE LENGTH

All LAN technologies are based on specifications that dictate the physical layer for the network, including the type of cable you can use and the maximum length of a cable segment. Cable length should be an important consideration from the very beginning of network planning. You must situate the components of your network so that the cables connecting them do not exceed the specified maximums.

You may therefore question why it is necessary to test the length of your cables if you have a plan that already accounts for their length. You may also deduce (correctly) that the maximum cable-length specifications are based, at least in part, on the need to avoid the signal degradation that can be caused by attenuation and crosstalk. If you are going to perform separate tests for these factors, why test the cable lengths, too?

You have several reasons. One is that if your network doesn't come close to exceeding the specifications for the protocol you plan to use, you may be able to double-check your cable lengths and omit other tests like those for crosstalk and attenuation. Another reason is that a cable-length test can also detect opens, shorts, and cable breaks in a connection. A third reason is that a length test measures the so-called electrical length of the wires inside the cable. Because the cable's wire pairs are twisted inside the outer jacket, the physical length of the wires is longer than the physical length of the cable.

Time-Domain Reflectometry

The length of a cable is typically tested in one of two ways: either by time-domain reflectometry or by measuring the cable's resistance.

A time-domain reflectometer (TDR) works much like radar, by transmitting a signal on a cable with the opposite end left open and measuring the amount of time that it takes for the signal's reflection to return to the transmitter, as shown in Figure 15.3. When you have this elapsed time measurement and you know the nominal velocity of propagation (NVP), you can calculate the length of the cable.

FIGURE 15.3
Time-domain reflectometry measures the time needed for a pulse to travel to the end of a cable and back.

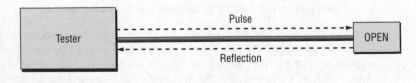

The NVP for a particular cable is usually provided by its manufacturer and expressed in relation to the speed of light. Some manufacturers provide the NVP as a percentage, such as 72 percent, whereas others express it as a decimal value multiplied by the speed of light (c), such as 0.72c. Many cable testers compute the length internally, based on the results of the TDR test and an NVP value that is either preprogrammed or that you specify for the cable you're testing.

When testing cable length, it's critically important that your tester uses the correct NVP value. The NVP values for various cables can range from 60 percent (0.6c) to 90 percent (0.9c), which creates a potential for error in the cable-length results ranging from 30 to 50 percent if the tester is using the wrong value. Time-domain reflectometry has other potential sources of inaccuracy as well. The NVP can vary as much as 4 to 6 percent between the different wire pairs in the same cable because of the deliberately varied twist intervals used to control crosstalk. The pulse generated by the TDR can be distorted from a square wave to one that is roughly sawtooth-shaped, causing a variance in the measured time delay of several nanoseconds, which converts to several feet of cable length.

Because of these possible sources of error, you should be careful when planning and constructing your network not to use cable lengths that closely approach or exceed the maximum recommended in your protocol specification.

Locating Cable Faults

Time-domain reflectometry has other applications in cable testing as well, such as the detection and location of cable breaks, shorts, and terminators. The reflection of the test pulse back to the transmitter is caused by a change in impedance on the cable. On a properly functioning cable, the open circuit at the opposite end produces the only major change in impedance. But if an open or short exists at some point midway in the cable run, it too will cause a reflection back to the transmitter. The size of the pulse reflected back is in direct proportion to the magnitude of the change in impedance, so a severe open or short will cause a larger reflection than a relatively minor fault, such as a kink, a frayed cable, or a loose connection. If there is no reflection at all, the cable has been terminated at the opposite end, which causes the pulse signal to be nullified before it can reflect back.

Cable testers use TDR to locate breaks and faults in cable by distinguishing between these various types of reflections. For example, a break located at 25′ in a cable run that should be at least 100′ long indicates that a fault in the cable exists and gives an indication of its approximate location (see Figure 15.4). The problem may be caused by a cable that has been entirely severed or by faulty or damaged wires inside the cable sheath. Sometimes you can't tell that a cable is faulty by examining it from the outside. This is why a test of each cable run during the installation process is so important.

FIGURE 15.4
TDRs are also used to locate breaks and other faults in a cable.

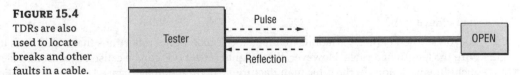

Resistance Measuring

The second method for determining the length of a cable is to measure its resistance using a digital multimeter (DMM). All conductors have a resistance specification, expressed in ohms per meter (or sometimes ohms per 100 meters or ohms per foot). If you know the resistance specification for the conductor per unit of length, you can measure the cable's resistance and divide the result by the manufacturer's specification to determine the cable's length. In the same way, if you already know the length of the cable from a TDR test, you can use the rating to determine the cable's total resistance.

Environmental factors can affect resistance, as can the cable's design and improper installation. Resistance increases with temperature, so your length calculations will suffer accordingly if you are measuring in a high- or low-temperature environment too far from the 20° C (68° F) temperature the resistance specification is based on. The twist intervals of the pairs also will influence your resistance measurement. Because the twists increase the actual length of the conductors, the resistance reading will be higher and result in a longer-than-actual cable length. And if the conductor was stretched during installation, high resistance readings will result, again producing longer-than-actual lengths for the cable.

PERFORMANCE TESTING

The tests we've discussed so far all relate to physical properties of the cable and ascertain if the cable has been terminated properly and is an acceptable length. They can be performed quickly and with relatively unsophisticated and inexpensive test devices. They are the basic, minimum levels of testing that should be performed to ensure your network will work.

But to properly characterize your cabling's performance, a battery of transmission tests must be administered; they determine the data-carrying capability of your cables and connectors. The following characteristics were all defined in Chapter 1, "Introduction to Data Cabling," so we won't explain them further here other than to note issues related to their testing.

All of the copper cable tests discussed in the following sections, except for propagation delay and delay skew, have formula-based performance requirements along a continuous-frequency spectrum. In the case of Category 5e, this range is 1MHz through 100MHz. For Category 6 and 6A, requirements at additional frequencies up to 250MHz and 500MHz are specified, respectively. If at any point along the spectrum the cable exceeds the specification limits, the cable fails. This is called *sweep* testing because the entire frequency range is being scanned.

Testing for transmission performance requires much more sophisticated equipment than that used for wire mapping, opens, shorts, and crosses—equipment that can cost several thousands of dollars per test set. However, the testing is essential for qualifying your cabling installation to a particular level of performance, such as Category 5e. If you can't afford such a set, either contract with an installation-and-testing company that has one, or rent an appropriate unit.

Impedance

As you learned earlier, variations in impedance cause signal reflections that a TDR uses to measure the length of a cable. However, these signal reflections can be caused by different factors, including variations in the cable manufacture, structural damage caused during installation, or connectors that are a poor match for the cable. The statistic that measures the uniformity of the cable's impedance is called its *structural return loss (SRL)*, which is measured in decibels (dB), with higher values indicating a better cable. Even when the SRL of a particular cable is acceptable, it is still possible for an installation of that cable to suffer from variations in impedance that cause signal reflections. When you construct a network to conform to a particular cabling specification, such as Category 5e UTP, to maintain a consistent level of impedance throughout the entire length of the cable run you have to use connectors and other hardware that have the same rating as the cable.

If, for example, during a twisted-pair installation, you fail to maintain the twist of the wire pairs up to a point no more than 0.5″ for Category 5e and 0.375″ for Category 6 and 6A from each connection, you run the risk of varying the impedance to the point at which a reflection occurs (as well as causing additional crosstalk). The cumulative amount of reflection caused by variations in impedance on a cable run is called its *return loss*, which, like impedance, is measured in ohms. If the return loss is too large, signal-transmission errors can occur at high transmission speeds. The worst-pair performance is reported at the frequency where the result came closest to the specified limits.

Attenuation

Attenuation is one of the most important specifications for high-speed networks; if it is too high, the signals can degrade prematurely and data can be lost. This is especially true if your network uses cable lengths that approach the maximum permitted by your networking protocol.

Testing the attenuation of a cable run requires a unit at both ends of the connection: one to transmit a calibrated signal and another to receive the signal and calculate how much it has degraded during the trip. Attenuation is measured in decibels (dB), and most good-quality cable testers include the secondary module needed to perform the test. The worst-case result is reported.

Near-End Crosstalk (NEXT)

Along with attenuation, near-end crosstalk (NEXT) is one of the major impediments to successfully installing and running a high-speed data network on twisted-pair cabling. Figure 15.5 shows NEXT. Testing for NEXT is a relatively simple process with today's sophisticated test sets. After terminating the far end of the cable run to prevent any reflections from interfering with the test, a signal is transmitted over one pair, and the magnitude of the crosstalk signal is measured on the other pairs (in decibels). For a complete assessment, you must test each wire pair against each of the three other pairs, for a total of six tests, and you must perform the six tests from both ends of the cable. The worst-case combination is reported as the cable's performance result.

FIGURE 15.5
Near-end crosstalk

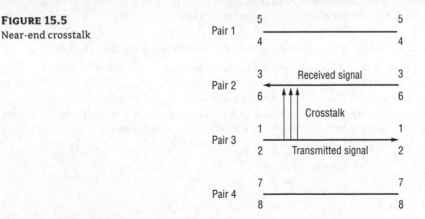

Power-Sum NEXT

Power-sum NEXT (sometimes called PS-NEXT) is a measurement of the cumulative effect of crosstalk on each wire pair when the other three pairs are transmitting data simultaneously. Figure 15.6 shows PS-NEXT. Each pair is tested separately, yielding four results. This test must also be performed at each end of the cable, and the worst-case result is reported.

FIGURE 15.6
Some high-speed protocols can generate excessive crosstalk by transmitting over two wire pairs at once.

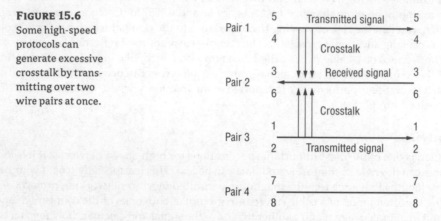

Attenuation to (Near-End) Crosstalk Ratio (ACR)

The attenuation-to-crosstalk ratio near end disturbances is a calculation used by cabling system manufacturers as a measure of the signal-to-noise ratio, with the noise being NEXT. ACR is the difference between the insertion loss for the disturbed cable run and the amount of NEXT it exhibits, both of which are measured in decibels. It is not specified in TIA 568-C, but the ACR is one of the best measurements of a cable run's overall quality, because it clearly indicates how robust the signal will appear in relation to the noise in the cable. Crosstalk varies at either end of the cable run, so you must run the test from both ends. The worst of the ACR measurements is the rating for the cable run.

Attenuation to Power-Sum (Near-End) Crosstalk Ratio (PSACR)

Another figure of merit is PSACR (Power-Sum ACR), which is the ratio of the signal to the PSNEXT noise. It is calculated the same way as ACR and it is not specified in ANSI/TIA-568-C. PSACR is another good measure of channel signal-to-noise ratio.

Far-End Crosstalk (FEXT) and Attenuation to Crosstalk Ratio Far End (ACRF)

Far-end crosstalk (FEXT) occurs when a signal crosses over to another wire pair as it approaches the far end of the cable, opposite the system that transmitted it. ANSI/TIA-568-C does not specify FEXT, but instead specifies ACRF, formerly known as Equal-Level Far-End Crosstalk (ELFEXT). The FEXT measurements are "equalized" for the amount of attenuation present, by subtracting the attenuation value from the FEXT value to achieve the "equal-level" FEXT (ELFEXT) or ACRF, which is the equivalent of the ACR for the far end of the cable.

Power-Sum FEXT (PSFEXT) and PSACRF

ANSI/TIA-568-C specifies the PS-ACRF (or PS-ELFEXT) test, which is a combined or power-sum measurement for all of the wire pairs in the cable, and a worst pair-to-pair ELFEXT test. In most cases, these measurements are not vital to an installation test, but some technologies, such as Gigabit Ethernet, require them. Testing must be done from both ends of the cable (each end of the cable is attached to a transceiver, so at some point each end is the far end), and the worst-case combinations are reported.

Propagation Delay and Delay Skew

The length of time required for a signal to travel from one end of a cable run to the other, usually measured in nanoseconds (ns), is its propagation delay. Because of the different twist rates used, the lengths of the wire pairs in a cable can vary. As a result, the propagation delay for each wire pair can be slightly different. When your network is running a protocol that uses only one pair of wires to transmit data, such as 100Base-TX Ethernet or Token Ring, these variations are not a problem. However, protocols that transmit over multiple pairs simultaneously, such as 100Base-T4, 1000Base-T, and 10GBase-T Ethernet, can lose data when signals traveling over the different pairs arrive too far apart in time.

To quantify this variation, some testers can measure a cable run's delay skew, which is the difference between the lowest and the highest propagation delay for the wire pairs within a cable. Propagation delay and delay skew are characteristics critical to some high-speed LAN applications, so they should be included in your battery of tests, especially for a network that will run one of the high-speed protocols that uses multiple pairs. For propagation delay, the worst pair is reported; for delay skew, it is the worst combination of any two pairs.

Noise

Most cable tests attempt to detect and quantify problems that result from the effects of the installation on the cable's own characteristics. However, environmental factors can also affect the functionality of the cable installation, and you should be sure to test each cable run for noise that emanates from outside sources. Outside noise is usually generated either by EMI, which is low-frequency, high-amplitude noise generated by AC power lines, electric motors, and fluorescent lights, or *radio frequency interference (RFI)*, which is high-frequency, low-amplitude noise created by radio and television sets and cell phones. Once again, this type of noise is usually not a problem on lower-speed networks, but it can be on protocols that run the network at 100MHz or more.

Testing for outside noise is a matter of shutting off or detaching all devices on the LAN and testing the cable for electrical activity. One of the most important elements of this kind of test is to perform it when all of the equipment at the site is operating as it normally does during work hours. For example, performing a noise test on an office network during the weekend, when most of the lights, copiers, coffee machines, air conditioners, and other equipment are shut down, will not give you an accurate reading.

Fiber-Optic Tests

Just as installing fiber-optic cable is completely different from installing copper cable, the testing processes also differ greatly. Much of the copper cable testing revolves around the various types of interference that can affect the performance of a network. Fiber-optic cable is completely immune from interference caused by crosstalk, EMI, and RFI, however, so tests for these are not needed. For a fiber-optic installation, you need to ensure that the signals arrive at their destinations with sufficient strength to be read and that the installation process has not degraded that strength.

Because of its superior signal-carrying capabilities, a fiber-optic cable installation can include various types of cable runs. The typical LAN arrangement consists of single-fiber links that connect a patch panel in a wiring closet or data center to wall plates or other individual equipment sites over relatively short distances, with patch cables at both ends to connect to a backbone network and to computers or other devices. Because of the limited number of connections they use, testing these types of links is fairly straightforward. However, fiber-optic cables can also support extremely long cable runs that require splices every two to four kilometers, which introduces a greater potential for connection problems.

To completely test a fiber-optic installation, you should perform your battery of tests three times. The first series of tests should be on the spooled cable before the installation to ensure that no damage occurred during shipping. The installation costs for fiber-optic cable can be high—often higher than the cost of the cable and other hardware—so it's worthwhile to test the cable before investing in its installation. Because excessive signal loss is caused mostly by the connections, simply testing the continuity of the cable at this stage is usually sufficient. This continuity testing is sometimes referred to as a "flashlight" test because it amounts to shining light in one end of the fiber strand and seeing if there is light at the other end.

The second series of tests should be performed on each separate cable segment as you install it, to ensure that the cable is not damaged during the installation and that each individual connector is installed correctly. By testing at this stage, you can localize problems immediately, rather than trying to track them down after the entire installation is completed.

Finally, you should test the entire end-to-end connection, including all patch cables and other hardware, to ensure that cumulative loss is within certified parameters.

For fiber-optic LAN installations, only two tests are generally required: optical power and signal loss. The following sections examine these tests and how you perform them. Other types of tests are used on long-distance fiber-optic links and in troubleshooting, which are much more complex and require more elaborate equipment. For more information on these, see the section "Cable Testing Tools" later in this chapter.

OPTICAL POWER

The most fundamental test of any fiber-optic-cable plant is the *optical-power test*, as defined in the EIA's FOTP-95 standard, which determines the strength of the signal passing through a cable run and is the basis for a loss-measurement (attenuation) test. The testing process involves connecting a fiber-optic power meter to one end of the cable and a light source to the other. The power meter uses a solid-state detector to measure the average optical power emanating from the end of the cable, measured in decibels. For data networks using multimode cable, you should perform optical-power tests at 850 and 1,300nm wavelengths; many testers run their tests at both settings automatically. Single-mode cables require a 1,300nm test and sometimes 1,550nm as well. The 1,550nm test determines whether the cable will support wavelength division multiplexing and can detect losses due to macrobending, which are not apparent at 1,300nm.

TIP Some people claim that you can use an optical time-domain reflectometer (OTDR) to test optical power and cable-plant loss. Although not necessary, the combination of a fiber-optic power meter and OTDR can provide other information such as splice and insertion loss at connections in addition to measuring optical power and signal loss.

Loss (Attenuation)

Loss testing, along with optical power, are the two most important tests for any fiber-optic cable installation. *Loss* is the term commonly used in the fiber-optic world for attenuation; it is the decrease in signal strength as light travels through the cable. The physics of optical transmission make it less susceptible to attenuation than any copper cable, which is why fiber cable segments can usually be much longer than copper ones. However, even if your network does not have extremely long fiber cable runs, there can be a significant amount of loss, not because of the cable, but because of the connections created during the installation. Loss testing verifies that the cables and connectors were installed correctly.

Measuring the loss on a cable run is similar in practice to measuring its optical power, except that you use a calibrated light source to generate the signal and a fiber-optic power meter to measure how much of that signal makes it to the other end. The combination of the light source and the power meter into one unit is called an *optical loss test set (OLTS)*. Because of the different applications that use fiber-optic cable, you should be sure to use test equipment that is designed for your particular type of network. For example, a light source might use either a laser or an LED to create the signal, and the wavelengths it uses may vary as well. For a fiber-optic LAN, you should choose a product that uses a light source at wavelengths the same as the ones your network equipment will use so that your tests generate the most accurate results possible.

The testing procedure begins with connecting the light source to one end of a *reference test cable* (also called the *launch cable*) and the power meter to the other end. The reference test cable functions as a baseline against which you measure the loss on your installed cable runs and should use the same type of cable as your network. After measuring the power of the light source over the reference test cable, you disconnect the power meter, connect the reference cable to the end of the cable you want to test, and connect the power meter to the other end. Some testers include a variety of adapters to accommodate various connector types. By taking another power reading and comparing it to the first one, you can calculate the loss for the cable run. As with the optical-power test, you should use both 850 and 1,300nm wavelengths for multimode fiber tests, and you should also test the cable from the other direction in the same way. When you have the results, compare them to the *optical loss budget (OLB)*, which is the maximum amount of signal loss permitted for your network and your application. (You may occasionally see *optical link budget*; though *optical loss budget* is preferred, the two terms are synonymous.)

WARNING Be sure to protect your reference test cables from dirt and damage. A faulty reference cable can produce false readings of high loss in your tests.

This type of test effectively measures the loss in the cable and in the connector to which the reference test cable is attached. The connection to the power meter at the other end introduces virtually no additional signal loss. To test the connectors at both ends of the cable run, you can add a second reference test cable to the far end, which is called a *receive cable*, and connect the power meter to it. This is known as a *double-ended loss test*. The type of test you perform depends on the type of cable run you are testing and the standard you're using as the model for your network.

The standard single-ended loss test is described in the FOTP-171 standard, which was developed by the EIA in the 1980s and intended for testing patch cables. The double-ended loss test for multimode cables is defined in the OFSTP-14 standard and is used to test an installed cable run. Another standard, the OFSTP-7, defines testing specifications for single-mode cables. The document describes two testing methods: the double-ended source/meter test from OFSTP-14 and an OTDR test, but when the results differ (and they usually will), the source/meter test is designated as definitive.

WARNING   Some older manuals recommend that you calibrate your power meter using both launch and receive cables, connected by a splice bushing, when you perform double-ended loss tests. This practice introduces additional attenuation into your baseline and can obscure the fact that one of your reference test cables is dirty or damaged. Always establish your testing baseline using a launch cable only.

Depending on the capabilities of your equipment, the loss-testing process might be substantially easier. Some power meters have a *zero loss reference* capability, meaning that you can set the meter to read 0dB while measuring the reference test cable. Then, when you test the installed cable run, the meter displays only the loss in decibels; no calculation is necessary.

Cable Plant Certification

So far in this chapter, you've learned about the types of tests you can perform on a cable installation but not about which tests you should perform for a particular type of network—or the results that you should expect from these tests. The tests you perform and the results you receive enable you to certify your cable installation as complying with a specific standard of performance. Many of the high-quality cable testers on the market perform their various tests automatically and provide you with a list of pass/fail results, but it is important to know not only what is being tested but also what results the tester is programmed to evaluate as passes and failures.

Changing standards and new technologies can affect the performance levels that you should expect and require from your network, and a tester that is only a year or two old may yield results that ostensibly pass muster but that are actually insufficient for the network protocol you plan to run. Always check to see what specifications a tester is using to evaluate your cable's performance. With some testers, the results that determine whether a cable passes or fails a test can be calibrated with whatever values you wish, whereas others are preprogrammed and cannot easily be changed. Obviously, the former is preferable, as it enables you to upgrade the tester to support changing standards.

The certification that you expect your network to achieve should be based not only on today's requirements but also on your expectation of future requirements. Professional consultants recommended that clients install Category 5e and 6 cable for many years, long before most of these clients even considered upgrading to Gigabit Ethernet or another technology that required Category 5e or 6. This was because the additional investment for a Category 5e or 6 installation was then minimal compared to the cost of completely recabling the network later on.

For the same reason, it may be a good idea for the cabling you install today to conform to the requirements for technologies you're not yet considering using. A few years ago, Gigabit Ethernet was a new and untried technology, yet now it is commonplace. It makes sense to assume that 10 Gigabit Ethernet will be just as common a few years from now. Installing Category 6A cable now and certifying it to conform to the highest standards currently available may not benefit you today, but in future years you may be proud of your foresight.

Creating a Testing Regimen

The level of performance that you require from a cable installation should specify which tests you have to perform during the installation process and what test results are acceptable. For example, a UTP installation intended only for voice-telephone traffic requires nothing more than a wire-mapping test to ensure that the appropriate connections have been made. Other factors will probably not affect the performance of the network sufficiently to warrant testing them. A data network, on the other hand, requires additional testing, and as you increase the speed at which data will travel over the network and the number of wire pairs used, the need for more extensive testing increases as well. Table 15.1 lists the most common Data Link layer protocols used on copper cable networks and the corresponding tests you should perform on a new cable installation.

TABLE 15.1   Minimum Cable Tests Required for Copper-Based Networking Protocols

NETWORK TYPE	TESTS REQUIRED
Voice telephone	Wire mapping
10Base-T Ethernet	Wire mapping, length, attenuation, NEXT
100Base-TX	Wire mapping, length, attenuation, NEXT, propagation delay, delay skew
Token Ring	Wire mapping, length, attenuation, NEXT
TP-PMD FDDI	Wire mapping, length, attenuation, NEXT
155Mbps ATM	Wire mapping, length, attenuation, NEXT
Gigabit Ethernet	Wire mapping, length, attenuation, NEXT, propagation delay, delay skew, PS-NEXT, ACR, PSACR, ACRF, PS-ACRF, return loss, insertion loss
10 Gigabit Ethernet	Wire mapping, length, attenuation, NEXT, propagation delay, delay skew, PS-NEXT, ACR, PSACR, ACRF, PS-ACRF, PSANEXT, PSAACRF, return loss, insertion loss

For fiber-optic cable installations, optical-power and loss testing is usually sufficient for all multimode fiber LANs. For 10GBase-SX installations where the optical loss budget is very tight, using an OTDR can be very helpful in locating where in the channel the light is being lost. For single-mode networks with long cable runs, OTDR testing is also recommended, as described in the section "Optical Time-Domain Reflectometers" later in this chapter.

Copper Cable Certification

The ANSI/TIA-568-C standard includes performance requirements for horizontal and backbone cabling. The standard defines two types of horizontal links for the purposes of testing. The *permanent link* refers to the permanently installed cable connection that typically runs from a wall plate at the equipment site to a patch panel in a wiring closet or data center. The *channel link* refers to the complete end-to-end cable run, including the basic link and the patch cables used to connect the equipment to the wall plate and the patch-panel jack to the hub or other device.

Table 15.2 and Table 15.3 summarize some performance levels required for copper cable testing, broken down by cable category at selected frequencies, for permanent link and channel link testing, respectively. For a full set of requirements, see ANSI/TIA-568-C.2.

TABLE 15.2   TIA Permanent-Link Testing Performance Standards

	CATEGORY 3	**CATEGORY 5E**	**CATEGORY 6**	**CATEGORY 6A**
Wire mapping	All pins properly connected	All pins properly connected	All pins properly connected	All pins properly connected
Length (in meters, not including tester cords)	< 90	< 90	< 90	< 90
NEXT (dB)				
@ 1 MHz	40.1	60.0	65.0	65.0
@ 10 MHz	24.3	48.5	57.8	57.8
@ 16 MHz	21.0	45.2	54.6	54.6
@ 100 MHz	N/A	32.3	41.8	41.8
@ 250 MHz	N/A	N/A	35.3	35.3
@ 500 MHz	N/A	N/A	N/A	26.7
PS-NEXT (dB)				
@ 1 MHz	N/A	57.0	62.0	62.0
@ 10 MHz	N/A	45.5	55.5	55.5
@ 100 MHz	N/A	29.3	39.3	39.3
@ 250 MHz	N/A	N/A	32.7	32.7
@ 500 MHz	N/A	N/A	N/A	23.8

TABLE 15.2   TIA Permanent-Link Testing Performance Standards   *(CONTINUED)*

	CATEGORY 3	CATEGORY 5E	CATEGORY 6	CATEGORY 6A
ACRF (dB)				
@ 1MHz	N/A	58.6	64.2	64.2
@ 10MHz	N/A	38.6	44.2	44.2
@ 100MHz	N/A	18.6	24.2	24.2
@ 250MHz	N/A	N/A	16.2	16.2
@ 500MHz	N/A	N/A	N/A	10.2
PS-ACRF (dB)				
@ 1MHz	N/A	55.6	61.2	61.2
@ 10MHz	N/A	35.6	41.2	41.2
@ 100MHz	N/A	15.6	21.2	21.2
@ 250MHz	N/A	N/A	13.2	13.2
@ 500MHz	N/A	N/A	N/A	7.2
PS-ANEXT (dB)				
@ 1MHz	N/A	N/A	N/A	67.0
@ 10MHz	N/A	N/A	N/A	67.0
@ 100MHz	N/A	N/A	N/A	60.0
@ 250MHz	N/A	N/A	N/A	54.0
@ 500MHz	N/A	N/A	N/A	49.5
PS-AACRF (dB)				
@ 1MHz	N/A	N/A	N/A	67.0
@ 10MHz	N/A	N/A	N/A	57.7
@ 100MHz	N/A	N/A	N/A	37.7
@ 250MHz	N/A	N/A	N/A	29.7
@ 500MHz	N/A	N/A	N/A	23.7

TABLE 15.2   TIA Permanent-Link Testing Performance Standards   *(CONTINUED)*

	CATEGORY 3	**CATEGORY 5E**	**CATEGORY 6**	**CATEGORY 6A**
Minimum Return Loss (dB)				
@ 1MHz	N/A	19.0	19.1	19.1
@ 10MHz	N/A	19.0	21.0	21.0
@ 100MHz	N/A	12.0	14.0	14.0
@ 250MHz	N/A	N/A	10.0	10.0
@ 500MHz	N/A	N/A	N/A	8.0
Maximum Insertion Loss (dB)				
@ 1MHz	3.5	2.1	1.9	1.9
@ 10MHz	9.9	6.2	5.5	5.5
@ 16MHz	13.0	7.9	7.0	7.0
@ 100MHz	N/A	21.0	18.6	18.0
@ 250MHz	N/A	N/A	31.1	29.5
@ 500MHz	N/A	N/A	N/A	43.8
Propagation delay @ 10MHz	< 498ns	< 498ns	< 498ns	< 498ns
Delay skew @ 10MHz	N/A	< 44ns	< 44ns	< 44ns

TABLE 15.3   TIA Channel-Link Testing Performance Standards

	CATEGORY 3	**CATEGORY 5E**	**CATEGORY 6**	**CATEGORY 6A**
Wire mapping	All pins properly connected	All pins properly connected	All pins properly connected	All pins properly connected
Length (in meters, not including tester cords)	< 100	< 100	< 100	< 100

TABLE 15.3   TIA Channel-Link Testing Performance Standards   *(CONTINUED)*

	CATEGORY 3	CATEGORY 5E	CATEGORY 6	CATEGORY 6A
NEXT (dB)				
@ 1MHz	39.1	60.0	65.0	65.0
@ 10MHz	22.7	47.0	56.6	56.6
@ 16MHz	19.3	43.6	53.2	53.2
@ 100MHz	N/A	30.1	39.9	39.9
@ 250MHz	N/A	N/A	33.1	33.1
@ 500MHz	N/A	N/A	N/A	26.1
PS-NEXT (dB)				
@ 1MHz	N/A	57.0	62.0	62.0
@ 10MHz	N/A	44.0	54.0	54.0
@ 100MHz	N/A	27.1	37.1	37.1
@ 250MHz	N/A	N/A	30.2	30.2
@ 500MHz	N/A	N/A	N/A	23.2
ACRF (dB)				
@ 1MHz	N/A	57.4	63.3	63.3
@ 10MHz	N/A	37.4	43.3	43.3
@ 100MHz	N/A	17.4	23.3	23.3
@ 250MHz	N/A	N/A	15.3	15.3
@ 500MHz	N/A	N/A	N/A	9.3
PS-ACRF (dB)				
@ 1MHz	N/A	54.4	60.3	60.3
@ 10MHz	N/A	34.4	40.3	40.3
@ 100MHz	N/A	14.4	20.3	20.3
@ 250MHz	N/A	N/A	12.3	12.3
@ 500MHz	N/A	N/A	N/A	6.3

TABLE 15.3 TIA Channel-Link Testing Performance Standards *(CONTINUED)*

	CATEGORY 3	CATEGORY 5E	CATEGORY 6	CATEGORY 6A
PS-ANEXT (dB)				
@ 1MHz	N/A	N/A	N/A	67.0
@ 10MHz	N/A	N/A	N/A	67.0
@ 100MHz	N/A	N/A	N/A	62.5
@ 250MHz	N/A	N/A	N/A	56.5
@ 500MHz	N/A	N/A	N/A	52.0
Minimum Return Loss (dB)				
@ 1MHz	N/A	17.0	19.0	19.0
@ 10MHz	N/A	17.0	19.0	19.0
@ 100MHz	N/A	10.0	12.0	12.0
@ 250MHz	N/A	N/A	8.0	8.0
@ 500MHz	N/A	N/A	N/A	6.0
Maximum Insertion Loss (dB)				
@ 1MHz	4.2	2.2	2.1	2.3
@ 10MHz	11.5	7.1	6.3	6.5
@ 16MHz	14.9	9.1	8.0	8.2
@ 100MHz	N/A	24.0	21.3	20.9
@ 250MHz	N/A	N/A	35.9	33.9
@ 500MHz	N/A	N/A	N/A	49.3
Propagation delay @ 10MHz	N/A	< 555ns	< 555ns	< 555ns
Delay skew @ 10MHz	N/A	<50ns	<50ns	<50ns

Fiber-Optic Certification

The main requirement of testing an installed fiber-optic cable link is ensuring that the insertion loss (IL) of the installed link meets the insertion loss budget (ILB) required for the intended application. The insertion loss is the total loss of the link comprising the loss of the connectors and splices, in the cable run, and the attenuation of the length of the cable in the link. The basic formula for computing insertion loss is as follows:

IL = connector loss + splice loss + cable loss

Test methods and equipment are available that will measure the insertion loss of the fiber-optic link (discussed in the "Cable Testing Tools" section). After testing the signal loss generated by a fiber-optic cable channel, you compare the results to the ILB for the channel to determine if the installation is within performance parameters.

The ILB depends on the application and is dictated by the application standards. Table 15.4 shows the ILBs for some key Ethernet applications.

TABLE 15.4 Insertion Loss Budgets for Key Ethernet Applications

APPLICATION	DATA RATE	DESIGNATION	STANDARD	INSERTION LOSS BUDGET (DB)
Ethernet	10Mbps	10BaseE-FL	IEEE 802.3	12.5
Fast Ethernet	100Mbps	100Base-FX	IEEE 802.3	11.0
Short Wavelength Fast Ethernet	10/100Mbps	100Base-SX	TIA/EIA-785	4.0
1 Gigabit Ethernet	1,000Mbps	1000Base-SX	IEEE 802.3z	3.56
10 Gigabit Ethernet	10,000Mbps	10GBase-SR	IEEE 802.3ae	2.60

Essentially, the ILB is calculated by the amount of acceptable loss for the length of the cable and for the number of splices and connectors. This is done by multiplying the rated length of cable and the number of splices and connectors by predefined loss coefficients. These coefficients vary according to the type of fiber cable you're using, the wavelength of the network, and the standard you adhere to. For the connectors, loss coefficients range from 0.5 to a maximum of 0.75dB. For the splices, loss coefficients are 0.2 or 0.3dB. For the cable-attenuation coefficient, use the values listed in Table 15.5. Using these coefficient values, the insertion loss can be approximated; however, the cable attenuation needs to be measured in the installed link and the results need to be compared to the requirements in Table 15.4.

TABLE 15.5   Cable Coefficients for Insertion Loss Calculations

FIBER TYPE	850NM	1,300NM	1,550NM
Multimode fiber	3 to 3.5dB/km	1 to 1.5dB/km	N/A
Single-mode fiber	N/A	0.4dB/km	0.3dB/km

For example, the ILB for 10GBase-SR was calculated based on a link comprising two connections in a 300 meter link. The following equation was used to set the insertion loss budget to 2.6dB:

ILB = (2 × 0.75db) + (0.3km × 3.5dB/km) = 2.6dB

CABLING @ WORK: LOSS BUDGETS GETTING MORE STRINGENT

As you can see in Table 15.4, the loss budgets for 850nm-based applications have been decreasing as network speeds have been increasing. Part of this is due to the rated cable length decreasing with increasing speed. However, as the budget decreases, the connection and splice loss component becomes a larger percentage of the total budget. We have seen many customers who have been having difficulty adding the desirable number of connections in 10Gbps 10GBase-SR links using OM3 multimode fiber.

Fiber cable manufacturers have found solutions. One such solution involves using a fiber that is rated for a greater distance than what would be required in the actual installed link. Interestingly, this creates *bandwidth headroom*. This headroom can be translated into an additional loss budget factor that can be added to the original budget of 2.6dB.

For example, many manufacturers promote a multimode fiber with an 850nm bandwidth of 4,700MHz-km (at the time of this writing, TIA is working on standardizing this fiber as an OM4 fiber) for use in 10GBase-SR links. These fibers are typically rated to 550 meters at 10Gbps. However, if this fiber is installed in a 300 meter 10GBase-SR link (normally requiring a 2,000MHz-km bandwidth OM3 multimode fiber), there is extra bandwidth headroom of about 2,700MHz-km. This can be converted to an additional ILB of 1.9dB. Adding this to the original 2.6dB ILB creates a total ILB of 4.5dB! This has allowed many network users to add additional connections or splice points in their 10Gbps fiber links (although, strictly speaking, it is not allowed by the standards yet).

Third-Party Certification

Testing your cable installation for compliance to a specific performance level is a great way to ensure that the cable plant will support the networking protocol you plan to run. If you have installed the cabling yourself, testing is needed to check your work, and if you have the cable installed by a third party, testing ensures that the job was done correctly. Handheld testers can perform a comprehensive battery of tests and provide results in a simple pass/fail format, but these results depend on what standards the device is configured to use.

In most cases, it isn't difficult to modify the parameters of the tests performed by these devices (either accidentally or deliberately) so that they produce false positive results. Improper use of these devices can also introduce inaccuracies into the testing process. As a general rule, it isn't a good idea to have the same people check the work that they performed. This is not necessarily an accusation of duplicity. It's simply a fact of human nature that intimate familiarity with something can make it difficult to recognize faults in it.

For these reasons, you may want to consider engaging a third-party testing-and-certification company to test your network after the cable installation is completed and certify its compliance with published standards. If your own people are installing the cable, then this is a good way to test their work thoroughly without having to purchase expensive testing equipment. If your cable will be installed by an outside contractor, adding a clause into the contract that states acceptance of the work is contingent on the results of an independent test is a good way of ensuring that you get a quality job, even if you accept the lowest bid. What contractor would be willing to risk having to reinstall an entire network?

Cable Testing Tools

The best method for addressing a faulty cable installation is to avoid the problems in the first place by purchasing high-quality components and installing them carefully. But no matter how careful you are, problems are bound to arise. This section covers the tools that you can use to test cables both at the time of their installation and afterward, when you're troubleshooting cable problems. Cable testing tools can range from simple, inexpensive, mechanical devices to elaborate electronic testers that automatically supply you with a litany of test results in an easy-to-read pass/fail format.

The following sections list the types of tools available for both copper and fiber-optic cable testing. This is not to say that you need all of the tools listed here. In fact, in some of the following sections, we attempt to steer you away from certain types of tools. In some cases, both high-tech and low-tech devices are available that perform roughly the same function, and you can choose which you prefer according to the requirements of your network, your operational budget, or your temperament. Some of the tools are extremely complicated and require extensive training to use effectively, whereas others are usable by anyone who can read.

You should select the types of tools you need based on the descriptions of cable tests given earlier in this chapter, the test results required by the standards you're using, and the capabilities of the workers—not to mention the amount of money you want to spend.

Wire-Map Testers

A *wire-map tester* transmits signals through each wire in a copper twisted-pair cable to determine if it is connected to the correct pin at each end. Wire mapping is the most basic test for twisted-pair cables because the eight separate wire connections involved in each cable run are a common source of installation errors. Wire-map testers detect transposed wires, opens (broken or unconnected wires), and shorts (wires or pins improperly connected to each other)—all problems that can render a cable run inoperable.

Wire-map testing is nearly always included in multifunction cable testers, but in some cases it may not be worth the expense to spend thousands of dollars on a comprehensive device.

Dedicated wire-map testers are relatively inexpensive and enable you to test your installation for the most common faults that occur during installation and afterward. If you are installing voice-grade cable, for example, a simple wire-mapping test may be all that's needed. Slightly more expensive devices do wire-map testing in addition to other basic functions, such as TDR length testing.

A wire-map tester consists of a remote unit that you attach to the far end of a connection and the battery-operated, handheld main unit that displays the results. Typically, the tester displays various codes to describe the type of faults it finds. In some cases, you can purchase a tester with multiple remote units that are numbered, so that one person can test several connections without constantly traveling back and forth from one end of the connections to the other to move the remote unit.

WARNING   The one wiring fault that is not detectable by a dedicated wire-map tester is a split pair, because even though the pinouts are incorrect, the cable is still wired straight through. To detect split pairs, you must use a device that tests the cable for the near-end crosstalk that split pairs cause.

Continuity Testers

A *continuity tester* is an even simpler and less expensive device than a wire-map tester. It is designed to check a copper cable connection for basic installation problems, such as opens, shorts, and crossed pairs. These devices usually cannot detect more complicated twisted-pair wiring faults such as split pairs, but they are sufficient for basic cable testing, especially for coaxial cables, which have only two conductors that are not easily confused by the installer. Like a wire-map tester, a continuity tester consists of two separate units that you connect to each end of the cable to be tested. In many cases, the two units can snap together for storage and easy testing of patch cables.

Tone Generators

The simplest type of copper cable tester is also a two-piece unit, a *tone generator and probe*, also sometimes called a *fox and hound* wire tracer. With a standard jack, you connect to the cable the unit that transmits a signal, or, with an alligator clip, you connect the unit to an individual wire. The other unit is an inductive amplifier, which is a penlike probe that emits an audible tone when touched to the other end of the conductor.

This type of device is most often used to locate a specific connection in a punch-down block. For example, some installers prefer to run all of the cables for a network to the central punch-down block without labeling them. Then they use a tone generator to identify which block is connected to which wall plate and label the punch-down block accordingly. You can also use the device to identify a particular cable at any point between the two ends. Because the probe can detect through the sheath the cable containing the tone signal, you can locate one specific cable out of a bundle in a ceiling conduit or other type of raceway. Connect the tone generator to one end and touch the probe to each cable in the bundle until you hear the tone.

In addition, by testing the continuity of individual wires using alligator clips, you can use a tone generator and probe to locate opens, shorts, and miswires. An open wire will produce no tone at the other end, a short will produce a tone on two or more wires at the other end, and an improperly connected wire will produce a tone on the wrong pin at the other end.

Using a tone generator is extremely time-consuming, however, and it's nearly as prone to errors as the cable installation. You either have to continually travel from one end of the cable to the other to move the tone generator unit or use a partner to test each connection, keeping in close contact using radios or some other means of communication. When you consider the time and effort involved, you will probably find that investing in a wire-map tester is a more practical solution.

Time-Domain Reflectometers

As described earlier in the section "Cable Length," a time domain-reflectometer (TDR) is the primary tool used to determine the length of a copper cable and to locate the impedance variations that are caused by opens, shorts, damaged cables, and interference with other systems. Two basic types of TDRs are available: those that display their results as a waveform on an LCD or CRT screen and those that use a numeric readout to indicate the distance to a source of impedance. The latter type of TDR provides less detail but is easy to use and relatively inexpensive. Many of the automated copper cable testers on the market have a TDR integrated into the unit. Waveform TDRs are not often used for field testing these days because they are much more expensive than the numeric type and require a great deal more expertise to use effectively.

You can use a TDR to test any kind of cable that uses metallic conductors, including the coaxial and twisted-pair cables used to construct LANs. A high-quality TDR can detect a large variety of cable faults, including open conductors; shorted conductors; loose connectors; sheath faults; water damage; crimped, cut, or smashed cables; and many other conditions. In addition, the TDR can measure the length of the cable and the distance to any of these faults. Many people also use the TDR as an inventory-management tool to ensure that a reel contains the length of cable advertised and to determine if a partially used reel contains enough cable for a particular job.

NOTE A special kind of TDR, called an *optical time-domain reflectometer (OTDR)*, is used to test fiber-optic cables. For more information, see the section "Optical Time-Domain Reflectometers" later in this chapter.

FAULT DETECTION

When a TDR transmits its signal pulse onto a cable, any extraordinary impedance that the signal encounters causes it to reflect back to the unit, where it can be detected by a receiver. The amount of impedance determines the magnitude of the reflected signal. The TDR registers the magnitude of the reflection and uses it to determine the source of the impedance. The TDR also measures the elapsed time between the transmission of the signal and the receipt of the reflection and, using the NVP that you supply for the cable, determines the location of the impedance. For example, on an unterminated cable with no faults, the only source of impedance is the end of the cable, which registers as an open, enabling the TDR to measure the overall length of the cable.

Faults in the cable return reflections of different magnitudes. A complete open caused by a broken cable prevents the signal from traveling any farther down the cable, so it appears as the last reflection. However, less serious faults enable the signal to continue on down the cable, possibly generating additional reflections. A waveform TDR displays the original test signal on an oscilloscope-like screen, as well as the individual reflections. An experienced operator can analyze the waveforms and determine what types of faults caused the reflections and where they are located.

Automated TDRs analyze the reflections internally and use a numerical display to show the results. Some of these devices are dedicated TDR units that can perform comprehensive cable-fault tests at a substantially lower price than a waveform TDR and are far easier to use. The unit displays the distance to the first fault located on the cable and may also display whether the reflection indicates a high impedance change (denoting an open) or a low impedance change (denoting a short). Some of these units even offer the ability to connect to a standard oscilloscope in order to display waveform results, if desired.

BLIND SPOTS

Some TDRs enable you to select from a range of pulse widths. The *pulse width* specifies the amount of energy the unit transmits as its test pulse. The larger the pulse width, the longer the signal travels on the cable, enabling the TDR to detect faults at greater distances. However, signals with larger pulse widths also take longer to transmit, and the TDR is all but incapable of detecting a fault during the time that it is transmitting. For example, because the signal pulse travels at approximately 3ns per meter, a 20ns pulse means that the beginning of the pulse will be about 6.6 meters from the transmitter when the end of the pulse leaves the unit. This time interval during which the pulse transmission tales place is known as a *blind spot*, and it can be a significant problem because faults often occur in the patch cables, wall plates, and other connectors near to the end of the cable run.

When you have a TDR with a variable pulse-width control, you should always begin your tests with the lowest setting so that you can detect faults that occur close to the near end of the cable. If no faults are detected, you can increase the setting to test for faults at greater distances. Larger pulse widths can also aid in detecting small faults that are relatively close. If a cable fault is very subtle and you use a low pulse-width setting, the attenuation of the cable may prevent the small reflection from being detected by the receiver. Larger pulse widths may produce a reflection that is more easily detected.

If your TDR uses a fixed-pulse width, you may want to connect an extra jumper cable between the unit and the cable run to be tested. This jumper cable should be at least as long as the blind spot and should use cable of the same impedance as the cable to be tested. It should also have high-quality connections to both the tester and the cable run. If you choose to do this, however, be sure to subtract the length of the jumper cable from all distances given in the test results.

INTEGRATED TDRs

Many of the combination cable testers on the market include TDR technology, primarily for determining the cable length, but they may not include the ability to detect subtle cable faults like the dedicated units can. Obviously, a severed cable is always detectable by the display of a shorter length than expected, but other faults may not appear. Some units are not even designed to display the cable length by default. Instead, they simply present a pass/fail result based on a selected network type. If, for example, you configure the unit to test a 100Base-T cable, any length less than 100 meters may receive a pass rating. For the experienced installer, a unit that can easily display the raw data in which the pass/fail results are based is preferable.

Another concern when selecting a TDR is its ability to test all four of the wire pairs in a twisted-pair cable. Some devices use time-domain reflectometry only to determine the length of the cable and are not intended for use as fault locators. So they might not test all the wire pairs, making it seem as though the cable is intact for its entire length when, in fact, opens or shorts could be on one or more pairs.

Fiber-Optic Power Meters

A *fiber-optic power meter* measures the intensity of the signal transmitted over a fiber-optic cable. The meter is similar in principle to a multimeter that measures electric current, except that it works with light instead of electricity. The meter uses a solid-state detector to measure the signal intensity and incorporates signal-conditioning circuitry and a digital display. Different meters are for different fiber-optic cables and applications. Meters for use on short-wavelength systems, up to 850nm, use a *silicon* detector, whereas long-wavelength systems need a meter with a germanium or indium gallium arsenide (InGaAs) detector that can support 850 to 1,550nm. In many cases, optical power meters are marketed as models intended for specific applications, such as CATV (cable television), telephone systems, and LANs.

Other, more expensive units can measure both long- and short-wavelength signals. Given that the cost of fiber-optic test equipment can be quite high, you should generally try to find products specifically suited for your network and application so that you're not paying for features you'll never use. A good optical power meter enables you to display results in various units of measure and signal resolutions, can be calibrated to different wavelengths, and measures power in the range of at least 0dBm to –50dBm. Some meters intended for special applications can measure signals as high as +20dBm to –70dBm. An optical power meter registers the average optical power over time, not the peak power, so it is sensitive to a signal source with a pulsed output. If you know the pulse cycle of the signal source, you can compute the peak power from the average power reading.

A fiber-optic power meter that has been properly calibrated to NIST (the United States' National Institute of Standards and Technology) standards typically has a ±5 percent margin for error, due primarily to variances introduced by the connection to the cable being tested, low-level noise generated by the detector, and the meter's signal-conditioning circuitry. These variances are typical for all optical power meters, regardless of their cost and sophistication.

The ability to connect the power meter to the cables you want to test is obviously important. Most units use modular adapters that enable you to connect to any of the dozens of connector styles used in the fiber-optic industry, although LC and SC connectors are most commonly used on LANs. The adapters may or may not be included with the unit, however, and reference test cables usually are not, so be sure to get all of the accessories you need to perform your tests.

Fiber-Optic Test Sources

To measure the strength of an optical signal, a signal source must be at the other end of the cable. Although you can use a fiber-optic power meter to measure the signal generated by your network equipment, accurately measuring the signal loss of a cable requires a consistent signal generated by a fiber-optic test source. A companion to the power meter in a fiber-optic tool kit, the test source is also designed for use with a particular type of network. Sources typically use LEDs (for multimode fiber) or lasers (for single-mode fiber) to generate a signal at a specific wavelength, and you should choose a unit that simulates the type of signals used by your network equipment.

Like power meters, test sources must be able to connect to the cable being tested. Some sources use modular adapters like those on power meters, but others, especially laser sources, use a fixed connector that requires you to supply a hybrid jumper cable that connects the light source to the test cable.

Like optical power meters, light sources are available in a wide range of models. LED sources are less expensive than laser sources, but beware of extremely inexpensive light sources. Some are intended only for identifying a particular cable in a bundle, using visible light. These devices are not suitable for testing signal loss in combination with a power meter.

Optical Loss Test Sets and Test Kits

In most cases, you need both an optical power meter and a light source in order to properly install and troubleshoot a fiber-optic network, and you can usually save a good deal of money and effort by purchasing the two together. You will thus be sure to purchase units that both support the wavelengths and power levels you need and that are calibrated for use together. You can purchase the devices together as a single combination unit called an optical loss test set (OLTS) or as separate units in a fiber-optic test kit.

An OLTS is generally not recommended for field testing, because it is a single unit. While useful in a lab or for testing patch cables, two separate devices would be needed to test a permanently installed link because you have to connect the light source to one end of the cable and the power meter to the other. However, for fiber-optic contractors involved in large installations, it may be practical to give workers their own OLTS set so that they can work with a partner and easily test each cable run in both directions.

Fiber-optic test kits are the preferable alternative for most fiber-optic technicians because they include a power meter and light source that are designed to work together, usually at a price that is lower than the cost of two separate products. Many test kits also include an assortment of accessories needed to test a particular type of network, such as adapters for various types of connectors, reference test cables, and a carrying case. Prices for test kits can range from $500 to $600 for basic functionality to as much as $5,000 for a comprehensive kit that can test virtually every type of fiber-optic cable.

TIP Communications can be a vital element of any cable installation in which two or more people are working together, especially when the two ends of the permanent cable runs can be a long distance apart, as on a fiber-optic network. Some test sets address this problem by incorporating voice communication devices into the power meter and light source, using the tested cable to carry the signals.

Optical Time-Domain Reflectometers

An *optical time domain reflectometer (OTDR)* is the fiber-optic equivalent of the TDR used to test copper cables. The OTDR transmits a calibrated signal pulse over the cable to be tested and monitors the signal that returns back to the unit. Instead of measuring signal reflections caused by electrical impedance as a TDR does, however, the OTDR measures the signal returned by *backscatter*, a phenomenon that affects all fiber-optic cables. Backscatter is caused by the back-reflection of photons due to normally occurring imperfections or defects in the glass, as shown in Figure 15.7. While the scatter occurs in all directions as shown, some of this reflected light will bounce all the way back to the transmitting end of the fiber, where it is detected by the OTDR. The scattered signal returned to the OTDR is much weaker than the original pulse, due to the attenuation of the outgoing pulse, the relatively small amount of signal that is scattered (called the *backscatter coefficient* of the cable), and the attenuation of the scattered signal on its way back to the source.

FIGURE 15.7
OTDRs detect the scattered photons on a fiber-optic cable that return to the transmitter.

As with a TDR, the condition of the cable causes variances in the amount of backscatter returned to the OTDR, which is displayed on an LCD or CRT screen as a waveform. By interpreting the signal returned, it's possible to identify cable faults of specific types and other conditions. An OTDR can locate splices and connectors and measure their performance, identify stress problems caused by improper cable installation, and locate cable breaks, manufacturing faults, and other weaknesses. Knowing the speed of the pulse as it travels down the cable, the OTDR can also use the elapsed time between the pulse's transmission and reception to pinpoint the location of specific conditions on the cable.

Loss is typically measured using a power meter and light source, which are designed to simulate the conditions of the network. Using an OTDR, it is possible to compute a cable's length based on the backscatter returned to the unit. The advantage to using an OTDR for this purpose is that you can test the cable from one end, whereas the traditional method requires that the light source be connected to one end and the power meter to the other.

OTDRs can have limited distance-resolution capabilities over short distances, making them quite difficult to use effectively in a LAN environment where the cables are only a few hundred feet long. OTDRs are used primarily on long-distance connections, such as those used by telephone and cable-television networks. As a result, you might find people who are experts at fiber-optic LAN applications that have never seen or used an OTDR. Other reasons could also account for why they may not have used an OTDR. One is that, as with TDRs, interpreting the waveforms generated by an OTDR takes a good deal of training and experience. Another reason is their jaw-dropping price. Full-featured OTDR units can cost $10,000 to $20,000. Smaller units (sometimes called mini-OTDRs) with fewer features can run from $5,000 to $10,000.

Fiber-Optic Inspection Microscopes

Splicing and attaching connectors to fiber-optic cables are tasks that require great precision, and the best way to inspect cleaved fiber ends and polished connection ferrules is with a microscope. Fiber-optic inspection microscopes are designed to hold cables and connectors in precisely the correct position for examination, enabling you to detect dirty, scratched, or cracked connectors and ensure that cables are cleaved properly in preparation for splicing. Good microscopes typically provide approximately 100-power magnification (although products range from 30 to 800 power), have a built-in light source (for illuminating the object under the scope), and are able to support various types of connectors using additional stages (platforms on which the specimen is placed), which may or may not be included.

Visual Fault Locators

The light that transmits data over fiber-optic cable is invisible to the naked eye, making it difficult to ensure without a formal test that installers have made the proper connections. A *visual fault locator* (sometimes called a *cable tracer*) is a quick and dirty way to test the continuity of a fiber-cable connection by sending visible light over a fiber-optic cable. A typical fault locator is essentially a flashlight that applies its LED or incandescent light source to one end of a cable, which is visible from the other end. A fault locator enables you to find a specific cable out of a bundle and ensure that a connection has been established.

More powerful units that use laser light sources can actually make points of high loss—such as breaks, kinks, and bad splices—visible to the naked eye, as long as the cable sheath is not completely opaque. For example, the yellow- or orange-colored sheaths commonly used on single-mode and multimode cables (respectively) usually admit enough of the light energy lost by major cable faults to make the losses detectable from outside. In a world of complex and costly testing tools, fault locators are one of the simplest and most inexpensive items in a fiber-optic toolkit. Their utility is limited when compared to some of the other tools described here, but they are a convenient means of finding a particular cable and locating major installation faults.

Multifunction Cable Scanners

The most heavily marketed cable-testing tools available today are the *multifunction cable scanners*, sometimes called *certification tools*. These devices are available for both copper and fiber-optic networks and perform a series of tests on a cable run, compare the results against either pre-programmed standards or parameters that you supply, and display the outcome as a series of pass or fail ratings. Most of these units perform the basic tests called for by the most commonly used standards, such as wire mapping, length, attenuation, and NEXT for copper cables, and optical power and signal loss for fiber optic. Many of the copper-cable scanners also go beyond the basics to perform a comprehensive battery of tests, including propagation delay, delay skew, PS-NEXT, ACRF, PS-ACRF, and return loss.

The primary advantage of a multifunction cable scanner is that anyone can use it. You simply connect the unit to a cable, press a button, and read off the results after a few seconds. Many units can store the results of many individual tests in memory, download them to a PC, or output them directly to a printer. This primary advantage, however, is also the primary disadvantage. The implication behind these products is that you don't really have to understand the tests being performed, the results of those tests, or the cabling standards used to evaluate them. The interface insulates you from the raw data, and you are supposed to trust the manufacturer implicitly and believe that a series of pass ratings means that your cables are installed correctly and functioning properly.

The fundamental problem with this process, however, is that the user must have sufficient knowledge of applicable specifications to ensure that appropriate parameters are being used by the test set. When evaluating products like these, it's important to choose units that are able to be upgraded or manually configured so that you can keep up with the constantly evolving requirements.

This configurability can lead to another problem, however. In many cases, it isn't difficult to modify the testing parameters of these units to make it easier for a cable to pass muster. For example, simply changing the NVP for a copper cable can make a faulty cable pass the unit's tests. An unscrupulous contractor can conceivably perform a shoddy installation using inferior cable and use his own carefully prepared tester to show the client a list of perfect "pass" test results. This should not preclude the use of these testers for certifying your cabling installation. Rather, you need to be vigilant in hiring reliable contractors.

As another example, some of the more elaborate (and more expensive) fiber-optic cable testers attempt to simplify the testing process by supplying main and remote units that both contain an integrated light source and semiconductor detector and by testing at the 850nm and 1,300nm wavelengths simultaneously. This type of device enables you to test the cable in both directions and at both wavelengths simply by connecting the two units to either end of a cable run. You needn't use reference test cables to swap the units to test the run from each direction or run a separate test for each wavelength.

However, these devices, apart from costing several times as much as a standard power meter/ light source combination, do not compare the test results to a baseline established with that equipment. Instead, they compare them to preprogrammed standards, which, when it comes to fiber-optic cables, can be defined somewhat loosely. So the device is designed primarily for people who really don't understand what they are testing and who will trust the device's pass-or-fail judgment without question—even when the standards used to gauge the test results are loose enough to permit faulty installations to receive a pass rating.

Multifunction test units are an extremely efficient means of testing and troubleshooting your network. But understand what they are testing and either examine the raw data gathered by the unit or verify that the requirements loaded to evaluate the results are valid. The prices of these products can be shocking, however. Both copper and fiber-optic units can easily run to several thousand dollars.

Troubleshooting Cabling Problems

Cabling problems account for a substantial number of network-support calls—some authorities say as many as 40 to 50 percent. Whether or not the figure is accurate, any network administrator will nevertheless experience network communication problems that can be attributed to no other cause than the network cabling. The type of cable your network uses and how it is installed will have a big effect on the frequency and severity of cabling problems.

For example, a coaxial thin Ethernet network allowed to run wild on floors and behind furniture is far more likely to experience problems than a 10Base-T network installed inside the walls and ceilings. This is true not only because the coaxial cables are exposed and more liable to be damaged but also because the bus topology is more sensitive to faults and the BNC connectors are more easily loosened. Once cabling is installed in the walls and verified for performance, very little is likely to go wrong with it. This goes to show that you can take steps toward minimizing the potential for cable problems by selecting the right products and installing them properly.

Establishing a Baseline

The symptoms of many cable problems are similar to symptoms of software problems, so it can often be difficult to determine when the cable causes a problem. The first step in simplifying the isolation of the source of network problems is to make sure that all of your cables are functioning properly at the outset. You do this by testing all of your cable runs as you install them, as described earlier in this chapter, and by documenting your network installation.

If you use a multifunction cable tester, you can usually store the results of your tests by retaining them in the tester's memory, copying them to a PC, or printing them out. You thus establish a performance baseline against which you can compare future test results. For example, by recording the lengths of all your cable runs at the time of the installation, you can tell if a cable break has occurred later by retesting and seeing if the length results are different than before. In the same way, you can compare the levels of crosstalk, outside noise, and other characteristics that may have changed since the cable was installed. Even if your tester does not have these data storage features, you should manually record the results for future reference.

Another good idea is to create and maintain a map of all your cable runs on a floor plan of your site. Sometimes cable problems can be the result of outside factors, such as interference from electrical equipment in the building. A problem that affects multiple cable runs might be traced to a particular location where new equipment was installed or existing equipment modified. When you install your cables inside walls and ceilings (and especially when outside contractors do it for you), it can be difficult to pinpoint the routes that individual cables take. A map serves as a permanent record of your installation, both for yourself and any future people working on the network.

Locating the Problem

Troubleshooting your network's cable plant requires many of the same common-sense skills as other troubleshooting. You try to isolate the cause of the problem by asking questions like the following:

◆ Has the cable ever worked properly?

◆ When did the malfunctions start?

◆ Do the malfunctions occur at specific times?

◆ What has changed since the cable functioned properly?

Once you've gathered all the answers to such questions, the troubleshooting consists of steps like the following:

1. Split the system into its logical elements.

2. Locate the element that is most likely the cause of the problem.

3. Test the element or install a substitute to verify it as the cause of the problem.

4. If the suspected element is not the cause, move on to the next likely element.

5. After locating the cause of the problem, repair or replace it.

You might begin troubleshooting by determining for sure that the cable run is the source of the problem. You can do this by connecting different devices to both ends of the cable to see if the problem continues to occur. Once you verify that the cable is at fault, you can logically break it down into its component elements. For example, a typical cable run might consist of two patch cables (one at each end), a wall plate, a punch-down block, and the permanently installed cable.

In this type of installation, it is easiest to test the patch cables, either by replacing them or testing them with a cable scanner. Replacing components can be a good troubleshooting method, as long as you know that the replacements are good. If, for example, you purchase a box of 100 cheap patch cables that are labeled Category 5e when they actually are Category 3 cable, replacing one with another won't do any good.

The most accurate method is to test the individual components with a cable scanner. If the patch cables pass, then proceed to test the permanent link. If you don't have a scanner available, you can examine the connectors at either end of the cable run and even reconnect or replace them to verify that they were installed correctly. However, there's little you can do if the problem is inside a wall or in some other inaccessible place. If you do have a scanner, the results of the tests should provide you with the information you need to proceed.

Resolving Specific Problems

Cable testers, no matter how elaborate, can't tell you what to do to resolve the problems they disclose. The following sections examine some of the courses of action you can take to address the most common cabling problems.

WIRE-MAP FAULTS

Wire-map faults are the result of an improper installation. When the wires within a twisted-pair cable are attached to the wrong pins, the cable is no longer wired straight through. If the pairs used to carry network data are involved, then signals won't reach their destination. In most cases, this fault occurs on a permanent link, although it is possible for a patch cable to be miswired.

The possible causes of wire-map faults are simple errors made during the installation or the use of different pinouts (T568-A and T568-B) at each end of the cable. Whatever the cause, however, the remedy is to rewire the connectors on one or both ends so that each pin at one end is connected to its equivalent pin at the other end.

EXCESSIVE LENGTH

Cable lengths should be carefully planned before network installation and tested immediately after installation to make sure that the cables are not longer than the recommended maximum according to the ANSI/TIA-568-C standard. Cable runs that are too long can cause problems like late collisions on an Ethernet network or excessive retransmissions due to attenuated signals. Most protocols have some leeway built into them that permit a little excess, so don't be overly concerned if the maximum allowable length for a cable segment is 95 meters and you have one run that is 96 meters long.

TIP   It's possible for a cable tester to generate incorrect length readings if the tester is improperly calibrated. If the cable length seems wrong, check to make sure that the nominal velocity of propagation (NVP) setting for the cable is correct. Also, some testers include the patch cable between the tester and the connection point in the calculated cable length.

To address the problem, you can start by using shorter patch cables, if possible. In some cases, you may find that an installer has left extra cable coiled in a ceiling or wall space that can be removed, and the end can be reconnected to the wall plate or punch-down block. Sometimes a more efficient cable route can enable you to rewire the run using less cable. If, however, you find that bad planning caused the problem and the wall plate is too far away from the punch-down block, you can still take actions.

The first and easiest action is to test the attenuation and NEXT on the cable run to see if they exceed the requirements for the protocol. These characteristics are the primary reasons for these maximum-length specifications. If you have installed cable that is of a higher quality than is required, you may be able to get away with the additional length, but if you are having network problems, chances are this isn't the answer.

OPENS AND SHORTS

Opens and shorts can be caused by improper installation, or they can occur later if a cable is damaged or severed. If the cable's length is correct but one or more wires are open or shorted, then a connector is likely faulty or has come loose and needs repairing or replacing. If all of the wires in a cable are reported as open in the same place or if the length of all the wires is suddenly shorter than it should be, the cable may have been accidentally cut by nearby equipment or by someone working in the area. Cables damaged but not completely severed may show up with drastically different lengths for the wire pairs or as shorts at some interim point.

Cable scanners usually display the distance to the open or short so that you can more easily locate and repair it. For cables installed in walls and ceilings, the cable map you (we hope) created during the installation can come in handy. If you don't know the cable's route, you can use a tone generator and probe to trace the cable to the point of the break.

It is tempting to try to splice the ends of the severed wires. *Don't do it*. You'll create a nexus for all sorts of potential transmission problems, including SRL (structural return loss) reflections, higher attenuation, and increased crosstalk. You must completely replace the permanent cable run. Broken or damaged patch cables should always be discarded.

EXCESSIVE ATTENUATION

A cable run can exhibit excessive attenuation for several different reasons, most of which are attributable to improper installation practices. The most obvious cause is excessive length. The longer the cable, the more the signals attenuate. Address this problem as you would any other excessive-length condition.

Another possible cause is that the cable used in the run is not suitable for the rate at which data will be transmitted. If, for example, you try to run a 100Base-TX network using Category 3 cable, one of the reasons it will fail is that the specified attenuation level for Category 3 allows the signal to decay more than 100Base-TX can handle. In this case, there is no other alternative than to replace the cable with the proper grade. Inferior or untwisted patch cables are a frequent cause of this type of problem. These are easily replaced, but if your permanent links are not of an appropriate performance grade, the only alternative is to replace the cabling.

Excessive attenuation can also be caused by other components that are of an inferior grade, such as connectors or punch-down blocks. Fortunately, these are generally easier to replace than the entire cable.

Environmental factors, such as a conductor stretched during installation or a high-heat environment, also cause excessive attenuation.

EXCESSIVE CROSSTALK

Crosstalk is a major problem that can have many different causes, including the following:

Inferior cable Cables not of the grade required for a protocol can produce excessive crosstalk levels. The only solution is to replace the cable with the appropriate grade.

Inferior components All the components of a cable run should be rated at the same grade, including all connectors. Using Category 3 connectors on a Category 5e network can introduce excessive crosstalk and other problems. Replace inferior components with those of the correct grade.

Improper patch cables   Replace inappropriate cables with twisted-pair patch cables that are rated the same as your permanent links. Silver-satin patch cables used for telephone systems may appear at first to work with data connections, but the wire pairs in these cables are not twisted, and the main reason for twisting conductors together in pairs is to minimize crosstalk.

Split pairs   Incorrect pinouts that cause data-carrying wires to be twisted together result in additional crosstalk, even when both ends are wired in the same way. Split pairs can be the result of mistakes during the installation or the use of the USOC pinouts. The solution is to reattach the connectors at both ends using either the T568-A or T568-B pinouts.

Couplers   Using couplers to join short lengths of cable generates more crosstalk than using a single cable segment of the appropriate length. Use one 12′ patch cable (for example) instead of two 6′ cables joined with a coupler. When repairing broken permanent links, pull a new length of cable rather than using couplers to join the broken ends together.

Twisting   The individual wire pairs of every Category 5e cable must remain twisted up to a point no farther than ½″ (0.5″) from any connector. A Category 6 pair must remain twisted to within ⅜″ (0.375″) of its termination. If the wires are too loosely twisted, reattach the connectors, making sure that all of the wire pairs are twisted tightly.

Sharing cables   Many network protocols use only two of the four wire pairs in a standard twisted-pair cable, so some people believe they can utilize the other two pairs for voice traffic or some other application. They shouldn't, however, because other signals running over the same cable can produce crosstalk. The problem may be difficult to diagnose in these cases because the crosstalk only occurs when the other application is using the other wire pairs, such as when the user is talking on the phone. If two pairs in a wire are used for another application, you must install new cabling for one application or the other so that they will no longer share a cable.

EXCESSIVE NOISE

The potential for noise generated by outside sources should be considered during the planning phase of a network installation. Cables should be routed away from AC power lines, light fixtures, electric motors, and other sources of EMI and RFI. Sometimes outside noise sources can be difficult to detect. You may, for example, test your cables immediately after you install them and detect no excess noise from outside sources and then find during later testing that your network performance is severely degraded by noise. It is entirely possible that a new source of interference has been introduced into the environment, but you also have to consider that your original tests may not have been valid.

If you installed and tested the cable plant during nights and weekends, your tests for outside noise may have generated all pass ratings because some sources of noise were not operating. When lights and machinery are turned on Monday morning, noise levels could be excessive. Always test your cable runs in the actual environmental conditions in which they'll be used.

If, after cable installation, a new source generates excessive noise levels, you must either move the cables and source away from each other or replace the UTP cables with fiber-optic cables.

The Bottom Line

Identify important cable plant certification guidelines.   The tests you perform and the results you receive enable you to certify your cable installation as complying with a specific standard of performance. Luckily the ANSI/TIA-568-C standards provide a very detailed set of requirements that will be likely be specified on behalf of the building owner. Although many high-quality cable testers on the market perform their various tests automatically and provide you with a list of pass/fail results, it is important to know not only what is being tested but also what results the tester is programmed to evaluate as passes and failures.

   Master It

   1. What is the difference between the permanent link and channel testing?

   2. If you have installed a Category 6A cable, what is the maximum NEXT that can be supported for a 500MHz channel?

   3. What is the insertion loss budget for a 300 meter link intended to operate at 10GBase-SR using OM3 fiber?

Identify cable testing tools.   Cable-testing tools can range from simple, inexpensive, mechanical devices to elaborate electronic testers that automatically supply you with a litany of test results in an easy-to-read pass/fail format. It's important to identify these and understand their benefits.

   Master It   You are asked to install a fiber-optic link that requires multiple connections and splice points. Based on the cable length, number of connections, and splice points, you have estimated the insertion loss to be very close to the insertion loss budget for the intended application. After installing the link, you find that the insertion loss is higher than the budget. It's now time to go and find the points in the link that could be causing unusually high loss. You suspect that it may be due to the splices in the system. Which measurement tool would you use and why?

Troubleshoot cabling problems.   Cabling problems account for a substantial number of network support calls. Since network uptime is a critical condition for maximizing an organization's business, troubleshooting problems quickly is very important. Understanding some basic steps will help you improve the speed of your process and repair of a customer's network.

   Master It   What are the basic recommended steps for troubleshooting a cable problem?

Chapter 16

Creating a Request for Proposal

All journeys begin with a single step. In the case of a telecommunications infrastructure and/or hardware project that is not performed in-house, that first step is the creation of the *request for proposal (RFP)*. The RFP is essential to the success of your telecommunications-infrastructure project.

Anyone who rushes into a project without a clear view of what he or she needs to accomplish is foolish. A vendor who accepts a job without a clear definition of the work to be performed is also foolish. The RFP is essential for setting the pace of a project that is going to involve both a client and an outside vendor. You may choose to write your own RFP, or you may choose to hand the entire cabling design project and RFP generation over to a specialized consulting company. Another option is to work with the consulting company but do much of the groundwork beforehand. Any of these three choices still requires that you have a good knowledge of generating an RFP.

The consulting companies that can perform the steps documented in this chapter are made up of experts in their field and can save you time and money. However, for installations smaller than a few hundred network points or locations, you may not need a consulting company to prepare an RFP.

In this chapter, you will learn to:

◆ Determine key aspects of writing an RFP

◆ Determine the components of the cabling infrastructure that should be included in the RFP

◆ Understand what to do after the RFP is written

What Is a Request for Proposal?

The request for proposal (RFP) is essential for defining what you want designed and built for the physical layer of your voice and data networks. An improperly constructed physical layer will contribute to poor reliability and poor performance.

NOTE   Though certainly medium- and large-scale projects will require an RFP, smaller projects (a few dozen cabling runs) on which you're working with a trusted vendor do not require one.

Setting the Tone for the Project

The RFP sets the tone for the entire cabling infrastructure project. The best way to think of an RFP is as a combination of a guidebook, map, and rulebook. It clearly articulates the project, goals, expectations, and terms of engagement between the parties. In addition, for it to serve your best interests, it must be designed to be fair to all parties involved. A well-thought-out and well-written RFP goes a very long way toward ensuring the success of the project. On the other hand, a poorly thought out and badly written RFP can make your project a nightmare.

Having been on both sides of the fence, we have seen the influence that the RFP has on both parties and on the overall success of an effort. One of the mistakes that we have seen made is that the buyer and vendor often see the RFP as a tool with which to take advantage of the other party. This is most unfortunate because it sets the stage for an adversarial relationship right from the beginning.

WARNING The RFP can ensure a successful relationship with your vendor and the successful completion of your cabling project. Unfortunately, some see it as a tool for taking advantage of the other party. We do *not* recommend using an RFP in this way!

The best way to prevent such a scenario from occurring is by making sure that the RFP clearly describes the scope of the project, the buyer's requirements and expectations of the vendor, and the responsibilities of all parties involved. Because it is to be used as a rule book, it must be designed to promote fairness.

To create a clear and fair RFP, you or your consulting company must do much legwork to ensure that many of the issues associated with the effort are identified, defined, addressed, and properly articulated in the RFP document. That preplanning often involves many people.

It is important to remember that although the project involves the installation of technology, it also involves many departments outside of the technology group, perhaps including management, finance, facility management, and legal departments as well as the departments getting the new network.

Before we get into some of the nuts-and-bolts aspects of creating the RFP, we'll talk about what the goal of the project should be.

The Goals of the RFP

The goal of every RFP should be the creation of an infrastructure that satisfies the needs of the organization today while being flexible enough to handle the emerging technologies of tomorrow. Everyone wants a system that they do not have to upgrade every time they need to install a faster piece of hardware or advanced application. In addition, no one wants to spend megabucks on their infrastructure, upgrading it every 1½ years to keep pace with industry advancements.

The goal of every RFP should also be to create an infrastructure that appears to be invisible. Wouldn't it be nice if an IT cabling infrastructure could be as invisible as electrical wiring? Think about your electrical wiring for just a second; when was the last time you had to upgrade it because you bought a new appliance? Or when did you have to add additional breakers to your electrical panel because you wanted to plug in another lamp or another computer?

Well, the good news is that with the proper planning and design, your communications infrastructure can become virtually invisible—thanks to the many advances within the infrastructure segment of the telecommunications and information industries within the last decade. It is possible to create cabling configurations that can and should become standard throughout every type of office in every site within your organization.

Perhaps the best part is that the potential system will not limit the types of data communications hardware purchased or the pool of infrastructure contractors you can invite to bid on the installation. It can also be flexible enough to satisfy everyone's needs. Sounds like a pipe dream, doesn't it? Well, it isn't. A well-designed, well-engineered, and well-installed infrastructure becomes the "enabler" for the rest of your applications and future technology requirements.

Developing a Request for Proposal

Developing the RFP involves quite a bit of work, depending on the size of both the project and your organization. The first part of the development process involves identifying exactly what your current and future needs are. Along with this, you must determine any restrictions and constraints that may be placed on the system you will install.

Once you know what your needs are and the factors that will constrain you, the next phase is to design the system and determine the components you will need. Once you have the system design and know what components will be necessary, you can proceed with putting together the RFP.

The Needs Analysis

Many people will be involved in the needs analysis, the most important step in creating an RFP, and you must be thorough. One of the reasons this step is important is because you can use this opportunity to establish "buy-in" from others in your organization.

For the sake of this discussion, we will assume that the infrastructure project being planned is at least of medium-sized scope. As one who is in charge of such things, your task may be to handle a simple 20-network node expansion. Then again, perhaps you have been given the task of overseeing and implementing an organization-wide infrastructure installation or upgrade involving hundreds of users located in multiple sites. In either scenario, the same basic approaches should be taken.

The objective of the needs analysis is to define the specific project. The needs analysis should involve anyone who will be affected by the installation of the system. Depending on the size of your organization, some of the people you will want to solicit advice from include the following:

- Those who are responsible for any type of information technology that will be affected by your project

- The people in facilities and/or facilities management

- The electrician or electrical contractor

- Managers who can help you gain a better understanding of the long-term goals of the organization as they relate to information technology and facilities

GETTING INPUT FROM KEY PLAYERS

It is important to get input from upper management and the strategic planners within your organization so you can understand the types of technology-dependent services, applications, and efficiencies that they may require or need to have deployed in order to realize the company's goals.

Through these meetings, those who are responsible will define the scope of the project, the intent, deadlines, payment terms, bonding, and insurance issues.

All these solicitations of input may sound like overkill, but we can assure you they are not. You may be saying to yourself, "Why would I want to make my life even more miserable than it already is by inviting all of these people's opinions into my project?" You may be surprised, but by doing so you will, in fact, be making your life much easier. Plus, you will save yourself a great deal of time, money, and aggravation in the not-too-distant future.

All the people from whom you will solicit opinions are going to have opinions anyway. Furthermore, if you don't find out these opinions before you start your project, it is safe to assume that these opinions will be communicated to you about two hours after it is too late to do anything reasonable about them. In addition, think of all the new friends you will make. (Most of these people are just dying to tell you what to do and how to do it.) By asking them for their opinions up-front, you are taking away their right to give you another one later. "Speak now or forever hold your peace" strictly applies here.

NOTE One IT director we know held "town hall" meetings with her company's managers when she began planning the infrastructure for their new location. The meetings often demonstrated how quickly the managers rushed to protect their own fiefdoms, but the combination of all the managers discussing the infrastructure needs also generated new ideas and requirements that she had not previously thought of.

Most important, though, is that scheduling meetings with these folks will help you understand the organization's overall needs from a variety of vantage points. The information and the understanding you receive will help you to get through the process. For instance, perhaps the facilities folks have information that, once conveyed to you and the wiring contractors, will lower the project cost and/or eliminate change orders and cost overruns. Or perhaps the in-house electrician is planning an installation in the same buildings during the same time frame, which would allow you to combine efforts and create efficiencies that would save time, money, and work. And finally, getting feedback from upper management about their long-range plans could prevent you from being hit with a new imaging-application rollout that will require "only" one more UTP circuit to be installed to every outlet location you just paid to have cabled!

You'd be surprised at some of the details that we have unearthed from employers and clients during the initial cabling system planning meetings. Some of these details would have caused time, money, and effort to be wasted if they had not been revealed, including:

◆ An entire wing of a building was to be renovated three months after the new cable was to have been installed.

◆ Major expansion was planned in six months, and an area that currently had only a few workstation locations would be accommodating several dozen additional salespeople.

◆ A departmental restructuring was taking place, and an entire group of people was going to be moved to a new location.

◆ The local building codes did not allow any data, voice, or electrical cabling in the plenum unless it was in conduit.

◆ Telecommunications infrastructure designs had to be approved by a registered, professional engineer, as required by the state in which the customer was located.

- A new phone-switch and voicemail system was being purchased and would also have to be cabled shortly after the data cabling project. Management mistakenly viewed this as a separate project.

- In a law firm, all the attorneys were going to be given a notebook computer in addition to their desktop computers. These notebook computers would require network access while the attorneys were in their offices, so each attorney's office would require two network connections.

- A new photocopier tracking system was to be installed that would require cabling to a tracking computer located in the computer room. In addition, these new photocopiers were going to function as network devices and would require their own Ethernet connections.

- Management wanted all offices of managers and senior project personnel, as well as conference rooms, to have cable TV hookups.

In each of these cases, prior to the initial meetings, we thought we had a pretty good idea of what the company was planning to do and what the regulations were. The additional information was helpful (and sometimes vital) to the successful installation of the new cabling installation.

THE BONUS: GETTING BUY-IN

In addition to gaining pertinent information, your wisely inclusive communication will get you "buy-in" on your project. In other words, if you show respect to the people involved by bringing them into the loop, many of them will feel a part of what is going on and (you hope) won't fight against it.

We must issue a warning here, however! You are the boss of the project and must remain so. If you let others control what you are responsible for, you have a recipe for disaster.

NOTE Responsibility for a project and the authority to make the final decision on a project need to go hand in hand.

If the one who has authority over the project is not held responsible for the outcome, be prepared for disaster. For any project to have a chance of succeeding, authority and responsibility must be welded together.

CABLING @ WORK: QUESTIONS TO ASK WHEN GATHERING INFORMATION

You have the key players in a room, and you have outlined what your cabling project will entail. They may look at you with a "So what am I doing here?" expression on their faces. What are some of the key questions that you may want to ask them? Here is a list of our favorites:

- How long is the company planning to occupy the space we are cabling?

- Will new technologies be implemented in the near or long term? These may include voice, data, video, network-attached PDAs, notebook computers, remote-controlled devices, security systems, and new photocopier technology.

◆ Will any new voice and data applications require fiber-optic cabling? What type of fiber-optic cable do these applications require?

◆ What electrical-code requirements and building requirements will influence the data communications cabling?

◆ Are the telecommunications rooms properly grounded?

◆ Are there installation time requirements? Will the areas that need to be worked in be accessible only during certain hours? Are they accessible on weekends and at night? Will the installation personnel have to work around existing office furniture and/or employees?

◆ Do building security requirements exist for contract personnel working in the areas in question? Are there places that contractors must be escorted?

◆ Does the contractor have to be unionized? Will any areas of the installation be affected by union rules, such as the loading dock and elevators?

◆ Are there plans to move to faster networking technologies, such as Gigabit Ethernet or 10 Gigabit Ethernet?

◆ How many work area outlets should be installed? (You may have your own opinions on this, but it is a good idea to hear others' thoughts on the matter.)

◆ Should the voice and data cabling infrastructures be combined, if they are not already?

◆ What insurance should the contractor carry? Should he or she be bonded? Licensed? Certified?

◆ Will construction or rewiring affect any deadlines?

◆ Will the contractor need to access any areas of the building that are not part of your leased space (such as entrance facilities and telecommunications rooms)?

◆ Will the company be providing parking and working space (such as storage space, office, telephone, and fax) for the contractor and his or her employees?

◆ If the organization spans multiple floors of a single building, is space available in the risers (conduits between floors) to accommodate additional cables? If not, who will have to give approval for new risers to be drilled?

◆ If the organization spans multiple buildings, how will the cable (usually fiber) be connected between buildings?

These are all questions that you will want to know the answers to if you are writing an RFP or if you are a contractor responding to an RFP. Your newfound friends may answer some of these questions, and you may have to answer others.

Designing the Project for the RFP

Once you have completed the needs analysis of your project, you should be prepared to enter the design phase. Although much of the design may be left up to the contractor whom you choose to install your system, many of the design-related questions should be answered in the RFP. This may seem a bit intimidating to the uninitiated, but we assure you that if you divide the project into small bite-sized (or "byte-sized," if you don't mind the pun) pieces, you will conquer the task. Even the largest and most intimidating projects become manageable when broken into small tasks.

COMPONENTS OF A CABLING INFRASTRUCTURE

The first step in dividing the project is to identify the four major subsystems of a cabling infrastructure:

- ◆ Backbone cabling
- ◆ Telecommunications rooms
- ◆ Horizontal cabling
- ◆ Work area components

These and other subsystems are described by the ANSI/TIA-568-C standard and are discussed in more detail in Chapter 2, "Cabling Specifications and Standards." Within each of these categories are several components.

NOTE   When designing cabling systems, you should conform to a known standard. In the United States, the standard you should use is the ANSI/TIA-C Commercial Building Telecommunications Cabling Standard. In Europe and other parts of the world, the ISO/IEC 11801 Ed. 2 Generic Cabling for Customer Premises Standard is the one to use. Most other countries in the world have adopted one of these standards as their own, but they may call it something different.

Backbone Cabling

The backbone (a.k.a. vertical, trunking, or riser) cabling connects the telecommunications rooms (TR) with the equipment room (ER); the equipment room is where the central phone and data equipment is located. Although the ANSI/TIA-568-C standard allows for an intermediate telecommunications room, we don't recommend them for data applications. Telecommunications rooms should be connected to the equipment room via backbone cabling in a hub-and-spoke manner.

Many RFPs will leave the determination of the number of cables and pairs up to the company that responds to the RFP; the responding company will figure out the number of pairs and feeder cables based on requirements you supply for voice and data applications. Other RFP authors will specify exactly how many cables and multipair cables must be used. The decision is up to you, but if you have little experience specifying backbone capacity, you may want to leave the decision to a professional. Some decisions that you may have to make with respect to backbone cable include the following:

- ◆ The number of multipair copper cables that must be used for voice applications.

- ◆ How many strands of fiber must be used between telecommunications rooms for data and voice applications.

- ◆ Whether single-mode or multimode fiber-optic cable will be used. Most data applications use 850nm optimized 50μm multimode fiber (a.k.a. OM3), though some newer voice and applications use single-mode fiber.

- ◆ Whether any four-pair UTP cable (Category 5e or 6) will be installed as backbone cabling.

NOTE Backbone sizing can be tricky. If you are not careful, you will incorrectly calculate the backbone capacity you need. If you are not sure of the exact capacity you require, leave it to the contractor to specify the right amount.

Telecommunications Rooms

Still more decisions must be made about the telecommunications room. Some of these decisions will be yours to make, and someone else in your organization will make other decisions. Here are some points about telecommunications rooms that you may need to think about:

◆ If the telecommunications room is to house electrical equipment, the room should be environmentally conditioned (temperature and humidity need to be maintained at acceptable levels).

◆ Appropriate grounding for racks and equipment has to exist. This is often the responsibility of the electrician. Don't ignore good grounding rules. Consult ANSI/TIA-607-B and the NEC for more information. If the rooms are not grounded properly, you need to know who will be responsible for installing grounding.

◆ Sufficient space and lighting need to be provided so that all necessary equipment can be installed and people can be in the room working on it.

◆ Backup power or uninterruptible power supplies (UPSs) should be installed in the telecommunications rooms.

◆ Proper access should be given to IT personnel. However, the rooms should be secure to prevent unwanted tampering or data theft.

The typical telecommunications room is going to include components such as the following:

◆ Punch-down blocks for voice. Some punch-down blocks can be used to cross-connect data circuits, but they are generally not recommended.

◆ Wall space, if you are going to use punch-down blocks (they are usually mounted on plywood that is mounted to the wall).

◆ Patch panels for copper and fiber circuits. It is a good practice to separate the patch panels used for data applications from the patch panels used for voice applications.

◆ Racks, cabinets, and enclosures for patch panels, telecommunications gear, UPSs, LAN hubs, switches, and so forth. Shelves for the racks and cabinets are often forgotten on RFPs and in the initial design. Don't forget extra mounting hardware for the racks, cabinets, and enclosures.

◆ Wire-management equipment used on the walls and on the racks. These are also often forgotten during the initial design phase. Good wire-management practice means that the telecommunications rooms will be cleaner and easier to troubleshoot.

◆ Patch cables for the copper and fiber-optic patch panels and hubs. These are the most commonly forgotten components on RFPs. Make sure that the patch cables match the category of cable that you use. ANSI/TIA-568-C specifies a maximum total length of 5 meters (16´) of patch cord per horizontal channel in the telecommunications room. If you don't need cables that long, you should use only the length necessary. You may want to order varying lengths of patch cables to keep things neat and untangled.

Horizontal Cabling

The horizontal cabling, sometimes called the distribution cabling, runs from the telecommunications room to the work area. The horizontal cabling is the component most often planned incorrectly. Cabling contractors know this and will often bid extremely low on the overall cost of a job so that they can get the follow-on work of moves, adds, and changes. This is because installing single runs of horizontal cable is far more costly than installing many runs of cable at once. To save yourself future unnecessary costs, make sure that you plan for a sufficient amount of horizontal cable.

Some of the components you will have to think about when planning your horizontal cable include:

◆ How much cable should be run between each work area and a telecommunications room? ANSI/TIA-568-C recommends that a minimum of two permanent links be provided to each work area: one UTP and one fiber cable, or two UTP cables. In an all-UTP environment, however, we recommend running four UTP cables to each work area.

◆ What category of UTP cable should be run? Most telephone applications today will use Category 3 cabling; 10Base-T Ethernet will also run on Category 3 cable. Faster Ethernet and other twisted-pair technologies require at least Category 5e cabling. A minimum of Category 5e is recommended.

◆ If using fiber-optic cable, what type of fiber cable should you use and how many pairs should you run to each work area? Typically, two pairs of multimode fiber-optic cable are used for horizontal cable, but this will depend on the applications in use and the number of data connections to be installed at each location.

◆ Per ANSI/TIA-568-C, the maximum distance that horizontal cable can extend (not including patch cables and cross-connections) is 90 meters (285′).

◆ Should you use some type of "shared-sheath" cabling for horizontal cabling? For example, since 10Base-T only uses two pairs of a four-pair cable, some network managers decide to use the other two pairs for an additional 10Base-T connection or a telephone connection. We *strongly discourage* the use of a shared sheath for data applications. All four pairs should be terminated in a single eight-position outlet for future applications.

Work Area

The final major area is the work area, which includes the wall plates, user patch cables, and user equipment. The work area can also include adapters such as baluns that modify the cable impedance. Many design issues relating to the work area will revolve around the choice of wall plates. Here are some points to think about relating to the work area:

◆ Know what type of wall plates you will use. The wall plate must have sufficient information outlets to accommodate the number of horizontal cables you use. Many varieties of modular wall plates on the market will accommodate fiber, UTP, video, audio, and coaxial modules in the same plates. See Chapter 9, "Wall Plates," for more information.

◆ For UTP cabling, the connecting hardware must also match the category of UTP cable you use.

◆ For fiber-optic cable, the wall plate and connector types must match the cable type you use and the requirements for the station cables (station patch cables) and fiber-optic connector types.

◆ Don't forget to estimate the number of patch cables you will need, and include this in the RFP. ANSI/TIA-568-C specifies a maximum length for patch cables of 5 meters (16′). UTP patch cables should be stranded copper cable and should match the category of cable used throughout the installation.

◆ Though not as common now as they were years ago, impedance-changing devices such as baluns might be necessary. Make sure that you have specified these in your cabling RFP if they are not being provided elsewhere.

How Much Is Enough?

Now that the categories and their components have been identified, the next order of business is to determine how many of these items will be needed. This is when you will begin to realize the benefits of the needs analysis that you performed. The size and components of your infrastructure are always based on the immediate needs of your organization with "realistic" future needs factored in.

Wall Plates and Information Outlets

When designing a cabling infrastructure, always start from the desktop and work backward. For instance, an accurate count of the number of people and their physical locations will determine the minimum number of information outlets that will be needed and where they will be installed.

KEY TERM *wall plates and information outlets* Depending on the design of the wall plate, a single wall plate can accommodate multiple *information outlets*. An information outlet can accommodate voice, data, or video applications.

Some IT and cabling professionals will automatically double this minimum number in order to give themselves room to grow. Our experience with information outlets is that, once you have your cabling system in place, you never seem to have enough. With wall plates in particular, there never seems to be one close to where you want to put phones and data equipment. Here are some ideas that may help you to plan information outlets:

◆ Don't forget locations such as network-printer locations and photocopier locations.

◆ In some larger offices, it may be helpful to install two wall plates at each location on opposite walls. This keeps the station patch-cable lengths to a minimum and also helps keep cable clutter to a minimum, as cables would not have to cross the entire length of a large office.

◆ Special-use rooms, such as conference rooms and war rooms, should be cabled with at least one wall plate identical to one in a typical workstation area.

◆ Training rooms should have at least four more information outlets than you anticipate needing.

◆ Use extreme caution when cabling to locations outside of your organization's office space, such as to a shared conference room; it may allow outsiders to access your data and voice systems. Although it may seem unlikely, we have seen it happen.

CABLING @ WORK: PUTTING DATA CABLES IN PLACES YOU WOULD NEVER IMAGINE

A few years ago during a convention center's remodeling and upgrading project, the convention center wired only the minimum locations required to install its new local area network and multi-media services. Later that year during a phone-system upgrade, the convention center had to rewire each location.

Shortly after that, this convention center decided to offer higher-speed video-teleconferencing/streaming and additional Ethernet connections. Each room had to have additional cabling installed. Now the convention center is again succumbing to the pressures of increasing bandwidth due to newer applications such as high-definition TV and is installing 1000Base-T Internet connections throughout their rooms. At the same time, it is wiring its restaurants and retail locations for Category 6 cabling because its new cash register system uses 1000Base-T network connections.

Though no precise figures have been calculated to see exactly how much would have been saved by doing the entire job at one time, estimates indicate that the convention center could have saved as much $200,000 by performing all the cabling infrastructure work at the same time.

Backbone and Horizontal Cabling

The number of information outlets required and their wall-plate locations will determine the sizing of your horizontal cables (fiber strands and copper pairs) as well as the size and placement of your main cross-connection (MC) and any required intermediate and horizontal cross-connections (IC and HC, respectively.). The applications to be run and accessed at the desktop determine the types of cables to be installed and the number of circuits needed at each wall plate location.

Some RFPs merely provide the numbers of wall plates and information outlets per wall plate and leave the rest of the calculations up to the cabling contractor. Other RFPs don't even get this detailed and expect the contractor to gather this information during his or her walk-through. Our preference is to have this information readily available to the contractor prior to the walk-through and site survey. The less ambiguous you are and the more you put into writing, the easier your job and your contractor's job will be. Information about wall plates and information outlets has to be gathered and documented by someone; you are the person who has to live with the results. Always remain open-minded to the contractor's suggestions for additional locations, though.

RULES FOR DESIGNING YOUR INFRASTRUCTURE

As you gather information and start planning the specifics of your cabling infrastructure, keep in mind our rules outlined here. Some of these result from our own experiences, and cabling and IT professionals have contributed others to us:

◆ Think "flexibility" and design accordingly. You will always find more uses and have more requirements of your infrastructure than you are seeing today. Technology changes and organizations change; be prepared.

◆ Create organizational standards and stick to them. Define the different outlet types that exist in your facility. For instance, those in the accounting department may need fewer circuits than those in operations, and operations may need a different configuration than those in sales. Once you determine the various types of configurations required, standardize them and commit to installing the same configuration throughout the particular department. On the other hand, you may decide that it makes sense to give *everyone* the same configuration. Some companies shift employees and departments around frequently, so this approach may be better suited to such an environment. Whatever you do, standardize one option or another and stick to it. Doing so will make troubleshooting, ordering of parts, and moves, adds, and changes (MACs) much less confusing.

◆ Use modular wall plates. Buy a wall plate that has more "openings" than circuits installed at the location. If you install two cables, buy wall plates that have three or four ports. The cost difference is minimal, and you will preserve your investment in the event extra cables are installed or activated.

◆ Never install any UTP cable that is not at least Category 5e–rated. For data applications, Category 3 is dead. You should even strongly consider installing Category 6, if your budget will allow. Three things in life never change: death, taxes, and the need for more bandwidth.

◆ Always install one more cable at each location than is going to be immediately used. If your budget is tight, you may choose not to terminate or test the circuit or not to add the necessary patch-panel ports, but do try to install the extra "pipe." Invariably, organizations find a use for that extra circuit. That cable will also enable you to quickly respond to any late special-connectivity needs with minimal cost and disruption.

◆ Whenever possible, leave a pull-string in the pathway to make future expansion easier.

◆ Use wire management above and below each patch panel. A neat patch panel begets a neat patch field. A messy patch field begets trouble.

◆ Make sure your connectors, patch cables, patch panels, and wall jacks are rated the same as your cable: Category 5e for Category 5e; Category 6 for Category 6. The same should hold true for installation practices.

◆ If you venture into the world beyond Category 6 cabling systems, remember that they are not standardized as of this writing. Further, vendor claims about the performance of these cables may be true only if you use all components from the same vendor or from a vendor alliance that includes cable and connectivity components.

◆ Label the circuits at the wall plate and at the patch panel. Although some people feel it is important to label the cable itself, do so only if it does not increase the cost of the cabling installation.

◆ Never underinstall fiber strands. *Never, ever, ever* install only two strands of fiber-optic cable between telecommunications rooms and the equipment room! Install only four strands if you have no money at all. You must try to install a minimum of six or eight strands. Much convergence is occurring in the low-voltage industries—alarm systems,

HVAC systems, and CCTV (closed-circuit television), CATV, and satellite video systems are all using digital information and running on fiber backbones. The installation of additional fiber beyond your current data needs could make you a hero the next time an alarm system or HVAC system is upgraded. Remember the movie *Field of Dreams*: Build it, and they will come.

◆ Include a few strands (four or eight) of single-mode fiber with your multimode fiber backbone. Even if you do not see the need for it today, put it in. To save money, you may choose not to terminate or test it, but it should be part of your backbone. Video applications and multimode's inability to handle some of the emerging higher bandwidth and faster-moving data applications makes this a very smart bet.

◆ Oversize your voice/copper backbone by a minimum of 10 to 25 percent if you can afford it. A safe way to size your voice copper trunk is to determine the maximum number of telephone stations you anticipate you will need. Determine the number of voice locations that will be fed from each room and then size your voice backbone to reflect 2.5 pairs per station fed. For example, in the case of a room that will feed a maximum of 100 telephone stations, you should install 250 pairs.

◆ Test and document all copper distribution. If you install Category 5e cable and components, insist upon 100 percent Category 5e compliance on all copper distribution circuits. All conductors of all circuits must pass. More sophisticated UTP cable testers provide printed test results; you should obtain and keep the test results from each location. Some testing software packages will allow you to keep these in a database. Tests should be reviewed prior to acceptance of the work. Note, though, that if you ask for Category 5e cable testing for each circuit, it may increase the cost of the overall installation.

◆ Test and document all fiber backbone cable. *Bidirectional attenuation testing* using a power meter is sufficient for LAN applications. (Bidirectional refers to testing each strand by shooting the fiber from the MC to the IC or HC and then shooting the fiber from the HC or IC to the MC.) Testing should be done at both 850nm and 1300nm on multimode fiber. Much has been made of the need to use an OTDR (optical time domain reflectometer) to test fiber; however, this is overkill. The key factor in the functionality of the fiber backbone is attenuation. The use of an OTDR increases the cost of testing significantly while providing nonessential additional information.

◆ Document the infrastructure on floor plans. Once this is done, maintain the documentation and keep it current. Accurate documentation is an invaluable troubleshooting and planning tool. Show the outlet location and use a legend to identify the outlet types. Differentiate between these:

   ◆ Data only, voice only, and voice/data locations

   ◆ Each circuit at that location (by number)

   ◆ All MC, IC, and HC locations

   ◆ Backbone cable routings

Although there is more to designing a telecommunications infrastructure system, the information in this section provides some basic guidelines that should help to remove some of the mystery.

Writing the RFP

If you have been successful at gathering information and asking the right questions, you are ready to start writing your RFP. Although no *exact* guideline exists for writing an RFP, this section provides a list of suggested guidelines. (We have also included a sample RFP at the end of this chapter.) By following these steps, you will be able to avoid many mistakes that could become very costly during the course of the project and/or the relationship.

Remember, though, regardless of how much ink is used to spell out expectations, the spirit of the agreement under which everyone operates really works to make a project successful.

TIP There is a terrific Word document template for a structured cabling RFP in the members-only section of the BICSI website (www.bicsi.org). If you are working with an RCDD, or someone in your company is a BICSI member, have them get it for you. It could save you a lot of work.

INCLUDING THE RIGHT CONTENT IN THE RFP

Will the RFP accurately specify what you want? To ensure that it does, take the following steps:

- Educate yourself about the components of the system to be specified and some of the options available to you.

- Evaluate specific desired features and functionality of the proposed system, required peripherals, interfaces, and expectations for the life cycle and warranty period.

- Solicit departmental/organizational input for desired features, requirements, and financial considerations.

- Determine the most cost-efficient solution for one-year, three-year, and even perhaps five-year projections.

- Evaluate unique applications (wireless, voice messaging, fiber, etc.) and their transmission requirements and departmental or operational requirements and restraints that could impact those applications.

- Analyze perceived versus actual needs for features and functionality, future applications, system upgradability, and so forth.

- Define contractor qualifications. Strongly consider requirements that call for vendors to be certified by the infrastructure component manufacturer whose products they are proposing to install. The same holds true for hardware bids. It is important that any vendor selling hardware be an authorized reseller for the hardware manufacturer. Call for proof of each certification and authorization as part of the initial bid submittal.

- Prepare a draft outline of selected requirements and acceptable timelines (which is subject to minor changes by mutual agreement).

◆ Define project milestones and completion dates. Milestones include bid conference dates, walk-through dates, dates to submit clarifications, the final bid due date, acceptance dates, project-start dates, and installation milestones.

◆ Include in the RFP sections on scope of work, testing acceptance, proposed payment schedules, liquidated damages (damages to be paid in the event of default), restoration, licensing, permits fees, and milestone dates.

◆ Call for all pricing to be in an itemized format, listing components and quantities to be installed and unit and extended prices.

◆ Request that costs to add or delete circuits—on a per circuit basis—be included in the response to the RFP.

◆ Include detailed language addressing the "intent" of the bid. Such language should articulate that the intent is to have a system installed that contains all of the components necessary to create a fully functional system. Language should be included that calls for the contractors to address at time of contract completion any omissions in the bid that would prohibit the system from being fully functional.

◆ Ask for references from similar jobs.

◆ Make sure to allow for adequate time for detailed site survey/estimating.

◆ Upon receipt of bids, narrow the field to three finalists and do the following:

  ◆ Correlate information and prioritize or rank the three remaining bids on cost versus performance. Don't get hung up on costs. If a bid seems too good to be true, it may be. Examine the vendor's qualifications and the materials they are specifying.

  ◆ Schedule meetings and/or additional surveys for best-and-final bids from remaining vendors.

  ◆ Specify that the RFP is the intellectual property of the client and should not be distributed. Though this won't stop an unscrupulous vendor from passing around information about your infrastructure, you have instructed them not to. One consulting company we know of actually assigns each vendor's copy of the RFP a separate number that appears on the footer of each page.

What Makes a Good RFP?

Does a good RFP have many pages (did you do in a few trees printing it)? Can you take advantage of the contractor? These are *not* good benchmarks for determining if your RFP will help to create a good working relationship between yourself and the company with which you contract. You may want to ask yourself the following questions about the RFP you are generating:

◆ Is it fair?

◆ Does it ensure that only competent bidders will meet the contractor qualifications?

- Is it nondiscriminatory?

- Does it communicate the objectives and the wishes of the client clearly and accurately?

- Does it provide protection to both the client and the service provider?

- Does it provide opportunities for dialogue between the parties (such as mandatory site walk-throughs and regular progress meetings)?

- Does it clearly state all deadlines?

- Does it define payment terms?

- Does it define the relationship of the parties?

- Does it address change-order procedures?

Distributing the RFP and Managing the Vendor-Selection Process

Once the RFP has been written, you may think you are home free. However, the next step is just as important as the creation of the RFP: you will then be ready to distribute the RFP to prospective vendors and begin the vendor-selection process.

Distributing RFPs to Prospective Vendors

If you worked with an infrastructure consultant on your RFP, that person may already have a list of contractors and vendors that you can use to fulfill your vendor needs. Many of these vendors may have already been tried and tested by your consultant. However, if you have developed your own RFP, you will need to find prospective vendors to whom you can distribute your RFP for bids.

How do you find the vendors? We suggest the following ways:

- Ask IT professionals from companies similar to yours for a list of vendors they have used for cabling.

- If you are a member of a users group or any type of professional organization, ask for vendor suggestions at your next group meeting.

- If you work with a systems-integration company, ask your contact at that company for one or more vendor recommendations. Chances are good that the contact has worked with vendors in the past that can respond to your RFP.

- Consult your phone-system (PBX) vendor. Many phone-system companies have a division that does cabling.

- If you have a contact at the telephone company, consult that person for suggestions.

As you distribute RFPs to potential vendors, be prepared to schedule vendor meetings and site inspections. For a cabling installation that involves approximately 500 to 2,000 nodes, you can expect to spend at least one full day in meetings and on-site inspections for each vendor to whom you submit the cabling RFP.

Vendor Selection

When reviewing the proposals you get, you may be tempted to simply pick the lowest-cost proposal. However, we recommend that you select a vendor based on criteria that include, but are not limited to, the following:

- Balance between cost and performance
- Engineering design and innovative solutions
- Proven expertise in projects of similar scope, size, and complexity
- Quality craftsmanship
- Conformance with all appropriate codes, ordinances, articles, and regulations

Check references. Ask not only about the quality of work but about the quality of a relationship the reference had with the specific vendor and whether the vendor completed all tasks on time.

Insist on a detailed warranty of a system's life cycle. Consider the ability to perform and any other requirements deemed necessary to execute the intent of contract.

Present a detailed description of work to be performed, payment agreements, and compliance with items contained in the RFP. Include this in the contract.

Identify key project personnel from both sides of the agreement, including the staff associated with accountability/responsibility for making decisions, and the authority to do so.

NOTE Once you have selected a vendor, promptly send letters or place phone calls to the rejected vendors. We agree that it is hard to tell vendors they have not been accepted, but it is worse if they hear about it through the grapevine or if they have to call you to find out.

Project Administration

After accepting the RFP, you are ready for the next phase of your installation challenge: the project-administration phase. This phase is no less critical than the others are.

Project Management Tips

Here are some tips we have found to be helpful during an infrastructure deployment:

- Schedule regular progress meetings. Progress reports should be submitted and compared to project milestones. Accountability should be assigned, with scheduled follow-up and resolution dates.

- Make sure that the contractor supervises 100 percent of the quality inspections of work performed. Cable-certification reports should be maintained and then submitted at progress meetings.

- Make sure that the contractor maintains and provides as-built documentation, the progress of which should be inspected at the regular progress meetings. The as-built documentation may include outlet locations, circuit numbers, telecommunications-room locations, and backbone and distribution routing. Particular care should be taken to make sure this documentation is done properly, as it tends to slip through the cracks.

Planning for the Cutover

If you install cabling in a location that you do not yet occupy, you do not have to plan for a cutover from an existing system. As long as the new cabling system is properly designed, it is relatively easy to move to the new system as you occupy the new location.

However, if you are supervising the installation for a location you do occupy, you will need to consider interoperability and the task of switching over to the new system. In a small system (fewer than 200 horizontal runs), the cutover may occur very quickly, but in medium to large systems, cutover can take days or weeks. Follow these guidelines:

◆ Cutover preparation should begin 5 to 15 days prior to the scheduled date, unless otherwise mutually agreed upon.

◆ Cutover personnel and backups should be designated and scheduled well in advance.

◆ Cutover personnel should have access to all records, diagrams, drawings, or other documentation prepared during the course of the project.

◆ Acceptance should begin at the completion of the cutover and could continue for a period of 5 to 10 working days prior to signing. The warranty should begin immediately upon signing of acceptance.

◆ Acceptance criteria should include 100 percent of all circuits installed. All circuits should pass specified performance tests, and that should be recorded in the project history file and cable management systems.

By following the guidelines, as appropriate to your particular situation, you can greatly reduce the chances of any aspect of your project spiraling out of control.

The final part of this chapter provides a sample of an RFP that has been successfully used in several projects in which we have been involved. Although we caution anyone against adopting an existing RFP without first doing a thorough analysis of his or her own specific needs, the following document can be a guide.

Technology Network Infrastructure Request for Proposal (A Sample RFP)

The following sample RFP, used for a school, may help you generate your own. It is suitable for small installations (fewer than 500 network points or locations). For larger installations, consider working with an infrastructure consultant. Remember, this document will probably not fit anyone's needs exactly.

General

The general section of this RFP includes contractor's requirements and defines the purpose of the RFP, the work that the RFP covers, and the RFP intent.

CONTRACTOR'S REQUIREMENTS

Picking the right contractor is a big part of ensuring the quality of the project meets expectations. In other words, picking the wrong contractor can create huge problems in quality. Here are some suggested requirements for contractors:

(a) The successful contractor must be a certified installer of the infrastructure components being provided and show proof thereof.

(b) The contractor must be an authorized reseller of the networking and infrastructure components quoted and show proof thereof.

(c) Work will be supervised by a Registered Communications Distribution Designer (RCDD) during all phases of the installation. An RCDD must be on site and available to technicians and installers any time work is being performed.

PURPOSE OF THIS RFP

It's important to be clear on the purpose of the RFP and roles and responsibilities of parties involved. Here are some examples:

(a) The purpose of the "Technology Network Infrastructure RFP" is to provide a functional specification for a comprehensive technology network system, including required network cabling and components and required network devices. The purpose of this is also to provide adequate details and criteria for the design of this technology network system.

(b) The contractor shall provide cables, network equipment, and components necessary to construct an integrated local area networking infrastructure.

(c) The contractor shall be responsible for the installation of the technology network systems as defined in the "Cable Plant."

This document provides specifications to be used to design the installation of a networking infrastructure and associated equipment. The contractor shall furnish all labor, materials, tools, equipment, and reasonable incidental services necessary to complete an acceptable installation of the horizontal and riser data communications cabling plant. This is to include, but is not necessarily limited to, faceplates, modular jacks, connectors, data patch panels, equipment racks, cable, and fiber optics.

WORK INCLUDED

Work shall include all components for both a horizontal and riser data cable plant from workstation outlet termination to wire-room terminations. All cable-plant components, such as outlets, wiring-termination blocks, racks, patch cables, intelligent-hub equipment, and so forth, will be furnished, installed, and tested by this contractor. The data cable plant is designed to support a 1000Mbps Ethernet computer network. The data cabling plant and components shall carry a manufacturer-supported 10-year performance warranty for data rates up to 1000Mbps. The bidder must provide such manufacturer guarantee for the above requirements as part of the bid submission.

The scope of work includes all activities needed to complete the wiring described in this document and the drawings that will be made available during the mandatory walk-through.

Any and all overtime or off-hours work required to complete the scope of work within the time frame specified are to be included in the contractor's bid. No additional overtime will be paid.

The awarded contractor must instruct the owner's representative in all the necessary procedures for satisfactory operation and maintenance of the plant relating to the work described in their specifications and provide complete maintenance manuals for all systems, components, and equipment specified. Maintenance manuals shall include complete wiring diagrams, parts lists, and so forth to enable the owner's representative to perform any and all servicing, maintenance, troubleshooting, inspection, testing, and so forth as may be necessary and/or requested.

The contractor shall respond to trouble calls within 24 hours after receipt of such a call considered not in need of critical service. Critical-service calls must be responded to on site, within four hours of receipt of a trouble call. The bidder must acknowledge their agreement to this requirement as part of the RFP response.

All basic electronic equipment shall be listed by Underwriters Laboratories, Inc. The contractor shall have supplied similar apparatuses to comparable installations, rendering satisfactory service for at least three years where applicable.

The installation shall be in accordance with the requirements of the National Electrical Code, state and local ordinances, and regulations of any other governing body having jurisdiction.

The cable system design is to be based on the ANSI/TIA-568-C Commercial Building Telecommunications Cabling Standard and Bulletins TSB-36 and TSB-40. No deviation from the standards and recommendations is permitted unless authorized in writing.

INTENT

This network cable system design will provide the connectivity of multiple microcomputers, printers, and/or terminals through a local area network environment. Each designated network interface outlet will have a capacity to support the available protocols, asynchronous 100- and 1000Mbps Ethernet, 4- and 16Mbps Token Ring, FDDI, and so forth through the network cabling and topology specified. The school may select one or any combination of the aforementioned media and access protocol methods; therefore, the design and installation shall have the versatility required to allow such combinations.

It is the intent of this document to describe the functional requirements of the computer network and components that comprise the "Technology Network System." Bid responses must include all of the above, materials, appliances, and services of every kind necessary to properly execute the work and to cover the terms and conditions of payment thereof and to establish minimum acceptable requirements for equipment design and construction and contract performance to assure fulfillment of the educational purpose.

Cable Plant

The following section covers the installation of horizontal cabling, backbone cabling, cable pathways, fire-code compliance, wire identification, and cross-connects.

HORIZONTAL CABLE

The following requirements apply for horizontal cabling:

(a) Each classroom shall have two quad-outlet wall plates installed. Each of the four information outlets shall be terminated with eight-pin modular jacks (RJ-45). The wall plates will be placed on opposite walls. There are a total of 37 classrooms.

(b) The computer skills classroom shall have 15 quad-outlet wall plates installed. Each will have four information outlets terminated using eight-pin modular jacks. Each wall plate will be located to correspond to a computer desk housing two computers. These locations will be marked on the blueprints supplied during the walk-through.

(c) Each administration office shall have one quad-outlet wall plate with four information outlets, each terminated using eight-pin modular jacks. There are 23 such office locations.

(d) Common administrative areas shall have one quad-outlet wall plate with four information outlets terminated with eight-pin modular jacks. There are 17 such common administrative areas.

(e) The school library shall have quad-outlet plates placed in each of the librarian work areas, the periodical desk, and the circulation desk. The student research area shall have two quad-outlet plates. Eight work areas total require quad outlets. The exact locations of where these are to be installed will be specified on the blueprints supplied during the walk-through.

(f) The school computer lab shall have 20 quad-outlet wall plates installed. The locations of these will be specified on the blueprints to be supplied during the walk-through.

(g) Horizontal cable shall never be open but rather will run through walls or be installed in the raceway if the cable cannot be installed in walls.

(h) The contractor is responsible for pulling, terminating, and testing all circuits being installed.

(i) The horizontal cable for the data network shall be twisted-pair wire specified as Category 5e by the ANSI/TIA-568-C standard and shall be UL-listed and verified. The cable shall meet all fire and smoke requirements of the latest edition of the NEC for the location of the installed cable.

(j) Testing for the distribution components will comply with ANSI/TIA-568-C Category 5e specifications and will certify 100 percent functionality of all conductors. All circuits must be tested and found to be in compliance. All testing results will be provided to the customer in a hard copy and electronic format that is a direct output of the test equipment. The contractor shall be responsible for applying for all available warranties and shall provide proof of warranty at the completion of the installation.

(k) The data-cable specifications are intended to describe the minimum standard for use in the "Technology Network System." The use of higher-grade data cabling is recommended if such can be provided in a cost-effective manner.

(l) Each cable shall be assigned a unique cable number.

(m) In the telecommunications room, the contractor shall install four separate color-coded patch panels. Each wall plate's information outlet shall use a different patch panel, and the wall-plate information outlets will be documented using the patch panel's color code and the patch-panel number.

(n) Wire management shall be employed in all telecommunications rooms and the equipment room.

DATA BACKBONE CABLING

The following specifications apply to the data backbone cabling:

(a) An ANSI/TIA-568-C–compliant 850nm laser-optimized 50/125 micron multimode (OM3) fiber-optic cable network is to be the backbone between the equipment room (the MC) and any telecommunications (wiring) rooms.

(b) All telecommunications rooms shall have 12 strands of multimode fiber-optic cable between the telecommunications room and the equipment room.

(c) All fiber must be FDDI- and 1000Base-SX–compatible.

(d) All fibers are to be terminated using SC-type connectors.

(e) All fiber is to be installed in an inner duct from rack to rack. A 15-foot coil of fiber is to be safely and securely coiled at each rack. The contractor will be responsible for any drilling or core holes and sleeving necessitated by national, state, and/or local codes.

(f) The fiber-optic patch panels are to be configured to the amount of strands terminated at each location. Fiber-optic panels shall be metallic, are to have a lockable slack storage drawer that can pull out, and shall occupy one rack position.

(g) Testing of fibers will be done using a power meter. The tests will be conducted at 850nm and 1300nm, bidirectionally. All test results will be provided to the customer in hard-copy format.

FIRE-CODE COMPLIANCE

All cabling installed in the riser and horizontal distribution shall meet or exceed all local fire codes. At a minimum, the requirements of the latest edition of the NEC shall be met, unless superseded by a local code.

WIRING PATHWAYS

The following are related to the installation of cable in plenum and other cable pathways:

(a) Cable pathway design should follow the TIA-569-B (Commercial Buildings Standards for Telecommunications Pathways and Spaces) standard.

(b) The methods used to run cable through walls, ceilings, and floors shall be subject to all state and local safety code and fire regulations. The contractor assumes all responsibility for ensuring that these regulations are observed.

(c) Cables shall be routed behind walls wherever possible. Surface-mount raceway shall be used where necessary.

(d) New cables shall be independently supported using horizontal ladders or other wire-suspension techniques. Cables shall not be allowed to lie on ceiling tiles or attached to electrical conduits.

(e) System layout shall restrict excessive cable lengths; therefore, routing of horizontal cables shall be in a manner as not to exceed 90 meters from device plate to patch panel located in the assigned wiring room. Each cable shall be home-run directly from its cross-connect to the wall plate.

(f) Cables shall be terminated at the rear of the patch panel within the telecommunications rooms and at the wall plates only. There shall be no splicing of any of the cables installed. Intermediate cross-connects and transition points are not allowed.

(g) The following are the minimum distances that Category 5e UTP shall be run from common sources of EMI:

EMI Source	Minimum Cable Separation
Fluorescent lighting	12 inches
Neon lighting	12 inches
Power cable 2 KVA or less	5 inches
Unshielded power cable over 2 KVA	39 inches
Transformers and motors	39 inches

WIRING IDENTIFICATION

All cables, wall jacks, and patch-panel ports shall be properly tagged in a manner to be determined at a later date. Each cable end must be identified within 6 inches from the termination point.

TELECOMMUNICATIONS ROOMS

The following are related to the installation of the telecommunications (wiring) rooms:

(a) The rooms to be used as the originating points for network cables that home-run to the room outlets are referred to as wire rooms or telecommunications rooms. All racks and their exact locations will be confirmed during the mandatory walk-through; their locations are specified on the blueprints that will be provided during the initial walk-through.

(b) Rack layout should provide enough space to accommodate the cabling, equipment racks, patch panels, and network-control equipment, as required. Additionally, the locations should provide for convenient access by operational personnel.

(c) All racks are to be configured as shown on the attached diagram, with all the fiber-optic cables at the top of the rack, the distribution below the fiber, and the hardware components mounted below the distribution patch panels.

(d) All racks, panels, and enclosures for mounting equipment shall meet 19-inch EIA mounting-width specifications. Each equipment rack should include two 19-inch rack shelves that can support the weight of a 50-pound uninterruptible power supply.

(e) Equipment racks shall be properly grounded to the nearest building ground and must be properly attached to the floor and supporting wall by means of a horizontal-rack bracket mount. All equipment racks must have a six-outlet 20-amp power strip with surge protection installed inside.

MC/IC Cable Management

The following relates to cable management for the main cross-connect (MC) and intermediate cross-connect (IC) in the equipment room, and horizontal cross-connects (HC) in telecommunications rooms:

(a) The contractor is required to install cable management on all racks installed. Cable management is to consist of horizontal management between each panel and vertical management on the sides of the rack.

(b) All cable management is to be of the "base-and-cover" style. Cable management is to be provided for the front of the rack only.

As-Built Diagrams

The contractor will provide as-built documentation within 15 days of completion of the project. These prints will include outlet locations, outlet numbers, MC/IC/HC locations, trunk-cable routing, and legends for all symbols.

Network Hardware Specifications

The networking hardware should be provided, installed, and serviced through a certified reseller/integrator or direct from the manufacturer.

Bidding Process

All work is to be completed based on the dates from the attached schedule. Dates on the attached schedule include walk-through dates, bid submission dates, and expected project-start and completion dates. Questions and comments are welcomed; prospective contractors are encouraged to submit these questions in writing.

Bid Submittals

The following are related to submittal of bids:

(a) All bids are to be submitted in triplicate.

(b) Each bid is to list all labor, material, and hardware costs in an itemized fashion. The detail is to include itemized unit pricing, cost per unit, and extended prices for each of the material and hardware components as well as the specific labor functions.

(c) A cost, per outlet, to add or delete outlet locations is to be included in the pricing format. This cost is not to include any changes in hardware or patch-panel quantities.

(d) There is also to be a scope of work provided that details all of the functions to be provided by the contractor for the project.

(e) Quote optional Category 5e patch cables and station cables on a per-unit cost basis. List pricing for 3-foot, 5-foot, 7-foot, 9-foot, and 14-foot patch cables.

(f) Quote optional network cutover assistance on a per-hour basis per technician.

MISCELLANEOUS

All data found in this RFP and associated documents is considered to be confidential information. Further, data gathered as a result of meetings and walk-through visits is considered to be confidential information. This confidential information shall not be distributed outside of organizations directly related to the contractor without expressed, written approval.

Further, all data submitted by prospected contractors will be treated as confidential and proprietary; it will not be shared outside of the vendor-evaluation committee.

The Bottom Line

Determine key aspects of writing an RFP.   The RFP is essential to the success of your telecommunications-infrastructure project. It is often referred to as a combination of a guidebook, map and rulebook. The guidebook quality ensures that the proper instructions are given for the design, installation, and testing phases of the project. The rulebook quality ensures that the roles and responsibilities of the relevant parties are established along with each of their liabilities. The map quality ensures that a direction has been set. This is a very important quality of the RFP and will help ensure the long term success of the project and many years of successful service after the project is completed.

> **Master It**   What analysis must be performed as the first part of the development of the RFP in order to define the customer's current and future needs, and key restrictions and constraints?

Determine the components of the cabling infrastructure that should be included in the RFP.   The design phase is an integral part of the RFP process. This may seem a bit intimidating, but even the largest and most intimidating projects become manageable when you divide the project into smaller sections.

> **Master It**   Identify the four major subsystems of the cabling infrastructure that should be included in the RFP.

Understand what to do after the RFP is written.   The steps following the writing of the RFP are just as important as the creation of the RFP. Once the guidebook, map, and rulebook are written, it is time to select the team and initiate the plan.

> **Master It**   What two steps should be taken after writing the RFP and obtaining approval by the customer in order to get started implementing the project?

Chapter 17

Cabling @ Work: Experience from the Field

Throughout the research phase for this book, cabling installers related to us their experiences, hints, tips, and stories from the field. There is no substitute for advice and stories from people who have been in the trenches. Although some of the topics mentioned in this chapter are also mentioned elsewhere in the book, we felt it was important to reiterate them through people's experiences.

Much of this chapter is targeted toward the professional cable installer, but anyone installing cabling will find some helpful information here. First, we'll give you some guidelines about the business of cable installation, and then we'll include a handful of case studies drawn from our experience in the industry.

In this chapter, you will learn to:

◆ Employ useful hints for planning the installation

◆ Understand tips for working safely

◆ Plan for contingencies

Hints and Guidelines

After a few years in the cabling business, you'll learn skills and approaches not specifically related to cabling technology, which will mark your work as professional. These skills and approaches, described in the following sections, will help you even if you're simply evaluating the work of others.

Know What You Are Doing

Purchasing and reading this book is a step in the right direction. You need to know more than just how to install cable, however. To design and implement a good cable plant for yourself or for a customer, you also need to know how networks are used and how they grow. Consider these factors:

Understand current technology.   Understand which technologies are appropriate for a given situation. UTP (Category 5e or greater) is the king of desktop cabling, for example, although optical fiber cabling has become the rule for campus and intra-building backbones. Wireless works great in mobile environments but over long distances introduces licensing issues. Read this book for information on cabling technologies and check other networking magazines and books for details on competing technologies.

Understand the standards that apply to your work.   The TIA publishes the ANSI/TIA-568-C standard, and the ISO/IEC publishes the ISO/IEC 11801 Ed. 2 standard. Both are discussed in Chapter 2, "Cabling Specifications and Standards." Professionals will be intimately familiar with one or the other of these standards (in the United States, it will be ANSI/TIA-568-C).

Know the limitations of the technology you use.   Don't try to run Category 5e twisted-pair cable for 500 meters, for example. Point-to-point lasers don't cope well with snow. Single-mode fiber can carry a signal farther than multimode fiber. You can pull only so many twisted-pair cables through a conduit. Outside plant cable is not the same as inside plant cable, which can be divided into plenum and nonplenum. (You do know the difference, don't you? Your fire marshal and building inspector certainly do.) This book tells you what you need to know about current network-cable technology.

Keep an eye on new developments.   Watch for changes in technology. Which advances in networking will make your cabling setup obsolete, and which will enhance it? We know people who were putting in 10Base-2 and 10Base-5 coaxial cable for networking in 1995, cable that was used for a year or two and then never used again because all local area networking moved to UTP cable. One of the interesting developments occurring as this book is being written is the use of high-density fiber-optic connectors that allow parallel optical transmission for speeds of 40 and 100Gbps. We suggest you subscribe to the cabling industry's trade magazines to keep abreast of the field; information about these magazines can be found in Appendix C, "Registered Communications Distribution Designer (RCDD) Certification."

Understand the business of cabling.   Even if you are just installing a network for your own company's use, you should strive to perform a professional job. After all, you are the customer in that instance, and you want to be pleased with the results. You should know how to plan the job, acquire materials, assemble a team, train and supervise the crew, oversee the installation, test, document, and sign off the job. Read the rest of this chapter for some hints on the business of cabling. Numerous industry periodicals can keep you up-to-date on the latest in cabling business. See Appendix B, "Cabling Resources," for more information.

Understand the business of business.   Though not related specifically to networks, if you are installing networks for others, you need to know how to run a business (or you need to hire people who will do it for you). That includes knowing about attracting work, bidding, developing and negotiating contracts, hiring, scheduling, billing, accounting, and so on. For more detailed information on the business of business, check out your local college's or university's business school.

Plan the Installation

Every well-executed job was at one point merely a well-planned job with realistic appraisals of the time, equipment, materials, and labor required. The following steps will help you develop that realistic plan:

Get the complete specification.   Obtain in writing, with detailed and accurate blueprints, exactly what sort of network the customer wants. Often it is up to you to plan the cable paths, but the customer usually has a good idea of where the network drops and patch panels should be located. Don't forget to confirm that the blueprints are accurate and up-to-date.

Perform a job walk.   Go to the site and walk through the job. Peer up into ceilings and look at conduits. Examine any walls that you'll have to penetrate. Make sure that the telecommunications room has sufficient space for your own racks and patch panels. Some areas are much easier to network than others—an office building that uses ceiling tiles and is still under construction, for example, is much easier to wire than an old brick building with plastered ceilings or an aircraft carrier with watertight bulkheads and convoluted cable paths.

Clarify inconsistencies and ambiguities.   If you don't see a way to get a cable from one location to another, point it out. Ask why the front desk doesn't have a drop planned. Doesn't the receptionist have a computer? Will a computer be placed there in the future? Questions you ask at this stage can save you from change orders later.

Calculate the lengths of network runs.   With an accurate blueprint, you can calculate lengths away from the site. Otherwise, you'll have to break out the measuring wheel and walk the path of the cables. You will have to use the measuring wheel for any outside cable runs (from one building to another, for example).

Plan for service loops and waste.   The last bit of cable you pull from the spool is always too short. Runs often have to go around unexpected obstacles. When you pull a group of cables to the same general area, some will need more length to get to their destination than others will. But you'll still have to pull the same amount of cable in the bundle for all the runs in an area and trim each cable to fit. You should trim the cable a little too long and push the extra back up into the wall or ceiling so that, if necessary, the jack location can be moved later without requiring the whole run to be pulled through again. That adds up to 10 to 30 percent more cable than a naive plan would indicate.

Evaluate your labor and skill level.   How many feet of cable can your installers pull in an hour? Do you have enough teams to pull groups of cable in different directions—and do you have supervisors for the teams? How many faceplates can each installer punch down in an hour? After you've gained some experience, you will be able to look at a job and know how long it will take your team. In the meantime, calculate it out.

Have the Right Equipment

The right tools indicate your commitment to doing the job right. Some tools are designed for the do-it-yourselfer, whereas others are designed for professionals who install cable every day. All experienced cable installers should carry punch-down tools, screwdrivers, snips, twine and fish tape, electrical tape, a measuring wheel, a cable tester, and patch-panel lights. See Chapter 6, "Tools of the Trade," for more information about tools.

Test and Document

Many cable installers view testing and documentation as a convenience to the customer and an annoyance to be avoided. We view the lack of testing and documentation as a threat to everyone's sanity. And, if you are thinking about offering a warranty that is backed up by the equipment manufacturer, you will need to submit test data and documentation for that warranty to be issued.

We can't count the number of times a customer has come to one of us and said, "This cable that you installed is bad. None of us can get any work done, and it is all your fault." If you kept the test documents (and you should *always* keep a copy of them for your own records), you can point to them and say, "But it passed with flying colors then, and you signed here. Of course, we stand by our work and we'll come out and fix it if it's broken, but you just might want to check your network adapter settings [or jumper cable or network equipment] before someone drives all the way out there…"

NOTE Don't discount the fact that damage can occur to a cable, jack, or patch-panel connection after the cabling system is installed.

Another common problem professional cable installers report is being called in to fix cabling problems left by another installer who didn't bother to test. Honest mistakes can be made in any cabling installation; the following are examples:

◆ Copper cables were routed past RF-noisy power lines.

◆ Cables had their jackets scraped off when pulled through narrow places or around corners.

◆ Installed cables, jacks, and/or patch panels were labeled incorrectly.

◆ Cables were bent in too tight a radius.

◆ Category 5e copper cables had their wires punched down in the wrong order.

◆ Cable was installed that exceeded the maximum length specified by the standards.

Testing your cable plant after you install it will pinpoint any of these problems. We are amazed that some installers simply assume that they've made no mistakes. Nobody's perfect—but you don't have to remind your customers of that.

Train Your Crew

You can get any group of enthusiastic people together and pull cable through a ceiling. Punching down the little colored strands of wire at the end of the cable into the faceplate is a different matter—show them how to do it first, give them some cable scraps, and have them punch down both ends. Then test those short cables. Until your crew gets a feel for punching down the cables correctly, you'll find crossed wires, marginal connections, and strands cut too short and too long. It takes practice to do it right, but the time you spend training your crew will be well worth a reduction in the number of problems to fix on the job site.

Terminating fiber-optic cable requires a different order of training altogether. Unless your installers have spent hours cutting, stripping, polishing, and terminating fiber-optic cable and then examining what they've done wrong in a microscope, they'll never get it done right. Have your installers make all their mistakes on your own property rather than on your customers' property.

Work Safely

Train your crew in safety as well as proper cabling methodology. If you take some basic steps to reduce the likelihood of accidents and your liability, you will sleep better at night.

Make sure that the safety lectures you give are themselves done safely, too. Once we had a contract to install fiber-optic cable in military hospitals. This was a retrofit situation, and we did not have precut holes in the drop locations; we had to cut the holes ourselves. A supervisor was showing how to properly wield a drill with a hole-saw bit installed and said, "And never chock the bit with your hand, like this," whereupon he grabbed the hole-saw bit with one hand and touched the trigger of the drill with the other to tighten the bit. Naturally, the drill whined, the bit spun, and blood dripped on the floor from the new gouge in the supervisor's hand. Fortunately, he did not drill a new hole through his hand. Because the incident happened in a wing of a hospital, a nurse took him away and bandaged his hand. He returned shortly thereafter and resumed his safety lecture. "And in the case of an on-the-job accident," he continued after looking at his watch, "you can take 15 minutes off." He was kidding, of course, but his unintentional example made an impression on the crew.

Network installers work quite a bit in ceilings, tight places, new-construction areas, dusty areas, and around all sorts of construction equipment. Make sure your employees know proper safety methods for handling ladders, wearing safety helmets, using dust masks, and so on.

Make It Pretty

The cable we install looks good. We are proud of our work, and so is the customer when we're done. Our telecommunications rooms look like something important happens there—huge bundles of cable swoop out of conduits and separate into neatly dressed branches that flow across to their requisite rack locations. The customer appreciates a telecommunications room that looks orderly, and an orderly wiring installation is easier to diagnose and fix network problems in. We needn't jiggle and pull cables to figure out which direction a bad line is running—when we look at it we know just by where it is. This saves a tremendous amount of time, both for us and for the customer, long after we're gone.

You can feel real pride when cabling systems you have installed are printed up in magazines. In one case, a cabling company was installing a fiber-to-the desk network for a local biotech company. This was when fiber was new and expensive (as opposed to fiber-network equipment now, which is simply expensive), and the supervisor was nervous about its installation and anxious to test it to make sure it all worked properly. Once the far ends of the cables had all been terminated, he terminated the fiber patch panels (which involves much cutting, polishing, gluing, and so forth and isn't something you just redo without a lot of expense). Unfortunately, he neglected to dress the cables first. Now, you can't untangle a knot once you've glued the ends together, which is essentially what he'd done. Fortunately, he'd left enough cable for a service loop (see the section "Plan for Contingencies" in a moment), and he used the extra length to push his new and permanent knot up into the ceiling where it couldn't be seen. With a junction box around the knot (look—cables go in, cables come out—never mind what's inside!), the plant was neat and ready for the photographers.

Look Good Yourself

It is important that you and your installation crew look appropriate for the job. The customer forms an impression about you and your company by how you walk, talk, and dress. The customer is reassured and happy when he sees professionals behaving in a professional manner. Even if you are installing cable for your own company, the way you and your crew carry yourselves will affect the work you do.

Every cabling company we have worked with has had a dress code of jeans and a T-shirt for their installers, and the company provides the T-shirts (with a company logo on the back, of course). The T-shirts identify the work crew on the site and provide free off-site advertisement as well (if the logo is not too terribly designed and the installers aren't embarrassed to wear them at home!).

Plan for Contingencies

No job ever goes exactly as planned. If you have only enough time, materials, and labor for the job as planned, you will inevitably come up short. Keep the following in mind as you plan:

Service loops   If you've read this far, you've already seen one reason for leaving service loops in installed cable (a service loop is an extra length—a few feet—of cable coiled up and left in the ceiling or wall). Another reason to leave service loops comes from a basic rule of cable: you can always cut it shorter, but you can't cut it longer. Inevitably, racks need to be moved, desks are reoriented, and partitions are shifted—often even before you're done with the job. If the cable is a bit longer than you originally needed, you can just pull it over to the new location and re-terminate rather than pulling a whole new cable to the new location. Also, if you determine while you're testing the plant (you *are* testing, aren't you?) that the cable has been punched down incorrectly, you will need an extra couple of inches to punch it down correctly.

Additional drops   Customers are always forgetting locations that they need network connectivity in ("Oh, you mean the printer requires a LAN connection too?"), so you should be prepared with additional cable, faceplates, and jacks for the inevitable change order. You can charge for the extra time and material required to install the extra drops, of course. Some companies bid low and expect to make their profit on exorbitant change-order costs, but we prefer to plan ahead and pleasantly surprise the customer with reasonable change-order prices. One installer that we know estimated that at least 75 percent of all cabling installations he has worked on required additional drops within the first month after installation.

Extra time required   What do you do when the installers punch down your entire network using Token Ring faceplates instead of Ethernet faceplates? (This actually happened once to someone we know, and although Ethernet and Token Ring use the same jack form factor, they connect different pins to the cable and the faceplates are not interchangeable. However, if the four-pair cable and either the T568-A or T568-B wiring patterns are used, the jack will support either type of connection.) What you do is tear out all the incorrect faceplates and re-terminate your network. If you haven't budgeted extra time for mix-ups like this, you risk serious disruption of your (and your customer's) schedule.

Labor shortages   People get sick. Competitors steal your best employees. The customer wants all four phases rolled out at once instead of week-by-week as you'd originally planned. Either you need to budget extra time for shortages in labor or you need to bring more installers on the job. Customers are always delighted when you beat your schedule, so it pays to pad the time budget a little bit (as long as you don't pad it so much you lose the bid).

Equipment and material shortages   When you plan a network installation, the amount of cable you allow for is always an estimate. Short of pulling a large amount of string from one location to another, you can't determine beforehand exactly how much cable an installation will require. The cable paths deviate from the planned ones due to interposed air ducts, inconvenient patch panels, already-full conduits, and so on. Typically, the amount of cable used exceeds the planned amount by 10 to 30 percent. Professionals are pretty good at estimating cable usage and will subconsciously add the "fudge factor" to their own calculations so that they seldom estimate more than 10 percent more than what will actually be required (sometimes to the installers' chagrin—see the section "Waste Not, Want Not," later in this chapter).

But cable isn't the only item you can run short of. Depending on the job, you may need innerduct tubing, racks, panels, faceplates, jacks, jumpers, rack screws, raceways, Velcro straps, string, T-shirts, and dust masks. It helps to have some spares on hand for when material is shipped to the wrong site, is held up in manufacturing, mysteriously disappears, gets used up too quickly, or gets stripped, scraped, burnt, painted, flooded, or stepped on. It pays especially to be sure of the arrival of any specialized equipment or peculiar material—for one job in particular, we were held up waiting for a manufacturer to actually *manufacture* the fiber-optic cable we needed.

Match Your Work to the Job

No two potential networks have exactly the same requirements. At first glance, two network jobs may appear to be identical—each may specify the same type of cable and indicate the same number of network drops to be installed in the same office environment with the same number of rack locations and so on. However, one job may take twice as long to execute as the other job. When a customer relaxes a constraint on how you install cable or when special considerations exist for a particular job, you should take advantage of the specific circumstances of that job.

When we install network cable, we find that some of our customers care about the order in which drop locations show up on the telecommunications room patch panels. Other customers do not care, as long as the patch panels are clearly labeled. This simple distinction has a huge impact on how long it takes to install the network cables.

One typical customer of ours (who, refreshingly, had a complete and comprehensive requirement for us to work from) specified the rack layout and cable arrangement all the way down to which location on each patch panel should correspond to which faceplate location in the building. To get the right cable to the right location to satisfy this requirement, we had to start in the telecommunications room, label one of each cable, pull them in groups to the general location of their destinations, and then cut and label the other ends. About 40 or 50 cable drops were in each general area, so we had to sort out the cables and pull them to their destinations from the general location that all the cables were originally pulled to. (We were lucky that none of the cable numbering had been rubbed off during the cable pulling!) We then cut them to the correct length and punched them down at the drop end. At the telecommunications room end of the cables the network technician also had the unwelcome job of sorting all those cables and punching them into the rack in correct order. All that writing and searching and sorting takes much time because you must not pull the wrong cable to the wrong location.

The other job we ran at the same time went far more quickly because the network administrator didn't care which patch-panel location went to which drop location—he was just going to plug them into a hub anyway. We just pulled cable from the telecommunications rooms to the drop locations, trimmed them, and terminated them in the wall sockets. The network technician in the telecommunications room simply punched that end of the cables down one by one, regardless of where the other end of the cable went.

Of course, we then had to determine which cable went where for testing purposes (*always* test, document, and label your work). However, we have some nifty devices that we plug into the patch panels that have LEDs that light up when a technician sticks a probe in the drop end of the cable. The technician calls out over the radio where that drop is, the patch-panel technician writes down the patch-panel location, they remove the LED unit and plug in the set, test the cable, and then move on to the next location. (If you are looking for these handy little gadgets, you can find them on the Internet at www.idealindustries.com.)

A job done this second way can take half the time (and therefore can be performed at a lower cost) of one done the hard way. But, of course, the easier method should only be done if the customer doesn't mind that the drop locations show up in random order in the patch panels.

Waste Not, Want Not

One cabling company owner found a good method for solving the tedious chore of cleaning up after a job. (As a crew installs cable, a good 10 to 30 percent of the cable ends up as trash. About half of the waste piles up next to the telecommunications room, and the rest lies scattered throughout the drop locations.) He observed that the scrap cable ends contained quite a bit of high-quality copper. Although his primary concern was to make sure the client had the best possible experience working with his company so that he'd get referrals, he was nevertheless bothered by the amount of trash he was generating.

He instituted a policy that all the leftover copper the installation technicians could scavenge from the cable belonged to them, and proceeds from the recycling of it would go toward a company party. All they had to do was gather it up, bag it, and place it in the back of his truck—one of the supervisors took it to the recycling plant for processing. The technicians were then meticulous about cleaning up—even the inch or two of a trimmed cable was picked up and stuck in the back pocket of a technician to go toward the fund. Of course, they had strict instructions not to pick up any of the other contractors' cable.

Customers were amazed at how clean this owner's job sites were at the final walk-through. He now has an excellent reputation in the industry because of his company's cleanliness and the way he handles other details of his business. And, finally, his employees enjoy working for him, have a reason to clean up their work, and have really cool parties.

Case Studies

To give you a better idea of what it is like to install network cable, here are a handful of case studies that are drawn from the experiences of real cabling installers and contractors in the field. The names have been changed to protect the innocent.

A Small Job

Recently, a medical website development company we'll call Quasicom decided to replace all the cables strewn through their hallways with a real network. They contacted a cabling contractor to whom they had been referred. The referring party said the company wouldn't have to worry about the quality of the contractor's work. A Quasicom representative asked the contractor to look at the company's problem and quote a price for installing the network.

That contractor took a hands-on approach, so he came down himself to perform the job walk. He talked with Quasicom's information services (IS) staff to determine the company's needs and got a written document that detailed where it wanted the network locations and the rack. With such a small network, the contractor didn't have much calculation to do, so he presented Quasicom with an offer that was accepted.

Two days later, after a contract was signed and exchanged, the contractor dispatched a team of two installers to the job site. Neither installer was a supervisor, but the senior member of the crew had enough experience to see that the job was done right. The contractor gave the plans to the crew at the job site, showing them where to put the rack and to run the cables and the wiring configuration to use when punching down the faceplates. He then left them to do the work (he had another job walk-through to do for a much larger customer).

Quasicom's premises were of typical modern office construction—a removable tile drop ceiling provided an easy way to run the cables from one location to another. (It is easy to drop network cables behind the drywall once you know how.) The crew expected no difficulties in installing the cable.

The two installers set up several boxes of spooled cable in the location where the rack was to be (in this case, it was not in a telecommunications room but in a server room where all the company's web hosts were hooked up to the Internet). The plan indicated that the biggest run they had was of eight cables to the eight drops in the front office area, so they set up all eight boxes of cable. Quasicom's IS staff wanted the patch-panel terminations of the wire to be in room-number order, so the installers marked the ends of the cable and the boxes they came in with indelible black marker.

The crew then pulled out the requisite number of feet of cable. (Normally, they would measure it, but in this case they just pulled it down the hall to its approximate drop point and made allowance for going up into the wall and down behind the drywall and added some service-loop extra.) They then used their snips to cut the cable off and marked the other end according to which box it came from.

They did the same with four boxes of cable for the quad run back to the back offices. That left two locations, each with a double-jack faceplate, which did not share a run with any other locations. They picked the two boxes that looked like they had just enough cable left in them to pull those runs from, and they drew those cable runs out as well.

Now it was time to put the cable up in the ceiling. The crew started by removing a few ceiling tiles so that they had access to the ceiling space. Then they tied the bundle of eight cables to the free end of a ball of twine and tossed the twine through the ceiling to a reasonably central area for all of the front-office drops. They used the twine to pull the cables through and down to that point. They performed the same operation for the other bundles of cables.

They had all the cables almost where they needed to be—just 20′ shy of the goal. They removed the ceiling tiles directly above the drop locations to determine that there would be no inconvenient obstacles (such as power conduits) and used drywall saws to cut holes in the wall for the faceplates. For each location, one member of the installation team found the corresponding cable and fed it across the ceiling to the other installer. That installer dropped it down the wall for the first installer to pull out of the hole.

Now all the cable was in place and merely needed to be terminated. Although pulling cable is a team process, terminating it is a solitary one. One member of the team began installing the boxes for the faceplates, stripping and punching down the faceplates (leaving enough cable length pushed up in the wall for a service loop), and screwing the faceplates into the box on the wall. The other team member went back to the server room to set up the rack and dress the cable for punching down on the rack.

This contractor always has his best installers perform the important job of punching down the racks. A small margin for error exists when terminating cable on a rack because so many cables feed into the same space-constrained location. If someone makes a mistake punching the cable down, he or she has to draw more cable out of the service loop (see the importance of a service loop?) to re-terminate. Doing so messes up the pretty swooping lines of cable tie-strapped to the rack and to the raceways. Installers can't terminate first and dress the cable later because they would end up with an ugly knot of cable in the ceiling.

That afternoon, the cabling contractor came back with his test equipment, and he and his crew verified that all of the drop locations passed a full Category 5e scan. They used a label gun to identify all of the drops and their corresponding rack locations and then had Quasicom's IS manager accompany them in a final walk-through. They provided Quasicom's manager with the documentation (the same plans he'd given them along with a printout of the test results), got his signature on an acceptance document, and it was all over but for the billing.

Job summary: Quasicom

Type of network: Interior Category 5e

Number of drops: 16

Number of telecommunications rooms: 1

Total cable length: 800 feet

Crew required: 2

Duration: 6 hours

A Large Job

The job walk-through that the contractor of the small job left to perform was for a defense contractor we'll call TechnoStuff, which had a new building under construction. For this job, the contractor was a subcontractor for another firm that had bid for all of the premises wiring for the new location. The Category 5e cable plant was to provide network access to a maze of cubicles and a handful of offices on two floors of the building. All but two of the telecommunications rooms were on the first floor.

Wiring cubicles is different from wiring offices because you have no walls to fish cable down. Also, interior-design consultants have an alarming tendency to move cubicles around. So instead of coming down the walls to faceplate locations, the runs to cubicle locations come up from locations in the floor. Sometimes the cables are terminated there (using *very* tough receptacles in the floor), and sometimes the customer wants the cables drawn up from the floor and along raceways in the cubicle walls to faceplates in the cubicles. In TechnoStuff's case, they wanted the cables terminated in the cubicles.

The contractor walked the site and examined the network plan. He talked with the foreman about construction schedules, when his crew would have access to the area (because he wanted to get in after the framing was done but before the drywall went up), who would be putting in the conduit and boxes, and so on. He emphasized that the conduit had to have string left in it (it's a pain to get cable through it otherwise). He also asked if he had to drill his own holes in the floor (typically the answer to this question is yes).

The next week, the contractor took his crew out and showed the supervisors what had to be done. At that point, they had the left wing of the first floor of TechnoStuff available to them, along with all of the telecommunications rooms. An office furniture contractor was setting up the cubicles in the right wing. The cabling contractor assigned one crew to set up the racks and pull the fiber-optic backbone from the periphery telecommunications rooms to the central one.

He directed the other two crews to pull the cable for that wing from the telecommunications rooms to the cubicle locations. He then left the supervisors in charge because he had a meeting with TechnoStuff regarding materials-delivery timetables and the pay schedule.

As soon as the supervisors had cable to the cubicle locations, one team began terminating the cable while the other continued pulling cable. TechnoStuff had no preference about which patch-panel location ended in what cubicle, so, to terminate for each location, the teams simply grabbed the next available cable in the area where they came up out of the floor. The Category 5e cables were pulled in bundles of 25 (any thicker bunches become unwieldy).

The contractor had evaluated the timetable and his available labor correctly. The crew just finished punching down the cable in the left wing when the right wing cubicles were all set up (the cable pullers had drawn their cables into the space and let the cubicles be set up around them). The first team had finished pulling and terminating the fiber-optic backbone and began terminating the Category 5e patch panels for the first floor.

The contractor came back at the end of the week to deal with some miscommunication about who was to provide the wall plates for the offices; he found to his disgust that he was responsible for doing so. The drywall had already gone up, and his crew would have to punch holes in it and dangle the cable down behind it instead of putting the cable in when the walls were open and easy to work with. In addition, TechnoStuff had decided at the last minute that it really did need network connections in its conference rooms, so extra cable had to be pulled to those locations as well. The contractor gave his supervisors the revised plans and additional instructions. He left to make sure that the additional cable and supplies would arrive in time.

With so much work in the open basement pulling cable from one place to another along the ceiling, one of the installers decided to "walk" his ladder from one hole to another while he was still on it instead of getting down and carrying it. Naturally, he overbalanced and fell, spraining his ankle. The supervisor sent him to get it checked, and the cable-pulling crew was down one member. He replaced him by pulling an installer off the punch-down crew, which was getting ahead of the cable pullers anyway.

At the end of two weeks, all of the cables were in place and terminated, and it was time to test and document. The contracting company discovered where each cable ended up by putting indicator lights in the patch panels and having an installer walk from location to location with a radio and a tool that would light up the indicator for that location when he plugged it in the faceplate. When the patch-panel location was found, they swapped the locator for the tester and measured the performance of that cable.

After all of the cables had been tested, labeled, and documented (and a handful of mistakes fixed), the supervisor took TechnoStuff's representative on a final walk-through and had him sign off to accept the job.

Job summary: TechnoStuff

Type of network: Interior Category 5e with fiber-optic backbone

Number of drops: 500 (times 4 jacks per location)

Number of telecommunications rooms: 6

Total cable length: 300,000 feet

Crew required: 12

Duration: 2 weeks

CABLING @ WORK: A Peculiar Job

One interesting job another contractor did a number of years ago required a great deal of ingenuity and adaptive thinking. A marine research institute had contracted with a local shipbuilding firm to construct from scratch a vessel designed for deep-sea exploration that we'll call (naturally) Cool Marine Institute Research Vessel—or CMIRV, for short. CMIRV wanted a ship designed for science, not just a fishing boat crudely adapted with generators, probes, computers, and insufficient living space for a bunch of scientists. The shipbuilding firm took the job but found itself at sea when it came to putting in a completely fiber-optic network with over 200 drops in a boat that was just 140´ long.

The cabling contracting company had put networks in ships before (for the U.S. Navy) and felt comfortable putting in a bid for this job. It was awarded the job. The customer then asked the contractor if his company could put in a telephone system, an alarm system, and a video system while it was at it! Of course, the contractor said yes.

The contracting company learned quickly that putting systems into a vessel under construction is a far different matter than retrofitting 10- and 20-year-old ships with a new network. First, it could not perform a job walk-through at the beginning because the boat hadn't even been built yet. Second, the pace for installing cable and equipment is slow but strict—the company would have to be prepared to run a cable when a path is accessible, and no holes could be drilled in bulkheads later. Also, CMIRV's main contractor required that the cabling contractor maintain a presence on the job site to resolve conflicts and ambiguities about cable and device placement, access requirements, and so on, regardless of whether any cabling work was planned.

The cabling contractor placed a reliable and steady employee at the job site and had him put in a portion of the network whenever it was possible. The company occasionally sent out additional crew when, for example, it was time to terminate a bunch of fiber-optic cable that had been pulled, to install and calibrate the cameras for the video system, or to hook up and program the telephone system.

> Job summary: CMIRV
>
> Type of network: Interior fiber optic
>
> Number of drops: 200
>
> Number of telecommunications rooms: 1
>
> Total cable length: 10,000 feet
>
> Crew required: 1+
>
> Duration: 1 year

An Inside Job

Don't think that good contractors and installers have always done cabling the right way. For many, the introduction to the art of network cabling installation came while performing a completely unrelated job. One person we know got her introduction to cabling while working for a university, managing the computer department for one of the colleges, which we'll call Budget

Nets College. It was finally time for Budget Nets to upgrade the campus network from ARCnet to Ethernet, and it had raised enough funds to upgrade internally rather than going through the traditional university appropriations channels. (Quite a bit of fuss was made about Budget Nets going outside the usual channels, but after the school got its LAN upgraded, the university's main IS group was more attentive to the other colleges' needs—probably because those in the group were afraid their jobs would become obsolete!)

Because Budget Nets had gone its own way to upgrade its network instead of waiting in queue for the university's dedicated networking crews to come in, it had to find someone else to pull the cable. The campus physical-plant department was happy to do that—all Budget Nets had to do was provide a requirement. The physical plant department would make a proposal and, if Budget Nets agreed to the cost, the physical plant department would pull the cable.

The physical plant's electrical workers knew everything about pulling wires through walls. They were even aware of the problems of RF interference and cable-length issues. They were not, however, experts in all the different kinds of jacks and jumpers that computer networking uses. Before Budget Nets knew it, the entire first floor of its building was wired with Token Ring instead of Ethernet faceplates.

Being a new administrator, our friend did not know what to do about the situation. The proper course of action would have been to call the electrical workers back in, make them take out the wrong faceplates, and have them put in the right ones. But meanwhile, the faculty and school staff needed to get back to work.

When used over twisted-pair cabling, both Ethernet and Token Ring use two pairs of the cable—one pair to receive and another pair to transmit. Unfortunately, Ethernet and Token Ring do not use the same pairs, so Token Ring faceplates can't be used in an Ethernet network—unless, of course, custom patch cables switch the pairs being used. So she made custom patch cables.

NOTE If the network had been wired to an ANSI/TIA-568-C recommended wiring pattern (T568-A or T568-B), the situation would not have occurred. If only the needed pin positions are wired, the cabling infrastructure will only support applications that use those particular pins.

Although the plan worked as a necessary quick fix for our administrator, we do not suggest you take this course of action in a similar situation, because it confuses people terribly when they find out that one cable will work in their outlets but another almost identical one will not. Eventually, she had to pull all those faceplates out and replace them. Those custom cables still show up on occasion and cause problems.

Job summary: Budget Nets College

Type of network: Interior Category 5e

Number of drops: 160

Number of telecommunications rooms: 2

Total cable length: 32,000 feet

Crew required: 4

Duration: 1 week

The Bottom Line

Employ useful hints for planning the installation. A well-thought-out and documented plan is critical for ensuring a well-executed job. Taking the time up-front to carefully plan the project will save a lot of time, materials, labor, and equipment.

Master It What are the recommended steps for developing a realistic plan?

Understand tips for working safely. Safety should be your top priority. Projects may need to be expedited or hurried if delayed. We will often need to work faster; but this may compromise our attention to safety. Please take the extra time to educate all of your employees of the importance and top priority of safety.

Master It What are some important aspects of safety training?

Plan for contingencies. Projects typically do not go as planned. No matter how good the plan is, there will likely be some things we may not have predicted. That's why every good plan has a set of contingency plans established in the event that things go wrong.

Master It What are the typical contingencies you should plan for?

Part II

Fiber-Optic Cabling and Components

Chapter 18

History of Fiber Optics

Like many technological achievements, fiber-optic communications grew out of a succession of quests, some of them apparently unrelated. It is important to study the history of fiber optics to understand that the technology as it exists today is relatively new and still evolving.

This chapter discusses the major accomplishments that led to the creation of high-quality optical fibers and their use in high-speed communications and data transfer, as well as their integration into existing communications networks.

In this chapter, you will learn to:

◆ Recognize the refraction of light

◆ Identify total internal reflection

◆ Detect cross-talk between two optical fibers

Evolution of Light in Communication

Hundreds of millions of years ago, the first bioluminescent creatures began attracting mates and luring food by starting and stopping chemical reactions in specialized cells. Over time, these animals began to develop distinctive binary, or on-off, patterns to distinguish one another and communicate intentions quickly and accurately. Some of them have evolved complex systems of flashing lights and colors to carry as much information as possible in a single glance. These creatures were the first to communicate with light, a feat instinctive to them but tantalizing and elusive to modern civilization until recently.

Early Forms of Light Communication

Some of the first human efforts to communicate with light consisted of signal fires lit on hill-tops or towers to warn of advancing armies, and lighthouses that marked dangerous coasts for ancient ships and gave them reference points in their journeys. To the creators of these signals, light's tremendous speed (approximately 300,000 kilometers per second) made its travel over great distances seem instantaneous.

An early advance in these primitive signals was the introduction of relay systems to extend their range. In some cases, towers were spread out over hundreds of kilometers, each one in the line of sight of the next. With this system, a beacon could be relayed in the time it took each tower guard to light a fire—a matter of minutes—while the fastest transportation might have taken days. Because each tower only needed in its line of sight the sending and receiving towers, the light, which normally travels in a straight line, could be guided around obstacles such as mountains as well as over the horizon. As early as the fourth century A.D., Empress Helena, the mother of Constantine, was believed to have sent a signal from Jerusalem to Constantinople in a single day using a relay system.

NOTE   The principle behind signal relay towers is still used today in the form of repeaters, which amplify signals attenuated by travel over long distances through optical fibers.

Early signal towers and lighthouses, for all their usefulness, were still able to convey only very simple messages. Generally, no light meant one state, while a light signaled a change in that state. The next advance needed was the ability to send more detailed information with the light. A simple but notable example is the signal that prompted Paul Revere's ride at the start of the American Revolution. By prearranged code, one light hung in the tower of Boston's Old North Church signaled a British attack by land, while two lights meant an invasion by sea. The two lamps that shone in the tower not only conveyed a change in state, but also provided a critical detail about that change.

The Quest for Data Transmission

Until the 1800s, light had proven to be a speedy way to transmit simple information across great distances, but until new technologies were available, its uses were limited. It took a series of seemingly unrelated discoveries and inventions to harness the properties of light through optical fibers.

The first of these discoveries was made by Willebrord van Roijen Snell, a Dutch mathematician who in 1621 wrote the formula for the principle of *refraction*, or the bending of light as it passes from one material into another. The phenomenon is easily observed by placing a stick into a glass of water. When viewed from above, the stick appears to bend because light travels more slowly through the water than through the air. Snell's formula, which was published 70 years after his death, stated that every transparent substance had a particular *index of refraction*, and that the amount that the light would bend was based on the relative refractive indices of the two materials through which the light was passing. Air has an approximate refractive index of 1 while water has a refractive index of 1.33.

The next breakthrough came from Jean-Daniel Colladon, a Swiss physicist, and Jacques Babinet, a French physicist. In 1840, Colladon and Babinet demonstrated that bright light could be guided through jets of water through the principle of *total internal reflection*. In their demonstration, light from an arc lamp was used to illuminate a container of water. Near the bottom of the container was a hole through which the water could escape. As the water poured out of the hole, the light shining into the container followed the stream of water. Their use of this discovery, however, was limited to illuminating decorative fountains and special effects in operas. It took John Tyndall, a natural philosopher and physicist from Ireland, to bring the phenomenon to greater attention. In 1854, Tyndall performed the demonstration before the British Royal Society and made it part of his published works in 1871, casting a shadow over the contribution of Colladon and Babinet. Tyndall is now widely credited with discovering total internal reflection, although Colladon and Babinet had demonstrated it 14 years previously.

Total internal reflection takes place when light passing through a material with a higher index of refraction (the water in the experiment) hits a boundary layer with a material that has a lower index of refraction (the air). When this takes place, the boundary layer becomes reflective, and the light bounces off the boundary layer, remaining contained within the material with the higher index of refraction.

Shortly after Tyndall, Colladon, and Babinet laid the groundwork for routing light through a curved material, another experiment took place that showed how light could be used to carry higher volumes of data.

In 1880, Alexander Graham Bell demonstrated his photophone, one of the first true attempts to carry complex signals with light. It was also the first device to transmit signals wirelessly. The photophone gathered sunlight onto a mirror attached to a mouthpiece that vibrated when a user spoke into it. The vibrating mirror reflected the light onto a receiver coated with selenium, which produced a *modulated* electrical signal that varied with the light coming from the sending device. The electrical signal went to headphones where the original voice input was reproduced.

Bell's invention suffered from the fact that outside influences such as dust or stray light confused the signals, and clouds or other obstructions to light rendered the device inoperable. Although Bell had succeeded in transmitting a modulated light signal nearly 200 meters, the photophone's limitations had already fated it to be eclipsed by Bell's earlier invention, the telephone. Until the light could be modulated and guided as well as electricity could, inventions such as the photophone would continue to enjoy only novelty status.

Evolution of Optical Fiber Manufacturing Technology

John Tyndall's experiment in total internal reflection had led to attempts to guide light with more control than could be achieved in a stream of water. One such effort by William Wheeler in 1880, the same year that Bell's photophone made its debut, used pipes with a reflective coating inside that guided light from a central arc lamp throughout a house. As with other efforts of the time, there was no attempt to send meaningful information through these conduits—merely to guide light for novelty or decorative purposes. The first determined efforts to use guided light to carry information came out of the medical industry.

Controlling the Course of Light

Doctors and researchers had long tried to create a device that would allow them to see inside the body with minimal intrusion. They had begun experimenting with bent glass and quartz rods, bringing them tantalizingly close to their goal. These tools could transmit light into the body, but they were extremely uncomfortable and sometimes dangerous for the patient, and there was no way yet to carry an image from the inside of the body out to doctors. What they needed was a flexible substance or *medium* that could carry whole images for about half a meter.

One such material was in fact pioneered for quite a different purpose. Charles Vernon Boys was a British physics teacher who needed extremely sensitive instruments for his continuing research in heat and gravity. In 1887, to provide the materials he needed, he began drawing fine fibers out of molten silica. Using an improvised miniature crossbow, he shot a needle that dragged the molten material out of a heat source at high speed. The resulting fiber—more like quartz in its crystalline structure than glass—was finer than any that had been made to date, and was also remarkably even in its thickness. Even though glass fibers had already been available for decades before this, Boys' ultra-fine fibers were the first to be designed for scientific purposes, and were also the strongest and smallest that had been made to date. He did not, however, pursue research into the optical qualities of his fibers.

Over the next four decades, attempts to use total internal reflection in the medical industry yielded some novel products, including glass rods designed by Viennese researchers Roth and

Reuss to illuminate internal organs in 1888, and an illuminated dental probe patented in 1898 by David Smith. A truly flexible system for illuminating or conveying images of the inside of the body remained elusive, however.

The next step forward in the optical use of fibers occurred in 1926. In that year, Clarence Weston Hansell, an electrical engineer doing research related to the development of television at RCA, filed a patent for a device that would use parallel quartz fibers to transmit a lighted image over a short distance. The device remained in the conceptual stage, however, until a German medical student, Heinrich Lamm, developed the idea independently in an attempt to form a flexible gastroscope. In 1930, Lamm bundled commercially produced fibers and managed to transmit a rough image through a short stretch of the first fiber-optic cable. The process had several problems, however, including the fact that the fiber ends were not arranged exactly, and they were not properly cut and polished. Another issue was to prove more daunting. The image quality suffered from the fact that the quartz fibers were bundled against each other. This meant that the individual fibers were no longer surrounded by a medium with a lower index of refraction. Much of the light from the image was lost to *crosstalk*. Crosstalk or optical coupling is the result of light leaking out of one fiber into another fiber.

The poor focus and resolution of Lamm's experimental image meant that a great deal more work would be needed, but Lamm was confident enough to write a paper on the experiment. The rise of the Nazis, however, forced Lamm, a Jew, to leave Germany and abandon his research. The dream of Hansell and Lamm languished until a way could be found to solve the problems that came with the materials available at the time.

Also in 1930, the chemical company DuPont invented a clear plastic material that it branded Lucite. This new material quickly replaced glass as the medium of choice for lighted medical probes. The ease of shaping Lucite pushed aside experiments with bundles of glass fiber, along with the efforts to solve the problems inherent in Lamm's probe.

The problems surfaced again twenty years later, when the Dutch government began looking for better periscopes for its submarines. They turned to Abraham van Heel, who was at the time the president of the International Commission of Optics and a professor of physics at the Technical University of Delft, the Netherlands. Van Heel and his assistant, William Brouwer, revived the idea of using fiber bundles as an image-transmission medium. Fiber bundles, Brouwer pointed out, had the added advantage of being able to scramble and then unscramble an image—an attractive feature to Dutch security officials.

When van Heel attempted to build his image carrier, however, he rediscovered the problem that Lamm had faced. The refractive index of adjacent fibers reduced a fiber's ability to achieve total internal reflection, and the system lost a great deal of light over a short distance. At one point, van Heel even tried coating the fibers with silver to improve their reflectivity, but the effort provided little benefit.

At his government's suggestion, van Heel approached Brian O'Brien, president of the Optical Society of America, in 1951. O'Brien suggested a procedure that is still the basis for fiber optics today: surrounding, or "cladding," the fiber with a layer of material with a lower refractive index.

Following O'Brien's suggestion, van Heel ran the fibers through a liquid plastic that coated them, and in April 1952, he succeeded in transmitting an image through a 400-fiber bundle over a distance of half a meter.

Van Heel's innovation, along with research performed by Narinder Kapany, who also coined the term fiber optics, and Harold Hopkins, helped make the 1950s the pivotal decade in the development of modern fiber optics.

Working in England, Kapany and Hopkins developed a method for ensuring that the fibers at each end of a cable were in precise alignment. They wound a single fine strand several thousand times in a figure-eight pattern and sealed a section in clear epoxy to bind the fibers together throughout the bundle. They then sawed the sealed portion in half, leaving the fiber ends bonded in exact alignment. The image transmitted with this arrangement was clearly an improvement, but the brightness degraded quickly since the fibers were unclad.

Extending Fiber's Reach

In January 1954, the British journal *Nature* chanced to publish papers on the findings of van Heel as well as Kapany and Hopkins in the same issue. Although their placement in the journal was apparently coincidental, the two advancements were precisely the right combination of ideas for Professor Basil Hirschowitz, a gastrosurgeon from South Africa who was working on a fellowship at the University of Michigan. Hirschowitz assembled a team to study the uses of these new findings as a way to finally build a flexible endoscope for peering inside the body. Assisting Hirschowitz were physicist C. Wilbur Peters and a young graduate student named Lawrence Curtiss.

Curtiss studied the work of Kapany and Hopkins and used their winding method to create a workable fiber bundle, but his first attempt at cladding used van Heel's suggestion of cladding glass fibers with plastic. The results were disappointing.

In 1956, Curtiss began working with a new type of glass from Corning, one with a lower refractive index than the glass he was using in his fibers. He placed a tube made of the new glass around a core made from the higher refractive index glass and melted the two together. The cladded glass fiber that he drew from this combination was a success. On December 8, 1956, Curtiss made a fiber with light-carrying ability far superior to that of any fiber before it. Even when he was 12 meters away from the glass furnace, he could see the glow of the fire inside the fiber that was being drawn from it. By early 1957, Hirschowitz and Curtiss had created a working endoscope, complete with lighting and optics. This event marked the first practical use of optical fibers to transmit complex information.

Curtiss' fibers were well suited for medical applications, but their ability to carry light was limited. Suffering a signal loss of one *decibel* per meter, the fibers were still not useful for long-distance communications. Many thought that glass was inherently unusable for communications, and research in this area remained at a minimal level for nearly a decade.

In the meantime, the electronic communications industry had been experimenting with methods of improving bandwidth for the higher volumes of traffic they expected to carry. The obvious choice for increasing the amount of information a signal could carry was to increase the frequency, and throughout the 1950s, researchers had pushed frequencies into the tens of *gigahertz*, which produced *wavelengths* of only a few millimeters. Frequencies in this range—just below the lowest infrared frequencies—required hollow pipes to be used as *waveguides*, because the signals were easily disturbed by atmospheric conditions such as fog or dust.

With the invention of the laser in 1960, the potential for increasing communication bandwidths literally increased exponentially. Wavelengths had been slashed from the millimeter range to the micrometer range, and true optical communications seemed within reach. The problems of atmospheric transmission remained, however, and waveguides used for lower frequencies were proving inadequate for optical wavelengths unless they were perfectly straight. Optical fibers, too, were all but ruled out as a transmission medium because at losses of 1000 decibels per kilometer, their *attenuation* was still too great.

One researcher did not give up on fiber, however. Charles K. Kao, working at Standard Telecommunications Laboratories, began studying the problems encountered in optical fibers. His conclusions revived interest in the medium after he announced in 1966 that signal losses in glass fibers were not caused by inherent deficiencies of the material, but by flaws in the manufacturing process. Kao proposed that improved manufacturing processes could lower attenuation to levels of 20 decibels per kilometer or better, while providing the ability to carry up to 200,000 telephone channels in a single fiber.

Kao's pronouncement sparked a race to find the lower limit of signal loss in optical fibers. In 1970, Corning used pure silica to create a fiber with a loss that achieved Kao's target of only 20 decibels per kilometer. That was just the beginning. Six years later, the threshold had dropped to just half a decibel per kilometer, and in 1979 the new low was 0.2 decibel per kilometer. Optical fiber had passed well into the realm of practicality for communications and could begin showing its promise as a superior medium to copper.

Evolution of Optical Fiber Integration and Application

Once signal losses in fiber dropped below Kao's projected figure of 20 decibels per kilometer, communications companies began looking seriously at fiber optics as a new transmission medium. The technology required for this fledgling medium was still expensive, however, and fiber-optic communications systems remained in closed-circuit, experimental stages until 1973. In that year, the U.S. Navy installed a fiber-optic telephone link aboard the USS Little Rock. Fiber optics had left the lab and started working in the field. Further military tests showed fiber's advantages over copper in weight and information-carrying capacity.

The first full-scale commercial application of fiber-optic communication systems occurred in 1977, when both AT&T and GTE began using fiber-optic telephone systems for commercial customers. During this period, the U.S. government breakup of the Bell Telephone system monopoly ushered in a boom time for smaller companies seeking to market long-distance service. A number of companies had positioned themselves to build microwave towers throughout the country to create high-speed long-distance networks. With the rise of fiber-optic technology, however, the towers were obsolete before they had even been built. Plans for the towers were scrapped in the early 1980s in favor of fiber-optic links between major cities. These links were then connected to local telephone companies that leased their capacity from the operators. The result was a bandwidth feeding frenzy. The fiber-optic links had such high capacities that extra bandwidth was leased to other local and long-distance carriers, which often undercut the owners of the lines, driving some out of business.

Following the success of fiber optics in the telecommunications industry, other sectors began taking advantage of this medium. During the 1990s, fiber-optic networks began to dominate in the fields of industrial controls, computers, and information systems. Improvements in lasers and fiber manufacturing continued to drive data rates higher and bring down operating costs.

Today, fiber optics have become commonplace in many areas as the technology continues to improve. Until recently, the transition to fiber optics was cost effective only for business and industry; equipment upgrades made it too expensive for telephone and cable companies to run fiber to every home. Manufacturing improvements have reduced costs, however, so that running fiber to the home is now an affordable alternative for telephone and cable companies.

In October 2004, the *Federal Communications Commission* (FCC) voted to further build on its policies that are designed to spark the development of new fiber-optic networks. These networks are now in place in some areas of the country and capable of providing advanced data, video, and voice services to consumers. *Fiber-to-the-home* (FTTH) is a reality and it is estimated that by 2014, 100 million homes in the United States will have fiber-optic connectivity.

The Bottom Line

Recognize the refraction of light.   Refraction is the bending of light as it passes from one material into another.

> **Master It**   You are fishing with a friend and in front of your boat; you see a large fish. You place your bait in front of the fish. Why does the fish not take the bait?

Identify total internal reflection.   In 1840, Colladon and Babinet demonstrated that bright light could be guided through jets of water through the principle of total internal reflection.

> **Master It**   You are swimming in a pool at night with a waterproof flashlight. When you point the light beam at the surface of the water, you see it bounce off the surface of the water and project on the bottom of the pool. Why does this happen?

Detect cross-talk between two optical fibers.   Crosstalk or optical coupling is the result of light leaking out of one fiber into another fiber.

> **Master It**   You have bundled two small glass rods together in an effort to make a fiber-optic scope that will allow you to look into the combustion chamber of a car engine when the spark plug is removed. After the engine cools, you insert your fiber-optic scope into the combustion chamber and are disappointed with the image you see. What is one possible cause for the poor performance of your fiber-optic scope?

Chapter 19

Principles of Fiber-Optic Transmission

Like Bell's photophone, the purpose of fiber optics is to convert a signal to light, move the light over distance, and then reconstruct the original signal from the light. The equipment used to do this job has to overcome all of the same problems that Bell encountered, while carrying more data over a much greater distance.

In this chapter, you will learn about the basic components that transmit, receive, and carry the optical signal. You will also learn some of the methods used to convert signals to light and light back to the original signals, as well as how the light is carried over the distances required.

In this chapter, you will learn to:

◆ Calculate the decibel value of a gain or loss in power

◆ Calculate the gain or loss in power from a known decibel value

◆ Convert dBm to a power measurement

The Fiber-Optic Link

A link is a transmission pathway between two points using some kind of generic cable. The pathway includes a means to couple the signal to the cable and a way to receive it at the other end in a useful way.

Any time we send a signal from one point to another over a wire, we are using a link. A simple intercom, for example, consists of the sending station (which converts voice into electrical signals), the wire over which the signals are transmitted, and the receiving station (which converts the electrical signal back into voice).

Links are often described in terms of their ability to send and receive signals as part of a communication system. When described in these terms, they are broken down into *simplex* and *duplex*. Simplex means that the link can only send at one end and receive at the other end. In other words, the signal goes only one way. An example is the signal from a radio station. Duplex means that the link has a transmitter and a receiver at each end. A *half-duplex* system allows signals to go only one way at a time—an example is an intercom system. A *full-duplex* system allows users to send and receive at the same time. A telephone is a common example of a full-duplex system.

A fiber-optic link, shown in Figure 19.1, is like any other link, except that it uses optical fiber instead of wire. A fiber-optic link consists of four basic components:

◆ Transmitter to convert a signal into light and send the light

◆ Receiver to capture the light and convert it back to a signal

◆ The optical fiber that carries the light

◆ The connectors that couple the optical fiber to the transmitter and receiver

FIGURE 19.1
The fiber-optic link

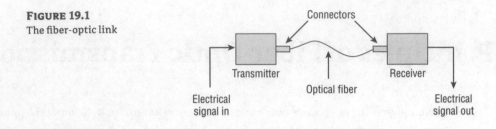

Now let's look at each component in a little more detail.

Transmitter

The transmitter, shown in Figure 19.2, converts an electrical signal into light energy to be carried through the optical fiber. The signal can be generated by many sources, such as a computer, a voice over a telephone, or data from an industrial sensor.

FIGURE 19.2
The fiber-optic transmitter

Receiver

The receiver is an electronic device that collects light energy from the optical fiber and converts it into electrical energy, which can then be converted into its original form, as shown in Figure 19.3. The receiver typically consists of a *photodiode* to convert the received light into electricity, and circuitry to amplify and process the signal.

FIGURE 19.3
The fiber-optic receiver

Optical Fibers

Optical fibers carry light energy from the transmitter to the receiver. An optical fiber may be made of glass or plastic, depending on the requirements of the job that it will perform. The advantage of light transmission through optical fiber as compared to the transmission of light through air is that the fiber can carry light around corners and over great distances.

Many fibers used in a fiber-optic link have a *core* between 8 and 62.5 microns (millionths of a meter) in diameter. For comparison, a typical human hair is about 100 microns in diameter. The *cladding* that surrounds the core is typically 125 microns in diameter.

The optical fiber's coating protects the cladding from abrasion. The thickness of the coating is typically half the diameter of the cladding, which increases the overall size of the optical fiber to 250 microns. Even with the additional thickness of the coating, optical fiber cabling is much smaller and lighter than copper cabling, as shown in Figure 19.4, and can carry many times the information.

FIGURE 19.4
Comparison of copper cable (top) and fiber cable (bottom)

Connectors

The connector is attached to the optical fiber and the fiber-optic cable. The connector allows the optical fiber to be mated to the transmitter or receiver. Transmitters or receivers typically have a *receptacle* that securely holds the connector and provides solid contact between the optical fiber and the optical subassembly of the device. The connector must align the fiber end precisely with the light source or photodiode to minimize signal loss.

The connectors, shown in Figure 19.5, could be considered the elements that make it possible for us to use fiber optics, because they allow large hands to handle the small, fragile fibers. They are also the only place in the link where the optical fiber is exposed.

FIGURE 19.5
Fiber-optic connectors

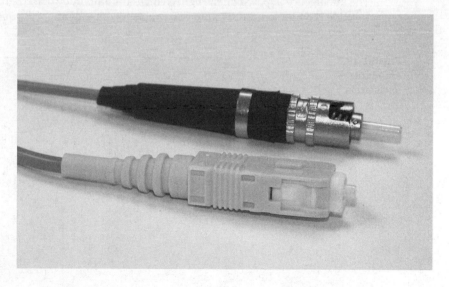

Now that you've seen the components required for a fiber-optic link, let's look at some of the methods that make it possible to transmit data with light.

Amplitude Modulation

One method used for converting electrical signals into light signals for transmission is *amplitude modulation (AM)*. Amplitude refers to the strength of a signal, represented by a waveform, as shown in Figure 19.6. In amplitude modulation, electrical energy with continuously varying voltage is converted into light with continuously varying brightness.

FIGURE 19.6
Amplitude on a
waveform

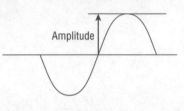

Amplitude modulation requires two components: a carrier and a signal that is imposed on the carrier—also known as the *intelligence*—to change it in some way. When we speak, we impose the intelligence created by the vibration of our vocal cords on air, which is the carrier. Similarly, Bell's photophone used sound to vibrate a mirror, which *modulated* the light reflected from it. At the receiving end, a similar arrangement worked in reverse to *demodulate* the light, retrieving the intelligence from it and creating the sound again.

To modulate the amplitude of the light in a fiber-optic transmitter, the intelligence is sent through a circuit that changes it to a continuously varying voltage. As the intelligence changes, the voltage controlling the light rises and falls, varying the light's intensity to match the intelligence. Figure 19.7 shows the basic process of amplitude modulation.

The intelligence imposed on the light changes the amplitude of the light, but not its wavelength. Amplitude modulation suffers from two problems that can affect the quality of the signal: attenuation and noise.

FIGURE 19.7
Amplitude
modulation

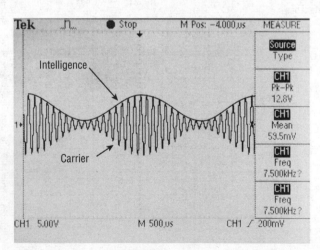

Attenuation is the loss of optical power as the signal passes through the optical fiber or interconnections. Attenuation in the optical fiber occurs as light is absorbed or scattered by impurities in the fiber. Attenuation from interconnections can be caused by several factors, which are covered in depth in later chapters.

As an amplitude-modulated signal is attenuated, its power decreases and small differences in amplitude become even smaller, or disappear entirely, as shown in Figure 19.8. When the light energy is converted back to electrical energy, these small differences are lost and cannot be reconstructed.

FIGURE 19.8
Attenuation of an
AM signal

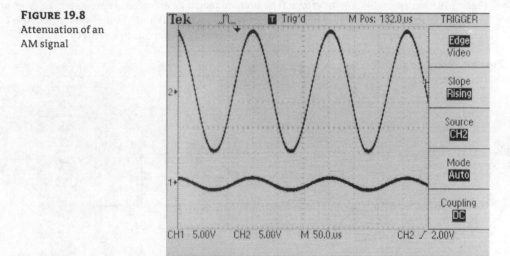

Noise is the introduction of unwanted energy into a signal. An example is static on an AM radio, especially when passing near high-voltage power lines. The unwanted energy changes the amplitude of the signal, sometimes rendering it unusable if the noise is great enough in comparison to the original signal.

Analog Transmission

Amplitude modulation is a form of *analog* transmission. An analog signal is one that varies continuously through time in response to an input. In addition, the response is infinitely variable within the specified range. In other words, a smooth change in the input will produce a smooth change in the signal.

A common example of an analog system is an electrical temperature sensor such as a thermocouple, which generates a small voltage that changes with the temperature. As the temperature rises, the thermocouple senses the temperature change, and the voltage increases. Because the relationship between temperature and voltage from the device is predictable, the thermocouple's output can be translated into a temperature reading. A reading of 3 millivolts (mV), for example, could indicate a temperature of 140° F.

When amplitude modulation is used with fiber optics, the amplitude of the optical transmission changes in relation to the strength of the incoming signal. Because of their infinitely variable response within a given range, analog signals are commonly used in RF-to-fiber applications.

Digital Data Transmission

In spite of the problems caused by noise, analog signals are still used in fiber-optic communications. If information is to be stored, carried, or manipulated by computers, however, it must be in a *digital* form—that is, represented by a series of on-off or high-low voltage readings. Figure 19.9 shows a digital waveform. The voltage readings are often represented as ones and zeros, with the high or on state being a one, and the low or off state being a zero. Because only two states—or digits—are used, the numbering system is referred to as *binary*.

FIGURE 19.9
Digital waveform

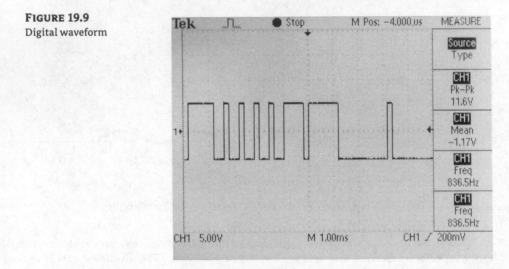

Recall that early signal fires, a form of digital communication, could announce a change in state by being lit, but could not communicate complex information. To make digital information more detailed, *binary digits*, or *bits*, are combined into eight-place sequences called *bytes*. A byte can be used to represent a single number in the same way that a voltage reading would be used in an analog transmission. For example, the above temperature reading of 140° F might be transmitted digitally as 10001100, the binary equivalent of 140.

Analog Data Transmission vs. Digital Data Transmission

One of the reasons that digital transmission is chosen over analog transmission is the fact that a digital signal is not affected by noise or attenuation the way an analog signal is.

Digital information can be stored and transmitted accurately because noise that would interfere with the analog data does not affect digital data. Each voltage in the sequence is either high or low, and voltages that do not match either the high or the low level do not change the meaning of the digital sequence.

The difference between the two is like the difference between a tape recording of a musical performance and a CD of the same performance. The analog recording may carry the same detail, but it would also contain a certain amount of hiss caused by electrical noise. The CD would be free of hiss, because the stray voltages do not register as either high or low signals.

More and more, digital transmissions are replacing analog transmissions, even in radio and television. Many radio stations now broadcast digital signals to receivers. In addition to carrying the regular programming as digital data, the broadcast can also carry digital data for display on the receiver, such as program details, announcers' names, and song titles. The Federal Communications Commission (FCC) requires all full-power television stations to transition from analog to digital by June 12, 2009. In fiber-optic transmission, digital signals make it possible to carry many thousands of conversations over a single fiber through the use of multiplexing, which will be explained later in this chapter.

Analog to Digital (A/D) Conversion

To transmit an analog signal such as a voice through a digital system, it is necessary to *digitize*, or *encode*, it. This is also known as analog to digital, or A/D, conversion.

In A/D conversion, the smooth, continuously variable analog signal is translated into a digital signal that carries the same information. To do this, the analog signal's voltage is "sampled" at regular intervals and converted into binary numbers that represent the voltage at each interval. In Figure 19.10, for example, each vertical line represents a sampling of the analog signal at a given time.

FIGURE 19.10
Sampling an
analog signal

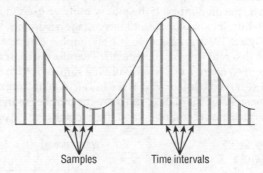

Samples   Time intervals

As with frequency measurements, the sample rate or sampling frequency is measured in terms of cycles per second, or hertz, so a rate of one sample per second would be designated 1Hz. A rate of 1,000 samples per second would be 1 kilohertz, or 1kHz.

Two factors affect the quality of the digital sample: sample rate and quantizing error.

Sample Rate

When an analog signal is digitized, any information between the samples is lost, so instead of a smooth transition over time, the digital information jumps from one voltage to the next in the signal. To smooth out the transitions and retain more of the information from the original analog signal, more samples must be taken over time. The higher the sampling rate, the more accurately the original analog signal can be digitized, as shown in Figure 19.11. Typically, audio signals for CDs and other digital music are sampled at 44.1kHz or 48kHz.

FIGURE 19.11
Low sample
rate vs. high
sample rate

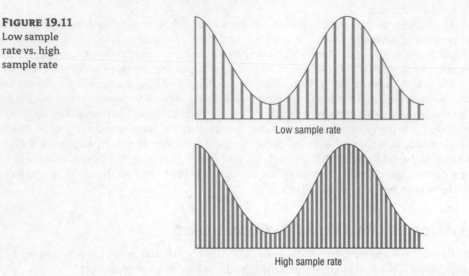

Low sample rate

High sample rate

Quantizing Error

The second factor affecting digital signal quality is called quantizing error. Quantizing error is caused by the inability of a binary number to capture the exact voltage of a digital sample.

Because an analog signal is infinitely variable, the sample's voltage could be any number within a specified range. If the binary number used to represent the voltage does not have enough bits, it cannot represent the voltage accurately.

In Figure 19.12, for example, a 4-bit number can represent 16 voltage levels—from 0 to 15 with 15 discrete steps or increments. Therefore, on a scale from 0V to +15V, each binary number represents a change of 1 whole volt. If a sample returned a reading of +1.5V, the binary number would still read 0001, or +1V. You can calculate the maximum error by dividing the number of discrete steps or the voltage range by the number of increments. In this case, $15 \div 15 = 1$, so you have a maximum error of 1V, and the average error is one-half of that, or 0.5V.

FIGURE 19.12
Sampling with a
4-bit number

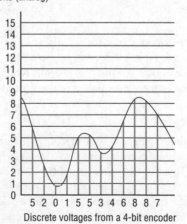

Volts (analog)

Discrete voltages from a 4-bit encoder

Increase the number of bits to eight, however, and you have 255 increments plus zero. The voltage between increments, and the maximum error, is now $15 \div 255 = 58.82 \times 10^{-3}$, or 58.82mV. Now, a reading of +1.5V is 0001 1001, or 25 steps from 0 instead of just one. Multiplying the number of increments by the voltage between them gives us $25 \times 58.82mV = 1.4705V$. This result is much closer to the analog reading of +1.5V.

As with the sample rate, the more bits used in encoding, the more accurate each sample can be. CD-quality audio signals are usually encoded at 16 bits, which means that there are 65,535 increments available, plus zero.

Digital-to-Analog (D/A) Conversion

When digital information is used to control analog devices such as temperature controls, or when analog information has been converted to digital data for transmission and must be converted back to analog data, digital-to-analog (D/A) conversion is used.

When digital data is converted to analog, two processes take place. First, a digital-to-analog converter converts each sequential binary sample to a proportional voltage. From our previous example of a 0V to +15V range represented by an 8-bit number, the binary sample 0001 1001 would be converted to +1.4705V. If the same binary sample were applied to a different voltage range, the result would be proportional to that range. The D/A converter outputs a stepped version of the analog signal, as shown in Figure 19.13. When reconstructing an encoded analog signal, the higher the sampling rate and the greater the number of bits in each sample, the more accurate the analog reconstruction can be.

FIGURE 19.13
Converting analog
to digital

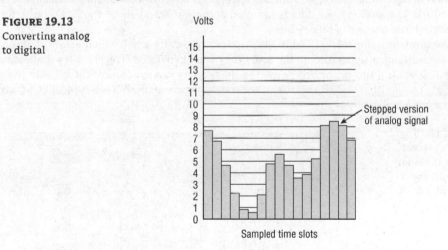

Next, the steps between each digital sample must be smoothed out to provide a transition from one voltage to another. No matter how many samples are used, the digital output will always produce a signal that jumps from one voltage to another, and then holds each voltage for the amount of time between samples. When a smooth analog signal is required, D/A converters have circuits that filter the stairstep voltage into a smooth waveform, as shown in Figure 19.14.

FIGURE 19.14
Filters convert
stairstep voltage
to a smooth
waveform.

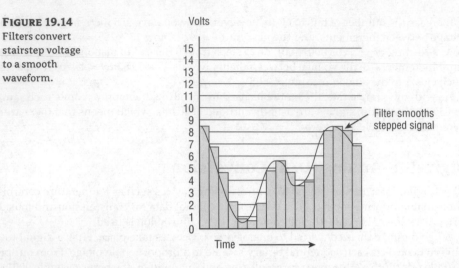

Pulse Code Modulation (PCM)

When an analog signal has been digitally encoded for transmission over a fiber-optic link, it has undergone a process known as *pulse code modulation (PCM)*. Pulse code modulation is a common method of digitizing analog data such as telephone conversations for transmission over a fiber-optic link. The analog voice data is sampled at regular intervals by the A/D converter and converted into a series of binary bits.

Data transmission using PCM in fiber optics is typically *serial*, which means that the binary bits are sent one after another in the order they were generated. The circuitry that converts the data also sends a timing, or *clock*, signal so the receiver can synchronize itself with the data that is being transmitted and reconstruct it accurately. Figure 19.15 shows a typical PCM sequence with a clock pulse burst.

FIGURE 19.15
In this PCM trans-
mission, the binary
numbers are sent
along with a clock
signal.

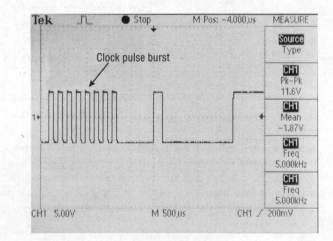

In order for pulse code modulation to be effective, an analog signal must be sampled at a rate that is at least twice the highest expected frequency. This number is referred to as the *Nyquist Minimum*. In practice, though, the sampling rate is usually closer to three times or more the highest expected frequency. This formula ensures that sampling will capture some portion of even the highest frequencies. For example, in a telephone conversation, the highest frequency encountered is about 4kHz. That means sampling must take place at the Nyquist Minimum of 8kHz to maintain a basic signal quality.

Multiplexing

Most fiber-optic data transmission systems can send data at rates that far exceed the requirements of a single stream of information. To take advantage of this fact, multiplexing can be used to carry several information channels, such as telephone conversations, nearly simultaneously. *Multiplexing* is the process of transmitting many channels of information over one link or circuit.

There are many multiplexing schemes or processes, which are discussed in detail in Chapter 29, "Passive Components and Multiplexers." Figure 19.16 shows a simple multiplexing process that may be used for the transmission of multiple telephone conversations. A *multiplexer* first divides each channel into several parts, each of which could represent a byte of voice data. In a process known as *interleaving*, the multiplexer sends the first part or byte of each channel, then the second part of each channel, continuing the process until all of the transmissions are completed. At the receiving end, a *demultiplexer* separates the transmissions into their individual channels and reassembles them in their proper order.

FIGURE 19.16
Multiplexing allows thousands of conversations to be carried in a single fiber.

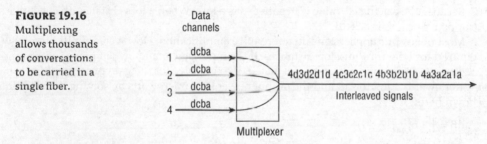

Decibels (dB)

As light travels away from its source through the link, it loses energy. Energy loss can be caused by several factors, such as the absorption or scattering of light by impurities in the fiber, or by light passing through the core and cladding and being absorbed in the coating.

It is important to be able to measure the amount of light energy lost in a fiber-optic link. Knowing the loss allows us to predict the strength of the light energy at the receiving equipment. The receiving equipment needs a minimum amount of light energy to accurately convert the light energy to the original signal. In addition, an understanding of how and where light is lost in a link can be very helpful when troubleshooting the link.

One of the more common terms used when discussing the quality of a signal in fiber optics is the *decibel (dB)*. The decibel was originally used to measure the strength of sounds as perceived by the human ear. Its name means "one-tenth of a bel."

NOTE A *bel* is a sound measurement named for telephone inventor Alexander Graham Bell.

Calculating dB Loss and Gain

The decibel is used to express gain or loss relative to a known value. In fiber optics, the decibel is most commonly used to describe signal loss through the link after the light has left the transmitter. Recall that in the 1960s, a signal loss of 20 decibels per kilometer was considered the goal for making fiber optics practical for communications. A 20 decibel loss means that of the original power put into the signal, only 1 percent remains.

To calculate the decibel value of a gain or loss in signal power, use the following equation:

$$dB = 10Log_{10}(P_{out} \div P_{in})$$

Let's apply this equation to the loss associated with a length of optical fiber. The transmitter couples 20 microwatts into the optical fiber and the power at the receiver is 3 microwatts. Since both values are in microwatts, we do not have to use microwatts in the equation. The equation would be written as:

$$dB = 10Log_{10}(3 \div 20)$$

$$dB = 10Log_{10} 0.15$$

$$dB = -8.24$$

$$Loss = 8.24dB$$

Because the calculated value is negative, it is a loss. When a loss is stated, the negative sign is dropped. The loss is 8.24dB.

Most fiber-optic applications do not require amplification. However, amplification may be required for long transmission distances. If the input power and output power of the amplifier are known, the gain in dB can be calculated. We can solve for the gain of an amplifier where the input power is 1 microwatt and the output power is 17 microwatts by solving the equation as shown here:

$$dB = 10Log_{10}(17 \div 1)$$

$$dB = 10Log_{10} 17$$

$$dB = 12.3$$

$$Gain = 12.3dB$$

If you know the decibel value and want to calculate the gain or loss, you will have to rearrange the equation as shown here:

$$(P_{out} \div P_{in}) = antilog(dB \div 10)$$

Let's apply this equation to the loss associated with a length of optical fiber. To solve for the gain or loss we do not need to know the input power or output power; we just need the gain or loss in dB. Remember the loss in dB is a ratio of power out divided by power in. This length of optical fiber has a loss of 7.5dB. The equation would be written as:

$$(P_{out} \div P_{in}) = antilog(dB \div 10)$$

$$(P_{out} \div P_{in}) = antilog(-7.5 \div 10)$$

$(P_{out} \div P_{in}) = $ antilog -0.75

$(P_{out} \div P_{in}) = 0.1778$

If one power is known, the other power can be calculated. In this example, the input power is 50mW. To calculate the output power the equation would be written as:

$(P_{out} \div 50mW) = 0.1778$

$P_{out} = 0.1778 \times 50mW$

$P_{out} = 8.89mW$

Remember that signals may be decreased, or attenuated, just about anywhere in the link. In addition to attenuation in the optical fiber itself, connectors, splices, and bends in the fiber-optic cable can also cause loss in the signal, sometimes considerable loss.

Expressing dB in Percentages

When measuring signal gain or loss, decibels are calculated relative to the original power, rather than as an absolute number. For example, a loss of 0.1 decibel means that the signal has 97.7 percent of its power remaining, and a loss of 3 decibels means that only 50 percent of the original power remains. This relationship is logarithmic, rather than linear, meaning that with each 10 decibels of loss, the power is 10 percent of what it was (as shown in Table 19.1)

TABLE 19.1   Decibel Losses Expressed in Percentages

LOSS IN dB	% POWER LOST	% POWER REMAINING
0.1	2.3%	97.7%
0.2	4.5%	95.5%
0.3	6.7%	93.3%
0.4	8.8%	91.2%
0.5	10.9%	89.1%
0.6	12.9%	87.1%
0.75	15.9%	84.1%
0.8	16.8%	83.2%
0.9	18.7%	81.3%
1	21%	79%
3	50%	50%
6	75%	25%

TABLE 19.1   Decibel Losses Expressed in Percentages   *(CONTINUED)*

LOSS IN dB	% POWER LOST	% POWER REMAINING
7	80%	20%
9	87%	13%
10	90%	10%
13	95.0%	5%
16	97%	3%
17	98.0%	2.0%
19	98.7%	1.3%
20	99.00%	1.00%
23	99.50%	0.50%
30	99.9%	0.1%
33	99.95%	0.05%
40	99.99%	0.01%
50	99.999%	0.001%
60	99.9999%	0.0001%
70	99.99999%	0.00001%

With each 10 decibels of gain, the power is 10 times what it was (as shown in Table 19.2).

TABLE 19.2   Decibel Gains Expressed in Percentages

GAIN IN dB	% POWER INCREASE	% TOTAL POWER
0.1	2.3%	102.3%
0.2	4.7%	104.7%
0.3	7.2%	107.2%
0.4	9.6%	109.6%
0.5	12.2%	112.2%
0.6	14.8%	114.8%

TABLE 19.2   Decibel Gains Expressed in Percentages   *(CONTINUED)*

GAIN IN dB	% POWER INCREASE	% TOTAL POWER
0.75	18.9%	118.9%
0.8	20.2%	120.2%
0.9	23.0%	123.0%
1	26%	126%
3	100%	200%
6	298%	398%
7	401%	501%
9	694%	794%
10	900%	1,000%
13	1,895%	1,995%
16	3,881%	3,981%
17	4,912%	5,012%
19	7,843%	7,943%
20	9,900%	10,000%
23	19,853%	19,953%
30	99,900%	100,000%
33	199,426%	199,526%
40	999,900%	1,000,000%
50	9,999,900%	10,000,000%
60	99,999,900%	100,000,000%
70	999,999,900%	1,000,000,000%

One of the advantages of using decibels when calculating gain and loss is their relative ease of use compared to percentages. When measuring loss through different components, you can algebraically add the decibel loss from each component and arrive at a total signal loss for the system. If you were to use percentages, you would have to calculate the remaining power after the signal passed through each component, then take another percentage off for the next component, and so on, until the loss through the entire system had been calculated.

The Rules of Thumb

When calculating gain or loss in a system, it is useful to remember the three rules of thumb shown in Table 19.3, which make it easier to perform some common decibel calculations. These rules help you calculate how much power you have after the indicated decibel gain or loss.

TABLE 19.3   Rules of Thumb

DECIBEL	LOSS	GAIN
3 dB	½	×2
7 dB	⅕	×5
10 dB	⅒	×10

For example, a loss of 3 decibels means that you have about ½ of the original power. A gain of 3 decibels means that you have about twice the original power. These rules are intended for rough calculations only, because a 3 decibel loss actually leaves 50.1187 percent of the original power, and a 7 decibel loss leaves 19.953 percent of the original power.

Because decibels can be algebraically added and subtracted, you can use combinations of the decibel values to determine total gains or losses. For example, the rules of thumb can be applied to find the power remaining after a 17dB reduction in signal strength. First, apply the 10dB rule and divide the original power value by 10. Then apply the 7dB rule by dividing the result of the first calculation by 5. That value is the power remaining.

Absolute Power

Taken by itself, the decibel is only a relative number, and it only has meaning when it is referenced to a known input or output power. In many cases, however, it is important to have an absolute value to use for comparison or for equipment specifications. When such a value is required in fiber optics, the decibel is referenced to 1 milliwatt (mW). When the decibel is referenced to 1mW, a lowercase "m" or sometimes "mW" is inserted after dB. The decibel referenced to 1mW is written as 0dBm.

When using dBm, power levels below 1mW are negative and power levels 1mW or greater are positive. The positive sign is typically implied and only the negative sign is used. As an example, a value of −10dBm means that 1mW of power has been attenuated by 10dB, so only 10 percent of its power, or 100 microwatts, remains. A value of 10dBm means that 1mW of power has been increased by 10dB, which has value of 10mW. Table 19.4 shows how dBm values convert to power measurements.

TABLE 19.4   Converting Power to dBm

OPTICAL POWER IN WATTS	OPTICAL POWER IN dBm
100mW	20dBm
20mW	13dBm
10mW	10dBm
8mW	9dBm
5mW	7dBm
4mW	6dBm
2mW	3 dBm
1mW	0dBm
500W	−3dBm
250W	−6dBm
200W	−7dBm
125W	−9dBm
100W	−10dBm
50W	−13dBm
25W	−16dBm
20W	−17dBm
12.5W	−19dBm
10W	−20dBm
5W	−23dBm
1W	−30dBm
500nW	−33dBm

The formula used to convert dBm to a power measurement is:

$$dBm = 10Log_{10}(P \div 1mW)$$

This equation is similar to the equation for finding a decibel value from an input power and an output power, except that in this case the input power is fixed at 1mW. Let's apply this equation to calculate optical power from a dBm value. In this example the power meter measures –15 dBm. To calculate the power in watts, the equation would be written as:

$$-15 = 10\mathrm{Log}_{10}\,(P \div 1\mathrm{mW})$$

$$-1.5 = \mathrm{Log}_{10}\,(P \div 1\mathrm{mW})$$

$$0.0316 = P \div 1\mathrm{mW}$$

$$P = 0.0316 \times 1\mathrm{mW}$$

$$P = 0.0316\mathrm{mW}$$

GETTING LOST IN LOSSES

You may find yourself interchanging dB and dBm; this happens to everyone from time to time. Remember that loss and gain are expressed in dB and absolute power is expressed in dBm. If you were to say that a link had a loss of 10dBm, you would actually be saying that the link has lost 10mW. However, you were probably trying to say the link lost 90 percent of its energy.

The Bottom Line

Calculate the decibel value of a gain or loss in power.   The decibel is used to express gain or loss relative to a known value. In fiber optics, the decibel is most commonly used to describe signal loss through the link after the light has left the transmitter.

Master It   The output power of a transmitter is 10mW and the power measured at the input to the receiver is 3mW. Calculate the loss in dB.

Calculate the gain or loss in power from a known decibel value.   If the gain or loss in power is described in dB, the gain or loss can be calculated from the dB value. If the input power is known, the output power can be calculated, and vice versa.

Master It   The loss for a length of optical fiber is 11dB. Calculate the loss in power from the optical fiber and the output power of the optical fiber for an input power of 3mW.

Convert dBm to a power measurement.   In fiber optics, dBm is referenced to 1mW of power.

Master It   A fiber-optic transmitter has an output power of –22dBm. Calculate the output power in watts.

Chapter 20

Basic Principles of Light

Even the simplest fiber-optic link is a triumph of innovative design, manufacturing precision, and technical creativity. One of the factors contributing to the high data rates and long transmission distances of fiber optics is knowledge of the principles of light. While the optical fiber is the transmission medium, light is the carrier on which the signals are imposed. Its small wavelengths and high transmission velocity make possible bandwidths that would be unimaginable using other forms of transmission. To get the best performance from every part of the fiber-optic link, you must understand the characteristics of light and the factors within the fiber-optic link that affect it.

This chapter describes the basic characteristics of light, especially the type of light used in fiber-optic communications. It also discusses the principles and materials that make fiber-optic communications possible, along with some of the problems that must be overcome.

In this chapter, you will learn to:

◆ Convert various wavelengths to corresponding frequencies

◆ Calculate the amount of energy in a photon using Planck's constant

◆ Calculate the speed of light through a transparent medium using its refractive index

◆ Use Snell's law to calculate the critical angle of incidence

◆ Calculate the loss in decibels from a Fresnel reflection

Light as Electromagnetic Energy

Whether it comes from the sun, an electric bulb, or a laser, all light is a form of electromagnetic energy. Electromagnetic energy is emitted by any object that has a temperature above absolute zero ($-273.15°$ C), which means that the atoms in the object are in motion. The electrons orbiting the atoms pick up energy from the motion, and the energy causes them to move to higher orbits, or energy levels. As they drop back to their original energy levels, they release the energy again. The energy takes two forms: an electrical field and a magnetic field, formed at right angles to each other and at right angles to their path of travel, as shown in Figure 20.1.

FIGURE 20.1
The three-dimensional nature of electro-magnetic energy

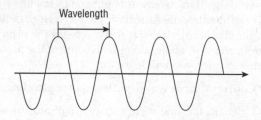

—— Electrical field
—— Magnetic field

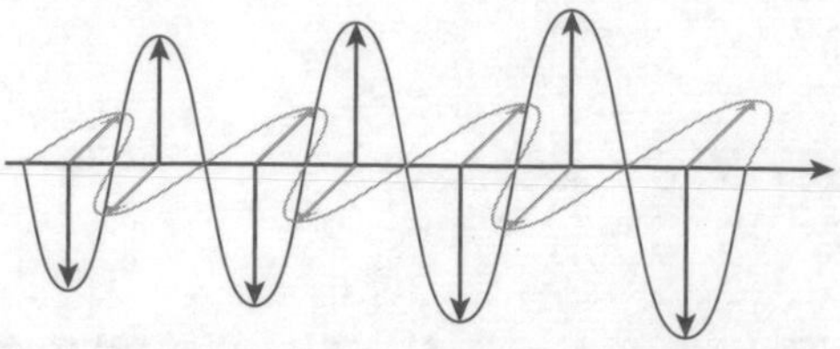

The combination of these electrical and magnetic fields is an electromagnetic wave, which travels through open space or air at approximately 300,000km/s (kilometers per second). Although the wave exists three-dimensionally, it is often represented as a two-dimensional sine wave, as shown in Figure 20.2. It is important to understand the three-dimensional nature of electromagnetic waves because different types of fiber-optic transmission take advantage of it, as will be discussed in later chapters.

FIGURE 20.2
Electromagnetic energy is often shown as a sine wave.

One of the most important characteristics of electromagnetic energy is its wavelength (λ). Wavelength is the distance between corresponding points on two consecutive waves. Depending on its wavelength, electromagnetic energy can occur as radio waves, light waves, X-rays, and more.

Wavelength is important because it allows us to calculate the electromagnetic energy's frequency (f), which is the number of waves that pass a given point in one second. Frequency is measured in cycles per second, or hertz (Hz).

A way to express the relationship between wavelength and frequency is that wavelength equals the velocity of the wave (v) divided by its frequency, or

$$\lambda = v \div f$$

Note that v is usually the speed of light in a vacuum or in open air, approximately 300,000km/s.

Likewise, you can calculate the frequency using the wavelength with the equation:

$$f = v \div \lambda$$

For example, one of the infrared light frequencies used in fiber optics has a wavelength of 850 nanometers (billionths of a meter), which translates to a frequency of 352.9 terahertz (THz), or 352.9 trillion hertz.

Note in Figure 20.3 that as the wavelength becomes smaller, more waves will occur in one second, which means that the frequency will increase as wavelength decreases.

FIGURE 20.3
As wavelength decreases, frequency increases.

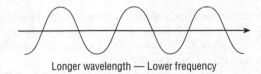

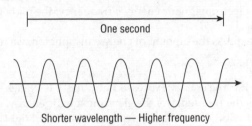

Longer wavelength — Lower frequency

One second

Shorter wavelength — Higher frequency

To illustrate this, the signal from an AM radio station transmitting at 790 kilohertz (kHz), or 790,000Hz, has a wavelength of 379.7 meters. If we choose a higher frequency, such as 850kHz, the wavelength goes down to 353.9 meters.

SOME USEFUL TERMS

Many of the terms used to express electromagnetic wavelengths and frequencies describe large multiples of cycles or small fractions of meters. To understand these terms, it is helpful to know the prefixes used with them.

Each prefix expresses a multiple of 10 or 1 divided by a multiple of 10 (as in ⅟₁₀₀₀) applied to the measurement unit, such as meters or cycles. Some terms may already be familiar, such as *kilo*meter, for a thousand meters, or *centi*meter, for a hundredth of a meter.

Here is a list of prefixes, along with their translations and mathematical equivalents, in descending order of magnitude:

Prefix	Meaning	Magnitude	Symbol
tera-	Trillion	10^{12}	T
giga-	Billion	10^{9}	G
mega-	Million	10^{6}	M
kilo-	Thousand	10^{3}	k
centi-	Hundredth	10^{-2}	c
milli-	Thousandth	10^{-3}	m
micro-	Millionth	10^{-6}	μ
nano-	Billionth	10^{-9}	n
pico-	Trillionth	10^{-12}	p

Electromagnetic radiation, like most radiated energy, has characteristics of both waves and particles. As a wave, it propagates through a medium and transfers energy without permanently displacing the medium. However, as a massless particle, called a *photon*, it travels in a wavelike pattern moving at the speed of light.

A photon, which is emitted from an electron as it changes energy levels, is the basic unit, or *quantum*, of electromagnetic energy. The amount of energy in each photon, however, depends on the electromagnetic energy's frequency: the higher the frequency, the more energy in the particle.

To express the amount of energy in a photon, we use the equation

$E = hf$

where E is the energy expressed in watts, h is Planck's constant, or 6.626×10^{-34} joule-seconds, and f is the frequency of the electromagnetic energy.

So, to find the energy of a photon of infrared light at 352.9 THz:

$$(3.529 \times 10^{14})(6.626 \times 10^{-34}) = 2.338 \times 10^{-19} \text{ W}$$

NOTE   Much of what we take for granted in the field of fiber optics comes from work done by pioneers in the field of physics in the late 1800s and early 1900s. The *joule* is a unit of energy named for James Prescott Joule, who studied the relationship between heat and mechanical work. One joule is equal to the amount of work required to produce 1 watt in 1 second. Planck's constant was defined in 1899 by Nobel Laureate Max Planck, who is regarded as the founder of quantum theory.

The Electromagnetic Spectrum

In 1964, scientists at a Bell Laboratories facility in New Jersey accidentally discovered electromagnetic radiation associated with the very beginnings of our universe. The radiation, predicted by certain cosmological theories, was emitted by hydrogen atoms at a temperature of just about 3° on the *Kelvin* (K) scale or –270.15° C, and is practical evidence that anything with a temperature above absolute zero puts out electromagnetic energy.

The wavelengths emitted by the cosmic hydrogen atoms, about 0.5mm to 5mm, are in the *microwave* range of the electromagnetic spectrum, just above radio waves in frequency, and just below infrared light.

As with sound, some characteristics of electromagnetic energy are dependent on wavelength. Longer wavelengths require less energy for propagation than shorter wavelengths of the same amplitude, making them useful for long distance communication. By the same rule, particles of higher frequency electromagnetic radiation have more energy than particles from lower frequency emissions of the same amplitude. This fact is important when determining which wavelengths to use in fiber-optic transmissions.

These transmissions consist of turning the carrier, or the light, on and off at high switching rates. At the higher rates desired for communications, higher frequencies of light are needed if an extremely short "on" cycle is to have enough energy to be detected.

Remember also that higher frequencies can carry more data, because more waves per second allow more bits per second to be carried. This fact extends the principle that had been applied first to radio and then to television, which relied on ever higher frequencies to carry greater amounts of information.

What we commonly call light is just one small part of the electromagnetic spectrum, shown in Figure 20.4. Visible light is an even smaller component, bordered by infrared, with longer wavelengths, and ultraviolet, with shorter wavelengths.

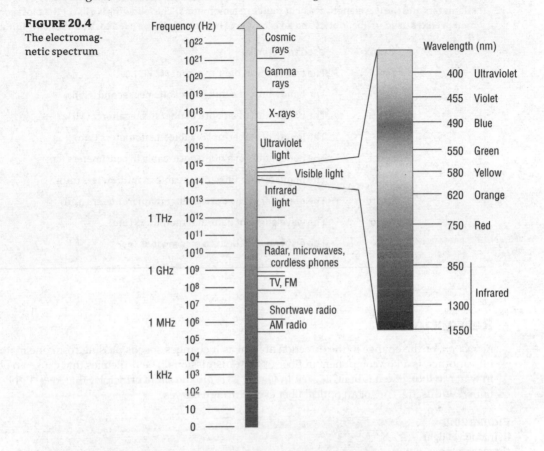

FIGURE 20.4
The electromagnetic spectrum

The wavelengths most commonly used for fiber optics are in the infrared range, at *windows* of 850 nanometers (nm), 1300nm, and 1550nm. The spectrum range of these wavelengths provides an important combination of characteristics: it is high enough to make high data rates possible, but low enough to require relatively low power for transmission over long distances.

NOTE   Windows were created as a way of standardizing useful wavelengths in fiber optics, making it easier to build interoperable equipment.

The specific wavelength windows have been selected because they provide the best possible characteristics for transmission. Even within the range between 850nm and 1550nm, certain regions have high losses due to materials in the fiber, such as stray water molecules, absorbing light at a wavelength of 1380nm. Other wavelengths, such as 1550nm, are favored because they have a low loss, allowing longer transmission distances. On the other hand, wavelengths near 1300nm suffer less from *dispersion*, which will be discussed at length in Chapter 22, "Optical Fiber Characteristics."

USING PREFIXES

Very wide ranges of numbers are used in fiber optics. You may be describing the length of a cable in kilometers and the wavelength of a light source in nanometers. The following is a list of the most common prefixes used in fiber optics and an example of how they may be used to describe something.

Prefix	Used to Describe
tera-	The frequency of light in terahertz (THz)
giga-	The bandwidth of a link in gigabits per second (Gbps)
mega-	The bandwidth of an optical fiber in megahertz (MHz)
kilo-	The length of a fiber-optic cable in kilometers (km)
centi-	The bend radius of a fiber-optic cable in centimeters (cm)
milli-	The diameter of a fiber-optic cable in millimeters (mm)
micro-	The diameter of the optical fiber in micrometers (µm)
nano-	The wavelength of light in nanometers (nm)
pico-	The dispersion of light in picoseconds (ps)

Refraction

Refraction, or the change in the direction of light as it changes speeds passing from one material into another, is a key component in fiber-optic transmissions. The principles that cause an object in water to look like it is bent, as seen in Figure 20.5, are the same principles that keep light contained within the core of an optical fiber even through curves.

FIGURE 20.5
Refraction of light through water

What Causes Refraction?

Refraction occurs when light waves change speed as they travel between two materials, each with a different refractive index or *index of refraction* (*n*). We commonly think of the speed of light as constant at about 300,000km/s (299,792.458km/s, to be precise), but that figure is actually the theoretical top speed of light, which only applies to its speed in a vacuum. In reality, light travels at lower velocities in various materials or media such as the earth's atmosphere, glass, plastic, and water. A *medium* is any material or space through which electromagnetic radiation can travel.

How quickly light travels through a medium is determined by its refractive index. Light travels more slowly in a material with a high refractive index and faster in a material with a low refractive index. It is important to note that even though light slows down when passing from a medium with a lower refractive index into a medium with a higher refractive index, it speeds up again when it passes into a medium with a lower refractive index (see Figure 20.6).

FIGURE 20.6
Light speeds up when it passes into a less dense medium.

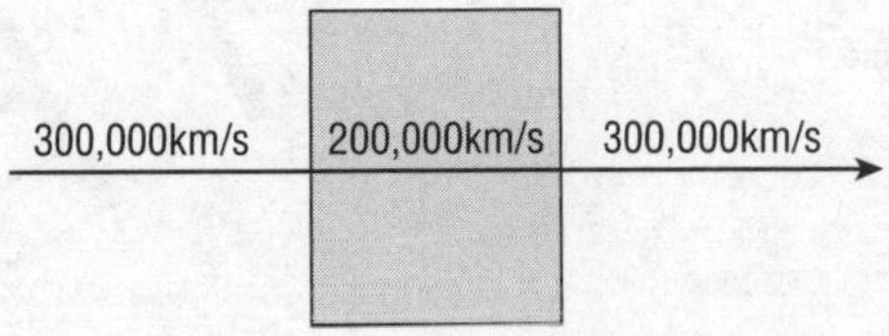

300,000km/s   200,000km/s   300,000km/s

When light changes velocity as it travels from one medium into another, refraction occurs. The amount of refraction is determined by the relative difference in the light's velocity in each of the media. The greater the difference, the greater the refraction.

In Figure 20.5, shown earlier, the portion of the oar in the water looks bent because the light rays reflected from it are bent as they pass through the boundary between the water and the air, as seen in Figure 20.7. Notice also that the portion of the oar that is underwater is distorted because the light rays reflected from it have been bent.

FIGURE 20.7
Light rays from the oar are refracted as they pass from water into the air.

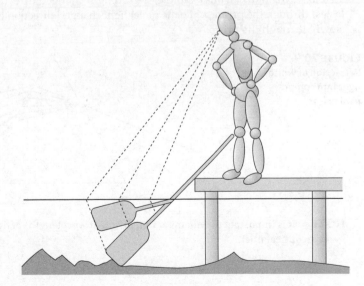

To explain refraction, we have to look at the wave nature of light. Recall that light consists of two perpendicular waves, and that the light travels in a path that forms right angles with both waves.

As the light wave meets the boundary between media with different refractive indexes, that portion of the light wave changes velocity or it experiences a *phase velocity* change, while the rest of the light wave maintains its original velocity. A change in the phase velocity of a light wave typically causes the light wave to change direction as described by Snell's law, which is discussed in detail later in this chapter. Figure 20.8 shows light waves changing direction because of a change in phase velocity as multiple light waves cross from a lower refractive index into a higher refractive index.

FIGURE 20.8
Light waves changing direction because of a change in phase velocity

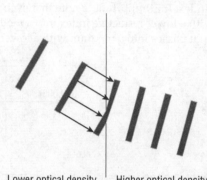

Lower optical density   Higher optical density

The velocity of light through different media such as glass depends on the light's wavelength. One of the most common ways to observe this is by seeing how white light is refracted through a prism and broken up into its component wavelengths, as shown in Figure 20.9. Notice that violet, which has the shortest visible wavelength, is refracted more than red, which has the longest visible wavelength. In this example, the red light changed direction the least and is traveling the fastest through the prism, while the violet light changed direction the most and is traveling the most slowly through the prism.

FIGURE 20.9
Refraction of white light into component colors

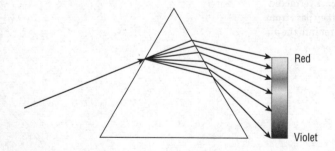

Red

Violet

NOTE   It is important to remember that different wavelengths of light travel at different speeds in an optical fiber.

Calculating the Index of Refraction

We know the velocity of light can change and its velocity depends on its wavelength and the index of refraction of the medium it is passing through. We also know that the higher the refractive index, the slower the light travels. In other words, the speed of light (c) in a medium is inversely proportional to its index of refraction.

It is important to be able to assign an index of refraction to different materials. Standard refractive index measurements are taken at a wavelength of 589nm. The refractive index of light wavelengths above or below 589nm will vary slightly. Table 20.1 lists the speed of light and index of refraction for some common materials.

TABLE 20.1   Refractive Indexes of Different Materials

MATERIAL	SPEED OF LIGHT TRAVELING THROUGH MATERIAL (km/s)	INDEX OF REFRACTION
Vacuum	300,000	1.0000
Air	300,000	1.0003
Water	225,056	1.333
Ethyl alcohol	220,588	1.36
Optical fiber cladding	205,479	1.46
Optical fiber core	204,082	1.47
Glass	200,000	1.5

The index of refraction is a relative value and it is based on the speed of light in a vacuum, which has an index of refraction of 1. We can calculate the index of refraction with the equation

$n = c/v$

where n is the index of refraction of the material, c is the velocity of light through vacuum, and v is the light's velocity through the material. So if light passes through a theoretical material at 210,000km/s, the material's index of refraction is

$n = 300000 \div 210000 = 1.43$

To calculate the amount of refraction that will take place when light passes from one material to another, we'll need a model and some basic terms. The model, shown in Figure 20.10, illustrates light passing from a medium with a lower index of refraction (n_1) to a medium with a higher index of refraction (n_2). The *interface* is represented by a horizontal line. A path perpendicular to the interface is known as the *normal*. Light traveling along the normal will change speed but will not change direction.

FIGURE 20.10
Model used to cal-
culate refraction

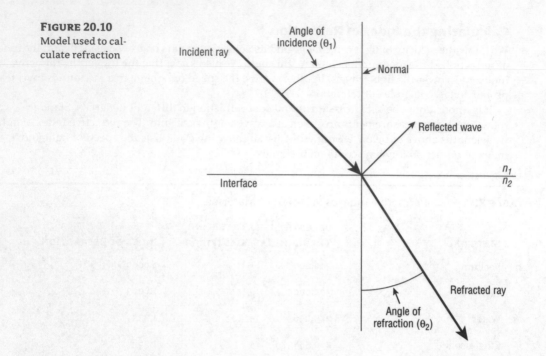

The *angle of incidence* represents the angle between the incident ray and normal. The *angle of refraction* represents the angle between the refracted ray and normal. Note that a small amount of light is also reflected from the interface at an angle equal to the angle of incidence.

Total Internal Reflection

As shown earlier in Figure 20.10, light passing from a lower index of refraction to a higher index of refraction refracts toward normal. This is illustrated by the angle of refraction (θ_2), which is smaller, then the angle of incidence (θ_1). When light passes from a higher index of refraction to a lower index of refraction, as shown in Figure 20.11, the light refracts away from normal.

We can calculate the amount of refraction using Snell's law, which shows the relationship between incident light and refracted light:

$$n_1\sin\theta_1 = n_2\sin\theta_2$$

where n_1 and n_2 are the index of refraction values of each material, θ_1 is the angle of incidence, and θ_2 is the angle of refraction.

Recall that the phenomenon that makes fiber-optic transmission possible, *total internal reflection* (TIR), is caused by the same principles that cause refraction. In the case of TIR, the light is passing from a higher index of refraction to a lower index of refraction at an angle that causes all of the light to be reflected. What has happened is that the angle of refraction has exceeded 90° from normal.

FIGURE 20.11
Light passing from higher index of refraction to lower index of refraction refracts away from normal.

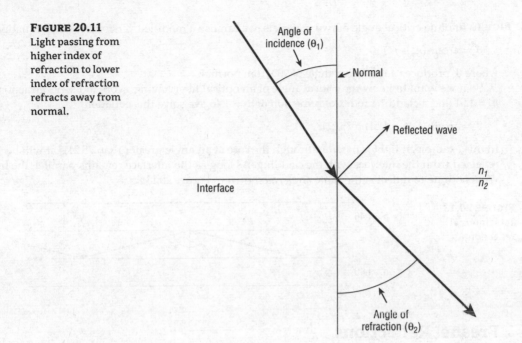

The incident angle required to produce a refracted angle of 90° is called the *critical angle.* As the incident ray moves from normal toward the critical angle, less and less of the incident ray's energy is carried into the refracted ray. At the critical angle all of the incident ray's energy is refracted along the interface. As the incident angle exceeds 90°, the light is reflected, as shown in Figure 20.12.

FIGURE 20.12
Reflection of light exceeding critical angle

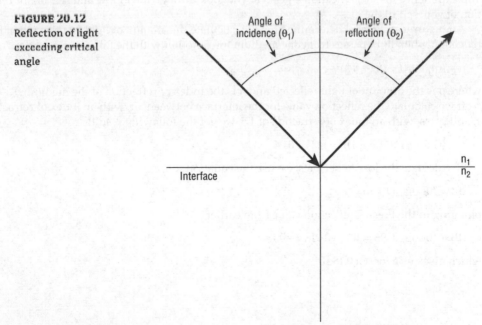

Angle of incidence > Critical angle

To find the critical angle of two materials, we can use a modified version of Snell's equation:

$$\theta_c = \arcsin(n_2 \div n_1)$$

where θ_c produces a refractive angle of 90° from normal.

So if we want to know the critical angle of an optical fiber having a core index of refraction of $n_1 = 1.51$ and a cladding index of refraction of $n_2 = 1.46$, we solve this equation:

$$\theta_c = \arcsin(1.46 \div 1.51) = 75.211°$$

In this example, if light is passing through the core at an angle greater than 75.211°, it will be reflected from the interface with the cladding. As long as the interface remains parallel, the light will continue to reflect at the same angle, as shown in Figure 20.13.

FIGURE 20.13
Total internal
reflection

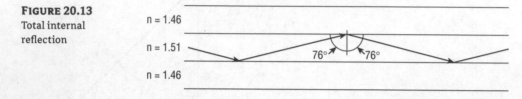

Fresnel Reflections

Recall that when light passes from one medium into another, refraction occurs as described by Snell's law. However, a reflection also occurs. This reflection is known as a *Fresnel reflection*. A Fresnel reflection occurs when light changes speed as it moves from a material with a given index of refraction into a material with a different index of refraction. The greater the difference in the index of refraction between the two materials, the greater the amount of light reflected. You experience Fresnel reflection whenever you look through a window and see a faint reflection of yourself in the glass.

Augustin-Jean Fresnel determined how to calculate the amount of light lost through a Fresnel reflection when light passes from one medium into another with the following equation:

$$\rho = ((n_1 - n_2) \div (n_1 + n_2))^2$$

where ρ is the amount of light reflected and n is the index of refraction of the medium.

If we calculate the reflection value for the interface between air, with an index of refraction of 1, and glass, with an index of refraction of 1.5, we get the following result:

$$\rho = ((1.5 - 1) \div (1.5 + 1))^2 = 0.2^2 = 0.04$$

To calculate the loss in decibels, we use the equation

$$dB = 10 \text{Log}_{10} (1 - \rho)$$

plugging in the Fresnel reflection value from earlier,

$$dB = 10 \text{Log}_{10}\, 0.96 = 10 \times -0.018 = -0.18$$

which gives us a loss of 0.18dB.

It is important to understand that a Fresnel reflection occurs whenever there is a velocity change. It does not matter whether the light is increasing speed or decreasing speed. Light traveling in the core of an optical fiber travels slower than it does in air. If there were an air gap at an interconnection, a Fresnel reflection would occur as light left the core of the optical fiber and entered the air. A Fresnel reflection of equal magnitude would also occur as light traveling in the air entered the core of the other optical fiber. The reflected light is lost and never makes it to its intended destination. Some of the reflected light may travel in the opposite direction in the core of the optical fiber back toward the source. Depending on the light source, this could cause a problem such as a shift in wavelength or a reduction in output power.

REDUCING FRESNEL REFLECTIONS AND MAKING THE INTERCONNECTION WORK

The connector endface is the part of the fiber-optic connector where the optical fiber is exposed. From time to time, the optical fiber at the connector endface is damaged, and typically the person causing the damage is unaware of it. The damage may occur because there was debris on one connector endface when it was mated with another connector. It may occur when a damaged connector is mated with an undamaged connector. Regardless of how it occurs, the end result is typically a fiber-optic link that does not function properly.

Critical fiber-optic links need to be restored as quickly as possible. This means that replacing the damaged connector(s) at this time may not be an option. Although it is not a common practice to use index matching gel in an interconnection, using it to reduce the Fresnel reflections can sometimes restore a damaged or poorly performing interconnection. This could mean the difference between a successful and a failed restoration. (*Index matching gel* is a substance that fills the air space within a connection with a material matching the refractive index of the fiber core.)

The Bottom Line

Convert various wavelengths to corresponding frequencies. The relationship between wavelength and frequency can be expressed with the formula $\lambda = v \div f$.

Master It If an electromagnetic wave has a frequency of 800kHz, what is its wavelength?

Calculate the amount of energy in a photon using Planck's constant. The amount of energy in each photon depends on the electromagnetic energy's frequency: the higher the frequency, the more energy in the particle. To express the amount of energy in a photon, we use the equation $E = hf$.

Master It Calculate the energy of a photon at a frequency of 375.25THz.

Calculate the speed of light through a transparent medium using its refractive index. The index of refraction is a relative value, and it is based on the speed of light in a vacuum, which has an index of refraction of 1. The index of refraction can be calculated using the equation $n = c/v$.

Master It If light is passing through a medium with a refractive index of 1.25, what will the velocity of the light through that medium be?

Use Snell's law to calculate the critical angle of incidence. The incident angle required to produce a refracted angle of 90° is called the critical angle. To find the critical angle of two materials, we can use a modified version of Snell's equation:

$$\theta_c = \arcsin(n_2 \div n_1)$$

Master It Calculate the critical angle for two materials where n_2 is 1.48 and n_1 is 1.52.

Calculate the loss in decibels from a Fresnel reflection. Augustin-Jean Fresnel determined how to calculate the amount of light lost from a Fresnel reflection using the equation $\rho = ((n_1 - n_2) \div (n_1 + n_2))^2$.

Master It Calculate the loss in decibels from a Fresnel reflection that is the result of light passing from a refractive index of 1.48 into air.

Chapter 21

Optical Fiber Construction and Theory

Optical fibers are called on to operate in a wide variety of conditions. Some of them must carry high volumes of data over many kilometers, while others carry smaller amounts of data inside an office building or aboard an aircraft. The type of job an optical fiber will do determines the type of fiber you'll choose to install. It is important to understand the types of fibers that are available and the ways in which they are built so that you can select and use them properly.

This chapter describes the construction of optical fibers and the components that make them up. We will discuss some of the important factors that must be considered in the manufacture and use of optical fibers, as well as the designs used to optimize them for different types of data transmission. Finally, the chapter will introduce you to some of the commonly used commercial optical fibers and describe their features.

In this chapter, you will learn to:

◆ Calculate the numerical aperture of an optical fiber

◆ Calculate the number of modes in an optical fiber

◆ Select the proper optical fiber coating for a harsh environment

Optical Fiber Components

Today's standard optical fiber is the product of precision manufacturing techniques and exacting standards. Make no mistake: even though it is found in almost any data or communications link, optical fiber is a finely tuned instrument requiring care in its production, handling, and installation.

As shown in Figure 21.1, a typical optical fiber comprises three main components: the core, which carries the light; the cladding, which surrounds the core with a lower refractive index and contains the light; and the coating, which protects the fragile fiber within.

Let's look at these components individually.

FIGURE 21.1
Optical fiber components include the core, cladding, and coating.

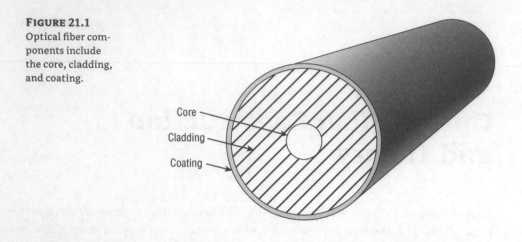

Core

Cladding

Coating

Core

The *core*, which carries the light, is the smallest part of the optical fiber. The optical fiber core is usually made of glass, although some are made of plastic. The glass used in the core is extremely pure silicon dioxide (SiO_2), a material so clear that you could look through 5 miles of it as though you were looking through a household window.

In the manufacturing process, dopants such as germania, phosphorous pentoxide, or alumina are used to raise the refractive index under controlled conditions.

Optical fiber cores are manufactured in different diameters for different applications. Typical glass cores range from as small as 3.7µm up to 200µm. Core sizes commonly used in telecommunications are 9µm, 50µm, and 62.5µm. Plastic optical fiber cores can be much larger than glass. A popular plastic core size is 980µm.

Cladding

Surrounding the core, and providing the lower refractive index to make the optical fiber work, is the *cladding*. When glass cladding is used, the cladding and the core are manufactured together from the same silicon dioxide–based material in a permanently fused state. The manufacturing process adds different amounts of dopants to the core and the cladding to maintain a difference in refractive indexes between them of about 1 percent. A typical core may have a refractive index of 1.49 at 1300nm while the cladding may have a refractive index of 1.47. These numbers, however, are wavelength dependent. The core of the same fiber will have a different refractive index at a different wavelength. At 850nm the core refractive index will increase slightly.

Like the core, cladding is manufactured in standard diameters. The two most commonly used diameters are 125µm and 140µm. The 125µm cladding typically supports core sizes of 9µm, 50µm, 62.5µm, and 85µm. The 140µm cladding typically has a 100µm core.

Coating

The *coating* is the true protective layer of the optical fiber. The coating absorbs the shocks, nicks, scrapes, and even moisture that could damage the cladding. Without the coating, the optical fiber is very fragile. A single microscopic nick in the cladding could cause the optical fiber to break when it's bent. Coating is essential for all-glass fibers, and they are not sold without it.

The coating is solely protective. It does not contribute to the light-carrying ability of the optical fiber in any way. The outside diameter of the coating is typically either 250μm or 500μm. The coating is typically colorless. In some applications, however, the coating is colored, as shown in Figure 21.2, so that individual optical fibers in a group of optical fibers can be identified.

FIGURE 21.2
The fiber coating can be color-coded to help identify it.

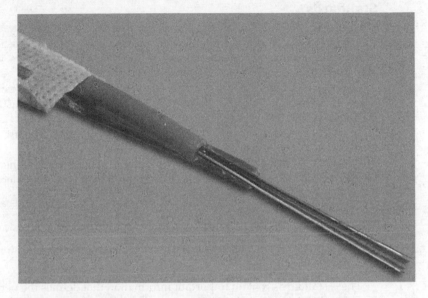

The coating found on an optical fiber is selected for a specific type of performance or environment. One of the most common types of coatings is *acrylate*. This coating is typically applied in two layers. The *primary coating* is applied directly on the cladding. This coating is soft and provides a cushion for the optical fiber when it is bent. The *secondary coating* is harder then the primary coating and provides a hard outer surface. Acrylate, however, is limited in temperature performance. A typical acrylate coating has a maximum operating temperature of 100° C while some high-temperature acrylates may perform at temperatures up to 125° C.

Silicone, carbon, and *polyimide* are coatings that may be found on optical fibers that are used in harsh environments such as those associated with avionics, aerospace, and space. They may also be used on optical fibers designed for mining, or oil and gas drilling.

Silicone is a soft material with a higher temperature capability then acrylate, typically 200° C. Like acrylate, silicone cushions the optical fiber during bending, but it is too soft to be the sole coating. It must be used in combination with a buffer or harder thermoplastic and is generally thicker than acrylate, with diameters up to 500μm. Silicone has other advantages besides a high-temperature capability; it has a high resistance to water absorption and low flammability.

Silicone or acrylate coatings must be removed prior to terminating the optical fiber. Both of these coatings can leave a residue, so extra care must be used when removing silicone. Terminating optical fibers is discussed in depth in Chapter 26, "Connectors."

Polyimide is a popular coating for aerospace applications because it can operate in temperatures up to 350° C. This coating is much thinner than acrylate or silicone, typically only 15μm thick. Beside a very high operational temperature, polyimide is resistant to abrasion and chemicals. However, polyimide is very difficult to remove. If it must be removed before termination, hot sulfuric acid or a portable polyimide-stripping machine is required.

Carbon is the thinnest coating of all, typically only a fraction of a micron thick. Carbon hermetically seals the glass surface, protecting it from moisture and increasing the optical fiber's fatigue resistance. The carbon coating does not have to be removed before termination.

Standards

While many combinations of core and cladding sizes are possible, standards are necessary to ensure that connectors and equipment can be matched properly. This is especially important when dealing with components as small as those used in fiber optics, where even slight misalignments can render the entire system useless.

Two organizations publish standards that define the performance of optical fibers used in the telecommunications industry; they are the Telecommunications Industry Association (TIA) and the International Telecommunications Union (ITU). While TIA and ITU publish many standards on optical fiber, the key standards that you should be familiar with are ANSI/TIA-568-C.3, ITU-T G.652, ITU-T G.655, and ITU-T G.657.

The ANSI/TIA-568-C.3 standard is applicable to *premises* optical fiber cabling components. The ITU standards are applicable to single-mode optical fiber and cable. Here are their descriptions:

- ITU-T G.652: Characteristics of a single-mode optical fiber and cable

- ITU-T G.655: Characteristics of a dispersion shifted single-mode optical fiber and cable

- ITU-T G.657: Characteristics of a non-zero dispersion-shifted single-mode optical fiber and cable

These standards contain important information that defines the performance of the optical fiber, fiber-optic cable, and components such as connectors and splices. Subsequent chapters will discuss these standards in detail.

Materials

Optical fibers are commonly made with a glass core and glass cladding, but other materials may be used if the fiber's performance must be balanced with the cost of installing the fiber, fitting it with connectors, and ensuring that it is properly protected from damage. In many cases, fibers must run only a short distance, and the benefits of high-quality all-glass fibers become less important than simply saving money. There are also circumstances in which the fibers are exposed to harsh conditions, such as vibration, extreme temperature, repeated handling, or constant movement. Different fiber classifications have evolved to suit different conditions, cost factors, and performance requirements. The major fiber classifications by material are:

Glass fibers   These have a glass core and glass cladding. They are used when high data rates, long transmission distances, or a combination of both are required. As you will see in later chapters, connectors for glass fibers must be built with a high degree of precision and can take time to apply correctly. Glass fibers are the most fragile of the various types available, and as a result they must be installed in environments where they will not be subjected to a great deal of abuse, or they must be protected by special cables or enclosures to ensure that they are not damaged.

Glass fibers are commonly found in long-distance data and interbuilding and interoffice networking applications.

Plastic-clad silica (PCS)   These fibers have a glass core and plastic cladding. The core is larger than all-glass fiber; typically, 200µm with a cladding thickness of 50µm. Like a silicone-coated glass optical fiber, the plastic coating of a PCS optical fiber is typically used with a thermoplastic buffer that surrounds the plastic cladding. A typical PCS fiber specification would be 200/300µm. The plastic cladding also serves as a protective layer for the glass core, so the coating normally found on all-glass fiber is not included on PCS fibers. PCS fibers are typically used for industrial sensing applications and medical/dental applications.

Hard-clad silica (HCS)   These fibers are similar to PCS fiber but they have a glass core with cladding made of a hard polymer or other material, typically stronger than other cladding materials. Hard-clad silica fiber is commonly used in locations where ruggedness is a prime consideration, such as manufacturing, factory automation, and other areas where shock and vibration would render standard glass fibers unreliable.

HCS optical fibers are typically much larger than glass optical fibers. A very popular size is 200/230µm.

Plastic fiber   These fibers have a plastic core and plastic cladding. They are selected for their low cost, ruggedness, and ease of use, and are installed where high bandwidth and long transmission distances are not required. While plastic fibers are unsuited for long-distance, high-performance transmissions, they can still carry signals with useful data rates over distances of less than 100m. Plastic fibers are much larger in diameter than glass fibers, as shown in the comparison in Figure 21.3. A very popular size is 980/1000µm. Plastic fiber is typically designed for visible wavelengths in the 650nm range. Some typical locations for plastic fiber include home entertainment systems, automotive, and manufacturing control systems. They may also be used in links between computers and peripherals and in medical equipment.

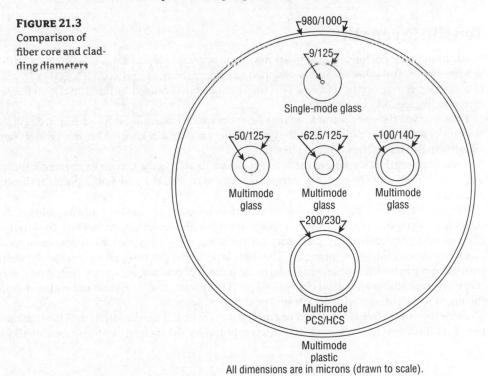

FIGURE 21.3
Comparison of fiber core and cladding diameters

Single-mode glass

Multimode glass

Multimode glass

Multimode glass

Multimode PCS/HCS

Multimode plastic

All dimensions are in microns (drawn to scale).

THE ADVANTAGES OF LARGE CORE PLASTIC OPTICAL FIBER

It is easy to get excited about the high bandwidth and long-distance transmission capabilities of glass optical fiber. It clearly outperforms any other medium. However, many applications do not require a high bandwidth over great distances. There are many applications for optical fiber in your home. You may already have a home entertainment system that uses plastic optical fiber, or you may own a car that uses plastic optical fiber to connect audio devices or a DVD changer. None of these applications requires high bandwidth over great distances. These applications are ideal for large core plastic optical fiber.

As you learned in this chapter, plastic optical fiber is typically designed to operate at a visible wavelength around the 650nm range. Being able to see the light as it exits the optical fiber has a significant advantage; no expensive test equipment is required. A power meter is needed to measure the light exiting a glass optical fiber operating in the infrared range. Power meters can cost more than your home entertainment system.

The large core of the plastic optical fiber has another advantage over small glass fibers: it is easy to align with another fiber or a light source or detector. Imagine aligning two human hairs so that the ends touch and are perfectly centered. Now imagine doing the same thing with two uncooked spaghetti noodles. Which do you think would be easier?

Plastic optical fiber is a great choice for audio and video electronics being integrated into homes and vehicles.

Tensile Strength

In addition to the coating, optical fibers may have several layers of protection such as buffers, jackets, armor, and other materials designed to protect the fiber and keep it together in cables. Like copper, though, optical fiber is still subject to hazards caused by handling, installation, careless digging, and bad weather.

One characteristic of optical fiber that deserves special attention is its tensile strength, or the amount of stress from pulling that the optical fiber can handle before it breaks. Optical fiber has a high tensile strength and resists stretching.

Tensile strength is important for several reasons. It affects the way fiber must be handled during installation, the amount of curvature it can take, and the way it will perform throughout its lifespan.

To understand optical fiber's tensile strength, let's look at a standard piece of plate glass. To cut the glass, all you have to do is scribe a sharp line through the surface layer. Once the strength of that layer is compromised, the glass snaps easily along the scribe, even if it follows a curve.

Optical fiber follows the same rule. The outer layer of the cladding provides much of the fiber's tensile strength, which is often measured in thousands of pounds per square inch (kpsi). A typical breaking strength for an optical fiber is 100kpsi. That means that a typical fiber with a cladding thickness of 125µm can withstand a pull of about 1.9 pounds.

Once the outer layer is scratched or cracked, however, the tensile strength is gone at that location. Like a scribed line, a scratch or crack compromises the integrity of the glass and allows the fiber to break more easily under stress. Scratching or cracking can occur due to mishandling, sharp blows, or bending beyond the fiber's *minimum bend radius*, especially if the bending takes place while the fiber is under tension. The minimum bend radius is the minimum radius the optical fiber can be bent without damaging it.

Manufacturing Optical Fiber

Optical fiber is manufactured to very high standards, because the core diameter and refractive indexes of the core and cladding must remain consistent over stretches of up to 80km.

Four methods are commonly used to make optical fiber:

◆ Modified chemical vapor deposition (MCVD)

◆ Outside vapor deposition (OVD)

◆ Vapor axial deposition (VAD)

◆ Plasma chemical vapor deposition (PCVD)

Each method uses variations on a process to create a *preform*, a short, thick glass rod that has a similar cross section to the final fiber. As shown in the block diagram in Figure 21.4, the preform is heated to 1900° C in a drawing tower until it begins to melt and a blob falls from the end, drawing a small thread of glass after it. The thread contains the core and cladding of the optical fiber, their relative thicknesses preset by the preform's construction.

FIGURE 21.4
Drawing the pre-
form into fiber

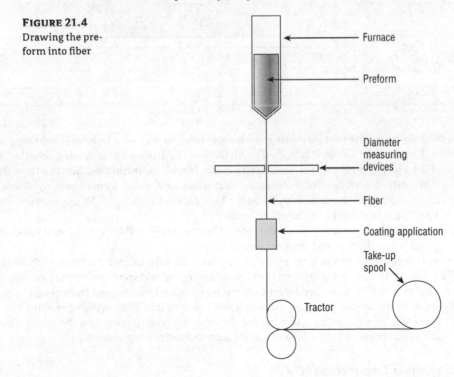

This thread is taken up by a pulling machine, or *tractor*, at a constant rate to maintain a consistent thickness. A thinner, longer fiber can be created by speeding up the draw rate. The entire process is closely monitored by laser measuring devices to ensure that the thickness remains consistent over the entire length of the fiber, which can be anywhere from 10 to 80km, depending on the size of the preform.

The fiber is then drawn through another process, which deposits and cures the coating. The coating is cured using ultraviolet lamps or thermal ovens. After the coating is applied, measurements are taken and the optical fiber is taken up on a spool.

Each of the preceding methods is best suited for different types of fiber, depending on the type of signals it will carry, the distance it will cover, and other factors, which will be discussed later in this chapter.

Let's look at the differences in the manufacturing methods.

Modified Chemical Vapor Deposition (MCVD)

Fibers manufactured using MCVD begin with a hollow glass tube about 2.5cm in diameter, as shown in Figure 21.5.

FIGURE 21.5
Modified chemical vapor deposition

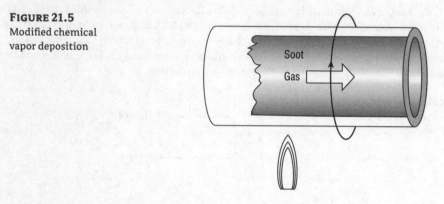

The cladding is created first by placing the hollow tube, or *bait*, on a lathe and spinning it rapidly. As the bait is spinning, it is heated by an oxygen/hydrogen torch passing lengthwise underneath as a gaseous mixture of vaporized silicon dioxide is introduced into the tube. The ultra-pure gas, mixed with carefully controlled impurities or *dopants*, forms a soot, which fuses to the inside of the bait in successive layers as the heat passes beneath it. The dopants are used to increase the refractive index of the fused material.

Every time the flame passes beneath the tube to heat it, another thin layer of soot adheres to the inside, building a thicker and thicker layer of glass.

The core is formed next by changing the gas/dopant mixture to create a vapor with a higher refractive index. This, too, is introduced into the spinning, heated tube, inside the first layers that were laid down. When the deposited material has reached the desired thickness, the tube is heated to an even higher temperature to consolidate the soot into glass without melting it, a process known as *sintering*. A drying gas is passed through the core to remove water contamination, and the tube is compressed into a preform and taken to the drawing tower.

Outside Vapor Deposition (OVD)

The OVD method is similar to MCVD, except that the fiber preform is built from the inside out.

In the OVD method, a glass target rod serves as the bait, as shown in Figure 21.6. The rod is placed on the lathe and spun as heat is applied to it. The core is laid down first by introducing a gas mixture between the heat and the rod. When the core layer is thick enough, the mixture is changed and the cladding is laid down.

FIGURE 21.6
Outside vapor
deposition

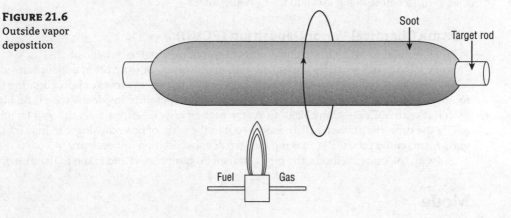

After the layers have been deposited, the rod is removed and the layers are dried and sintered, and then collapsed into the preform.

Vapor Axial Deposition (VAD)

As with MCVD and OVD, the VAD process uses a heated glass bait to collect a soot of silicon dioxide and dopant. With VAD, however, the glass rod is suspended vertically and the heat source is at the lower end, as shown in Figure 21.7.

FIGURE 21.7
Vapor axial
deposition

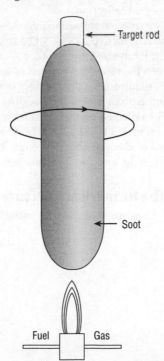

The gas is introduced between the end of the rod and the heat source, and the soot builds up in a radial pattern. This method gives the manufacturer a great deal of control over not only the refractive index of the various layers, but also the pattern in which they are laid down.

As with OVD, the rod is then removed, and the layers are dried, sintered, and collapsed into the preform before being taken to the drawing tower.

Plasma Chemical Vapor Deposition (PCVD)

PCVD is similar to MCVD, but allows a finer layer of material to be deposited.

In PCVD, the gas particles are heated with microwaves, causing plasma to form inside the tube. The plasma contains electrons with extremely high energy levels, approximating those found in a gas heated to 60,000° C, even though the temperature inside the tube is no higher than it is with MCVD—about 1200° C. As the high-energy electrons meet the soot forming inside the tube, they transfer their energy to it in the form of heat, causing it to fuse into the final glass form on the tube walls, making the step of consolidation unnecessary.

As with the other methods, the tube can then be compressed and taken to the drawing tower.

Mode

One of the most important characteristics used to distinguish types of fiber is the number of potential paths light can take through it. It may seem that light would go straight through the fiber core, following all of its curves, until it comes out the other end. However, the light itself is a complex combination of electrical and magnetic waves, and their wavelengths can be many times smaller than the core of the fiber. Like a rubber ball shot through a sewer pipe, the light actually has more than one path or *mode*. There are many potential paths or modes it can follow, depending on the size of the core, wavelength of the light source, and *numerical aperture*.

To understand modes in optical fibers, let's take another look at the basis for fiber-optic transmission: total internal reflection.

Recall that total internal reflection depends on the principle that light passing through a medium of a given refractive index will be reflected off the interface with a medium of a lower refractive index if it hits the interface at or above the critical angle, which is determined by the difference between the refractive indexes of the two media.

To satisfy this requirement, optical fiber is manufactured with a core having a refractive index slightly higher than that of the cladding surrounding it. As a result, light entering the end of the fiber will be reflected off the interface with the cladding, and guided through the fiber. By definition, each reflection will be at the same angle as the angle of incidence.

The number of modes possible in a length of fiber depends on the diameter of the core, the wavelength of the light, and the core's numerical aperture.

Calculating the Numerical Aperture and Modes

You can find the numerical aperture using the following formula:

$$\text{NA} = \sqrt{n^2{}_{core} - n^2{}_{cladding}}$$

For example, if the core has a refractive index of 1.48 and the cladding has a refractive index of 1.46, we can calculate:

$$NA = \sqrt{1.48^2 - 1.46^2} = 0.242$$

Note that the numerical aperture has no dimension. It is intended as a relative indication of the light-gathering capacity of the fiber core.

To find the number of modes, we can use the equation

$$M = (2D \times \pi \times NA/\lambda)^2/2$$

where D is the diameter of the core and λ is the wavelength of the light. For example, let's use our previously derived value for the numerical aperture with a core diameter of 50μm and a light wavelength of 1300nm:

$$M = (100 \times 10^{-6} \times \pi \times 0.242/1300 \times 10^{-9})^2/2 = 1710.07$$

Because light cannot have part of a mode, we must round down to the nearest whole number if we come up with a decimal. So the answer is 1,710 modes, or potential paths for the light to follow.

Refractive Index Profiles

What happens when the light rays entering a fiber take slightly different paths, some entering at sharper angles, some at shallower angles? As shown in Figure 21.8, the light can follow modes ranging from a straight line through the fiber (*zero-order* mode) to a low number of reflections (*low-order* mode) to a high number of reflections (*high-order* mode). Depending on a fiber's construction and the wavelength of the light source, the fiber may support one mode or more than 1,000,000 modes (in fibers with large cores such as those found in plastic optical fibers). The path or paths that light takes through an optical fiber can be understood by examining the fiber's *refractive index profile*.

FIGURE 21.8
The three types of
light modes

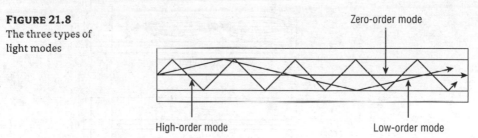

The refractive index profile graphically represents the relationship between the refractive index of the core and that of the cladding. Several common profiles are found on optical fibers used in telecommunications, avionics, aerospace, and space applications:

◆ Step index or matched clad

◆ Graded index

◆ Depressed clad

These refractive index profiles are shown in Figure 21.9.

FIGURE 21.9
Refractive index
profiles

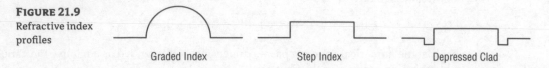

Graded Index   Step Index   Depressed Clad

So far in this text, we have described a simple relationship in which the core has one refractive index and the cladding another, and there is a single interface between the two. This is known as *step-index* fiber. If a step index fiber has a small core and only permits one path for light, it is referred to as a *single-mode* optical fiber. However, if the fiber has a large enough core diameter with room for the light to take a number of different possible paths, or modes, by way of reflection, it is referred to as *multimode* fiber.

MULTIMODE STEP-INDEX FIBER

Also known simply as step-index fiber, *multimode step-index fiber*, as already discussed, has a core with a single refractive index and a cladding with another, slightly lower refractive index. These are separated by a single interface, which reflects light that hits it at any angle greater than its critical angle. The large core of a step index multimode optical fiber allows a ray of light to take many possible paths, as shown in Figure 21.10.

FIGURE 21.10
Light propagation
profiles

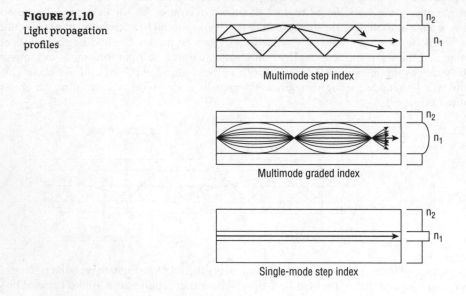

Multimode step index

Multimode graded index

Single-mode step index

The result is that even though all of the rays pass through the same length of fiber, the reflections create a longer path for the light to follow. The more reflections, the longer the path. You would get the same result if you and a friend walked down the same road at the same rate of speed, with you walking down the middle of the road and your friend zigzagging from one side of the road to the other. Even if your friend started out before you did, you would eventually pass him, because he is taking a longer path. As a result, you would arrive at the end of the path before your friend, even though you started out together.

The same effect occurs as light is reflected through the fiber. Even though light rays enter the fiber in the order 1, 2, 3, they may arrive in the order 3, 1, 2, or even overlap each other as they arrive, causing the information they carry to become muddled and useless. This effect is called *modal dispersion*, and is an important factor limiting bandwidth in optical fiber.

Typically, modal dispersion in step-index fiber offsets light rays by about 15 to 30ns/km, depending on the diameter of the fiber core and the wavelength of the light. In other words, if one ray takes the most direct path through the fiber, and another ray takes the longest possible path by reflecting off the fiber walls, the ray taking the longest path will follow the ray taking the shortest path by anywhere from 15 to 30 nanoseconds (15 to 30 billionths of a second) for every kilometer of fiber length. That means that for every kilometer of fiber length, each bit would have to be separated from the one before it by at least 30ns, or roughly 6m, to ensure that the bits would not overlap. This results in a data rate limit of 33.33MHz for a 1km fiber, 16.67MHz for a 2km fiber, 11.11MHz for a 3km fiber, and so on—far below the capabilities of transmitters, receivers, and processors, and completely unsuited for long-distance fiber-optics transmission. For this reason multimode step index optical fibers are not used in telecommunications and their performance is not addressed in any of the standards discussed in this text. PCS, HCS, and plastic optical fibers are typically step-index.

Two methods used to reduce modal dispersion are multimode graded-index fiber and single-mode step-index fiber.

MULTIMODE GRADED-INDEX FIBER

Multimode graded-index fiber tackles the problem of modal dispersion by increasing the speed of the high-order mode light rays, allowing them to keep up with the low-order mode rays.

The fiber accomplishes this by using the very principles upon which fiber optics is based: the laws of refraction.

Remember that when light passes from a higher RI medium to a lower RI medium, it gains velocity. A *graded-index* fiber core actually consists of many concentric glass layers with refractive indexes that decrease with the distance from the center. Viewed on end, these layers resemble the rings of a tree.

Any light that passes straight through the fiber travels at a constant speed. If a light ray enters at an angle, however, it passes through the graded layers. As it does, two things happen.

First, the light is refracted away from normal, because the refractive index has decreased. Second, the light propagation velocity increases. This happens in every new layer the light traverses, until it has reached the cladding or has been bent to the critical angle for one of the layers. At this point, the light is reflected and begins its new direction, this time refracting toward normal and slowing down until it reaches the highest refractive index, at the center of the optical fiber core. It then passes through the center and begins the cycle again on the opposite side of the fiber core. The path resembles a segmented sine wave, as shown in Figure 21.10.

Note that as the light follows a longer path, its average velocity increases, offsetting the extra distance it must travel. The time it takes to move from one end of the fiber to the other is now much closer to that of light following a straight path through the fiber. In fact, modal dispersion in graded-index fiber can be reduced to as little as 1ns/km.

SINGLE-MODE STEP-INDEX FIBER

Single-mode step-index fiber uses another approach to reduce modal dispersion. It doesn't give the light enough space to follow anything but a single path through the fiber.

Single-mode fiber uses a core so small that light can only travel in one mode. The diameter of the single-mode fiber is typically only about 8 to 10µm, but its performance also depends on the wavelength of the light.

For example, if a fiber is designed to carry a wavelength of 1310nm in only one mode, there is room for light at 850nm to travel in several modes, so the fiber has become a multimode fiber at the smaller wavelength. The smallest wavelength at which the fiber carries only one mode is called the *cutoff wavelength*. The cutoff wavelength for a fiber designed to carry 1310nm in single mode is about 1260nm.

Even in single-mode fiber, light does not follow a straight path. Because of the way light propagates, it actually follows a corkscrew-like path through the fiber core and propagates in a portion of the cladding, as shown in Figure 21.11. As a result, manufacturers must make single-mode fiber with cladding that will carry the light with less attenuation.

FIGURE 21.11

Light propagation in single-mode fiber

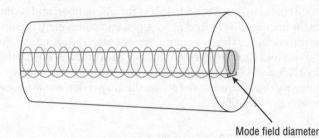

Mode field diameter

The unique light propagation characteristics of single-mode fiber also make it necessary to take the light propagated in the cladding into account when matching fiber sizes for connections, so the light in the cladding is not lost. Single-mode fibers are typically specified by their *mode field diameter* rather than their core and cladding size. As you can see in Figure 21.11, the mode field diameter is the real estate used by the light within the core and the cladding as it propagates.

SINGLE-MODE DEPRESSED-CLAD FIBER

Single-mode depressed-clad optical fiber has three levels of refractive indexes. Like all optical fibers, the core has the highest refractive index. However, the cladding has two refractive indexes. The refractive index of the cladding immediately adjacent to the core is lower than the outer cladding. This refractive index profile resists attenuation from bends with a small radius better than step index.

As described in Chapter 1, "Introduction to Data Cabling," fiber to the home is a reality. Bringing fiber to the home or an apartment complex can require the cable to be bent sharply. Bending an optical fiber can cause loss, and single-mode optical fibers can experience considerable loss at longer wavelengths such as 1550nm.

To reduce attenuation from bending, fiber manufacturers created single-mode optical fibers that have very little loss when bent sharply. These optical fibers feature proprietary refractive index profiles similar to the depressed clad shown in Figure 21.11. These fibers are beyond the scope of this book.

GETTING TO KNOW THE STANDARDS

If you are a sports fan, you know the statistics of your favorite team and players. If you plan to be the person everyone turns to when there is a question about optical fiber, you need to know the standards just as well as the sports fan who recites the statistics of their favorite team and players. The nice thing about standards, however, is they do not change nearly as fast as your favorite team's record does. Standards typically take years to develop and just as long to change.

The Bottom Line

Calculate the numerical aperture of an optical fiber.   To find the numerical aperture of an optical fiber, you need to know the refractive index of the core and the cladding.

Master It   The core of an optical fiber has a refractive index of 1.49 and the cladding has a refractive index of 1.47. Calculate the numerical aperture.

Calculate the numbers of modes in an optical fiber.   To find the number of modes in an optical fiber, you need to know the diameter of the core, the wavelength of the light source, and the numerical aperture.

Master It   Calculate the number of possible modes in an optical fiber with the following specifications:

Core diameter = 62.5μm

Core refractive index = 1.50

Cladding refractive index = 1.46

Wavelength of the light source = 1300nm

Select the proper optical fiber coating for a harsh environment.   The coating is the true protective layer of the optical fiber; it absorbs the shocks, nicks, scrapes, and even moisture that could damage the cladding. Without the coating, the optical fiber is very fragile. The coating found on an optical fiber is selected for a specific type of performance or environment.

Master It   Choose a coating for an optical fiber that must operate in temperatures as high as 250° C.

Chapter 22

Optical Fiber Characteristics

In addition to factors such as construction, materials, and size as discussed in the previous chapter, optical fibers have certain performance or operational characteristics that define them. These characteristics may describe limitations or features of the fiber with regard to its light-carrying ability under various conditions, and are generally affected by its physical properties.

In this chapter, we describe the characteristics of optical fiber that affect the way it is selected, handled, installed, and used. We'll cover in detail how these characteristics change a fiber's ability to carry light, as well as the methods used to take advantage of some characteristics while minimizing the effects of others.

In this chapter, you will learn to:

♦ Calculate the attenuation in dB for a length of optical fiber

♦ Calculate the usable bandwidth for a length of optical fiber

♦ Calculate the macrobending loss in a single-mode optical fiber

♦ Calculate the acceptance for an optical fiber

It All Adds Up

If you've ever looked at your bank balance and wondered where it all went, you have some idea of why it's important to understand optical fiber characteristics in great detail.

Chances are, when you took another look at your account, you remembered that some of it went to necessities, some to entertainment, some to service charges, and so on. It didn't all go away at once: it was spent little by little, in many different places. Any single expenditure may have gone almost unnoticed, but over time all of those expenses can add up to a significant amount.

This is typically the fate of light as it passes through an optical fiber. By the time it reaches the other end, it is diminished in several ways—most notably in its power and its ability to carry a signal. In spite of the purity of the materials that go into fibers, they are still not perfect—in part by design, and in part because there are factors that simply cannot be overcome with current technology.

Going back to your bank account, if you know where the money goes, you can take steps to make it last longer or make sure that when it is spent, you get more use out of it.

The same goes for our light beam. If you understand the characteristics in an optical fiber that take away from its ability to carry a signal at high speed and over long distances, it is possible to overcome them, or use them to advantage. Remember, though, that propagation of light within a fiber is a complex mix of influences, and each of the characteristics that we'll be discussing affects different portions of this mix.

Some aspects of these characteristics have been covered in previous chapters to describe why fiber is constructed in certain ways. Now we are going to look at them more closely.

Dispersion

In general, *dispersion* is the spreading of light as it travels away from its source. The light spreads because different components of it travel at slightly different velocities or take slightly different paths, depending on the conditions in the medium through which it is traveling and the wavelengths that make up the light. There are different kinds of dispersion, however, and the kind that is taking place depends on several factors in the optical fiber and in the light itself.

The greatest effect of dispersion is that as the light spreads, it can degrade or destroy the distinct pulses of the digital signals in the light by making them overlap each other, as shown in Figure 22.1, blurring and blending them to the point that they are unusable. The effect grows more pronounced as the distance the light travels increases.

FIGURE 22.1
The effects of dispersion on a signal

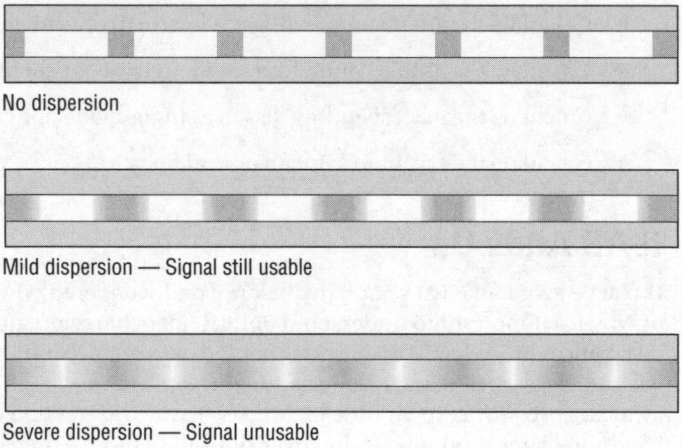

No dispersion

Mild dispersion — Signal still usable

Severe dispersion — Signal unusable

The effect is similar to looking into a hallway through a frosted glass window. If people are moving through the hallway close together, the glass spreads their images so much that they merge with one another and look like a single mass rather than individuals. If they spread out far enough from each other, however, you can see each person moving past the window. The images are still spread out, but the space between each person is great enough to see.

To prevent signal degradation due to dispersion, it is necessary to keep the pulses far enough apart to ensure that they do not overlap. This limits the signals to a bit rate that is low enough to be only minimally affected.

Dispersion limits the optical fiber's *bandwidth*, or the amount of information it can carry. Fortunately, optical fiber manufacturers and standards organizations provide information about the bandwidth or the information-carrying capacity of optical fibers at specific wavelengths.

The types of dispersion that affect optical fiber performance are:

- Modal dispersion
- Material dispersion
- Waveguide dispersion
- Chromatic dispersion
- Polarization-mode dispersion

Let's look at these types of dispersion more closely.

Modal Dispersion

We mentioned modal dispersion in the previous chapter to explain why fibers are classified as multimode or single-mode. It will help to review some of the important points.

Modal dispersion results from light taking different paths, or modes, as it passes through the optical fiber. The number of potential modes the light can take is determined by the diameter of the optical fiber core, the refractive indices of the fiber core and cladding, and the wavelength of the light.

A mode can be a straight line through the fiber, or the light can follow an angular path, resulting in reflections every time the light meets the interface between the core and the cladding. The more reflections, the longer the path through the fiber, and the longer the light takes to pass through it.

Depending on the mode, some parts of the light will pass through the fiber in less time than others. The difference in travel time can cause parts of the light pulses to overlap each other, or in extreme cases to arrive in a different order from the order they were transmitted. If this happens, the signal is no longer usable.

Methods for overcoming modal dispersion include:

Lower bit rate   Lowering the bit rate increases the gap between bits in the signal. While dispersion will still affect the bits, they will not overlap one another, and the signal will still be usable.

Graded index fiber   *Graded index fiber* gradually reduces the refractive index of the fiber core from the center toward the cladding, allowing the light that follows a more angled path to speed up as it leaves the center and causing it to slow down again as it reaches the center. This effect reduces the difference in travel time between modes and allows greater bandwidths and faster data transmission. Graded index fiber is a moderately priced solution that allows wider bandwidths than multimode step index fiber. In addition, the gradual change of indexes as the light heads for the cladding causes the light to curve back into the core of the fiber before it has a chance to approach the cladding at a penetrating angle and be lost or reflected with a destructive time delay.

Reducing the core size or increasing the wavelength   In the previous chapter, we calculated the number of potential modes in a multimode optical fiber. For a given wavelength, a smaller core optical fiber will have fewer modes than a larger core optical fiber. For a given core size, a multimode optical fiber will have fewer modes at a longer wavelength than a shorter wavelength.

Single-mode fiber   Single-mode fiber has a core that is small enough for only one mode to propagate, eliminating the problems caused by multiple modes. This type of fiber requires more expensive connectors and equipment because of the small core size and is typically used when very wide bandwidth requirements over long distances justify the cost.

Material Dispersion

Material dispersion occurs when different wavelengths of light travel at different velocities in the optical fiber. Even the light from a laser is made up of different wavelengths within a narrow range called the *spectral width,* which varies depending on the light source. Depending on the wavelength, a *light-emitting diode* (*LED*) may have a spectral width as narrow as 20nm or as wide as 170nm. A *laser's* spectral width is much narrower than that of the LED. Depending on the laser type, the spectral width can range from 0.1nm to 3nm.

Recall the formula for determining the refractive index of a material:

$$n = c/v$$

where n is the refractive index, c is the speed of light in a vacuum, and v is the speed of the wavelength of light through the material. In this equation, n changes with the wavelength of the light passing through the material. Remember that this is why white light breaks into its component colors in a prism.

When the different wavelengths travel at different velocities, the slower wavelengths begin to lag behind as the light travels down the optical fiber core, causing the light to spread as shown in Figure 22.2. If the light pulses must travel a great distance, the lag in the slower wavelengths can cause them to overlap the faster wavelengths of the bits following them. As with modal dispersion, these overlaps can degrade and ultimately destroy the signal.

FIGURE 22.2
Material dispersion in fiber causes some wavelengths to travel more slowly than others.

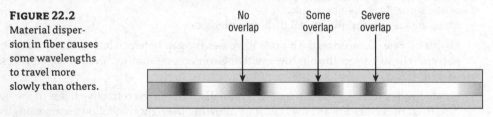

Because the wavelengths used in fiber-optic transmissions have a narrow spectral width, material dispersion takes place on a much smaller scale than modal dispersion. Its effects in a fiber are measured in picoseconds per nanometer of spectral width per kilometer (ps/nm/km), and are insignificant in a multimode fiber when compared to the effects of modal dispersion.

Material dispersion only becomes a problem when modal dispersion is overcome with single-mode fiber and higher data rates are used over long distances.

Waveguide Dispersion

Waveguide dispersion occurs in single-mode fiber as the light passes through not only the core, but also part of the cladding, as shown in Figure 22.3. By design, the core has a higher refractive index than the cladding, so the light will travel more slowly through the core than through the cladding.

FIGURE 22.3
Waveguide disper-
sion in optical fiber

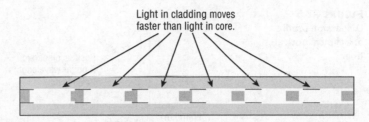

Light in cladding moves
faster than light in core.

While the difference in refractive indexes of single-mode fiber core and cladding are minus-
cule, they can still become a factor over great distances. In addition, waveguide dispersion can
combine with material dispersion to create another problem for single-mode fiber: chromatic
dispersion (discussed in the next section).

Various tweaks in the design of single-mode fiber can be used to overcome waveguide dispersion, and manufacturers are constantly refining their processes to reduce its effects.

Chromatic Dispersion

Chromatic dispersion occurs in single-mode optical fiber, and results from the combination of
effects from material dispersion and waveguide dispersion.

When chromatic dispersion occurs, the effects of material dispersion, as shown in
Figure 22.4, compound the effects of waveguide dispersion. At lower data rates and in multimode fiber, the effects of chromatic dispersion are so small as to be unnoticed, especially when
buried under modal dispersion. It is mostly a problem in single-mode fiber carrying high bit
rates over long distances, where the detrimental effects build up.

FIGURE 22.4
Waveguide disper-
sion and material
dispersion com-
bine to create chro-
matic dispersion.

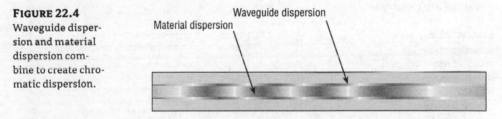

Material dispersion

Waveguide dispersion

One way to reduce chromatic dispersion is to take advantage of the fact that the relationship
between wavelength, refractive index, and velocity is not linear. In the infrared range of most
fiber-optic transmissions, the light's velocity through the medium drops as the wavelength
increases until it reaches the range between 1300nm and 1550nm. At wavelengths greater than
1550nm, the longer wavelengths have a higher velocity. Somewhere in the 1300nm to 1550nm
range there is a crossover where, depending on the specific composition of the fiber, the refractive index is the same for the wavelengths within the narrow spectral width of the light being
transmitted. In other words, as shown in Figure 22.5, as the wavelength approaches this range,
dispersion drops to zero. This *zero-dispersion point* normally occurs at 1300nm in a standard
single-mode fiber. Unfortunately, other characteristics of the optical fiber attenuate the signal at
this wavelength, making it unusable for long-distance runs.

FIGURE 22.5
Dispersion profile
of a typical optical
fiber

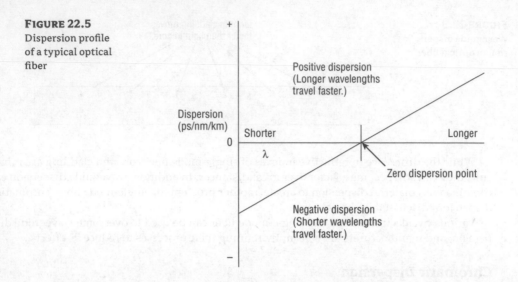

There are two approaches to reduce chromatic dispersion. One approach is to use *dispersion-shifted fiber* and the other is to reduce the spectral width of the light source. The spectral width of the light source is discussed in Chapter 27, "Fiber-Optic Light Sources."

Dispersion-shifted fiber, sometimes called *zero-dispersion-shifted fiber*, is made so that the zero dispersion point is shifted from 1300nm to 1550nm, where the attenuation is lowest, as shown in Figure 22.6. Shifting the zero dispersion point to 1550nm allows longer transmission distances because of less attenuation. Dispersion-shifted fiber is used when high data rates over long distances are required.

FIGURE 22.6
Zero dispersion
points of non-
shifted, disper-
sion-shifted, and
non-zero-disper-
sion-shifted fibers

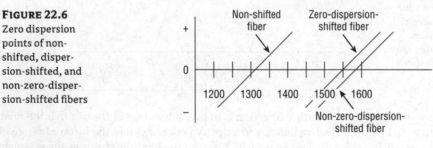

A variation of the dispersion-shifted fiber, *non-zero-dispersion-shifted fiber (NZ-DSF),* is used when multiple frequencies are being used to send more than one channel through the fiber, a process known as *wavelength division multiplexing (WDM).* When WDM takes place at the zero dispersion point, an effect known as *four-wave mixing* occurs, in which multiple wavelengths combine to form new wavelengths. These new wavelengths can interfere with the signal carrying wavelengths, reducing their power and introducing noise into the system.

To reduce four-wave mixing, NZ-DSF is used to move the dispersion point just off the 1550nm mark. The slight amount of dispersion introduced minimizes four-wave mixing while preserving most of the benefits of dispersion-shifted fiber.

The refractive index profile of dispersion-shifted fiber is shown in Figure 22.7. The peak in the center of the profile reveals an inner core that has its highest refractive index at the center. The refractive index gradually decreases toward a thin inner layer of cladding. The smaller peaks represent a ring of glass with a higher refractive index surrounding the inner cladding, slowing the light that would normally increase its velocity in the cladding. This effect reduces waveguide dispersion, which in turn reduces chromatic dispersion.

FIGURE 22.7
Refractive index profile of dispersion-shifted fiber

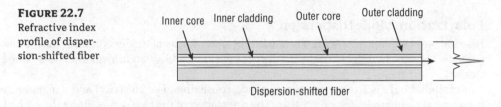

Dispersion-shifted fiber

MIXING IT UP—THE PROBLEM OF FOUR-WAVE MIXING

It seems that nobody can leave a good thing alone. Once engineers overcame the problem of chromatic dispersion in single-mode optical fiber using dispersion shifting, they decided to squeeze all the use they could out of it by piling on different wavelengths to create multiple transmission channels. The idea behind this is that different wavelengths can actually occupy the same space but remain distinct from one another until they are sorted out at the other end of the fiber link.

It seemed to make very good sense, but then another problem cropped up. The wavelengths used in the multiple channels must stay near the zero-dispersion range of 1550nm, so you end up with individual channels only 2nm apart, typically at 1546, 1548, 1550, and1552nm, for example. It's difficult for any two things to be only 2nm apart and not interact, so interact they do. In fact, they create new wavelengths that can interfere with the wavelengths that are part of the transmission.

The problem gets exponentially worse as the number of wavelengths being transmitted increases. The formula for predicting the number of new waves created is:

$$FWM = (n^2 (n - 1))/2$$

where *FWM* is the number of waves created and *n* is the number of wavelengths being transmitted through the fiber. So if 2 wavelengths are being used, an extra 2 wavelengths will appear. That's not too bad in itself. But if you are transmitting 4 original wavelengths, 24 extra wavelengths are created:

$$FWM = (16 \times 3)/2 = 24$$

And 8 original wavelengths will produce 224 extra wavelengths:

$$FWM = (64 \times 7)/2 = 224$$

The solution to four-wave mixing actually involves creating just enough dispersion in the fiber to render the newly created wavelengths harmless to the signals while leaving the original signal clear enough to use. The fiber created for this purpose is *non-zero-dispersion-shifted fiber (NZ-DSF)*. Non-zero-dispersion shifting moves the zero-dispersion point slightly away from the wavelength used for the transmission, usually about 10nm above or below the transmission wavelength, so there is sufficient dispersion to keep the effects of four-wave mixing to minimum.

REDUCED SPECTRAL WIDTH

Because material dispersion is caused by an overabundance of wavelengths in the optical signal, the simplest solution is to reduce the number of wavelengths by reducing the spectral width. Recall that dispersion is expressed as picoseconds per nanometer of spectral width per kilometer of fiber, so any reduction in spectral width will have a significant effect on the amount of material dispersion.

Polarization-Mode Dispersion

Polarization-mode dispersion (PMD) is masked by other forms of dispersion until the bit rate exceeds 2.5Gbps. In order to understand PMD, we must look at an information pulse more closely.

Recall that light is an electromagnetic wave, consisting of an electrical and a magnetic wave traveling at right angles to one another. The orientation of the two waves along the path of propagation determines the light's *polarization mode*, or *polarity*.

As shown in Figure 22.8, it is possible to have different polarities of light traveling through the fiber in a signal, occupying different parts of the fiber as they pass through it. Because no fiber is perfect, there will be obstacles in one part of the fiber that are not present in another. As a result, the light having one polarity may pass an area without interference, while light having another polarity may pass through a defective region, slowing it down.

FIGURE 22.8
Polarized light shown in a cross section of optical fiber

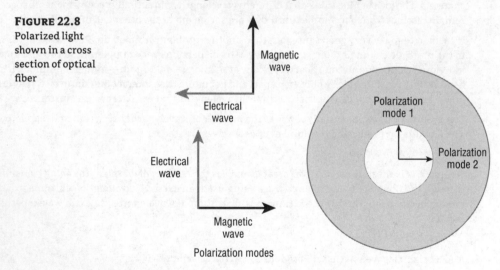

Polarization-mode dispersion is not so much a function of the fiber's overall characteristics as it is a result of irregularities, damage, or environmental conditions such as temperature. Small areas of damage called *microbends* can cause PMD, as can fiber that is not perfectly round or concentric.

Because PMD is caused by specific conditions within the fiber, and not the fiber's overall characteristics, it is difficult to assign a consistent PMD value to a length of fiber. The exact amount of PMD changes with external conditions, the physical condition of the fiber, and the polarization state of the light passing through it at any given moment. For this reason, PMD is measured in terms of the total difference in the travel time between the two polarization states, referred to as the *differential group delay (DGD)* and measured in picoseconds (ps).

How Dispersion Affects Bandwidth

While different types of dispersion have different causes and are measured different ways (ns/km, ps/nm/km), all of them have one effect in common: they place a limit on the bandwidth of optical fibers. The bandwidth of a specific brand of optical fiber can be found in the manufacturer's data sheet. The bandwidth of a type of optical fiber can also be found in an industry standard. Remember, an industry standard defines the minimum performance levels; optical fibers are available that exceed industry standards.

The product of bandwidth and length (MHz · km) expresses the information carrying capacity of a multimode optical fiber. Bandwidth is measured in megahertz (MHz) and the length is measured in kilometers. The MHz/km figure expresses how much bandwidth the fiber can carry per kilometer of its length. The fiber's designation must always be greater than or equal to the product of the bandwidth and the length of the optical fiber.

The minimum bandwidth-length product for a 50/125μm multimode optical fiber as defined in ANSI/TIA-568-C.3 is 500MHz · km. To find the usable bandwidth for this optical fiber at a length of 1.7km you must change the equation around and solve as shown here:

Bandwidth-length product = 500MHz · km

Optical fiber length = 1.7km

Bandwidth in MHz = 500 ÷ 1.7

Bandwidth = 294.1MHz

As you can see, the bandwidth decreases as the optical fiber length increases. However, the opposite is true when the length of the optical fiber decreases as shown here:

Bandwidth-length product = 500MHz · km

Optical fiber length = 0.25km

Bandwidth in MHz = 500 ÷ 0.25

Bandwidth = 2,000MHz

Bandwidth-length product is defined in ANSI/TIA-568-C.3 for multimode optical fiber, but not for single-mode optical fiber. The bandwidth for a length of a single-mode optical fiber is specified by dispersion for a specific wavelength. The spectral width of the light source will affect the bandwidth. For a given length of optical fiber, a laser with a wide spectral width will have less bandwidth than a laser with a narrower bandwidth. Spectral width can vary tremendously among lasers. Laser sources are covered in detail in Chapter 28, "Fiber-Optic Detectors and Receivers."

Attenuation

The *attenuation* of a fiber-optic signal is the loss of optical power as the signal travels through the fiber. There are two types of attenuation: intrinsic and extrinsic. *Intrinsic attenuation* occurs because no manufacturing process can produce a perfectly pure fiber. Either by accident or by design, the fiber will always have some characteristics that attenuate the signal passing through it. *Extrinsic attenuation* is caused by external mechanisms that bend the optical fiber. These bends are broken into two categories: *microbends* and *macrobends*, which are discussed in detail later in this chapter.

The wavelength of the light passing through the fiber also affects attenuation. In general, attenuation decreases as wavelength increases, but there are certain wavelengths that are more easily absorbed in plastic and glass fibers than others. One of the reasons for establishing standard operating wavelengths of 850nm, 1300nm, and 1550nm in glass fiber and in the visible range of 650nm for plastic is because the wavelengths in between are considered high-loss regions. Specifically, these wavelengths for glass optical fiber are in the ranges of 730, 950, 1250, and 1380nm.

NOTE   Attenuation provides a good example of the superiority of fiber over copper for carrying signals. When an electrical signal is carried through copper wire, attenuation increases with the data rate of the signal, requiring an increase in transmission power or, more often, the use of repeaters. Attenuation per unit length in an optical signal for an optical fiber of a given type is constant no matter what the data rate.

Remember from Chapter 19, "Principles of Fiber-Optic Transmission," that attenuation is measured in *decibels* (dB). Decibels help us account for the loss of power when we are measuring attenuation in an optical fiber. In glass optical fiber, attenuation is lowest at long wavelengths and highest at the short wavelengths. This is not true for plastic optical fiber. Plastic optical fiber is typically designed for visible light sources with a center wavelength of 650nm. Attenuation increases dramatically above or below this wavelength.

Attenuation values for a length of optical fiber can be calculated using the attenuation coefficient for a specific type of optical fiber. This information can be found in the manufacturer's data sheet or in a standard. The "Fiber Specifications" section, later in this chapter, provides attenuation information for multimode and single-mode optical fibers.

You can calculate the maximum attenuation for a length of optical fiber using this formula:

Attenuation coefficient = 3.5dB/km

Optical fiber length = 0.75km

Attenuation for the length of optical fiber = $0.75 \times 3.5 = 2.625$

Maximum attenuation for 0.75km of optical fiber is 2.625dB

Remember that the attenuation is wavelength dependent. The shorter wavelengths will attenuate more and the longer wavelengths less. Also remember that the attenuation coefficient is a "not to exceed" number. In other words, the attenuation should be no greater than that. The measured attenuation will typically be less.

While decibels are useful in measuring *total attenuation*, we can also divide attenuation into two types: *absorption* and *scattering*.

Absorption

The absorption of light is a form of intrinsic attenuation. Absorption accounts for a very small percentage of attenuation in an optical fiber—typically between 3 and 5 percent.

All materials, even the clearest glass, absorb some light. The amount of absorption depends on the type of material and the wavelength of the light passing through it. You can see absorption easily in sunglasses. Even on the brightest days, only a fraction of the light energy passes through the tinted lenses. The wavelengths that do not pass through are mostly absorbed by impurities that have been placed in, or coated on, the lens material.

In an optical fiber, absorption occurs when impurities such as water or ions of materials such as copper or chromium absorb certain wavelengths, as shown in Figure 22.9. By keeping these impurities as low as possible, manufacturers can produce fibers with a minimum of attenuation.

FIGURE 22.9
Absorption in
optical fiber

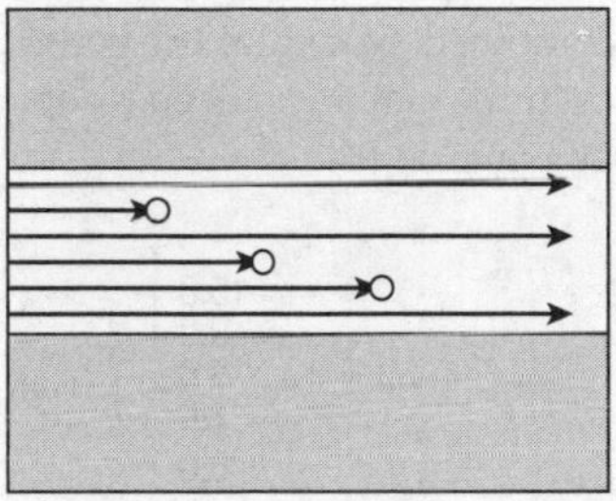

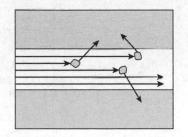

Scattering

The scattering of light is a form of intrinsic attenuation. Scattering accounts for the greatest percentage of attenuation in an optical fiber—typically between 95 to 97 percent.

Scattering is caused by atomic structures and particles in the fiber redirecting light that hits them, as shown in Figure 22.10. The process is called *Rayleigh scattering*, for Lord Rayleigh, a British physicist who first described the phenomenon in the late nineteenth century.

FIGURE 22.10
Scattering in
optical fiber

Rayleigh scattering is also the answer to the age-old question "Why is the sky blue?" The blue that we see is actually the more prevalent blue wavelengths of light from the sun being scattered by particles in the atmosphere. As the sun moves toward the horizon and the light must pass through more of the atmosphere, the scattering increases to the point where the blue light is almost completely attenuated, leaving the red wavelengths, which are less affected by the scattering for reasons that we'll see shortly.

Rayleigh scattering depends on the relationship between wavelength and the size of the structures in the fiber. Scattering decreases as the wavelength of the light approaches the size of the structures, which means that in fiber-optic communications, shorter wavelengths are more likely to be scattered than the longer wavelengths. This is one of the main reasons that infrared wavelengths are used in fiber optics. The relatively long wavelengths of infrared are less subject to scattering than visible wavelengths. It also explains why the sun turns red on the horizon. The shorter blue wavelengths are more likely to be scattered by the similarly sized particles in the atmosphere than are the red wavelengths.

Total Attenuation

Total attenuation is the combination of the intrinsic effects of absorption and scattering in a fiber. Figure 22.11 shows a typical attenuation curve for an optical fiber with the effects of absorption and scattering combined. Note that the general curve is caused by the effects of scattering, while the irregularities in the plot are caused by specific impurities, such as hydroxyl molecules, absorbing light in those wavelengths.

FIGURE 22.11
An optical fiber's attenuation curve

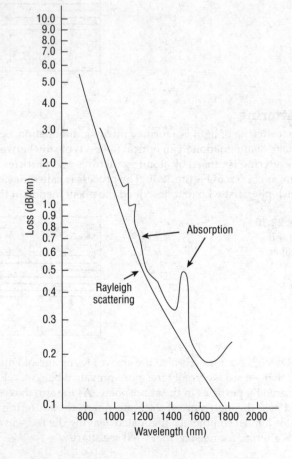

Note also the windows at the 850, 1300, and 1550nm ranges. Remember that while the 1300nm range is better in terms of dispersion, it still has a higher attenuation than the 1550nm range, which is the reason for dispersion-shifted fiber.

Bending Losses

In addition to intrinsic losses within the optical fiber, the actual condition of the fiber can lead to losses as well. These losses are referred to as extrinsic losses and are caused by bends in the optical fiber. These bends, depending on their diameter, are referred to as microbends or macrobends.

Microbends

Microbends are small distortions of the boundary layer between the core and cladding caused by crushing or pressure. Microbends are very small and may not be visible when looking at the fiber-optic cable.

Microbends change the angle of incidence within the fiber, as shown in Figure 22.12. Changing the angle of incidence forces high-order light rays to reflect at angles that prevent further reflection, causing them to be lost in the cladding and absorbed.

FIGURE 22.12
Losses caused by
microbending and
macrobending

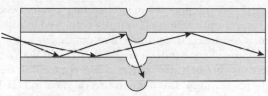

Microbending loss

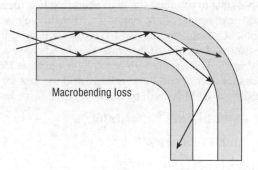

Macrobending loss

Microbends can be caused by the coating, cabling, packaging, and installation of the cable. Remember from Chapter 21, "Optical Fiber Construction and Theory," that one of the most common types of coatings is *acrylate*. Acrylate is limited in temperature performance. Exceeding the operating temperature range of the acrylate coating can cause microbending which can result in significant attenuation.

Macrobends

Compared to a microbend, a macrobend has a much larger radius. Macrobends occur when the fiber is bent around a radius that can be measured in millimeters. As shown in Figure 22.12, these tight radii change the angle of incidence within the fiber, causing some of the light rays to reflect outside of the fiber and, as with microbending, be lost in the cladding and absorbed.

At long wavelengths such as 1550nm to 1625nm, single-mode optical fiber can experience severe attenuation from macrobends. This is not the case with multimode optical fiber. A multimode fiber can be wrapped around a pencil several times and experience very little loss while a single-mode optical fiber would experience considerable loss.

As you will learn in the "Fiber Specifications" section later in this chapter, single-mode optical fiber can be broken into two groups: bend sensitive and bend insensitive. There is a considerable difference in the bending attenuation of the two fiber types. ITU-T G.652, ITU-T G.655, and ITU-T G.657 define macrobending loss for each fiber type.

The macrobending loss is defined for a number of turns around a mandrel at a specified radius and wavelength. ITU-T G.652 allows for a maximum attenuation of 0.1dB at 1550nm with 100 turns around a 30mm radius mandrel. The 30mm radius of the mandrel is slightly less then the radius of a tennis ball. Reducing the radius of the bend will increase the attenuation. Optical fiber with a macrobending specification like this is not a good choice for installations that may require several sharp bends.

ITU-T G.657 defines the macrobending loss for bend-insensitive single-mode optical fiber. This type of optical fiber is typically employed in *passive optical networks* (*PONs*) that run to your home or residence. Installing a fiber-optic cable in the wall of a home or apartment may require one or more sharp bends. A single sharp bend with an optical fiber that is not bend insensitive could cause enough attenuation to prevent the link from functioning.

Macrobending loss is defined several ways in ITU-T G.657. Tables 22.5 and 22.6 in the section "Fiber Specifications" list the parameters. In Table 22.6, the maximum macrobending loss for one turn around a 10mm radius mandrel at a wavelength of 1550nm is 0.1dB. With this information, it is possible to estimate the macrobending loss for a typical installation. If the number of bends and the diameter of each bend are known, the macrobending losses can be calculated.

Let's look at a single-mode ITU-T G.657 class B bend-insensitive fiber-optic cable installed in a home with three small radius bends. Two bends have a radius of 10mm and one bend has a radius of 7.5mm. The wavelength of the light source is 1550nm. Referring to Table 22.6, we see that the loss for each 10mm bend is 0.1dB and the loss for the 7.5mm bend is 0.5dB. The total macrobending loss (MBL) will be the sum of the three losses as shown here:

MBL 1 + MBL 2 + MBL 3 = Total MBL

0.1dB + 0.1dB + 0.5dB = 0.7dB

Total macrobending loss = 0.7dB

Numerical Aperture

As we saw in Chapter 21, many of a fiber's characteristics are determined by the relationship between the core and the cladding. The *numerical aperture (NA)* expresses the light-gathering ability of the fiber. The NA is a dimensionless number, meaning that it is to be used as a variable in determining other characteristics of the fiber, or as a means of comparing two fibers.

As discussed earlier, the numerical aperture is determined by the refractive indexes of the core and the cladding:

$$NA = \sqrt{n^2_{core} - n^2_{cladding}}$$

where n_1 is the refractive index of the core and n_2 is the refractive index of the cladding.

Recall that in order for light to be contained within a multimode fiber it must stay above the *critical angle*, or the angle at which it reflects off the boundary between the core and the cladding, rather than penetrating the boundary and refracting through the cladding.

In order to maintain the critical angle, light must enter within a specified range defined by the *acceptance cone* or *cone of acceptance*. As shown in Figure 22.13, this region is defined by a cone extending outside the optical fiber's core. Light entering the core from outside of the cone will either miss the core or enter at an angle that will allow it to pass through the boundary with the cladding and be lost. The *acceptance angle* defines the acceptance cone. Light entering the core of the optical fiber at an angle greater than the acceptance angle will not propagate the length of the optical fiber. For light to propagate the length of the optical fiber, it must enter the core at an angle that does not exceed the acceptance angle.

FIGURE 22.13
The cone of acceptance

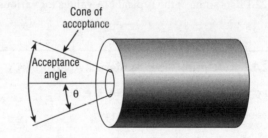

The cone of acceptance is determined using the numerical aperture:

$$NA = \sin\theta$$

where θ is the angle defining the cone of acceptance.

To determine the acceptance angle of a fiber with a core refractive index of 1.48 and a cladding refractive index of 1.47, we must determine the NA of the fiber:

$$NA = \sqrt{n^2_{core} - n^2_{cladding}} = \sqrt{(2.19 - 2.16)} = \sqrt{0.03} = 0.17$$

Next, we can use the NA to determine the acceptance angle. Remember that the acceptance angle is the value of θ.

$$NA = \sin\theta$$

So

$$\theta = \arcsin NA = \arcsin 0.17 = 9.97$$

The acceptance angle is 9.97°.

The acceptance angle is also used to determine how light emerges from a fiber. The light that comes out of a fiber end is the light that has not been absorbed or lost in the cladding, so with the exception of a small percentage of light that has propagated in the boundary between the core and the cladding, what emerges is coming out at an angle equal to or greater than the critical angle, as shown in Figure 22.14.

FIGURE 22.14
Light emerges from the fiber in the cone of acceptance.

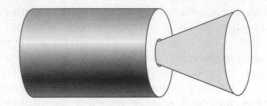

The numerical aperture varies depending on the optical fiber type. Single-mode optical fibers have the smallest NA and narrowest acceptance cone while large core plastic optical fibers have the largest NA and widest acceptance cone. NA is not defined in ANSI/TIA-568-C.3, ITU-T G.652, ITU-T G.655, or ITU-T G.657. However, it is defined in IEC 60793-2 for multimode optical fibers. Table 22.1 lists some of the typical NA values for various single-mode and multimode optical fibers.

TABLE 22.1   Optical Fiber Numerical Aperture (NA)

FIBER TYPE	NA
Single-mode	0.14
Multimode 50/125µm	0.20
Multimode 62.5/125µm	0.275
Multimode 200/230µm	0.37
Multimode 980/1000µm	0.47

Equilibrium Mode Distribution

Equilibrium mode distribution (EMD) is a condition in which the light traveling through the fiber populates the available modes in an orderly way.

When light energy first enters a fiber, it does not automatically fill every available mode with an equal amount of energy. Some modes will carry a high amount of light energy, while others may carry nothing at all. Over distance, this condition balances out as light transfers between modes because of imperfections or bends in the fiber. Until EMD is achieved, however, light may be traveling in inefficient modes that will eventually lose energy. These modes may include paths that carry the light through the cladding, or high-order modes that produce a high number of reflections through the fiber core.

Let's back up to where the light enters the fiber and see how EMD is reached.

A light source rarely, if ever, perfectly matches the numerical aperture of the fiber. Typically, the beam will either be larger or smaller than the NA. If it is larger, as is the case with most LEDs, the fiber will be overfilled. This means that the light energy will occupy most of the modes, including those that are less efficient and are doomed to loss somewhere down the length of the fiber. The *overfilled launch* is discussed in detail in Chapter 28.

If the source has a smaller beam, as with most lasers, the fiber will be underfilled. The light energy will occupy only a few of the available modes until twists, turns, and imperfections in the fiber distribute some of the light energy into higher-order modes.

The effect is similar to traffic entering a highway. If eight lanes of cars are trying to get onto a six-lane highway, you'll see cars jockeying for position in the available lanes, and possibly even a few driving on the shoulder until they are attenuated by the highway patrol. This condition will persist until the drivers settle down into their new pattern. On the other hand, if a single lane of cars enters the same six-lane highway, they will first occupy only a single lane, but eventually spread out to a more evenly distributed pattern.

Even when EMD is achieved, the effect is short-lived due to minute changes in the fiber characteristics and the effects of connectors. It does not take much to throw the modes out of equilibrium.

The distance required to achieve EMD depends on the fiber material. Light passing through plastic fibers reaches EMD in a few meters, but light passing through glass fibers may reach equilibrium only after several kilometers.

It is important to understand EMD because it affects the measurement of light energy in a fiber. In an overfilled fiber, light that has not yet reached EMD still has much of the energy traveling in the inefficient or high-order modes, and this energy can be measured over a short distance. Over a longer distance, however, once EMD has been reached, the energy in the less efficient modes has been lost, and the energy reading may drop significantly.

Once EMD has been reached, however, energy loss drops off, because the light is traveling in more efficient modes and is less likely to be lost. For this reason, loss before EMD is proportional to the length of the fiber, and loss after EMD has been reached is proportional to the square root of the fiber length. In other words, if the energy loss before EMD is 0.2dB/km, the loss over 2km would be 0.4dB. After reaching EMD, however, the loss would be $0.2 \times \sqrt{2}$.

Fiber Specifications

In order for optical fibers to work properly within a system or network, they must meet certain minimum standards established by standards organizations. In Chapter 21 we learned about the Telecommunications Industry Association (TIA) and the International Telecommunications Union (ITU). These organizations publish many different standards on optical fiber, fiber-optic cable, and testing.

Chapter 21 introduced ANSI/TIA-568-C.3, ITU-T G.652, ITU-T G.655, and ITU-T G.657. ANSI/TIA-568-C.3 is applicable to *premises* optical fiber cabling components while the ITU standards are applicable to single-mode optical fiber and cable. These standards define the minimum performance requirements for multimode and single-mode optical fibers. They help ensure compatibility with other components within the system, even if the manufacturers vary.

Section 4 of ANSI/TIA-568-C.3 contains the performance specifications for optical fiber cables recognized in premises cabling standards. Throughout this book, different chapters focus on different sections of ANSI/TIA-568-C.3. This chapter focuses on optical fiber cable transmission performance parameters as defined in Section 4.2 of ANSI/TIA-568-C.3.

Table 22.2 describes the attenuation and bandwidth characteristics for multimode and single-mode optical fibers recognized in ANSI/TIA-568-C.3. This table introduces a laser-optimized multimode optical fiber that was not previously described in ANSI/TIA-568-B.3. It also defines new bandwidth requirements for multimode optical fiber. Bandwidth is now defined with an overfilled launch (OFL) and a restricted mode launch (RML).

TABLE 22.2   Characteristics of ANSI/TIA-568-C.3-Recognized Optical Fibers

OPTICAL FIBER CABLE TYPE	WAVELENGTH (NM)	MAXIMUM ATTENUATION (DB/KM)	MINIMUM OVERFILLED MODAL BANDWIDTH-LENGTH PRODUCT (MHZ · KM)	MINIMUM EFFECTIVE MODAL BANDWIDTH-LENGTH PRODUCT (MHZ · KM)
62.5/125μm Multimode	850	3.5	200	Not Required
	1300	1.5	500	Not Required
50/125μm Multimode	850	3.5	500	Not Required
	1300	1.5	500	Not Required
850nm Laser-Optimized 50/125μm Multimode	850	3.5	1500	2000
	1300	1.5	500	Not Required
Single-mode Indoor-Outdoor	1310	0.5	N/A	N/A
	1550	0.5	N/A	N/A
Single-mode inside plant cable	1310	1.0	N/A	N/A
	1550	1.0	N/A	N/A
Single-mode outside plant cable	1310	0.5	N/A	N/A
	1550	0.5	N/A	N/A

Up until 1996 when TIA formed the TIA FO-2.2.1 Task Group on Modal Dependence of Bandwidth to study the interaction between multimode optical fiber and a laser source, multimode optical fiber bandwidth was defined with an overfilled launch. However, as laser technology improved and IEEE began developing a Gigabit Ethernet Standard (IEEE 802.3z) for high-speed local area networks (LANs) using lasers, a better indicator of bandwidth performance in a laser-based system was required.

Since 1996 the FO-2.2.1 Task Group has developed two Fiber Optic Test Procedures (FOTPs). FOTP 203 defines a standard procedure for measuring the launch power distribution of a laser-based multimode fiber-optic transmitter. FOTP 204 describes the methods used to measure the information-carrying capacity of a multimode optical fiber using an overfilled launch (OFL) or restricted mode launch (RML). As we learned earlier bandwidth-length product (MHz · km) is used to define the information-carrying capacity of a multimode optical fiber.

As you can see in Table 22.2, bandwidth-length product is not defined for single-mode optical fiber. As we learned earlier, the bandwidth of a single-mode optical fiber depends on the dispersion for a given length at a given wavelength. It is also impacted by the spectral width of the light source. A light source with a narrow spectral width will provide a greater bandwidth over a given distance than a light source with a wide spectral width.

ITU-T G.652, ITU-T G.655, and ITU-T G.657 are three standards typically used to describe the performance characteristics of single-mode optical fiber:

◆ ITU-T G.652 defines the characteristics of a single-mode optical fiber and cable.

◆ ITU-T G.655 defines the characteristics of a non-zero-dispersion-shifted single-mode optical fiber and cable.

◆ ITU-T G.657 defines the characteristics of a bending loss-insensitive single-mode optical fiber and cable for the access network.

ITU-T G.652 contains several tables that list the performance characteristics of single-mode optical fiber and cable. The values in each table are identical for many of the fiber and cable attributes. Table 22.3 lists commonly used optical and geometric characteristics for single-mode optical fiber and cable.

TABLE 22.3   ITU-T G.652, Single-Mode Optical Fiber and Cable

FIBER ATTRIBUTES, G.652.A, G.652.B, G.652.C, AND G.652.D		
Mode field diameter	Wavelength	1310nm
	Range of nominal values	8.6–9.5µm
	Tolerance	±0.6µm
Cladding diameter	Nominal	125.0µm
	Tolerance	±1µm
Core concentricity error	Maximum	0.6µm
Cladding noncircularity	Maximum	1%
Cable cut-off wavelength	Maximum	1260nm
Macrobend loss	Radius	30mm
	Number of turns	100
	Maximum at 1550nm or 1625nm	0.1dB

TABLE 22.3 ITU-T G.652, Single-Mode Optical Fiber and Cable *(CONTINUED)*

Cable attributes, G.652.A		
Attenuation coefficient	Maximum at 1310nm	0.5dB/km
	Maximum at 1550nm	0.4dB/km
Cable attributes, G.652.B		
Attenuation coefficient	Maximum at 1310nm	0.4dB/km
	Maximum at 1550nm	0.35dB/km
	Maximum at 1625nm	0.4dB/km
Cable attributes, G.652.C		
Attenuation coefficient	Maximum from 1310nm to 1625nm	0.4dB/km
	Maximum at 1550nm	0.3dB/km
Cable attributes, G.652.D		
Attenuation coefficient	Maximum from 1310nm to 1625nm	0.4dB/km
	Maximum at 1550nm	0.3dB/km

ITU-T G.655 contains several tables that list the characteristics of a non-zero-dispersion-shifted single-mode optical fiber and cable. Unlike ITU-T G.652, several of the fiber attributes are not similar among the cable types. Table 22.4 lists commonly used optical and geometric characteristics for non-zero-dispersion-shifted single-mode optical fiber and cable.

TABLE 22.4 ITU-T G.655.C, G.655.D, and G.655.E, Non-Zero-Dispersion-Shifted Single-Mode Optical Fiber and Cable

FIBER ATTRIBUTES			
		G.655.C	G.655.D and G.655.E
Mode field diameter	Wavelength	1550nm	1550nm
	Range of nominal values	8–11μm	8–11μm
	Tolerance	±0.7μm	±0.6μm
Cladding diameter	Nominal	125.0μm	125.0μm
	Tolerance	±1μm	±1μm

TABLE 22.4 ITU-T G.655.C, G.655.D, and G.655.E, Non-Zero-Dispersion-Shifted Single-Mode Optical Fiber and Cable   *(CONTINUED)*

FIBER ATTRIBUTES			
		G.655.C	G.655.D and G.655.E
Core concentricity error	Maximum	0.8µm	0.6µm
Cladding noncircularity	Maximum	2%	1%
Cable cut-off wavelength	Maximum	1450nm	1450nm
Macrobend loss	Radius	30mm	30mm
	Number of turns	100	100
	Maximum at 1625nm	0.5dB	0.1dB
Cable attributes, G.655.C, G.655.D, and G.655.E			
Attenuation coefficient	Maximum at 1550nm	0.35dB/km	
	Maximum at 1625nm	0.4dB/km	

ITU-T G.657 contains two tables that list the characteristics of a bending loss-insensitive single-mode optical fiber and cable for the access network. Like ITU-T G.655, several of the fiber attributes are not similar among the cable types, primarily in the area of bend radius. Table 22.5 and Table 22.6 list commonly used optical and geometric characteristics for bending loss-insensitive single-mode optical fiber and cable for the access network.

TABLE 22.5 ITU-T G.657, Specifications for G.657 Class A Bending Loss-Insensitive Single-Mode Optical Fiber and Cable

FIBER ATTRIBUTES, G.657 CLASS A		
Mode field diameter	Wavelength	1310nm
	Range of nominal values	8.6–9.5µm
	Tolerance	±0.4µm
Cladding diameter	Nominal	125.0µm
	Tolerance	±0.7µm
Core concentricity error	Maximum	0.5µm

TABLE 22.5 ITU-T G.657, Specifications for G.657 Class A Bending Loss-Insensitive Single-Mode Optical Fiber and Cable *(CONTINUED)*

FIBER ATTRIBUTES, G.657 CLASS A			
Cladding noncircularity	Maximum	1%	
Cable cut-off wavelength	Maximum	1260nm	
Macrobend loss	Radius	15mm	10mm
	Number of turns	10	1
	Maximum at 1550nm	0.25dB	0.75dB
	Maximum at 1625nm	1.0dB	1.5dB
Cable Attributes			
Attenuation coefficient	Maximum from 1310 to 1625nm	0.4dB/km	
	Maximum at 1550nm	0.3dB/km	

TABLE 22.6 ITU-T G.657, Specifications for G.657 Class B Bending Loss-Insensitive Single-Mode Optical Fiber and Cable

FIBER ATTRIBUTES, G.657 CLASS B			
Mode field diameter	Wavelength	1310nm	
	Range of nominal values	6.3–9.5μm	
	Tolerance	±0.4μm	
Cladding diameter	Nominal	125.0μm	
	Tolerance	±0.7μm	
Core concentricity error	Maximum	0.5μm	
Cladding noncircularity	Maximum	1%	
Cable cut-off wavelength	Maximum	1260nm	
Macrobend loss	Radius (mm)	15	10	7.5
	Number of turns	10	1	1
	Maximum at 1550nm	0.03dB	0.1dB	0.5dB
	Maximum at 1625nm	0.1dB	0.2dB	1.0dB

TABLE 22.6 ITU-T G.657, Specifications for G.657 Class B Bending Loss-Insensitive Single-Mode Optical Fiber and Cable *(CONTINUED)*

Cable Attributes		
Attenuation coefficient	Maximum at 1310nm	0.5dB/km
	Maximum at 1550nm	0.3dB/km
	Maximum at 1625nm	0.4dB/km

Table 22.7 describes the typical attenuation and bandwidth characteristics for plastic fibers and HCS (hard-clad silica) and PCS (plastic-clad silica) fibers. The TIA or ITU standards covered in this text do not address the performance of these optical fibers. The data contained in this table was obtained from manufacturers' data sheets and can vary from manufacturer to manufacturer.

TABLE 22.7 Characteristics of HCS/PCS and Plastic Fibers

OPTICAL FIBER CABLE TYPE	WAVELENGTH (NM)	MAXIMUM ATTENUATION	MAXIMUM INFORMATION TRANSMISSION CAPACITY FOR OVERFILLED LAUNCH (MHZ · KM)
200/230μm HCS/PCS	650	10dB/km	17
	850	8dB/km	20
1mm plastic fiber	650	0.19dB/m	1.5

The Bottom Line

Calculate the attenuation in dB for a length of optical fiber. The attenuation values for a length of optical fiber can be calculated using the attenuation coefficient for a specific type of optical fiber. This information can be found in the manufacturer's data sheet or in a standard.

Master It Calculate the maximum attenuation for a 15km ITU-T G.657 optical fiber cable at 1550nm.

Calculate the usable bandwidth for a length of optical fiber. The product of bandwidth and length (MHz · km) expresses the information carrying capacity of a multimode optical fiber. Bandwidth is measured in megahertz (MHz) and the length is measured in kilometers. The MHz · km figure expresses how much bandwidth the fiber can carry per kilometer of its length. The fiber's designation must always be greater than or equal to the product of the bandwidth and the length of the optical fiber.

Master It Refer to Table 22.2 and calculate the usable bandwidth for 700 meters of 50/125 μm multimode optical fiber at a wavelength of 850nm.

Calculate the macrobending loss in a single-mode optical fiber. The macrobending loss is defined for a number of turns around a mandrel at a specified radius and wavelength.

Master It A single-mode ITU-T G.657 class B bend-insensitive fiber-optic cable is installed in a home with three small radius bends. Two bends have a radius of 7.5mm and one bend has a radius of 10mm. The wavelength of the light source is 1625nm.

Calculate the acceptance angle for an optical fiber. The acceptance angle defines the acceptance cone. Light entering the core of the optical fiber at an angle greater than the acceptance angle will not propagate the length of the optical fiber. For light to propagate the length of the optical fiber, it must be at an angle that does not exceed the acceptance angle.

Master It Determine the acceptance angle of an optical fiber with a core refractive index of 1.48 and a cladding refractive index of 1.45.

Chapter 23

Safety

Whether you work as a technician or an installer, your work with fiber optics can expose you to several workplace hazards that are defined and regulated by the Occupational Safety and Health Administration (OSHA). OSHA has published numerous regulations on workplace hazards ranging from laser light sources to ladders, and employers are required to be familiar with these regulations and follow them to keep the workplace safe.

You are responsible for your own safety as well as for the safety of your co-workers. It is up to you to know and incorporate safe work practices in everything you do.

This chapter describes the types of hazards that you will encounter as you work with fiber optics. Some of the hazards are unique to fiber optics work, but others are more common. This chapter discusses the dangers that these hazards create, and informs you of different methods of working safely around them.

This chapter provides a general overview on safety; it is not the catch-all for safety. Each company and individual is responsible for maintaining a safe environment. The availability of safety information is also the responsibility of the company and individual.

In this chapter, you will learn to:

◆ Classify a light source based on optical output power and wavelength

◆ Identify the symptoms of exposure to solvents

◆ Calculate the proper distance the base of a ladder should be from a wall

Basic Safety

Whenever you work in a hazardous environment, such as a construction site, a lab, or a production facility, you must always be aware of the potential dangers you face. Your workplace is required by law to provide you with equipment and facilities that meet standards set by OSHA, but you will have to be an active participant in your own safety.

You can use three lines of defense to help you get through the day safely: engineering controls, personal protective equipment (PPE), and good work habits.

Engineering Controls

Engineering controls are the mechanisms that your facility has established to make a hazardous situation safer. They may include ventilation in the form of exhaust fans or hoods, special cabinets for storing flammables, or workstations that minimize the hazards of specialized work, such as cutting optical fibers.

Do not ignore or try to get around the engineering controls set up in your workplace. By doing so, you only endanger yourself and others. Make sure that fans and ventilation systems are working properly. If they are not, report any problems to your facility supervisor immediately.

Do not try to alter or modify the engineering controls unless the modifications have been approved by your safety officer. Improper modifications could reduce the effectiveness of the controls and create a greater hazard.

Personal Protective Equipment (PPE)

Personal protective equipment consists of anything that you would wear to protect yourself from materials or situations. PPE can include protective gloves and eyewear for cutting and grinding operations, respirators for working with chemicals that put out harmful vapors, and specialized goggles for working with lasers.

Your PPE protects you not only from short-term accidents, such as cuts or flying shards of glass, but also from damage that can build up over time. Such damage may include dust from construction operations such as drywall sanding that can build up over time in your lungs and cause diseases such as silicosis, or exposure to chemicals such as solvents that can have harmful long-term or chronic effects as well as harmful short-term effects.

Whenever you use PPE, inspect it carefully to ensure that it is in good condition. Look for cuts, tears, or other signs of damage in protective outerwear such as gloves or aprons. Inspect eyewear for cracks or pitting. If you use goggles designed to protect you from certain light wavelengths, make sure they are clean and free from scratches that could reduce their effectiveness.

If you wear contact lenses, be sure your facility allows them in your work area. If you work with adhesives or solvents, you should avoid wearing them anyway, because splashed chemicals could be trapped in the lens and be more difficult to wash out. You may be able to obtain safety goggles with prescription lenses if you have to use them on a regular basis.

If you work with a respirator, test it every time you put it on. Cover the canisters with your hands and try to inhale, then cover the exhaust port and try to exhale. The respirator should form a good seal with your face, and no air should leak through the canisters or exhaust.

Some construction areas may require hardhats. Do not take these warnings lightly. Even a small hand tool dropped from a few feet can injure or kill you if you are not protected. Hardhats are designed to absorb the shock from falling objects so your head doesn't have to. To make sure the hardhat fits properly, adjust the inner band so that it fits snugly against your forehead and does not allow the hat to move around on your head. Make sure there is enough room between the suspension and the hardhat shell to absorb any blows.

Good Work Habits

Good work habits are in some ways the simplest and most effective means to working safely. Good work habits can help you prevent accidents and spot potential problems in time to correct them.

Here are some general rules for working safely:

◆ Keep a clean workspace. Clean up at the end of your work day and store tools properly. A "rat's nest" can hide problems and add to confusion.

◆ Observe your surroundings. Look up from what you are doing once in a while to make sure everything around you is the way it should be.

◆ Use tools for the job they were designed to perform. Misuse of tools is one of the most common causes of accidents in the workplace.

◆ Do not eat or drink in the work area. In addition to accidentally drinking from the wrong bottle, you could accidentally ingest glass fiber or other dangerous materials that might get mixed in with your food.

◆ Report problems or injuries immediately. Let your facility supervisors know about hazards so they can correct them as soon as possible.

◆ Know how to reach emergency personnel. Have emergency numbers posted by the nearest telephone so you don't waste time fumbling through a directory in an emergency.

◆ Put your emergency contact information in your cell phone.

Let's look at some of the hazards directly related to your work with fiber optics.

Light Sources

Even though most lasers and LEDs used in fiber optics operate in the near-infrared and infrared (IR) wavelengths and are invisible to the eye, they can still cause damage if they are delivered at high intensity, or if the exposure is long enough. The possibility of damage is even greater because you cannot see the beam, and in many cases, the damage is done before you know it.

A laser can be especially dangerous because it can concentrate a great amount of power into a small beam of coherent light. Many lasers used in fiber optics operate below dangerous levels, but some, such as those used for transmission over long distances, put out enough power to cause damage in a very short time.

Injuries from the infrared wavelengths output by lasers used in fiber-optic systems and test equipment include but are not limited to cataracts and corneal and retinal burns.

Federal Regulations and International Standards

Federal regulations and international standards have been created to prevent injuries from laser radiation.

In Chapter 21, "Optical Fiber Construction and Theory," TIA and ITU were introduced as organizations that publish standards on the performance of optical fibers used in the telecommunications industry. There is a difference between a standard and a federal regulation. A federal regulation is a law. If there are federal regulations for a specific type of product sold in the United States, such as a laser, that product must meet federal regulations. However, it does not need to meet any standards. A standard only provides guidance; it is not a law.

FEDERAL REGULATIONS

The U.S. Food and Drug Administration (FDA) Center for Devices and Radiological Health (CDRH) is responsible for eliminating unnecessary human exposure to man-made radiation from medical, occupational, and consumer products. The FDA has had performance standards for light-emitting products since 1976. These performance standards are described in the Code of Federal Regulations 21 CFR, Subchapter J.

Fiber-optic equipment sold in the United States that contains a laser must meet 21 CFR, Subchapter J. However, it is not required by 21 CFR, Subchapter J to meet any standard. Many fiber-optic products that contain lasers only meet 21 CFR, Subchapter J.

21 CFR, Subchapter J encompasses all consumer products that contain a laser. In this chapter only the areas of 21 CFR, Subchapter J typically required for fiber-optic communication and test equipment will be addressed; detailed information can be obtained from the FDA website, `http://www.accessdata.fda.gov/scripts/cdrh/cfdocs/cfcfr/CFRSearch.cfm?FR=1040.10`.

The laser classifications described in 21 CFR, Subchapter J are based on emission duration and emission limits for specific wavelengths. There are two wavelength windows that apply to fiber-optic communication and test equipment:

>400nm and ≤ 1400nm

>1400nm and ≤ 2500nm

CFR, Subchapter J divides laser products into classes and defines the labeling requirements based on the optical output power:

◆ Class I products do not require a label.

◆ Class II products require a "Caution" label bearing the warning logotype.

◆ Class IIIa, Class IIIb, and Class IV laser products must have a "Danger" label affixed similar to the labels, as shown in Figure 23.1.

FIGURE 23.1
Warning placards
for lasers

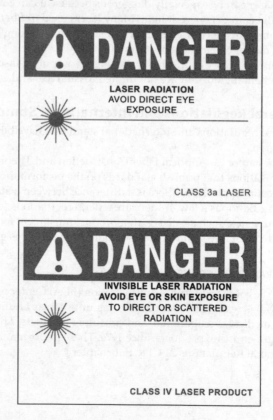

The classification tables for accessible emission limits for laser radiation found in 21 CFR, Subchapter J are very detailed and cover all consumer laser products. The following is a list of the hazards associated with each class. Many of the products used in fiber-optic communication systems and test equipment are Class I; however, some equipment may be Class IV.

Class I   These levels of laser radiation are not considered to be hazardous.

Class IIa   These levels of laser radiation are not considered to be hazardous if viewed for any period of time less than or equal to 1,000 seconds but are considered to be a chronic viewing hazard for any period of time greater than 1,000 seconds.

Class II   These levels of laser radiation are considered to be a chronic viewing hazard. A caution label is required that states "DO NOT STARE INTO THE BEAM."

Class IIIa   These levels of laser radiation are considered to be, depending on the irradiance, either an acute intrabeam viewing hazard or chronic viewing hazard, and an acute viewing hazard if viewed directly with optical instruments. A caution label is required with an irradiance less than or equal to 2.5mW. A danger label is required with an irradiance greater than 2.5mW. The caution label should state "DO NOT STARE INTO THE BEAM OR VIEW DIRECTLY WITH OPTICAL INSTRUMENTS." The danger label should state "AVOID DIRECT EYE EXPOSURE."

Class IIIb   These levels of laser radiation are considered to be an acute hazard to the skin and eyes from direct radiation. A danger label is required that states "AVOID DIRECT EXPOSURE TO BEAM."

Class IV   These levels of laser radiation are considered to be an acute hazard to the skin and eyes from direct and scattered radiation. A danger label is required that states "AVOID EYE OR SKIN EXPOSURE TO DIRECT OR SCATTERED RADIATION."

STANDARDS

Two organizations have published recognized laser safety standards: the International Electrotechnical Commission (IEC) and the American National Standards Institute (ANSI). ANSI Z136.2 addresses the safe use of optical fiber communication systems utilizing laser diode and LED sources. IEC 60825-2 addresses the safety of optical fiber communication systems. The objective of both of these standards is to protect people from optical radiation released by an optical communication system.

These organizations have also published other laser standards. ANSI Z136.1-200 addresses the safe use of lasers in general and is not specific to optical fiber communication systems. IEC 60825-1 addresses the safety of laser products in general, not specific to optical fiber communication systems.

Test equipment or an optical fiber communication system (OFCS) containing a laser sold in the United States should have an FDA classification and the manufacturer may advertise compliance to an ANSI or IEC standard. Any OFCS or test equipment sold outside the United States may or may not have an FDA classification and may or may not comply with an ANSI or IEC standard.

OSHA Standards for OFCS Service Groups

OSHA also classifies an OFCS. Section III, Chapter 6 of the OSHA Technical Manual describes the Optical Fiber Service Group (SG) Designations. It is based on ANSI Z136.2. These SG designations relate to the potential for ocular hazards to occur only when the OFCS is being serviced. "Being serviced" can involve something as simple as removing a connector from a receptacle. The SGs outlined in the OSHA Technical Manual are described next:

SG1   An OFCS in this SG has a total output power that is less than the accessible emission limit (AEL) for Class I (400nW) and there is no risk of exceeding the maximum permissible irradiance (MPI) when viewing the end of a fiber with a microscope, an eye loupe, or the unaided eye.

SG2   An OFCS is in this SG only if wavelengths between 400nm and 700nm are emitted. Such lasers are potentially hazardous if viewed for more than 0.25 second.

SG3A   An OFCS in this SG is not hazardous when viewed with the unaided eye and is hazardous only when viewed with a microscope or eye loupe.

SG3B   An OFCS in this SG does not meet any of the previous criteria.

OSHA Laser Hazard Classes

If the total power for an OFCS is at or above 0.5W, it does not meet the criteria for an optical fiber SG designation and should be treated as a standard laser system. The OSHA laser hazard classes are based on ANSI Z136.1 and summarized next:

Class 1   These lasers cannot emit laser radiation at known hazard levels, typically 400nW for a continuous wave at visible wavelengths. Users of these lasers are generally exempt from radiation hazard controls during operation and maintenance, but not necessarily during service.

Class II   These lasers only emit visible wavelengths above Class I but below 1mW. These lasers can cause damage if you look directly at them for more than 0.25 second, but have limited requirements for protection. It is assumed that the normal aversion to pain will cause anyone looking at the bright light to turn away or close their eyes before any damage can take place.

Class IIIA   These intermediate power lasers have a continuous wave output from 1.0 to 5.0mW. Directly viewing the beam can damage the eye.

Class IIIB   These moderate power lasers have a continuous wave output in the 5–500mW range (up to +27dBm) and can cause damage to the eye, even if the beam is reflected.

Class IV   These high-power lasers have a continuous wave output above 500mW, or +27dBm, and can burn almost any living tissue they contact under any viewing condition. They also pose a fire hazard.

IEC Laser Hazard Classifications

IEC standard 60825-1 addresses the safety of laser products and is not specific to an OFCS. However, components in the OFCS may be classified by this standard. Classifications that appear in this standard are summarized here:

Class 1   These levels of laser radiation are safe under normal operating conditions including the use of a microscope or eye loupe.

Class 1M   These levels of laser radiation are safe under normal operating conditions; however, they may be hazardous when viewed with a microscope or eye loupe.

Class 2   These levels of laser radiation at wavelengths between 400nm and 700nm are not hazardous if viewed for less than 0.25 second. It is assumed that the normal aversion to pain will cause anyone looking at the bright light to turn away or close his or her eyes (blink).

Class 2M   These levels of laser radiation at wavelengths between 400nm and 700nm are not hazardous if viewed for less than 0.25 second. It is assumed that the normal aversion to pain will cause anyone looking at the bright light to turn away or close his or her eyes (blink). However, they may be hazardous when viewed with a microscope or eye loupe.

Class 3R   These levels of laser radiation are potentially hazardous when viewed directly.

Class 3B   These levels of laser radiation are normally hazardous when viewed directly.

Class 4   These levels of laser radiation are considered to be an acute hazard to the skin and eyes from direct beam viewing or diffuse reflections.

IEC Laser Hazard Levels for OFCS

IEC 60825-2 addresses the safety of an OFCS. This standard describes hazard levels associated with an OFCS. It also defines labeling requirements and the location type for each hazard level. There are three location types:

Unrestricted   This applies to domestic premises, industrial or business premises, and any public areas.

Restricted   This applies to secured areas within industrial and business/commercial premises not open to the public; general areas within switching centers; delimited areas not open to the public on trains, ships, or other vehicles; overhead fiber-optic cables and cable drops to a building; and optical test sets.

Controlled   This applies to cable ducts, street cabinets, manholes, and dedicated and delimited areas of network operator distribution centers.

The hazard level labeling requirements for each location type that appear in this standard are summarized here:

Hazard level 1   There are no labeling requirements for unrestricted, restricted, or controlled location types.

Hazard level 1M   Labels are required for unrestricted, restricted, or controlled location types.

Hazard level 2 and 2M   Labels are required for unrestricted, restricted, or controlled location types.

Hazard level 3R   This hazard level is not permitted in an unrestricted location type. Labels are required for restricted or controlled location types.

Hazard level 3B   These hazard levels are not permitted in an unrestricted or restricted location type. A label is required for a controlled location type.

Hazard level 4   These hazard levels are not permitted in unrestricted, restricted, or controlled location types.

Laser Safety

Because most lasers used in fiber-optic systems emit IR radiation, you cannot see the beam, no matter how powerful it is. As a result, you will not be able to tell if the system is powered, especially if you are working on a piece of fiber far from the transmitter.

You should treat an optical fiber coupled to a laser with the same caution that you would treat electrical cables connected to a breaker panel. Do not assume that the system is turned off, especially if you have to use a microscope to look at the fiber end. Do not take anyone else's word that the system is off, or that the fiber is uncoupled from the laser. You will have to endure the results for the rest of your life.

Unless you can be sure that the fiber is not coupled to a laser, do not look at the fiber end without some kind of protection. Use filters and protective eyewear that block out the specific wavelengths used by the lasers. Use a video microscope when available instead of a handheld optical microscope to view the end face of a connector.

Hazardous laser areas should be clearly identified with warning placards and signs stating that access is limited to personnel with proper safety gear and authorized access, as shown in Figure 23.1. Do not ignore these signs or think that they don't apply to you. They are there for your protection and for the safety of those working inside the restricted areas.

If the lab has a separate door and a hazardous laser is operating inside, the door should have interlocks to kill the laser before the door is opened. Some of these doors may have separate combination locks to prevent unauthorized entry.

Handling Fiber

In spite of optical fiber's flexibility, remember that it is glass. In short pieces, it is stiff enough to pierce your skin or eyes and cause discomfort, pain, or damage. If the pieces become airborne, you may even accidentally inhale or swallow them, risking damage to your throat or respiratory system. You can protect yourself with correct procedures and the right PPE.

When you cut, cleave, scribe, or accidentally break optical fibers, the ends can get lost easily, either by becoming airborne or by rolling along a surface. These ends can have extremely sharp edges, and if they are mishandled, they can lodge in your skin or eyes. If they are not removed immediately, the pieces can work themselves in further, increasing the risk of damage or infection. As shown in Figure 23.2, always work over a nonreflective black surface, which makes it easier to keep track of cut fiber ends. Also, keep a separate labeled container with a lid nearby for cut fiber ends.

To prevent injury to your hands, always handle cut pieces of fiber with tweezers. To prevent eye injury, always wear proper eye protection. It takes only one piece of glass to damage your vision permanently. If you have been handling fiber, do not rub your eyes or put your hands near them until you have washed your hands. If you do get a piece of fiber in your skin, remove it immediately with a pair of tweezers or seek medical attention.

You may not always have the convenience of a laboratory or workshop environment for your fiber work. Work areas for splicing, building connectors, or other tasks may include basements, crawlspaces, underground vaults, an attic, or the back of a van. Don't take shortcuts just because you don't have the luxury of a full workshop at your disposal. Make sure you have an appropriate work surface and the proper tools and safety equipment before you start working.

FIGURE 23.2
A nonreflective black surface, a fiber waste container, and safety glasses can help prevent injury from fiber ends.

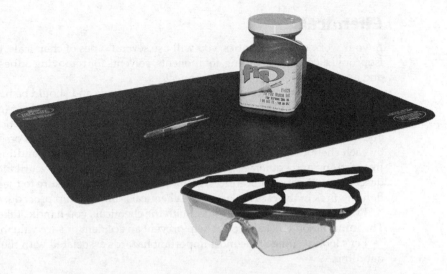

REMOVING FIBER FROM YOUR DIET

Whenever you are cleaving fiber, you are going to end up with ends. No matter how short or long they are, rest assured that they will be hard to spot when it's time to clean up. Your first line of defense against these hard-to-spot fibers is immediate response. In other words, as soon as you cleave, pick up and dispose of the end.

If this discipline doesn't work for you, or if you find yourself with stray ends anyway, you'll still have to round them up and dispose of them properly. There are some low-tech and high-tech strategies to help you, as long as you can prepare them beforehand.

First, to prevent fiber ends from traveling beyond your work area, create a barrier around your work surface. This will keep bouncers and fliers confined to the work surface, where you can find them more easily.

If you want to make sure you've collected all of your strays, or if you know that you created more ends than you have collected, use a bright light source, such as an LED flashlight. To find the fibers, place the light at the level of the work surface and rotate along the surface. The fiber ends will pick up the light and Fresnel reflection will cause them to shine brightly.

As a last, low-tech measure, double up a piece of adhesive tape and pat it all over the work surface. This will pick up any minute fiber particles that you could not spot or retrieve using other methods.

Chemicals

In your work with fiber optics, you will use several types of chemicals, including 99 percent isopropyl alcohol for cleaning components, solvents for removing adhesives and other materials, and anaerobic epoxy for making connectors.

Each of these chemicals poses a number of hazards and should be handled carefully. Each chemical that you use is accompanied by a Material Safety Data Sheet (MSDS), which provides important information on the chemical's properties, characteristics such as appearance and odor, and common uses. The MSDS also gives you information on specific hazards posed by each chemical and ways to protect yourself through specific handling procedures, protective clothing and equipment, and engineering controls such as ventilation. Finally, the MSDS describes emergency procedures, including first aid for exposure to the chemical, methods for fighting fires in the case of flammable chemicals, and cleanup procedures for spills.

Even if you think you're familiar with the chemicals you handle, take time to read the MSDS. The information you gain could help prevent an accident or save valuable time in an emergency.

Let's look at some of the most important hazards associated with the chemicals you'll be handling.

Isopropyl Alcohol

Even though alcohol is commonly used in the home and the lab, its hazards should not be ignored or taken lightly. Alcohol vapors escape into the air easily, and can cause damage to your liver and kidneys if they are inhaled. The vapors are also highly flammable and can ignite if exposed to a spark or flame in high enough concentrations.

Alcohol can also cause irritation to your eyes, skin, and mucous membranes (nose and mouth) if it comes in direct contact with them.

Always use alcohol in a proper dispenser, shown in Figure 23.3. Store and transfer it carefully to avoid spills and excess evaporation. If you do spill any alcohol, clean it up with a dry cloth and dispose of the cloth in a container designed for flammable waste materials. As with all flammables, do not use alcohol in areas where sparks, open flames, or other heat sources will be present.

FIGURE 23.3
An alcohol dispenser helps reduce the risk of spills and fire.

It's important to remember that alcohol flames are almost invisible, and spills could lead to a mad scramble as you try to dodge flames you cannot see. Alcohol fires can be extinguished with water or a Class A fire extinguisher.

Solvents

Many solvents have similar properties that require that you handle them with great care. Like alcohol, solvents are very volatile and sometimes flammable. Their primary danger, however, is their hazard to your health.

One of the health hazards posed by solvents comes from the fact that they can cause excessive drying in your skin and mucous membranes, and the resultant cracking of the surface layers can leave you open to infection. The hazard is more serious if you inhale the solvent vapors, as they can damage your lungs and respiratory system.

Solvents can also cause organ damage if inhaled or ingested. The molecules that make up most solvents can take the place of oxygen in the bloodstream and find their way to the brain and other organs. As the organs are starved of oxygen, they can become permanently damaged. One of the first signs that this kind of damage is taking place is dizziness or a reaction similar to intoxication. If you feel these symptoms, get to fresh air immediately.

Solvents may come in glass or plastic containers. Make sure that the container you are using is properly marked for the solvent it contains. Do not leave solvent in an unmarked container.

If you are carrying solvent in a glass bottle, use a rubber cradle to carry the bottle. The cradle protects the bottle from breaking if it falls.

Never leave a solvent container open. Keep the top off just long enough to transfer the amount necessary for the job, and replace it firmly to prevent the vapors from escaping.

Anaerobic Epoxy

The two-part epoxy used for making connectors is typically used in small quantities and does not present any immediate health hazards. If you are working in an enclosed space, however, such as the back of a van or an access area, vapors from the adhesive portion can irritate your eyes, nose, and throat. If you feel any of these symptoms, get to an open space immediately. The adhesive can also irritate your skin or eyes on contact. If you get any of the adhesive on you, wash the area immediately. If any material splashes in your eyes, flush them for 15 minutes at an eyewash station or sink.

Use caution when working with the primer portion of these adhesives. With a flash point of $-18°$ C, it is highly flammable. Do not use anaerobic epoxy where there is an open spark or flame, or where heating components or elements are being used.

As with solvents, do not leave epoxy containers open any longer than necessary to dispense the amount you are using.

Site Safety

Many of the locations for fiber-optic components may be in areas that require special safety precautions. These may include construction sites, enclosed areas, locations near high-voltage power lines, or areas requiring access by ladder or scaffold.

Always follow the on-site safety requirements and observe all warning signs. Here are some general safety rules to help you.

Electrical

When fiber-optic systems run through the same area as electrical wiring or cabinets, use extreme caution with tools and ladders. One wrong move can send enough voltage through your body to kill you. Remember that electrical fields can exist beyond a cable's insulation if high voltages are present, so use wooden ladders to reduce the possibility of exposure to induced voltages. Use care with cutters and other tools to avoid accidental contact with electrical wires, and report any hazardous conditions that may exist.

Remember that high voltage causes most of its damage by making muscles seize up, including the heart and lungs. The greatest chance of damage comes when voltage passes through your heart to get the ground, such as when you touch a wire with your hand and your opposite leg provides the path to ground, or when the current passes from one hand to the other.

If you accidentally grab a live wire, the voltage may keep your hand clenched, making it nearly impossible to release the wire. If you see someone who is in this situation, do not try to pull them away with your hands. You may be caught up in the circuit as well. Instead, use a nonconducting stick, such as a wooden or fiberglass broom handle, to knock the victim away from the voltage source.

If the victim is not breathing, artificial respiration may be necessary to get the heart and lungs operating again.

Ladders

You may often find that you need a ladder to reach a work area. When choosing a ladder, make sure you select one that matches your requirements. Self-supporting ladders, such as stepladders, should be used only if the work area is not near a vertical support such as a wall, and the floor beneath the work area is even and firm.

Non-self-supporting ladders, such as extension ladders, are useful when there is a firm vertical support near the work area and there is a stable, nonslip surface on which to rest the ladder. When setting up a non-self-supporting ladder, it is important to place it at an angle of 75½° to support your weight and be stable. A good rule for finding the proper angle is to divide the working height of the ladder (the length of the ladder from the feet to the top support) by four, and place the feet of the ladder that distance from the wall. For example, if the ladder is 12 feet tall, the bottom should be 3 feet from the wall.

Be sure that the ladder you select can carry your weight along with the weight of any tools and equipment you are carrying. Read the labels and warnings on the ladder you select to make sure that it is the right one for the work you are performing.

If you are working near live electrical systems, be sure to use a nonconducting ladder of wood or fiberglass. If you are working near high heat, select an aluminum ladder to avoid scorching or melting your only means of support.

In the work area, place the ladder so that you can work comfortably without having to reach too high or too far to the side. Overreaching can cause you to lose your balance and fall, or place too much weight above or to the side of the ladder, causing the entire ladder to come down. If your work area extends beyond a comfortable reach, climb down the ladder and move it. Do not try to "walk" the ladder to a new work area.

Inspect ladders before using them. Make sure the rungs and rails are in good shape, and are not split, broken, or bent. Make sure all fittings and fasteners are secure, and that all locking mechanisms are working properly.

When carrying a ladder to the worksite, reach through the rail and balance it on your shoulder. Be aware of obstacles and corners as you carry it, and make sure others are aware that you have an awkward load, especially if you are walking through hallways or other limited visibility areas.

Trenches

If you are working on a fiber system in a trench, be sure the trench is properly dug and shored before entering it. Never work in a trench without someone else around, in case of a collapse.

If you have never worked in a trench before, learn the proper way to enter and exit a trench. Always use a ladder. Never jump into a trench or try to climb down the sides. You could trigger a collapse.

If you witness a collapse and others are trapped but not in immediate danger, do not try to dig them out yourself. You risk making the problem worse. Get help immediately. Special training is required to recover victims from a trench collapse, so leave it to the experts.

Emergencies

It takes only one slip-up to create an emergency. It could come from a moment of carelessness, an attempt at taking a shortcut, or ignorance of the proper procedures. Whatever the cause, the first response is always the same. Remain calm. Panic can cause even more damage and complicate matters beyond repair.

The best way to handle emergencies is to accept the fact that they will occur, and be ready for them. Make sure you know what can go wrong with the materials and chemicals you handle, and what you can do to minimize the damage.

Injury

Injuries can be caused by misuse of tools, fibers penetrating your skin or eyes, burns, falls, or any number of other mishaps. Make sure that you and your co-workers know the location of first-aid kits in your work area. Also, make sure you know how to reach emergency personnel. If you are on a new job site, make it a priority to familiarize yourself with emergency procedures and contact information.

Chemical Exposure

Accidental chemical exposure can result in anything from temporary discomfort to permanent injury or death. The first few seconds of an emergency involving chemical exposure can be critical in the victim's recovery.

If the chemical is splashed on the skin or in the eyes, flush the affected area with clean water immediately, using a shower or eyewash station if available. Continue flushing the area for at least 10 minutes. This washes the chemical away, but also dilutes its effects if it has been absorbed by the skin or eyes.

In case of inhalation, move the victim to fresh air immediately and call for medical attention. If a chemical has been accidentally swallowed, induce vomiting unless the chemical is corrosive and could damage the esophagus and throat as it comes back up. Use a neutralizing liquid such as milk to dilute corrosive chemicals if they have been swallowed, and seek medical attention immediately.

Fire

If a fire breaks out in your work area, it may be small enough for you to handle alone. If it is small, you can smother it with a damp cloth. If it is larger, but contained in a trash can or other enclosure, use the appropriate fire extinguisher for the material that is burning.

To use a fire extinguisher properly, remember the acronym PASS as you use the following procedure:

Pull Pull the pin from the fire extinguisher trigger.

Aim Aim the extinguisher at the base of the fire.

Squeeze Squeeze the handle firmly to activate the extinguisher.

Sweep Sweep the extinguisher discharge at the base of the fire until the flames are out.

Do not give the fire a chance to trap you. If you think that the extinguisher will not put out the fire completely and there is a risk that your exit will be cut off, leave immediately and call for help. You can do more good with a phone call than you can in a failed attempt at being a hero.

The Bottom Line

Classify a light source based on optical output power and wavelength. The OSHA Technical Manual describes the Optical Fiber Service Group (SG) Designations based on ANSI Z136.2. These SG designations relate to the potential for ocular hazards to occur only when the OFCS is being serviced. OSHA also classifies standard laser systems.

Master It An OFCS does not fall into any of the SG designations defined in the OSHA Technical Manual because during servicing it is possible to be exposed to laser emissions greater than 550mW. What classification would be assigned to this OFCS based on the OSHA Technical Manual?

Identify the symptoms of exposure to solvents. Solvents can cause organ damage if inhaled or ingested. The molecules that make up most solvents can take the place of oxygen in the bloodstream and find their way to the brain and other organs. As the organs are starved of oxygen, they can become permanently damaged.

Master It Your co-worker spilled a bottle of liquid several minutes ago and now appears impaired. What has your co-worker potentially been exposed to?

Calculate the proper distance the base of a ladder should be from a wall. Remember non-self-supporting ladders, such as extension ladders, require a firm vertical support near the work area and a stable, nonslip surface on which to rest the ladder. When setting up a non-self-supporting ladder, it is important to place it at an angle of 75½° to support your weight and be stable.

Master It A non-self-supporting ladder is 16′ in length. How far should the feet of the ladder be from the wall?

Chapter 24

Fiber-Optic Cables

So far we have studied the principles and characteristics of individual optical fibers. While they are certainly adequate for the job of carrying signals from one place to another, they are not rugged enough to withstand the rigors of handling, transportation, and installation. In addition, some installations require multiple optical fibers for sending and receiving or for routing to a number of locations.

For an optical fiber to be suitable for everyday use, it must be incorporated into cables that provide standardized fiber groupings, protection from the environment, and suitable size for handling.

In this chapter, we will describe standard and harsh environment fiber-optic cables used in many types of installations. We will detail different types of fiber-optic cables and the uses for which they were designed. We will also describe some of the basic requirements for handling and installation.

In this chapter, you will learn to:

◆ Determine the cable type from the NEC markings

◆ Identify the fiber number from the color code

◆ Identify the optical fiber type from the color code

◆ Determine the cable length using sequential markings

Basic Cable

You may already be familiar with cables used for electrical wiring. These cables typically consist of two or more insulated wires bundled together and surrounded by a protective outer covering called the *jacket* or *sheath*. In addition to holding the wires in place, the jacket or sheath also protects the wires from the environment and damage that may occur during handling or installation.

Some of the largest cables, used for telephone transmissions, can be several inches in diameter and contain hundreds of wires, as shown in Figure 24.1. These cables are very heavy, difficult to bend, and expensive.

FIGURE 24.1
Cable with hundreds of wires

Fiber-optic cables like the one shown in Figure 24.2 do not need to be as large as the cable shown in Figure 24.1 because the bandwidth of an optical fiber over a long distance is many times greater than the bandwidth of a wire. Greater bandwidth means fewer optical fibers are required to carry the same information the larger cable is carrying. This results in a small, low-cost fiber-optic cable that is much easier to handle and terminate than the cable shown in Figure 24.1. Chapter 32, "Fiber-Optic System Design Considerations," compares the performance of fiber-optic and copper cables.

FIGURE 24.2
Typical fiber-optic cable

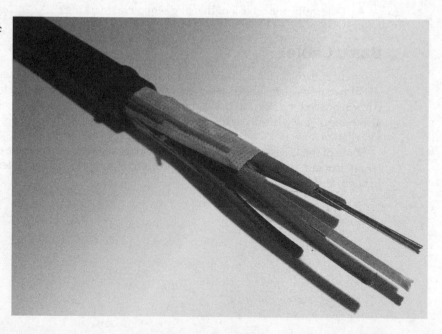

Optical fibers are used in many different configurations and environments; manufacturers have created a wide variety of cable types to meet the needs of almost any application. The type of signal being carried and the number of optical fibers required are just two of the many considerations when selecting the right cable for an application. Other factors include:

◆ Tensile strength

◆ Temperature range

◆ Bend radius

◆ Flammability

◆ Buffer type

◆ Jacket type

◆ Weight

◆ Armor

◆ Crush resistance

The exact combination of these factors varies, depending on the application and operational environment. A cable installed inside an office building, for example, will be subject to less extreme temperatures than one installed outdoors or in an aircraft. Cabling installed in a manufacturing facility may be exposed to abrasive dusts, corrosive chemicals, or hotwork, such as welding, requiring special protection. Some cables may be buried underground, where they are exposed to burrowing or chewing animals, while others may be suspended between poles, subject to their own weight plus the weight of animals and birds who think the cables were put there for them.

Cable Components

Whether a cable contains a single optical fiber, several optical fibers, or hundreds of optical fibers, it has a basic structure in common with other cables. As shown in Figure 24.3, a typical fiber-optic cable consists of the optical fiber (made up of the core, cladding, and coating), a buffer, a strength member, and an outer protective jacket or sheath. Let's look at these components individually.

FIGURE 24.3
Fiber-optic cable
components

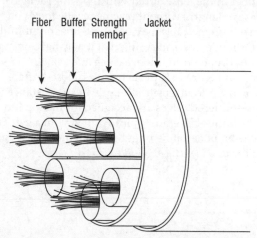

Fiber   Buffer   Strength   Jacket
                  member

Buffer

As you learned in Chapter 21, "Optical Fiber Construction and Theory," all optical fiber has a coating. The coating is the optical fiber's first protective layer. The protective layer placed around the coating is called the buffer. There are two buffer types: *loose buffer* and *tight buffer*

LOOSE BUFFER

Loose buffer is also referred to as loose tube buffer or loose buffer tube. A loose buffer consists of a buffer layer or tube that has an inner diameter much larger than the diameter of the coated optical fiber, as shown in Figure 24.4.

FIGURE 24.4
Loose-buffered cable has a buffer diameter greater than the fiber diameter.

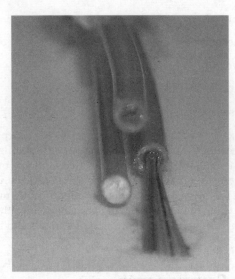

The loose buffer provides room for additional optical fiber when the cable is manufactured. This extra room also allows the optical fiber to move independently of the buffer and the rest of the cable components. This is an important factor when the cable is subjected to temperature extremes. A variety of materials are used to make up a fiber-optic cable. These materials may not expand or contract the same way as the optical fiber during temperature changes. The additional fiber in the loose buffer will allow the cable to expand (increase in length) or contract (decrease in length) without placing tensile forces on the optical fiber.

Loose-buffered cable may be single-fiber or multifiber (not to be confused with single-mode and multimode fiber), meaning that it may have one or many optical fibers running through each buffer tube. In addition, a cable may contain a number of loose buffers grouped together, as shown in Figure 24.5. In such cables, loose-buffered tubes are grouped around a central core that provides added tensile strength and a resistance to bending.

Loose-buffered cables are designed for indoor and outdoor applications. For outdoor applications, the buffer tubes are typically filled with a gel. The gel displaces or blocks water and prevents it from penetrating or getting into the cable. A gel-filled loose-buffered cable is typically referred to as a *loose tube, gel-filled (LTGF) cable*.

Loose-buffered cables are ideal for direct burial and aerial installations. They are also very popular for indoor/outdoor applications.

FIGURE 24.5
Loose-buffered
tubes in a cable

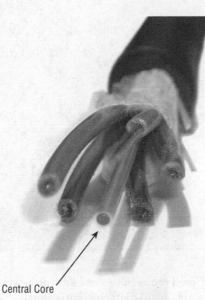

Central Core

TIGHT BUFFER

Tight-buffered cable is typically used in more controlled environments where the cable is not subjected to extreme temperatures or water. In short, tight-buffered cable is generally used for indoor applications rather than outdoor applications.

As shown in Figure 24.6, tight-buffered cable begins with a 250μm coated optical fiber. The buffer is typically 900μm in diameter and is applied directly to the outer coating layer of the optical fiber. In this way, it resembles a conventional insulated copper wire. The buffer may have additional strength members running around it for greater resistance to stretching.

FIGURE 24.6
Tight-buffered
cable uses a buf-
fer attached to the
fiber coating.

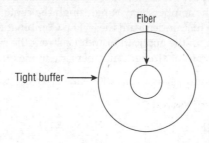

Fiber

Tight buffer

One of the benefits of tight-buffered cable is that the buffered optical fiber can be run outside of the larger cable assembly for short distances, as shown in Figure 24.7. This makes it easier to attach connectors. The standard diameter of the buffer allows many different types of connectors to be applied.

FIGURE 24.7
Tight-buffered
optical fibers
extending out of
the cable assembly

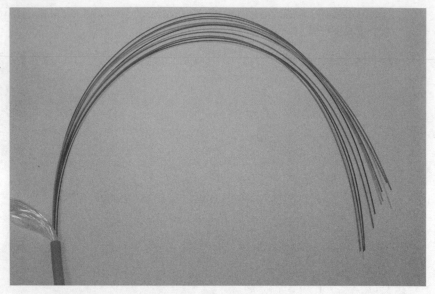

Tight-buffered cables have some advantages over loose-buffered cables. Tight-buffered cables are generally smaller in diameter than comparable loose-buffered cables. The minimum bend radius of a tight-buffered cable is typically smaller than a comparable loose-buffered cable. These two advantages make tight-buffered cable the choice for indoor installations where the cable is routed through walls and other areas requiring tight bends.

Strength Members

We've already learned a bit about how strength members help increase a cable's tensile strength. The primary importance of these members is to ensure that no tensile stress is ever placed on the optical fiber. The installation of a fiber-optic cable may require a great deal of pulling, which places considerable tensile stress on the strength members. The strength members need to be able to support the installation tensile stresses and the operational tensile stresses.

Strength members may run through the center of a fiber-optic cable or they may surround the buffers just underneath the jacket. Combinations of strength members may also be used depending on the application and the stress the cable is designed to endure. Some of the most common types of strength members are made of:

◆ Aramid yarns, usually Kevlar

◆ Fiberglass rods

◆ Steel

Aramid yarns, as shown in Figure 24.8, are useful when the entire cable must be flexible. These fine, yellowish or gold yarns are made of the same material used in high-performance sails, bulletproof vests, and fire protection gear. They have the advantages of being light, flexible, and quite strong. Aramid yarns may be used in cable subgroups if they will be bundled into larger cables.

FIGURE 24.8
Aramid (Kevlar)
yarns used as
strength members

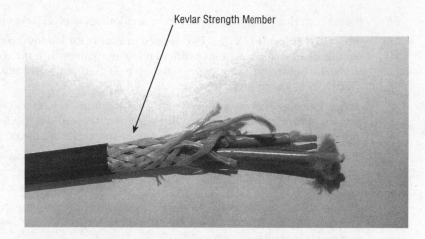

Kevlar Strength Member

Larger cables and most loose-buffered cables typically have a central member of either fiber-glass or steel added, as shown in Figure 24.9. This central member provides tensile strength and a resistance to bending. This resistance to bending prevents the cable from being over-bent and kinking the loose-buffered tubes.

FIGURE 24.9
A fiberglass cen-tral member adds tensile strength and a resistance to bending

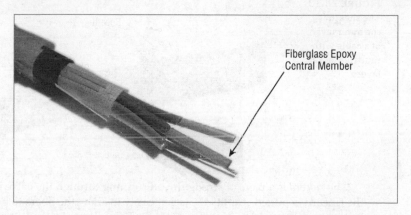

Fiberglass Epoxy
Central Member

Although the tensile strength of steel is greater than the tensile strength of fiberglass, many applications require a *dielectric*, or nonconductive cable. Fiberglass is nonconductive while steel is not. Steel running through a cable outdoors makes an excellent lightning rod.

Jacket

The jacket is the cable's outer protective layer. It protects the internal components from the out-side world. The jacket may be subject to many factors such as sunlight, ice, animals, equipment accidents, and ham-handed installers. The jacket must also provide protection from abrasion, oil, corrosives, solvents, and other chemicals that could destroy the components in the cable.

Jacket materials vary depending on the application. Typical jacket materials include:

Polyvinyl chloride (PVC)   PVC is used primarily for indoor cable runs. It is fire-retardant and flexible, and is available in different grades to meet different conditions. PVC is water resistant; however, PVC does not stand up well to solvents. In addition, it loses much of its flexibility at low temperatures.

Polyethylene   This material is typically used outdoors. It offers excellent weather and sun resistance in addition to excellent water resistance and flexibility in low temperatures. Specially formulated polyethylene offers low-smoke, zero-halogen performance.

Polyvinyl difluoride (PVDF)   This material is chosen for its low-smoke and fire-retardant properties for use in cables that run through airways or *plenums* in a building. PVDF cables are not as flexible as other types, so their fire safety properties are their primary draw.

Polytetrafluoroethylene (PTFE)   This material is primarily used in aerospace fiber-optic cables. It has an extremely low coefficient of friction and can be used in environments where temperatures reach as high as 260° C.

Some cables contain multiple jackets and strength members, as shown in Figure 24.10. In such cables, the outermost jacket may be referred to as the sheath, while the inner protective layers are still called jackets.

FIGURE 24.10
The sheath is the outer layer of a cable with multiple jacket layers.

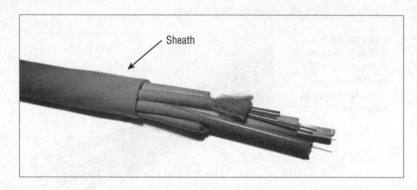

Sheath

The *ripcord* is a piece of strong thread running through the cable just under the jacket. When the ripcord is pulled, it splits the jacket easily to allow the fibers to be separated for connector-ization or splicing. The ripcord reduces the risk incurred when cutters or knives are used to split the jacket.

Cable Types

As uses for optical fiber have become more varied, manufacturers have begun producing cables to meet specific needs. Cable configurations vary based on the type of use, the location, and future expansion needs, and it is likely that more will be created as future applications emerge.

Bear in mind that different cable arrangements are variations on a theme. Different combinations of buffer type, strength members, and jackets can be used to create cables to meet the needs of a wide variety of industries and users.

Let's look at some of the commonly available optical-fiber cables.

Cordage

The simplest types of cables are actually called *cordage*, and are used in connections to equipment and patch panels. They are typically made into patch cords or jumpers. The major difference between cordage and cables is that cordage only has one optical-fiber/buffer combination in a jacket, whereas cables may have multiple optical fibers inside a jacket or sheath.

The two common types of cordage are *simplex* and *duplex*.

SIMPLEX CORDAGE

Simplex cordage, shown in Figure 24.11, consists of a single optical fiber with a tight buffer, an aramid yarn strength member, and a jacket.

FIGURE 24.11
Simplex cordage

Simplex cordage gets its name from the fact that, because it is a single fiber, it is typically used for one-way, or simplex, transmission, although bidirectional communications are possible using a single fiber.

DUPLEX CORDAGE

Duplex cordage, also known as *zipcord*, is similar in appearance to household electrical cords, as you can see in Figure 24.12. Duplex cordage is a convenient way to combine two simplex cords to achieve duplex, or two-way, transmissions without individual cords getting tangled or switched around accidentally.

FIGURE 24.12
Duplex cordage

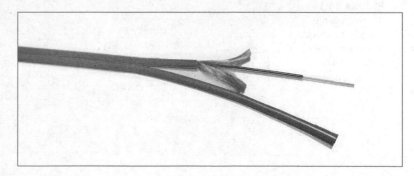

Distribution Cable

When it is necessary to run a large number of optical fibers through a building, distribution cable is often used. Distribution cable consists of multiple tight-buffered fibers bundled in a jacket with a strength member. These cables, like the one shown in Figure 24.13, may also feature a dielectric central member to increase tensile strength, resist bending, and prevent the cable from being kinked during installation.

FIGURE 24.13
Distribution cable
with dielectric
central member

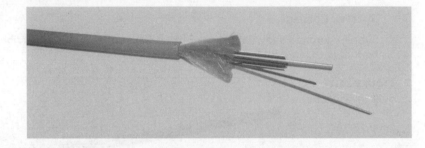

Distribution cables are ideal for inter-building routing. Depending on the jacket type they may be routed through plenum areas or riser shafts to telecommunications rooms, wiring closets, and workstations. The tight-buffered optical fibers are not meant to be handled much beyond the initial installation, because they do not have a strength member and jacket.

Distribution cables may carry up to 144 individual tight-buffered optical fibers, many of which may not be used immediately but allow for future expansion.

Breakout Cable

Breakout cables are used to carry optical fibers that will have direct termination to the equipment, rather than being connected to a patch panel.

Breakout cables consist of two or more simplex cables bundled with a strength member and/or central member covered with an outer jacket, as shown in Figure 24.14. These cables are ideal for routing in exposed trays or any application requiring an extra rugged cable that can be directly connected to the equipment.

FIGURE 24.14
Breakout cable
containing simplex
cables bundled
with a central
strength member

Armored Cable

Armored cable can be used for indoor applications and outdoor applications. An armored cable typically has two jackets. The inner jacket is surrounded by the armor and the outer jacket or sheath surrounds the armor.

An armored cable used for outdoor applications, shown in Figure 24.15, is typically a loose tube construction designed for direct burial applications. The armor is typically a corrugated steel tape surrounded by an outer polyethylene jacket. This combination of outer jacket and armor protects the optical fibers from gnawing animals and the damage that can occur during direct burial installations.

FIGURE 24.15
Armored cable for outdoor applications

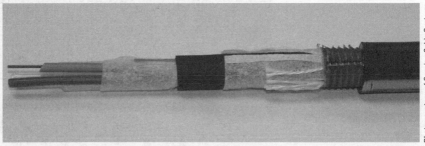

Photo courtesy of Corning Cable Systems

Armored cable used for indoor applications may feature tight-buffered or loose-buffered optical fibers, strength member(s), and an inner jacket. The inner jacket is typically surrounded by a spirally wrapped interlocking metal tape armor. This type of armor, shown in Figure 24.16, is rugged and provides crush resistance. These cables are used in heavy traffic areas and installations that require extra protection, including protection from rodents.

FIGURE 24.16
Armored cable for indoor applications

Messenger Cable

When a fiber-optic cable must be suspended between two poles or other structures, the strength members alone are typically not enough to support the weight of the cable and any additional forces that may be placed on the cable. For aerial installations a messenger wire is required. The messenger wire can be external to the fiber-optic cable or integrated into the cable.

When the messenger wire is integrated into the fiber-optic cable, the cable is typically referred to as a *messenger cable*. These cables typically feature a 0.25″ stranded steel messenger wire. The messenger wire, sometimes referred to simply as the messenger, is integrated into the outer jack of the cable, as shown in Figure 24.17.

FIGURE 24.17
Messenger cable
used for aerial
installations

Also called *figure 8 cable* for the appearance of its cross section, messenger cable greatly speeds up an aerial installation because it eliminates the need to lash the fiber-optic cable to a pre-run messenger wire.

If a messenger wire is not incorporated into the cable assembly used, the cable will have to be lashed to messenger wire. A messenger wire is a steel or aluminum wire that supports the fiber-optic cable in an aerial installation. Either way, cables in aerial installations must be able to withstand loading from high winds, ice, birds and climbing animals, and even windblown debris such as branches.

Ribbon Cable

Ribbon cable is a convenient solution for space and weight problems. The cable contains fiber ribbons, which are actually coated optical fibers placed side by side, encapsulated in Mylar tape (see Figure 24.18), similar to a miniature version of wire ribbons used in computer wiring. A single ribbon may contain 4, 8, or 12 optical fibers. These ribbons can be stacked up to 22 high.

FIGURE 24.18
Ribbon cables
consist of parallel
fibers held together
with Mylar tape.

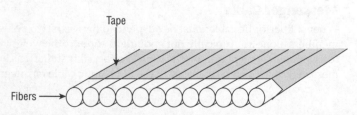

Tape

Fibers →

Because the ribbon contains only coated optical fibers, this type of cable takes up much less space than individually buffered optical fibers. As a result, ribbon cables are denser than any other cable design. They are ideal for applications where limited space is available, such as in an existing conduit that has very little room left for an additional cable.

As shown in Figure 24.19, ribbon cables come in two basic arrangements. In the *loose tube ribbon cable*, fiber ribbons are stacked on top of one another inside a loose-buffered tube. This type of arrangement can hold several hundred fibers in close quarters. The buffer, strength members, and cable jacket carry any strain while the fiber ribbons move freely inside the buffer tube.

The *jacketed ribbon cable* looks like a regular tight-buffered cable, but it is elongated to contain a fiber ribbon. This type of cable typically features a small amount of strength member and a ripcord to tear through the jacket.

While ribbon fiber provides definite size and weight savings, it does require special equipment and training to take advantage of those benefits. Connectors, strippers, cleavers, and fusion splicers must all be tailored to the ribbon fiber. For these reasons, ribbon fiber may not be the best solution in all situations.

Submarine Cable

Submarine cable is specially designed for carrying optical fiber underwater. Not all submarine cable is the same, however. Depending on the distance it will span and the type of service it will provide, submarine cable can take many different forms.

Submarine cable may be laid in trenches under the bottom of waterways where shipping or fishing activities threaten to snag or damage the cable, or they may be laid directly on the bottom of less-traveled waterways, or on the deep ocean floor where such activities do not penetrate. Figure 24.20 shows a single-optical-fiber submarine cable with armor.

FIGURE 24.19
Armored loose tube ribbon cable (bottom) and jacketed ribbon cable (top)

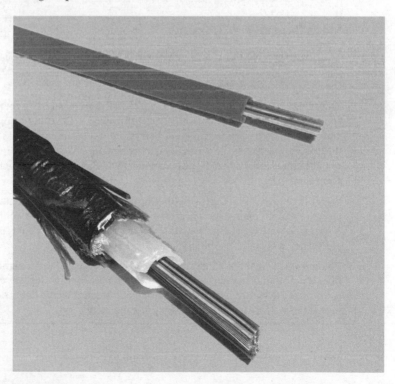

FIGURE 24.20
Submarine cable
with armor

Jacket   Armor   Fiber

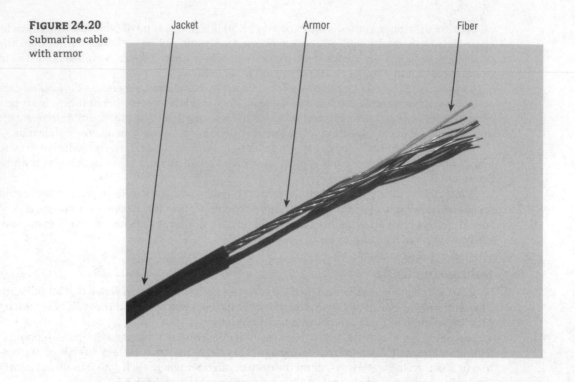

MORE UNDERWATER HAZARDS

While fishing operations, anchoring, and other hazards pose a threat to underwater cables of all types, some fiber-optic cables running in the Grand Canary Islands encountered a special danger soon after they were laid.

The cables were part of a fiber-optic test system run by AT&T in 1985. Only three weeks after the cables were laid about ¾ of a mile below the surface, the cable stopped operating and was out of commission for a week. Technicians first thought that the cable had separated because of abrasion due to scraping across the ocean floor, but when the cable was examined more closely, it was found to have small shark's teeth embedded in it. The shark attacks were repeated twice over the next several months.

Why the repeated shark bites on fiber-optic cable? Sharks are known to be drawn to prey by their electromagnetic emissions, and attacks on standard coaxial cable would have been much more likely if the cable were not shielded against electromagnetic interference. But the fiber did not carry any electricity, only light, so the sharks should have ignored it.

The answer lay in the fact that light does have an electromagnetic signature, but because the optical fiber is not subject to electromagnetic interference, no shielding was applied to the cable. The small sharks were drawn to the weak electrical fields put out by the light traveling through the fiber, and in the dark water attacked it, thinking it was a meal.

Aerospace Cable

Aerospace cables are designed to be installed in aircraft and spacecraft. These cables are designed to operate in extreme temperature environments. In addition, they must protect the optical fiber from the shock and vibration associated with aircraft and spacecraft. While the optical fiber used in many aerospace cables may be the same optical fiber used in other cable types; the coating, buffer, and jacket are typically different.

Aerospace fiber-optic cables come in a variety of types to be used in different locations on the aircraft or spacecraft. Cables designed to be used inside avionics boxes or cabinets typically have a temperature range of –40° C to +85° C. Cables designed to be used between cabinets typically have a temperature range of –65° C to +125° C, while those cables used by the engines have a temperature range of –65° C to +260° C.

Figure 24.21 shows the components of a typical aerospace fiber-optic cable. This cable features a primary and secondary buffer. The primary buffer is the coating applied directly over the cladding of the optical fiber. Aerospace cables may use a variety of coatings depending on the temperature range required. A typical acrylate coating has a maximum operating temperature of 100° C while some high-temperature acrylates may perform at temperatures up to 125° C. Silicone, carbon, and polyimide coatings have greater temperature ranges than acrylate. Detailed information on optical fiber coatings can be found in Chapter 21.

Secondary Buffer
Cable Jacket
Cladding
Barrier
Core
Primary Buffer
Bondable Layer
Strength Member
Drawing courtesy of Carlisle Interconnect Technologies

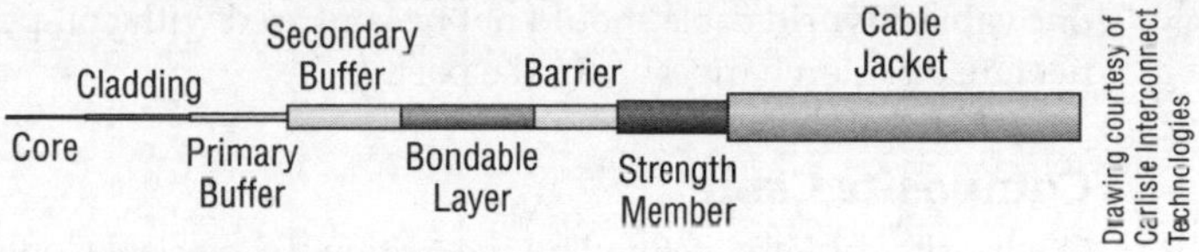

FIGURE 24.21 Typical aerospace fiber-optic cable.

The secondary buffer shown in Figure 24.22 is polytetrafluoroethylene (PTFE). The bondable layer surrounding the secondary layer is a polyimide tape while the barrier layer is unsealed PTFE. The strength member is braided Kevlar while the jacket shown in Figure 24.23 is extruded FEP (fluorinated ethylene propylene). This jacket material has a service temp range of –100° C to +200° C as well as excellent chemical and UV resistance properties, and is also flame retardant.

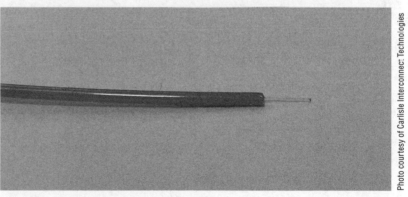

Photo courtesy of Carlisle Interconnect Technologies

FIGURE 24.22 PTFE secondary buffer

FIGURE 24.23
FEP jacket and
braided Kevlar
strength member

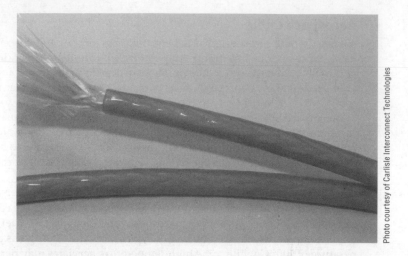

Hybrid Cable

Hybrid cable, as applied to fiber optics, combines multimode and single-mode optical fibers in one cable. Hybrid cable should not be confused with composite cable, although the terms have been used interchangeably in the past.

Composite Cable

Composite cable, as defined by the National Electrical Code (NEC), is designed to carry both optical-fiber and current-carrying electrical conductors in the same run. As shown in Figure 24.24, this composite cable consists of optical fibers along with twisted-pair wiring typical of telephone wiring. This arrangement is convenient for networks that carry fiber-optic data and conventional telephone wiring to the same user. Composite cable also provides installers with a way to communicate during fiber installation and provides electrical power to remote equipment, such as repeaters, along the fiber's route.

FIGURE 24.24
Composite cable
carries fiber and
wiring in the
same run.

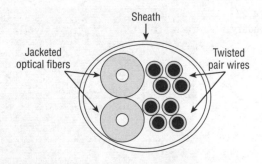

Sheath

Jacketed
optical fibers

Twisted
pair wires

Cable Duty Specifications

The various combinations of strength members, jacket materials, and fiber arrangements are determined by the specific requirements of an installation. Among the factors considered are the amount of handling a cable will take, the amount of stress the cable must endure in normal use, and the locations where it will run.

Cable duty specifications are typically divided into two basic types:

Light-duty cables   These are designed for basic protection of the fiber within and minimal handling. A good example of a light-duty fiber cable is a simplex cord, which consists of a tight-buffered fiber with a jacket. Simplex cordage is not engineered to withstand excessive pulling forces, and the jacket is engineered for flexibility and ease of handling, not harsh environments.

Heavy-duty cables   These are designed for more and rougher handling, with additional strength members and jacketing around the fiber. They are made for harder pulling during installation, and they protect the fiber within from damage in exposed or extreme environments.

Cable Termination Methods

Some fibers, such as those found in simplex and duplex cords and breakout cable, are already set up to receive connectors and can be handled easily. Others, including loose-buffered cables, must be prepared for connectors and handling with special kits.

These kits, known as *fanout kits* and *breakout kits*, are designed to adapt groups of coated fibers for connectors by separating them and adding a tight buffer (fanout kit) or a tight buffer, strength member, and jacket (breakout kit) to each one.

Fanout Kit

The fanout kit, shown in Figure 24.25, converts loose-buffered fibers into tight-buffered fibers ready for connectors. A typical fanout kit contains an enclosure sometimes called a *furcation unit*. The furcation unit attaches to the loose-buffered tube. Hollow tight-buffered tubes 900μm in diameter are applied over the optical fibers and passed into the furcation unit, which is then closed, locking the tight buffers in place on the fibers. After the fibers have the buffers applied, connectors can be attached for use in a patch panel or other protected enclosure.

FIGURE 24.25
A fanout kit adds a tight buffer to individual fibers.

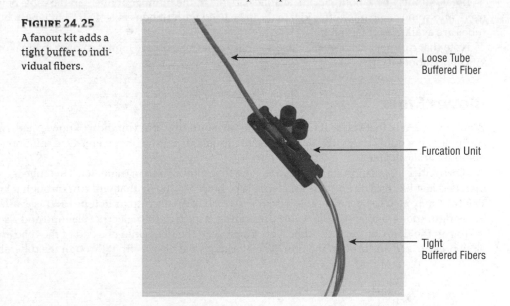

Loose Tube
Buffered Fiber

Furcation Unit

Tight
Buffered Fibers

Breakout Kit

The breakout kit, shown in Figure 24.26, is similar to the fanout kit in that it spreads the fibers from the loose-buffered tube through a furcation kit and provides 900μm tight buffers to be applied over the optical fiber. The breakout kit, however, is designed to allow the optical fiber to be connected directly to equipment with standard connectors.

FIGURE 24.26
A breakout kit adds a tight buffer, strength member, and a jacket to individual optical fibers.

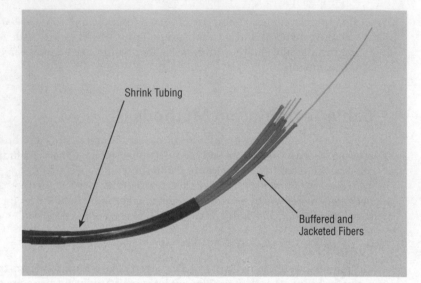

Shrink Tubing

Buffered and Jacketed Fibers

In addition to buffer material, the breakout kit provides a 3mm diameter jacket with an aramid strength member that slips over the optical fiber. Heat-shrink tubing and epoxy is used to join the individual jackets to the cable.

Breakout kits and fanout kits are available to match the number of fibers in the cable being used, and some companies offer kits that can be used with ribbon cable. Different length buffer tubes are available for fanout kits.

Note that different vendors may use a variety of terms or trade names to refer to breakout kits and fanout kits.

Blown Fiber

What if you knew that you wanted to run fiber in a building, but you didn't know exactly what equipment it would be serving, or what the requirements for the fiber were? One solution to this dilemma is *blown fiber*.

Blown fiber installation starts with a hollow tube about 5mm in diameter. This tube is installed just like a cable and acts as a loose tube buffer for fibers that will run through it later. The tube may also be part of a cable assembly, which may carry up to 19 tubes and is available in configurations similar to other cables, including armored, all-dielectric, plenum, and riser.

To run the fibers through the tube, you simply lead the fibers into one end of the tube and blow pressurized air through the tube. The air carries the fibers with it through the tube, similar to the way that pneumatic carriers at drive-through tellers shoot your deposits into the bank building.

Blown fiber can be used, as already described, when the fiber needs in a building are not yet known, or when fiber needs to be repaired or upgraded. Blown fiber is also useful when a company cannot afford to install all of the fiber that it might potentially use. As a business grows and the new fiber can be economically justified, it can be blown in as needed.

One advantage of blown-in fiber, even when it is installed all at once, is that cuts or damaged sections can be repaired quickly. In case of an accidental cut from power equipment, for example, the damaged section can be cut out, a new section spliced in, and new fiber blown in. This can be a great time-savings over splicing multiple broken fibers in a cable. Blown fiber can be run 1,000′ or more through a building, around curves, and uphill, in a very short time.

Blown fiber is covered in detail in Chapter 31, "Cable Installation and Hardware."

NEC Standards for Fiber-Optic Cables and Raceways

The National Electrical Code (NEC) is published by the National Fire Protection Association (NFPA). It provides specific guidance for running fiber-optic cable within buildings. The code requires cables to be tested for fire resistance and smoke characteristics, determines which types of cables may be run in different areas of a building, and specifies the types of raceways that can be used with each type of cable.

Be sure you stay current with the latest NEC guidelines. The NEC is updated every three years. The information is this chapter can be found in NEC Article 770, "Optical Fiber Cables and Raceways."

NEC Fiber-Optic Cable Types

The NEC recognizes three types of fiber-optic cables in Article 770:

Nonconductive Cables containing no metallic members or other electrically conductive materials.

Conductive Cables containing non-current-carrying conductive members, including strength members, armor, or sheath.

Composite Cables containing optical-fiber and current-carrying electrical conductors. These cables may also contain non-current-carrying conductive members.

The NEC classifies fiber-optic cables according to their electrical and fire safety characteristics and contains rules for the use of each cable type to minimize hazards. Any fiber-optic cabling run indoors must be listed with the NEC as suitable for the purpose. Any outdoor non-listed fiber-optic cabling that is brought into a structure must be terminated after no more than 50′ and connectorized or spliced to a cable listed as being resistant to the spread of fire.

Fire Resistance

Optical-fiber cables installed within buildings shall be listed as being resistant to the spread of fire. These cables are also required to be marked. Three types of cables are recognized in Article 770 for indoor installations: plenum cable, riser cable, and general-purpose cable.

PLENUM CABLE

Plenum cables, whether conductive or nonconductive, are suitable for use in ducts, plenums, and other space used for environmental air. These cables will have fire resistance and low smoke-producing characteristics.

RISER CABLE

Riser cables, whether conductive or nonconductive, are suitable for a vertical run in a shaft or from floor to floor. These cables will have fire-resistance characteristics capable of preventing the carrying of a fire from floor to floor.

GENERAL-PURPOSE CABLE

General-purpose cables, whether conductive or nonconductive, are resistant to the spread of fire. However, these cables are not suitable for plenum or riser applications.

Table 24.1 shows the cable types along with the markings each cable will contain. Markings should be clearly visible on the cable, as shown in Figure 24.27, and no cable should be used indoors unless it contains the NEC cable type on it.

TABLE 24.1   NEC Cable Types and Description

MARKING	TYPE	LOCATION	PERMITTED SUBSTITUTIONS
OFNP	Nonconductive optical-fiber plenum cable	Ducts, plenums, other air spaces	None
OFCP	Conductive optical-fiber plenum cable	Ducts, plenums, other air spaces	OFNP
OFNR	Nonconductive optical-fiber riser cable	Risers, vertical runs	OFNP
OFCR	Conductive optical-fiber riser cable	Risers, vertical runs	OFNP, OFCP, OFNR
OFNG	Nonconductive optical-fiber general-purpose cable	General-purpose use except for risers and plenums	OFNP, OFNR
OFCG	Conductive optical-fiber general-purpose cable	General-purpose use except for risers and plenums	OFNP, OFCP, OFNR, OFCR, OFNG, OFN
OFN	Nonconductive optical-fiber general-purpose cable	General-purpose use except for risers, plenums, and spaces used for environmental air	OFNP, OFNR
OFC	Conductive optical-fiber general-purpose cable	General-purpose use except for risers, plenums, and spaces used for environmental air	OFNP, OFCP, OFNR, OFCR, OFNG, OFN

FIGURE 24.27
NEC cable type
marking

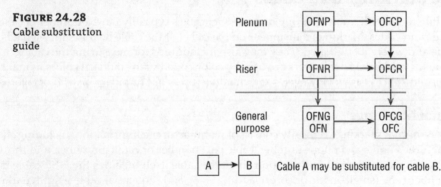

All indoor cables must be resistant to the spread of fire, and those listed as suitable for environmental air spaces such as plenums must also have low smoke-producing characteristics. Note in the substitutions list that nonconductive cables may always be substituted for conductive cables of an equal or lower rating, but conductive cables may never be substituted for nonconductive cables. Figure 24.28 shows the order in which cables may be substituted for one another.

FIGURE 24.28
Cable substitution
guide

	Plenum	OFNP → OFCP
	Riser	OFNR → OFCR
	General purpose	OFNG OFN → OFCG OFC

A → B   Cable A may be substituted for cable B.

NOTE   The NFPA 262 describes test methods for flame travel and smoke of cables for use in air-handling spaces.

The NEC also describes the *raceways* and *cable trays* that can be used with fiber-optic cables and the conditions under which each type of cable may be used. The code prohibits, for example, conductive fibers being run in the same cable tray or raceway with wiring for power, lights, and other electrical power systems. They can, however be run in the same cable tray or raceway as lower-power communication circuits and control systems.

NEC-Listed Raceways

The NEC defines a raceway as an "enclosed channel of metal or nonmetallic materials designed expressly for holding wires, cables, or busbars."

This definition may include metal or nonmetallic conduit and tubing along with raceways running through concrete or under floors. Where optical-fiber cables are concerned, however, the NEC describes three types of raceways:

Plenum optical-fiber raceways   These run through spaces that carry environmental air. The NEC requires these raceways to be fire-resistant and have low smoke-producing characteristics.

Riser optical-fiber raceways   These run through vertical spaces, often between floors, where flammable materials would give fire a chance to spread quickly. A riser raceway is required to have fire-resistant characteristics capable of preventing the carrying of fire from floor to floor.

General-purpose optical-fiber raceways   These are used where fire-resistance and low smoke-producing characteristics of the other raceway types are not as critical, including non-environmental air spaces. The NEC calls for these raceways to be resistant to the spread of fire as defined in UL 2024, "Standards for Optical Fiber Cable Raceways."

Additional information on the NEC can be found in Chapter 31.

Cable Markings and Codes

In addition to the NEC cable marking, optical-fiber cables typically have a number of other markings and codes. Markings that appear on the jacket of the cable help identify what is inside the cable and where it originated. They may also include aids for measuring the cable.

Inside the cable, the fiber buffers are usually color-coded with standard colors to make connections and splices easier. However, some manufacturers use numbers instead of colors.

External Markings

A cable's external markings typically consist of manufacturer's information, including the manufacturer's name and phone number, cable part number or catalog number, and the date the cable was manufactured. Information about the cable itself includes the NEC cable marking, the fiber type (single-mode or core/cladding size), and sequential cable length markings in meters or feet.

Color Codes

As with copper wiring, optical fibers running in cables must have some way of being distinguished from one another so that they can be connected properly at each end. TIA-598-C provides color-coding schemes for premises jackets and optical fibers within a fiber-optic cable. Table 24.2 shows the color-coding scheme for individual fibers bundled in a cable.

TABLE 24.2   TIA-598-C Fiber Color Code

FIBER NUMBER	BASE COLOR/TRACER	FIBER NUMBER	BASE COLOR/TRACER
1	Blue	19	Red/Black
2	Orange	20	Black/Yellow
3	Green	21	Yellow/Black
4	Brown	22	Violet/Black
5	Slate	23	Rose/Black
6	White	24	Aqua/Black
7	Red	25	Blue/Double Black
8	Black	26	Orange/Double Black
9	Yellow	27	Green/Double Black
10	Violet	28	Brown/Double Black
11	Rose	29	Slate/Double Black
12	Aqua	30	White/Double Black
13	Blue/Black	31	Red/Double Black
14	Orange/Black	32	Black/Double Yellow
15	Green/Black	33	Yellow/Double Black
16	Brown/Black	34	Violet/Double Black
17	Slate/Black	35	Rose/Double Black
18	White/Black	36	Aqua/Double Black

Table 24.3 shows the color coding used on premises cable jackets to indicate the type of fiber they contain, if that is the only type of fiber they contain.

TABLE 24.3   TIA-598-C Premises Cable Jacket Colors

Fiber Type	Jacket Color for Nonmilitary Applications	Jacket Color for Military Applications
Multimode (50/125µm)	Orange	Orange
Multimode (50/125µm) Laser-optimized	Aqua	—
Multimode (62.5/125µm)	Orange	Slate
Multimode (100/140µm)	Orange	Green
Single-mode (NZDS)	Yellow	Yellow
Polarized maintaining Single-mode	Blue	—

Sequential Markings

A cable's external markings may include sequential markings, as shown in Figure 24.29. Sequential markings are numbers that appear every 2′ or 1m. These markings are useful in determining how much cable is left on a reel, measuring off large runs of cable, or simply determining the length of a piece of cable without pulling out a tape measure.

FIGURE 24.29
Sequential marking in feet on a fiber-optic cable

The numbers themselves indicate cable length, not optical-fiber length. To measure the length of the cable using the sequential markings, first determine the measurement standard that is being used. Next, subtract the number at the low end from the number at the high end. The difference between the two is the length.

Because some measurements dealing with fiber optics use meters rather than feet, you may need to convert any measurements in feet to the metric system. The formula for converting length is:

1 foot = 0.3048 meter

Let's practice measuring out a length of cable using sequential markings. To find the length in meters of a cable that has sequential markings of 9,846 at one end and 12,218 at the other end, and the markings are measuring the cable in feet, you first determine the distance between the markings using this formula:

12,218 − 9,846 = 2,372 feet

Now, convert feet to meters:

2,372 × 0.3048 = 722.99 meters

Remember that the length of the cable is not necessarily the length of the optical fiber inside of it. In loose tube cable, the fiber is actually slightly longer than the cable. This fact becomes important if fault location procedures tell you that there is a fault at a specific distance from the end of the fiber. That distance could be short of the measured distance on the outside of the cable since the fiber wanders inside of the buffer tube.

Bend Radius Specifications

Throughout the installation process, optical fiber's light-carrying abilities are threatened by poor handling, damage from tools or accidents, and improper installation procedures.

One of the installation hazards that can cause attenuation or damage the optical fiber is extreme bending. When an optical fiber is bent too far, the light inside no longer reflects off the boundary between the core and the cladding, but passes through to the coating.

To reduce the risk of excessive bending during installation, manufacturers specify a minimum installation and operational bend radius for their optical-fiber cables. Following the manufacturer's guidelines reduces the risk of damage to the cable and optical fiber during and after installation. Chapter 31 provides detailed information on cable bend radius.

The Bottom Line

Determine the cable type from the NEC markings. Article 770 of the NEC states that optical-fiber cables installed within buildings shall be listed as being resistant to the spread of fire. These cables are also required to be marked.

Master It You have been asked to install the two cables shown here in a plenum space. From the markings on the cables, determine the cable types and determine if each cable can be installed in a plenum space.

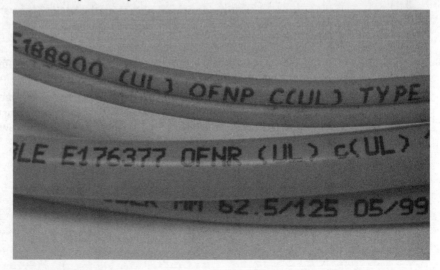

Identify the fiber number from the color code. TIA-598-C defines a color code for optical fibers within a cable assembly.

Master It You have been asked to identify the first and last tight-buffered optical fibers in the 12-fiber cable shown here. There are no numbers on the tight buffers; however, each buffer is colored as defined in TIA-598-C. What are the colors of the first and last tight-buffered optical fibers?

Identify the optical-fiber type from the color code. Premises cable jacket colors are defined in TIA-598-C.

Master It It is your first day on the job and your supervisor has asked you to get some laser-optimized multimode optical fiber from the back of the van. The van contains several different cables, each with a different jacket color. What is the jacket color of the laser-optimized cable?

Determine the cable length using sequential markings. Many markings are found on a cable. Sequential markings typically appear every 2 feet or every 1m.

Master It You have been asked to determine the length of fiber-optic cable coiled up on the floor. The sequential markings for this cable are meters. The marking on one end of the cable is 1,235 and the marking on the other end is 2,485. How long is this cable?

Chapter 25

Splicing

In this chapter, we will discuss fiber-optic splices and examine the factors that affect splice performance. We will also describe different tools and methods used to splice optical fibers as well as standards used to gauge splice performance.

In this chapter, you will learn to:

◆ Determine if the splice loss is from an intrinsic factor

◆ Determine if the splice loss is from an extrinsic factor

◆ Calculate the potential splice loss from a core diameter mismatch

Why Splice?

Up to this point, we have not addressed how to connect or mate two optical fibers together. Nor have we discussed why we would want to do this. This chapter discusses splicing. Splicing is one way to join two optical fibers together so the light energy from one optical fiber can be transferred into another optical fiber.

A splice is a permanent connection of two optical fibers. Once the two optical fibers are joined with a splice, they cannot be taken apart and put back together as they can if you join them using connectors. A splice is typically employed for one of three reasons: to repair a damaged cable, to extend the length of a cable, or to join two different cable types.

Splice Performance

How well a splice performs depends on many variables. These variables can be broken into two groups: *intrinsic factors* and *extrinsic factors*.

As you have learned in earlier chapters, optical fibers are not perfect and variations between optical fibers can affect splice performance. These variations are referred to as intrinsic factors.

The performance of a splice can also be affected by alignment and optical fiber mating issues that have nothing to do with the optical fiber. The factors that affect the alignment and/or mating of the optical fibers are referred to as extrinsic factors.

Intrinsic Factors

Even when fibers are manufactured within specified tolerances, there are still slight variations from one optical fiber to another. These variations can affect the performance of the splice even though the optical fibers are perfectly aligned when mated. The variations between two optical fibers that affect splice performance are referred to as intrinsic factors.

Let's look at the most common types of variations.

NUMERICAL APERTURE (NA) MISMATCH

A numerical aperture (NA) mismatch occurs when the NA of one optical fiber is different from the NA of the other optical fiber. If the NA of the transmitting fiber is larger than the NA of the receiving optical fiber, a loss may occur. However, a loss will not occur if the NA of the transmitting optical fiber is less than the NA receiving optical fiber. NA mismatch loss is typically only a concern with multimode optical fiber.

In Chapter 22, "Optical Fiber Characteristics," we showed you how to calculate the NA of an optical fiber using the refractive indexes of the core and cladding. Recall that in order for light to be contained within a multimode optical fiber it must reflect off the boundary between the core and the cladding, rather than penetrating the boundary and refracting through the cladding. Light must enter within a specified range defined by the acceptance cone or cone of acceptance.

Light entering the core of an optical fiber from outside of the acceptance cone will either miss the core or enter at an angle that will allow it to pass through the boundary with the cladding and be lost as shown in Figure 25.1. Light entering the core of the optical fiber at an angle greater than the acceptance angle will not propagate the length of the optical fiber. For light to propagate the length of the optical fiber, it must enter at an angle that does not exceed the acceptance angle.

FIGURE 25.1

When NA mismatch loss occurs, the receiving optical fiber cannot gather all of the light emitted by the transmitting fiber.

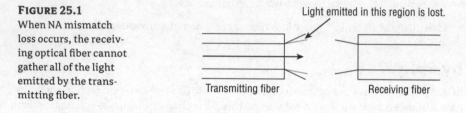

Light emitted in this region is lost.

Transmitting fiber   Receiving fiber

The exact loss from an NA mismatch is difficult to calculate. Factors such as light source type, light source launch condition, optical fiber length, and bends in the optical fiber all affect the potential loss. It is possible to have an NA mismatch between two optical fibers and no loss resulting from the mismatch.

CORE DIAMETER MISMATCH

Core diameter mismatch occurs when there is a difference in the core diameters of the two optical fibers. A core diameter mismatch loss results when the core diameter of the transmitting optical fiber is greater than the core diameter of the receiving optical fiber, as shown in Figure 25.2. A loss occurs when light at the outer edge of the transmitting optical fiber core falls outside the diameter of the receiving optical fiber core. This light is lost in the cladding of the receiving optical fiber. Core diameter mismatch loss is typically only a concern with multimode optical fiber.

FIGURE 25.2

Core diameter mismatch loss is the result of the transmitting optical fiber having a larger core diameter than the receiving optical fiber.

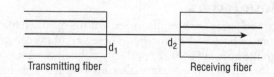

Transmitting fiber   Receiving fiber

It is not uncommon for two multimode optical fibers with different core diameters to be spliced together. As you saw in Chapter 22, ANSI/TIA-568-C.3 recognizes multimode optical fibers with 50 µm cores and 62.5 µm cores. We can calculate the worst-case loss percentage for a splice that joins a 50 µm core and 62.5 µm core using this formula:

$$Loss = [(d_1)^2 - (d_2)^2] / (d_2)^2$$

where d_1 is the diameter of the transmitting core and d_2 is the diameter of the receiving core.

$$62.5^2 - 50^2 = 1406.25$$

$$1406.25/50^2 = 0.5625 \text{ or } 56.25\%$$

Using the following formula, the decibel loss can be calculated:

$$dB = 10Log_{10}(P_{out} / P_{in})$$

Note that the decibel formula is a ratio where the power output (P_{out}) is divided by the power input (P_{in}). In this case, the power input is the power being put into the receiving optical fiber. We know that the receiving optical fiber will not accept 56.25 percent of the light from the transmitting optical fiber. Subtracting 56.25 percent from 100 percent, as shown next, results in a difference of 43.75 percent, which can be written as 0.4375.

$$100\% - 56.25\% = 43.75\% \text{ or } 0.4375$$

$$10Log_{10} (0.4375 / 1) = -3.59dB$$

$$Loss = 3.59dB$$

The 3.59dB loss that we just calculated is the maximum loss possible from the core diameter mismatch. The actual loss may vary depending on factors such as light source type, light source launch condition, optical fiber length, and bends in the optical fiber.

Mode Field Diameter Mismatch

A mode field diameter mismatch occurs when there is a difference in the mode field diameters of two single-mode optical fibers. A mode field diameter mismatch loss results when the mode field diameter of the transmitting optical fiber is greater than the mode field diameter of the receiving optical fiber, as shown in Figure 25.3. A loss occurs when optical fiber with the smaller mode field diameter will not accept all of the light from the optical fiber with the larger mode field diameter.

FIGURE 25.3

Mode field diameter mismatch loss is the result of the transmitting optical fiber having a larger mode field diameter then the receiving optical fiber.

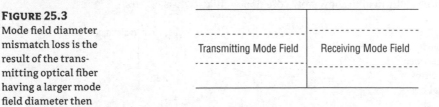

Transmitting Mode Field | Receiving Mode Field

CLADDING DIAMETER MISMATCH

Cladding diameter mismatch occurs when the cladding diameters of the transmit and receive optical fibers are not the same. Cladding diameter mismatch loss occurs when the cores of the optical fiber are not aligned because of the cladding diameter mismatch, as shown in Figure 25.4. A cladding diameter mismatch can cause the light exiting the core of the transmitting optical fiber to enter the cladding of the receiving optical fiber. The light entering the cladding is lost, causing attenuation.

FIGURE 25.4
Cladding diameter mismatch loss results from differing cladding diameters.

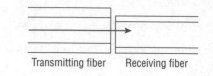

Transmitting fiber   Receiving fiber

CONCENTRICITY

Ideally, the core and cladding of an optical fiber are perfectly round and *concentric*, which means that they share a common geometric center. However, optical fibers are not perfect and there will be concentricity variations. These concentricity variations can cause the optical fiber cores to misalign, as shown in Figure 25.5, causing a loss when the light exiting the core of the transmitting optical fiber enters the core of the receiving optical fiber.

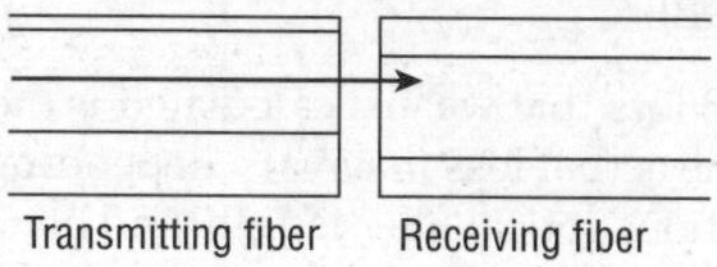

FIGURE 25.5
Off-center fiber cores cause concentricity loss.

Transmitting fiber   Receiving fiber

The illustration in Figure 25.5 is greatly exaggerated to clearly show how a concentricity loss may occur. The core and cladding concentricity differences are typically less than 1μm.

NONCIRCULARITY

Just as the core and cladding of an optical fiber may not be perfectly concentric, they may also not be perfectly circular. The noncircularity of the core will cause a loss when light from the core of the transmitting optical fiber enters the cladding of the receiving optical fiber. In Figure 25.6, the core is elliptical to show how core noncircularity loss occurs.

FIGURE 25.6
Noncircularity loss takes place when the ellipticities of two optical fibers do not match exactly.

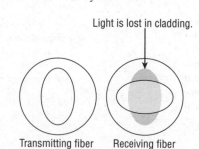

Light is lost in cladding.

Transmitting fiber   Receiving fiber

Cladding noncircularity may cause loss when it causes part of the core of the transmitting optical fiber to align with the cladding of the receiving optical fiber. Any light that enters the cladding of the receiving optical fiber will be lost, causing attenuation.

Note that the amount of loss depends on the alignment of the ellipticities of the two cores. Maximum loss occurs when the long or major axes of the cores are at right angles to one another, and minimum loss occurs when the axes of the cores are aligned. After reading about all the possible intrinsic losses, you may find it hard to believe that two optical fibers can be spliced together with virtually no loss. Optical fiber manufacturing produces optical fibers with little variation, as shown in Tables 25.1, 25.2, and 25.3. These tables list the attributes and tolerances for different single-mode optical fibers as defined in ITU standards ITU-T G.652 and ITU-T G.655.

TABLE 25.1   ITU-T G.652, Single-Mode Optical Fiber Attributes

FIBER ATTRIBUTES, G.652.A, G.652.B, G.652.C, AND G.652.D		
Mode field diameter	Wavelength	1310nm
	Range of nominal values	8.6–9.5µm
	Tolerance	±0.6µm
Cladding diameter	Nominal	125.0µm
	Tolerance	±1µm
Core concentricity error	Maximum	0.6µm
Cladding noncircularity	Maximum	1%

TABLE 25.2   ITU-T G.655.C, Non-Zero-Dispersion-Shifted Single-Mode Optical Fiber and Cable

FIBER ATTRIBUTES		
Mode field diameter	Wavelength	1550nm
	Range of nominal values	8–11µm
	Tolerance	±0.7µm
Cladding diameter	Nominal	125.0µm
	Tolerance	±1µm
Core concentricity error	Maximum	0.8µm
Cladding noncircularity	Maximum	2%

TABLE 25.3   ITU-T G.655.D, Non-Zero-Dispersion-Shifted Single-Mode Optical Fiber and Cable

FIBER ATTRIBUTES		
Mode field diameter	Wavelength	1550nm
	Range of nominal values	8–11µm
	Tolerance	±0.6µm
Cladding diameter	Nominal	125.0µm
	Tolerance	±1µm
Core concentricity error	Maximum	0.6µm
Cladding noncircularity	Maximum	1%

Extrinsic Factors

Extrinsic factors that affect optical fiber splice performance are factors related to the condition of the splice itself, external to the optical fiber. In an ideal splice, the optical fibers are identical and they are aligned so that cores are perfectly centered on each other and the core axes are perpendicular to the endfaces being joined, as shown in Figure 25.7. However, there is no such thing as an ideal splice, only a real splice. In a real splice, intrinsic and extrinsic factors affect splice performance.

FIGURE 25.7
Conditions for an
ideal splice

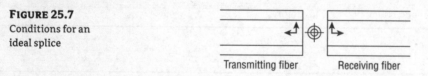

Transmitting fiber   Receiving fiber

In this section of the chapter, we will examine common extrinsic factors that affect splice performance. As you read about these extrinsic factors, keep in mind that many times they are caused by dirt and contamination. Microscopic particles of dirt can cause the misalignment of one or both optical fibers, creating a high-loss splice.

LATERAL MISALIGNMENT

Lateral misalignment occurs when the two optical fibers are offset, as shown in Figure 25.8. Lateral misalignment loss occurs when light from the core of the transmitting optical fiber enters the cladding of the receiving optical fiber, creating a loss. As the lateral misalignment increases, less light from the core of the transmitting optical fiber makes its way into the core of the receiving optical fiber, increasing the loss of the splice.

FIGURE 25.8
Lateral misalign-
ment of the optical
fibers

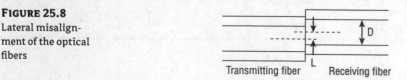

Transmitting fiber   Receiving fiber

END SEPARATION

Even if the optical fibers are perfectly aligned, the splice still may still experience loss from *end separation*. End separation is simply a gap between the transmitting and receiving optical fibers, as shown in Figure 25.9.

FIGURE 25.9
End separation is a gap between fiber ends in a mechanical splice.

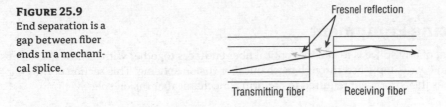

Fresnel reflection

Transmitting fiber   Receiving fiber

Two different types of losses can be generated from end separation. The first is through Fresnel reflection, which takes place when light passes from the higher refractive index of the core in the transmitting optical fiber into the lower refractive index of the air, and then back into the core of the receiving optical fiber. Each change in the refractive index causes a certain amount of light to be reflected and therefore lost.

One way to overcome the effects of Fresnel reflections in separated optical fibers is to use an *index matching gel*, which is a transparent gel having a refractive index close to that of the core of the optical fibers being spliced. The gel fills the gap and reduces or eliminates the Fresnel reflection. Index matching gel is typically used in all mechanical splices.

End separation also causes loss because when light exits the transmitting optical fiber it spreads out like the light from a flashlight. Some of the light leaving the core of the transmitting optical fiber may enter the cladding of the receiving optical fiber, causing a loss. How much the light spreads out depends on several variables, including the distance between the optical fibers, light source type, launch type, length of the transmitting optical fiber, and the bends in the transmitting optical fiber.

ANGULAR MISALIGNMENT

If the optical fibers in a splice meet each other at an angle, a loss from *angular misalignment* may occur, as shown in Figure 25.10. The amount of loss depends the severity of the angular misalignment and the acceptance cones of the transmitting and receiving optical fibers. Because the NA of a multimode optical fiber is greater then the NA of a single-mode optical fiber, multimode splices tolerate angular misalignment better than single-mode splices.

FIGURE 25.10
Angular misalignment results when fiber ends are not perpendicular to fibers.

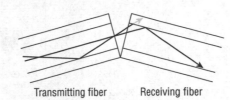

Transmitting fiber   Receiving fiber

The loss from angular misalignment occurs when light from the core of transmitting optical fiber enters the cladding of the receiving optical fiber or enters the core of receiving optical fiber at an angle exceeding the acceptance angle. Light entering the core of the receiving optical fiber at an angle exceeding the acceptance angle will not propagate the length of the receiving optical fiber.

Angular misalignment also prevents the endfaces from contacting each other, resulting in Fresnel reflections exactly like those described in the "End Separation" section earlier in this chapter. Again, this is why index-matching gel is used with every mechanical splice.

Splicing Equipment

The goal of any splice is to join two optical fiber endfaces together with as little loss as possible. This can be accomplished through mechanical or fusion splicing. This section of the chapter describes the equipment required to perform a mechanical or fusion splice.

Cleaning Materials

One of the keys to performing a successful fiber-optic splice is cleanliness. The optical fiber needs to be properly cleaned prior to cleaving and splicing. In addition, the splicing area should be kept as clean as possible. As discussed earlier in the chapter, a microscopic piece of dirt could cause lateral displacement, end separation, or angular misalignment.

Prior to starting any splice, ensure you have lint-free wipes designed for optical fiber and a solvent designed to clean and remove the coating residue from the optical fiber. Lint-free wipes like those shown in Figure 25.11 are engineered to lift away oils, grime, and dust from the surface of the optical fiber.

FIGURE 25.11
Lint-free wipes
engineered for
cleaning optical
fibers

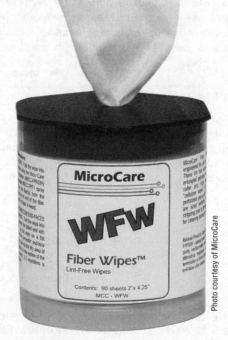

When handling a lint-free wipe, be sure to only touch one side of the wipe because the wipe will absorb the oil from your skin. Use the side of the wipe that you did not touch for cleaning the optical fiber. You do not want to contaminate the optical fiber during the cleaning process. Cleaning of the optical fiber is covered in depth later in the splicing procedures section.

When you select an optical fiber cleaning solvent, ensure it has been engineered to clean optical fibers. The goal is to select a solvent that leaves little or no residue on the surface of the optical fiber. If you choose isopropyl alcohol, make sure that it is virtually free of water and has been filtered to remove impurities, like the isopropyl alcohol shown in Figure 25.12. Do not use the isopropyl alcohol commonly available at your local pharmacy; this alcohol contains a large percentage of water and will leave a residue on the optical fiber.

FIGURE 25.12
Isopropyl alcohol for cleaning optical fibers

Photo courtesy of MicroCare

Isopropyl alcohol poses a problem for air travel because of its flammability. If the splicing job requires air travel, you will not be able to bring isopropyl alcohol on the aircraft. For jobs requiring air travel, you need to select a cleaning fluid that is approved for air travel. Remember from Chapter 23, "Safety," that isopropyl alcohol vapors are highly flammable and can ignite if exposed to a spark or flame in high enough concentrations.

Cleaning fluids like the one shown in Figure 25.13 are approved for air travel and engineered specifically to clean the optical fibers. These cleaning fluids typically contain little or no isopropyl alcohol and have been filtered to remove microscopic contaminates.

MicroCare, Corning Cable Systems, and AFL Telecommunications manufacture some of the most commonly used cleaning materials.

MicroCare www.microcare.com

Corning www.corningcablesystems.com

AFL Telecommunications www.afltele.com

FIGURE 25.13
Optical fiber clean-
ing fluid approved
for air travel

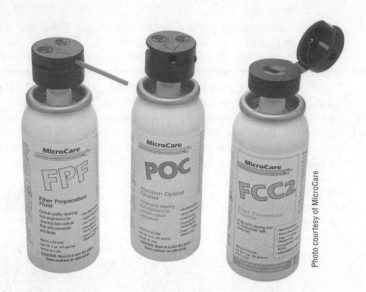

Photo courtesy of MicroCare

Cleavers

A cleaver is required for all splice types. Cleavers are available for single optical fibers, like the one shown in Figure 25.14, and ribbon fiber, like the one shown in Figure 25.15.

FIGURE 25.14
Single optical
fiber cleaver

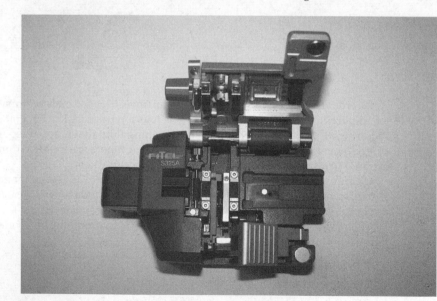

Photo courtesy of KITCO Fiber Optics

Photo courtesy of KITCO Fiber Optics

FIGURE 25.15
Ribbon fiber
cleaver

Cleaving is typically a two-step process, and it takes place after the optical fiber has been properly prepared and cleaned. The first step in the process is scoring the optical fiber. This is accomplished with a scoring blade. The scoring blade should be held perpendicular to the optical fiber, as shown in Figure 25.16. When the cleaver is operated, the scoring blade lightly contacts the optical fiber, creating a small surface flaw.

FIGURE 25.16
Scoring blade
ready to score the
optical fiber

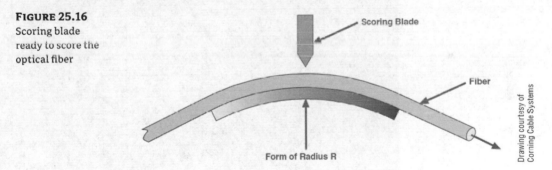

Drawing courtesy of
Corning Cable Systems

After the optical fiber is scored, the next step is to bend the optical fiber. The optical fiber is bent slightly, as shown in Figure 25.16. Slightly bending the optical fiber places tensile stress on the surface of the optical fiber where it has been scored. This causes the small surface flaw to open and split the fiber.

After the optical fiber is cleaved, the endface should be perpendicular to the optical fiber without any surface defects and the cleave angle should not exceed $1.0°$. Cleave angles can only be evaluated with a microscope. This is typically only done when a fusion splice is performed. Many fusion splicers will measure the cleave angle of the optical fiber prior to splicing.

Figure 25.17 is a photograph of a cleaved optical fiber with defects. In an ideal cleave, the mirror area would cover the entire endface. However, in this example the mirror area only represents a small portion of the endface. The rest of the endface looks like it was hacked with a sharp object many times. These irregular defects are commonly referred to as hackles.

FIGURE 25.17
Cleaved endface
with hackles

Mechanical Splice

Many manufacturers offer mechanical splices. A mechanical splice like the one shown in Figure 25.18 is typically permanent. It aligns the cleaved optical fibers and holds them in place. Index matching gel inside the mechanical splice reduces or eliminates Fresnel reflections.

FIGURE 25.18
Mechanical splice

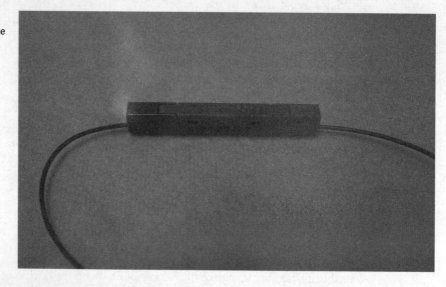

Mechanical splices can be used for both multimode and single-mode optical fiber. Mechanical splices do not outperform fusion splices. However, they will outperform mated connector pairs. The key advantage to a mechanical splice over a fusion splice is the low cost of the equipment required to perform the mechanical splice.

The assembly tool that holds the optical fibers and the mechanical splice is relatively inexpensive when compared to the price of a fusion splicer. However, the actual mechanical splice shown in Figure 25.18 costs considerably more than the protective sleeve required for a fusion splice.

3M and Corning Cable Systems manufacture some of the most commonly used mechanical splices.

3M www.3m.com

Corning www.corningcablesystems.com

Fusion Splice

A fusion splicer creates a permanent splice by melting the optical fiber endfaces to one another with an electric arc between two electrodes, as shown in Figure 25.19. The fusion splice is then enclosed in heat-shrink tubing, as shown in Figure 25.20. An oven built into the splicer heats the tubing, shrinking it around the fusion splice. The heat-shrink tubing also contains a metal rod that prevents the splice from bending.

FIGURE 25.19
The electrodes of a fusion splicer

Electrodes

FIGURE 25.20
A fusion splice covered with heat shrink held rigid with a metal rod.

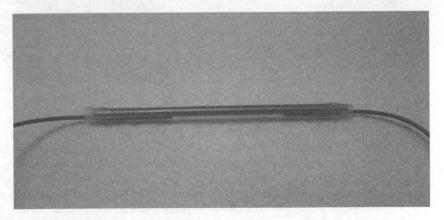

Fusion splicers are more expensive than the assembly tool required for a mechanical splice. However, they provide the lowest-loss splice possible. In addition, fusion splices do not produce Fresnel reflections.

Fusion splicing is the most accurate and durable method for joining two optical fibers. After the optical fibers are stripped and cleaved, they are placed into the fusion splicer where the optical fibers are aligned between two electrodes, as shown in Figure 25.21. Alignment may be accomplished several ways depending on the type of fusion splicer used.

FIGURE 25.21
Optical fibers aligned between the fusion splicer electrodes

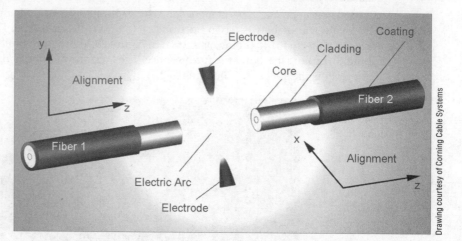

Drawing courtesy of Corning Cable Systems

There are several alignment techniques used to align the optical fibers. A very common alignment technique is to use a fixed V-groove, as shown in Figure 25.22. Each optical fiber is placed in a precision-machined V-groove. The V-grooves align the cladding of the optical fibers. The fusion splicer moves the optical fibers axially into the electric arc, melting the optical fibers together.

FIGURE 25.22
Optical fibers in a V-groove aligned between the fusion splicer electrodes

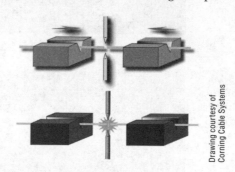

Drawing courtesy of Corning Cable Systems

Fusion splicers with this type of alignment system have the lowest cost and are typically the smallest and lightest. However, they align the cladding of the optical fiber instead of the core and can create higher loss splices than fusion splicers that align the cores of the optical fibers.

A fusion splicer that aligns the cores of the optical fibers will produce the lowest-loss splice. However, aligning the cores of the optical fibers is more difficult than aligning two optical fibers placed in V-grooves. To align the cores, the fusion splicer must be able to detect the cores and have the ability to move the optical fibers on each axis.

There are several patented technologies used to detect the cores of optical fibers. Because they are patented, they will not be discussed in detail in this chapter.

Fusion splicers on the market today typically feature a display that allows you to see the optical fibers on two different axes. The cameras in the fusion splicer magnify the optical fibers so that the endfaces can be evaluated. You cannot look at an optical fiber with the naked eye and evaluate the endface. Without magnification, optical fiber with a perfect cleave will look just like an optical fiber that has been broken. Figure 25.23 is a photograph of a fusion splicer display showing a two-axis view of two broken optical fibers. This photograph clearly shows that a broken optical fiber can have a jagged endface.

FIGURE 25.23
Optical fibers with jagged endfaces viewed on the display of a fusion splicer

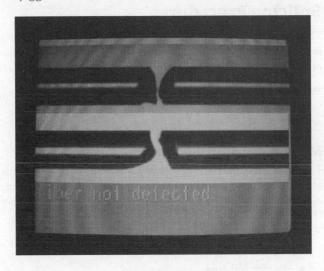

Most fusion splicers available today also have the ability to approximate the loss for the splice after it is completed. There are also fusion splicers available that have the ability to measure the loss for the splice after it is completed. Figure 25.24 is a photograph of a fusion splicer display showing the estimated splice loss. The splice loss is displayed on the bottom of the screen, and different axis views of the spliced optical fiber are shown at the top of the screen. Note that the spliced optical fiber appears as a single optical fiber with no flaws.

FIGURE 25.24
Fusion splicer displaying the estimated splice loss

You can find out more about specific features of fusion splicers through their manufacturers' websites. Some of the major manufacturers of fusion splicers are:

Corning www.corningcablesystems.com

AFL Telecommunications www.afltele.com

Fitel www.fitel.com

Next let's look at the procedures used with each type of splicer.

Splicing Procedures

This section will familiarize you with the basic procedures used for mechanical or fusion splicing. While manufacturers' models will vary, many of the same principles and requirements apply. Be sure to read and follow the directions for your particular equipment carefully.

WARNING Whenever handling optical fiber, remember to follow the safety precautions described in Chapter 23.

Mechanical Splicing Procedure

To prepare for mechanical splicing, make sure that the work area is clean, dry, and well lit. Do all your work over a fiber-optic mat and place any scrap optical fibers in their proper container, as described in Chapter 23. Assemble the following tools before you begin:

- Mechanical splice assembly tool
- Mechanical splice
- Buffer and coating removal tool
- Optical fiber cleaning fluid
- Lint-free wipes
- Cleaver

Once your materials are assembled, proceed with the following steps:

1. Remove the mechanical splice from its protective packaging and load the splice into the assembly tool by pressing firmly at the ends of the splice. Do not depress the raised section on the mechanical splice.

2. Strip approximately 3cm of buffer and/or coating from the optical fiber using a stripper, as shown in Figure 25.25.

3. Clean the optical fiber by pulling the fiber through a lint-free wipe soaked in optical fiber cleaning fluid, as shown in Figure 25.26.

FIGURE 25.25
Stripping the buffer and coating from an optical fiber

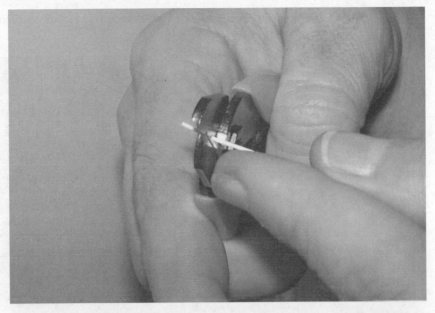

FIGURE 25.26
Clean the optical fiber by pulling it through the lint-free wipe soaked in optical fiber cleaning fluid.

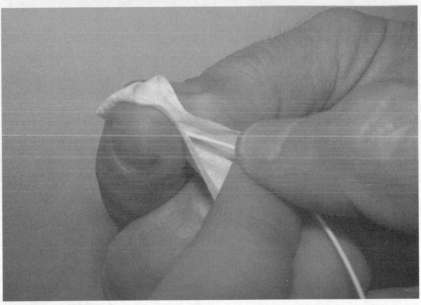

4. Place the optical fiber in the cleaver, as shown in Figure 25.27, to the length specified by the mechanical splice manufacturer.

Photo courtesy of KITCO Fiber Optics

FIGURE 25.27
Place the optical fiber in the cleaver at the length specified by the manufacturer.

5. Cleave the optical fiber.

6. If a gauge is provided, check the cleave length with the gauge.

7. Repeat steps 2 through 6 for the other fiber end to be spliced.

8. Push one cleaved optical fiber into one end of the mechanical splice until it stops moving.

9. Push the other cleaved optical fiber into the other end of the mechanical splice until it stops moving.

10. Place both optical fibers in the clamping mechanisms on the opposite sides of the splicing tool, forming a modified loop in the optical fiber, as shown in Figure 25.28.

FIGURE 25.28
Mechanical splice tool properly set up to perform a mechanical splice

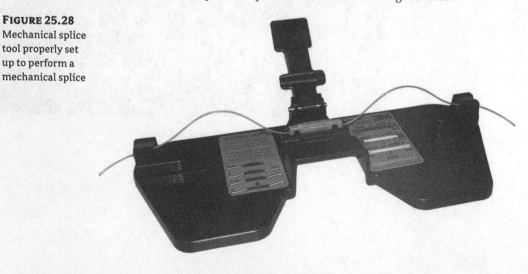

11. Pivot the splicing tool handle down until it contacts the top of the splice, then squeeze the tool handle to complete the assembly and lock the spliced ends in place.

12. Remove the optical fibers from the clamping mechanisms and lift the mechanical splice from the tool.

CAN'T AFFORD A FUSION SPLICER?

If you want the joy of fusion splicing without the pain of the purchase price, you can always rent or lease a fusion splicer. Companies such as Fiber Instrument Sales, Inc. (www.fiberinstrumentsales .com) provide rental arrangements for fusion splicers and other high-end fiber equipment if you know you won't need it on a long-term or recurring basis.

Rental rates typically run ¹⁄₁₀ to ¹⁄₁₂ of the purchase price of a new fusion splicer per month. Bear in mind, though, that the rental period starts the day that the splicer is shipped to you and ends the day that it is received back at the rental office. In other words, you're going to pay for the days that it is in transit.

If you still want to buy a fusion splicer but don't want to pay full price, consider a preowned or reconditioned unit. These are occasionally available through distributors and still have years of life left in them.

Fusion Splicing Procedure

The equipment required for fusion splicing is far more complex and expensive than the equipment required for mechanical splicing. However, many of the steps required to perform a fusion splice are identical to the steps required to perform a mechanical splice.

Many fusion splicers contain a feature that automatically positions the fiber ends in proper relationship with each other and with the electrodes for the best possible splice. All that is required of the operator is to prepare the fibers properly and place them in the fusion splicer as outlined by the manufacturer.

The fusion-splicing procedures outlined in this chapter are general procedures designed to list the steps required to fusion-splice single optical fibers and ribbon optical fibers.

To prepare for fusion splicing, as with mechanical splicing, make sure that the work area is clean, dry, and well lit. Do all your work over a fiber-optic mat and place any scrap optical fibers in their proper container, as described in Chapter 23. Assemble the following tools before you begin:

- Fusion splicer
- Buffer and/or coating removal tool
- Optical fiber cleaning fluid
- Lint-free wipes
- Cleaver
- Heat-shrink protective covering

Once your materials are assembled, proceed with the following steps. First we'll look at single optical fiber fusion splicing:

1. Power on the fusion splicer and select the appropriate splicing program for the optical fiber you will be splicing.

2. Slide the protective heat-shrink tubing over one optical fiber end and move it far enough up the optical fiber to place it out of the way.

3. Strip approximately 3cm of buffer and/or coating from the optical fiber using a stripper, as shown in Figure 25.29.

FIGURE 25.29
Stripping the buffer and coating from an optical fiber

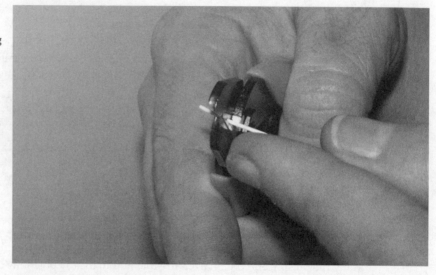

4. Clean the optical fiber by pulling the fiber through a lint-free wipe soaked in optical fiber cleaning fluid, as shown in Figure 25.30.

FIGURE 25.30
Clean the optical fiber by pulling it through the lint-free wipe soaked in optical fiber cleaning fluid.

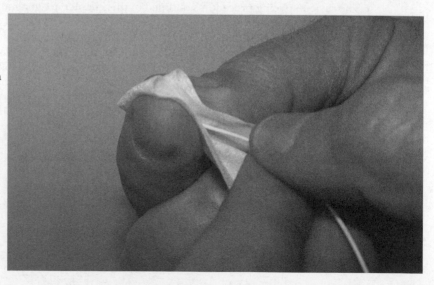

5. Place the optical fiber in the cleaver, as shown in Figure 25.31, to the length specified by the mechanical splice manufacturer.

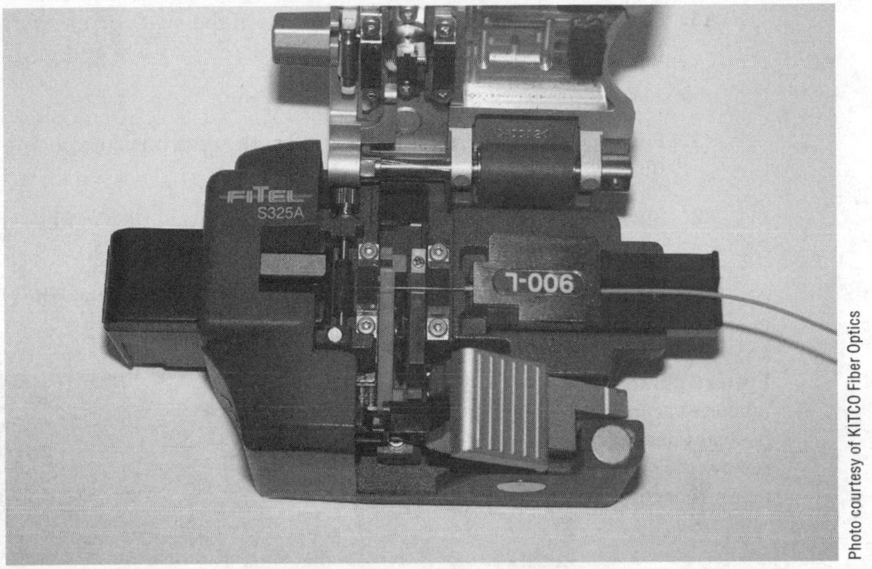
FITEL
S325A
900-L
Photo courtesy of KITCO Fiber Optics

FIGURE 25.31 Place the optical fiber in the cleaver at the length specified by the manufacturer.

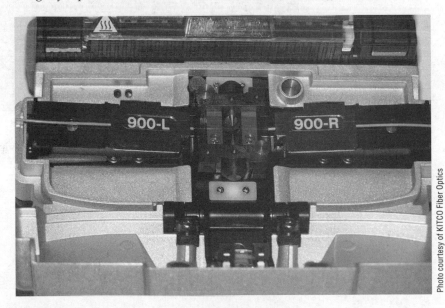

6. Cleave the optical fiber.
7. Place the optical fiber in the fusion splicer following the manufacturer's instructions. Position the endface of the optical fiber between the electrodes.
8. Repeat steps 3 through 7 for the other fiber end to be spliced. The properly placed fibers should be slightly separated between the electrodes, as shown in Figure 25.32.

900-L
900-R
Photo courtesy of KITCO Fiber Optics

FIGURE 25.32 Properly placed fibers almost touch each other between the electrodes.

9. Close the electrode cover.

10. Begin the fusion-splicing process.

11. Carefully remove the splice and position the heat-shrink tubing from step 2 over it. Place the splice and tubing in the heat-shrink oven to seal and protect the splice.

Next we'll look at ribbon fiber fusion splicing:

1. Power on the fusion splicer and select the appropriate splicing program for the optical fiber you will be splicing.

2. Slide the protective heat-shrink tubing over one ribbon fiber end and move it far enough up the optical fiber to place it out of the way.

3. Strip approximately 3cm of Mylar tape and coating from the ribbon fiber, as shown in Figure 25.33.

FIGURE 25.33
Ribbon fiber stripper using heat to remove the Mylar tape and coating

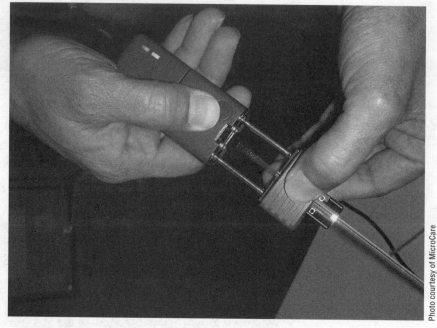

Photo courtesy of MicroCare

4. Clean the ribbon fiber by pulling the fiber through a lint-free wipe soaked in optical fiber cleaning fluid, as shown in Figure 25.34.

FIGURE 25.34
Ribbon fiber being
pulled through
a lint free wipe
soaked on optical
cleaning fluid

Photo courtesy of MicroCare

5. Place the ribbon fiber in the cleaver, as shown in Figure 25.35, to the length specified by the manufacturer of splicer you are using.

6. Cleave the ribbon fiber.

FIGURE 25.35
Ribbon fiber placed
in the cleaver for
cleaving

Photo courtesy of MicroCare

7. Place the ribbon fiber in the fusion splicer, as shown in Figure 25.36. Position the ribbon fiber as described in the manufacturer's data sheet.

8. Lock the ribbon fiber in place as described in the manufacturer's data sheet.

9. Repeat steps 3 through 8 for the other fiber end to be spliced.

10. Close the electrode cover.

11. Begin the fusion-splicing process.

12. Carefully remove the splice and position the heat-shrink tubing from step 2 over it. Place the splice and tubing in the heat-shrink oven to seal and protect the splice.

Photo courtesy of MicroCare

FIGURE 25.36
Ribbon fiber placed in the fusion splicer for splicing

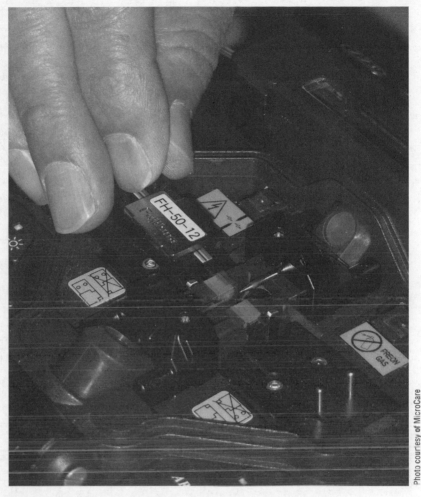

Splice Requirements

As with connectors, standards specify a maximum permissible loss for splices, depending on their location. For inside plant splices, ANSI/TIA-568-C.3 states that optical fiber splices, whether fusion or mechanical, will have a maximum attenuation of 0.3dB.

For outside plant splices, ANSI/TIA-758-A states that the splice insertion loss shall not exceed 0.1dB mean, with a maximum of 0.3dB, as measured with an optical time-domain reflectometer (described in Chapter 33, "Test Equipment and Link/Cable Testing").

FUSION OR MECHANICAL?

It may seem like an easy choice between mechanical and fusion splicing: fusion splices are superior to mechanical splices because they essentially make one optical fiber out of two. The cost per fusion splice is virtually nil, while mechanical splices typically cost from $10 to $15 per splice. In short, if you have a fusion splicer, there is really no need to even look at mechanical splicing.

So why bother with mechanical splices at all?

The one key factor that favors mechanical splicing is the initial setup cost. Mechanical splicing and fusion splicing both require a cleaver, hand tools, and cleaning supplies. However, only a mechanical splice assembly tool is required to assemble the mechanical splice. This tool typically costs around $100 while many fusion splicers cost more than $10,000.

If your application only requires a few splices, mechanical splicing is the most economical way to go. However, if your application requires many splices, fusion splicing may be the most economical way to go.

The Bottom Line

Determine if the splice loss is from an intrinsic factor. Even when fibers are manufactured within specified tolerances, there are still slight variations from one optical fiber to another. These variations can affect the performance of the splice even though the optical fibers are perfectly aligned when mated.

> **Master It** Two optical fibers are about to be fusion spliced together, as shown here. Determine what type of problem exists with the optical fibers about to be spliced.

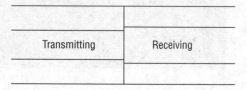

Determine the splice loss is from an extrinsic factor. Extrinsic factors that affect optical fiber splice performance are factors related to the condition of the splice itself, external to the optical fiber.

> **Master It** Two optical fibers are about to be fusion spliced together, as shown here. Determine what type of problem exists with the optical fibers about to be spliced.

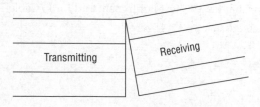

Calculate the potential splice loss from a core diameter mismatch.   A core diameter mismatch loss results when the core diameter of the transmitting optical fiber is greater than the core diameter of the receiving optical fiber.

Master It   Calculate the worst-case loss for a splice where the transmitting optical fiber has a core diameter of 51μm and the receiving optical fiber has a core diameter of 49.5μm.

Chapter 26

Connectors

The ideal traveling environment for a pulse of light is an unbroken optical fiber. At some point, however, that optical fiber must connect to a piece of equipment or join another optical fiber in order to extend its length or change the type of cable being used.

One of the most common methods for *terminating* an optical fiber, or making its end useful, is to use a connector. A connector is a device that protects the end of the optical fiber while allowing it to be quickly and reliably joined to equipment or other optical fibers.

Connectors are often used instead of splices to join two optical fibers together because they allow the optical fibers to be disconnected and reconnected easily. Splices, on the other hand, are permanent connections between two optical fibers. Connectors can be useful when network assignments must be changed, when equipment must be removed/replaced, or when expansion is anticipated.

This chapter describes several common connectors used in optical fiber termination. It describes the factors that affect connector performance and methods used to improve performance. The chapter also discusses methods used to install connectors so they meet industry performance standards.

In this chapter, you will learn to:

- Evaluate connector endface polishing

- Evaluate connector endface cleaning

- Evaluate connector endface geometry

- Identify an optical fiber type from the color of the connector strain relief

The Fiber-Optic Connector

The job of a fiber-optic connector is to couple an optical fiber end mechanically to a piece of equipment or to another optical fiber so that the cores line up accurately and produce the smallest amount of loss. Inherent in this requirement is the need for the connector to protect the fiber from repeated handling during connection and disconnection, align the fiber end precisely with its counterpart in the interconnection, and prevent strain on the fiber itself. The strength members in the fiber-optic cable normally provide strain relief, but once the fiber itself is attached to the connector, the connector must perform that job.

Although there are several different types of connectors recognized by industry standards, they all contain common components, shown in Figure 26.1.

FIGURE 26.1
Fiber-optic con-
nector components

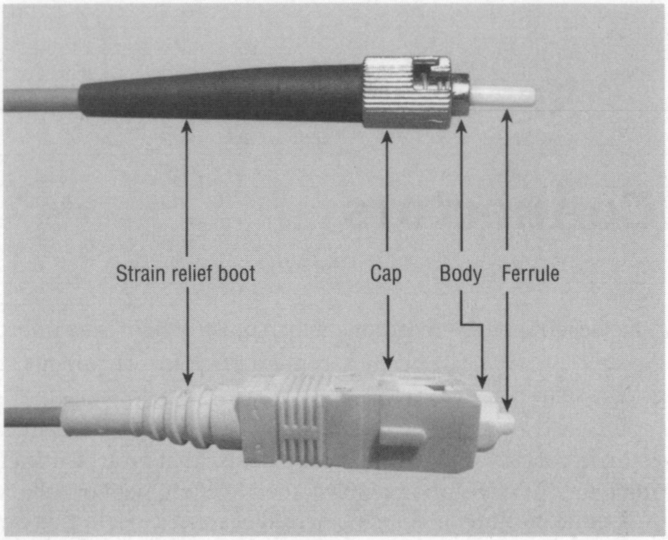

Strain relief boot   Cap   Body   Ferrule

Beginning at the working end of the connector, the *ferrule* holds the fiber in place. The ferrule must hold the fiber exactly centered in its endface for the best possible connection, so its construction is critical. Not only must the hole for the optical fiber be accurately placed; it must also be sized precisely to receive the exact diameter of the optical fiber cladding. The optical fiber's coating is stripped away prior to inserting the optical fiber into the ferrule.

Ferrules typically are made of metal, ceramic, or plastic with a selection of hole or bore diameters ranging from slightly larger than the optical fiber diameter to slightly smaller to allow for minute variations in the manufactured optical fiber cladding diameters. For example, for a 125µm optical fiber, ferrules are available with hole sizes ranging from 124µm to 127µm. Because the ferrule must align the optical fiber end precisely, it must meet several important criteria:

◆ The ferrule must be strong enough to withstand many cycles of connection and disconnection without bending, cracking, or breaking.

◆ The ferrule must maintain dimensional stability to ensure proper alignment of the optical fiber.

◆ The ferrule must be of the right shape and have material properties to ensure a low-loss interconnection.

Ceramic materials such as aluminum oxide and zirconium oxide are among the best materials for ferrules, offering the best combination of characteristics. They are hard enough to protect the fiber end, and their *coefficient of thermal expansion*, the measure of how much a material expands and contracts with temperature changes, is about the same as the optical fiber itself.

Metal ferrules, typically made of stainless steel, are stronger than ceramic, but they are less dimensionally stable. Plastic ferrules are less expensive than metal or ceramic, but they are neither as strong nor as stable as the other materials.

When a connector is built, which we'll describe in detail later in this chapter, the optical fiber is typically epoxied into the ferrule with the end protruding slightly beyond the endface of the ferrule. The optical fiber end is later trimmed and polished with the ferrule endface for a precise fit.

The ferrule fits inside the next component, the body of the connector. The body, which can be either metal or plastic, holds the ferrule, the optical fiber, and the cable in place and transfers any strain placed on the connector to the cable rather than the optical fiber.

The cap, sometimes called the coupling nut, fits over the body of the connector and provides a means to secure the connector, as shown in Figure 26.2. The cap can be a locking mechanism, a threaded ring, or a snap attachment, depending on the type of connector.

FIGURE 26.2
Connector
cap types

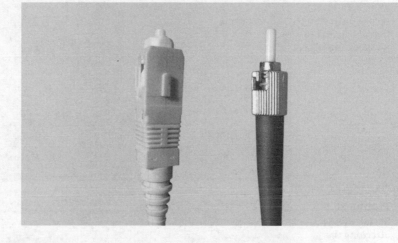

The strength member of the fiber-optic cable is secured to the connector. This allows all the tensile stresses placed on the connector to be handled by the cable's strength member instead of the optical fiber. There are several ways to accomplish this; crimping a ring or band around the strength member, securing it to the connector body, is a very popular method. Figure 26.3 shows the band of an LC connector securely attaching the strength member to the connector body.

FIGURE 26.3
Crimped band of
an LC connector
securely attaching
the cable strength
member to the
connector body

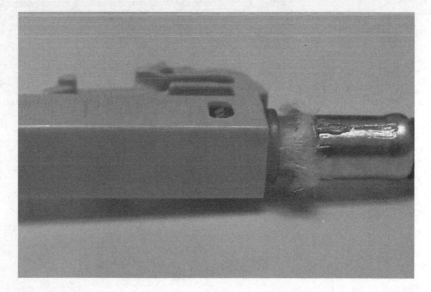

Depending on the connector type, the cable jacket may or may not be securely fastened to the connector body. Sometimes a band or ring is crimped around the cable jacket to secure it to the connector body, as shown in Figure 26.4. Other times a piece of shrink tubing is used to secure the jacket to the connector body, as you can see in Figure 26.5.

FIGURE 26.4
Crimped band of an SC connector securely attaching the cable jacket to the connector body

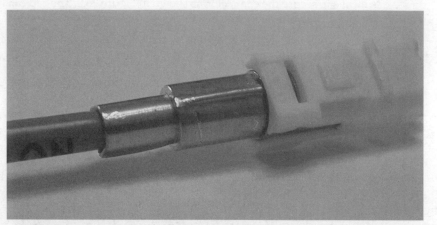

FIGURE 26.5
Shrink tubing attaching the cable jacket to an LC connector body

A strain relief or boot is typically placed over the cable jacket and secured to the connector body. The boot is generally made of an elastic material that slides over the connector body. Friction between the boot and the connector body holds the boot in place. The boot prevents the cable from being pulled at too great an angle against the connector, as shown in Figure 26.6.

FIGURE 26.6
Boot or strain relief preventing the fiber-optic cable from being overbent

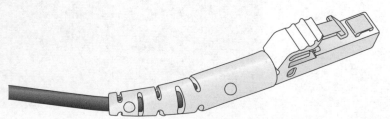

Fiber-optic connectors, commonly referred to as *plugs*, mate with other connectors or receptacles. The ferrule of the connector is the plug, and when two connectors mate, their ferrules are aligned with a sleeve that is typically called a *mating sleeve* or an *alignment sleeve*. Figure 26.7 is a photograph of an LC alignment sleeve that mates two LC connectors.

FIGURE 26.7
LC mating or alignment sleeve with connectors

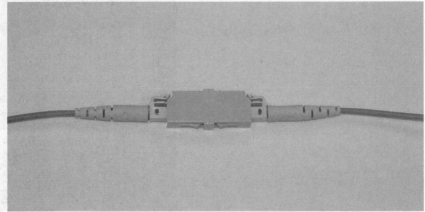

Connectors also mate with equipment or device receptacles. The device may be a transmitter or receiver, as discussed in Chapter 27, "Fiber-Optic Light Sources," and Chapter 28, "Fiber-Optic Detectors and Receivers." Or it may be a passive device like one of those described in Chapter 29, "Passive Components and Multiplexers." The receptacle is very similar to one half of a mating sleeve. Figure 26.8 shows an ST mating sleeve and Figure 26.9 shows ST transceiver receptacles. Looking at these two photographs, you can see how closely they match. The transceiver receptacles look just like one half of the mating sleeve.

FIGURE 26.8
ST mating or alignment sleeve without connectors

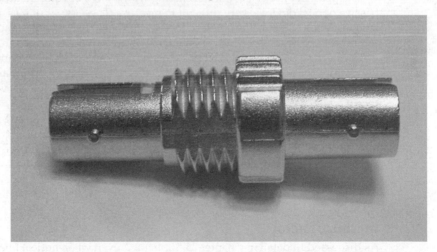

FIGURE 26.9
ST transceiver
receptacles

Connection Performance

The performance of a single connector cannot be measured. As described earlier in this chapter, the connector mates with either another connector or the equipment. The performance of two connectors mated together, also referred to as a mated connector pair, can be measured. This is described in detail in Chapter 33, "Test Equipment and Link/Cable Testing."

The performance of a connector mating with a piece of equipment such as a transceiver or passive optical device is very difficult to measure. These types of measurement require specialized test equipment that is not described in this book. Because of the specialized equipment required to perform these measurements, they are rarely performed in the field.

Connection performance describes the performance of an interconnection. In this section of the chapter, we are going to examine the factors that affect the performance of an interconnection. This interconnection may consist of two connectors and a mating sleeve or a connector and a receptacle.

Connection performance is dependent on several factors, including intrinsic factors, extrinsic factors, geometry, endface finish, and cleanliness. Intrinsic factors, extrinsic factors, and cleanliness apply to the connector, mating sleeve, and the receptacle. Endface finish and geometry only apply to the connector. Endface finish and the cleaning of the connector endface are discussed at the end of the chapter after connector assembly. Cleaning and inspection of the mating sleeve and receptacle is discussed in detail in Chapter 34, "Troubleshooting and Restoration."

Intrinsic Factors

As discussed in detail in Chapter 25, "Splicing," even when optical fibers are manufactured within specified tolerances, there are still slight variations from one optical fiber to another. The same is true for the mating sleeve and receptacle. Variations in these components can affect the performance of the connection.

Let's briefly review the most common types of intrinsic variations:

NA mismatch   A numerical aperture (NA) mismatch occurs when the NA of one optical fiber is different from the NA of the other optical fiber. A loss may occur if the NA of the transmitting optical fiber is larger than the NA of the receiving optical fiber.

Core diameter mismatch   Core diameter mismatch occurs when there is a difference in the core diameters of the two optical fibers. A loss may occur when the core diameter of the transmitting optical fiber is greater than the core diameter of the receiving optical fiber.

Mode field diameter mismatch   A mode field diameter mismatch occurs when there is a difference in the mode field diameters of two single-mode optical fibers. A loss may occur when the mode field diameter of the transmitting optical fiber is greater than the mode field diameter of the receiving optical fiber.

Cladding diameter mismatch   Cladding diameter mismatch occurs when the cladding diameters of the two optical fibers are not the same. A loss may occur when the cores of the optical fibers are not aligned because of the cladding diameter mismatch.

Ferrule bore diameter mismatch   Ferrule bore diameter mismatch occurs when the bore or hole diameter of the transmit ferrule is different from the bore diameter of the receive ferrule. A loss may occur if the bore diameter mismatch prevents the cores of the optical fibers from properly aligning.

Mating sleeve diameter mismatch   Mating sleeve diameter mismatch occurs when the diameters of the transmit side of the mating sleeve and the receive side are not the same. A loss may occur when the cores of the optical fibers are not aligned because of the mating sleeve diameter mismatch.

Optical fiber concentricity   The core and cladding of an optical fiber should be round and concentric, which means that they share a common geometric center. Concentricity variations can cause the optical fiber cores to misalign, creating a loss when the light exiting the core of the transmitting optical fiber enters the cladding of the receiving optical fiber.

Ferrule concentricity   Concentricity variations in the ferrule can also cause a loss. The ferrule and the bore or hole through the ferrule that accepts the optical fiber should have the same geometric center just like the core and cladding of the optical fiber. Variations in this geometric center can cause the optical fiber cores to misalign when two connectors are mated, creating a loss when light from the core of the transmit optical fiber enters the cladding of the receive optical fiber.

Mating sleeve concentricity   Concentricity variations in the transmit side and receive side of the mating sleeve can also cause a loss. The transmit and receive sides of the mating sleeve should have the same geometric center just like the core and cladding of the optical fiber. Variations in the geometric center of the two sides of the sleeve can cause the optical fiber cores to misalign when two connectors are mated, which may create a loss when light from the core of the transmit optical fiber enters the cladding of the receive optical fiber.

Optical fiber noncircularity   In an ideal optical fiber, the core and the cladding are perfectly circular. Noncircularity of the core or cladding can cause the cores of the transmit and receive fibers to misalign. This misalignment will cause a loss when light from the core of the transmitting optical fiber enters the cladding of the receiving optical fiber.

Ferrule noncircularity   Just as the core and cladding of an optical fiber may not be perfectly circular, the ferrule and the bore through the ferrule may also not be perfectly circular. The noncircularity of the ferrule or the bore through the ferrule may cause the core of the transmit and receive optical fibers to misalign. A loss will occur when light from the core of the transmitting optical fiber enters the cladding of the receiving optical fiber.

Mating sleeve noncircularity Noncircularity in either side of the mating sleeve may prevent the transmit and receive ferrules from properly aligning, which may cause the cores of the optical fibers to misalign. A loss will occur if the misalignment from the noncircularity causes light from the core of the transmitting optical fiber to enter the cladding of the receiving optical fiber.

Extrinsic Factors

Extrinsic factors that affect connection performance are factors related to the condition of the interconnection itself or factors that affect the interconnection. An ideal interconnection would have identical optical fibers, ferrules, and mating sleeve ends. These components would allow the cores of the transmitting and receiving optical fibers to align perfectly, resulting in a low-loss interconnection. However, there is no such thing as a perfect interconnection, only an imperfect interconnection. In a real interconnection, intrinsic and extrinsic factors affect connection performance.

Extrinsic factors and cleanliness go hand in hand. Dirt or contamination can cause any or all of the extrinsic factors described in this chapter. This section of the chapter briefly reviews the common extrinsic factors that affect connection performance. Cleaning is covered later in the chapter and extrinsic factors are covered in detail in Chapter 25.

Lateral misalignment Lateral misalignment occurs when the two optical fibers are offset laterally. A loss occurs when the lateral misalignment causes light from the core of the transmitting optical fiber to enter the cladding of the receiving optical fiber. As the lateral misalignment increases, the loss increases because less light from the core of the transmitting optical fiber makes its way into the core of the receiving optical fiber.

In an interconnection, lateral misalignment can be caused by intrinsic variations in the optical fiber, ferrule, or mating sleeve. It may also be the result of dirt, contamination, or wear.

End separation End separation is simply an unintended air gap between the endfaces of the transmitting and receiving optical fibers. The air gap will cause a loss and produce a Fresnel reflection. End separation may be caused by dirt, contamination, an improperly polished connector endface, or wear. However, it is more often the result of not fully inserting and securing the connector in the receptacle or mating sleeve.

Figure 26.10 shows two ST connectors in a mating sleeve. The connector on the right has been inserted into the mating sleeve; however, it has not been latched like the connector on the left of the mating sleeve. This happens often and will cause end separation.

FIGURE 26.10
ST connectors in a mating or alignment sleeve: the connector on the left is latched and the connector on the right is not latched.

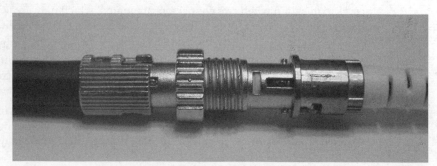

Angular misalignment   Angular misalignment occurs when the optical fibers in an interconnection meet each other at an angle. Loss from angular misalignment occurs when light from the core of transmit optical fiber enters the cladding of the receive optical fiber or enters the core of receive optical fiber at an angle exceeding the acceptance angle. Light entering the core of the receive optical fiber at an angle exceeding the acceptance angle does not propagate the length of the receive optical fiber.

As with end separation and lateral misalignment, angular misalignment may be caused by dirt, contamination, or wear.

Geometry

Geometry refers to the shape of the ferrule endface. It seems at first that the ideal shape would be a flat surface that would mate with the flat surface of another ferrule, but this geometry actually presents the most potential problems.

Flat endfaces like the one shown in Figure 26.11 will always have some polishing irregularities; they will never be perfectly flat across. When two precision flat metal pieces such as the head and cylinder block of a car engine are brought together, a gasket is required. The gasket fills the small voids between the head and engine block, compensating for the irregularities of the metal surfaces. Without the gasket, the car would not run properly.

The slightest variation in the flat endface surface will keep the fiber ends far enough apart to cause end separation loss due to Fresnel reflection. In addition, some of the reflected light may return down the core of the optical fiber to the transmitter as *return reflection*, which can interfere with the operation of the light source. Light interference is discussed in detail in Chapter 27.

To ensure *physical contact (PC)* between optical fiber ends, the best endface geometry is a convex curve. This curve, or *PC finish*, shown in Figure 26.11, ensures that the highest feature or *apex* on the endface is the center of the optical fiber end. When the optical fiber ends are in direct physical contact, the light behaves as if the connected ends are a continuous piece of optical fiber and passes through with very little loss and back reflection.

FIGURE 26.11
Endface geometry
configurations

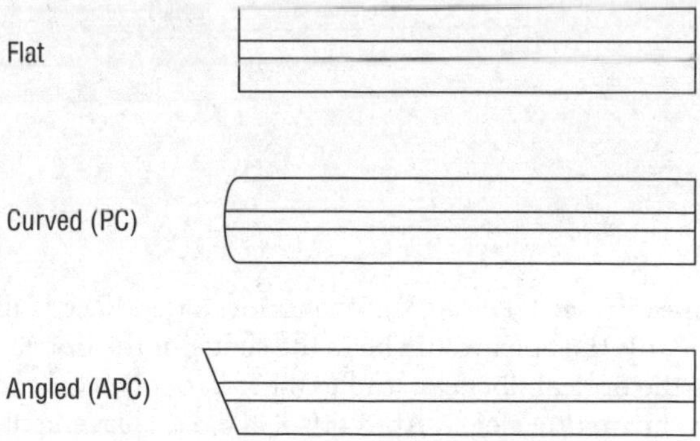

Flat

Curved (PC)

Angled (APC)

Another method of reducing return reflection is with an *angled PC (APC) finish*. This type of finish, as shown in Figure 26.11, puts an angle of about 8° on the endface, with the intent that it will mate with a similarly angled endface when properly aligned. With the endface and the fiber end angled, any light that is reflected is sent into the cladding and absorbed by the coating, rather than traveling back through the core of the optical fiber toward the light source.

Interferometer

An interferometer is a device that uses light to take precision measurements of the fiber-optic connector endface. The key measurements provided by the interferometer are *radius of curvature, apex offset,* and *fiber undercut or protrusion.* These critical parameters are required by Telcordia GR-326 to evaluate connector endface geometry for single-mode connectors and jumper assemblies. Telcordia GR-326 is an industry standard that defines generic requirements for single-mode connectors and jumpers.

When two connectors with a PC finish are brought together in a mating sleeve, springs are used to provide compressive forces that force the endfaces to touch under pressure. These compressive forces maintain glass on glass contact, producing a low-loss and low-reflection interconnection. The geometry of the connector endface is critical to ensure a high-performance interconnection.

Radius of curvature   The radius of curvature describes the roundness of the connector endface, and it is measured from the center axis of the connector ferrule, as shown in Figure 26.12. The minimum and maximum values for the radius of curvature as defined in Telcordia GR-326 are 7 and 22mm. Values below or above this range increase the risk of optical fiber damage. They also increase the possibility of reflection and insertion loss.

FIGURE 26.12
Connector radius
of curvature

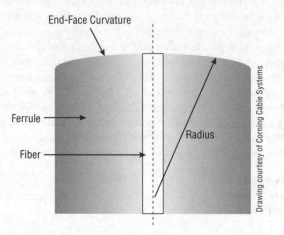

Apex offset   The apex of the connector endface is the highest point of the rounded endface. Ideally, the apex would be in the center of the optical fiber. Not having the apex in the center of the optical fiber can lead to the lack of physical contact between the two optical fiber cores in the mating sleeve. Apex offset is the displacement between the apex of the endface and the center of the optical fiber core, as shown in Figure 26.13. Telcordia GR-326 states that apex offset cannot exceed 50µm.

FIGURE 26.13
Connector
apex offset

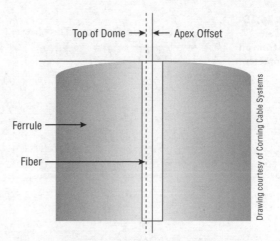

Fiber undercut or protrusion Optical fiber undercut or protrusion describes the distance the optical fiber endface is below or above the rounded connector endface, as shown in Figure 26.14. When the optical fiber protrudes too far from the ferrule endface, the compressive forces on the optical fiber increase and so does the chance that the optical fiber will be damaged. When the optical fiber sits too far below the ferrule endface, an air gap will result, increasing the loss and reflection. Telcordia GR-326 states that an undercut value cannot exceed ±55nm.

FIGURE 26.14
Connector
undercut

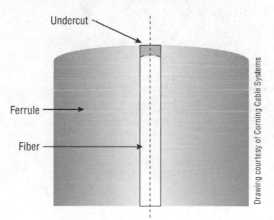

A connector endface is like a snowflake—no two are exactly alike. It is for this reason that Telcordia GR-326 provides ranges for critical geometric values. An interferometer will typically provide a surface contour display, surface-fitting display, 3D-mesh display, and a live image to evaluate the connector endface, as shown in Figure 26.15. It will also provide numerical data for the key geometric values, as shown at the bottom of Figure 26.15.

Courtesy of KITCO Fiber Optics

FIGURE 26.15
Views of an endface from an interferometer

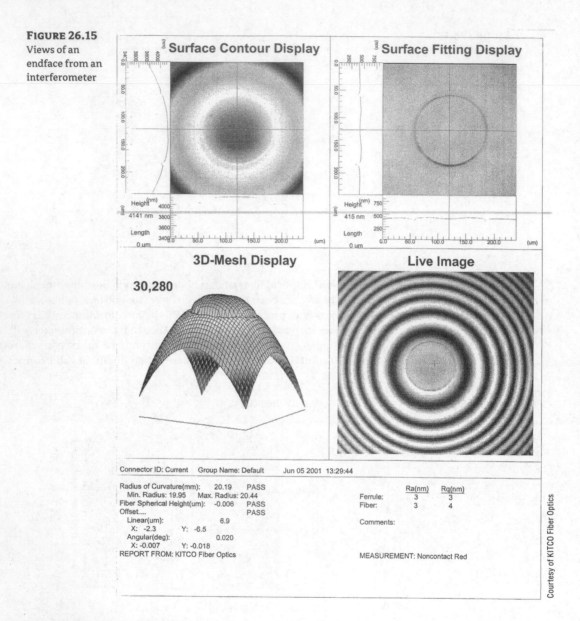

Connector Types

The purpose of all fiber-optic connectors is the same—to align a terminated optical fiber end with another terminated optical fiber end or a piece of equipment and hold it firmly in place. The manner in which this has been accomplished, however, has continued to evolve as technology improves and as requirements change. As a result, manufacturers have created many different types of connectors. Most of the connectors are still in use, though some have become more common while others exist only in older or legacy systems.

Connectors can be divided into two main groups: single-fiber connectors and multifiber connectors. Single-fiber connectors are designed for only one optical fiber, but they may be used to break out multiple optical fibers from a cable for individual termination in patch panels or hardware. Multifiber connectors may be used to connect paired fibers in duplex systems, or to connect entire ribbons of 12 optical fibers or more with one connector.

Because there are so many different types of connectors, the Telecommunications Industry Association (TIA) has not attempted to recognize or create standards for all of the available types. In the ANSI/TIA-568-C standard, however, they recognized performance standards for connectors, which will be described at the end of this chapter.

Let's look at the different types of connectors.

Single-Fiber Connectors

Single-fiber connectors have a wide variety of connection methods. Some, including the earliest types, are engaged by pushing and twisting or by using a threaded sleeve to draw the connector tight. Newer forms, however, are square or rectangular snap-in connectors. These are engaged by a simple push that engages a locking mechanism.

SC CONNECTORS

The SC (subscriber connector), shown in Figure 26.16, is among the most widely used connectors. Originally developed by Nippon Telephone and Telegraph (NTT), the SC has a standard-sized 2.5mm ferrule and a snap-in connector that was created as an alternative to connectors that required turning or twisting to keep them in place. In addition, SC connectors can be installed in only one orientation, making them suitable for an APC endface.

FIGURE 26.16
The SC connector is one of the most common types in use.

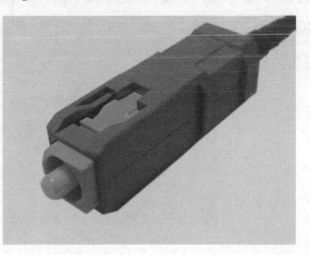

ST CONNECTORS

The ST (straight tip) connector, shown in Figure 26.17, was developed by AT&T as a variation on a design used with copper coaxial cables. This connector has a metal connector cap that must be twisted to lock into place. The ST is considered a legacy connector, as it has been around for quite some time and can still be found in many installations today.

FIGURE 26.17
An ST connector must have its connector cap twisted to lock in place.

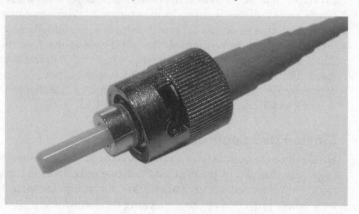

FC CONNECTORS

An FC (face contact) connector is a rugged metal connector with a screw-on connector cap and a 2.5mm ferrule. Like the SC connector, the FC, shown in Figure 26.18, is used in connections where proper polarization must be maintained. Because the connector is cylindrical, it must be aligned with a built-in key. Note, however, that there are several different standards for the size of the key, meaning that the connector must be properly matched with its adapter.

FIGURE 26.18
FC connector with screw-on connector cap

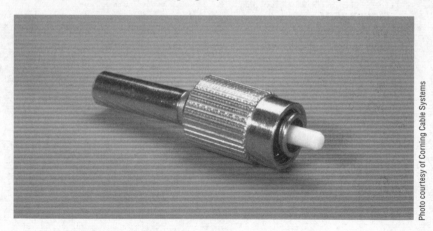

Photo courtesy of Corning Cable Systems

LC CONNECTORS

The LC connector, shown in Figure 26.19, is a *small form factor* connector. Developed by Lucent Technologies, this snap-in connector is considered to be a smaller version of the SC connector and is sometimes referred to as a mini-SC. The small form factor connector has a 1.25mm ferrule, half the size of an SC connector ferrule.

FIGURE 26.19
The LC connector is a small form factor version of the SC connector.

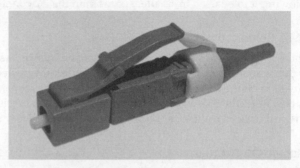

The smaller form factor allows two LC connectors to fit into roughly the same size space as a single SC connector. Figure 26.20 shows two LC connectors on one side of an SC mating sleeve and an SC connector plugged into the other side of the mating sleeve. The two LC connectors require no more room than a single SC.

FIGURE 26.20
SC mating sleeve with an SC connector on the left side and two LC connectors on the right side.

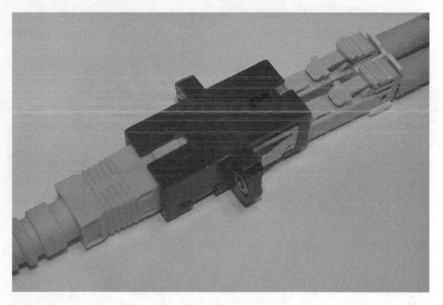

D4 CONNECTORS

The D4 connector, also known as a DIN connector, is an older style heavy-duty metal connector with a 2.5mm ferrule and a threaded connector cap that must be screwed on to secure the connector. It was developed by Siemens and is mostly used by Deutsche Telecom, the parent company of T-Mobile Wireless. It is similar in function to the FC connector, but its profile is slightly smaller, allowing it to be used in smaller spaces than the FC connector.

SMA CONNECTORS

The SMA connector was developed by Amphenol from its line of microwave connectors known as SubMiniature A. The connector has a 3mm stainless steel ferrule and a connector cap that is threaded on the inside. Because it was originally developed before the invention of single-mode optical fiber, the SMA connector does not provide as precise a connection as more recent designs. It is still used in military applications and in the delivery of high-power laser light.

BICONIC CONNECTORS

The biconic connector features a stainless steel or ceramic cone-shaped ferrule that helps in aligning fiber ends during connection. Biconic connectors are threaded on the outside for screw-in placement, and although they are considered legacy connectors, they are still used in military applications because of their robustness. The biconic connector is also incorporated into the tactical fiber-optic cable assembly (TFOCA). The biconic plug and receptacle used in the TFOCA are shown in Figure 26.21.

FIGURE 26.21
The biconic connector has a cone-shaped ferrule.

Plug

Receptacle

Photo courtesy of Amphenol Fiber Systems

MINI-BNC CONNECTORS

The mini-BNC, shown in Figure 26.22, is similar in appearance to its counterpart made for copper coaxial cables. The mini-BNC (short for either Bayonet Nut Connector or Bayonet Neill-Concelman, the inventor of the original BNC) is a metal twist-lock connector with a 2.5mm ferrule. The mini-BNC is considered a legacy connector, so while it is rarely installed in new systems, you may find it on older installations.

FIGURE 26.22
The mini BNC connector resembles the BNC connector used in RF applications.

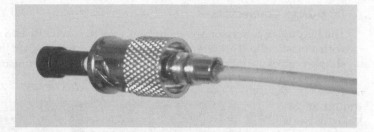

Multiple-Fiber Connectors

Multiple-fiber connectors are mostly designed to support duplex operations and ribbon fibers, allowing a greater number of connections in a smaller space. Some duplex connectors resemble larger or ganged versions of single-fiber connectors.

FDDI CONNECTORS

The FDDI connector was developed by the American National Standards Institute (ANSI) for use in Fiber-Distributed Data Interface (FDDI) networks. This plastic duplex connector is also called FDDI MIC for *medium interface connector*, a connector that is used to link electronics and fiber transmission systems. It features protective shields around the two 2.5mm ferrules to prevent damage. The connector locks into place with latches on the sides and can be keyed to ensure proper orientation of the two fiber ends.

ESCON CONNECTORS

The ESCON connector was developed for IBM's ESCON (Enterprise Systems Connection) architecture in the early 1990s. It is similar to the FDDI, but it has a shroud that retracts from the ferrules when the connector is engaged.

SC DUPLEX CONNECTORS

The SC duplex connector, shown in Figure 26.23, is actually two single SC connectors joined with a plastic clip. This arrangement can be extended into even more plugs, if necessary, but is most commonly applied in the duplex configuration.

FIGURE 26.23
The SC duplex is made of two single SC connectors joined together with a clip.

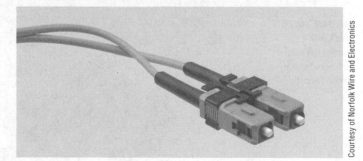

Courtesy of Norfolk Wire and Electronics

LC DUPLEX CONNECTORS

The LC duplex connector, shown in Figure 26.24, is actually two single LC connectors joined with a plastic clip. This arrangement is typically applied when the LC connectors are attached to simplex cables as opposed to duplex cordage. It keeps the transmit and receive optical fibers for a single channel together.

FIGURE 26.24
The LC duplex is made of two single LC connectors joined together with a clip.

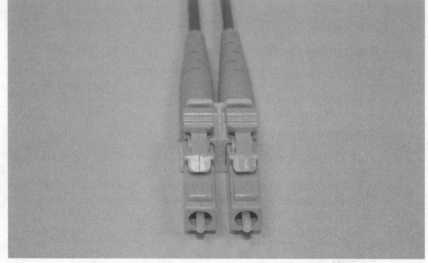

MPO CONNECTORS

The MPO (Multifiber Push On) connector, shown in Figure 26.25, is built on the MT-style ferrule, designed by NTT. The MT (mechanical transfer) ferrule is designed to hold up to 12 fibers in a ferrule 7mm wide and is ideally suited for ribbon fiber connections. In addition, precision-machined guide pins maintain the close alignment necessary for connecting 12 fibers at once. These guide pins can be arranged as necessary between the mating connectors depending on the way they will be used. Connectors designed for multiple fibers are also known as *array connectors*.

The MPO connector has a plastic body that is spring-loaded to keep the connectors together.

FIGURE 26.25
The MPO connector packs 12 fibers into a single ferrule.

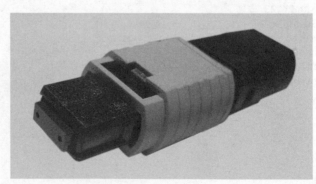

MTP CONNECTORS

The MTP connector shown in Figure 26.26 is designed by US Conec and built around the MT ferrule. It is designed as a high-performance version of the MPO and will interconnect with MPO connectors.

FIGURE 26.26
This MTP connector is a high-performance version of the MPO.

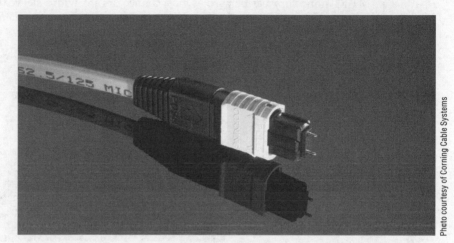

Photo courtesy of Corning Cable Systems

MT-RJ CONNECTORS

The MT-RJ connector, shown in Figure 26.27, was developed to emulate the functionality of the RJ-45 modular connector, which is used in most office networking. It provides a snap-in, duplex connection in a housing that resembles the familiar modular network plug. In fact, the connector is designed to fit in the same physical opening as the RJ-45, so many networking fixtures and wall plates can be retained after the hardware inside them has been upgraded for fiber.

FIGURE 26.27
The MT-RJ connector is designed to emulate the RJ-45 modular plug in fit and ease of use.

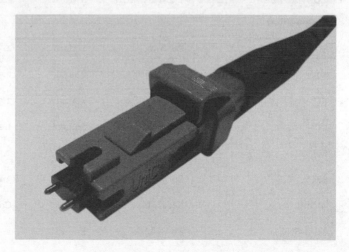

The MT-RJ's single ferrule holds two fibers in a housing smaller than most single-fiber connectors, so it is attractive in many applications where size matters. The connector is built like a mechanical splice, so the fibers are inserted into the ferrule and the plug is closed to hold the fibers in place without the need for epoxy.

MT-RJ connectors may be used as most connectors are, with a sleeve that joins two connectors together, as shown in Figure 26.28, or in a true plug-and-jack arrangement with bare fibers or a transceiver installed in the jack.

FIGURE 26.28
Two MT-RJ connectors and a mating sleeve

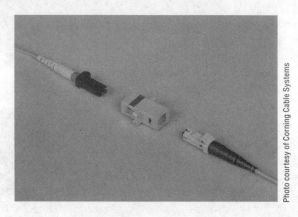

Photo courtesy of Corning Cable Systems

There are many other connector types in use, and more are being added as the technology to align and secure them improves. Table 26.1 gives you a quick reference to the connectors we've discussed and their key features.

TABLE 26.1   Quick Reference for Fiber-Optic Connectors

CONNECTOR	NO. OF FIBERS	CONNECTION	SHAPE	FERRULE SIZE
Single-Fiber				
SC	1	Push-in	Square	2.5mm
ST	1	Twist-lock	Round	2.5mm
FC	1	Screw-on	Round	2.5mm
LC	1	Push-in	Square	1.25mm
D4 (DIN)	1	Screw-on	Round	2.5mm
SMA	1	Screw-on	Round	3mm
Biconic	1	Screw-on	Round	Conical
Mini-BNC	1	Twist-lock	Round	2.5mm

TABLE 26.1   Quick Reference for Fiber-Optic Connectors   *(CONTINUED)*

Connector	No. of Fibers	Connection	Shape	Ferrule Size
Multifiber				
FDDI	2	Push-in	Rectangular	(2) 2.5mm
ESCON	2	Push-in	Rectangular	(2) 2.5mm
SC Duplex	2	Push-in	Rectangular	(2) 2.5mm
MPO	12	Push-in	Rectangular	7mm
US Conec MTP	12	Push-in	Rectangular	7mm
MT-RJ	2	Push-in	Rectangular	5mm

PIGTAILS

Pigtails are fiber ends with connectors factory-attached for future splicing into a system, as shown in Figure 26.29. Typically, a pigtail starts as a manufactured patch cord or jumper with a connector at each end. You can then cut the jumper in half and have two pigtails ready to splice.

FIGURE 26.29
Use a pigtail to
splice a connector
to an existing fiber.

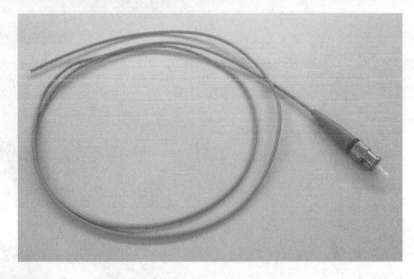

Pigtails are available with a variety of connectors, depending on your needs. These products have advantages over field terminations because the connectors are factory-installed and polished to exacting standards. This can save time over attaching a connector to the end of an optical fiber, and it typically produces a better connection. They are especially useful in situations where many connectors have to be added to cables in a relatively short time, or in a location where it is easier to make a splice than it is to add a connector.

On the downside, pigtails require hardware to protect the splice and investment in a fusion or mechanical splicer, in addition to the cost of the connector and hardware itself.

SPECIALIZED CONNECTORS

While standard commercial connectors are used in most networks and communications links, fiber-optic connectors are also available in specialized configurations that feature durable cases, heavy-duty connector bodies, and multiple-fiber connections, such as the TFOCA Gen X connector shown in Figure 26.30. This connector is often found in military applications where heavy use and destructive environments are common.

FIGURE 26.30
The TFOCA Gen X connector is a specialized connector for military applications.

Photo courtesy of STRAN Technologies

The connectors shown in Figure 26.31 are typically used in aerospace applications. This is a military qualified MIL-DTL-38999 four-optical-fiber connector cabled for an aerospace application.

FIGURE 26.31
Two MIL-DTL-38999 connectors cabled for an aerospace application

Photo courtesy of Carlisle Interconnect Technologies

Connector Termination

One of the most exacting jobs you will encounter in working with fiber optics is connector termination. If you do the job properly, an optical fiber the size of a human hair will mate with another optical fiber or a piece of hardware and transfer a large percentage of the light passing through it. If you do the job wrong, the light could stop at the connector and go no further, or so little of the light could transfer that it would essentially be useless. With those encouraging words urging us forward, let's look at three ways in which an optical fiber can be terminated with a connector.

NOTE There are many mechanical precision tools used for processing connections, but the actual process is largely an art that must be practiced. Do not make the mistake of making your first connection in a real installation; practice before every installation.

Tools

Regardless of the method that you choose to terminate the optical fiber, tools will be required. In this section, we discuss three different termination methods. Each of these methods requires some of the tools discussed in this section.

Shears Even though cabling materials are fairly easy to cut, the aramid or Kevlar strength member is strong, fibrous, and loose fitting, making it difficult to cut. You will need a good pair of sharp shears like the pair shown in Figure 26.32 to cut through the strength member quickly and cleanly.

FIGURE 26.32
Shears designed to cut the Kevlar strength member

Photo courtesy of KITCO Fiber Optics

Stripper A stripper is used to remove the outer jacket, tight buffer, and coating so that only the optical fiber itself is exposed. Some strippers are designed to only remove the tight buffer and coating, like the one shown in Figure 26.33. Other strippers may be designed to remove the jacket, tight buffer, and coating, like the one shown in Figure 26.34. Some are adjustable, like the one shown in Figure 26.35, and can be used to remove the jacket, tight buffer, and coating. However, they need to be adjusted for each application.

FIGURE 26.33
No-Nik stripper designed to remove only the tight buffer and coating

FIGURE 26.34
Stripper designed to remove the jacket, tight buffer, and coating

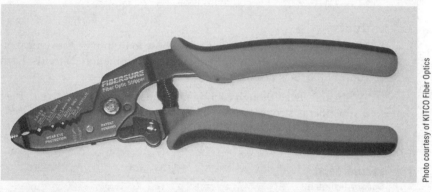

FIGURE 26.35
Adjustable stripper that can be set to remove the jacket, tight buffer, and coating

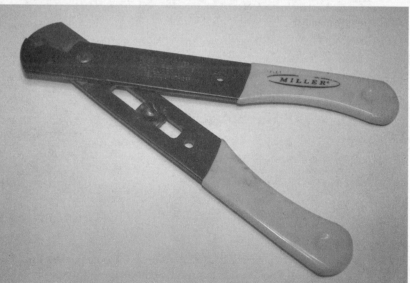

Photo courtesy of KITCO Fiber Optics

Scribe   The scribe, shown in Figure 26.36, is used for precision work in removing the fiber end once it has been adhered inside the ferrule. To use it, you only need to score or nick one side of the optical fiber lightly. This breaks the surface of the cladding, which provides the fiber's tensile strength. You'll need some practice with a scribe to get the right results, since it requires just the right touch to keep from damaging or breaking the optical fiber.

FIGURE 26.36
Scribe used to score or nick the optical fiber

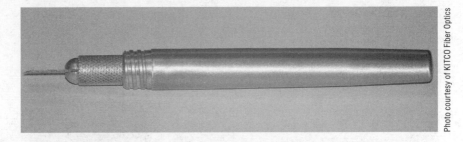

Photo courtesy of KITCO Fiber Optics

Polishing fixture (puck)   The polishing fixture, or puck as it is typically referred to, is shown in Figure 26.37. The puck is used to ensure that the ferrule stays perpendicular to the polishing film during the polishing process. (See Figure 26.38.) Even the slightest variation in the polishing angle can affect the endface finish of the connector, which, in turn, could impact the performance of the connector.

FIGURE 26.37
Polishing fixture (puck) for a 2.5mm ferrule

Photo courtesy of KITCO Fiber Optics

FIGURE 26.38
Polishing fixture (puck) holding an LC connector perpendicular to the polishing film

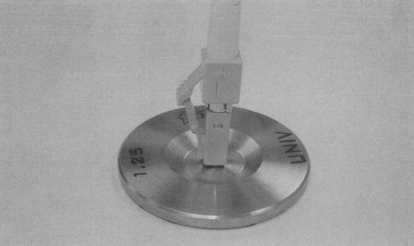

Lint-free wipes Lint-free wipes like those shown in Figure 26.39 are engineered to lift away oils, grime, and dust from the surface of the optical fiber. When handling a lint-free wipe, only touch one side of the wipe because the wipe will absorb the oil from your skin. Use the side of the wipe that you did not touch for cleaning the optical fiber. You do not want to contaminate the optical fiber during the cleaning process.

Photo courtesy of MicroCare

FIGURE 26.39
Lint-free wipes engineered for cleaning optical fibers

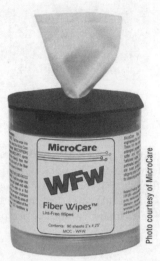

Cleaning fluid Cleaning fluids like those shown in Figure 26.40 are approved for air travel and engineered specifically to clean the optical fibers without leaving a residue. These cleaning fluids typically contain little or no isopropyl alcohol and have been filtered to remove microscopic contaminates.

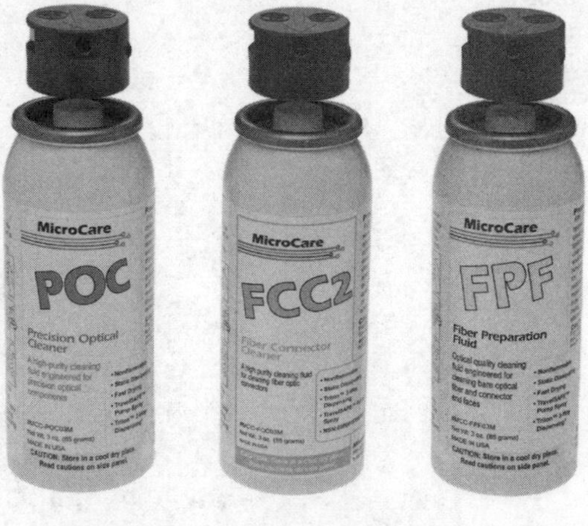

Photo courtesy of MicroCare

FIGURE 26. 40
Optical fiber cleaning fluids approved for air travel

Alcohol   Isopropyl alcohol is used to remove the residual coating from the optical fiber prior to termination. While it is widely used, it does leave a residue and you may want to consider a cleaning fluid that does not leave a residue. If you choose isopropyl alcohol, make sure that it is virtually free of water and has been filtered to remove impurities, like the isopropyl alcohol shown in Figure 26.41. Do not use the isopropyl alcohol commonly available at your local pharmacy; this alcohol contains a large percentage of water and will leave a residue on the optical fiber.

FIGURE 26.41
Isopropyl alcohol
for cleaning optical
fibers

Cleaver   A cleaver is only required for pre-polished connectors. Cleavers like the one shown in Figure 26.42 score and break the optical fiber, leaving a perpendicular finish. Cleaving is typically a two-step process and it takes place after the optical fiber has been properly prepared and cleaned. Cleaving is described in detail in Chapter 25.

FIGURE 26.42
Single optical fiber
cleaver

Nonreflective mat   Always work over a nonreflective black surface like the one shown in Figure 26.43. The nonreflective black mat makes it easier to keep track of cut optical fiber ends.

FIGURE 26.43
A nonreflective black mat makes it easier to see the bare optical fiber.

Optical fiber disposal container   Always use a labeled container with a lid like the one shown in Figure 26.44 to dispose of bare optical fibers.

FIGURE 26.44
Container designed and labeled for optical fiber waste

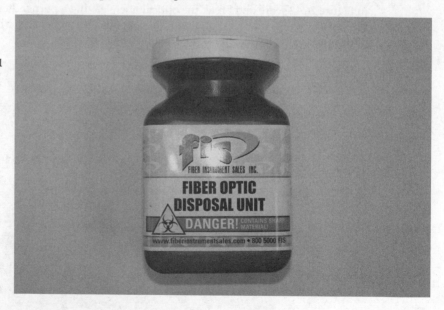

Safety glasses To prevent eye injury, always wear proper eye protection like the safety glasses shown in Figure 26.45.

FIGURE 26.45
Safety glasses designed to prevent eye injuries.

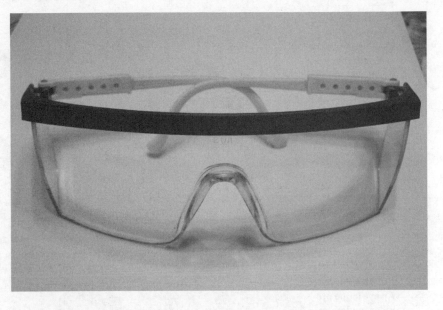

Tweezers To prevent injury to your hands from the optical fiber, you may want to handle pieces of bare optical fiber with tweezers, like the pair shown in Figure 26.46.

FIGURE 26.46
Tweezers can be used to pick up the pieces of optical fiber.

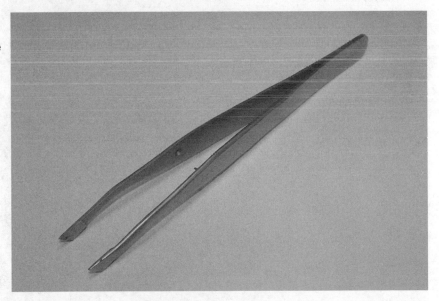

Crimper The crimper like the one shown in Figure 26.47 is typically used to secure the strength member and jacket to the connector.

FIGURE 26.47
Crimpers are used to secure the strength member and jacket to the connector.

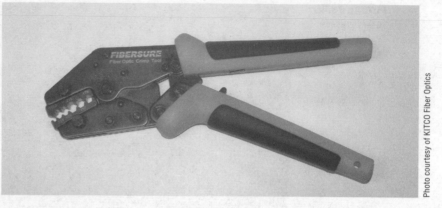

Photo courtesy of KITCO Fiber Optics

Eye loupe An eye loupe like the one shown in Figure 26.48 provides moderate magnification and is used to view the end of the connector ferrule after the scribing process.

FIGURE 26.48
An eye loupe provides moderate magnification for viewing the end-face of the connector ferrule.

Photo courtesy of KITCO Fiber Optics

Rubber pad The rubber pad like the one shown in Figure 26.49 is placed under the polishing film. It provides a cushioned surface for PC connectors.

FIGURE 26.49
The rubber pad provides a cushioned surface for the polishing film.

Photo courtesy of KITCO Fiber Optics

Cure adapter The cure adapter like the one shown in Figure 26.50 is placed over the ferrule of the connector. It helps transfer heat from the curing oven and protects the optical fiber during the curing process.

FIGURE 26.50
The cure adapter slides over the connector ferrule and helps transfer heat from the curing oven to the connector.

Photo courtesy of KITCO Fiber Optics

Curing oven The curing oven is only required for epoxies that need to be heated to a specific temperature to cure. The curing oven like the one shown in Figure 26.51 accepts the cure adapter.

FIGURE 26.51
Curing oven used to heat the connector and cure the epoxy

Photo courtesy of W.R. Systems

Epoxy

This section of the chapter discusses oven-cured and anaerobic epoxies. There are advantages and disadvantages to each type. The epoxy should be chosen based on the application. This section provides a general overview on epoxy. You should always consult the epoxy data sheet for specific information. If you have specific questions about the type of epoxy to use for your application, consult the manufacturer of the connector or epoxy.

OVEN-CURED EPOXY

Oven-cured epoxy is probably the best type you can use, but it is also the most cumbersome. The epoxy itself consists of a resin and a hardener, which must be mixed in the right proportions. Once it is mixed, it has a limited *pot life* before it begins to harden, and any epoxy that is unused must be discarded. On the other hand, once a batch is mixed, several connectors can be assembled at the same time, so it is useful in making a large number of terminations at one location.

Once the connector has been assembled, you have to insert it into a specially built oven to cure the epoxy and then let the assembly cool before you can begin polishing it. The oven will hold a number of connectors, so as each one is built it can be inserted for curing.

While oven-cured epoxy is time-consuming and equipment-intensive, it produces a hard, fully cured bead around the base of the optical fiber, as shown in Figure 26.52. This bead reduces the risk that the fiber will break inside the ferrule during cleaving, which could ruin the connector.

FIGURE 26.52
The bead of oven-cured epoxy protects the fiber during cleaving.

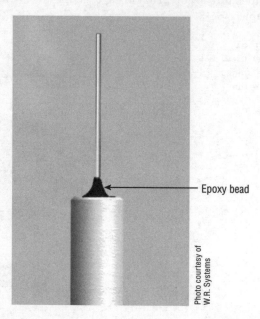

Epoxy bead

Photo courtesy of
W.R. Systems

ANAEROBIC EPOXY

Anaerobic epoxy is typically used where there is no power available or when time is at a premium. While the adhesive still comes in two parts, each part is applied separately to the fiber and the ferrule. When the two are joined, the epoxy hardens and cures in about 10 seconds.

While this may seem to be an ideal solution for all situations, there are some drawbacks to anaerobic epoxy. For one or two connectors, there is a significant time-savings. However, because of its quick hardening time, anaerobic epoxy does not lend itself to be used in batches as oven-cured epoxy does. You typically prepare and work on only one connector at a time. In addition, this epoxy does not form a hard bead the way oven-cured epoxy does, so greater skill is required for cleaving and polishing the optical fiber. Finally, this epoxy is not approved for air travel.

Abrasives

Abrasives are used to polish the optical fiber end. Depending on the application, different types of abrasives are needed. Essentially, they are like very fine sandpaper, but the abrasive material is adhered or fixed to a Mylar film backing instead of paper. Two popular abrasive materials are aluminum oxide and diamond.

ALUMINUM OXIDE

Aluminum oxide is an abrasive that is harder than the optical fiber but softer than the ceramic ferrule. Because of this, it will polish down the optical fiber but not the ferrule. While this abrasive prevents the person polishing the connector from altering the endface geometry of the ferrule, it does not produce a very attractive finish when compared to diamond abrasive, and it chips the optical fiber at the epoxy ring. (The epoxy ring is the area between the optical fiber and the ferrule where the epoxy rests.) This abrasive will produce an endface similar to the one shown in Figure 26.53.

FIGURE 26.53
Multimode end-face polished with aluminum oxide polishing film.

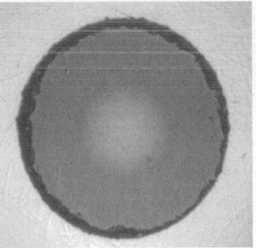

Photo courtesy of W.R. Systems

While the cosmetic finish of an endface polished with aluminum oxide may not look as attractive as an endface polished with diamond, like the one shown in Figure 26.54, aluminum oxide polishing films are significantly less expensive than diamond. The cosmetic finish does not always tell the whole story when it comes to connector performance. The endface shown in Figure 26.53 will offer insertion loss performance that exceeds the ANSI/TIA-568-C.3 requirements for multimode interconnections.

FIGURE 26.54
Single-mode
endface polished
with diamond
polishing film

Photo courtesy of MicroCare

DIAMOND

Diamond is an abrasive that is harder than both the optical fiber and the ceramic ferrule. Because of this, it will polish down the optical fiber and the ceramic ferrule. This abrasive does not prevent the person polishing the connector from altering the ferrule's endface geometry. However, it does produce a very attractive finish and it does not chip the optical fiber at the epoxy ring. This abrasive will produce an endface like the one shown in Figure 26.54.

While the cosmetic finish of an endface polished with diamond may be very attractive, its geometry may also be altered. When polishing with a diamond abrasive, you have to be very careful not to overpolish. As discussed earlier in this chapter, an interferometer can be used to evaluate the geometry of the connector endface. However, an interferometer is expensive and not always available.

You can use the inspection microscope to determine if you are polishing the ferrule endface and possibly changing its geometry. When you look at Figure 26.53, which shows a connector endface polished with aluminum oxide, you can see some texture markings on the ferrule. The same type of texture markings exist on the unpolished connector ferrule shown in Figure 26.55. This clearly shows that the aluminum oxide abrasive did not polish down the ferrule endface. However, when you look at the endface polished with a diamond abrasive shown in Figure 26.54, you see there are no texture markings on the ferrule because they have been removed with the diamond abrasive.

The risk associated with polishing the ferrule is that you may alter the shape of the ferrule endface. When you buy a connector, the endface meets the geometric specifications you selected. Polishing the connector with a diamond abrasive may alter those specifications. It is easier to control the geometry with a polishing machine than with hand polishing.

FIGURE 26.55
An unpolished
connector endface
without an optical
fiber.

Photo courtesy of W.R. Systems

Hand Polishing

While some of the procedures involved may vary, the main task is the same: you must insert a bare optical fiber into a hole in the ferrule so that it sticks out slightly beyond the ferrule endface and secure it there; then you must *score*, or *cleave*, the excess optical fiber and polish the ferrule endface and the optical fiber together to achieve the proper profile and finish.

While this task sounds simple, it requires some careful planning and execution. Remember that the hole in the ferrule is almost microscopic, since it must be only slightly larger than the optical fiber itself.

There are many different fiber-optic connectors on the market from a variety of manufacturers. Each manufacturer has its own polishing process. This section discusses two basic connector assembly and polishing processes using oven-cured epoxy and anaerobic epoxy. The connectors will be terminating simplex cordage.

Assembling the Connector

Assemble your tools before you start so you don't have to look for them in the middle of the process. It may seem like a fussy detail, but your work will proceed more smoothly and efficiently if you have prepared all of your tools and materials and laid them out where you can reach them easily when you need them.

CABLE AND OPTICAL FIBER PREPARATION

The following steps describe how to prepare simplex cordage for connectorization.

1. Cut the cable 2″ longer than you'll need.

2. Install the strain relief boot on the cable end, small diameter first. Important: The strain relief boot will not fit over the connector, so it must be placed on the cable first.

3. Install the crimp sleeve if one was provided with the connector.

4. Locate the appropriate strip chart from the connector manufacturer.

5. Mark the cable jacket as shown on the strip chart, cut the jacket with the stripper, and then twist the jacket and pull it away from the cable to remove it.

6. Pull the strength member back; it will be cut to length in the last step.

7. Strip away the tight-buffer and coating to the length shown on the strip chart in ¼″ increments to reduce the chance of breaking the optical fiber.

8. Clean the fiber with a lint-free wipe soaked in optical fiber cleaning fluid; this will remove the coating residue so that the epoxy in the ferrule can bond with the cladding surface.

9. Using the shears, cut the strength member to the length shown on the strip chart, and then distribute it evenly around the tight buffer.

OVEN-CURED EPOXY APPLICATION, CONNECTOR ASSEMBLY, AND POLISHING

The following procedures apply to oven-cured epoxies. Remember these are general procedures and you should always refer to the epoxy and connector manufacturer data sheets for detailed information.

Oven-Cured Epoxy Application and Connector Assembly

This procedure describes how to prepare the epoxy, inject it into the connector, and assemble the connector for polishing:

1. Turn on the curing oven and set the temperature as recommended by the epoxy manufacturer.

2. Mix the epoxy according to the manufacturer's instructions.

3. Secure the needle to the syringe. Load the syringe with epoxy and ensure that the syringe is free of air pockets.

4. Insert the needle tip into the back of the connector until it bottoms out against the ferrule. Maintain pressure and slowly inject the epoxy until a bead appears on the end of the ferrule tip. Continue to inject epoxy until the bead covers about a third of the ferrule diameter end.

5. Release the pressure on the plunger, wait 5 seconds, and then remove the needle.

6. If required, place the cure adapter over the connector ferrule.

7. Gently feed the optical fiber through the connector, as shown in Figure 26.56, making sure the strength member fans back as you push the optical fiber forward.

8. Crimp the connector using the crimper as described in the manufacturer's data sheet.

FIGURE 26.56
Feed the fiber through the connector.

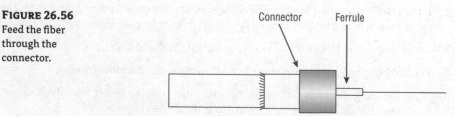

Connector   Ferrule

9. Place the connector in the curing oven and cure the epoxy according to the cure manufacturers cure schedule.

10. After the epoxy has cured, carefully remove the connector and allow it to cool. Then proceed to polishing the connector, described in the next section.

Oven-Cured Epoxy Connector Polishing

This procedure describes how to polish the endface of the connector after the epoxy has cured and the connector has cooled. This polishing process is only for PC finishes. The same process can be applied for a flat finish by substituting a glass plate for the rubber pad.

The following procedure is a general polishing procedure. You should always refer to the epoxy and connector manufacturer data sheets for detailed information.

1. Score the fiber with the scribe where it exits the epoxy bead, and then gently pull the optical fiber, lifting it from the connector endface. A small optical fiber nub should be protruding from the connector endface. You can use the eye loupe to view the connector endface.

2. Polish down or de-burr the fiber end by holding the connector so that it is facing up. Arch a piece of 5μm polishing film over it. Lightly rub the film in a circle over the connector until you no longer feel the fiber end "grab" the film. Then stop.

3. Place a clean piece of 5μm aluminum oxide polishing film over the rubber pad.

4. Insert the connector into the puck and place the puck on the polishing film.

5. Slowly move the puck in a figure-8 pattern over the polishing film to begin removing the epoxy bead. After about five or ten figure-8 strokes, you should start to feel less resistance as the epoxy bead polishes down. Stop polishing before you have removed all of the epoxy.

6. Remove the 5μm polishing film and place a 1μm polishing film on the rubber pad. Slowly move the puck in a figure-8 pattern over the polishing film to begin removing what is left of the epoxy bead. After about five or ten figure-8 strokes, the polishing puck should glide smoothly over the polishing film. Stop polishing and go to the next step for single-mode applications.

7. For single-mode applications, remove the 1μm polishing film and place a 0.3 or 0.1μm diamond polishing film on the rubber pad. Slowly move the puck in a figure-8 pattern over the polishing film and stop after ten figure-8 strokes. Polishing is complete.

NOTE An aluminum oxide abrasive works well for most non-laser-optimized multimode applications; however, a diamond abrasive is recommended for single-mode and laser-optimized multimode applications. Do not overpolish when using a diamond abrasive.

ANAEROBIC EPOXY APPLICATION, CONNECTOR ASSEMBLY, AND POLISHING

The following procedure applies to anaerobic epoxies. Remember this is a general procedure and you should always refer to the epoxy and connector manufacturer data sheets for detailed information.

Anaerobic Epoxy Application and Connector Assembly

This procedure describes how to inject the adhesive into the connector, apply the primer to the optical fiber, and assemble the connector:

1. Place a needle on top of the anaerobic adhesive and insert the needle tip into the back of the connector until it bottoms out against the ferrule. Maintain pressure and slowly inject the epoxy until a bead appears on the end of the ferrule tip. Continue to inject epoxy until the bead covers about a third of the ferrule diameter end.

2. Open the primer. Using the small brush on the cap of the primer, coat the bottom two-thirds of the exposed optical fiber with the primer. Ensure that you place the lid back on the primer immediately after use; the primer evaporates quickly and will melt the nonreflective mat if spilled on it.

3. Gently feed the optical fiber through the connector, as shown in Figure 26.57, making sure the strength member fans back as you push the optical fiber forward. Wait 10 seconds for the epoxy to cure before going to the next step.

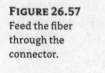

FIGURE 26.57
Feed the fiber through the connector.

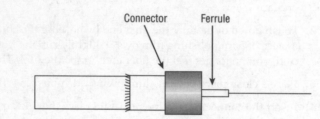

Connector   Ferrule

WARNING   Do not pull back on the optical fiber; this may break the optical fiber.

4. Crimp the connector as described in the manufacturer's data sheet, and then proceed to the next section.

Anaerobic Epoxy Connector Polishing

This procedure describes how to polish the endface of the connector after the anaerobic epoxy has cured. This polishing process is only for PC finishes. The same process can be applied for a flat finish by substituting a glass plate for the rubber pad.

The following procedure is a general polishing procedure. You should always refer to the connector manufacturer data sheet for detailed information.

NOTE   The anaerobic epoxy bead is not hard like the oven-cured epoxy bead and does not offer any support to the optical fiber.

1. Score the optical fiber with the scribe where it exits the connector endface, and then gently pull the optical fiber, lifting it from the connector endface. A small optical fiber nub should be protruding from the connector endface. You can use the eye loupe to view the connector endface.

2. Polish down or de-burr the optical fiber end by holding the connector so that it is facing up. Arch a piece of 5μm polishing film over it. Lightly rub the film in a circle over the connector until you no longer feel the fiber end "grab" the film. Then stop.

3. Place a clean piece of 1μm polishing film on the rubber pad.

NOTE   An aluminum oxide abrasive works well for most non-laser-optimized multimode applications; however, a diamond abrasive is recommended for single-mode and laser-optimized multimode applications. Do not overpolish when using a diamond abrasive.

4. Insert the ferrule into the puck and place the puck on the polishing film.

5. Slowly move the puck in a figure-8 pattern over the polishing film to begin polishing down the optical fiber. After about five or ten figure-8 strokes, you should start to feel less resistance and the puck should glide across the polishing film. Stop polishing and go to the next step for single-mode applications.

6. For single-mode applications, remove the 1μm polishing film and place a 0.3 or 0.1μm diamond polishing film on the rubber pad. Slowly move the puck in a figure-8 pattern over the polishing film and stop after ten figure-8 strokes. Polishing is complete.

Machine Polishing

Machine polishing is typically performed at the factory. Depending on the polishing machine, one or more connectors can be polished at a time. Machine polishing produces the best connector endface and produces more consistent results than hand polishing. Most single-mode connectors are polished with a machine.

The same steps required to assemble a connector for hand polishing are also required to assemble a connector for machine polishing. The polishing machine is substituted for the hand polishing. Connectors are inserted into the polishing machine after they have been cleaved and de-burred. When the polishing process is complete the connectors are removed and inspected.

Pre-polished Connectors

Pre-polished connectors have been around for many years and do not require hand or machine polishing. A pre-polished connector is very similar to a mechanical splice because a cleaved piece of optical fiber is inserted into the rear of the connector and mated with a small piece of optical fiber that was bonded to the connector ferrule at the factory. The gap between the two optical fibers is filled with index matching gel to reduce Fresnel reflections. The small piece of optical fiber bonded to the connector ferrule has a cleaved perpendicular finish on the side mating with the other optical fiber. The other end of the optical fiber is polished with a machine to produce an exacting endface.

Pre-polished connectors are available for single or multiple fiber applications. When using a pre-polished connector, you must ensure that the optical fiber in the connector is the same size and type of optical fiber you are terminating. You cannot terminate a multimode optical fiber that has a 50μm core with a pre-polished connector that has a 62.5μm core and vice versa. You should not terminate a non-laser-optimized multimode optical fiber that has a 50μm core with a pre-polished connector that has a laser-optimized 50μm core and vice versa.

Two of the most popular brands of pre-polished connectors are the UniCam from Corning Cable Systems and the OptiCam from Panduit. Both manufacturers offer many different versions of these connectors and kits. They have trained personnel who can answer any questions you may have about your application. Shown in Figure 26.58 is the UniCam LANscape kit; Figure 26.59 shows the Panduit OptiCam kit.

FIGURE 26.58
Corning Cable
Systems' UniCam
LANscape kit

FIGURE 26.59
Panduit
OptiCam kit

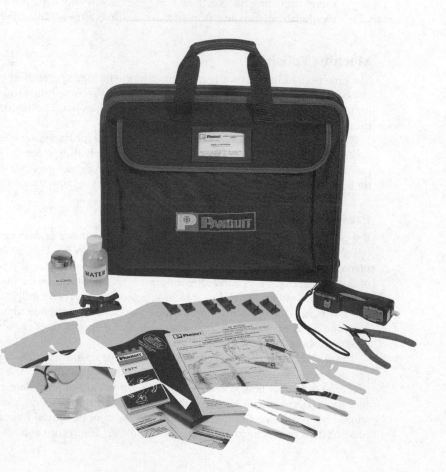

Endface Cleaning and Inspection

Good cleaning and inspection skills are essential for low-loss interconnections. Often the cause of a problem is very basic and can be discovered thorough inspection and repaired by cleaning.

Endface Cleaning

After a connector is polished or before it is inserted into a receptacle or mating sleeve, the endface should be cleaned and inspected. Mating a clean connector with a dirty connector will result in two dirty connectors. Figure 26.60 shows what a formerly clean connector endface looks like after it has been mated with a connector that has skin oil on the endface. You can tell that this connector had been mated because the oil is distributed in a crater-like pattern.

FIGURE 26.60
Skin oil distributed in a crater-like pattern, the result of mating a clean and dirty connector

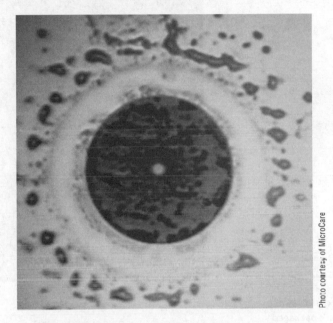

Photo courtesy of MicroCare

Cleaning the endface of a connector is a very simple process; however, it does require cleaning products engineered for that purpose. Only two basic products are required: lint-free wipes and a cleaning fluid engineered to clean the optical fiber endface.

Often a cleaning fluid is not required to clean connector endface. The connector endface can be cleaned with a dry, lint-free wipe, as shown in Figure 26.61. This wipe has been engineered to lift oils, dirt, and contamination from the endface of the connector. This wipe is stored in a sealed foil package until it is ready to be used. The sealed package prevents the wipe from being contaminated prior to use. Notice that the wipe is glued to the foil on one side; this allows you to handle the wipe without getting oil and dirt from your hand on the wipe.

Lint-free wipes are also available in bulk packaging, as shown in Figure 26.62. These are the same wipes found in the sealed foil package. When using these wipes, be careful to touch only one side of the wipe with your hand. Only use the side that did not touch your hand to clean the connector endface.

FIGURE 26.61
Lint-free wipe that
was sealed in a foil
package being used
to clean a connec-
tor endface

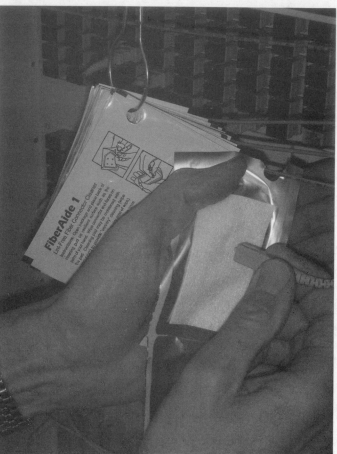

Photo courtesy of MicroCare

FIGURE 26.62
Bulk-packaged
lint-free wipes

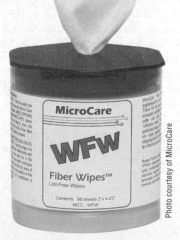

Photo courtesy of MicroCare

To clean the endface with a lint-free wipe, place the wipe in the palm of your hand, as shown in Figure 26.63, and move the connector in a straight line from top to bottom several times. Be sure to lift the connector from the wipe before it reaches the end of the wipe. If the connector comes in contact with your skin or hair, it will pick up the oil, as shown in Figure 26.64.

FIGURE 26.63
Cleaning the connector endface with a lint-free wipe that was sealed in a foil package

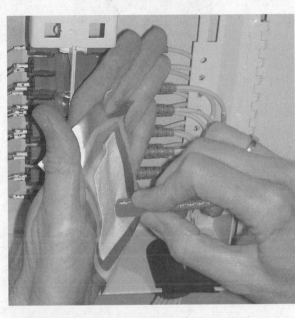

Photo courtesy of MicroCare

FIGURE 26.64
Oil on the endface of a connector

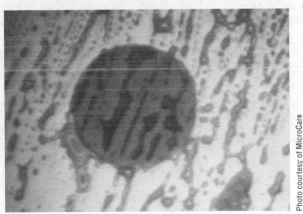

Photo courtesy of MicroCare

If cleaning the endface of the connector with a dry, lint-free wipe does not remove the contamination, wet the wipe with an optical fiber cleaning fluid. To do this, fold the wipe over, place it in the cleaning fluid dispenser, and pump the dispenser, as shown in Figure 26.65. Remove the wipe and clean the connector endface. Be careful not to touch the cleaning surface of the wipe with your skin.

FIGURE 26.65
Wetting the lint-free wipe with a cleaning fluid

A connector that has been cleaned properly will be free of any dirt or contamination, as shown in Figure 26.66. Only after the connector has been cleaned and inspected should it be mated with another connector or receptacle.

FIGURE 26.66
Properly cleaned single-mode connector endface

Endface Inspection

TIA-455-57B provides a guideline for examination of an optical fiber endface. This standard is covered in depth in Chapter 34, "Troubleshooting and Restoration." This section will discuss the tools required to inspect the connector endface and examine common defects and their causes.

INSPECTION MICROSCOPES

To evaluate the endface of a fiber-optic connector, you need a microscope designed for this application. This type of microscope is typically referred to as an *inspection microscope*. There are many different inspection microscopes available, and they can be grouped into two basic categories: optical microscopes and video microscopes. Video microscopes can be further broken into two groups: bench top and video probe.

An optical inspection microscope allows the direct viewing of the connector endface.

WARNING   Care should be taken to ensure that there is no light source on the other end of the connector endface being viewed. As you have learned in this book, fiber-optic light sources are infrared and are not visible to the naked eye. Viewing a connector endface that is radiating infrared light with an optical microscope can temporarily or permanently damage your eye.

Optical inspection microscopes are typically handheld and are available with magnification levels from 100X to 400X. Three different light sources are also available to illuminate the endface: incandescent lamp, LED, and laser. The 400X handheld inspection microscope shown in Figure 26.67 uses a laser to illuminate the connector endface, and the 400X model shown in Figure 26.68 uses an LED.

FIGURE 26.67
400X microscope with a laser for illumination

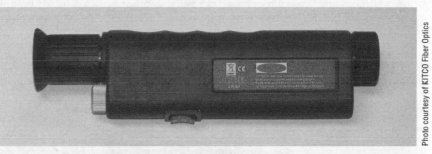

Photo courtesy of KITCO Fiber Optics

FIGURE 26.68
400X microscope with an LED for illumination and an LC connector inserted for inspection

The key advantage to the LED light source is long battery life, and the LED should never need replacing. The laser light source requires more power than the LED; therefore battery life is reduced. However, the laser does make it easier to identify light scratches and defects in the optical fiber endface, and it too should never need replacing. Incandescent lamps are typically found in lower-cost 100X microscopes like the one shown in Figure 26.69.

FIGURE 26.69
100X microscope with an incandescent lamp for illumination

Video inspection microscopes allow you to view the connector endface indirectly. This eliminates any possibility of your eye being damaged from infrared light while inspecting the endface with the microscope. Figure 26.70 is a photograph of a 200X video probe microscope.

FIGURE 26.70
200X video probe microscope

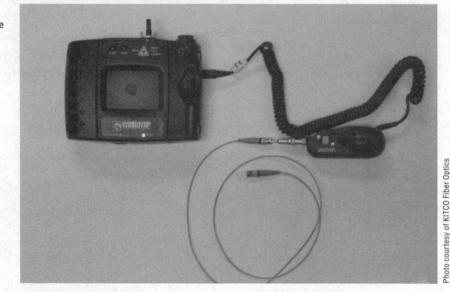

Photo courtesy of KITCO Fiber Optics

Typically, a multimode connector can be evaluated with a 100X microscope, while a single-mode connector requires a minimum of 200X magnification. A 400X microscope, of course, works great for both multimode and single-mode.

EXAMPLES OF ENDFACE INSPECTION

This section of the chapter will examine different endface finishes. Some finishes will be a hand polish with an aluminum oxide abrasive, and some will be a machine polish with a diamond abrasive. Each endface will be evaluated with a 400X video microscope.

Good Quality, Clean Endfaces

The multimode endface shown in Figure 26.71 was hand polished with a 5µm aluminum oxide abrasive to remove the epoxy, followed with a 1µm aluminum oxide. The light scratches on the core and the cladding are typical of a 1µm aluminum oxide abrasive. The chipping and pitting around the epoxy ring is also typical for this abrasive. This endface is an example of a good hand polish with an aluminum oxide abrasive.

FIGURE 26.71

Good hand polish with an aluminum oxide abrasive viewed at 400X

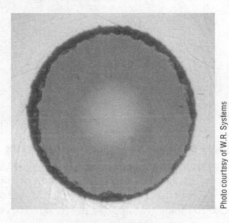

Photo courtesy of W.R. Systems

The single-mode endface shown in Figure 26.72 was machine polished. The final polishing step used a 0.1µm diamond abrasive. This endface is cosmetically perfect; there are no visible scratches on the optical fiber and no chipping or pitting near the epoxy ring. This endface is an example of a good machine polish with a diamond abrasive.

FIGURE 26.72

Good machine polish with a diamond abrasive viewed at 400X

Photo courtesy of MicroCare

Good Quality, Dirty Endfaces

The multimode endface shown in Figure 26.73 was hand polished with a 5μm aluminum oxide abrasive to remove the epoxy, followed by a 1μm aluminum oxide. The light scratches on the core and the cladding are typical of a 1μm aluminum oxide abrasive. Note that the endface is covered with skin oil and needs to be cleaned.

FIGURE 26.73
Skin oil covering a good hand polish with an aluminum oxide abrasive viewed at 400X

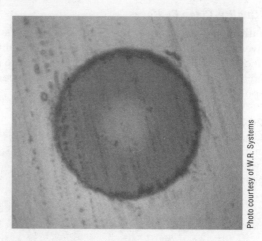

Photo courtesy of W.R. Systems

The single-mode endface shown in Figure 26.74 was machine polished. The final polishing step used a 0.1μm diamond abrasive. This endface is cosmetically perfect; however, it is covered with hair oil. You can typically tell the difference between hair oil and skin oil by the size of the oil deposits. Skin oil as shown in Figure 25.73 typically leaves smaller deposits than hair oil, and the oil tends to look it was applied in streaks. This connector needs to be cleaned.

FIGURE 26.74
Hair oil covering a good single-mode machine polish with a diamond abrasive viewed at 400X

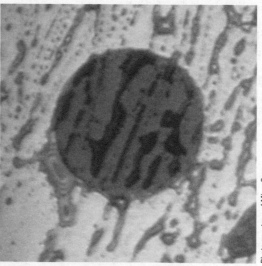

Photo courtesy of MicroCare

The single-mode endface shown in Figure 26.75 was machine polished. The final polishing step used a 0.1μm diamond abrasive. This endface is cosmetically perfect; however, it was cleaned with a shirttail instead of a lint-free wipe. Debris from the shirttail was deposited on the endface of the connector. This connector needs to be cleaned with a lint-free wipe.

FIGURE 26.75
Debris from a shirttail covering a good single-mode machine polish with a diamond abrasive viewed at 200X

Photo courtesy of MicroCare

Broken or Damaged Endfaces

It is impossible to tell how a connector was broken or damaged if it is only inspected after polishing is complete. If damage happens more than once, stop after every step and inspect the endface to determine which step is causing the problem.

The multimode endface shown in Figure 26.76 was hand polished with a 5μm aluminum oxide abrasive to remove the epoxy, followed by a 1μm aluminum oxide. Somewhere in the polishing process, this connector was broken.

FIGURE 26.76
Broken multimode endface viewed at 200X

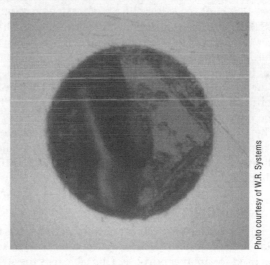

Photo courtesy of W.R. Systems

The multimode endface shown in Figure 26.77 was hand polished with a 5μm aluminum oxide abrasive to remove the epoxy, followed by a 1μm aluminum oxide. Somewhere in the polishing process, this connector was damaged.

FIGURE 26.77
Damaged mul-
timode endface
viewed at 400X

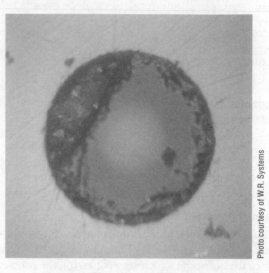

Photo courtesy of W.R. Systems

Connector Performance

If you have assembled, polished, cleaned, and inspected the connector properly, there is a good chance that your connector will provide a low-loss interconnection. Appendix A of ANSI/TIA-568-C.3 defines fiber-optic connector performance. The maximum insertion loss allowed by ANSI/TIA-568-C.3 for a multimode or single-mode mated connector pair is 0.75dB.

Return loss as defined in ANSI/TIA-568-C.3 is the ratio of the power of the outgoing optical signal to the power of the reflected signal. This ratio is expressed in dB. The minimum return loss for a multimode interconnection is 20dB. This means that light energy from the light source traveling through the interconnection is 20dB greater then the light energy being reflected back by the interconnection toward the source.

The minimum return loss for a single-mode interconnection is 26dB unless that interconnection is part of a broadband analog video (CATV) application. Interconnections used in CATV applications must have a minimum return loss of 55dB. Optical return loss testing is discussed in Chapter 33.

Connector Color Code

Multimode and single-mode connectors and adapters can be identified using the color code in section 5.2.3 of ANSI/TIA-568-C.3. The color code in Table 26.2 should be used unless color coding is used for another purpose. This table lists the strain relief and adapter housing color for different optical fiber types.

TABLE 26.2   Multimode and Single-Mode Connector and Adapter Identification

OPTICAL FIBER TYPE OR CONNECTOR TYPE	STRAIN RELIEF AND ADAPTER HOUSING COLOR
850nm laser-optimized 50/125µm optical fiber	Aqua
50/125µm optical fiber	Black
62.5/125µm optical fiber	Beige
Single-mode optical fiber	Blue
Angled contact ferrule single-mode connectors	Green

The Bottom Line

Evaluate connector endface polishing.   To evaluate the endface of a fiber-optic connector, you need a microscope designed for this application.

Master It   You are inspecting the multimode endface shown here with a 200X inspection microscope. Is there anything wrong with this connector endface?

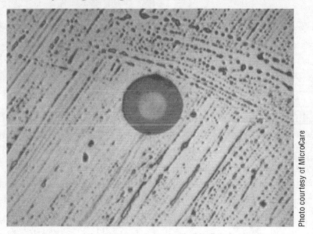

Photo courtesy of MicroCare

Evaluate connector endface cleaning.   A connector that has been cleaned properly will be free of any dirt or contamination.

> **Master It**   You are inspecting the connector endfaces shown in these two graphics. Which of these connectors has oil on the endface?

Photo courtesy of MicroCare

Evaluate connector endface geometry.   The key measurements provided by the interferometer are radius of curvature, apex offset, and fiber undercut or protrusion. These critical parameters are required by Telcordia GR-326 to evaluate connector endface geometry.

> **Master It**   You are evaluating a machine polished endface with the interferometer. The radius of curvature measured by the interferometer is 25mm. Does this endface radius of curvature fall within the range specified by Telcordia GR-326?

Identify an optical fiber type from the color of the connector strain relief.   Multimode and single-mode connectors and adapters can be identified using the color code in ANSI/TIA-568-C.3.

> **Master It**   You need to mate two 850nm laser-optimized 50/125µm fiber-optic cables together with a jumper at a patch panel. What color strain reliefs would the correct jumper have?

Chapter 27

Fiber-Optic Light Sources

As discussed in Chapter 18, "History of Fiber Optics," the idea of transmitting information with light is not new—only the technology that makes it easily possible. Like optical fiber, light source technology has improved rapidly over the decades. These technological advances have greatly increased data transmission rates and reduced costs. Fiber-optic transmitters are available to support every standardized network with a variety of connector choices.

This chapter discusses current fiber-optic light source and transmitter technology as it applies to common telecommunication network standards, industrial control systems, and general-purpose systems. The performance standards that we discuss in this chapter do not represent the highest levels achievable in the lab or on the test bench. They represent commonly available parts for standardized networks. All performance values were obtained from manufacturers' data sheets or networking standards.

In this chapter, you will learn to:

♦ Determine the minimum optical output power for an LED transmitter

♦ Determine the maximum optical output power for an LED transmitter

♦ Determine the minimum optical output power for a laser transmitter

♦ Determine the maximum optical output power for a laser transmitter

Semiconductor Light Sources

The light sources used in fiber-optic communication systems are far different from the light sources used to illuminate your home or office. Fiber-optic light sources must be able to turn on and off millions to billions of times per second while projecting a near-microscopic beam of light into an optical fiber. On top of this performance, they must be reasonably priced, highly reliable, easy to use, and available in a small package.

Semiconductor light source technology has made all this possible. Today's fiber-optic communication systems use *light-emitting diodes (LEDs)* and *laser diodes* (from this point forward, the laser diode will be referred to as the *laser*) exclusively. These semiconductor light sources are packaged to support virtually every fiber-optic communication system imaginable.

LED Sources

A basic LED light source is a semiconductor diode with a *p* region and an *n* region. When the LED is *forward biased* (a positive voltage is applied to the *p* region and a negative voltage to the *n* region), current flows through the LED. As current flows through the LED, the junction where the *p* and *n* regions meet emits random photons. This process is referred to as *spontaneous emission*. Figure 27.1 shows a forward-biased LED in a basic electric circuit.

FIGURE 27.1
Forward-
biased LED

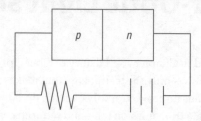

Photons emitted from the junction where the *p* and *n* regions meet are not in phase, nor are they launched in the same direction. These out-of-phase photons are called *incoherent light*. This incoherent light cannot be focused so that each photon traverses down the optical fiber. Because of this, only a small percentage of the photons emitted will be coupled into the optical fiber. Figure 27.2 shows the out-of-phase photons being spontaneously emitted from the LED.

FIGURE 27.2
Radiating forward-
biased LED

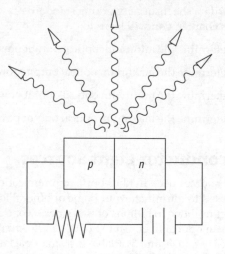

Two types of LEDs, the *surface-emitting LED* and the *edge-emitting LED*, are commonly used in fiber-optic communication systems. Surface-emitting LEDs are a *homojunction structure*, which means that a single semiconductor material is used to form the *pn* junction. Incoherent photons radiate from all points along the *pn* junction, as shown in Figure 27.3.

Edge-emitting LEDs are a *heterojunction structure*, which means that the *pn* junction is formed from similar materials with different refractive indexes. The different refractive indexes are used to guide the light and create a directional output. Light is emitted through an etched opening in the edge of the LED, as shown in Figure 27.4.

FIGURE 27.3
Radiating surface-emitting LED

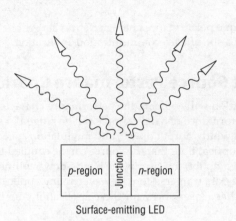

Surface-emitting LED

FIGURE 27.4
Radiating edge-emitting LED

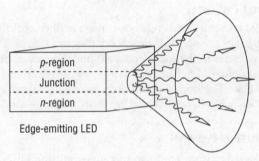

Edge-emitting LED

Laser Sources

The term *laser* is actually an acronym that stands for *light amplification by stimulated emission of radiation*. Like the LED, the laser is a semiconductor diode with a *p* region and an *n* region. Unlike the LED, the laser has an optical cavity that contains the emitted photons with reflecting mirrors on each end of the diode. One of the reflecting mirrors is only partially reflective. This mirror allows some of the photons to escape the optical cavity, as shown in Figure 27.5.

FIGURE 27.5
Radiating laser

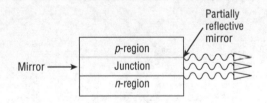

Every photon that escapes the optical cavity is a duplicate of the first photon to escape. These photons have the same wavelength, phase relationship, and direction as the first photon. This process of generating light energy is called *stimulated emission*. The photons radiated from the laser have a fixed relationship that is referred to as *coherent light* or coherent radiation. Figure 27.5 shows the in-phase light waves emitted from the laser.

There are three families of lasers used in fiber-optic communication systems: *Fabry-Pérot*, *distributed feedback* (*DFB*), and the *vertical-cavity surface-emitting laser* (*VCSEL*). Each laser family has unique performance characteristics that will be discussed in this chapter and is designed to support a specific telecommunication application.

Light Source Performance Characteristics

This section will compare the performance characteristics of the LED and laser light sources. The performance of a light source can be judged in several areas: output pattern, wavelength, spectral width, output power, and modulation speed. These performance areas determine the type of optical fiber that the source can be coupled to, transmission distance, and data rate. Without a doubt, the laser is the hands-down winner when it comes to ultra high-speed long-distance data transmission. However, many applications require only a fraction of the performance that a laser offers; for these applications, an LED is used.

Output Pattern

The LED and laser semiconductors used in fiber-optic light sources are packaged to couple as much light as possible into the core of the optical fiber. The output pattern or NA of the light source directly relates to the energy coupled into the core of the optical fiber. The LED has a wide output pattern compared to a laser and does not couple all its light energy into the core of a multimode optical fiber. The output pattern of a laser light source is very narrow, allowing a majority of the light energy to be coupled into the core of a single-mode or multimode optical fiber.

LED OUTPUT PATTERN

Unpackaged surface-emitting and edge-emitting LEDs have wide output patterns, as shown earlier in Figures 27.3 and 27.4. An LED light source must be assembled in a package to mate with a specific connector type and optical fiber. However, occasionally there may be a requirement for an LED to be packaged as a *pigtail*. A pigtail is a short length of tight-buffered optical fiber permanently bonded to the light source package, as shown in Figure 27.6.

FIGURE 27.6
Pigtailed light source

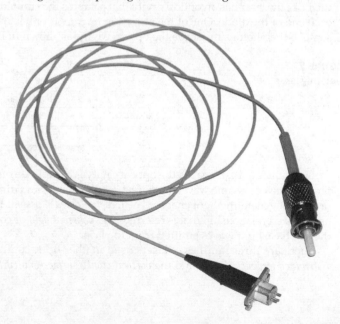

To couple as much light as possible into the core of the optical fiber, the manufacturer typically mounts a micro lens in the shape of a sphere directly on top of the LED. The manufacturer may also use an additional lens to further control the output pattern. Figure 27.7 shows a surface-emitting LED with a series of lenses directing light into the optical fiber.

FIGURE 27.7
Packaged surface-emitting LED

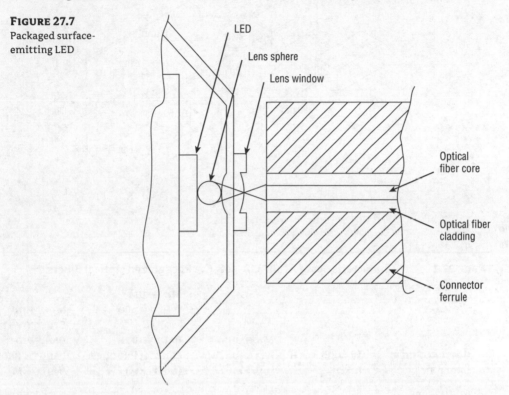

LED

Lens sphere

Lens window

Optical fiber core

Optical fiber cladding

Connector ferrule

Even the best lens can't direct all the light energy into the core of the optical fiber. The typical LED light source overfills the optical fiber, allowing light energy to enter the cladding and the core at angles exceeding the NA. These high-order or marginal modes are attenuated over a short distance. Light energy that enters the core exceeding the acceptance angle will not propagate to the end of the optical fiber. This light energy will refract into the cladding. Figure 27.8 shows a typical LED source overfilling an optical fiber, which is illustrated by the light spot that the source projects.

The process of an LED transmitter overfilling a multimode optical fiber is referred to as an *overfilled launch (OFL)*. An OFL is used to define minimum effective modal bandwidth length product of a multimode optical fiber. Multimode OFL optical fiber bandwidth requirements are defined in ANSI/TIA-568-C.3, as shown in Table 27.1.

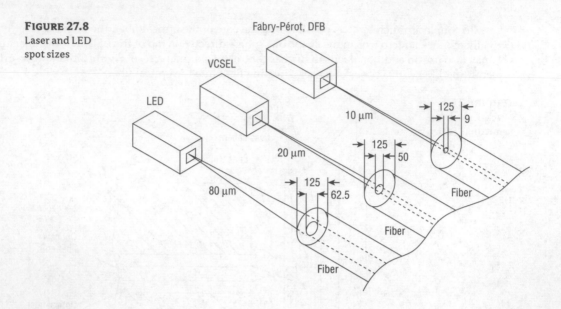

FIGURE 27.8
Laser and LED
spot sizes

TABLE 27.1   Characteristics of ANSI/TIA-568-C.3 Recognized Optical Fibers

OPTICAL FIBER CABLE TYPE	WAVELENGTH (NM)	MAXIMUM ATTENUATION (DB/KM)	MINIMUM OVERFILLED MODAL BANDWIDTH-LENGTH PRODUCT (MHZ · KM)	MINIMUM EFFECTIVE MODAL BANDWIDTH-LENGTH PRODUCT (MHZ · KM)
62.5/125μm multimode	850	3.5	200	Not Required
	1300	1.5	500	
50/125μm multimode	850	3.5	500	Not Required
	1300	1.5	500	
850 nm laser-optimized 50/125μm multimode	850	3.5	1500	2000
	1300	1.5	500	Not Required

Laser Output Pattern

Unlike the LED, the laser light source has a narrow output pattern or NA. Like the LED, the laser must be packaged to align the source with the optical fiber and couple as much energy as possible into the core. Lasers can be packaged for a specific connector or bonded to a pigtail.

Fabry-Pérot and DFB laser light sources are designed for either multimode or single-mode applications. VCSELs are currently designed only for multimode optical fiber applications. The VCSEL emits a wider output pattern and larger spot size than the Fabry-Pérot or DFB lasers. The output of the VCSEL fills only the center of the core of the multimode optical fiber, as shown in Figure 27.8.

Fabry-Pérot and DFB lasers are designed for single-mode and multimode optical fiber applications. The output pattern and spot size of these lasers for single-mode applications are narrower and smaller than the VCSEL. Figure 27.8 shows the typical output pattern of the Fabry-Pérot, DFB, and VCSEL laser.

As bandwidth requirements increase, LED transmitters are being replaced with laser transmitters. However, the multimode optical fiber is not being replaced with single-mode. A laser transmitter, unlike an LED transmitter, does not overfill the multimode optical fiber. Power from a laser transmitter is distributed in the narrow center region of the multimode optical fiber core. This is very different from an LED transmitter, where power is distributed to 100 percent of the core.

Multimode optical fiber has a greater bandwidth potential with a laser transmitter than an LED transmitter. The launch from a laser transmitter results in far fewer modes than the overfilled launch from an LED transmitter. Fewer modes mean less modal dispersion, resulting in greater potential bandwidth.

While all of the laser types—Fabry-Pérot, DFB, and VCSEL—are used with multimode optical fiber, only the VCSEL is used for 850nm applications. The VCSEL is not currently available at a wavelength of 1300nm. For 1300nm or 1310nm applications, the Fabry-Pérot or DFB are used.

Because the launch conditions for the VCSEL transmitter are different from those for the LED, ANSI/TIA-568-C.3 now defines the bandwidth of a multimode optical fiber both with an OFL and with a *restricted mode launch (RML)*. The minimum effective modal bandwidth length product of a multimode optical fiber at 850nm is defined with a restricted mode launch condition. RML only applies to laser-optimized multimode optical fiber, as shown in Table 27.1.

Source Wavelengths

The performance of a fiber-optic communication system is dependent on many factors. One key factor is the wavelength of the light source. Short wavelengths (650nm through 850nm) have more modes and greater attenuation than longer wavelengths (1300nm through 1600nm). The wavelength of the light source can determine system bandwidth and transmission distance.

LED Wavelengths

LED light sources are manufactured from various semiconductor materials. The output wavelength of the LED depends on the semiconductor material that it's manufactured from. Table 27.2 breaks down the semiconductor materials and their associated wavelengths. LED wavelengths are chosen for fiber-optic communication systems based on their application.

Visible wavelengths (650nm) are typically used in short-distance, low-data-rate systems with large-core-diameter optical fiber. These systems are typically found in industrial control systems. Infrared wavelengths (820nm, 850nm, and 1300nm) are typically used for longer-distance, higher-data-rate systems with smaller-core optical fiber.

TABLE 27.2   LED Semiconductor Materials

WAVELENGTH	SEMICONDUCTOR MATERIALS
650nm	Aluminum (Al), gallium (Ga), arsenic (As)
820nm	Aluminum (Al), gallium (Ga), arsenic (As)
850nm	Aluminum (Al), gallium (Ga), arsenic (As)
1300nm	Indium (In), gallium (Ga), arsenic (As), phosphorus (P)

LASER WAVELENGTHS

Just like the LED, a laser light source output wavelength is dependent on the semiconductor material that it's manufactured from, as shown in Table 27.3. As you learned in Chapter 22, "Optical Fiber Characteristics," single-mode optical fiber performance can be defined at 1310nm, 1550nm, or 1625nm. Fabry-Pérot and DFB lasers can be manufactured to operate at or near these wavelengths.

TABLE 27.3   Laser Semiconductor Materials

WAVELENGTH	SEMICONDUCTOR MATERIALS
850nm	Aluminum (Al), gallium (Ga), arsenic (As)
1310nm	Indium (In), gallium (Ga), arsenic (As), phosphorus (P)
1550nm	Indium (In), gallium (Ga), arsenic (As), phosphorus (P)

NOTE   As VCSEL technology improves, it will begin to replace the Fabry-Pérot for 1310nm and 1550nm applications.

The DFB laser is unique because it can be tuned to a specific wavelength. This allows many lasers to operate around a center wavelength in a multichannel system. Multichannel systems will be discussed in greater detail in Chapter 29, "Passive Components and Multiplexers."

The DFB laser is virtually a single-wavelength laser and has the least dispersion of laser fiber-optic sources. Because of this, the DFB is the choice for long-distance, high-speed applications. The 1550nm DFB laser can support a 2.5Gbps data rate at a transmission distance of 80km.

Source Spectral Output

The spectral output of a fiber-optic light source can have a significant impact on the bandwidth of a fiber-optic communication system. Different wavelengths of light travel at different speeds through an optical fiber. Because of this, a light pulse made up of more than one wavelength will disperse as each wavelength travels at a different velocity down the optical fiber. This dispersal limits the bandwidth of the system. To achieve the highest bandwidth possible, a fiber-optic light source must output the narrowest range of wavelengths possible. The range of wavelengths that a light source outputs is described by its *spectral width*.

LED SPECTRAL WIDTH

The spectral width of an LED light source is much wider than the spectral width of a laser. LED spectral width is typically described by the term Full Width, Half Maximum (FWHM). When the spectral width of an LED is displayed graphically, FWHM is measured at one-half the maximum intensity across the full width of the curve, as shown in Figure 27.9.

FIGURE 27.9
Typical FWHM spectral width of 650nm, 850nm, and 1300nm LEDs

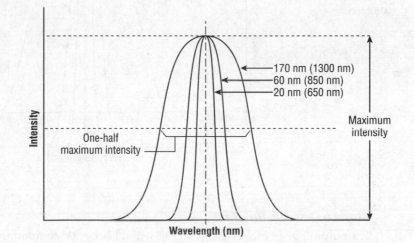

LED spectral widths vary tremendously. A short wavelength 650nm visible LED may have a spectral width as narrow as 20nm. The spectral width of a long wavelength 1300nm infrared LED may be as wide as 170nm. Because an LED is used only for multimode applications, the wide spectral width has no effect on the bandwidth of the communication system. Figure 27.9 compares the spectral width of various LED wavelengths.

LASER SPECTRAL WIDTH

Laser light sources offer the highest level of performance for a fiber-optic communication system. Unlike the LED, lasers output a very narrow spectrum of light. The manufacturer may describe the spectral output width of a laser several ways. These include FWHM, Root Mean Squared (RMS), and 20dB below peak spectral density.

All laser light sources output a narrow spectrum of light. However, there are significant differences in the spectral widths of each laser family. The spectral width of the Fabry-Pérot laser may span as much as 5nm, causing dispersion problems in high-speed or long transmission distance systems. VCSELs have shorter cavities than Fabry-Pérot lasers, which reduce their spectral output width. The DFB laser is virtually a one-frequency laser with spectral widths less than 0.1nm.

The DFB laser achieves this narrow spectral output width by using a series of corrugated ridges on the semiconductor substrate to diffract unwanted wavelengths back into the laser cavity. These corrugated ridges form an internal diffraction grating, which is discussed in detail in Chapter 29. The laser manufacturer can set the desired output wavelength with the spacing of the corrugated ridges on the internal diffraction grating. The DFB laser is used in the highest performance systems. Figure 27.10 compares the spectral widths of the three laser families.

FIGURE 27.10
Typical –20dB spectral width of Fabry-Pérot, VCSEL, and DFB laser

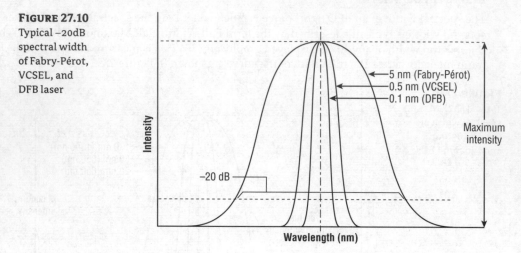

Source Output Power

The output power of the light sources used in fiber-optic communication systems varies dramatically depending on the application. LED light sources are typically designed to support transmission distances up to 2km while laser light sources may support transmission distances in excess of 80km. Laser optical power output levels can exceed LED optical output power levels by more than 20dB.

LED Output Power

Today's LED technology seems to be changing daily. High-power LED light sources are beginning to replace many incandescent light sources. As of this writing, single-lamp 5-watt LED white light flashlights are available with beams so powerful that viewing them directly is like viewing a flashbulb. However, the LED light sources used in fiber-optic communication output considerably less power than that flashlight.

The optical output power of a typical LED light source used in fiber-optic communications is 10dB or more below the lowest power laser light source. Output power is typically expressed by the manufacturer at the Beginning of Life (BOL) and the End of Life (EOL). An LED light source will lose some output power over its usable lifetime. The industry convention allows for a degradation of 1.5dB from BOL to EOL.

LED light sources couple only a small portion of their light energy into the core of the optical fiber. The amount of energy coupled into the optical fiber depends on the optical fiber core size and NA. The larger the core size and greater the NA, the more energy coupled into the core.

Off-the-shelf LED light sources are available to support a wide range of glass, plastic, and PCS optical fiber. Most LED light sources used in telecommunication applications are designed to work with either 50/125µm or 62.5/125µm optical fiber. The light energy coupled into the core of a 50/125µm optical fiber with an NA of 0.20 is approximately 3.5dB less than the energy coupled into the core of a 62.5/125µm optical fiber with an NA of 0.275. Because the same LED light source may be used with different sizes of optical fiber, the manufacturer typically measures the output power at the end of one meter of optical fiber with the cladding modes removed. The cladding modes can be removed by a mode filter, which is discussed in detail in Chapter 33, "Test Equipment and Link/Cable Testing."

LASER OUTPUT POWER

Laser light sources offer the highest output power available for fiber-optic communication systems. The higher output power of the laser allows greater transmission distances than with an LED. Laser output power varies depending on the application. VCSELs used in multimode applications have the lowest output power. Fabry-Pérot and DFB lasers used in single-mode applications have the greatest output power.

Like the LED, the output power of the laser will diminish over its usable lifetime. Laser manufacturers typically provide BOL and EOL minimum optical power levels. The industry convention for lasers allows for a 1.5dB reduction in power from BOL to EOL. Optical output power levels are normally expressed as the amount of light coupled into a one meter optical fiber.

Source Modulation Speed

There are many advantages to using optical fiber as a communications medium. High bandwidth over long distances is one of them. What limits the bandwidth of today's fiber-optic communication systems is not the optical fiber but the light source. The modulation speed of a light source is just one factor that can limit the performance of a fiber-optic communication system.

LED MODULATION SPEED

Today's LED light sources can be modulated to support data rates greater than 400Mbps. However, most LED light sources are designed to support network standards that do not require a data rate that high. An example of this is the IEEE 802.3 Ethernet standard, which establishes network data rates at 10Mbps, 100Mbps, 1Gbps, and 10Gbps. The LED can easily support the 10Mbps and 100Mbps data. However, current LED technology does not support the 1Gbps or greater data rate.

LASER MODULATION SPEED

Laser light sources are constantly evolving. Laser manufacturers have been able to modulate all three-laser families at data rates up to 10Gbps. However, many applications for laser light sources do not require data rates in the 10Gbps range. Laser light sources are used in network applications with data rates as low as 10Mbps when transmission distances greater than 2,000 meters are required.

Transmitter Performance Characteristics

Up to this point, we have discussed only the characteristics of the unpacked and packaged LED and laser light source with no mention of the electronics required to drive these light sources. The incredible data rates discussed in the previous text would not be possible without integrated electronics packaged into the transmitter.

For most users, the LED or laser transmitter is a black box. The manufacturer has neatly integrated everything required to convert the electrical input signal into light energy in the smallest package possible. However, it is important to be able to interpret the manufacturer data sheet and verify the optical output of the transmitter.

LED Transmitter Performance Characteristics

LED transmitters are designed to support multimode optical fibers with core sizes ranging from as small as 50µm to as large as 1mm. LED transmitters are directly modulated. (Direct modulation is when the drive current through the LED is varied.) The output power of the LED is directly proportional to the current flow through the LED, as shown in Figure 27.11. In a digital application, the drive current is switched on and off. In an analog application, the drive current is varied.

FIGURE 27.11
LED and laser optical output power versus drive current

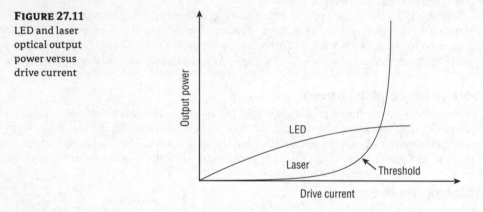

A typical LED transmitter contains an electrical subassembly, optical subassembly, and receptacle, as shown in Figure 27.12. The electrical subassembly amplifies the electrical input signal with the driver integrated circuit (IC). The driver IC provides the current to drive the LED in the optical subassembly. The optical subassembly mates with the receptacle to direct light into the optical fiber.

Because LED fiber-optic links are simplex, they require a transmit optical fiber and a receive optical fiber. Typically, the LED transmitter is packaged with the receiver section. Packaging the transmitter and receiver together reduces the overall space required, simplifies circuit board design, and reduces cost. Figure 27.13 is a photograph of a 1300nm transceiver with an ST receptacle.

FIGURE 27.12
Block diagram
of a typical LED
transmitter

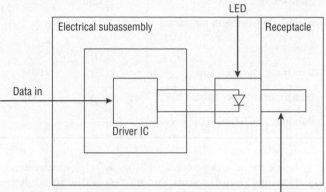

FIGURE 27.13
A 100Mbps
1300nm LED trans-
ceiver with an ST
receptacle

The performance characteristics of the LED transmitter are typically broken up into four groups: recommended operating conditions, electrical characteristics, optical characteristics, and data rate. The recommended operating conditions describe maximum and minimum temperature and voltage ranges that the device can operate in without damage. Table 27.4 shows the typical recommended operating conditions for a 1300nm 100Mbps LED transmitter.

TABLE 27.4   LED Transmitter Recommended Operating Conditions

PARAMETER	SYMBOL	MIN.	TYP.	MAX.	UNIT
Ambient Operating Temperature	T_A	0		70	°C
Supply Voltage	V_{CC}	4.75		5.25	V
Data Input Voltage—Low	$V_{IL}-V_{CC}$	−1.810		−1.475	V
Data Input Voltage—High	$V_{IH}-V_{CC}$	−1.165		−0.880	V
Data and Signal Detect Output Load	R_L		50		Ω

The electrical characteristics of the LED transmitter describe the supply current requirements, the data input requirements, and the power dissipated by the device. Table 27.5 shows the typical electrical characteristics for a 1300nm 100Mbps LED transmitter.

TABLE 27.5   LED Transmitter Electrical Characteristics

PARAMETER	SYMBOL	MIN.	TYP.	MAX.	UNIT
Supply Current	I_{CC}		145	185	mA
Power Dissipation	P_{DISS}		0.76	0.97	W
Data Input Current—Low	I_{IL}	−350	0		µA
Data Input Current—High	I_{IH}		14	350	µA

The optical characteristics of the LED transmitter at a minimum include output power, center wavelength, and spectral width. Table 27.6 shows the typical optical characteristics for a 1300nm 100Mbps LED transmitter.

TABLE 27.6   LED Transmitter Optical Characteristics

PARAMETER	SYMBOL	MIN.	TYP.	MAX.	UNIT
Optical Output Power: BOL62.5/125µm NA = 0.275 Fiber: EOL	P_o	−19 −20	−16.8	−14	dBm avg.
Optical Output Power: BOL50/125µm, NA = 0.20 Fiber: EOL	P_o	−22.5 −23.5	−20.3	−14	dBm avg.
Optical Extinction Ratio			0.001 −50	0.03 −35	% dB
Optical Output Power at Logic "0" State	P_o("0")			−45	dBm avg.

TABLE 27.6   LED Transmitter Optical Characteristics   *(CONTINUED)*

PARAMETER	SYMBOL	MIN.	TYP.	MAX.	UNIT
Center Wavelength	λ_c	1270	1308	1380	nm
Spectral Width—FWHM	Δl		137	170	nm
Optical Rise Time	t_r	0.6	1.0	3.0	ns
Optical Fall Time	t_f	0.6	2.1	3.0	ns
Duty Cycle Distortion Contributed by the Transmitter	DCD		0.02	0.6	ns_{P-P}
Data-Dependent Jitter Contributed by the Transmitter	DDJ		0.02	0.6	ns_{P-P}

As you can see in Table 27.6, there is a minimum, maximum, and typical value for the optical output power of an LED transmitter. These values are defined for 62.5/125µm optical fiber and for 50/125µm optical fiber. As you learned in Chapter 22, the NA of a 62.5/125µm optical fiber is typically 0.275 and the NA of a 50/125µm optical fiber is 0.20.

The amount of optical output power coupled into a multimode optical fiber depends on the core diameter and the NA of the fiber. As you saw earlier in this chapter, the same LED transmitter will typically couple 3.5dB more power into the 62.5/125µm optical fiber than into the 50/125µm. The 3.5dB difference is shown in Table 27.6. There is a 3.5dB difference between the minimum optical output power for the 62.5/125µm and 50/125µm optical fiber.

Earlier in this chapter, we stated that light sources age, and as they age, their output power decreases. The industry convention allows for 1.5dB of aging. However, the difference between the BOL value and EOL for the LED transmitter described in Table 27.6 is only 1dB. The difference between BOL and EOL output power will vary from manufacturer to manufacturer. Always obtain this data from the manufacturer's data sheet.

As you will learn in Chapter 32, "Fiber-Optic System Design Considerations," the optical output power of a fiber-optic transmitter is required to create a power budget for a fiber-optic link. The output power of most fiber-optic transmitters is very close to the typical value. However, the output power can vary as defined by the minimum and maximum values in Table 27.6.

To find the minimum optical output power for a specific optical fiber, locate the Optical Output Power row for that optical fiber in Table 27.6. In the Min. column, you will see a BOL and an EOL value. The higher number is the BOL value, and the lower number is the EOL value.

To find the maximum optical output power for a specific optical fiber, locate the Optical Output Power row for that optical fiber in Table 27.6. The number in the Max. column is the value for the maximum optical output power. The optical output of the fiber-optic transmitter should never exceed this value. There is no BOL or EOL value for the maximum optical output power

LED Transmitter Applications

LED transmitters are often designed to support one or more network standards. A 100Mbps 1300nm transmitter could support 100Base-FX applications or 100Mbps ATM applications. Table 27.7 lists the data rates, wavelengths, and optical fiber types for various LED transmitters used in IEEE 802.3 Ethernet communication systems.

TABLE 27.7	LED Transmitters for Ethernet Applications	
NETWORK	**DATA RATE**	**WAVELENGTH AND MEDIA**
10Base-FL	10Mbps	850nm, multimode fiber
100Base-FX	100Mbps	1300nm, multimode fiber
100Base-SX	10/100Mbps	850nm, multimode fiber

Laser Transmitter Performance Characteristics

Laser transmitters are designed to support either single-mode optical fiber systems or multi-mode optical fiber systems. The single-mode transmitter is designed to interface with 8–11µm optical fiber cores. Until recently, all single-mode transmitters were from the Fabry-Pérot or DFB families. However, manufacturers keep improving VCSEL technology and the VCSEL is beginning to emerge as an alternative 1300nm single-mode optical fiber transmitter.

PROBLEMS ASSOCIATED WITH CORE DIAMETER MISMATCH

LED transmitters are typically designed to support different core size multimode optical fiber. For example, a 100Base-SX transmitter is designed to operate with 50/125µm or 62.5/125µm optical fiber, as shown in Table 27.6. If the LED transmitter receptacle is designed to accept an SC connector, any SC connector regardless of optical fiber size or type can be plugged into the transmitter.

So what happens when a 62.5/125µm patch cord from the LED transmitter is mated with a 50µm horizontal cable? Back reflections and attenuation. As you may remember from the "LED Output Power" section, we mentioned that an LED light source couples roughly 3.5dB more light energy into a 62.5µm core than a 50µm core. So when you take the 62.5µm core from the transmitter and mate it with the 50µm core horizontal cable, there is roughly a 3.5dB loss at that connection. As mentioned in Chapter 19, "Principles of Fiber-Optic Transmission," a loss of 3dB means that half of the power is lost. So if less than half the photons from a light pulse get coupled into the horizontal cable, where do the rest of the photons go?

Well, some of the photons will be absorbed by the coating of the 50µm optical fiber. Others will be reflected back into the cladding of the 62.5µm optical fiber, and absorbed by the coating. However, some of the photons will be reflected back into the core of the 62.5µm optical fiber. Many of those photons will travel back to the transmitter where they will be reflected off the transmitter window or lens assembly back into the core. These photons will travel the length of the optical fiber to the receiver. There is a good possibility that the receiver will detect these photons and convert their light energy into electrical energy. Now the circuit decoding the electrical energy is being bombarded with good and bad data that prevents the system from operating.

The fix for this problem is simple: use a 50µm patch cord. The bottom line is that you should pay careful attention to core diameter to prevent core diameter mismatch. You can't look at only the color of the patch cord to determine the core diameter. There are orange patch cords with a 62.5µm core and there are orange patch cords with a 50µm core. You must read the cable markings from the manufacturer to find out the actual core diameter.

Like the LED transmitter, the laser transmitter can be broken into three sections: the electrical subassembly, optical subassembly, and receptacle or pigtail, as shown in Figure 27.14. The electrical subassembly amplifies the input signal and provides the drive current for the laser. Unlike the LED, the laser emits very few photons until the drive current passes a threshold level, as shown earlier in Figure 27.11. A feedback circuit that monitors laser output intensity typically controls the drive current provided to the laser to ensure a constant output level.

FIGURE 27.14
Block diagram of a typical laser transmitter

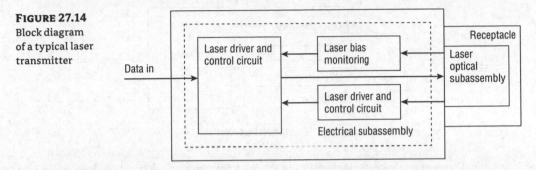

The optical subassembly contains a photodiode that receives part of the energy output by the laser. The electrical output of the photodiode is monitored in the electrical subassembly to help control drive current. The optical subassembly may mate with an external modulator, receptacle, or pigtail.

Laser transmitters may or may not be packaged with the receiver section. How the transmitter is packaged depends on the application. Laser transmitters operating up to 2.5Gbps typically do not require coolers and are often packaged with a receiver section. Laser transmitters operating at 10Gbps or greater produce enough heat to require an integrated cooler to help the laser maintain a constant temperature. Temperature variations will affect the laser's threshold current, output power, and frequency.

The VCSEL is used in multimode transmitters. Currently VCSEL multimode transmitters support only 850nm operation with 50/125µm or 62.5/125µm optical fiber. The VCSEL can support data rates as high as 10Gbps through multimode optical fiber.

The VCSEL, Fabry-Pérot or DFB laser is capable of direct modulation up to 10Gbps. To achieve data rates greater than 10Gbps, the laser must typically be indirectly modulated. Indirect modulation does not vary the drive current as direct modulation does. Maintaining a constant drive current prevents the output wavelength from changing as the electron density in the semiconductor material changes. This is commonly referred to as *laser chirp*, which causes dispersion and limits bandwidth over long transmission distances.

Indirect modulation requires a constant drive current to the laser. The laser light source outputs a continuous wave (CW) that is modulated by an external device. The external modulation device is typically integrated into the laser transmitter package. Indirect modulation of a laser allows higher modulation rates (up to 40Gbps) and less dispersion than direct modulation. Figure 27.15 shows a functional block diagram of direct and indirect modulation.

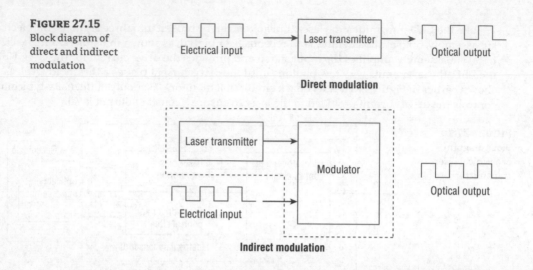

FIGURE 27.15
Block diagram of
direct and indirect
modulation

Direct modulation

Indirect modulation

The performance characteristics of the laser transmitter are typically broken up into four groups: recommended operating conditions, electrical characteristics, optical characteristics, and data rate. The recommended operating conditions describe the maximum and minimum temperatures and voltage ranges that the device can operate in without damage. Table 27.8 shows the typical recommended operating conditions for a 1300nm, 2.5Gbps laser transmitter.

TABLE 27.8   Laser Transmitter Recommended Operating Conditions

PARAMETER	SYMBOL	MIN.	TYP.	MAX.	UNIT
Ambient Operating Temperature	T_A	0		+70	°C
Supply Voltage	V_{CC}	3.1		3.5	V
Power Supply Rejection	PSR		100		mV_{PP}
Transmitter Differential Input Voltage	V_D	0.3		2.4	V
Data Output Load	R_{DL}		50		Ω
TTL Signal Detect Output Current—Low	I_{OL}			1.0	mA
TTL Signal Detect Output Current—High	I_{OH}	−400			μA
Transmit Disable Input Voltage—Low	T_{DIS}			0.6	V
Transmit Disable Input Voltage—High	T_{DIS}	2.2			V
Transmit Disable Assert Time	T_{ASSERT}			10	μS
Transmit Disable Deassert Time	$T_{DEASSERT}$			50	μS

The electrical characteristics of the laser transmitter describe the supply current requirements, the data input requirements, and the power dissipated by the device. Table 27.9 shows the typical electrical characteristics for a 1300nm, 2.5Gbps laser transmitter.

TABLE 27.9   Laser Transmitter Electrical Characteristics

PARAMETER	SYMBOL	MIN.	TYP.	MAX.	UNIT
Supply Current	I_{CCR}		115	140	mA
Power Dissipation	P_{DISS}		0.38	0.49	W
Data Output Voltage Swing (single-ended)	$V_{OH} - V_{OL}$	575		930	mV
Data Output Rise Time	t_r		125	150	ps
Data Output Fall Time	t_f		125	150	ps
Signal Detect Output Voltage—Low	V_{OL}			0.8	V
Signal Detect Output Voltage—High	V_{OH}	2.0			V
Signal Detect Assert Time (OFF to ON)	AS_{MAX}			100	µs
Signal Detect Deassert Time (ON to OFF)	ANS_{MAX}			100	µs

The optical characteristics of the laser transmitter at a minimum include output power, center wavelength, spectral width, and back-reflection sensitivity. Excessive back reflections can interfere with the operation of the laser. Table 27.10 shows the typical optical characteristics for a 1300nm, 2.5Gbps laser transmitter.

TABLE 27.10   Laser Transmitter Optical Characteristics

PARAMETER	SYMBOL	MIN.	TYP.	MAX.	UNIT
Optical Output Power 9µm SMF	P_{OUT}	10	−6	−3	dBm
Center Wavelength	λ_c	1260		1360	nm
Spectral Width—rms	ΔLOW		1.8	4	nm rms
Optical Rise Time	t_r		30	70	ps
Optical Fall Time	t_f		150	225	ps
Extinction Ratio	E_R	8.2	12		dB

TABLE 27.10   Laser Transmitter Optical Characteristics   *(CONTINUED)*

PARAMETER	SYMBOL	MIN.	TYP.	MAX.	UNIT
Optical Output Eye	Compliant with eye mask Telecordia GR-253-CORE				
Back-Reflection Sensitivity				−8.5	dB
Jitter Generation	pk to pk			70	mUI
	RMS			7	mUI

As you can see in Table 27.10, there is a minimum, maximum, and typical value for the optical output power of a laser transmitter. These values are defined for a single-mode optical fiber with a core diameter of 9µm. However, the NA is not stated as it was in the LED transmitter data sheet. Recall from Chapter 22 that single-mode optical fibers have the smallest NA and narrowest acceptance cone. A typical NA for a single-mode optical fiber is 0.14.

Like the LED, the laser ages and its output power decreases over time. In Table 27.10, there is no BOL or EOL value. If the manufacturer does not provide these values, use the industry convention of 1.5dB and assume that the values stated for optical output power are the BOL values.

To find the minimum optical output power, locate the Optical Output Power row in Table 27.10. In the Min. column, there is a value. Since BOL and EOL are not stated, assume this value is the BOL value. To approximate the EOL value, subtract 1.5dB.

To find the maximum optical output power, locate the Optical Output Power row in Table 27.10. The number in the Max. column is the value for the maximum optical output power. The optical output of the fiber-optic transmitter should never exceed this value. There is no BOL or EOL value for the maximum optical output power.

Laser Transmitter Applications

Laser transmitters can typically support multiple data rate operation. The same transmitter may be used in a 155Mbps ATM switch or a 2.5Gbps SONET OC-48 switch. Table 27.11 lists the data rates, wavelengths, and optical fiber type for various laser transmitters used in IEEE 802.3 Ethernet communication systems.

TABLE 27.11   Laser Transmitters for Ethernet Applications

NETWORK	DATA RATE	WAVELENGTH AND MEDIA
100Base-LX	10/100Mbps	1300nm single-mode fiber
100Base-LH	10/100Mbps	1300nm single-mode fiber
100Base-LH	10/100Mbps	1550nm single-mode fiber
1000Base-SX	1,000Mbps	850nm VCSEL, multimode fiber

TABLE 27.11   Laser Transmitters for Ethernet Applications   *(CONTINUED)*

NETWORK	DATA RATE	WAVELENGTH AND MEDIA
1000Base-LX	1,000Mbps	1310nm FP, multimode fiber
10GBase-SR	10,000Mbps	850nm VCSEL, multimode fiber
10GBase-LX4	10,000Mbps	1310nm DFB, multimode fiber

Light Source Safety

The light sources used in fiber-optic communication systems operate at very low power levels. Unlike an incandescent lamp, their light energy is distributed over a narrow spectrum, typically the infrared spectrum. This narrow spectrum light energy is often invisible, and it can be a danger to the fiber-optic installer or fiber-optic technician. The level of danger depends on the classification of the light source and the amount of energy coupled into the optical fiber. Laser light sources pose the greatest risk because of their narrow spectral width coherent light.

Classifications

We have learned that fiber-optic light sources are very different from the lightbulbs used to illuminate our home or office. We shop for lightbulbs by their wattage. The wattage shown on the package of the lightbulb tells us how much energy the bulb requires to operate. A 100 watt incandescent lightbulb does not output 100 watts of light energy; it requires 100 watts of power to operate. Much of the energy required to operate the bulb is actually emitted as heat, not visible light. We all know how hot a 100 watt incandescent lightbulb can get.

That same 100 watt lightbulb emits a very broad spectrum of visible light. In other words, the bulb emits many different wavelengths of visible light that combine together to create white light. Each wavelength of light radiating from the bulb represents only a small fraction of the overall light energy output by the bulb. Unlike the laser, the visible light that radiates from the bulb is incoherent. Because the broad visible spectrum of light radiating from the bulb is incoherent and the light energy is spread out over many wavelengths, the light from the bulb will not damage your eye. The coherent light from the laser and its lack of divergence is what can cause eye damage. An ordinary incandescent lightbulb is not an eye hazard like a laser. However, this doesn't mean that you should stare at 100 watt light bulbs.

Fiber-optic light sources are classified by their ability to cause damage to your eye. As you learned in Chapter 23, "Safety," there are standards and federal regulations that classify the light sources used in fiber-optic transmitters. Every laser transmitter sold in the United States must be classified as defined in 21CFR, Chapter 1, Subchapter J. Only these classifications are covered in this chapter. Listed here are the hazards associated with each class:

Class I   These levels of laser radiation are not considered to be hazardous.

Class IIa   These levels of laser radiation are not considered to be hazardous if viewed for any period of time less than or equal to 1,000 seconds, but are considered to be a chronic viewing hazard for any period of time greater than 1,000 seconds.

Class II   These levels of laser radiation are considered to be a chronic viewing hazard.

Class IIIa These levels of laser radiation are considered to be, depending on the irradiance, either an acute intrabeam viewing hazard or chronic viewing hazard, and an acute viewing hazard if viewed directly with optical instruments.

Class IIIb These levels of laser radiation are considered to be an acute hazard to the skin and eyes from direct radiation.

Class IV These levels of laser radiation are considered to be an acute hazard to the skin and eyes from direct and scattered radiation.

Safe Handling Precautions

There are some fundamental precautions that you should always observe when working with a fiber-optic communication system.

◆ Assume that the fiber-optic cable assembly you are handling is energized.

◆ Never directly view the end of an optical fiber or the end face of a fiber-optic connector without verifying that the optical fiber is dark.

◆ View the end face of a fiber-optic connector from at least 6″ away when testing continuity or using a visible fault locator.

The Bottom Line

Determine the minimum optical output power for an LED transmitter. The minimum optical output power of an LED transmitter should be defined in the manufacturer's data sheet.

Master It Refer to the following table to determine the BOL and EOL values for the minimum optical output power of a 50/125µm optical fiber.

Parameter	Symbol	Min.	Typ.	Max.	Unit
Optical Output Power: BOL62.5/125µm, NA = 0.275	P_0	−22	−19.8	−17	dBm avg.
Fiber: EOL		−23			
Optical Output Power: BOL50/125µm, NA = 0.20	P_0	−25.5	−23.3	−17	dBm avg.
Fiber: EOL		−26.5			
Optical Extinction Ratio			0.001	0.03	%
			−50	−35	dB
Optical Output Power at Logic "0" State	P_0 ("0")			−45	dBm avg.
Center Wavelength	λ_c	1270	1308	1380	nm

Parameter	Symbol	Min.	Typ.	Max.	Unit
Spectral Width—FWHM	ΔI		137	170	nm
Optical Rise Time	t_r	0.6	1.0	3.0	ns
Optical Fall Time	t_f	0.6	2.1	3.0	ns
Duty Cycle Distortion Contributed by the Transmitter	DCD		0.02	0.6	ns_{P-P}
Data-Dependent Jitter Contributed by the Transmitter	DDJ		0.02	0.6	ns_{P-P}

Determine the maximum optical output power for an LED transmitter. The manufacturer's data sheet should contain the maximum value for the optical output power of an LED transmitter.

Master It Refer to the following table to determine the maximum optical output power of a 62.5/125µm optical fiber.

Parameter	Symbol	Min.	Typ.	Max.	Unit
Optical Output Power: BOL62.5/125µm NA − 0.275	P_0	−24	−21.8	−19	dBm avg.
Fiber: EOL		−25			
Optical Output Power: BOL50/125µm, NA = 0.20	P_0	−27.5	−25.3	−19	dBm avg.
Fiber: EOL		−28.5			
			0.001	0.03	%
			−50	−35	dB
Optical Output Power at Logic "0" State	P_0 ("0")			−45	dBm avg.
Center Wavelength	λ_c	1270	1308	1380	nm
Spectral Width—FWHM	ΔI		137	170	nm
Optical Rise Time	t_r	0.6	1.0	3.0	ns
Optical Fall Time	t_f	0.6	2.1	3.0	ns
Duty Cycle Distortion Contributed by the Transmitter	DCD		0.02	0.6	ns_{P-P}
Data-Dependent Jitter Contributed by the Transmitter	DDJ		0.02	0.6	ns_{P-P}

Determine the minimum optical output power for a laser transmitter. The manufacturer's data sheet should contain the minimum, maximum, and typical value for the optical output power of a laser transmitter.

Master It Refer to the following table to determine the BOL and EOL values for the minimum optical output power.

Parameter	Symbol	Min.	Typ.	Max.	Unit
Optical Output Power 9μm SMF	P_{OUT}	−6	0	+3	dBm
Center Wavelength	λ_c	1260		1360	nm
Spectral Width—rms	ΔLOW		1.8	4	nm rms
Optical Rise Time	t_r		30	70	ps
Optical Fall Time	t_f		150	225	ps
Extinction Ratio	E_R	8.2	12		dB
Optical Output Eye	Compliant with eye mask Telecordia GR-253-CORE				
Back-Reflection Sensitivity				−8.5	dB
Jitter Generation	pk to pk			70	mUI
	RMS			7	mUI

Determine the maximum optical output power for a laser transmitter. The manufacturer's data sheet should contain the maximum value for the optical output power of a laser transmitter.

Master It Refer to the following table to determine the maximum optical output power.

Parameter	Symbol	Min.	Typ.	Max.	Unit
Optical Output Power 9μm SMF	P_{OUT}	−12	−8	−5	dBm
Center Wavelength	λ_c	1260		1360	nm
Spectral Width—rms	ΔLOW		1.8	4	nm rms
Optical Rise Time	t_r		30	70	ps
Optical Fall Time	t_f		150	225	ps
Extinction Ratio	E_R	8.2	12		dB

Parameter	Symbol	Min.	Typ.	Max.	Unit
Optical Output Eye	Compliant with eye mask Telecordia GR-253-CORE				
Back-Reflection Sensitivity				−8.5	dB
Jitter Generation	pk to pk			70	mUI
	RMS			7	mUI

Chapter 28

Fiber-Optic Detectors and Receivers

In Chapter 18, "History of Fiber Optics," we introduced you to the fiber-optic receiver. The job of the receiver is to take light energy from the optical fiber and convert it to electrical energy. In this chapter, we will explain the basic components that make up the fiber-optic receiver, starting with the photodiode. You will learn about the effects of optical input power on the performance of the receiver and the performance characteristics of LED and laser receivers.

In this chapter, you will learn to:

◆ Calculate the dynamic range for an LED receiver

◆ Calculate the dynamic range for a laser receiver

Photodiode Fundamentals

A *photodiode* in a fiber-optic receiver is like the tire on your car. The photodiode is where the rubber meets the road. Light energy from the optical fiber stops at the photodiode. It's the job of the photodiode to convert the light energy received from the optical fiber into electrical energy. Just as there are different performance-level tires that you can put on your car, so there are different performance-level photodiodes that can be incorporated into a receiver. This section of the chapter discusses the fundamentals of basic photodiode operation and the different types of photodiodes that may be used in a receiver.

The best way to imagine a photodiode is to think about a *solar cell*. Most of us have seen the exhibits at museums where a solar cell or a group of solar cells powers a small boat or car with the light energy from a light bulb. Maybe you own a solar charger for your boat battery or have decorative outdoor lighting that uses solar cells to recharge the batteries.

The solar cell takes the light energy it receives and converts it into electrical energy. In other words, the photons absorbed by the solar cell cause electrons to flow within the solar cell. This electron flow is called a *current*. The current from the solar cell flowing through the motor of the small boat or car causes the motor to rotate. The more current, the faster the motor rotates and the faster the boat or car travels.

Like the LED we learned about in Chapter 27, "Fiber-Optic Light Sources," the basic photodiode is a semiconductor diode with a *p* region and an *n* region, as shown in Figure 28.1. Photons absorbed by the photodiode excite electrons within the photodiode in a process called *intrinsic absorption*. When stimulated with an outside bias voltage, these electrons produce a current flow through the photodiode and the external circuit providing the bias voltage.

FIGURE 28.1
PN photodiode

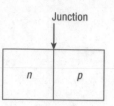

The *PN photodiode* is *reverse biased* when used in an electrical circuit. This is the opposite of how the LED is used in an electrical circuit. Reverse bias means that the *n* region of the photodiode is connected to a positive electrical potential and the *p* region is connected to a negative electrical potential.

In an electrical circuit, as shown in Figure 28.2, light that is absorbed by the photodiode produces current flow through the entire external circuit. As current flows through the *resistor*, it produces a voltage drop across the resistor. This voltage drop is input to an amplifier for amplification.

FIGURE 28.2
PN photodiode
in an electrical
circuit

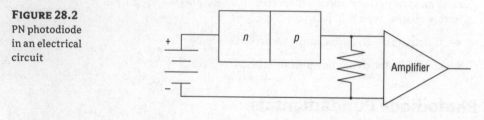

You may be wondering why the output of the photodiode needs to be amplified, since there is no amplification used with a solar cell. The solar cell is typically supplied with photons by a powerful light source such as the sun or a very bright lamp. The photodiode used in a fiber-optic receiver gets all of its light energy from the optical fiber connected to it. As we have learned in this book, the core of an optical fiber is extremely small and carries very little light energy in comparison to the energy that a solar cell receives. A photodiode that is stimulated by the light energy in an optical fiber does not produce a great amount of electrical current flow. This is why amplification is required. It's also the reason why different photodiodes have been developed for various fiber-optic receiver applications.

Other Types of Photodiode

The basic photodiode is called a PN photodiode, but there are two other types: the PIN photodiode and the avalanche photodiode. This section describes how they work.

PIN Photodiode

The *PIN photodiode* works like a PN photodiode; however, it is manufactured to offer better performance. The better performance comes in the form of improved efficiency and greater speed. Improved efficiency means that it has a better photon-to-carrier conversion ratio. If the same amount of light energy hit a PN photodiode and a PIN photodiode, the PIN photodiode would generate more current flow through an external circuit.

Greater speed means that the diode can turn on and off faster. Remember that in fiber optics the light pulses being sent by the transmitter happen at a very fast rate. The photodiode needs to be able to stop and start electron flow fast enough to keep up with the incoming light pulses.

The PIN photodiode shown in Figure 28.3 is constructed a little differently than the PN photodiode. An *intrinsic* layer is used to separate the *p* region and the *n* region. This creates a large depletion region that absorbs the photons with improved efficiency when compared to the PN photodiode.

FIGURE 28.3
PIN photodiode

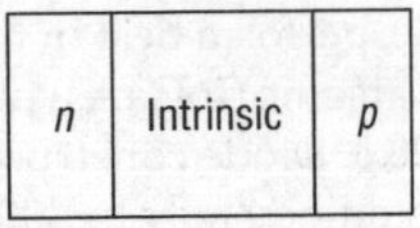

n	Intrinsic	p

Avalanche Photodiode

The *avalanche photodiode (APD)* works just as its name suggests. On a snow-covered mountain, a small vibration can trigger an avalanche of snow. With the APD, a small bundle of photons can trigger an avalanche of electrons. The APD accomplishes this through a process called *photomultiplication*.

The APD is constructed with one more *p* region than the PIN photodiode, as shown in Figure 28.4. When the APD is biased very close to its breakdown voltage, it acts like an amplifier with a multiplication factor, or *gain*. An APD with a multiplication factor of 50 sets free on the average 50 electrons for each photon absorbed. The free electrons produce current flow through the electrical circuit connected to the APD.

FIGURE 28.4
Avalanche
photodiode

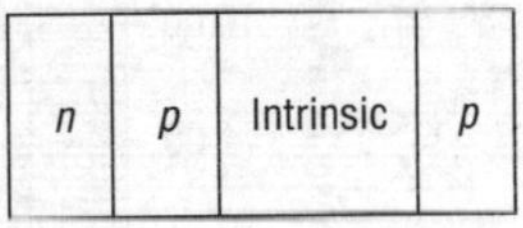

n	p	Intrinsic	p

Photodiode Responsivity, Efficiency, and Speed

This section describes the factors determining how responsive a photodiode is.

Responsivity

The *responsivity* of the photodiode describes how well the photodiode converts a wavelength or a range of wavelengths of optical energy into electrical current. It's the ratio of the photodiode electrical output current to its optical input power. The greater the responsivity, the greater the electrical current output for a given amount of optical input power.

Responsivity is described in amperes/watt (A/W). However, a photodiode in a fiber-optic receiver will never generate an ampere of electrical current. That's not to say that a photodiode *couldn't* be built to generate an ampere of electrical current, but remember that a photodiode in a fiber-optic receiver receives a very small amount of light energy. A typical LED receiver works well with an optical input power as low as one microwatt. One microwatt is one millionth of a watt.

The overall responsivity of a photodiode depends on three factors:

- Semiconductor material makeup

- Wavelength

- Diode construction

Photodiodes are constructed for specific wavelengths. Some semiconductor materials perform better at longer wavelengths and some perform better at shorter wavelengths. Silicon photodiodes, for example, perform best in the visible and short infrared wavelengths. Germanium and indium gallium arsenic (InGaAs) photodiodes perform best at long infrared wavelengths, as shown in Figure 28.5. Diode construction also plays a large role in responsivity. The responsivity of an APD photodiode may be 100 or more times greater than a PIN photodiode.

FIGURE 28.5
Photodiode
semiconductor
responsivity

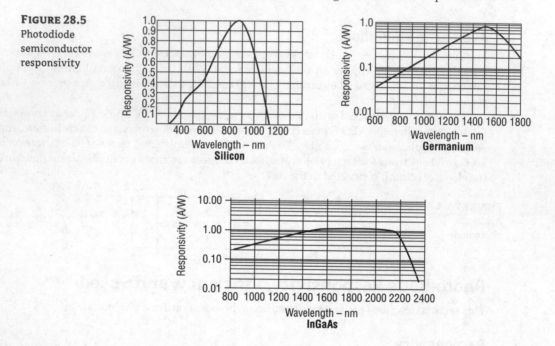

Quantum Efficiency

The responsivity of a photodiode depends on its quantum efficiency. Quantum efficiency describes how efficiently a photodiode converts light energy into electrical energy—that is, photons into free electrons. It is typically expressed as a percentage. A quantum efficiency of 47 percent means that for every 100 photons absorbed, 47 current-generating electrons will be created.

Switching Speed

It has been stressed many times in this book that fiber-optic systems transmit and receive millions or billions of light pulses per second. When data is being moved at this rate, there is very little time for the photodiode to switch on and off. Switching speed depends on physical size, construction, and electrical biasing. Photodiodes in fiber-optic receivers are very small in size and biased to produce the fastest possible switching time.

Fiber-Optic Receiver

Many different fiber-optic receiver designs are in use today. The complexity of these receivers varies by application. The job of the receiver is to take the light energy from the optical fiber and convert it into electrical energy. The output of the receiver is designed to interface with electronics that handle the information after the light-to-electricity conversion. A typical receiver can be broken into three subassemblies: electrical subassembly, optical subassembly, and *receptacle*, as shown in Figure 28.6.

FIGURE 28.6
Block diagram of a typical LED receiver

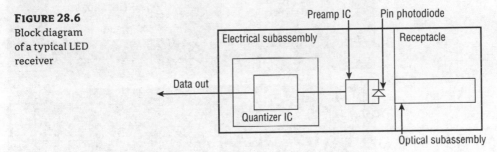

Packaging

As you learned in Chapter 27, fiber-optic receivers are typically packaged with the transmitter. Together, the receiver and transmitter form a *transceiver*. Engineers are finding more and more applications for fiber optics in different types of environments. Because of this, fiber-optic transceivers are being developed in many different form factors or packages.

In this section, we will look at two different types of receiver construction or packaging:

Transceiver package soldered to a circuit board   One receiver type is part of a transceiver package that is typically soldered to a circuit board. The receptacle that accepts the fiber-optic connector is part of the transceiver package. The receptacles found on these receivers are not designed for a harsh environment. However, the circuit board containing the transceiver package may be installed in an enclosure that is designed for harsh environments. These receivers accept connectors such as the ST, SC, or LC, like the transceivers soldered to the circuit board in Figure 28.7.

Photo courtesy of Protokraft

FIGURE 28.7
LED transceivers soldered to circuit board

Transceiver package contained within a receptacle   The other receiver type is part of a transceiver package that is contained within a harsh environment receptacle like those covered in Chapter 26, "Connectors." The electrical subassembly and optical subassembly are integrated into the receptacle. They may be integrated into the receptacle two different ways depending on the size of the receptacle cavity. The electrical subassembly and optical subassembly can be incorporated into a size 8 cavity insert. The size 8 insert shown in Figure 28.8 is a Gigabit Ethernet receiver.

FIGURE 28.8
Size 8 cavity inserts with Gigabit Ethernet receiver

The electrical subassembly and optical subassembly can also be integrated into the receptacle, as shown in Figure 28.9. The electrical subassembly of this transceiver receiver is soldered directly to the circuit board.

FIGURE 28.9
Transceiver receptacle soldered to the circuit board

Photo courtesy of Protokraft

Integrating the optical and electrical subassemblies into the harsh environment receptacle eliminates an interconnection and the loss associated with that interconnection. If the transceiver shown in Figure 28.7 were to be used in a harsh environment, it would need to be mounted in an enclosure that protects the transceiver and interconnection. An interconnection would be required at the enclosure bulkhead to bring the optical signal into the enclosure. A short length of fiber-optic cable would be required inside the enclosure to bring the optical signal to the receiver. Having the optical and electrical subassemblies in the receptacle eliminates one interconnection and the short fiber-optic cable.

Receptacle

Regardless of the receiver packaging, the receptacle is the part of the fiber-optic receiver that accepts the connector. Its job is to provide alignment for the ferrule or terminus so that the optical fiber is perpendicular to the photodiode in the optical subassembly.

Optical Subassembly

The *optical subassembly* guides the light energy from the optical fiber to the photodiode. Depending on the receiver packaging, a window or terminus endface in the optical subassembly makes contact with the optical fiber endface. The window may be shaped like a lens to focus the light energy onto the photodiode, as shown in Figure 28.10, or a lens may be placed between the window and the photodiode.

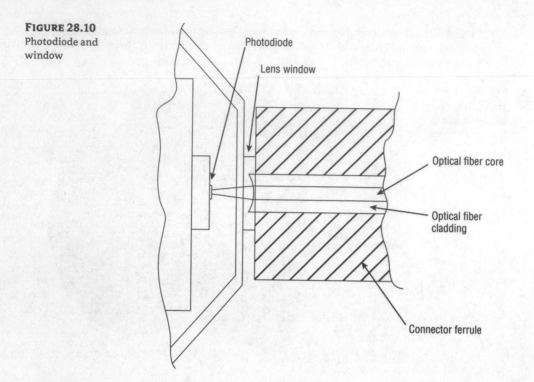

FIGURE 28.10
Photodiode and window

In many receivers, the photodiode and preamplifier are housed in the optical subassembly. Sometimes the optical subassembly is referred to as the photodiode preamplifier subassembly. As we learned earlier in this chapter, the photodiode used in a receiver does not produce a substantial electrical signal. Mounting the preamplifier as physically close as possible to the photodiode allows the preamplifier to receive the strongest electrical signal.

Electrical Subassembly

The electrical subassembly is typically built on a multilayer printed circuit board. This printed circuit contains the *quantizer IC* and other passive components required to complete the electrical circuit. The photodiode and preamplifier contained in the optical subassembly are electrically connected to the printed circuit board. A metal shield, shown in Figure 28.11, is typically placed over the electrical subassembly. The shield serves two purposes; it protects the electrical subassembly from external electromagnetic radiation and reduces the amount of electromagnetic radiation from the electrical subassembly.

When the receiver is incorporated into a cavity insert, the metal housing acts as a shield. The metal receptacle also provides additional shielding.

The photodiode converts the light pulses from the optical fiber to electrical current pulses, as discussed earlier in this chapter. The *transimpedance amplifier*, or preamplifier, amplifies the electrical current pulses from the photodiode and outputs voltage pulses to the quantizer IC. The *limiting amplifier* in the quantizer IC amplifies the voltage pulses and provides a binary decision. It determines whether the electrical pulses received represent a binary 1 or a binary 0.

FIGURE 28.11
Transceiver
module with EMI
shield exposed

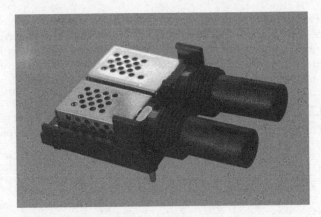

The quantizer IC also measures the received optical energy. It sets the signal detect line when there is adequate signal strength to convert the light energy into electrical energy. This prevents the electronics from trying to decode a weak signal or noise.

Receiver Performance Characteristics

This section examines several key performance characteristics of a fiber-optic receiver. It generalizes *dynamic range* and *operating wavelength*. The specific performance characteristics of LED and laser receivers are covered in detail. The data used in this section was extracted from manufacturer data sheets and represents typical performance characteristics of readily available fiber-optic receivers.

Dynamic Range

Fiber-optic receivers are limited in the amount of optical input power they can receive. Too much optical input power will saturate the photodiode. Saturating the photodiode prevents the photodiode from turning off after the light pulse has been absorbed. This can cause electrical output pulses of the receiver to overlap, creating a *bit error*.

The dynamic range of the receiver is measured in decibels. It is the difference between the maximum and minimum optical input power that the receiver can accept. If the maximum optical input power is –14dBm and the minimum optical input power is –32dBm, the dynamic range would be the difference between the two values, or 18dB.

The receiver will generate a minimum number of errors when the optical input power is kept within the minimum and maximum values. All receivers generate errors. Error generation is typically described by the receiver's *bit error rate*, or BER. A typical receiver may have a BER of one error in a billion to one error in a trillion. Typically this would be written as 10^{-9} or 10^{-12}, respectively.

Operating Wavelength

Fiber-optic receivers are designed to operate within a range of wavelengths. A typical 1300nm receiver may have an operating wavelength range from 1270nm to 1380nm. This is because fiber-optic transmitters output optical energy within a wavelength range. The receiver must be able to accept a 1300nm transmitter that has a center wavelength of 1275nm or 1375nm. A receiver designed for 1300nm may not perform well or may not perform at all when connected to an 850nm or 1550nm transmitter because the responsivity of the photodiode changes over a range of wavelengths.

LED Receiver Performance Characteristics

The performance characteristics of the LED receiver are typically broken up into four groups:

♦ Recommended operating conditions

♦ Electrical characteristics

♦ Optical characteristics

♦ Data rate

The recommended operating conditions describe the maximum and minimum temperature and voltage ranges that the device can operate in without damage. Table 28.1 shows the typical recommended operating conditions for a 1300nm 100Mbps LED receiver.

TABLE 28.1   LED Receiver Recommended Operating Conditions

PARAMETER	SYMBOL	MIN.	TYP.	MAX.	UNIT
Ambient operating temperature	T_A	0		70	°C
Supply voltage	V_{CC}	4.75		5.25	V
Data input voltage—low	$V_{IL} - V_{CC}$	−1.810		−1.475	V
Data input voltage—high	$V_{IH} - V_{CC}$	−1.165		−0.880	V
Data and signal detect output load	R_L		50		Ω

The electrical characteristics of the LED receiver describe the supply current requirements, data output voltages, signal detect output voltages, rise/fall times, and power dissipated by the device. Table 28.2 shows the typical electrical characteristics for a 1300nm 100Mbps LED receiver.

TABLE 28.2   LED Receiver Electrical Characteristics

PARAMETER	SYMBOL	MIN.	TYP.	MAX.	UNIT
Supply current	I_{CC}		82	145	mA
Power dissipation	P_{DISS}		0.3	0.5	W
Data output voltage—low	$V_{OL} - V_{CC}$	−1.840		−1.620	V

TABLE 28.2   LED Receiver Electrical Characteristics   *(CONTINUED)*

PARAMETER	SYMBOL	MIN.	TYP.	MAX.	UNIT
Data output voltage—high	$V_{OH} - V_{CC}$	−1.045		−0.880	V
Data output rise time	t_r	0.35		2.2	ns
Data output fall time	t_f	0.35		2.2	ns
Signal detect output voltage—low	$V_{OL} - V_{CC}$	−1.840		−1.620	V
Signal detect output voltage—high	$V_{OH} - V_{CC}$	−1.045		−0.880	V
Signal detect output rise time	t_r	0.35		2.2	ns
Signal detect output fall time	t_f	0.35		2.2	ns

The optical characteristics of the LED receiver at a minimum include minimum optical input power, maximum optical input power, and operating wavelength. Table 28.3 shows the typical optical characteristics for a 1300nm 100Mbps LED receiver.

TABLE 28.3   Receiver Optical Characteristics

PARAMETER	SYMBOL	MIN.	TYP.	MAX.	UNIT
Optical input power minimum at window edge	$P_{IN\,Min.}$ (W)		−33.5	−31	dBm avg.
Optical input power maximum	$P_{IN\,Max.}$	−14	−11.8		dBm avg.
Operating wavelength	λ	1270		1380	nm

If the LED receiver optical characteristics are known, the dynamic range for the receiver can be calculated. The dynamic range is the difference between the minimum value for the maximum optical input power and the maximum value for the minimum optical input. The minimum value for the maximum optical input power in Table 28.3 is –14dBm. The maximum value for the minimum optical input power is –31dBm. (Remember that this information should be obtained from the manufacturer's data sheet.)

To calculate the dynamic range, subtract the minimum value for the maximum optical input power ($P_{IN\,Max}$) from the maximum value for the minimum optical input power ($P_{IN\,Min}$), as shown in the following formula:

$P_{IN\,Max} - P_{IN\,Min}$ = Dynamic range

–14dBm – –31dBm = 17dB

Dynamic range = 17dB

LED receivers are typically designed to support one or more network standards. A 100Mbps 1300nm receiver could support a 100Base-FX Ethernet application or a 100Mbps ATM application. Table 28.4 lists the data rates, wavelengths, and optical fiber type for various LED receivers used in IEEE 802.3 Ethernet communication systems.

TABLE 28.4   LED Receivers for Ethernet Applications

NETWORK	DATA RATE	WAVELENGTH AND MEDIA
10Base-FL	10Mbps	850nm, multimode fiber
100Base-FX	100Mbps	1300nm, multimode fiber
100Base-SX	10/100Mbps	850nm, multimode fiber

Laser Receiver Performance Characteristics

The performance characteristics of the laser receiver, like the LED receiver, are typically broken up into four groups:

◆ Recommended operating conditions

◆ Electrical characteristics

◆ Optical characteristics

◆ Data rate

The recommended operating conditions describe maximum and minimum temperature and voltage ranges that the device can operate in without damage. Table 28.5 shows the typical recommended operating conditions for a 1300nm 2.5Gbps laser transmitter.

TABLE 28.5   Laser Receiver Recommended Operating Conditions

PARAMETER	SYMBOL	MIN.	TYP.	MAX.	UNIT
Ambient operating temperature	T_A	0		+70	°C
Supply voltage	V_{CC}	3.1		3.5	V
Power supply rejection	PSR		100		mV_{P-P}
Transmitter differential input voltage	V_D	0.3		2.4	V

TABLE 28.5   Laser Receiver Recommended Operating Characteristics   *(CONTINUED)*

PARAMETER	SYMBOL	MIN.	TYP.	MAX.	UNIT
Data output load	R_{OL}		50		Ω
TTL signal detect output current—low	I_{OL}			1.0	mA
TTL signal detect output current—high	I_{OH}	−400			µa
Transmit disable input voltage—low	T_{DIS}			0.6	V
Transmit disable input voltage—high	T_{DIS}	2.2			V

The electrical characteristics of the laser receiver describe the supply current requirements, data output characteristics, and power dissipated by the device. Table 28.6 shows the typical electrical characteristics for a 1300nm 2.5Gbps laser receiver.

TABLE 28.6   Laser Receiver Electrical Characteristics

PARAMETER	SYMBOL	MIN.	TYP.	MAX.	UNIT
Supply current	I_{CC}		115	140	mA
Power dissipation	P_{DISS}		0.38	0.49	W
Data output voltage swing (single-ended)	$V_{OH} - V_{OL}$	575		930	mV
Data output rise time	t_r		125	150	ps
Data output fall time	t_f		125	150	ps
Signal detect output voltage—low	V_{OL}			0.8	V
Signal detect output voltage—high	V_{OH}	2.0			V
Signal detect assert time (OFF to ON)	AS_{MAX}			100	µS
Signal detect deassert time (ON to OFF)	ANS_{MAX}			100	µS

The optical characteristics of the laser receiver at a minimum include minimum optical input power, maximum optical input power, center wavelength, spectral width, and reflectance. *Reflectance* is the ratio of reflected power to incident power, where incident power is the light energy exiting the optical fiber into the receiver and reflected power is the light energy reflected from the receiver into the core of the optical fiber traveling toward the light source. Table 28.7 shows the typical optical characteristics for a 1300nm 2.5Gbps laser receiver.

TABLE 28.7   Laser Receiver Optical Characteristics

PARAMETER	SYMBOL	MIN.	TYP.	MAX.	UNIT
Receiver sensitivity	$P_{IN\,Min}$		−23	−19	dBm avg.
Receiver overload	$P_{IN\,Max}$	−3	+1		dBm avg.
Input operating wavelength	λ	1260		1570	nm
Signal detect—asserted	P_A		−27.3	−19.5	dBm avg.
Signal detect—deasserted	P_D	−35	−28.7		dBm avg.
Signal detect—hysteresis	P_H	0.5	1.4	4	dB
Reflectance			−35	−27	dB

If the laser receiver optical characteristics are known, the dynamic range for the receiver can be calculated. As with the LED receiver, the dynamic range for the laser receiver is the difference between the minimum value for the maximum optical input power and the maximum value for the minimum optical input. The minimum value for the maximum optical input power in Table 28. 7 is −3dBm. The maximum value for the minimum optical input power is −19dBm. (Remember that this information should be obtained from the manufacturer's data sheet.)

To calculate the dynamic range, subtract the minimum value for the maximum optical input power ($P_{IN\,Max}$) from the maximum value for the minimum optical input power ($P_{IN\,Min}$) as shown in this formula:

$P_{IN\,Max} - P_{IN\,Min}$ = dynamic range

−3dBm − −19 dBm = 16dB

Dynamic range = 16dB

Laser transmitters can typically support multiple data rate operations. The same transmitter may be used with data rates ranging from 155Mbps in an ATM switch to 2.5Gbps for a SONET OC-48 switch. Table 28.8 lists the data rates, wavelengths, and optical fiber type for various laser transmitters used in IEEE 802.3 Ethernet communication systems.

TABLE 28.8   Laser Receivers for Ethernet Applications

NETWORK	DATA RATE	WAVELENGTH AND MEDIA
100Base-LX	10/100Mbps	1300nm single-mode fiber
100Base-LH	10/100Mbps	1300nm single-mode fiber

TABLE 28.8 Laser Receivers for Ethernet Applications *(CONTINUED)*

NETWORK	DATA RATE	WAVELENGTH AND MEDIA
100Base-LH	10/100Mbps	1550nm single-mode fiber
1000Base-SX	1,000Mbps	850nm VCSEL, multimode fiber
1000Base-LX	1,000Mbps	1310nm FP, multimode fiber
10GBase-SR	10,000Mbps	850nm VCSEL, multimode fiber
10GBase-LX4	10,000Mbps	1310nm DFB, multimode fiber

IS IT THE COMPUTER HARDWARE OR SOFTWARE?

We were troubleshooting a communication problem between a piece of rack-mounted computing equipment and a router. The equipment and the router were communicating over 50/125μm multimode optical fiber. Everything would work for a while, and then the equipment and router would stop communicating.

Typically, when things like this happen, the programmers blame the failure on the hardware and we hardware engineers blame the failure on the software. Because we had worked together on the design of the network switch in the computing equipment, we immediately became involved when the communication failure occurred. The next phase was troubleshooting to prove that the hardware was not the problem.

During the troubleshooting process, someone questioned whether the receiver on the router was receiving any light energy from the switch's transmitter. Because the two pieces of equipment communicate at a wavelength of 1300nm, the light is not visible. To quickly answer that question, a power meter was used to measure the optical output power from the transmitters on both pieces of equipment. A mode filter was used on the 1 meter jumper from the transmitter to the power meter.

The measurements obtained were compared to the manufacturer's optical characteristics for both the transmitter and the receiver. The optical output power for each transmitter was within specifications. About the same time that we determined the hardware was functioning properly, the programmers made some minor changes and the problem was resolved.

Being able to measure the transmitter's optical output power and determine that it's within the acceptable range for the receiver is a fundamental skill that will be needed in various troubleshooting scenarios.

The Bottom Line

Calculate the dynamic range for an LED receiver. The dynamic range of the LED receiver is the difference between the minimum value for the maximum optical input power and the maximum value for the minimum optical input.

Master It Refer to the following table and calculate the dynamic range for the LED receiver.

Parameter	Symbol	Min.	Typ.	Max.	Unit
Optical input power minimum at window edge	$P_{IN\ Min.}$ (W)		−30.5	−28	dBm avg.
Optical input power maximum	$P_{IN\ Max.}$	−16	−13.8		dBm avg.
Operating wavelength	λ	1270		1380	nm

Calculate the dynamic range for a laser receiver. The dynamic range of the laser receiver is the difference between the minimum value for the maximum optical input power and the maximum value for the minimum optical input.

Master It Refer to the following table and calculate the dynamic range for the laser receiver.

Parameter	Symbol	Min.	Typ.	Max.	Unit
Receiver sensitivity	$P_{IN\ Min}$		−20	−16	dBm avg.
Receiver overload	$P_{IN\ Max}$	−1	+3		dBm avg.
Input operating wavelength	λ	1260		1570	nm
Signal detect—asserted	P_A		−24.3	−16.5	dBm avg.
Signal detect—deasserted	P_D	−32	−25.7		dBm avg.
Signal detect—hysteresis	P_H	0.5	1.4	4	dB
Reflectance			−35	−27	dB

Chapter 29

Passive Components and Multiplexers

The objective of this chapter is to help you gain an understanding of fiber-optic *passive components* and *multiplexers*. This chapter covers not only particular devices and their applications, but also the reasons why the components were chosen and when they should be used.

Fiber-optic passive components and multiplexers are elementary items but necessary in all applications that require the transmission, combining, or distribution of optical signals. Passive components are components that do not require an external energy source. Multiplexers are devices that are used to combine two or more signals into a single output. The term *multiplexing* is used to refer to the process by which the signals are combined.

Some of the optical devices we cover in this chapter are *couplers*, *switches*, *attenuators*, *isolators*, *amplifiers*, and *filters*. We will also examine multiplexers and their associated processes, in particular *wavelength division multiplexing* and *dense wavelength division multiplexing*.

In this chapter, you will learn to:

- ◆ Calculate the output power of a real tee coupler port
- ◆ Calculate the output power of a real star coupler
- ◆ Calculate attenuator values

Standards

There are many standards for passive components and multiplexers. Passive components and multiplexers play a key role in bringing fiber optics to your home or business. They are the core of a *passive optical network (PON)*, and the application of many of the devices in this chapter is covered in Chapter 30, "Passive Optical Networks."

This chapter provides an overview of many different passive components and their operation. For each of these components, one or more standards define performance and testing requirements. The International Electrotechnical Commission (IEC) has developed many of these standards.

IEC has two technical subcommittees—86B and 86C—under Technical Committee 86. The subcommittees focus on standards for many of the components in this chapter. Subcommittee 86B focuses on fiber-optic interconnecting devices and passive components. These components include attenuators, switches, wavelength multiplexers/demultiplexers, couplers, and isolators. This subcommittee has developed numerous performance and testing standards for interconnecting devices and passive components.

IEC Subcommittee 86C focuses on fiber-optic systems and active devices, including terminology, test and measurement methods, functional interfaces, and mechanical, optical, environmental, and electrical requirements. The documents developed by this subcommittee help to ensure the interoperability of components and systems, which leads to reliable system performance.

Parameters

This chapter discusses many different passive devices and some of the common parameters that apply to each device. When working with passive components it is important to have a basic understanding of common parameters. Some of the parameters that you need to be familiar with are optical fiber type, connector type, *center wavelength, bandwidth, insertion loss, excess loss, polarization-dependent loss (PDL), return loss, crosstalk, uniformity, power handling*, and *operating temperature*.

Connector types and optical fiber types   Many passive devices are available with receptacles or pigtails. The pigtails may or may not be terminated with a connector. If the device is available with a receptacle or connector, the type of receptacle or connector needs to be specified when ordered. You should also note the type of optical fiber used by the manufacturer of the device to ensure it is compatible with the optical fiber used for your application.

Center wavelength and bandwidth   Passive devices also have a center wavelength and bandwidth or bandpass. The center wavelength is the nominal operating wavelength of the passive device. The bandwidth and bandpass are the range of wavelengths over which the manufacturer guarantees the performance of the device. Some manufacturers will list an operating wavelength range instead.

Types of loss   Insertion loss, excess loss, and PDL are all measured in decibels.

Insertion loss   This is the optical power loss caused by the insertion of a component into the fiber-optic system. When working with passive devices, you need to be aware of the insertion loss for the device and the insertion loss for an interconnection. Insertion loss as stated by the manufacturer typically takes into account all other losses, including excess and PDL. Insertion loss is the most useful parameter when designing a system.

Excess loss   This may or may not be defined by the manufacturer. Excess loss, associated with optical couplers, is the amount of light lost in the coupler in excess of the light lost from splitting the signal. In other words, when a coupler splits a signal, the sum of the power at the output ports does not equal the power at the input port; some optical energy is lost in the coupler. Excess loss is the amount of optical energy lost in the coupler. This loss is typically measured at the specified center wavelength for the device.

Polarization-dependent loss (PDL)   This is only a concern for single-mode passive devices. It is often the smallest value loss, and it varies as the polarization state of the propagating light wave changes. Manufacturers typically provide a range for PDL or define a not-to-exceed number.

Return loss   When a passive device is inserted, some of the optical energy from the source is going to be reflected back toward the source. This is typically described as return loss, or reflection loss. Return loss is expressed in decibels and is the negative quotient of the power received divided by the power transmitted. Return loss is covered in detail in Chapter 33, "Test Equipment and Link/Cable Testing."

Crosstalk   Crosstalk in an optical device describes the amount of light energy that leaks from one optical conductor to another. Crosstalk is not a concern in a device where there is a single input and multiple outputs. However, it is a concern with a device that has multiple inputs and a single output, such as an optical switch. Crosstalk is expressed in decibels, where the value defines the difference between the optical power of one conductor and the amount of leakage into another conductor. In an optical switch with a minimum crosstalk of 60dB, there is a 60dB difference between the optical power of one conductor and the amount of light that leaked from that conductor into another conductor.

Uniformity   Uniformity is expressed in decibels. It is a measure of how evenly optical power is distributed within the device. For example, if a device is splitting an optical signal evenly into four outputs, how much those outputs could vary from one another is defined by uniformity. Uniformity is typically defined over the operating wavelength range for the device.

Power handing   Power handling describes the maximum optical power at which the device can operate while meeting all the performance specifications defined by the manufacturer. Power handling may be defined in milliwatts or decibels, where 0dBm is equal to 1mW.

Operating temperature   Operating temperature describes the range of temperatures that the device is designed to operate in. This can vary significantly between devices, because some devices are only intended for indoor applications while others may be used outdoors or in other harsh environments.

Couplers

In many applications, it may not be possible to have a design of many point-to-point connections. In these cases, optical couplers are used. A fiber-optic coupler is a device that combines or splits optical signals. A coupling device may combine two or more optical signals into a single output, or the coupler may be used to take a single optical input and distribute it to two or more separate outputs. Figure 29.1 is an example of a basic four-port coupler.

FIGURE 29.1
Four-port coupler

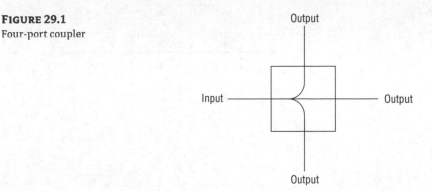

Many couplers are designed bidirectionally, which enables the same coupler to be used to combine signals or split signals. A coupler being used to split a signal may be referred to as a splitter.

Couplers are available with a wide range of input and output ports. A basic coupler may have only one input port and two output ports. Today's technology supports couplers with up to 64 input and 64 output ports, as shown in Figure 29.2.

FIGURE 29.2
128-port coupler

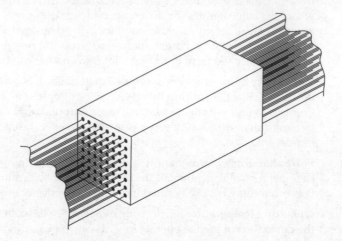

There are many different types of couplers, and the number of input and output ports is dependent on the intended usage. Some of the types of *optical couplers* are *optical combiners*, *Y couplers*, *star couplers*, *tee couplers*, *tree couplers*, and *optical splitters*; in this chapter, we will only focus on the tee coupler and the star coupler.

The Tee Coupler

A *tee coupler* is a three-port optical coupling device that has one input port and two output ports, as shown in Figure 29.3.

FIGURE 29.3
Tee coupler

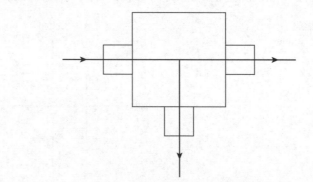

The tee coupler is a passive device that splits the optical power from the input port into two output ports. The tee coupler is in essence an optical splitter. The uniqueness of the tee coupler is that this type of coupler typically distributes most of the optical input power to one output and only a small amount of power to the secondary output. Note that when the outputs are evenly distributed, the coupler is called a Y coupler. The tee coupler is also referred to as an optical tap, due to the nature of the device. A majority of the power continues forward, but a portion of the signal (determined by the splitting ratio) is tapped to be used for an output port.

DISTRIBUTION AND LOSS OF OUTPUT POWER IN A TEE COUPLER

The tee coupler is a 1 × 2 coupler, meaning that it has one input port (or connection) and two output ports. As previously stated, the optical output power of the two output ports is typically not evenly distributed. Common splitting ratios are 90:10, 80:20, 70:30, 60:40, and 50:50 (a Y coupler). Not all manufacturers follow the convention of placing the larger value to the left of the colon and the smaller to the right. Some manufacturers simply reverse this and place the smaller value to the left of the colon and the larger to the right.

A typical use for a tee coupler would be to supply optical signals to a bus type network of in-line terminals. Assuming ideal conditions and a 90:10 split on the tee coupler, the first terminal would receive 10 percent of the optical signal and 90 percent of the optical signal would go forward to the next tee coupler, and so on, as shown in Figure 29.4.

FIGURE 29.4

Ideal tee couplers in a bus type network

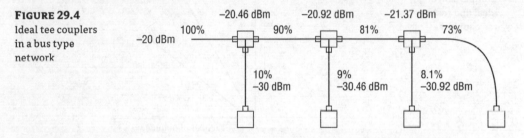

It's easy to see how the optical signal power levels decrease from one terminal to the next. Keep in mind that we did not account for any losses; this is an ideal system. If we were to take losses into account, the results (shown in Figure 29.5) would be much different.

FIGURE 29.5

Real tee couplers in a bus type network

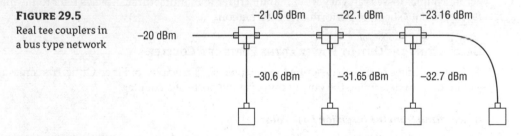

Now let's look at some of the common losses associated with couplers and their effects on the network of interconnections. To be realistic, we have to take into account the insertion loss for the device and the insertion loss for each individual interconnection, whether a splice or a mated connector pair.

METHODS OF DESCRIBING INSERTION LOSS

The insertion loss for the coupler may be described two ways. One manufacturer may provide an insertion loss value for a device and you will have to calculate the power at each output port based on the splitting ratio. However, another manufacturer may provide the insertion loss combined with the loss from the splitting ratio. In this section, you will learn how to calculate the power at an output port using the insertion loss values provided by the manufacturer.

In Figure 29.5, the insertion loss for each coupler and each interconnection is identical; each interconnection had an insertion loss of 0.3dB and each coupler had an insertion loss of 0.3dB. These losses add up and affect the optical output power of each port. Each coupler adds additional insertion losses; these losses reduce the number of taps a real tee coupler can support when compared to an ideal tee coupler. Figure 29.6 graphically compares the performance differences between ideal tee couplers and real tee couplers.

FIGURE 29.6
Comparison of ideal and real tee couplers

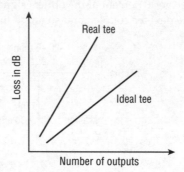

As shown by the previous example, taking into account the losses of the device and the interconnections can make a large difference in anticipated optical output power. Remember that actual coupler losses can vary from manufacturer to manufacturer. Always refer to the manufacturer's data sheet for performance information.

CALCULATING THE OUTPUT POWER AT THE PORTS OF A COUPLER

The decibel rules of thumb described in Chapter 19, "Principles of Fiber-Optic Transmission," can be used to determine the output power at the ports of a coupler.

Calculating from the Insertion Loss Value

To determine the output power at both output ports of a tee coupler with a 50:50 splitting ratio using the decibel rules of thumb, we will assume that the insertion loss for each interconnection is the same and that the manufacturer only provided the insertion loss for the coupler and did not take into account the loss for the splitting ratio. In this case, the total loss is the sum of the insertion losses for the interconnections (LI) and the insertion loss for the device (LC) as shown in this formula:

$LI_1 + LI_2 + LI_3 = LI_T$

$LI_T + LC = Total Loss$

For this example, let's assume that the insertion loss for each interconnection is 0.3dB and the insertion loss for the coupler is 0.5dB. The total coupler loss would be 1.4dB, as shown here:

0.3dB + .03dB + 0.3dB = 0.9dB

0.9dB + 0.5dB = 1.4dB

Remember from earlier in the chapter that uniformity is a measure of how evenly optical power is distributed within the device. In the next step, we will calculate the optical power at each output port without taking into account uniformity. To find the optical output power at each port, we must know the input power.

Assume that the input power to the coupler is –20dBm. The first step is to subtract the total coupler losses (1.4dB) from the input power (–20dBm). As shown here, the remaining power available to the ports is –21.4dBm:

–20dBm – 1.4dB = –21.4dBm

The second step is to split the remaining power using the decibel rules of thumb covered in Chapter 19.

Because this coupler has a 50:50 splitting ratio, each output port should receive 50 percent of the energy. A loss of 50 percent is a change of 3dB. The output power at each port will be 3dB less then the power remaining after all the insertion losses have been accounted for as shown here:

–21.4dBm – 3dB = –24.4dBm

Each output port should have a power output of –24.4dBm. However, couplers are not perfect and the amount of power actually available will vary depending on the uniformity. Assume that the uniformity for this coupler is 0.2dB. This means that the difference between the highest and lowest insertion loss for the coupler output ports will not exceed 0.2dB within the bandpass. In this example, the difference between the output power in each port could vary as much as 0.2dB.

Calculating with the Insertion Loss and Excess Loss

In the next example, the manufacturer provided the insertion loss combined with the loss from the splitting ratio. This tee coupler has a splitting ratio of 80:20. The decibel rules of thumb will not be required to calculate the power available at each output port because the manufacturer states the insertion loss in decibels as 1.5/7.6. This means that the power at the 80 percent port will be 1.5dB less than the input power to the coupler and the power at the 20 percent port will be 7.6dB less than the input power to the coupler.

If the input power to the coupler is –20dBm, as in the previous example, the power at each port can be calculated by subtracting the insertion loss for each port defined by the manufacturer and subtracting the interconnection insertion losses from -20dBm. Assume that insertion loss for each interconnection is 0.25dB. As shown here, the output power at the 80 percent port is –22.0dBm and the output power at the 20 percent port is –28.1dBm.

–20dBm – 1.5dB – 0.25dB – 0.25dB = –22dBm

–20dBm – 7.6dB – 0.25dB – 0.25dB = –28.1dBm

The Star Coupler

The star coupler is used in applications that require multiple ports—input and/or output. The star coupler will distribute optical power equally from one or more input ports to two or more output ports. Figure 29.7 shows a basic star coupler with four input ports and four output ports. Star couplers are available in 1×64 up to 64×64 dimensions.

FIGURE 29.7
Eight-port star
coupler

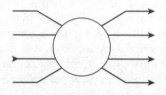

A special version of the star coupler, called a *tree coupler*, is used when there is one input port and multiple output ports or when there are multiple input ports and one output port.

Star couplers are frequently used in network applications when there are a large number of output terminals. In our tee coupler example, we had to account for interconnection insertion loss and coupler insertion loss at each tee connection. However, with the star coupler there is only one coupler insertion loss regardless of the number of ports. With only one coupler insertion loss, a multiple port star coupler is more efficient than a series of tee couplers, as shown in Figures 29.4 and 29.5. So the larger the network is, the more efficient the star coupler becomes.

Two types of star couplers are commonly used: the reflective star and the transmissive star. Couplers are typically considered to be a black box—that is, only the manufacturer knows what's inside. However, many star couplers are made of fused optical fibers.

The reflective star, shown in Figure 29.8, is defined as a coupler that distributes optical power to all input and output ports when a signal is applied to only one port.

The transmissive star, shown in Figure 29.9, is defined as a coupler that distributes optical power to all the output ports when a signal is applied to any of the input ports.

FIGURE 29.8
Fused reflective
star coupler

FIGURE 29.9
Fused transmis-
sive star coupler

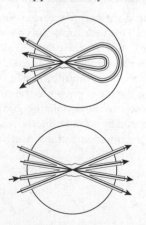

CALCULATING THE LOSSES WITH A STAR COUPLER

In our previous example, we looked at the losses in a network of four terminals with tee couplers. Let's compare the losses for the same number of terminals using a star coupler. Figure 29.10 shows the power delivered to each terminal of a four-port star coupler using the same values as in the previous example, 0.3dB interconnection insertion losses and 0.3dB coupler insertion loss (due to only one coupler). This example shows a tree coupler with four output ports and one input port, similar to the sequential tee-coupled workstation network shown in Figure 29.5.

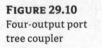

FIGURE 29.10
Four-output port
tree coupler

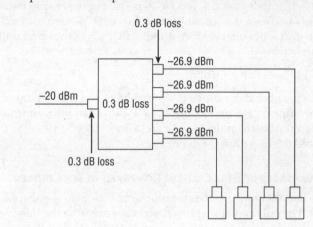

Let's determine the output power at each output port of a five-output port star coupler using the decibel rules of thumb. (Note that this exercise does not use the coupler shown in Figure 29.10.)

In this example, we will assume that the insertion loss for each interconnection is the same. Unlike the tee coupler, manufacturers typically only provided the insertion loss for the coupler and do not take into account the loss for the splitting ratio. In this case, the total loss is the sum of the insertion losses for the interconnections (LI) and the insertion loss for the coupler (LC) as shown here:

$LI_1 + LI_2 + LI_3 + LI_4 + LI_5 = LI_T$

$LI_T + LC =$ Total Loss

For this example, let's assume that the insertion loss for each interconnection is 0.3dB and the insertion loss for the coupler is 1.2dB. This coupler has five output ports and one input port. The total coupler loss would be 3dB, as shown here:

0.3dB + .03dB + 0.3dB + 0.3dB + 0.3dB + 0.3dB = 1.8dB

or

0.3dB × 6 = 1.8dB

1.8dB + 1.2dB = 3dB

Assume that the input power to the coupler is –7dBm. The first step is to account for the star coupler insertion losses and the interconnection losses. This is done by subtracting 3dB from the input power of –7dBm. The remaining power available to the ports is –10dBm, as shown here:

–7dBm – 3dB = –10dBm

The second step is to split the remaining power using the decibel rules of thumb covered in Chapter 19.

Each output port will receive 20 percent of the energy because the energy is distributed evenly between the output ports. A loss of 80 percent is a change of 7dB. The output power at each port will equal –10dBm minus 7dB. Each output port will have a power output of –17dBm, as shown here:

–10dBm – 7dB = –17dBm

Remember that couplers are not perfect and the amount of power actually available will vary depending on the uniformity. If we assume that the uniformity for this coupler is 0.3dB, the difference between the highest and lowest insertion loss for the coupler output ports will not exceed 0.3dB within the bandpass.

ADVANTAGES OF STAR COUPLER COMPARED TO TEE COUPLER

The key advantage to the star coupler is that there is only one insertion loss caused by the coupler. The only remaining insertion losses are from the interconnections. The advantage of the star coupler becomes very apparent as the number of ports increases. A simple loss-comparison chart, as shown in Figure 29.11, can reveal the significance in the number of terminals versus loss for the tee and star couplers.

FIGURE 29.11
Real tee coupler vs.
real star coupler
comparison chart

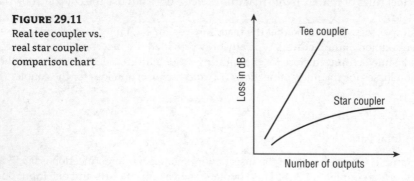

A star coupler has another advantage over a series of tee couplers. If one of the tee couplers in the series is disconnected, none of the other terminals down the line will receive an optical signal. However, disconnecting a terminal from the star coupler will not impact the operation of the other terminals.

Optical Switches

The next device we will be looking at is an optical switch. The fiber-optic switch can be a mechanical, optomechanical, or electronic device that opens or closes an optical circuit. The switch can be used to complete or break an optical path. Passive fiber-optic switches will route an optical signal without electro-optical or optoelectrical conversion. However, a passive optical switch may use an electromechanical device to physically position the switch. An optical switch may have one or more input ports and two or more output ports. Figure 29.12 shows a basic optical switch with one input port and four output ports.

FIGURE 29.12
Basic optical
switch

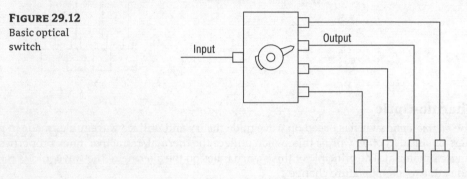

As with any other type of switch, the optical switch has many uses, depending on the complexity of the design. In essence, the switch is the control for making, breaking, or changing the connections within an optical circuit. This definition can be expanded to incorporate the concept of the switch as the control that interconnects or transfers connections from one optical circuit to another.

It is important to be aware of the basic switch parameters for an optical switch. Some of the performance parameters to consider are the number of input and output ports (required size of the switch), optical fiber type, connector type, center wavelength, bandwidth, losses, crosstalk, switching speed, durability (number of switching cycles), power handling, and repeatability (the amount of change in output power each time the switch changes state).

Optomechanical Switch

An *optomechanical* switch redirects an optical signal by moving fiber or bulk optic elements by means of mechanical devices. These types of switches are typically stepper motor driven. The stepper will move a mirror that directs the light from the input to the desired output, as shown in Figure 29.13. Although optomechanical switches are inherently slow due to the actual physical movement of the optical elements, their reliability, low insertion losses, and minimal crosstalk make them a widely deployed type of switch.

The optomechanical switch works on the premise that the input and output light beams are collimated within the fiber and "matched" within the switching device—the beams are moved within the device to ensure the switched connection from the inputs to the outputs. The optomechanical switch can be physically larger than alternative switches, but there are many micromechanical fiber-optic switches becoming available.

FIGURE 29.13
Optomechanical
switch

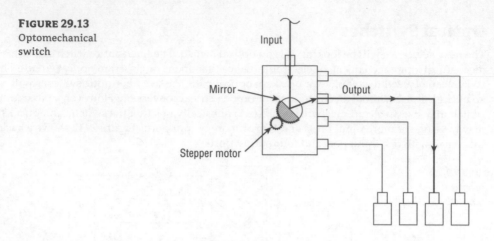

Thermo-Optic

The *thermo-optic* switch is based on waveguide theory and utilizes waveguides made in polymers or silica. In other words, this switch utilizes the thermal/refractive index properties of the device's material. The principle of this switch relies on the altering of the waveguide's refractive index due to a temperature change.

The temperature change can be accomplished in many ways, but generally the device is heated by using a resistive heater, which has the effect of slowing down light in one of the paths. The device then combines the light in the two paths in a constructive or destructive effect, making it possible to attenuate or switch the signal. This type of switch is inherently slow due to the time it takes to heat the waveguide. It's like a burner on an electric stove: it takes a while to heat up and a while to cool down.

This type of device typically has less optical loss than the optomechanical switch. Thermo-optic switches are attractive for several reasons: they work well in low optical power applications, are small in size, and have the potential to be integrated with a number of devices based on silicon wafer theory.

Electro-Optic

Electro-optic refers to a variety of phenomena that occur when an electromagnetic wave in the optical spectrum travels through a material under the stress of an electric field. An electro-optic switch is based on the changing of the refractive index of a waveguide by using an electric field. This device is semiconductor-based and therefore boasts high speed and low optical power loss similar to that of the thermo-optic devices. This device is still in the research stage; however, the technology is rapidly advancing.

In summary, optical switches can be used in a variety of applications, large and small. The use of a fiber-optic switch allows data to be routed where and when it is needed.

Optical Attenuators

An *optical attenuator* is a passive device that is used to reduce the power level of an optical signal. The attenuator circuit will allow a known source of power to be reduced by a predetermined factor, which is usually expressed as decibels. Optical attenuators are generally used in single-mode long-haul applications to prevent optical overload at the receiver.

Optical attenuators typically come in two forms of packaging. The bulkhead optical attenuator shown in Figure 29.14 can be plugged into the receiver receptacle. The inline attenuator resembles a patch cord and is typically used between the patch panel and the receiver.

FIGURE 29.14
Bulkhead optical
attenuator

Principles of Optical Attenuators

Optical attenuators use several different principles in order to accomplish the desired power reduction. Attenuators may use the gap-loss, absorptive, or reflective technique to achieve the desired signal loss. The types of attenuators generally used are fixed, stepwise variable, and continuously variable.

GAP-LOSS PRINCIPLE

The principle of gap-loss is used in optical attenuators to reduce the optical power level by inserting the device in the fiber path using an in-line configuration. Gap-loss attenuators are used to prevent the saturation of the receiver and are placed close to the transmitter. Gap-loss attenuators use a longitudinal gap between two optical fibers so that the optical signal passed from one optical fiber to another is attenuated. This principle allows the light from the transmitting optical fiber to spread out as it leaves the optical fiber. When the light gets to the receiving optical fiber, some of the light will be lost in the cladding because of gap and the spreading that has occurred. The gap-loss principle is shown in Figure 29.15.

FIGURE 29.15
Gap-loss principle
attenuator

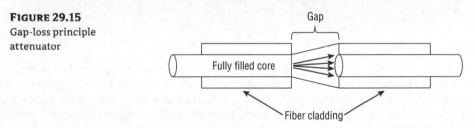

The gap-loss attenuator will only induce an accurate reduction of power when placed directly after the transmitter. These attenuators are very sensitive to modal distribution ahead of the transmitter, which is another reason for keeping the device close to the transmitter to keep the loss at the desired level. The farther away the gap-loss attenuator is placed from the transmitter, the less effective the attenuator is, and the desired loss will not be obtained. To attenuate a signal farther down the fiber path, an optical attenuator using absorptive or reflective techniques should be used.

Keep in mind that the air gap will produce a Fresnel reflection, which could cause a problem for the transmitter.

USING A BULKHEAD ATTENUATOR TO TEST RECEIVER SENSITIVITY

In Chapter 27, "Fiber-Optic Light Sources," you learned about the output parameters of a fiber-optic transmitter, and in Chapter 28, "Fiber-Optic Detectors and Receivers," you learned about the input parameters of a fiber-optic receiver. In Chapter 33, "Test Equipment and Link/Cable Troubleshooting," you will learn how to apply this information to analyze the performance of a fiber-optic link. Knowing how to test the sensitivity of a fiber-optic receiver is an important skill.

A fiber-optic receiver provides optimal performance when the optical input power is within a certain range. But how do you test the receiver to see if it will provide optimal performance at the lowest optical input powers? One way is to use optical attenuators, such as bulkhead attenuators. Typically only a couple of values are required to complete your testing. This process involves three steps:

1. Measure the optical output power of the fiber-optic transmitter with the power meter. Remember from Chapters 27 and 28 that industry standards define transmitter optical output power and receiver optical input power for a particular network standard. If you are testing a 100BaseFX receiver, you should be using a 100BaseFX transmitter. The optical output power of the transmitter should be within the range defined by the manufacturer's data sheet.

2. Connect the transmitter to the receiver and verify proper operation at the maximum optical output power that the transmitter can provide. You need to test the receiver at the minimum optical input power that the receiver can accept while still providing optimal performance. To do this, you need to obtain the lowest optical input power level value from the manufacturer's data sheet.

3. Calculate the attenuation level required for the test. Let's say that the transmitter's optical output power is –17dBm and the minimum optical power level for the receiver is –33dBm. The difference between –17dBm and –33dBm is 16dB. You would use a 16dB bulkhead attenuator at the input of the receiver and retest the receiver. If the receiver still operates properly, it's within specifications.

ABSORPTIVE PRINCIPLE

The absorptive principle, or absorption, accounts for a percentage of power loss in optical fiber. This loss is realized because of imperfections in the optical fiber that absorb optical energy and convert it to heat. (See Chapter 22, "Optical Fiber Characteristics," for a detailed discussion of the subject.) This principle can be employed in the design of an optical attenuator to insert a known reduction of power.

The absorptive principle uses the material in the optical path to absorb optical energy. The principle is simple, but can be an effective way to reduce the power being transmitted and/or received. Figure 29.16 shows the principle of the absorption of light.

FIGURE 29.16
Absorptive prin-
ciple attenuator

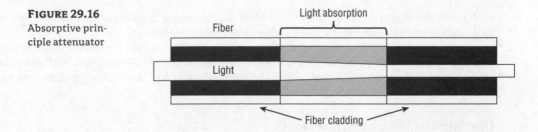

REFLECTIVE PRINCIPLE

The reflective principle, or scattering, accounts for the majority of power loss in optical fiber and again is due to imperfections in the optical fiber, which in this case cause the signal to scatter. This topic is also discussed in detail in Chapter 22. The scattered light causes interference in the optical fiber, thereby reducing the amount of transmitted and/or received light. This principle can be employed in the planned attenuation of a signal. The material used in the attenuator is manufactured to reflect a known quantity of the signal, thus allowing only the desired portion of the signal to be propagated. This principle is shown in Figure 29.17.

FIGURE 29.17
Reflective prin-
ciple attenuator

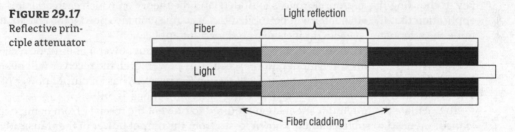

Now that we have looked at the principles behind the attenuator theories, we will discuss some of the types of attenuators. We will examine fixed, stepwise variable, and continuously variable attenuators and when they should be used.

Types of Attenuators

The types of attenuators generally used are fixed, stepwise variable, and continuously variable.

FIXED ATTENUATORS

Fixed attenuators are designed to have an unchanging level of attenuation. They can theoreti-cally be designed to provide any amount of attenuation that is desired. The output signal is attenuated relative to the input signal. Fixed attenuators are typically used for single-mode applications.

STEPWISE VARIABLE ATTENUATORS

A stepwise variable attenuator is a device that changes the attenuation of the signal in known steps such as 0.1dB, 0.5dB, or 1dB. The stepwise attenuator may be used in applications dealing with multiple optical power sources—for example, if there are three inputs available, there may be a need to attenuate the signal at a different level for each of the inputs.

Conversely, the stepwise attenuator may also be used in situations where the input signal is steady, yet the output requirements change depending on the device that the signal is output to.

The stepwise attenuator should be used in applications where the inputs, outputs, and operational configurations are known.

CONTINUOUSLY VARIABLE ATTENUATORS

A continuously variable attenuator is an attenuator that can be changed on demand. These attenuators generally have a device in place that allows the attenuation of the signal to change as required. A continuously variable attenuator is used in uncontrolled environments where the input characteristics and/or output needs continually change. This allows the operator to adjust the attenuator to accommodate the changes required quickly and precisely without any interruption to the circuit.

Calculating the Attenuation Value

In summary, there are many types of attenuators and many principles on which they work. The key to choosing the appropriate one is to understand the theory on which each operates and the application that the attenuator will be applied to. Of course, you also need to be able to determine the attenuator value in decibels required for your application.

In this example let's assume that the maximum optical input power a fiber-optic receiver can operate with is –6dBm. If the input power exceeds this power level, the receiver will be overloaded. The transmitter, which is located 10km from the receiver, has an output power of 3dBm. The loss for the 10km of optical fiber, including interconnections, is 5dB.

To calculate the minimum attenuation required to prevent the receiver from being overloaded, we need to subtract all the known losses from the output power of the transmitter as shown here:

Transmitter power (TP) = 3dBm

Receiver maximum optical input power (MP) = –6dBm

Total losses (TL) = 5dB

Minimum attenuation required = MP + TL – TP

–6dBm + 5dB – 3dBm = –4dB

At a minimum, a 4dB attenuator is required. However, an attenuator with a larger value could be used as long as it did not over-attenuate the signal. Refer to Chapter 28 to determine the dynamic range of a receiver.

Optical Isolator

Many laser-based transmitters and optical amplifiers use an optical isolator because the components that make up the optical circuit are not perfect. Connectors and other types of optical devices on the output of the transmitter may cause reflection, absorption, or scattering of the optical signal. These effects on the light beam may cause light energy to be reflected back at the source and interfere with source operation. To reduce the effects of the interference, an *optical isolator* is used.

The optical isolator comprises elements that will permit only forward transmission of the light; it does not allow for any return beams in the fiber transmission routes or in the amplifiers. There are a variety of optical isolator types, such as polarized (dependent and independent), composite, and magnetic.

Polarized Optical Isolator

As mentioned, the polarized optical isolator transmits light in one direction only. This is accomplished by using the polarization axis of the linearly polarized light. The incident light is transformed to linearly polarized light by traveling through the first polarizer. The light then goes through a Faraday rotator; this takes the linearly polarized light and rotates the polarization 45 degrees, then the light passes through the exit polarizer. The exit polarizer is oriented at the same 45 degrees relative to the first polarizer as the Faraday rotator is. With this technique, the light is passed through the second polarizer without any attenuation. This technique allows the light to propagate forward with no changes, but any light traveling backward is extinguished entirely.

The loss of backward-traveling light occurs because when the backward light passes through the second polarizer, it is shifted again by 45 degrees. The light then passes through the rotator and again is rotated by 45 degrees in the same direction as the initial tilt. So when the light reaches the first polarizer, it is polarized at 90 degrees. And when light is polarized by 90 degrees, it will be "shut out." Figure 29.18 shows the forward-transmitted light in a dependent polarized optical isolator.

FIGURE 29.18
Forward-transmitted light through a polarized optical isolator

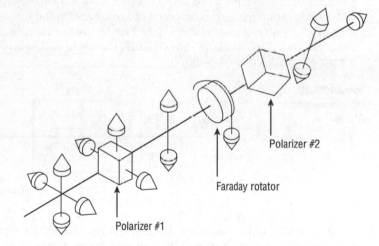

Figure 29.19 shows the backward-traveling light in a dependent polarized optical isolator.

FIGURE 29.19
Reverse-transmit-
ted light through a
polarized optical
isolator

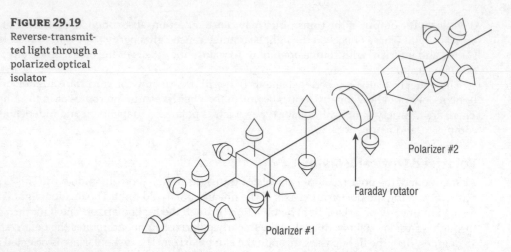

It should be noted that these figures depict the dependent type of polarized optical isolator. There is also an independent polarized optical isolator. The independent device allows all polarized light to pass through, not just the light polarized in a specific direction. The principle of operation is roughly the same as the dependent type, just slightly more complicated. The independent optical isolators are frequently used in optical fiber amplifiers.

Magnetic Optical Isolator

Magnetic optical isolators are another name for polarized isolators. The magnetic portion of any isolator is of great importance. As mentioned, there is a Faraday rotator in the optical isolators. The Faraday rotator is a rod composed of a magnetic crystal having a Faraday effect and operated in a very strong magnetic field. The Faraday rotator ensures that the polarized light is in the correctly polarized plane, thus ensuring that there will be no power loss. Figure 29.20 shows a basic magnetic optical isolator.

FIGURE 29.20
Magnetic optical
isolator

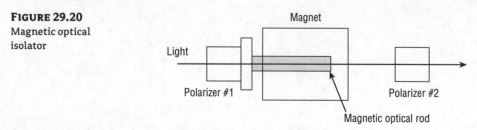

In summary, optical isolators are used to ensure stabilization of laser transmitters and amplifiers as well as to maintain good transmission performance.

Wavelength Division Multiplexing

Wavelength division multiplexing (WDM) is the combining of different optical wavelengths from two or more optical fibers into just one optical fiber. This combining or coupling of the wavelengths can be very useful in increasing the bandwidth of a fiber-optic system. WDM multiplexers are used in pairs: one at the beginning of the fiber to couple the inputs and one at the end of the fiber to decouple and then route the separated wavelengths into separate fibers. A WDM multiplexer can be thought of as an optical fiber highway; the highway can support a very large bandwidth, thus increasing the system's capacity.

Each channel in a WDM multiplexer is designed to transmit a specific optical wavelength. The multiplexer operates very much like a coupler at the beginning of the optical fiber and as a filter at the end of the optical fiber. For example, an eight-channel multiplexer would have the ability to combine eight different channels or wavelengths from separate optical fibers onto one optical fiber. Again, to take advantage of the enormous bandwidth at the end of the optical fiber, another multiplexer will recover the separate wavelengths.

Figure 29.21 shows a simple WDM system composed of multiple light sources, a multiplexer or combiner that combines the wavelengths into one optical fiber, and a *demultiplexer* or splitter that separates the wavelengths to their respective receivers.

FIGURE 29.21
Simple WDM
system

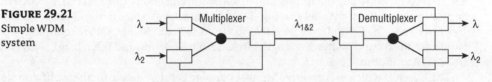

WDM multiplexers are available in a variety of sizes, but will most commonly be found with 2-, 4-, 8-, 16-, 32-, and 64-channel configurations. The types of multiplexers are wideband (or crossband), narrowband, and dense. Wideband or crossband multiplexers are devices that combine a broad range of wavelengths, such as 1310nm and 1550nm. A narrowband multiplexer will combine multiple wavelengths with 1000GHz channel spacing. A dense multiplexer combines wavelengths with 100GHz channel spacing. Figure 29.22 shows a basic wideband or crossband WDM system.

FIGURE 29.22
Basic wideband or
crossband WDM
system

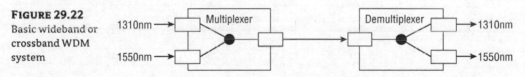

Narrowband WDM systems have channels spaced 1000GHz, or approximately 8nm, apart. Table 29.1 shows the wavelength and frequency data for a narrowband WDM system.

TABLE 29.1   Narrowband WDM Channel Spacing

λ (NM)	F (THz)
1531.90	195.7
1539.77	194.7
1547.72	193.7
1555.75	192.7

Figure 29.23 shows a basic narrowband WDM system.

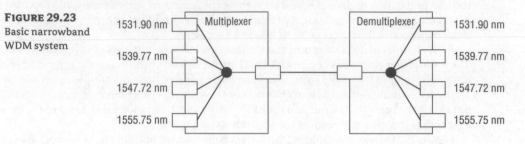

FIGURE 29.23
Basic narrowband
WDM system

The industry standard on the dense wavelength division multiplexing (DWDM) multiplexers, as recommended by the International Telecommunications Union (ITU), is 100GHz, or approximately 0.8nm, channel spacing. Table 29.2 displays values in the C band, which is the 1550nm band. The C band uses wavelengths from 1530 to 1565nm. Other commonly-referred-to bands are the S band, with wavelengths from 1525 to 1538nm, and the L band, with wavelengths from 1570 to 1610nm.

The C band, which is currently the most popular band in use, is further split into short (blue) and long (red) bands, as shown in Table 29.2.

TABLE 29.2   DWDM 100GHz Channel Spacing

	SHORT/BLUE BAND			LONG/RED BAND	
Channel	λ (nm)	F (THz)	Channel	λ (nm)	F (THz)
65	1525.66	196.5	40	1545.32	194.0
64	1526.44	196.4	39	1546.12	193.9
63	1527.21	196.3	38	1546.92	193.8
62	1527.99	196.2	37	1547.72	193.7
61	1528.77	196.1	36	1548.51	193.6
60	1529.55	196.0	35	1549.32	193.5
59	1530.33	195.9	34	1550.12	193.4
58	1531.12	195.8	33	1550.92	193.3
57	1531.90	195.7	32	1551.72	193.2

TABLE 29.2   DWDM 100GHz Channel Spacing   *(CONTINUED)*

SHORT/BLUE BAND			LONG/RED BAND		
Channel	λ (nm)	F (THz)	Channel	λ (nm)	F (THz)
56	1532.68	195.6	31	1552.52	193.1
55	1533.47	195.5	30	1553.33	193.0
54	1534.25	195.4	29	1554.13	192.9
53	1535.04	195.3	28	1554.94	192.8
52	1535.82	195.2	27	1555.75	192.7
51	1536.61	195.1	26	1556.55	192.6
50	1537.40	195.0	25	1557.36	192.5
49	1538.19	194.9	24	1558.17	192.4
48	1538.98	194.8	23	1558.98	192.3
47	1539.77	194.7	22	1559.79	192.2
46	1540.56	194.6	21	1560.61	192.1
45	1541.35	194.5	20	1561.42	192.0
44	1542.14	194.4	19	1562.23	191.9
43	1542.94	194.3	18	1563.05	191.8
42	1543.73	194.2	17	1563.86	191.7
41	1544.53	194.1	16	1564.68	191.6

As shown in Table 29.2, the closer the channels are spaced together, the higher the number of channels that can be inserted into a band. Currently a spacing of 50GHz is available. It is important to note that as the spacing or the width of each channel decreases, the smaller the spectral width becomes. This is relevant because the wavelength must be stable or sustainable long enough not to drift into an adjacent channel.

Now let's look at a different view of channel spacing. Figure 29.24 shows a four-channel narrowband WDM spectrum using DFB laser transmitters with a spectral width of 1.0nm measured at –20dB.

Figure 29.25 shows a 32-channel dense WDM spectrum using distributed feedback (DFB) laser transmitters with a spectral width of 0.3nm measured at −20dB.

FIGURE 29.24
Four-channel narrowband WDM spectrum

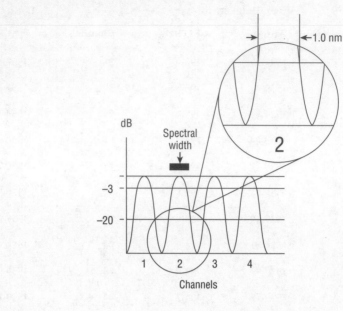

FIGURE 29.25
32-channel dense WDM spectrum

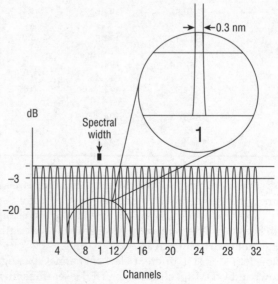

You can quickly see that as the channel spacing decreases, the laser transmitter spectral width must also decrease. To achieve 50GHz channel spacing, the laser transmitter spectral width needs to be very narrow—or about as narrow as today's laser technology permits.

Besides having a very narrow spectral width, the laser transmitter cannot drift—it must output the same wavelength at all times. If the laser transmitter's output wavelength changes even a few tenths of a nanometer, it could drift into the next channel and cause interference problems.

There are different configurations of WDM multiplexers. Everything we have covered up to this point describes a unidirectional WDM system. The unidirectional WDM multiplexers are configured so that the multiplexer only connects to optical transmitters or receivers. In other words, they allow the light to travel in only one direction and they provide only simplex communication over a single optical fiber. Therefore, full-duplex communications require two optical fibers.

A WDM multiplexer that is designed to connect with both transmitters and receivers is called bidirectional; in essence, the multiplexer is designed for optical transmission in both directions using only one optical fiber. Two channels will support one full-duplex communication link. Figure 29.26 shows two bidirectional WDM multiplexers communicating over a single optical fiber.

FIGURE 29.26
Two-channel
bidirectional
WDM system

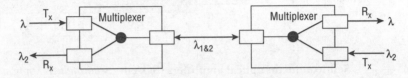

As with any other device that is added to a fiber-optic network, there are factors that must be considered. As mentioned earlier in the chapter, losses are a factor that must be taken into account. When using WDM multiplexers, remember that the greater the number of channels, the greater the insertion losses. Other specifications to keep in mind when using WDM multiplexers are isolation, PMD, and the spectral bandwidth.

In summary, WDM multiplexers are widely used devices. They provide a way to utilize the enormous bandwidth capacity of optical fiber without the expense of using the fastest laser transmitters and receivers. Just think about it: an eight-channel WDM system using directly modulated 2.5Gbps laser transmitters carries twice as much data as a single indirectly modulated 10Gbps laser transmitter. WDM systems allow designers to combine modest performance parts and create an ultra-performance system. WDM systems deliver the most bang for the buck!

Optical Amplifier

As optical signals travel through an optical fiber, they are attenuated. In long-haul applications, the signal is attenuated to the point where re-amplification is required. Traditionally, a device commonly referred to as a repeater accomplished this re-amplification.

A repeater is basically a receiver and transmitter combined in one package. The receiver converts the incoming optical energy into electrical energy. The electrical output of the receiver drives the electrical input of the transmitter. The optical output of the transmitter represents an amplified version of the optical input signal plus noise.

The technology available today eliminates the need for repeaters. Passive optical amplifiers are now used instead of repeaters. A passive optical amplifier amplifies the signal directly without the need for optical-to-electric and electric-to-optical conversion. There are several different techniques with which to passively amplify an optical signal: erbium doped fiber amplifiers, semiconductor optical amplifiers, and Raman amplification, all of which use a technique called laser pumping.

Erbium doped fiber amplifiers (EDFAs)   *Erbium doped fiber amplifiers (EDFAs)* are generally used for very long fiber links such as undersea cabling. The EDFAs use a fiber that has been treated or "doped" with erbium, and this is used as the amplification medium. The pump lasers operate at wavelengths below the wavelengths that are to be amplified. The doped fiber is energized with the laser pump. As the optical signal is passed through this doped fiber, the erbium atoms transfer their energy to the signal, thereby increasing the energy or the strength of the signal as it passes. With this technique, it is common for the signal to be up to 50 times or 17dB stronger leaving the EDFA than it was when it entered.

An example of an EDFA is shown in Figure 29.27. EDFAs may also be used in series to further increase the gain of the signal. Two EDFAs used in series may increase the input signal as much as 34dB.

FIGURE 29.27
Erbium doped fiber
amplifier

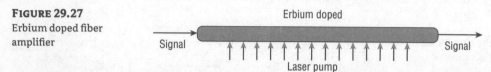

Semiconductor optical amplifiers (SOAs)   *Semiconductor optical amplifiers (SOAs)* use a technique similar to that of EDFAs but without doping the optical fiber. Unlike the EDFA, which is energized with a laser pump, the SOA is energized with electrical current. The SOAs use an optical waveguide and a direct bandgap semiconductor that is basically a Fabry-Pérot laser to inject light energy into the signal, as shown in Figure 29.28. This technique, however, does not offer the high amplification that the EDFAs do. SOAs are typically used in shorter fiber links such as *metropolitan area networks (MANs)*.

FIGURE 29.28
Semiconductor
optical amplifier

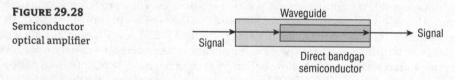

One problem with SOAs is that the gain is very hard to control. By using the semiconductor technique and a waveguide, the signal may deplete the gain of a signal at another wavelength. This can introduce crosstalk among channels by allowing the signal at one wavelength to modulate another.

Raman amplification   *Raman amplification* is a method that uses pump lasers to donate energy to the signal for amplification. However, unlike EDFAs, this technique does not use doped fiber, just a high-powered pumping laser, as shown in Figure 29.29. The laser is operated at wavelengths 60nm to 100nm below the desired wavelength of the signal. The laser signal energy and the photons of the transmitted signal are coupled, thereby increasing the signal strength.

FIGURE 29.29
Raman
amplification

Raman amplification does not amplify as much as the EDFAs, but it does have an advantage in that it generates much less noise.

These techniques can be combined to take advantage of their amplification characteristics. In some cases, Raman and EDF amplifiers are combined in long-haul fiber links to ensure high amplification and decreased noise levels.

In summary, each amplification technique has advantages and disadvantages. Remember to keep in mind the amplification that the amplifier is being used in. For example, if a signal needed amplification but noise was an issue, a Raman amplifier would most likely be the best choice. If the signal needed to be amplified by just a small amount, the SOA might be best.

All of these amplification methods have one big advantage: optical amplifiers will amplify all signals on a fiber at the same time. Thus, it is possible to simultaneously amplify multiple wavelengths. But it is important to keep in mind that the power levels must be monitored carefully because the amplifiers can become saturated, thereby causing incorrect operation.

Optical Filter

An optical filter is a device that selectively permits transmission or blocks a range of wavelengths. Optical filters are typically bandpass or band-reject.

A bandpass optical filter allows a certain range of optical wavelengths to pass and attenuates the rest, as shown in Figure 29.30.

FIGURE 29.30
Optical bandpass
filter response

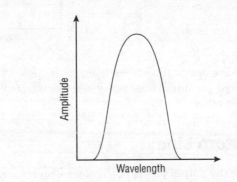

A band-reject optical filter attenuates a band of optical wavelengths and allows the others to pass, as shown in Figure 29.31.

FIGURE 29.31
Optical band-reject
filter response

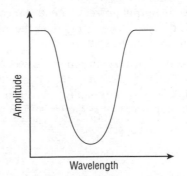

An example of a basic optical filter would be the optical filter used on a traffic light. A typical traffic light contains three optical filters, one red, one yellow, and one green. The bulb behind each optical filter is the same and emits a wide range of visible wavelengths. The optical filters allow only a certain range of wavelengths to pass, creating the red, yellow, or green light.

Bandpass optical filters are designed to transmit a specific waveband or wavelength range. A wideband optical filter may allow wavelengths plus and minus 20nm off the center frequency to pass. This type of optical filter would be used when signals are separated by several hundred nanometers, such as with a 1310nm and 1550nm source.

A narrowband optical filter allows only a very narrow range of optical energy to pass, as shown in Figure 29.32. The bandwidth of a narrowband optical filter may be less than one nanometer. The narrowband optical filter would be used in a DWDM application to reject adjacent optical channels.

FIGURE 29.32
Narrowband optical filter response

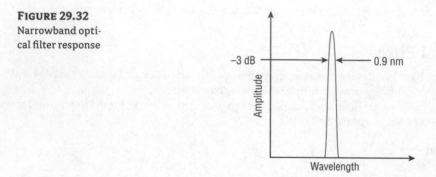

Each of these optical filter types is simple in theory yet is a vital part of some fiber-optic systems. As stated, an optical filter is a device that selects the wavelengths it will allow to pass and will reject the others.

The Bottom Line

Calculate the output power of a real tee coupler.   Remember that manufacturers may or may not provide the insertion loss combined with the splitting ratio.

    Master It   The manufacturer provided the insertion loss combined with the loss from the splitting ratio for a 90:10 tee coupler. Calculate the power available at each output port.

Calculate the output power of a real star coupler.   Remember that unlike the tee coupler, a manufacturer typically only provides the insertion loss for the coupler and does not take into account the loss for the splitting ratio.

    Master It   Calculate the output power at each port of a star coupler with one input port and 10 output ports. Take into account the interconnection insertion losses. Assume the insertion loss for the coupler is 1.7dB and each interconnection insertion loss is 0.25dB. The input power to the coupler is –5dBm.

Calculate attenuator values. Remember that if the maximum optical input power for a fiber-optic receiver is exceeded, the receiver will not operate properly.

Master It Calculate the minimum attenuation required to prevent the receiver from being overloaded. Assume the transmitter power is –3dBm, the total losses are 6dB, and the maximum optical input power the receiver can tolerate is –14dBm.

Chapter 30

Passive Optical Networks

Today we have access to more information than ever before. We live in a digital world and bandwidth is what makes a digital world happen. Aging copper networks are being taxed by residential and business customers. However, passive optical networks (PON) such as fiber-to-the-home (FTTH) are increasingly being deployed to meet the current and future bandwidth needs that the aging copper networks cannot support.

The fastest-growing global broadband technology today is FTTH. Telecommunications service providers are moving quickly to maximize the number of services that can be offered to a residential or commercial customer. Today's networks provide more bandwidth than has ever been available and tomorrow's networks will provide twice that bandwidth.

This chapter discusses the fundamentals of a PON including fiber-to-the-home, fiber-to-the-building, fiber-to-the-curb, and fiber-to-the-node. It also introduces the outside plant hardware and equipment required for these passive optical networks.

In this chapter, you will learn to:

◆ Identify the different PON configurations

◆ Identify the cables used in a PON

◆ Identify different access points in a PON

Passive and Active Network Fundamentals

There are many types of networks carrying many different types of information. However, all these individual networks can be placed in one of two categories: passive or active. A *passive network* does not use electrically powered equipment or components to get the signal from one place to another. An *active network* uses electrically powered equipment or components to route the signal from one place to another.

Passive Copper Network

There are many types of passive copper networks. However, the one virtually everyone is familiar with is their home cable TV network. In a copper cable TV network, the cable provider supplies the signal to the home over a coaxial cable.

In the most basic network, the cable enters the home and is routed to a single television. However, few homes have a single television. For homes with multiple TVs, the signal from the cable provider must be split for each television to receive the signal. The splitting is usually accomplished with an inexpensive device commonly referred to as a *splitter*. The splitter requires no electrical power. It will typically have a single input and may have two, three, four, or more outputs. Figure 30.1 is an example of a splitter that has a single input and four outputs. An individual cable is routed from the splitter to each television.

FIGURE 30.1
Splitter with one
input and four
outputs

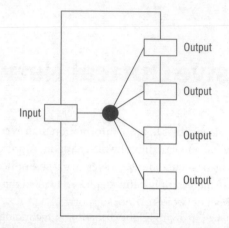

One of the problems with this type of network is loss of signal strength. As the signal from the cable provider is split and routed to multiple televisions, the signal strength to each television is reduced. Adding too many televisions can reduce the signal strength to the point where none of the televisions receives adequate signal strength to operate properly. When this happens, it is time to look at installing an active cable TV network.

Active Copper Network

Just as there are many types of passive copper networks, there are also many types of active copper networks. The previous section focused on a passive home cable TV network and pointed out that you can only connect a limited number of televisions to this type of network. To have adequate signal strength for multiple televisions—for example, one in each room—an active network is required.

In one example of an active home cable TV network, one cable enters the home and is routed to a distribution amplifier. The distribution amplifier boosts or amplifies and splits the signal from the cable provider. Each output of the distribution amplifier has a signal strength approximately equal to the signal strength on the input cable from the cable provider. An individual cable is routed from the distribution amplifier to each television.

This type of active network overcomes the signal strength problem associated with a passive network. However, it does add a level of complexity and requires power. If the distribution amplifier were to fail, all the televisions would lose their signals. The same would be true if the distribution amplifier were accidentally unplugged: every television in the house would be without a signal.

Passive Optical Network

There are many types of passive optical networks (PON). One of the most common types is very similar to the passive cable TV network previously described. However, optical fiber is used instead of instead of coaxial cable. In Chapter 29, "Passive Components and Multiplexers," you learned about many different passive devices that are available to support different types of physical network topologies. Couplers are the core of any PON. A coupler may combine two or more optical signals into a single output, or the coupler may take a single optical input and distribute it to two or more separate outputs.

Figure 30.2 is an example of a seven-port coupler. The coupler is splitting a single input signal into six outputs.

FIGURE 30.2
Seven-port coupler

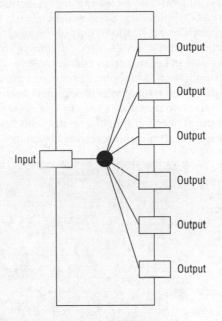

Many couplers are designed for bidirectional operation, which enables the same coupler to be used either to combine signals or to split signals. In a bidirectional coupler, therefore, each port can be either an input or an output. However, for a PON application, a coupler being used to split a signal may be referred to as a splitter.

In a PON, the input to the coupler in Figure 30.2 would be split equally between the six outputs. Data going into the coupler would be sent to each output just as the signal from the cable TV provider is sent to each TV in the passive copper network. Although each output will carry the same information as the input, the signal strength will be reduced based on the number of outputs. There is a finite limit on the number of outputs for a PON application; typically, the limit is 32. However, some applications may support more.

Active Optical Network

An active optical network is very similar to the active home cable TV network previously described. One optical fiber connects to a switch instead of a distribution amplifier. The switch rebroadcasts the data to each individual user. A separate cable is routed from the switch to each individual user.

This type of active network overcomes the signal strength problem associated with a passive network. However, it does add a level of complexity and requires power. If the switch were to fail, all the users would lose access to incoming data. The same would be true if the switch lost power: data would stop flowing.

Fiber to the X

This section of the chapter looks at different PON configurations. These different configurations can be grouped into what is known as *Fiber to the X*, or *FTTX*. FTTX can be used to describe any optical fiber network that replaces all or part of a copper network.

FTTX is different from a traditional fiber-optic network that would be used for a local area network (LAN) application. One key difference is the number of optical fibers required for each user. In most FTTX applications, only one optical fiber is used. That single optical fiber passes data in both directions. This is very different from a LAN application where the transmit optical fiber sends data in one direction and the receive optical fiber sends data in the other direction. In a LAN application, both optical fibers can have data passing through them at the same time.

We've discussed transceivers in several chapters in this book. A transceiver is typically a device that has two receptacles like the one shown in Figure 30.3. One receptacle mates with the transmit optical fiber and the other mates with the receive optical fiber. This allows the transceiver to be transmitting and receiving simultaneously. This is known as *full-duplex* operation.

FIGURE 30.3
Transceiver with a transmit and a receive receptacle

In an FTTX single optical fiber application, full-duplex operation is typically not possible; usually only half-duplex operation takes place. This means that part of the time the optical fiber is carrying a signal in one direction, and the rest of the time, it is carrying a signal in the other direction. The key to making this system work is timing. Data is sent downstream for a predetermined amount of time and then data is sent upstream for a predetermined amount of time. This process is also known as *time division multiplexing*.

FTTX systems typically use multiple wavelengths. The downstream laser is always a different wavelength than the upstream laser. The downstream laser is typically the longer wavelength, such as 1480nm or 1550nm (or both), and the upstream laser is typically 1310nm.

FTTX is possible with optical fiber distances up to 20km because optical fiber is capable of transmitting information with a very low level of loss. The typical loss for an FTTX optical fiber at 1550nm is 0.25dB/km and 0.35dB/km at 1310nm.

Fiber to the Home

A fiber-to-the-home (FTTH) PON uses optical fiber from the central office to the home; there are no active electronics helping with the transmission of data in between the two locations. The central office is a communications switching facility. It houses a large number of complex switches that establish temporary connections between subscriber lines that terminate at the central office.

At the home, a converter box changes the optical signal from the optical fiber into electrical signals. The converter box interfaces with existing home cabling such as coaxial cabling for cable TV, twisted-pair cabling for telephone, and Category 5e or 6 cabling for Internet connectivity.

Fiber to the Building

A fiber-to-the-building (FTTB) PON is very similar to an FTTH PON. It uses optical fiber from the central office to the building and there are no electronics helping with transmission in between. The optical signal from the optical fiber is converted into electrical signals in a converter box at the building. The converter box interfaces with existing cabling such as coaxial cabling for cable TV, twisted-pair cabling for telephone, and Category 5e or 6 cabling for Internet connectivity.

Fiber to the Curb

In a fiber-to-the-curb (FTTC) PON, optical fiber runs from the central office and stops at the curb. The "curb" may be right in front of the house or some distance down the block. The converter box is located where the optical fiber stops, and it changes the optical signal from the optical fiber into electrical signals. These electrical signals are typically brought into the home through some existing copper cabling. The electrical signals may need to be processed by another converter box inside the house to interface with existing cabling such as coaxial cabling for cable TV, twisted-pair cabling for telephone, and Category 5e or 6 cabling for Internet connectivity.

Fiber to the Node

Fiber to the node (FTTN) is sometimes referred to as fiber to the neighborhood. A FTTN PON only has optical fiber from the central office to the node. The node is typically a telecommunications cabinet that serves a neighborhood or section of a neighborhood. The optical signal from the optical fiber is converted into electrical signals inside the telecommunications cabinet. These electrical signals are distributed throughout the neighborhood through existing copper cables to the houses.

Outside Plant Components

This section will discuss the major outside plant components for an FTTX PON. The outside plant components make up the PON infrastructure and are all designed for installation exterior to buildings. The cables connect different access points in the PON. Everything is initiated from the central office or central switching point.

Cables

Several different types of cables are employed in an FTTX PON, including feeder, distribution, and drop cables.

FEEDER CABLES

Feeder cables run from the central switching point to the local convergence point. These cables typically contain multiple ribbons of 12 single-mode optical fibers each. A common *feeder cable* will contain 18 ribbons for a total of 216 single-mode optical fibers.

DISTRIBUTION CABLES

Distribution cables run from the local convergence point to the network access point. Like the feeder cables, they typically contain multiple ribbons of 12 single-mode optical fibers. However, distribution cables do not contain as many optical fibers as feeder cables. A *distribution cable* can have as few as 12 optical fibers or as many as 144. A typical distribution cable has 72 optical fibers.

DROP CABLES

A *drop cable* is a single optical fiber cable that is terminated at the factory, typically with SC connectors on both ends. The cable is environmentally sealed and the connectors are sealed when they are mated. Drop cables, like the one shown in Figure 30.4, are typically available in 15′ increments in lengths from 90′ to 180′ and in 50′ increments in lengths from 200′ up.

FIGURE 30.4
Single optical fiber drop cable with SC connector

Photo courtesy of Corning Cable Systems

Drop cables run from the network access point to the residence or building. They are designed to minimize installation cost and to provide years of trouble-free service.

Local Convergence Point

The *local convergence point (LCP)* is the access point where the feeder cables are broken out into multiple distribution cables. It is typically located in a field-rated cabinet like the 432–optical fiber cabinet shown in Figure 30.5 or the 864–optical fiber cabinet shown in Figure 30.6. A local convergence point services a neighborhood or business park.

FIGURE 30.5
432–optical fiber field rated local convergence cabinet

Photo courtesy of Corning Cable Systems

FIGURE 30.6
864–optical fiber field rated local convergence cabinet

Photo courtesy of Corning Cable Systems

Depending on the architecture of the PON, the local convergence point may or may not be the place where the optical signals are split. The optical signals may be split at the network access point with a splitter like the one shown in Figure 30.7. This splitter distributes the optical signal to 32 individual optical fibers. Each optical fiber is terminated with a connector for easy installation and configuration.

FIGURE 30.7
32-optical fiber splitter

Photo courtesy of Corning Cable Systems

Network Access Point

The network access point (NAP) is located close to the homes or buildings it services. This is the point where a distribution cable is broken out into multiple drop cables. The NAP is a terminal that serves as a connection point for drop cables. It may be installed in an aerial installation, in a pedestal, or in a hand hole.

Depending on the architecture of the PON, the NAP may or may not house the optical splitter. Figure 30.8 is a photograph of a system terminal. A system terminal serves as the NAP for some FTTX installations. Figure 30.9 is a photograph of an aerial system terminal serving as a NAP.

FIGURE 30.8
System terminal that serves as a network access point

Photo courtesy of KITCO Fiber Optics

Photo courtesy of Corning Cable Systems

FIGURE 30.9
Aerial system terminal serving as a network access point

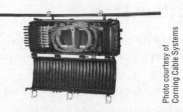

Network Interface Device

The drop cable runs from the NAP to the network interface device (NID). The NID is typically mounted to the outside of the house or building. It is an all-plastic enclosure designed to house the electronics that support the network. The SC connector on the end of the drop cable mates with the connector in the NID.

The passive optical network ends at the NID. The electronics in the NID will interface with existing cabling for television, telephone, and internet connectivity. Figure 30.10 is a photograph of a NID. You can see the cabling entering the NID from the bottom.

FIGURE 30.10
Network interface device with the cabling entering from the bottom

Photo courtesy of KITCO Fiber Optics

The Bottom Line

Identify the different PON configurations.   FTTX can be used to describe any optical fiber network that replaces all or part of a copper network.

> **Master It**   In this PON, the optical signals are converted right in front of the house or some distance down the block. What type of PON is this?

Identify the cables used in a PON.   Several different types of cables are employed in an FTTX PON.

> **Master It**   This cable type runs from the central switching point to the local convergence point. What type of cable is it?

Identify different access points in a PON.   Cables connect the different access points in the PON.

> **Master It**   What do you call the point where a distribution cable is broken out into multiple drop cables?

Chapter 31

Cable Installation and Hardware

Under the right conditions, an optical fiber will carry light great distances almost instantly. If the cable is not installed properly, though, the light it is supposed to carry may not even travel from one part of an office building to another.

Proper installation depends on a thorough knowledge of the strengths and limits of a fiber-optic cable, as well as the methods for protecting the cable from damage both during the installation and over its lifetime. The cable must endure the pulling force that is necessary for it to be put in place; environmental conditions that threaten to freeze, soak, or otherwise damage it; and the daily stresses that result from its location or position.

The rules governing fiber-optic cable installation are designed to minimize the short-term and long-term stresses on the cable as well as ensure that the installation conforms to codes governing fire safety. Some of these requirements were discussed in Chapter 24, "Fiber-Optic Cables," with regard to the structure of the cable itself. The rules governing cable construction can only be effective, though, if the cable is installed properly.

This chapter describes the requirements for a successful fiber-optic cable installation. We discuss the conditions affecting fiber-optic cable and the methods for installing the cable so that the effects of those conditions are minimized. We describe regulations for electrical safety in cable installations as well as methods for routing cables in different situations. The chapter also examines the ways in which cables are enclosed and terminated, including proper labeling methods.

In this chapter, you will learn to:

◆ Determine the minimum cable bend radius

◆ Determine the maximum cable tensile rating

◆ Determine the fill ratio for a multiple-cable conduit installation

Installation Specifications

Many fiber-optic cables are created for specific types of service and must be installed in accordance with their manufacturer's specifications. While some specifications are concerned with the particular duty or job a cable will perform, others apply to traits shared by all cables. Two of these specifications are *bend radius* and *tensile strength* or *pull strength*. Both specifications apply to conditions faced by fiber-optic cable while it is being installed and, once it has been installed, to its normal working conditions.

As we have learned in previous chapters, Section 4 of ANSI/TIA-568-C.3 defines performance standards for optical fiber cables recognized in premises cabling standards. Section 4.3

defines the physical requirements for optical fiber cables. Four cable types are covered in this section: *inside plant*, *indoor-outdoor*, *outside plant*, and *drop cable*.

Inside plant cables   Designed for installation in the building interior.

Indoor-outdoor cables   Designed for installation in the building interior or exterior to the building. These cables are not designed for long-haul outdoor applications. They are typically used to provide connectivity between two buildings. Unlike an outside plant cable, these cables resist the spread of fire as defined in Article 770 of the National Electrical Code (NEC).

Outside plant cables   Designed for outdoor installations. These cables typically do not resist the spread of fire as defined in Article 770 of the NEC. Because of this, the length of outside plant cable within a building cannot exceed 50′ (15.2m) from the point of entrance into the building. To comply with Article 770 of the NEC, these cables must also be terminated in an enclosure. This is discussed in detail in the section "Fire Resistance and Grounding" later in this chapter.

Drop cables   Designed to link a drop terminal to a premise terminal. A terminal is a device that is capable of sending and/or receiving information over a communications channel that in this case uses an optical fiber. The service provider typically owns the drop terminal.

The bend radius and pull load requirements for these cables as defined in ANSI/TIA-568-C.3 will be discussed in detail in this chapter.

Bend Radius

As we have seen in previous chapters, optical fiber depends on the maintenance of total internal reflection to carry an optical signal. Macrobends, or very small radius bends in the optical fiber, can change the light's angle of incidence enough to cause some or a majority of the light to pass into the cladding, severely attenuating the signal or cutting it off completely.

Manufacturers often specify a minimum installation and operational bend radius for their optical fiber cables. The minimum installation bend radius is the short-term bend radius, and the minimum operational bend radius is the long-term bend radius. The minimum is actually the larger of the two because of the tensile stresses that may be placed on the cable during installation.

Following the manufacturer's guidelines reduces the risk of damage to the cable and optical fiber during and after installation. It also reduces the risk of macrobends. Keep in mind that bends approaching the minimum bend radius might cause some attenuation. However, how much attenuation occurs depends on multiple factors, including cable construction, optical fiber type, and the wavelength of the light source.

Another, more basic reason for the minimum bend radius is that the optical fibers, while flexible, are not indestructible. Bending beyond the minimum radius may cause the fiber to break inside the cable, which would require repair at the very least and replacement of the entire cable as a worst-case scenario.

Section 4.3 of ANSI/TIA-568-C.3 defines the bend radius requirements for inside plant, indoor-outdoor, outside plant, and drop cables.

Inside plant cables are broken into three groups:

◆ Cables with four or fewer optical fibers intended for Cabling Subsystem 1

◆ Cables with four or fewer fibers intended to be pulled through *pathways* during installation

◆ All other inside plant cables

Inside plant cables intended for Cabling Subsystem 1 run from an *equipment outlet* to a *distributor* in the *hierarchical star topology*. This distributor may be an optional connection facility, intermediate connection facility, or central connection facility. These cables must support a bend radius of 25mm (1″) when not subject to a tensile load. In other words, 25mm (1″) is the minimum bend radius for this cable type.

Inside plant cables with four or fewer fibers intended to be pulled through pathways during installation must support a bend radius of 50mm (2″) while under a pull load of 220N (50 pound-force [lbf]). For all other inside plant cables, the bend radius is based on the cable diameter and the loading conditions. Under no tensile load or pull load, these cables must support a bend radius of 10 times the cable's outside diameter. At the maximum tensile load rating or pull load defined by the manufacturer, these cables must support a bend radius of 20 times the cable's outside diameter.

NOTE While ANSI/TIA-568-C.3 provides bend radius requirements, not all cables comply with these requirements. You should always refer to the manufacturer's data sheet for the bend radius and pull load performance for each cable being installed.

Tensile Rating

Recall that there are different types of strength members in fiber-optic cables. The job of the strength member is to ensure that no tensile stress is placed on the optical fiber during and after installation. The strength member does have physical limitations. Tensile forces that exceed the physical limitations the cable was designed to handle can damage the cable and possibly break the optical fiber. Excessive tensile loading may also create macrobends, causing the optical fiber to attenuate a signal.

There are two types of tensile loads:

◆ A *static load* is a tensile load on a cable that does not change. A static load is often referred to as the *operational load*.

◆ A *dynamic load* is a changing tensile load. Dynamic load is referred to as the *installation load*.

ANSI/TIA-568-C.3 Section 4.3 defines the minimum pull strength requirements for inside plant, indoor-outdoor, outside plant, and drop cables. *Pull load, pull strength,* and *tensile load* are all the same type of load on a cable, and the terms may be interchanged.

Inside plant cables   As you learned in the section "Bend Radius," earlier in the chapter, inside plant cables are broken into three groups. The minimum pull load requirements only apply to cables with four or fewer fibers intended to be pulled through pathways during installation. The minimum pull load for these cables is 220N (50lbf).

Indoor-outdoor cables   Indoor-outdoor cables with 12 or fewer fibers must have a minimum pull strength or tensile load rating of 1,335N (300lbf). Cables with more than 12 fibers must have a minimum tensile load rating or pull strength of 2,670N (600lbf).

Outside plant cables   The minimum tensile load rating or pull strength for an outside plant cable is not dependent on the fiber count. All outside plant cables must have a minimum pull strength of 2,670N (600lbf).

Drop cables The minimum pull strength for a drop cable depends on how the cable is installed. A drop cable that is directly buried, placed by trenching, or blown into a duct must have a minimum pull strength 440N (100lbf). If the cable is installed by pulling, the minimum pull strength is 1,335N (300lbf).

When installing any cable type, always refer to the manufacturer's data sheet for physical requirement specifications. Some cables may or may not meet the minimum physical requirements defined in ANSI/TIA-568-C.3. Other cables may exceed these requirements.

Let's examine the physical requirements specification for two fiber-optic cables. Table 31.1 lists the tensile load and bend radius specifications from a manufacturer's data sheet for two fiber-optic cables. This table defines the minimum bend radius and the maximum tensile load or pull load for each cable. These values are defined for short- and long-term operation.

The cables described in Table 31.1 are inside plant cables. The single-fiber cable is shown in Figure 31.1.

FIGURE 31.1
Single fiber cable
with ST connectors

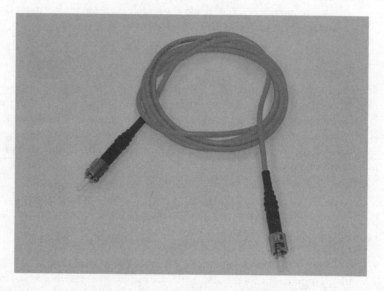

TABLE 31.1: Inside Plant Cable Physical Requirements Specification

CABLE TYPE	DIAMETER	WEIGHT	MINIMUM BEND RADIUS: SHORT TERM	MINIMUM BEND RADIUS: LONG TERM	MAXIMUM LOAD: SHORT TERM	MAXIMUM LOAD: LONG TERM
Single-fiber cable	0.114″, 2.9mm	6lb/kft, 9kg/km	1.8″, 4.5cm	1.2″, 3.0cm	50lb., 220N	15lb., 66N
12-fiber cable	0.26″, 6.6mm	27lb/kft, 40kg/km	3.9″, 9.9cm	2.6″, 6.6cm	300lb., 1,334N	90lb., 396N

The 12-fiber cable is shown in Figure 31.2. Both cables utilize a tight-buffered fiber surrounded by aramid yarn strength members, with a flexible jacket.

FIGURE 31.2
12-fiber cable

Note that the weight of the each cable over distance is included in Table 31.1. The reason for including the weight is that it can add to the tensile load if the fiber is to be suspended or hung vertically for long distances. In the single-fiber cable, for example, the long-term load limit is 15 pounds, while its weight is 6 pounds per thousand feet. If the cable must support itself for a length of 500′, it already has 3 pounds of tensile loading on it.

Note also that the manufacturer's specified bend radius for installation is about 15 times the cable diameter, while the long-term bend radius is about 10 times the cable diameter. This greater minimum bend radius for installation allows a greater tensile load—about three times as much as the long-term tensile load—to be used for pulling or wrestling the cable into position. Remember that the figures used in this instance are for a cable that is installed inside a building. Cables that will be installed outside of a building will typically have a larger minimum bend radius.

Installation Hardware

Fiber-optic cables and the optical fibers within them have a number of specialized requirements for their installation and termination. From cable-pulling tools to cable-protecting enclosures, installation hardware has been specially designed and built to meet the needs of fiber-optic cables in almost any environment and situation.

Let's take a look at some of the hardware commonly used in fiber-optic installation.

Pulling Eye

Sooner or later, you will need to run a cable through a wall, conduit, or other inaccessible space. An indispensable tool for this job is the pulling eye, shown in Figure 31.3. This device is specially designed to attach to the cable's strength member at one end and a pulling line at the other. The pulling line is fed through the space to be occupied by the cable. The line is then used to pull the eye through the space with the cable attached.

FIGURE 31.3
The pulling eye is used to pull cable through the conduit.

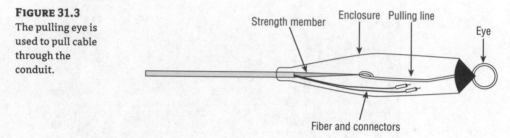

The pulling eye also uses a sheath that encloses the fiber ends to protect them from damage while the cable is being pulled.

Pull Box

Optical fiber is typically small enough, light enough, and flexible enough to be relatively easy to pull through conduits. However, friction is always a concern. Friction on the cable as it is pulled through the conduit will increase the tensile loading. Increases in friction as the cable is pulled through a turn or several turns can cause the tensile load to exceed its maximum.

To make the cable easier to pull and to ease the tensile load on it, pull boxes are installed at intervals in the conduit. Typically, pull boxes are installed after long straight runs and every time a set of turns totals 180° or more. The purpose of a pull box is to create an intermediate opening for pulling the cable to reduce the length that is being pulled through the conduit and to reduce the number of turns through which the cable must be pulled at any one time.

To use a pull box, pull the fiber-optic cable through the box and out the large opening. Be very conscious of how much the cable is being bent while you are doing this. Also be very careful not to place a bend in the cable that has a radius less than the minimum bend radius defined by the manufacturer. Once the cable has been pulled as far as it will go, feed it into the other side of the box and down the conduit, again paying close attention to the bend radius. This process minimizes the stress placed on the cable.

There are two types of pull boxes, as shown in Figure 31.4: straight and corner.

FIGURE 31.4
Straight and corner pull boxes ease tensile loading on fiber during installation.

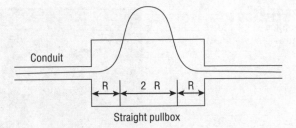

R = Manufacturer's specified minimum bend radius

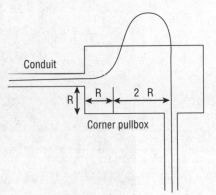

Straight pull box   Straight pull boxes are installed in-line in the conduit and should have an opening at least four times the minimum bend radius of the cable being pulled. This length prevents the cable from exceeding its minimum bend radius as the last of it is pulled through the box.

Corner pull box   Corner pull boxes are installed at angles in the conduit and typically require a length of three times the cable's bending radius and a depth that equals or exceeds the bending radius. This requirement prevents the cable from dragging against a sharp turn when it is pulled through.

Splice Enclosures

Any time you have a splice in an optical fiber, whether it is a mechanical or a fusion splice, you must protect it from exposure and strain. *Splice enclosures* take many forms, depending on their location and specific application. Some have been adapted from electrical splice enclosures used in the telecommunications industry for aerial and underground cable, while others are designed specifically for optical fibers and are used for indoor installations.

Splice enclosures can typically be placed into two categories:

Radial splice enclosure   With a radial splice enclosure, cables enter and exit the enclosure through the same side, as shown in Figure 31.5.

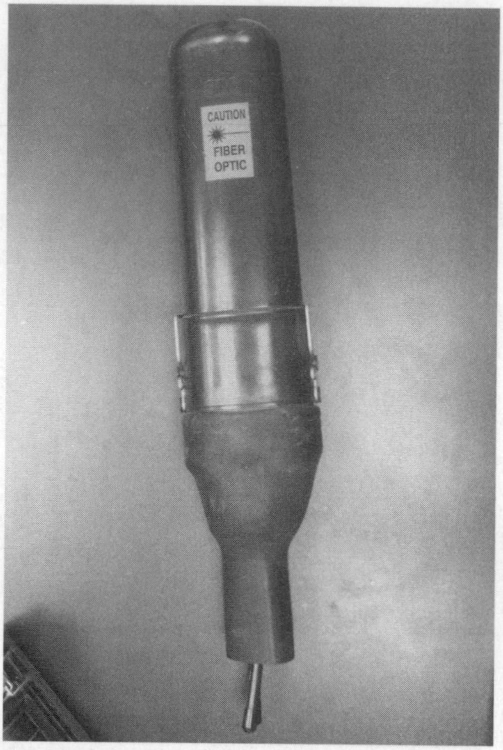

FIGURE 31.5
Radial splice
enclosure

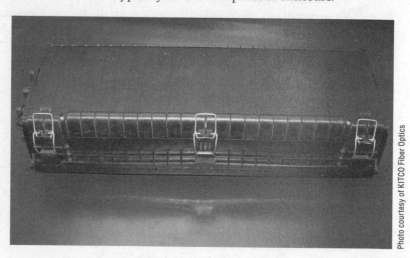

Photo courtesy of KITCO Fiber Optics

Axial splice enclosure   With an axial splice enclosure, cables enter and exit the enclosure at opposite ends, as shown in Figure 31.6. Both types can be used for an aerial or direct burial installation. However, the radial is typically used with a pedestal enclosure.

FIGURE 31.6
Axial splice
enclosure

Photo courtesy of KITCO Fiber Optics

The *pedestal enclosure* shown in Figure 31.7 is used when the fiber-optic cable has been buried underground. The pedestal enclosure is placed over the cables as they enter or exit the ground.

FIGURE 31.7
Fiber-optic ped-
estal

Photo courtesy of KITCO Fiber Optics

Splice enclosures designed for outdoor applications like those shown in Figures 31.5 and 31.6 are environmentally sealed. However, indoor splice enclosures may or may not be environmentally sealed. The rack-mounted splice enclosure shown in Figure 31.8 is for indoor applications and is not environmentally sealed.

FIGURE 31.8
Rack-mounted
splice enclosure

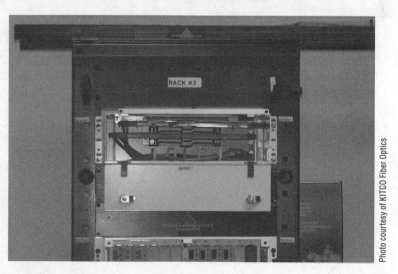

Photo courtesy of KITCO Fiber Optics

Typically, splice enclosures will incorporate the following features:

♦ A strain-relief system that ensures the strength member will carry all of the tensile loading.

♦ Clips incorporated into a panel or tray, or bonded to the enclosure. The clips hold the actual splices in an orderly fashion.

♦ Space for looping the extra optical fiber required to perform the splice outside of the enclosure.

The splice enclosure shown in Figure 31.9 has three mechanical splices and three fusion splices. The clips that hold the splices are bonded to the enclosure. The mechanical splices are held in the top three clips and the fusion splices in the bottom three. Notice that this splice enclosure allows sufficient space for the optical fiber to be bent without exceeding its minimum bend radius.

FIGURE 31.9
A splice enclosure with three mechanical and three fusion splices

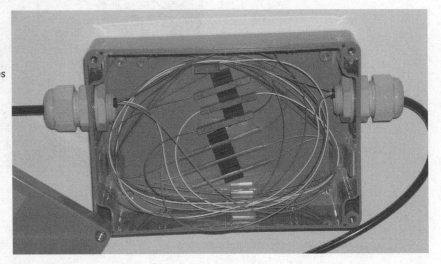

Patch Panels

A *patch panel* is an interconnection point for fiber-optic cables. They allow signals to be routed from one cable to another with a patch cord or jumper. Patch panels are available in many different shapes and sizes. Patch panels are often mounted in a rack, as shown in Figure 31.10.

However, patch panels may be contained within an enclosure, as shown in Figure 31.11.

The patch panel contained within the wall-mounted enclosure shown in Figure 31.11 allows the interconnection of the four fiber-optic cables entering the box from the top. Signals from one cable can be routed to another simply by making a connection between the two optical fibers with a jumper. The jumpers to the right of the patch panel provide an optical path from one fiber-optic cable to another.

FIGURE 31.10
A rack-mounted
patch panel

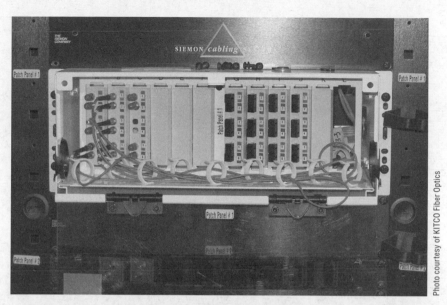

FIGURE 31.11
An enclosure-
mounted
patch panel

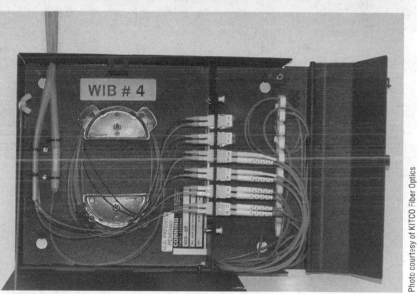

Patch panels use a bulkhead-mounted *mating sleeve* to make the interconnection. These mating sleeves can feature identical connector receptacles or different receptacles. Figure 31.12 shows an ST-to-ST mating sleeve with connectors on each end.

FIGURE 31.12
ST-to-ST mating sleeve with connectors

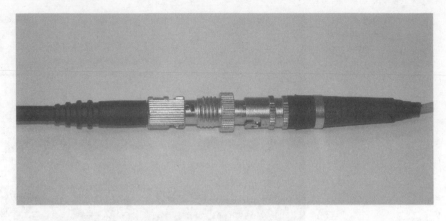

Figure 31.13 shows a mating sleeve that mates an ST connector with an SC connector. Mating sleeves with different connector receptacles are often referred to as *hybrid adapters*.

FIGURE 31.13
ST-to-SC mating sleeve or hybrid adapter with connectors

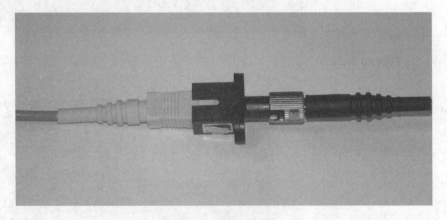

Like connectors, receptacles have dust caps to prevent contamination. Figure 31.14 shows an ST-to-SC hybrid adapter with the dust caps in place.

FIGURE 31.14
ST-to-SC hybrid adapter with dust caps installed

Installation Methods

Optical fiber has already reached into most of the places that once only knew copper cable. As technology, regulations, and pricing permit, fiber will ultimately replace copper for most signal-carrying applications, even into the home.

Many fiber installations resemble those used for copper wiring and have built on the lessons learned from it. There are some installation requirements and methods unique to fiber, however, that are required to protect the cable and ensure the highest-quality transmission.

Let's look at a typical application for optical fiber that uses a variety of installation methods. In our example, shown in Figure 31.15, a manufacturing plant is using fiber to carry instrumentation and control signals between the production building and another building several hundred meters away (called the control building in the diagram). The cable carrying the signals must be collected in a central area and routed into a cable that runs through trays and ductwork to the outside. The cable then runs from the building and is strung across several poles until it reaches a road. There, the cable runs underground until it enters the next building and is distributed to data collection and control systems.

FIGURE 31.15
A sample fiber
installation
scenario

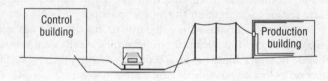

Let's look at installation for each of these situations.

Tray and Duct

Tray and duct installation is used inside structures and is similar to installation methods used for electrical wiring. Because many optical fiber cables are nonconductive, some of the requirements and restrictions for copper cables do not apply to fiber.

When the optical fiber cable rests in trays or horizontal ductwork, as shown in Figure 31.16, the weight of the cable is usually not a factor as long as the runs remain on the same vertical level. If optical fiber cabling is run vertically, however, the cable will have to support itself or be secured using either cable clamps or hangers. Be sure to follow the cable manufacturer's specifications for vertical cable rise.

FIGURE 31.16
Tray and duct
installation

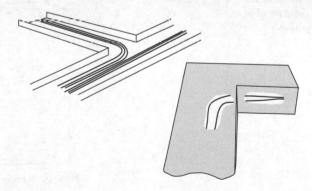

There are two methods for clamping cables for vertical runs:

♦ One type of clamp, which can also be used to secure cables horizontally, secures the cable directly to the surface by placing pressure against it. These should be installed carefully to prevent crushing the cable with excessive force.

♦ The second type of vertical clamp is a tight wire mesh that wraps around the cable and is secured to a hanger. This method has the advantage of reducing pressure against the cable itself while still supporting it. It also allows the cable to be removed more easily, since the mesh wrap is slipped over the hanger, and not permanently attached.

When installing cable in a tray, be aware of whether the tray will also be occupied by other cables, especially electrical cables. Aside from safety considerations surrounding the use of conductive or hybrid optical fiber cables, the weight of the much larger copper cables can cause macrobends if they are piled on top of the fiber. If there is a risk of this happening, choose an optical fiber cable that will protect optical fiber inside from the crushing weight of other cables.

If you are laying the cable into the tray, rather than pulling it, you can meet the manufacturer's minimum bend radius specification. If you will be pulling it, however, be sure to leave some extra radius to account for the tensile loading.

Whether you are installing in a tray or in a duct, do not allow the fiber to contact any sharp edges or exceed the minimum bend radius. Ideally, you should be working with components that do not have these hazards built in; however, if these hazards exist you should keep cable tension as low as possible during installation.

Conduit

Conduit installation uses dedicated conduits for the cable runs, which may be installed by feeding pulling lines through the conduit with a fish tape, attaching the cable's strength member to a pulling eye, and then pulling the cable through the conduit.

Conduit may be run inside structures or underground, and in many cases conduits may already be in place for other applications, such as power or telephone lines. When you are installing cable in conduits within a structure, be sure that you have allowed enough room in the conduit for the cable or cables you are installing, as shown in Figure 31.17. If possible, account for future expansion by installing a cable with extra optical fibers.

FIGURE 31.17
The conduit must
leave room for fiber
to be pulled.

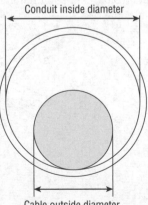

Conduit inside diameter

Cable outside diameter

When preparing for a conduit installation, you must be aware of the *fill ratio* for the conduit. If you are installing cable in existing conduit already occupied by one or more cables containing electrical conductors, you will need to determine the fill ratio of the conduit prior to selecting your cable. The NEC specifies the following maximum fill ratios by cross-sectional area for conduit:

1 cable: 53 percent

2 cables: 31 percent

3 or more cables: 40 percent

To determine the fill ratio, use the following formula:

Fill Ratio = $(OD^2\text{cable}_1 + OD^2\text{cable}_2 \dots) / ID^2\text{conduit}$

where *OD* is the outside diameter of the cable and *ID* is the inside diameter of the conduit.

After you have determined the fill ratio of the existing cables in the conduit, select your cable. You need to ensure that the cable you select does not cause the NEC fill ratios to be exceeded. Fortunately, optical fiber is available in many different cable types. If the existing fill ratio will only permit a fiber-optic cable with a very small cross-sectional area, use a ribbon cable.

Let's determine the conduit fill ratio with four existing cables in place. The inside diameter of the conduit is 1.5″. The cable to be installed has an outside diameter of 0.35″. The outside diameters of the existing cables are 0.45″, 0.37″, 0.25″, and 0.55″. To determine the fill ratio of the conduit before the fiber-optic cable is installed, use the following formula:

Fill Ratio = $(OD^2\text{cable}_1 + OD^2\text{cable}_2 + OD^2\text{cable}_3 + OD^2\text{cable}_4) / ID^2\text{conduit}$

Fill Ratio = $(0.45^2 + 0.37^2 + 0.25^2 + 0.55^2) / 1.5^2$

Fill Ratio = $(0.2025 + 0.1369 + 0.0625 + 0.3025) / 2.25$

Fill Ratio = $0.7044 / 2.25$

Fill Ratio = 0.313 or 31.3%

The next step is to determine the conduit fill ratio with the addition of the fiber-optic cable. To determine that fill ratio, use the following formula:

Fill Ratio = $(OD^2\text{cable}_1 + OD^2\text{cable}_2 + OD^2\text{cable}_3 + OD^2\text{cable}_4 + OD^2\text{cable}_5) / ID^2\text{conduit}$

Fill Ratio = $(0.45^2 + 0.37^2 + 0.25^2 + 0.55^2 + 0.35^2) / 1.5^2$

Fill Ratio = $(0.2025 + 0.1369 + 0.0625 + 0.3025 + 0.1225) / 2.25$

Fill Ratio = $0.8269 / 2.25$

Fill Ratio = 0.3675 or 36.75%

The calculated fill ratio is less than 40 percent, so the conduit is large enough as defined by the NEC.

If a large conduit is already in place and contains other cables, or if it is likely that other cables will be run through the conduit in the future, you may want to consider installing an *innerduct*, a conduit sized for the optical fiber cable that will protect it from other activities in the larger conduit.

Direct Burial

To run an outdoor fiber-optic cable out of sight, we can install it by *direct burial*. As the name implies, this method can be as simple as placing a suitable cable directly in the ground. Direct burial methods also include placing a cable within a protective pipe or conduit and burying it.

Direct burial has some advantages. Cables that are buried underground are not visible and do not obstruct the scenery like cables that span telephone poles or buildings. Burying a cable keeps it out of the wind, rain, snow, and ice. The wind from a tornado or hurricane will not damage a buried cable. Unlike aerial cables, a buried cable will not break from the excessive weight of ice from an ice storm.

However, direct burial of fiber-optic cables also has some disadvantages. Fiber-optic cables buried underground are difficult to locate since they do not emit any electromagnetic energy the way a copper cable such as a power line does. Animals that burrow can damage cables buried underground. A direct burial cable typically features an armor layer that protects it from burrowing animals and damage that may be caused when rock is used to cover the cable as is placed in the ground.

To place a direct burial cable in the ground, a backhoe or shovel can be used to dig a trench in which the cable is laid and then covered up. This approach is typically used for short runs. For longer runs that require more efficient methods, a cable-laying plow is available. This device is designed to open a trench, lay the cable, and cover it up again while on the move. It is a more complicated machine, but it is useful when longer distances must be covered.

When using direct burial methods, be sure to dig the trench deep enough to be below the frost line. In some areas, this can be as much as 30″ deep.

Remember that you must contact the local utilities before you begin digging or trenching operations. The location of existing underground cables must be known before you dig.

Aerial

As our cable leaves the production building, it's going to be spliced onto a fiber that is part of a messenger cable, which will be strung along a series of poles in an aerial installation. These poles already carry power lines, but they will not affect the data on the optical fiber itself because it is an insulator.

If a messenger wire is not incorporated into the cable assembly used, the cable will have to be lashed to messenger wire. A messenger wire is a steel or aluminum wire running between the poles to support the fiber-optic cable. Either way, cables in aerial installations must be able to withstand loading from high winds, ice, birds and climbing animals, and even windblown debris such as branches.

Blown Fiber

Blown fiber is an alternative to installing cables. In new construction or renovation, it is a good idea to consider blown fiber as an alternative to traditional methods.

There are three steps to installing blown fiber:

1. Install the special tubing or conduit.

2. Blow the optical fiber through the tubes from location to location.

3. Terminate the optical fiber.

The optical fiber used in a blown fiber installation is the same optical fiber found in fiber-optic cables. However, the strand or strands of optical fiber will be coated with a protective layer that creates friction as dry gas passes over it. The friction generated by the dry gas as it flows through the tubing pulls the optical fiber or optical fiber bundle through the tubing. During this process no tensile stress is placed on the optical fiber.

The tubing is about 5mm in diameter. Typically, a jacket is applied around two or more tubes, forming what looks like a cable. This cable-like bundle of tubes is installed without the optical fiber. The optical fiber is blown through these tubes after the installation. Figure 31.18 shows two different bundles of tubes and a single tube that is spliced in the center.

FIGURE 31.18
Blown fiber tubing assemblies

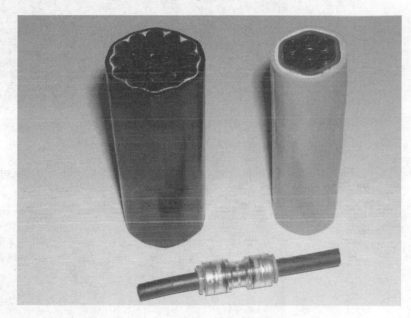

Blown optical fiber is used in buildings, between buildings, and on board ships. The U.S. Navy uses blown optical fiber on aircraft carriers. The life of a ship can exceed 50 years. Removing and installing cables on board a ship is expensive, and there is always the risk that this process may cause damage to other cables or equipment.

The tubing installed for blown optical fiber never needs to be removed because old optical fiber can be pulled out and new fiber blown in. The process of pulling out the old or damaged fiber and blowing in new fiber takes very little time in comparison to the time it would take to remove and reinstall a new cable. Figure 31.19 is a photograph of shipboard enclosure where blown optical tubes are interconnected with a splice.

FIGURE 31.19
Interconnected
blown fiber tubes

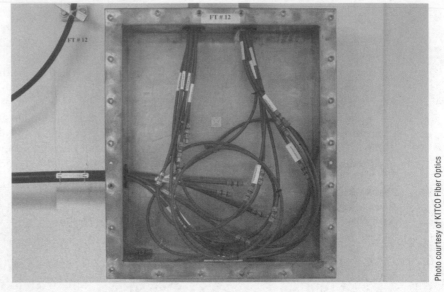

Photo courtesy of KITCO Fiber Optics

Figure 31.20 is a photograph of a shipboard enclosure where the optical fibers in the tubes are terminated.

FIGURE 31.20
Terminated blown
optical fibers

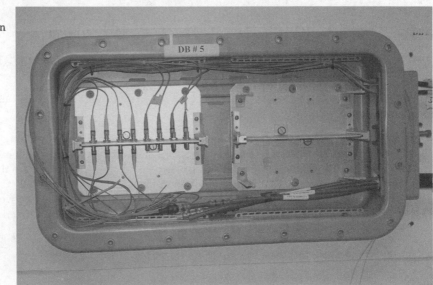

Photo courtesy of KITCO Fiber Optics

LEAVING SLACK FOR REPAIRS

No matter where you install cable, you'll want to leave some slack for the inevitable repairs that will have to be performed. How much slack should you leave? That depends on the situation. Look at the location; the number of turns that the fiber takes as it passes through conduits, trays, or ductwork; and the likelihood that the fiber will be disturbed after you have installed it. Consider the following cautionary tale:

During the construction of a strip mall in Virginia, cable installers ran cable underground in an area where construction was ongoing. After the cable was laid, backhoe operators accidentally pulled up the conduit carrying the cable seven times, requiring the cable to be spliced and re-laid in the ground. After the sixth event, all of the slack in the cable had been used up in repairs and the cable had to be replaced.

For the record, the location of the cable was clearly marked.

Cable Slack

How much slack an installation should have and where the slack should be stored depends on the installation. However, there are some basic guidelines for slack.

When a cable is brought into a building from outside and the cable is not a listed indoor-outdoor type cable, the slack should be outside the building. Typically, 10 meters of slack is sufficient for most installations.

When a cable is brought into a building from outside and the cable is a listed indoor-outdoor type cable, the slack should be inside the building. Typically, 10 meters of slack is sufficient for most installations.

Cables that span the inside of a building between cross-connects should have slack on each end of the cable. As with the other installations, 10 meters of slack is typically sufficient.

All cable slack should be stored in the form of a loop. For an indoor installation, it should either be hung from the wall or placed under a subfloor. For an outdoor installation, it should be placed in the closest suitable outdoor access point.

Fire Resistance and Grounding

As you learned in Chapter 24, Article 770 of the NEC places requirements on fiber-optic cables and their installation within buildings. One requirement is that the cable must be able to resist the spread of fire. This requirement applies to all fiber-optic cables installed in a building. Other requirements address the metallic members in some cable types.

This section will address cable selection to meet fire resistance requirements defined in Article 770 of the NEC. It will also address the installation of fiber-optic cables that contain metallic members.

NOTE This chapter provides an overview of Article 770 and is not a substitute for the NEC. The National Fire Protection Association (NFPA) updates the NEC every three years. The person performing the installation is responsible for compliance with the current version of the NEC. In addition, the installer needs to be aware of local, state, and federal codes that apply to the installation.

Fire Resistance

Article 770 of the NEC states that optical fiber cables installed within buildings shall be listed as being resistant to the spread of fire. These cables are also required to be marked in accordance with Table 31.2.

TABLE 31.2: NEC Cable Markings

CABLE MARKING	CABLE TYPE
OFNP	Nonconductive optical fiber plenum cable
OFCP	Conductive optical fiber plenum cable
OFNR	Nonconductive optical fiber riser cable
OFCR	Conductive optical fiber riser cable
OFNG	Nonconductive optical fiber general-purpose cable
OFCG	Conductive optical fiber general-purpose cable
OFN	Nonconductive optical fiber general-purpose cable
OFC	Conductive optical fiber general-purpose cable

There are three exceptions to this:

◆ An optical fiber cable is not required to be listed and marked as long as the length of the cable within the building, measured from its point of entrance, does not exceed 50′ and the cable enters the building from the outside and is terminated in an enclosure.

◆ A conductive optical fiber cable is not required to be listed or marked where the cable enters the building from the outside and is run in rigid metal conduit or intermediate metal conduit and the conduit is grounded to an electrode in accordance with Chapter 8 of the NEC.

◆ A nonconductive optical fiber cable is not required to be listed and marked where the cable enters the building from the outside and is run in a raceway installed in compliance with Chapter 3 of the NEC.

Plenum cable　Plenum cables, whether conductive or nonconductive, are suitable for use in ducts, plenums, and other space used for environmental air. These cables will have fire-resistance and low smoke-producing characteristics.

Riser cable　Riser cables, whether conductive or nonconductive, are suitable for a vertical run in a shaft or from floor to floor. These cables will have fire-resistance characteristics capable of preventing the carrying of a fire from floor to floor.

General-purpose cable　General-purpose cables, whether conductive or nonconductive, are resistant to the spread of fire. However, these cables are not suitable for plenum or riser applications.

Grounding

Even though optical fiber cable itself does not carry electrical power, there may be some circumstances in which you will have to contend with electricity. If a cable contains conductive components such as armor or metal strength members, and the cable is likely to come in contact with electrical circuits, there is a chance that, in an accident, the metal could become a path for an electrical current, potentially leading to fire or personal injury.

In the event that a conductive component in a cable comes in contact with an electrical current, that current is going to seek the path of the least resistance. If you happen to touch the cable or anything to which the cable is connected, that path could be through you. Depending on the voltage and current involved, you could face severe injury or even death.

For this reason, Article 770 places a grounding or isolation requirement on any fiber-optic cable entering a building that contains electrically conductive materials and that may be exposed to or come in contact with electrical light or power conductors. The metallic member of the cable must be grounded as close to the point of entrance as practicable. An alternative to grounding is placing an insulating joint or equivalent device as close to the point of entrance as practicable. The insulating joint interrupts the flow of current through the conductive materials in the cable.

Cable Types

Article 770 of the NEC groups fiber-optic cables into three types:

◆ Nonconductive cables have no metallic members or electrically conductive materials.

◆ Conductive cables have conductive components such as metallic strength members, vapor barriers, or armor. These conductive components were not designed into the cable to conduct current and are thus referred to as *non-current-carrying conductive members*.

◆ Composite cables contain optical fibers and electrical conductors designed to carry current. These cables may also have metallic members that were not designed to carry current such as metallic strength members, vapor barriers, or armor.

Hardware Management

Eventually, fiber-optic cables end up in some kind of cabinet, panel, rack, or enclosure. Whether it's a patch panel, rack, splice enclosure, or wall outlet, the preparation of these items is crucial to the performance of the entire network. In these locations, the fiber or cable is terminated and connected in some way, to another fiber, a connector, or a piece of hardware, and this process involves a risk of mistakes and mismanagement.

Poor hardware management can lead to confusion, excess strain on cables and fibers, and inefficient troubleshooting. In addition, a poorly organized hardware space reflects badly on the installer.

In this section we will look at some guidelines for good hardware management.

Cleanliness

A clean working environment is essential to fiber work. Whether you are splicing fibers, terminating cables, or mounting cables in cabinets, it's important to keep dust, trash, water, and other contaminants away from your work area.

If possible, block off the space in which you are working to prevent exposure to contaminants. Use air filters to draw dust and grit out of the air, and try to limit the amount of nonessential work traffic in the area where you are working. Crews have been known to park their work trucks over manholes to keep dirt from the street from falling inside while they work. In other areas, it may be enough to use good housekeeping practices to ensure that the work area won't be contaminated.

Organization

Once a cable enters the telecommunications room, it is likely to be joined by anywhere from one to dozens of other cables. If these cables are not organized from the very beginning, they can quickly become a rat's nest of tangled, interwoven lines.

Many manufacturers offer cable management devices. These devices will help you organize cables for a neater, more user-friendly layout. Often the shortest distance between two points is not the best way to route a cable. Cables should only be routed vertically and horizontally. A cable should never be routed diagonally.

Secure cables in place when required. This helps relieve strain on the cable and keeps it from assuming new and interesting configurations after the cabinet is closed. Figure 31.21 shows cables routed to the top of a 19″ rack. These cables are secured to minimize stress and maintain a configuration.

FIGURE 31.21
Secured cables
routed to a 19″ rack

Photo courtesy of KITCO Fiber Optics

Organization and neatness are also essential for efficient troubleshooting. It's much easier to trace connections and links if the cables are organized and neatly bundled than if they are spread all over the cabinet and tangled together.

Clamps and Cable Ties

Sometimes the best-planned installation can be destroyed by a simple cable tie or clamp. Earlier in the chapter, we discussed the bend radius of a fiber-optic cable. We learned that exceeding the bend radius can cause attenuation or damage the cable or optical fiber. A cable tie or clamp, if improperly applied, can place a sharp bend in a cable. This sharp bend can produce a bend radius small enough to cause attenuation or damage.

When you are installing a cable tie or clamp, be very conscious of the pressure that is being placed on the cable. Look closely at the jacket of the cable to see if you are compressing or crushing it. You do not have to be very strong to damage a cable with a clamp or cable tie. Clamps and cable ties should be installed so they do not place unnecessary pressure on the cable jacket.

It is not always easy to tell that you have placed too much pressure on the jacket with a clamp or cable tie. One way to avoid problems is to test a piece of cable with a light source on one end of the cable and a power meter on the other. Apply the clamp or cable tie to the center of the cable under test and watch the power meter to see if there are any changes in power. If the power meter shows a reduction in power as you apply the clamp or cable tie, the pressure is too great. Testing with a light source and power meter is covered in detail in Chapter 33, "Test Equipment and Link/Cable Testing."

Another way to test, without a light source and power meter, is to apply the clamp or cable tie to the desired pressure. Let it sit for several minutes, and then remove it. Carefully look at the cable after the clamp or cable tie is removed to see if an imprint was left in the jacket. If no imprint was left in the jacket, the pressure was not too great. If an imprint was left, the pressure should be reduced.

Labeling

Labeling, as we will see in the section "Labeling Requirements," later in this chapter, is key to a well-run network. Labeling helps you find the right port or cable quickly and easily, and it makes troubleshooting more efficient.

All cables must be labeled within 300mm (about 12″) of their ends. The labeling must be done in a consistent and easily readable format, which will be described shortly.

It's also important to mark ports and outlets with labels that will not come off easily. In large systems, you can lose a great deal of time just trying to remember which outlet is connected to which panel in a room at the other end of the building.

Documentation

Even after the cable is properly organized and labeled, you'll need to keep good records of the network, down to the designations and locations of each port and cable. Good documentation can speed up troubleshooting and system repair and modification, and even improve safety, because it reduces the amount of time spent climbing around in spaces just to find out where a cable leads.

Documentation, like labeling, needs to follow a consistent and easily readable format, which is described in ANSI/TIA-606-A. It should identify not only the cables and hardware but also the spaces occupied by the network and the locations of outlets and type of installation used.

Labeling Requirements

Even the simplest fiber-optic network can instantly become confusing if it is improperly or insufficiently labeled to show where cables originate, how they are connected, and where they go.

It may be tempting to come up with a handy system of your own for labeling, but unless you can predict all of the changes that will take place as the system grows, you will soon find yourself with a cumbersome mess on your hands. In addition, if you don't keep track of your labeling system, it will be difficult if not impossible for others to decipher it.

The ANSI/TIA standard for labeling is contained in ANSI/TIA-606-A. This document contains detailed information on conventions for labeling optical fiber connections within buildings and between buildings that share a communication network. The standard specifies the methods used to determine how fiber links are identified and a numbering/lettering system that is standardized according to a top-down classification.

ANSI/TIA-606-A requires that a label be attached to each end of the cable assembly. These labels need to be clearly visible and durable. The labels should be designed to last as long as the cable assembly and must be printed with a mechanical device designed for this purpose. Handwritten labels are not acceptable.

If the patch panel has provisions for inserting labels between the wiring blocks, these labels are to be color-coded. The color of the label identifies the origin of the cables. Labels may also contain additional information about the connection such as the cabinet number and shelf number. Table 31.3 lists the termination type color-coding scheme.

TABLE 31.3: ANSI/TIA-606-A Patch Panel Color Coding Scheme

TERMINATION TYPE	LABEL COLOR	TYPICAL APPLICATION TYPE
Demarcation point	Orange	Central office or entrance facility terminations
Network connection	Green	User side of central office or entrance facility connection, equipment room
Common equipment	Purple	Connections to a mainframe, LAN, or multiplexer
Key system	Red	Connections to key telephone systems
First level backbone	White	Main cross-connect to intermediate cross-connect terminations
Second level backbone	Gray	Intermediate cross-connect to telecommunications room cable terminations
Campus backbone	Brown	Inter-building backbone terminations
Horizontal	Blue	Horizontal cable terminations
Miscellaneous	Yellow	Alarms, security, or energy management

As an example of labeling according to ANSI/TIA-606-A, imagine working with a wall outlet that has a port labeled:

1A-B002

This particular number, known as a horizontal link identifier, immediately communicates the following information:

1: The fiber originates in a telecommunications room (TR) on the first floor of the building.

A: The TR is designated as A.

B: The fiber originates in patch panel B in TR A.

002: The fiber originates in port 002 of the patch panel.

Using this labeling system, we can very quickly trace the fiber to a specific connection point, even in a very large building.

While the previous example is fairly simple, it shows that by using a consistent labeling system, it is possible to communicate fiber routings accurately. Other numbers may be used to show *backbones*, or links between telecommunication spaces, interbuilding cabling, and other major connections.

The Bottom Line

Determine the minimum cable bend radius.   Remember that cable manufacturers typically define the minimum bend radius when the cable is stressed and unstressed. These numbers are typically in compliance with ANSI/TIA-568-C.3.

Master It   Refer to the following table and determine the minimum installation bend radius and the minimum operational bend radius for the 24-fiber cable.

Cable Type	Diameter	Weight	Minimum Bend Radius: Short Term	Minimum Bend Radius: Long Term	Maximum Load: Short Term	Maximum Load: Long Term
12-fiber cable	0.26″, 6.6mm	27lb/kft, 40kg/km	3.9″, 9.9cm	2.6″, 6.6cm	300lb., 1334N	90lb., 396N
24-fiber cable	0.375″, 9.5mm	38.2lb/kft, 56.6kg/km	5.5″, 14cm	3.7″, 9.4cm	424lb., 1887N	127lb., 560N

Determine the maximum cable tensile rating. Remember that cable manufacturers typically define a dynamic and static pull load.

Master It Refer to the following table and determine the maximum dynamic tensile load and the maximum static tensile load for the 12-fiber cable.

Cable Type	Diameter	Weight	Minimum Bend Radius: Short Term	Minimum Bend Radius: Long Term	Maximum Load: Short Term	Maximum Load: Long Term
12-fiber cable	0.26″, 6.6mm	27lb/kft, 40kg/km	3.9″, 9.9cm	2.6″, 6.6cm	300lb., 1334N	90lb., 396N
18-fiber cable	0.325″, 8.3mm	33.8lb/kft, 50kg/km	4.9″, 12.4cm	3.25″, 8.3cm	375lb., 1668N	112lb., 495N

Determine the fill ratio for a multiple-cable installation. When preparing for a multiple cable conduit installation, you must be aware of the fill ratio for the conduit. If you are installing cable in existing conduit already occupied by one or more cables that contain electrical conductors, you will need to determine the fill ratio of the conduit prior to selecting your cable.

Master It You need to install one cable into a conduit that has three existing electrical cables. The inside diameter of the conduit is 1″. The cable to be installed has an outside diameter of 0.25″. The outside diameters of the existing cables are 0.28″, 0.32″, and 0.38″. Determine the fill ratio of the conduit before and after the fiber-optic cable is installed and verify that this installation will not overfill the conduit as define by the NEC.

Chapter 32

Fiber-Optic System Design Considerations

Up to this point, you have learned about transmitters, receivers, couplers, attenuators, connectors, splices, and fiber-optic cable. These pieces are some of the building blocks of a fiber-optic system. A basic fiber-optic system contains a transmitter, receiver, fiber-optic cable, and connectors.

There are many ways to approach fiber-optic system design, and there are many different fiber-optic systems. Fortunately, there are industry standards that simplify fiber-optic system design. This chapter focuses on the basic design considerations for a fiber-optic system, compares optical fiber to copper, explains how to break down a fiber-optic link to analyze performance, and shows how to prepare a power budget.

In this chapter, you will learn to:

♦ Calculate the bandwidth for a length of optical fiber

♦ Calculate the attenuation for a length of optical fiber

♦ Calculate the power budget and headroom for a multimode fiber-optic link

♦ Calculate the power budget and headroom for a single-mode fiber-optic link

Basic Fiber-Optic System Design Considerations

Before beginning a basic fiber-optic system design, three questions need to be answered:

♦ How much data needs to be moved today?

♦ How much data will need to be moved in the future?

♦ What is the transmission distance?

Throughout this book, you have learned how the physical properties of light and of the optical fiber determine bandwidth and transmission distance. Now let's take some of these lessons learned and apply them in the design of a basic fiber-optic system.

The release of a laser-optimized multimode optical fiber standard and the improvements in VCSEL bandwidth are changing the traditional approaches to designing a fiber-optic system. In the past, high data rates such 1Gbps or 10Gbps were typically accomplished using single-mode optical fiber. However, the VCSEL, when used with laser optimized multimode optical fiber, is capable of these types of data rates over several hundred meters.

Many local area network (LAN) applications do not require transmission distances greater than several hundred meters. This means the entire network could be designed using laser-optimized multimode optical fiber. As you have learned in earlier chapters, multimode optical fiber has a much larger core than single-mode optical fiber. The larger multimode core is more forgiving than the smaller single-mode core at interconnections. A small misalignment at a single-mode interconnection could prevent the system from operating whereas that same amount of misalignment would have little or no effect on the multimode interconnection.

We mentioned earlier that before you begin a design, the current data rate, future data rate, and transmission distance must be known. Let's look at three rules of thumb that will help you design a link to meet your current data rate, future data rate, and transmission distance expectations.

Rule of thumb number 1   For current and future data rates up to 1Gbps with a transmission distance no greater than 860m, choose laser-optimized multimode optical fiber.

Rule of thumb number 2   For current and future data rates greater than 1Gbps up to 10Gbps and transmission distances no greater than 300m, choose laser-optimized multimode optical fiber.

Rule of thumb number 3   For all other applications, choose a laser transmitter and single-mode optical fiber.

You may remember from Chapter 27, "Fiber-Optic Light Sources," that LED transmitters are typically not designed for data rates exceeding 155Mbps, whereas laser transmitters that support data rates from 10Mbps through 10Gbps are readily available.

The three rules of thumb just discussed are not all-encompassing. They are intended to be a general guideline and are based on IEEE 802.3 Ethernet standards, Fibre Channel standards, and SONET standards for a single laser and single optical fiber. There are applications where these rules of thumb do not apply; however, those applications are not basic systems and may require more than one laser or optical fiber.

The Advantages of Optical Fiber over Copper

So far in this chapter, you have learned the rules of thumb for a basic system design without hearing about any of the factors that affect system performance. This section will look at seven different performance areas within a system and compare the performance of optical fiber to Category 5e and coaxial copper cable.

Performance in this comparison will be evaluated in the areas of bandwidth, attenuation, *electromagnetic immunity*, size, weight, safety, and security. The fiber-optic system will operate at an 850nm wavelength with a VCSEL source. System performance data for the optical fiber is derived from ANSI/TIA-568-C.3. Laser-optimized optical fiber will be compared to Category 5e copper cable. Performance data for the Category 5e copper cable will be derived from Commercial Building Telecommunication Cabling Standard, ANSI/TIA-568-B-2001.

At the time this chapter was written, TIA was in the process of updating the 568 standards. ANSI/TIA-568-C.3 was available for purchase while ANSI/TIA-568-C.0, C.1, and C.2 were not. All the comparisons in this chapter are based on latest standards that were available for purchase at the time the chapter was written.

Bandwidth

Bandwidth is a very popular buzzword these days. We are bombarded with commercials advertising high-speed downloads from a *cable modem,* satellite, or fiber-to-the-home (FTTH). The competition to sell us bandwidth is fierce. As we learned in Chapter 18, "History of Fiber Optics," by 2014 some 100 million homes should have fiber-optic connectivity.

In earlier chapters, we looked at the physical properties of the optical fiber and fiber-optic light source that limit bandwidth. You learned that single-mode systems with laser transmitters offer the greatest bandwidth over distance and that multimode systems with LED transmitters offer the least bandwidth and are limited in transmission distance. You also learned that the bandwidth of the optical fiber is inversely proportional to its length. In other words, as the length of the optical fiber increases, the bandwidth of the optical fiber decreases.

So how does length affect the bandwidth of a copper cable? Well, when it comes to cable length, copper suffers just like optical fiber does. As the length of a copper cable is increased, the bandwidth for that cable decreases. So if both copper and optical fiber lose bandwidth over distance, why is optical fiber superior to copper? To explain that, we will look at the minimum bandwidth requirements defined in ANSI/TIA-568-C.3 for multimode optical fiber and ANSI/TIA-568-B.1 for *unshielded twisted-pair (UTP) Category 5e* cable. Remember that the values defined in ANSI/TIA-568-B.1 and ANSI/TIA-568-C.3 are the minimum values that a manufacturer needs to achieve. There are many manufacturers offering optical fiber and copper cables that greatly exceed these minimum requirements. However, for this comparison, only values defined in these two standards will be used.

One of the problems in doing this comparison is cable length. ANSI/TIA-568-B.1 defines Category 5e performance at a maximum physical length of 100m. ANSI/TIA-568-C.3 does not define a maximum bandwidth at a maximum optical fiber length. As you learned in Chapter 22, "Optical Fiber Characteristics," ANSI/TIA-568-C.3 defines multimode optical fiber bandwidth two ways depending on the type of light source. In this comparison the 850nm VCSEL is used with 50.5/125µm optical fiber. The minimum effective modal bandwidth-length product in MHz · km is 2,000 for this combination, as shown in Table 32.1.

TABLE 32.1   Characteristics of ANSI/TIA-568-C.3 Recognized Optical Fibers

Optical Fiber Cable Type	Wavelength (nm)	Maximum Attenuation (dB/km)	Minimum Overfilled Modal Bandwidth-Length Product (MHz · km)	Minimum Effective Modal Bandwidth-Length Product (MHz · km)
62.5/125µm Multimode	850	3.5	200	Not required
	1300	1.5	500	
50/125µm Multimode	850	3.5	500	Not required
	1300	1.5	500	

TABLE 32.1   Characteristics of ANSI/TIA-568-C.3 Recognized Optical Fibers   *(CONTINUED)*

Optical Fiber Cable Type	Wavelength (nm)	Maximum Attenuation (dB/km)	Minimum Overfilled Modal Bandwidth-Length Product (MHz · km)	Minimum Effective Modal Bandwidth-Length Product (MHz · km)
850nm laser-optimized 50/125μm multimode	850 1300	3.5 1.5	1500 500	2000 Not required
Single-mode Indoor-outdoor	1310 1550	0.5 0.5	N/A N/A	N/A N/A
Single-mode inside plant cable	1310 1550	1.0 1.0	N/A N/A	N/A N/A
Single-mode outside plant cable	1310 1550	0.5 0.5	N/A N/A	N/A N/A

NOTE   ANSI/TIA-568-B.3, section 5.5.1 states that Category 5e multipair balanced 100Ω backbone cabling supporting data applications up to 100MHz should be limited to a total distance of 90m. The 90m distance allows for an additional 5m at each end for equipment cables or patch cords. The total combined length cannot exceed 100m.

To do this comparison, we will need to calculate the bandwidth limitations for 100m of laser-optimized multimode optical fiber as shown here:

Bandwidth-length product = 2,000 MHz · km

Optical fiber length = 0.1km

Bandwidth in MHz = 2,000 ÷ 0.1

Bandwidth = 20,000MHz

The 100m laser-optimized multimode optical fiber has a bandwidth 200 times greater than the Category 5e UTP cable. This clearly shows that optical fiber offers incredible bandwidth advantages over Category 5e UTP cable.

NOTE   How do we get from Gbps to MHz? Remember that in Chapter 19, "Principles of Fiber-Optic Transmission," we learned about bits and symbols. Baud rate is the number of symbols per second. For most applications, the baud rate equals the bit rate. In other words, a 1Gbps transmission would transmit one billion symbols per second. As you learned in Chapter 19, one complete cycle of a square wave contains two symbols, a high symbol and a low symbol. So a 500MHz square wave transmits one billion symbols per second, or 1Gbps.

Now that we have seen how the bandwidth of laser-optimized multimode optical fiber with an 850nm source greatly exceeds that of Category 5e cable, let's do a comparison with RG6 *coaxial cable*. This cable has been chosen because it is widely used in homes and buildings for video distribution. In this comparison, the performance data for the RG6 coaxial cable is taken from the average of several manufacturers' data sheets. ANSI/TIA-568-B does not define performance parameters for RG6 coaxial cable.

An RG6 coaxial cable with a transmission frequency of 1GHz has roughly the same 100m transmission distance characteristics as a Category 5e cable. We know from the previous comparison that a 100m length of 50/125μm multimode optical fiber will support transmission frequencies up to 20GHz over that same distance with an 850nm light source. In this comparison, the laser-optimized multimode optical fiber offers a bandwidth advantage 20 times greater than the RG6 coaxial cable.

These comparisons have demonstrated the bandwidth advantages of optical fiber over copper cable. The comparisons were done at very short distances because at this point we have not addressed how attenuation in a copper cable changes with the transmission frequency, whereas in an optical fiber, attenuation is constant regardless of the transmission frequency.

Attenuation

All transmission mediums lose signal strength over distance. As you know, this loss of signal strength is called attenuation and is typically measured in decibels. Optical fiber systems measure attenuation using optical power. Copper cable systems typically use voltage drop across a defined load at various transmission frequencies to measure attenuation. The key difference here is not that optical fiber uses power and copper uses voltage. The key difference is that attenuation in copper cables is measured at different transmission frequencies. This is not the case with optical fiber, where attenuation is measured with a continuous wave light source.

The attenuation in a copper cable increases as the transmission frequency increases. Table 32.2 shows the maximum worst pair insertion loss for a 100m horizontal Category 5e and Category 3 cable as defined in ANSI/TIA-568-B.2, section 4.3.4.7. This table clearly shows the effects that transmission frequency has on a copper cable.

TABLE 32.2   Horizontal Cable Insertion Loss, Worst Pair*

FREQUENCY (MHZ)	CATEGORY 3 (DB)	CATEGORY 5E (DB)
0.772	2.2	1.8
1.0	2.6	2.0
4.0	5.6	4.1
8.0	8.5	5.8
10.0	9.7	6.5
16.0	13.1	8.2
20.0		9.3

TABLE 32.2   Horizontal Cable Insertion Loss, Worst Pair*   *(CONTINUED)*

FREQUENCY (MHz)	CATEGORY 3 (DB)	CATEGORY 5E (DB)
25.0		10.4
31.25		11.7
62.5		17.0
100.0		22.0

For a length of 100m (328´)

The maximum allowable attenuation in an optical fiber is defined in section 4.2 of ANSI/TIA-568-C.3. Table 32.3 shows the attenuation portion of the optical fiber cable transmission performance parameters. This table defines attenuation for both multimode and single-mode optical fibers. You will notice that there is no column for transmission frequency in this table. That is because optical fiber does not attenuate as transmission frequency increases like copper cable does.

TABLE 32.3   Optical Fiber Cable Attenuation Performance Parameters

OPTICAL FIBER CABLE TYPE	WAVELENGTH (NM)	MAXIMUM ATTENUATION (DB/KM)
50/125µm multimode	850	3.5
	1300	1.5
62.5/125µm multimode	850	3.5
	1300	1.5
Single-mode inside plant cable	1310	1.0
	1550	1.0
Single-mode outside plant cable	1310	0.5
	1550	0.5

Now that we know much about how optical fiber and copper cable attenuate, let's do a comparison. The first comparison will put optical fiber up against Category 5e cable. The distance will be 100m and the transmission frequency will be 100MHz.

Table 32.2 lists 22dB for the worst-case attenuation for the Category 5e cable at 100MHz. This means the Category 5e cable loses 99.37 percent of its energy over a distance of 100m at that transmission frequency. As you learned in Chapter 22, the attenuation of an optical fiber is different for each wavelength. Table 32.3 lists the maximum attenuation in dB/km. We can solve for the maximum attenuation at both wavelengths as shown here:

Attenuation coefficient at 850nm = 3.5dB/km

Optical fiber length = 0.1km

Attenuation for the length of optical fiber = 0.1 × 3.5 = 0.35

Maximum attenuation at 850nm for 100m of optical fiber is 0.35dB

Attenuation coefficient at 1300nm = 1.5dB/km

Optical fiber length = 0.1km

Attenuation for the length of optical fiber = 0.1 × 1.5 = 0.15

Maximum attenuation at 1300nm for 100m of optical fiber is 0.15dB

The equations show that multimode optical fiber loses 7.7 percent of the light energy from the transmitter over the 100m distance at 850nm. At 1300nm, the loss is only 3.4 percent. However, the Category 5e cable loses 99.4 percent of the transmitted signal over that distance. This means that the Category 5e cable loses roughly 150 times more energy than the optical fiber.

This comparison clearly shows the attenuation advantages of optical fiber over Category 5e cable.

Now let's compare the same optical fiber at the same wavelengths to an RG6 coaxial cable. For this comparison, the RG6 attenuation characteristics are taken from the average of several different manufacturers' published data sheets, as shown in Table 32.4.

TABLE 32.4   RG6 Cable Insertion Loss*

FREQUENCY (MHz)	RG6 (DB)
1.0	1.0
5.0	1.8
10.0	2.3
20.0	3.3
50.0	5.0
100.0	6.2
200.0	9.2
300.0	11.0
400.0	12.5
500.0	14.0
1000.0	19.4
3000.0	35.1

*For a length of 100m (328′)

As you look at Table 32.4, you can see that RG6 coaxial cable easily outperforms Category 5e cable. However, it does not even begin to approach the multimode optical fiber operating at 850nm or 1300nm. The RG6 coaxial cable has 6.2dB of attenuation at a transmission frequency of 100MHz over a distance of 100m. Whereas the Category 5e cable lost 99.4 percent of its signal strength, in the previous comparison the RG6 coaxial cable lost only 76 percent of its energy over the same distance.

These comparisons should make it clear that optical fiber has an enormous attenuation advantage over copper cable. The comparisons were done only with multimode optical fiber. Outside plant single-mode optical fiber greatly outperforms multimode optical fiber. Single-mode optical fiber links are capable of transmission distances greater than 80km without re-amplification. Based on the previous comparisons, Category 5e cable would require re-amplification every 100m at a transmission frequency of 100MHz. The RG6 coaxial cable would only require re-amplification every 350m at a transmission frequency of 100MHz.

An 80km RG6 coaxial cable link transmitting a frequency of 100 MHz would require roughly 228 amplifiers to re-amplify the signal. However, the single-mode optical fiber link would not require re-amplification. Optical fiber links are unsurpassed in transmission distance.

Electromagnetic Immunity

Electromagnetic interference (EMI) is electromagnetic energy, sometimes referred to as *noise*, that causes undesirable responses, degradation, or complete system failure. Systems using copper cable are vulnerable to the effects of EMI because a changing electromagnetic field will induce current flow in a copper conductor. Optical fiber is a dielectric or an insulator, and current does not flow through insulators. What this means is that EMI has no effect on the operation of an optical fiber.

Let's take a look at some examples of EMI-induced problems in a copper system. A Category 5e cable has four pairs of twisted conductors. The conductors are twisted to keep the impedance uniform along the length of the cable and to decrease the effects of EMI by canceling out opposing fields. Two of the conductor pairs are used to transmit and two of the conductor pairs are used to receive.

For this example, only one pair is transmitting and one pair is receiving. The other two pairs in the cable are not used. Think of each pair of conductors as an antenna. The transmitting pair is the broadcasting antenna and the receiving pair is picking up that transmission just like your car radio antenna.

The data on the transmitting pair is broadcast and picked up by the receiving pair. This is called *crosstalk*. If enough current is induced into the receiving pair, the operation of the system can be affected. This is one reason why it is so critical to maintain the twists in a Category 5e cable.

Would we have this problem with an optical fiber? Does crosstalk exist in optical fiber? Regardless of the number of transmitting and receiving optical fibers in a cable assembly, crosstalk does not occur. To have crosstalk in an optical fiber cable assembly, light would have to leave one optical fiber and enter one of the other optical fibers in the cable assembly. Because of total internal reflection, under normal operating conditions light never leaves the optical fiber. Therefore, crosstalk does not exist in optical fiber cable assemblies.

Now let's take a look at another EMI scenario. Copper cables are being routed through a manufacturing plant. This manufacturing plant houses large-scale electromechanical equipment that generates a considerable amount of EMI, which creates an EMI-rich environment.

Routing cables through an EMI-rich environment can be difficult. Placing the cables too close to the EMI-generating source can induce unwanted electrical signals strong enough to cause systems to function poorly or stop operating altogether. Copper cables used in EMI-rich environments typically require electrical shielding to help reduce the unwanted electrical signals. In addition to the electrical shielding, the installer must be aware of the EMI-generating sources and ensure that the copper cables are routed as far as possible from these sources.

Routing copper cables through an EMI-rich environment can be challenging, time-consuming, and expensive. However, optical fiber cables can be routed through an EMI-rich environment with no impact on system performance. The fiber-optic installer is free to route the optical fiber as efficiently as possible. Optical fiber is very attractive to every industry because it is immune to EMI.

Size and Weight

The size of any cable must always be taken into consideration when preparing for an installation. Often fiber-optic cables will be run through existing conduits or raceways that are partially or almost completely filled with copper cable. This is another area where small fiber-optic cable has an advantage over copper cable.

Let's do a comparison and try to determine the reduced-size advantage that fiber-optic cable has over copper cable. As you learned in Chapter 21, "Optical Fiber Construction and Theory," a coated optical fiber is typically 250μm in diameter. You learned that ribbon fiber-optic cables sandwich up to 12 coated optical fibers between two layers of Mylar tape. Twelve of these ribbons stacked on top of each other form a cube roughly 3mm by 3mm. This cube can be placed inside a buffer and surrounded by a strength member and jacket to form a cable. The overall diameter of this cable would be only slightly larger than an RG6 coaxial cable or a bundle of four Category 5e cables.

So how large would a copper cable have to be to offer the same performance as the 144 optical fiber ribbon cable? That would depend on transmission distance and the optical fiber data rate. Because we have already discussed Category 5e performance, let's place a bundle of Category 5e cables up against the 144 optical fiber ribbon cable operating at a modest 2.5Gbps data rate over a distance of just 100m.

A Category 5e cable contains four conductor pairs and as defined in ANSI/TIA-568-B.2 is 0.25″ in diameter. Each pair is capable of a 100MHz transmission over 100m. As you learned earlier in this chapter, a 100MHz transmission carries 200 million symbols per second. If each symbol is a bit, the 100MHz Category 5e cable is capable of a 200Mbps transmission rate. When the performance of each pair is combined, a single Category 5e cable is capable of an 800Mbps transmission rate over a distance of 100m.

Now let's see how many Category 5e cables will be required to provide the same performance as the 144 optical fiber ribbon cable. The 144 optical fiber ribbon cable has a combined data transmission rate of 360Gbps. When we divide 360Gbps by 800Mbps, we see that 450 Category 5e cables are required to equal the performance of this modest fiber-optic system.

When 450 Category 5e cables are bundled together, they are roughly 5.3 inches in diameter. As noted earlier in this chapter, the 144 optical fiber ribbon cable is approximately the size of four Category 5e cables bundled together. The Category 5e bundle thus has a volume roughly 112.5 times greater than the 144 optical fiber ribbon cable. In other words, Category 5e bundles need 112.5 times more space in the conduit than the 144 optical fiber ribbon cable.

The comparison we just performed is very conservative. The distance we used was kept very short and the transmission rate for the optical fiber was kept low. We can get even a better appreciation for the cable size reduction fiber-optic cable offers if we increase the transmission distance and the data rate.

In this comparison, let's increase the transmission distance to 1,000m and the data transmission rate to 10Gbps. The bandwidth of a copper cable decreases as distance increases, just as with fiber-optic cables. Because we have increased the transmission distance by a factor of 10, it's fair to say that the Category 5e cable bandwidth will decrease by a factor of 10 over 1000m.

With a reduction in bandwidth by a factor of 10, we will need ten times more Category 5e cables to equal the old 2.5Gbps performance. In other words, we need 4,500 Category 5e cables bundled together. In this comparison, however, the bandwidth has been increased from 2.5Gbps to 10Gbps. This means we have to quadruple the number of Category 5e cables to meet the bandwidth requirement. We now need 18,000 Category 5e cables bundled together. Imagine how many cables we would need if the transmission distance increased to 80,000m. We would need a whopping 1,440,000 Category 5e cables bundled together.

These comparisons vividly illustrate the size advantage that fiber-optic cables have over copper cables. The advantage becomes even more apparent as distances increase. The enormous capacity of such a small cable is exactly what is needed to install high-bandwidth systems in buildings where the conduits and raceways are almost fully populated with copper cables.

Now that we have calculated the size advantages of optical fiber over Category 5e cable, let's look at the weight advantages. It is pretty easy to see that thousands, tens of thousands, or millions of Category 5e cable bundled together will outweigh a ribbon fiber-optic cable roughly one half of an inch in diameter. It's difficult to state exactly how much less a fiber-optic cable would weigh than a copper cable performing the same job—there are just too many variables in transmission distance and data rate. However, it's not difficult to imagine the weight savings that fiber-optic cables offer over copper cables. These weight savings are being employed in commercial aircraft, military aircraft, and the automotive industries, just to mention a few.

Security

We know that optical fiber is a dielectric and because of that it is immune to EMI. So why is an optical fiber secure and virtually impossible to tap? Because of total internal reflection, optical fiber does not radiate.

In Chapter 22, you learned about macrobends. Excessive bending on an optical fiber will cause some of the light energy to escape the core and cladding. This light might penetrate the coating, buffer, strength member, and jacket. This energy is detectable by means of a *fiber identifier*.

Fiber identifiers detect light traveling through an optical fiber by inserting a macrobend. Photodiodes are placed against the jacket or buffer of the fiber-optic cable on opposite sides of the macrobend. The photodiodes detect the light that escapes from the fiber-optic cable. The light energy detected by the photodiodes is analyzed by the electronics in the fiber identifier. The fiber identifier can typically determine the presence and direction of travel of the light.

If the fiber identifier can insert a simple macrobend and detect the presence and direction of light, why is fiber secure? Detecting the presence of light and determining the source of the light does not require much optical energy. However, as you learned earlier in this chapter, a fiber-optic receiver typically has a relatively small window of operation. In other words, the fiber-optic receiver typically needs at least 10 percent of the energy from the transmitter to accurately decode the signal on the optical fiber. Inserting a macrobend in a fiber-optic cable and directing 10 percent of the light energy into a receiver is virtually impossible. A macrobend this severe would also be very easy to detect with an *optical time-domain reflectometer* (OTDR). There is no transmission medium more secure than optical fiber.

NOTE   The fiber identifier and OTDR are covered in detail in Chapter 33, "Test Equipment and Link/Cable Testing."

Safety

Electrical safety is always a concern when working with copper cables. Electrical current flowing through copper cable poses shock, spark, and fire hazards. Optical fiber is a dielectric that cannot carry electrical current; therefore, it presents no shock, spark, or fire hazard.

Because optical fiber is a dielectric, it also provides electrical isolation between electrical equipment. Electrical isolation eliminates ground loops, eliminates the potential shock hazard when two pieces of equipment at different potentials are connected together, and eliminates the shock hazard when one piece of equipment is connected to another with a ground fault.

Ground loops are typically not a safety problem; they are usually an equipment operational problem. They create unwanted noise that can interfere with equipment operation. A common example of a ground loop is the hum or buzz you hear when an electric guitar is plugged into an amplifier with a defective copper cable or electrical connection. Connecting two pieces of equipment together with optical fiber removes any path for current flow, which eliminates the ground loop.

Using copper cable to connect two pieces of equipment that are at different electrical potentials poses a shock hazard. It's not uncommon for two grounded pieces of electrical equipment separated by distance to be at different electrical potentials. Connecting these same two pieces of equipment together with optical fiber poses no electrical shock hazard.

If two pieces of electrical equipment are connected together with copper cable and one develops a ground fault, there is now a potential shock hazard at both pieces of equipment. Everyone is likely to experience, or hear of someone experiencing, a ground fault at least once in their life. A common example is when you touch an appliance such as an electric range or washing machine and experience a substantial electrical shock. If the piece of equipment that shocked you was connected to another piece of equipment with a copper cable, there is a possibility that someone touching the other piece of equipment would also be shocked. If the two pieces of equipment were connected with optical fiber, the shock hazard would exist only at the faulty piece of equipment.

Nonconductive fiber-optic cables offer some other advantages, too. They do not attract lightning any more than any other dielectric. They can be run through areas where faulty copper cables could pose a fire or explosion hazard. The only safety requirement that Article 770 of the NEC places on nonconductive fiber-optic cables addresses the type of jacket material. When electrical safety, spark, or explosion hazards are a concern, there is no better solution than optical fiber.

NOTE   Article 770 of the NEC is discussed in detail in Chapter 24, "Fiber-Optic Cables."

Link Performance Analysis

Now that we have discussed fiber-optic link design considerations and the advantages of optical fiber over copper cable, let's look at how to analyze the performance of a fiber-optic link. This section focuses on link performance analysis using the ANSI/TIA-568-C.3 Optical Fiber Cabling Components Standard.

ANSI/TIA-568-C.3 defines optical fiber components' performance. We will use three parts of this standard to analyze the performance of a link:

- Section 4.2, cable transmission performance

- Section 5.3, optical fiber splice

- Annex A, optical fiber connector performance specifications

When analyzing link performance using ANSI/TIA-568-C.3, you are performing a worst-case analysis. In other words, your link should perform no worse than the performance levels defined in the standard. Typically, a link will outperform the standard or greatly exceed the standard. The standard sets the minimum requirements for cable transmission performance, splice performance, and connector performance.

There are typically three different methodologies that can be used to develop a loss budget or power budget:

Worst-case method   The worst-case method that will be discussed in this chapter is commonly viewed as the most conservative engineering approach. When using this method, you use worst-case maximum value loss for each passive component. The minimum transmitter output power and minimum receiver sensitivity are also used.

Statistical method   The statistical method is less conservative than the worst-case method. Instead of using worst-case values for each component, the statistical method distributes the properties for each component in some manner around the mean value for that component type. The goal of this method is to more accurately quantify the predicted performance of the fiber-optic link.

Numerical method   The numerical method is the most calculation-intensive method and typically requires the use of a computer program to accomplish. This method looks at the environmental conditions of each component and the performance is calculated based on these conditions.

Cable Transmission Performance

As you saw earlier in this chapter, ANSI/TIA-568-C.3, section 4.2 addresses the performance of 50/125µm multimode optical fiber, 62.5/125µm multimode optical fiber, and single-mode inside and outside plant optical fiber. Maximum attenuation and minimum bandwidth-length product is defined for each fiber type by wavelength, as shown in Table 32.5.

TABLE 32.5   Characteristics of ANSI/TIA-568-C.3 Recognized Optical Fibers

Optical Fiber Cable Type	Wavelength (nm)	Maximum Attenuation (dB/km)	Minimum Overfilled Modal Bandwidth-Length Product (MHz · km)	Minimum Effective Modal Bandwidth-Length Product (MHz · km)
62.5/125µm multimode	850 1300	3.5 1.5	200 500	Not required
50/125µm multimode	850 1300	3.5 1.5	500 500	Not required
850nm laser-optimized 50/125µm multimode	850 1300	3.5 1.5	1500 500	2000 Not required
Single-mode indoor-outdoor	1310 1550	0.5 0.5	N/A N/A	N/A N/A
Single-mode inside plant cable	1310 1550	1.0 1.0	N/A N/A	N/A N/A
Single-mode outside plant cable	1310 1550	0.5 0.5	N/A N/A	N/A N/A

Splice and Connector Performance

ANSI/TIA-568-C.3, section 5.3 addresses the performance of fusion or mechanical optical fiber splices. The standard states that a fusion or mechanical splice shall not exceed a maximum optical insertion loss of 0.3dB when measured in accordance with ANSI/EIA/TIA-455-34-A, Method A (factory testing), or ANSI/TIA-455-78-B (field testing).

This section also defines the minimum *return loss* for mechanical or fusion optical fiber splices. Multimode mechanical or fusion splices shall have a minimum return loss of 20dB while single-mode mechanical or fusion splices shall have a minimum return loss of 26dB when measured in accordance with ANSI/TIA/EIA-455-107-A. The minimum single-mode return loss for broadband analog video CATV applications is 55dB when measured in accordance with ANSI/TIA/EIA-455-107-A.

ANSI/TIA-568-C.3, Annex A describes the optical fiber connector performance requirements. Section A.4 states that all multimode connectors, adapters, and cable assemblies shall meet the requirements at both 850nm and 1300nm ± 30nm wavelengths. All single-mode connectors, adapters, and cable assemblies shall meet the requirement at both 1310nm and 1550nm ± 30nm wavelengths.

Section A.4.2 states the maximum insertion loss for a mated connector pair. The maximum insertion loss of a mated multimode or single-mode connector pair is 0.75dB. Multimode-mated pairs shall be tested in accordance with FOTP-171 methods A1 or D1, or FOTP-34 method A2. Single-mode mated pairs shall be tested in accordance with FOTP-171 methods A3 or D3, or FOTP-34 method B.

USING A POWER BUDGET TO TROUBLESHOOT A FIBER-OPTIC SYSTEM

This chapter explains how to calculate a power budget for a multimode fiber-optic link and a single-mode fiber-optic link. A power budget tells the fiber-optic installer or technician the maximum allowable loss for every component in the fiber-optic link. The sum of these losses equals the maximum acceptable loss for the link per the ANSI/TIA-568-C.3 standard.

The maximum acceptable loss for the link is the worst-case scenario. Every component in the fiber-optic link should outperform the ANSI/TIA-568-C.3 standard. Links with a total loss just under the maximum acceptable loss may not function properly. This is especially true with short links.

To illustrate, we'll tell a story about a short fiber-optic link on a submarine. Of course, all fiber-optic links on a submarine are short. This link, when tested with the light source and power meter, had a total loss under the maximum allowable value. However, when the link was connected to the receiver and transmitter, the fiber-optic system did not function properly.

Because the link was short, the loss for the optical fiber was negligible. The calculated loss came from the multiple interconnections. The maximum allowable loss for the interconnections was 3.0dB. The measured loss for the link was 2.1dB. This is 0.9dB below the maximum allowable. The typical performance for a link with this number of interconnections would have been less than 1.0dB.

One of the interconnections in the link had a loss much greater than the maximum allowable. This bad interconnection was reflecting roughly 20 percent of the light energy from the transmitter back at the transmitter. This back-reflected energy was reflected toward the receiver at the transmitter interconnection. The receiver detected the back-reflected energy from the transmitter interconnection, causing the system to malfunction.

Just because the total loss for a fiber-optic link is below the maximum allowable doesn't mean your system will perform without problems. If your system is having problems, start by evaluating each component in the fiber-optic link.

Power Budget

Now that the link performance requirements have been identified, let's put them to use in a power budget. A power budget as defined in IEEE Standard 802.3 is the minimum optical power available to overcome the sum of attenuation plus power penalties of the optical path between the transmitter and receiver. These losses are calculated as the difference between the minimum transmitter launch power and the minimum receive power.

MULTIMODE LINK ANALYSIS

Table 32.6 lists some of the typical optical characteristics for a 1300nm LED transmitter. In this table, we see that manufacturers typically list three values for optical output power. There is a typical value and there are the maximum and minimum values. The minimum value represents the least amount of power that the transmitter should ever output. This is the output power level we will use in calculating our power budget.

TABLE 32.6   LED Transmitter Optical Characteristics

PARAMETER		SYMBOL	MIN.	TYP.	MAX.	UNIT
Optical output power	BOL	P_0	−19	−16.8	−14	dBm avg.
62.5/125μm, NA = 0.275 fiber	EOL		−20			
Optical output power	BOL	P_0	−22.5	−20.3	−14	dBm avg.
50/125μm, NA = 0.20 fiber	EOL		−23.5			
Optical extinction ratio				0.001	0.03	%
				−50	−35	dB
Optical output power at logic "0" state		P_0 ("0")			−45	dBm avg.
Center wavelength		λ_c	1270	1300	1380	nm
Spectral width—FWHM		Δ		137	170	nm
Optical rise time		t_r	0.6	1.0	3.0	ns
Optical fall time		t_f	0.6	2.1	3.0	ns
Duty cycle distortion contributed by the transmitter		DCD		0.02	0.6	ns_{P-P}
Duty cycle jitter contributed by the transmitter		DDJ		0.02	0.6	ns_{P-P}

NOTE   Some of the parameters in this table are outside the scope of this book. You do not have to understand all aspects of the datasheet, just the information you need.

Table 32.7 lists some of the typical optical characteristics for a 1300nm receiver. In this table, the manufacturers list the maximum and minimum optical input power, and under each of those, there is a typical value and a minimum or maximum value. The receiver will perform best when the input power is between the minimum value for maximum optical input power and maximum value for minimum optical input power.

TABLE 32.7   LED Receiver Optical Characteristics

PARAMETER	SYMBOL	MIN.	TYP.	MAX.	UNIT
Optical input power minimum at window edge	$P_{IN\,Min.}$ (W)		−33.5	−31	dBm avg.
Optical input power maximum	$P_{IN\,Max.}$	−14	−11.8		dBm avg.
Operating wavelength	λ_c	1270		1380	nm

NOTE   Remember from Chapter 27 that the manufacturer-stated optical output power of a typical laser transmitter is measured after 1m of optical fiber. Therefore, we do not have to account for the connector loss at the transmitter. Remember from Chapter 28, "Fiber-Optic Detectors and Receivers," that the manufacturer typically states minimum optical input power for the receiver at the window edge. This means we do not have to account for the connector loss at the receiver.

To determine the power budget for the transmitter defined in Table 32.6 and the receiver defined in Table 32.7, we need to choose an optical fiber. In this example we will use a 62.5/125μm multimode optical fiber. We will also use the EOL value for transmitter output power to account for the aging of the light source.

The minimum EOL value for the optical output power is –20dBm for a 62.5/125μm multimode optical fiber. The maximum value for the minimum optical input power is –31dBm. The difference is 11dB. This is the window of operation where the receiver will provide the best performance. This is also our power budget.

To better understand a power budget, let's take a look a basic fiber-optic link as shown in Figure 32.1 and calculate the maximum loss allowable according to ANSI/TIA-568-C.3. After we have calculated the maximum allowable loss, we can compare maximum loss to the power budget and determine the minimum power to the receiver. If the minimum power to the receiver falls within the window of operation, the link should support low bit error rate data transmission.

FIGURE 32.1
Basic
fiber-optic link

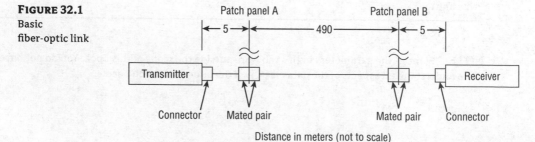

Distance in meters (not to scale)

The link shown in Figure 32.1 has a 490m span of 62.5/125µm multimode optical fiber. Each end of the optical fiber is connectorized and plugged into a patch panel. On the other side of each patch panel, there is a 5m patch cord. One patch cord connects to the transmitter and the other to the receiver. The total link length is 500m.

You saw earlier in Table 32.3 that ANSI/TIA-568-C.3 defines the maximum attenuation for a 62.5/125µm optical fiber at both 850nm and 1300nm. Because the transmitter selected for this example has a 1300nm output, we will only evaluate link performance at 1300nm. The maximum allowable attenuation per ANSI/TIA-568-B.3 is 1.5dB per km of 62.5/125µm optical fiber. The total link length was 500m with a maximum allowable attenuation of 0.75dB, as shown here:

Attenuation coefficient at 1300nm = 1.5dB/km

Optical fiber length = 0.5 km

Attenuation for the length of optical fiber = 0.5 × 1.5 = 0.75

Maximum attenuation for 500m of optical fiber is 0.75dB

The link shown in Figure 32.1 has six connectors. One connector is plugged into the transmitter, one is plugged into the receiver, and the remaining four are mated together in pairs at each patch panel. When evaluating connector loss, only the mated pairs are accounted for. The loss at the transmitter and receiver is ignored.

Because the transmitter and receiver manufacturers have accounted for connector loss, we only have to account for mated pair loss. In the link shown in Figure 32.1, there are two mated pairs, one at each patch panel. As you learned earlier in this chapter in the "Splice and Connector Performance" section, ANSI/TIA-568-C.3 states that the maximum allowable insertion loss for a mated connector pair is 0.75dB. Because there are two mated pairs on our link, the maximum interconnection loss for our link is 1.5dB.

At this point, we have identified the maximum allowable loss for the optical fiber and the interconnections. The maximum allowable loss for the link per ANSI/TIA-568-C.3 is the sum of these two losses. The maximum allowed loss for the link in Figure 32.1 is 2.25dB.

Now that all the performance parameters of the link have been identified, let's determine the minimum power that should be available to the receiver. The minimum power available to the receiver is equal to the minimum optical output power of the LED transmitter minus the loss of the link. The minimum optical output power for the LED transmitter is –20dBm, the link loss is 2.25dB, and the minimum power available at the receiver should be –22.25dBm. The power budget for the link in Figure 32.1 was 11dB. The maximum loss allowable for this link per ANSI/TIA-568-B.3 is 2.25dB. The loss for this link is small and provides 8.75dB of headroom for the life of the LED transmitter.

The link we are going to examine in Figure 32.2 is a 50/125µm multimode optical fiber link with a splice. The transmitter for this link as an 850nm VCSEL. The minimum EOL optical output power for the 850nm VCSEL transmitter is –8dBm. The minimum optical input power for the receiver is –17dBm.

Looking at Figure 32.2, we see two patch panels with 800m of 50/125µm multimode optical fiber between them. There is a mechanical splice 400m away from patch panel A. Each end of the optical fiber is connectorized and plugged into the patch panel. Patch cords 5m in length each are used to connect the VCSEL transmitter to one patch panel and the receiver to the other patch panel.

FIGURE 32.2
Multimode
fiber-optic link
with splice

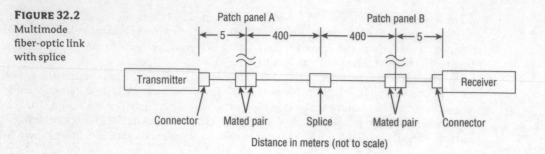

TABLE 32.8   Blank Multimode ANSI/TIA-568-C.3 Power Budget Calculation Table

STEP NO.	DESCRIPTION	VALUE
1	Minimum EOL optical output power.	dBm
2	Minimum optical input power.	dBm
3	Subtract line 2 from line 1 to calculate the power budget.	dB
4	km of optical fiber.	
5	Number of interconnections.	
6	Number of splices.	
7	Multiply line 4 × 3.5.	
8	Multiply line 4 × 1.5.	
9	Multiply line 5 × 0.75.	
10	Multiply line 6 × 0.3.	
11	Add lines 7, 9, and 10 for total link loss at 850nm.	dB
12	Add lines 8, 9, and 10 for total link loss at 1300nm.	dB
13	Subtract line 11 from line 3 for headroom at 850nm.	dB
14	Subtract line 12 from line 3 for headroom at 1300nm.	dB

Table 32.8 is a blank power budget calculation table. This table can be used to calculate the power budget for a link, sum all the losses in the link, and calculate headroom.

Table 32.9 has all the values for the link filled in. The power budget for the link is 9dB. The loss for the link is approximately 4.6dB and the headroom for the link is approximately 4.4dB.

TABLE 32.9 Completed Multimode ANSI/TIA-568-C.3 Calculation Table for an 850nm VCSEL Transmitter

STEP NO.	DESCRIPTION	VALUE
1	Minimum EOL optical output power.	–8dBm
2	Minimum optical input power.	–17dBm
3	Subtract line 2 from line 1 to calculate the power budget.	9.0dB
4	km of optical fiber.	0.810
5	Number of interconnections.	2.0
6	Number of splices.	1.0
7	Multiply line 4 × 3.5.	2.8
8	Multiply line 4 × 1.5.	
9	Multiply line 5 × 0.75.	1.5
10	Multiply line 6 × 0.3.	0.3
11	Add lines 7, 9, and 10 for total link loss at 850nm.	4.6dB
12	Add lines 8, 9, and 10 for total link loss at 1300nm.	
13	Subtract line 11 from line 3 for headroom at 850nm.	4.4dB
14	Subtract line 12 from line 3 for headroom at 1300nm.	

Up to this point, we have only looked at calculating the power budget for a link using a table. However, the power budget can be plotted graphically. The graphical representation of the losses in the link very much resembles the trace of an OTDR. The OTDR is covered in detail in Chapters 33 and 34.

Figure 32.3 is the graphical representation of the link we analyzed in Figure 32.2. When analyzing a link graphically, power gain or loss is noted on the vertical scale and distance is noted on the horizontal scale. Power gain and loss may be drawn to scale; however, there is not usually sufficient room on your paper to draw distance to scale.

The first step in setting up the graphical representation of the link is to assign values for the vertical axis. As shown in Table 32.6, the maximum value assigned is the minimum EOL optical output power, –8dBm. The minimum value assigned is the minimum optical input power for the receiver, –17dBm.

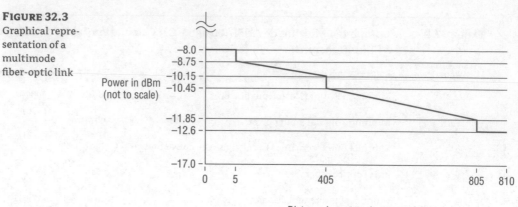

FIGURE 32.3
Graphical representation of a multimode fiber-optic link

The next step is to assign values to the vertical axis that represent the before and after power levels for the interconnections. Working from the transmitter to the receiver, the first interconnection is a mated connector pair at patch panel A. There is a 5m patch cord between the transmitter and patch panel A. At a wavelength of 850nm, the maximum loss for the 5m patch cord is only 0.0175dB. This very small value represents only a fraction of a percent of the loss for this link and can be ignored in the graphical representation. Typically you can ignore all patch cord loss when generating a graphical representation of a link. This simplifies the drawing and reduces the time required to generate the drawing.

The amount of light energy entering the first mated pair is –8dBm. The maximum loss allowed by ANSI/TIA-568-C.3 for a mated pair is 0.75dB. Using this value, the power exiting the mated pair should be no less than –8.75dBm. In a linear fashion working from top to bottom, the next value recorded on the vertical axis is –8.75dBm.

The next value on the vertical axis represents the light energy entering the mechanical splice. There is 400m of optical fiber from the first mated pair to the mechanical splice. The loss for the 400m of optical fiber is 1.4dB. The light energy entering the mechanical splice is –10.15dBm and is recorded on the vertical axis. The maximum loss allowed by ANSI/TIA-568-C.3 for a mechanical splice is 0.3dB. Using this value, the power exiting the mechanical splice should be no less than –10.45dBm, which is recorded on the vertical axis.

Between the mechanical splice and patch panel B is 400m of optical fiber. The loss of the 400m of optical fiber is 1.4dB. The light energy entering the mated pair at patch panel B is –11.85dBm, which should be recorded on the vertical axis. As we already know, ANSI/TIA-568-C.3 allows a maximum loss of only 0.75dB for a mated pair. The minimum amount of light energy exiting the mated pair should be –12.6dBm. If we ignore the loss for the 5m patch cord between patch panel B and the receiver, the power available to the receiver is –12.6dBm. This is the last value recorded on the vertical axis.

The next step in this process is to record the horizontal axis values. Working from left to right starting at the transmitter, the first value recorded on the horizontal axis is zero. It is from this point that we will record the successive segments of optical fiber.

The first segment of optical fiber is the 5m patch cord. With an overall length of 810m, it's not practical to draw the 5m patch cord to scale unless you have a very large sheet of paper. Looking at Figure 32.3, you can see that the first segment on the horizontal axis is long enough to clarify that the 5m patch cord is a separate segment of optical fiber. The second value recorded on the horizontal axis is 5m.

The second segment is a 400m span of optical fiber from patch panel A to the mechanical splice. The next value recorded on the horizontal axis is 405m.

The third segment of optical fiber is a 400m span from the mechanical splice to patch panel B. The next value recorded on the horizontal axis is 805m.

The last segment of optical fiber is the 5m patch cord that connects the receiver to patch panel B. The last value recorded on the horizontal axis is 810m.

With all the values for the vertical and horizontal axes recorded, the last step is to draw a series of lines that graphically represent the recorded values. (You may want to use a straight edge for this.) The first line segment will be a horizontal line from –8.0dBm on the vertical axis to the 5m point on the horizontal axis. This line represents the patch cord that connects the transmitter to patch panel A. The line is horizontal because the patch cord loss is insignificant. At the end of this, a vertical line will be drawn that runs from –8.0dBm to –8.75dBm; it represents the loss for the mated pair at patch panel A.

The third line will extend from the 5m point on the horizontal axis to the 405m point. It will be a sloping line from left to right that goes from –8.75dBm on the vertical axis to –10.15dBm. This line represents the loss for the 400m section of optical fiber from patch panel A to the mechanical splice. At the end of this line a vertical line is drawn from –10.15dBm to –10.45dBm to represent the loss for the mechanical splice.

The fifth line will extend horizontally from the 405m point to the 805m point. This line will slope from –10.45dBm on the vertical axis to –11.85dBm; it represents the loss for the 400m section of optical fiber from the mechanical splice to patch panel B. At the end of this line, a vertical line will be drawn from –11.85dBm to –12.6dBm, representing the loss for the mated pair at patch panel B.

The last line drawn is a horizontal line from the 805m point on the horizontal axis to the 810m point. This line represents the patch cord from patch panel B to the receiver. There is no slope to this line because the loss for the patch cord is insignificant. If everything has been drawn correctly, your drawing should look like Figure 32.3.

SINGLE-MODE LINK ANALYSIS

As with multimode optical fiber, ANSI/TIA-568-C.3 defines performance requirements for single-mode optical fiber, connectors, and splices. However, ANSI/TIA-568-C.3 only addresses *inside plant* splice performance. *Outside plant* splice performance is addressed in ANSI/TIA-758-A. ANSI/TIA-758-A states that splice insertion loss shall not exceed 0.1dB mean (0.3dB maximum) when measured with OTDR testing.

Figure 32.4 is a basic single-mode optical fiber link with 4510m of optical fiber spliced in three locations between two patch panels located in separate buildings. The transmitter is connected to patch panel A with a 5m patch cord and the receiver is connected to patch panel B with a 5m patch cord. There is an inside plant splice 50m from patch panel A, another 25m from patch panel B, and an outside plant splice 2000m from patch panel A.

FIGURE 32.4
Single-mode fiber-optic link

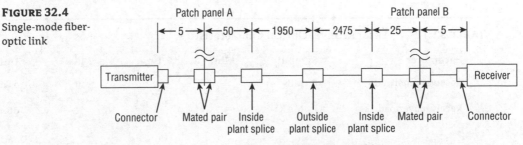

Distance in meters (not to scale)

Before we can analyze the link and do a power budget, we need to look at how manufacturers typically describe the performance characteristics for laser transmitters and receivers. Table 32.10 lists some of the typical optical power characteristics for a 1310nm laser transmitter. In this table, we see that manufacturers usually list three values for optical output power. There is a typical value, and there are maximum and minimum values. The minimum value represents the least amount of power the transmitter should ever output. This is the output power level we will use in calculating our power budget.

TABLE 32.10   Laser Transmitter Optical Characteristics

PARAMETER	SYMBOL	MIN.	TYP.	MAX.	UNIT
Optical output Power 9µm SMF	P_{OUT}	−10	−6	−3	dBm
Center wavelength	λ_c	1260		1360	nm
Spectral width—rms	Δ_c		1.8	4	nm rms
Optical rise time	t_r		30	70	ps
Optical fall time	t_f		150	225	ps
Extinction ratio	E_R	8.2	12		dB
Optical output eye	Compliant with eye mask Telecordia GR-253-CORE				
Back-reflection sensitivity				−8.5	dB
Jitter generation	pk to pk			70	mUI
	RMS			7	mUI

Table 32.11 lists some of the typical optical power characteristics for a 1310nm receiver. In this table, the manufacturers list the maximum and minimum optical input power, and under each of those there is a typical and a minimum or maximum value. The receiver will perform best when the input power is between the minimum value for maximum optical input power and maximum value for minimum optical input power.

TABLE 32.11   Laser Receiver Optical Characteristics

Parameter	Symbol	Min.	Typ.	Max.	Unit
Receiver sensitivity	P_{IN} MIN		−23	−19	dBm avg.
Receiver overload	P_{IN} MAX	−3	+1		dBm avg.

TABLE 32.11   Laser Receiver Optical Characteristics   *(CONTINUED)*

Parameter	Symbol	Min.	Typ.	Max.	Unit
Input operating wavelength	λ	1260		1550	nm
Signal detect — asserted	P_A		−27.3	−19.5	dBm avg.
Signal detect — deasserted	P_D	−35	−28.7		dBm avg.
Signal detect — hysteresis	P_H	0.5	1.4	4	dB
Reflectance			−35	−27	dB

   With all the transmitter and receiver values defined, we can calculate the power budget using the power budget calculation table shown in Table 32.12. The minimum value for the maximum optical input power is –10dBm. The maximum value for the minimum optical input power is –19dBm. The difference is 9dB; however, the manufacturer of the transmitter did not state an EOL value. Since the EOL value was not provided we should assume the transmitter output power will decay 1.5dB. This value is defined as the industry convention (see Chapter 27). With transmitter output power decay added in, the power budget is 7.5dB for this transmitter-receiver combination.

TABLE 32.12   Blank Single-Mode ANSI/TIA-568-C.3, ANSI/TIA-758-A Power Budget Calculation Table

STEP NO.	DESCRIPTION	VALUE
1	Minimum EOL optical output power.	dBm
2	Minimum optical input power.	dBm
3	Subtract line 2 from line 1 to calculate the power budget.	dB
4	km of inside plant optical fiber.	
5	km of outside plant optical fiber.	
6	Number of interconnections.	
7	Number of inside plant splices.	
8	Number of outside plant splices.	
9	Multiply line 4 × 1.0.	
10	Multiply line 5 × 0.5.	

TABLE 32.12   Blank Single-Mode ANSI/TIA-568-C.3, ANSI/TIA-758-A Power Budget
Calculation Table   *(CONTINUED)*

STEP NO.	DESCRIPTION	VALUE
11	Multiply line 6 × 0.75.	
12	Multiply line 7 × 0.3.	
13	Multiply line 8 × 0.1.	
14	Add lines 9, 10, 11, 12, and 13 for total link loss.	dB
15	Subtract line 14 from line 3 for headroom.	dB

Now that the power budget has been calculated, the power budget calculation table can be completed. The link shown in Figure 32.4 contains two mated pairs, two inside plant splices, one outside plant splice, and 4510m of optical fiber. The maximum allowable loss for the two mated pairs is 1.5dB. The maximum allowable loss for the two inside plant splices is 0.6dB, and for the outside plant splice 0.1dB.

The 4510m of optical fiber is broken up into inside plant fiber and outside plant fiber. The 4425m of optical fiber between the two inside plant splices will be evaluated as outside plant and the remaining 85m will be evaluated as inside plant. The maximum allowable loss for the outside plant optical fiber is rounded down to 2.2dB. The maximum allowable loss of the inside plant optical fiber is 0.085, which is rounded up to 0.1dB. The maximum allowable loss for this link is rounded up to 4.5dB. Our power budget was 7.5dB; that leaves us 3.0dB of headroom. Table 32.13 is the completed power budget calculation table with the rounded values.

TABLE 32.13   Completed Single-Mode ANSI/TIA-568-C.3, ANSI/TIA-758-A Power Budget
Calculation Table

STEP NO.	DESCRIPTION	VALUE
1	Minimum EOL optical output power.	−11.5dBm
2	Minimum optical input power.	−19.0dBm
3	Subtract line 2 from line 1 to calculate the power budget.	7.5dB
4	km of inside plant optical fiber.	0.085
5	km of outside plant optical fiber.	4.425
6	Number of interconnections.	2
7	Number of inside plant splices.	2

TABLE 32.13   Completed Single-Mode ANSI/TIA-568-C.3, ANSI/TIA-758-A Power Budget Calculation Table   *(CONTINUED)*

STEP NO.	DESCRIPTION	VALUE
8	Number of outside plant splices.	1
9	Multiply line 4 × 1.0.	0.1
10	Multiply line 5 × 0.5.	2.2
11	Multiply line 6 × 0.75.	1.5
12	Multiply line 7 × 0.3.	0.6
13	Multiply line 8 × 0.1.	0.1
14	Add lines 9, 10, 11, 12, and 13 for total link loss.	4.5dB
15	Subtract line 14 from line 3 for headroom.	3.0dB

With the power budget calculation table completed, you should be able to quickly sketch out the graphical representation of the link. Figure 32.5 is the completed graphical representation for the link we just discussed without rounding.

FIGURE 32.5
Graphical representation of a single-mode fiber-optic link

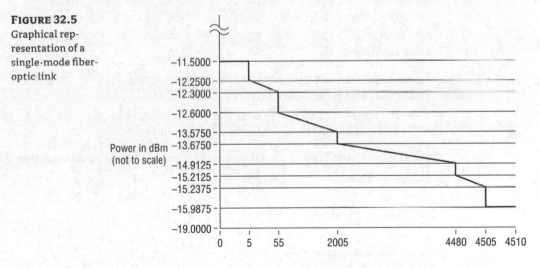

Power in dBm (not to scale)

Distance in meters (not to scale)

The Bottom Line

Calculate the bandwidth for a length of optical fiber. The minimum effective bandwidth-length product in MHz · km is in TIA-568-C.3.

Master It Refer to the following table and determine the minimum bandwidth for 700 meters of 850nm laser-optimized 50/125μm multimode optical fiber.

Optical Fiber Cable Type	Wavelength (nm)	Maximum Attenuation (dB/km)	Minimum Overfilled Modal Bandwidth-Length Product (MHz · km)	Minimum Effective Modal Bandwidth-Length Product (MHz · km)
62.5/125μm Multimode	850	3.5	200	Not required
	1300	1.5	500	
50/125μm Multimode	850	3.5	500	Not required
	1300	1.5	500	
850nm laser-optimized 50/125μm multimode	850	3.5	1500	2000
	1300	1.5	500	Not required
Single-mode indoor-outdoor	1310	0.5	N/A	N/A
	1550	0.5	N/A	N/A
Single-mode inside plant cable	1310	1.0	N/A	N/A
	1550	1.0	N/A	N/A
Single-mode outside plant cable	1310	0.5	N/A	N/A
	1550	0.5	N/A	N/A

Calculate the attenuation for a length of optical fiber. The maximum allowable attenuation in an optical fiber is defined in section 4.2 of ANSI/TIA-568-C.3.

Master It Refer to the following table and determine the maximum allowable attenuation for 5.6km of outside plant single-mode optical fiber.

Optical Fiber Cable Type	Wavelength (nm)	Maximum Attenuation (dB/km)
50/125μm multimode	850	3.5
	1300	1.5
62.5/125μm multimode	850	3.5
	1300	1.5
Single-mode inside plant cable	1310	1.0
	1550	1.0
Single-mode outside plant cable	1310	0.5
	1550	0.5

Calculate the power budget and headroom for a multimode fiber-optic link. Before we can analyze the link and do a power budget, we need to know the optical performance characteristics for transmitter and receiver.

Master It Refer to the following graphic and two tables to calculate the power budget and headroom for the multimode fiber-optic link with 50/125µm multimode optical fiber.

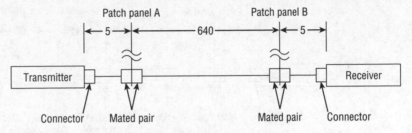

Distance in meters (not to scale)

Parameter		Symbol	Min.	Typ.	Max.	Unit
Optical output power	BOL	P_0	–19	–16.8	–14	dBm avg.
62.5/125µm, NA = 0.275 fiber	EOL			–20		
Optical output power	BOL	P_0	–22.5	–20.3	–14	dBm avg.
50/125µm, NA = 0.20 fiber	EOL			–23.5		
Optical extinction ratio				0.001	0.03	%
				–50	–35	dB
Optical output power at logic "0" state		P_0 ("0")			–45	dBm avg.
Center wavelength		λ_c	1270	1300	1380	nm
Spectral width—FWHM		Δ		137	170	nm
Optical rise time		t_r	0.6	1.0	3.0	ns
Optical fall time		t_f	0.6	2.1	3.0	ns
Duty cycle distortion contributed by the transmitter		DCD		0.02	0.6	ns$_{P\text{-}P}$
Duty cycle jitter contributed by the transmitter		DDJ		0.02	0.6	ns$_{P\text{-}P}$

Parameter	Symbol	Min.	Typ.	Max.	Unit
Optical input power minimum at window edge	$P_{IN\,Min.}$ (W)		–33.5	–31	dBm avg.
Optical input power maximum	$P_{IN\,Max.}$	–14	–11.8		dBm avg.
Operating wavelength	λ_c	1270		1380	nm

Calculate the power budget and headroom for a single-mode fiber-optic link. Before we can analyze the link and do a power budget, we need to know the optical performance characteristics for transmitter and receiver.

Master It Refer to the following graphic and two tables to calculate the power budget and headroom for the single-mode fiber-optic link.

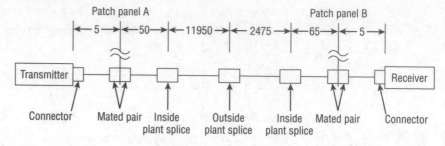

Distance in meters (not to scale)

Parameter	Symbol	Min.	Typ.	Max.	Unit
Optical output Power 9μm SMF, EOL	P_{OUT}	−5	−2	0	dBm
Center wavelength	λ_c	1260		1360	nm
Spectral width—rms	Δ_c		1.8	4	nm rms
Optical rise time	t_r		30	70	ps
Optical fall time	t_f		150	225	ps
Extinction ratio	E_R	8.2	12		dB
Optical output eye	Compliant with eye mask Telecordia GR-253-CORE				
Back-reflection Sensitivity				−8.5	dB
Jitter generation	pk to pk			70	mUI
	RMS			7	mUI

Parameter	Symbol	Min.	Typ.	Max.	Unit
Receiver sensitivity	P_{IN} MIN		−23	−19	dBm avg.
Receiver overload	P_{IN} MAX	0	+1		dBm avg.
Input operating wavelength	λ	1260		1550	nm
Signal detect — asserted	P_A		−27.3	−19.5	dBm avg.
Signal detect — deasserted	P_D	−35	−28.7		dBm avg.
Signal detect — hysteresis	P_H	0.5	1.4	4	dB
Reflectance			−35	−27	dB

Chapter 33

Test Equipment and Link/Cable Testing

This chapter explains the basic operation and application of essential fiber-optic test equipment. The basic theory, operation, and application of each piece of test equipment will be explained. Many of the test methods described in this chapter are based on current industry standards. How to test to these standards is described in detail.

At the time of this writing, the ANSI/TIA-568 standard was under revision. ANSI/TIA-568-B.3 had been replaced with ANSI/TIA-568-C.3. However, ANSI/TIA-B.1 and B.2 had not been replaced. This chapter focuses on the ANSI/TIA-568 standards that were available for purchase.

In this chapter, you will learn to:

- ◆ Perform optical loss measurement testing of a cable plant
- ◆ Determine the distance to a break in the optical fiber with an OTDR
- ◆ Measure the loss of a cable segment with an OTDR
- ◆ Measure the loss of an interconnection with an OTDR
- ◆ Measure the loss of a fusion splice or macrobend with an OTDR

Continuity Tester

The continuity tester is a basic and essential tool for every fiber-optic toolkit. It is also one of the least expensive tools in your toolkit. This low-cost tool will allow you to quickly verify the continuity of an optical fiber.

Many of the continuity testers available today are just modified flashlights. Some flashlights, such as the one pictured in Figure 33.1, have been modified to use an LED instead of an incandescent lamp. Others may just use a brighter incandescent lamp, like the tester pictured in Figure 33.2.

FIGURE 33.1
LED continuity
tester

FIGURE 33.2
Incandescent
continuity tester

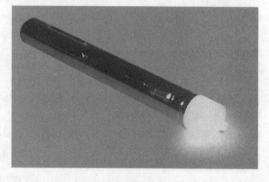

The job of the continuity tester is to project light from the LED or incandescent lamp into the core of the optical fiber. This is usually accomplished by attaching a receptacle to the lamp end of the flashlight. The receptacle is designed to center and hold the connector ferrule directly above the LED or incandescent lamp. When the connector ferrule is inserted into the receptacle, the endface is typically just above the light source. This approach eliminates the need for a lens to direct light into the core of the optical fiber. However, it directs only a fraction of the light emitted by the lamp or LED into the core of the optical fiber.

Because there is no lens used to focus light energy into the optical fiber, the continuity tester works best with multimode optical fiber. The measured optical output power of an LED continuity tester is typically less than –36dBm when measured at the end of one meter of 62.5/125μm optical fiber. The optical output power is reduced another 3–4dB when used with 50/125μm optical fiber and 16–20dB when used with single-mode optical fiber. The continuity tester works best with multimode optical fiber; however, it can be used with single-mode optical fiber. For best results with single-mode optical fiber, dim the lights in the test area if possible.

The low output power of the continuity tester combined with the high attenuation of visible light by an optical fiber limit the length of optical fiber that can be tested. A multimode optical fiber will attenuate a 650nm light source roughly 7dB per kilometer. This high attenuation limits the use of the continuity tester to multimode optical fibers no greater than 2km in length.

LED continuity testers have a couple of advantages over incandescent lamp testers. They typically feature a red (635–650nm) LED that is easy to see. They require far less power from the batteries than an incandescent lamp. An LED continuity tester may provide 10 or more times longer battery life compared to an incandescent lamp. The LED lamp will never need replacing, unlike an incandescent lamp, which may last only 10 hours. LED lamps are also shock resistant.

Whether you are using an LED or an incandescent lamp continuity tester, operation and testing are identical. The continuity tester can test only for breaks in the optical fiber. It does not have sufficient power to identify the location of a break.

The first step when using the continuity tester is to clean and visually inspect the endface of the connector before inserting it into the continuity tester. You need to visually inspect the connector to verify that there is no endface damage. A shattered endface will significantly reduce the light coupled into the core of the optical fiber under test.

After the connector has been cleaned and inspected, you need to verify that the continuity tester is operating properly. Turn the continuity tester on and verify that it is emitting light. This could save you from embarrassment. You don't want to tell the customer that there is a break in their fiber-optic cable when the real problem is dead batteries in your continuity tester.

Depending on where the other end of the fiber-optic cable to be tested is located, you may need an assistant to help you. With the continuity tester turned on, insert the ferrule of the connector under test into the receptacle. If light is being emitted from the other end of the optical fiber, there is good continuity. This means only that there are no breaks in the optical fiber. This does not mean that there are no macrobends or high-loss interconnections in the fiber-optic cable or link under test.

The continuity tester is often used to verify that there are no breaks in a reel of fiber-optic cable before it is installed. There are a couple of ways you could approach testing the reel. One way would be to install a connector on either end of the cable. The other end of the cable should have the jacket and strength member stripped back so that the buffer is exposed. You should remove a small amount of buffer to expose the optical fiber under test. This will allow you to clearly see the light from the continuity tester, ensuring accurate results.

Another approach is to use a pigtail with a mechanical splice or alignment sleeve. The pigtail would have a connector on one end that will mate with the continuity tester receptacle. The other end should have the jacket, strength member, and buffer stripped back so that 10 to 15mm of optical fiber is exposed, as shown in Figure 33.3. The optical fiber should have a perpendicular cleave.

FIGURE 33.3
Pigtail prepared
for temporary
mechanical splice

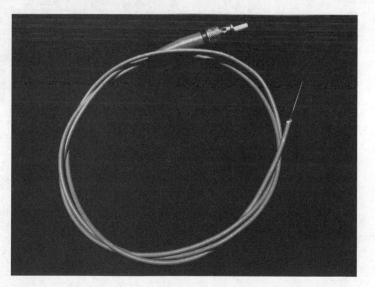

One end of the cable under test should be prepared just like the nonconnectorized end of the pigtail. The other end should have the jacket and strength member stripped back so that the buffer is exposed. A small amount of buffer should be removed to expose the optical fiber under test.

To test for continuity, insert the pigtail connector into the continuity tester receptacle. Turn on the continuity tester and verify that light is being emitted from the exposed optical fiber on the opposite end of the pigtail. Insert the exposed optical fiber from the pigtail into one side of a mechanical splice or alignment sleeve. Insert the optical fiber from the cleaved end of the cable into the other side of the mechanical splice or alignment sleeve and check for continuity.

Visual Fault Locator

Like the continuity tester, the *visual fault locator (VFL)* is an essential tool for every fiber-optic toolkit. Unlike the continuity tester, it is not one of the least expensive tools in your toolkit. The VFL will allow you to quickly identify breaks or macrobends in the optical fiber, and identify a poor fusion splice in multimode or single-mode optical fiber.

The big difference between the continuity tester and the VFL is the light source and optical output power of the light source. The VFL typically uses a red (635–650nm) laser light source. The optical output power of the laser is typically 1mW or less. Because of the high optical output power, you should never view the output of the VFL directly.

The VFL is available in different shapes and sizes. Some may look like a pen, others may be built into an optical time-domain reflectometer (OTDR), and some may look like a small test equipment box. Figure 33.4 shows a VFL with an LC receptacle.

FIGURE 33.4
VFL with LC receptacle

The VFL fills the core of the optical fiber with light from the laser. The light from the laser escapes the optical fiber at a break or macrobend. The light escaping from the optical fiber will typically illuminate the buffer surrounding the optical fiber. Macrobends are not always visible through the jacket but are typically visible through the buffer. Breaks may be visible through the jacket of the fiber-optic cable depending on jacket color, thickness, number of optical fibers in the cable, and amount of strength member.

Unlike the continuity tester, the VFL is not limited to testing multimode optical fibers 2km or less in length. The VFL can be used to verify continuity of multimode or single-mode optical fiber longer than 2km. Due to attenuation of the 635–650nm laser light source by the optical fiber, macrobends may not be detectable beyond 1km in multimode optical fiber and 500 meters in single-mode optical fiber. The same holds true for finding breaks in the optical fiber through the jacket of the fiber-optic cable.

The first step before using the VFL for testing an optical fiber is to measure the output power of the VFL. This can be done with a multimode test jumper and an optical power meter. Connect the output of the VFL to the test jumper and the other end of the test jumper to the power meter input. Set the power meter at the shortest wavelength possible. This is typically 850nm. The measured optical output power of the VFL should be no less than –7dBm with a 62.5/125µm optical fiber. If the measured optical output power is less than –7dBm, replace the battery and retest. Most VFLs will have a measured optical output power greater than –3dBm with a fresh battery. Optical output power below –7dBm will limit the effectiveness of the VFL to identify breaks and macrobends.

The second step before using the VFL is to clean and visually inspect the endface of the connector on the optical fiber to be tested before inserting it into the VFL. You need to visually inspect the connector to verify that there is no endface damage. A shattered endface will not allow light to be properly coupled into the optical fiber under test. A shattered endface may also damage the VFL. Some VFLs use a short pigtail from the laser to the receptacle. The other end of the pigtail has a connector that will mate with the receptacle; this is referred to as a contact VFL. A shattered endface can damage the optical fiber in the connector on the end of the pigtail.

NOTE There are two types of VFLs: contact and non-contact. With a contact VFL, the optical fiber under test will make contact with the VFL. However, with a non-contact VFL the optical fiber under test will not touch the VFL.

To identify a macrobend in the optical fiber, visually inspect the length of the cable under test. If you locate a bend in the cable that is glowing, this is a macrobend. Figure 33.5 shows simplex cordage with a macrobend.

FIGURE 33.5
The light from a VFL identifying a macrobend in simplex cordage

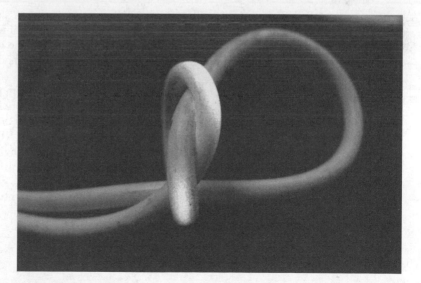

To identify a break in the optical fiber, visually inspect the length of the cable under test. If you locate a small spot on the cable that is emitting red light, this is a break. Figure 33.6 shows multimode cordage with a break.

FIGURE 33.6
The light from a VFL identifying a break in simplex cordage

The VFL can be used to test the continuity of an optical fiber in the same manner described in the "Continuity Tester" section earlier in this chapter. The VFL, however, will couple 1,000 times (30dB) or more light energy into the optical fiber than the LED continuity tester. You should never view the output of the VFL or the endface of a connector being tested by the VFL directly.

Using the VFL to test for continuity can produce false results. A single break in an optical fiber does not stop all the light from the VFL cold. Some of the light will typically reach the end of the optical fiber. In most cases, so much light reaches the end of the optical fiber that the person viewing the light might believe there is good continuity when in fact the optical fiber is broken.

A break in the optical fiber will typically attenuate the light from the VFL by 20–30dB. As you learned earlier in the chapter, the VFL will couple 30dB or more light energy into the optical fiber than the LED continuity tester. The maximum loss from a break in an optical fiber typically never exceeds 30dB. Therefore, the light energy from the VFL after passing through a single break with 30dB of loss will be equal to or greater than the light energy from an LED continuity tester. While the VFL can be used to test continuity, the continuity tester is typically the best tool for that job.

The VFL is often used in conjunction with an OTDR to help identify the actual location of the fault. Chapter 34, "Troubleshooting and Restoration," describes how to use the VFL to troubleshoot macrobends or breaks in an optical fiber.

Fiber Identifier

The fiber identifier acts as the fiber-optic installer or technician's infrared eyes. By placing a slight macrobend in an optical fiber or fiber-optic cable, it can detect infrared light traveling through the optical fiber and determine the direction of light travel. Some fiber identifiers can also detect test pulses from an infrared (800–1700nm) light source.

The fiber identifier typically contains two photodiodes that are used to detect the infrared light. The photodiodes are mounted so that they will be on opposite ends of the macrobend of the optical fiber or fiber-optic cable being tested. Figure 33.7 shows the location of the photodiode assemblies in the fiber identifier. The photodiode assemblies look like two small glass lenses.

FIGURE 33.7
Photodiode assemblies in the fiber identifier

The fiber identifier can typically be used with coated optical fiber, tight-buffered optical fiber, a single optical fiber cable, or a ribbon cable. Each of these must be placed in the center of the photodiodes during testing. Selecting the correct attachment for the optical fiber or optical-fiber cable type under test typically does this. Figure 33.8 shows three attachments.

FIGURE 33.8
Fiber identifier optical fiber and fiber-optic cable attachments

To test an optical fiber or fiber-optic cable, select the correct attachment and install it on the fiber identifier. Figure 33.9 shows the fiber identifier ready to test a 900µm tight-buffered optical fiber. Center the tight-buffered optical fiber over the photodiode assemblies and insert a macrobend into the tight-buffered optical fiber by pushing the slide into the tight-buffered optical fiber and compressing it, as shown in Figure 33.10.

FIGURE 33.9
Fiber identifier
ready to test a
900μm tight-buff-
ered optical fiber

FIGURE 33.10
Fiber identifier
compressing a
tight-buffered
optical fiber

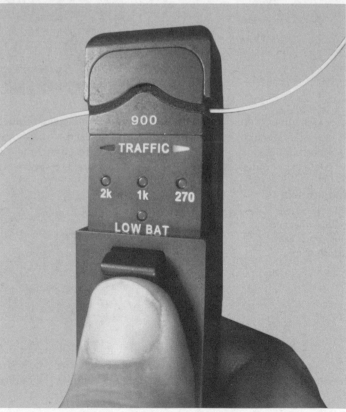

If there is sufficient infrared light energy traveling through the optical fiber, one of the directional arrows on the fiber identifier should illuminate. The directional arrow is pointing in the direction the light is traveling. If a test pulse is being transmitted through the optical fiber, the directional arrow and test pulse indicator should illuminate.

The fiber identifier works best with coated optical fiber or tight-buffered optical fiber. Like the VFL, the fiber identifier does not always work with single optical-fiber cables. How well it works with a cable depends on the amount of strength member within the cable, jacket thickness, jacket color, and amount of light energy available from the optical fiber.

The fiber identifier can be used by itself or in conjunction with an OTDR. It can be used to identify traffic in a working fiber-optic link or can be used to help identify the location of a fault. Fault location techniques are discussed in detail in Chapter 34.

Optical Return Loss Test Set

Optical return loss (ORL) testing is performed with an optical return loss test set. ORL testing measures the amount of optical light energy that is reflected back to the transmit end of the fiber-optic cable. The energy being reflected back, or back reflections, comes from mechanical interconnections, passive devices, fiber ends, and Rayleigh scattering caused by impurities within the optical fiber.

Besides reducing the amount of light transmitted, back reflections can cause various laser light source problems. They can cause the laser's output wavelength to vary. They can also cause fluctuations in the laser's optical output power and possibly permanently damage the laser light source.

The impact that back reflections have on the laser light source can cause problems in analog and digital systems. In digital systems, they can increase the BER. In analog systems, they reduce the *signal-to-noise ratio (SNR)*.

The ORL test set measures return loss using an *optical continuous wave reflectometer (OCWR)*. A light source within the ORL test set continuously transmits light through a directional coupler, as shown in Figure 33.11. Light energy returned from the optical fiber is directed to the photodiode of a power meter. The light energy measured by the power meter is the return loss.

FIGURE 33.11
Directional coupler in an ORL test set

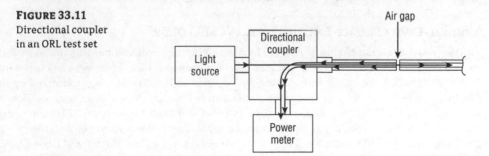

ORL measurements of a fiber-optic link should be taken with all patch cords and equipment cords in place. All system equipment should be turned off. The receive connector should remain plugged into the equipment receiver. The transmit connector should be unplugged from the equipment transmitter and plugged into the ORL test set after the test set has been calibrated, as shown in Figure 33.12. The test set should be calibrated as described in the manufacturer's operation manual. Prior to performing ORL testing, clean and inspect all connectors. Allow sufficient time for the ORL test set to warm up and stabilize.

FIGURE 33.12
ORL test set connected to fiber-optic link and system equipment

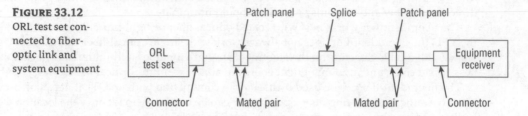

ORL test sets are available for both multimode and single-mode optical fiber. The multimode ORL test uses an LED light source with a typical output power of –20dBm. The return loss measurement range is usually between 10 and 45dB. Accuracy is normally within 0.5dB.

The single-mode ORL test set uses a laser source with a typical output power of –10dBm. The return loss measurement range is usually between 0 and 50dB or 0 and 60dB. Accuracy in the 0 to 50dB range is normally 0.5dB with a decrease in the 0 to 60dB range of 1dB.

Light Source and Optical Power Meter

The light source and optical power meter work hand in hand with each other. They are typically referred to as an *optical loss test set (OLTS)*. Many different OLTSs are available. Some support only multimode testing, some single-mode, and a few can be used for both. The multimode OLTS is typically the lowest priced, followed by the single-mode and then the combination multimode and single-mode.

This section explains the basic operation of the multimode and single-mode OLTS.

Multimode OLTS

The multimode OLTS is available with either a VCSEL or an LED light source. This section discusses the differences between applications of these test sets.

VERTICAL-CAVITY SURFACE-EMITTING LASER (VCSEL) OLTS

As discussed in earlier chapters, ANSI/TIA-568-C.3 defines optical-fiber transmission performance parameters. This standard includes laser-optimized optical fiber, which was not included in ANSI/TIA/EIA-568-B.3. It would seem that the best source to test laser-optimized optical fiber would be the VCSEL, and several manufacturers do offer VCSEL light sources. However, multimode optical-fiber test standards such as TIA-TSB-140 still specify an LED light source.

ANSI/TIA-568-C.3 defines laser-optimized optical-fiber transmission performance parameters with both an LED and a VCSEL light source. As discussed in Chapter 27, "Fiber-Optic Light Sources," the LED light source overfills the multimode optical fiber while the VCSEL only fills the center section of the core. Since the VCSEL only fills part of the core, it does not allow the entire core to be tested. The VCSEL tester thus might not detect the misalignment of a mated connector pair.

Currently the LED is the preferred light source for multimode optical-fiber testing. While VCSEL light sources are available, this chapter focuses on the LED light sources, which are required by multiple industry standards.

LED OLTS

The multimode OLTS consists of an LED-based light source and an optical power meter. The typical light source consists of an 850nm LED and a 1300nm LED, like the one shown in Figure 33.13. The optical power meter is typically selectable for 850nm, 1300nm, and 1550nm, like the one shown in Figure 33.14.

FIGURE 33.13

Multimode 850nm and 1300nm light source

FIGURE 33.14

Multimode 850nm, 1300nm, and 1550nm optical power meter

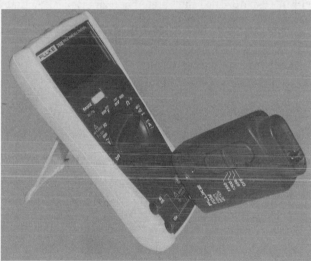

Two types of LED light sources are available: filtered and unfiltered. A filtered light source does not overfill the optical fiber. An unfiltered light source does overfill the optical fiber with high-order modes. Multimode *insertion loss* measurements should be performed with a light source that meets the launch requirements of ANSI/TIA-455-78-B. This launch requirement can be achieved within the test equipment, which is the case with the filtered light source or with an external mandrel wrap on the transmit test jumper. Most LED light sources in use today are unfiltered and require the mandrel wrap. The mandrel wrap is discussed in great detail later in the "Mode Filter" section of this chapter

The optical output power of an unfiltered LED light source is typically –20dBm for both the 850nm and 1300nm LEDs when measured with a 62.5/125μm optical fiber. The optical output power increases roughly 4dB when used with 100/140μm optical fiber and decreases roughly 3dB when used with a 50/125μm optical fiber.

NOTE Remember that ANSI/TIA-568-C.3 recognizes only 50/125μm or 62.5/125μm multimode optical fiber.

Most multimode light sources are designed to operate between the temperatures of 0° to +40° C or 0° to +50° C. Within these temperature ranges, they should provide a stable optical output power that typically varies less than ±0.2dB over an eight-hour period. To achieve this performance, the light source should be allowed to warm up before beginning testing. The manufacturer specifies warm-up times.

The multimode power meter typically uses a single photodiode with a switch to select the proper wavelength. The selector switch compensates for the responsivity of the photodiode at the different wavelengths. The output of the optical power meter is in decibels (dB) referenced to one milliwatt (0 dBm), where 0dBm is equal to 1mW of optical power.

Many power meters will support both multimode and single-mode testing. Acceptable optical-fiber types may range from 9/125μm through 100/140μm. The optical input power range is typically +3dBm to −50dBm.

As with the light sources, most optical power meters are designed to operate between the temperatures of 0° and +40° C or 0° and +50° C. When operated within this temperature range, absolute accuracy is typically within ±0.25dB, relative accuracy within ±0.15dB, and repeatability within ±0.04dB. The values stated here are typical. Actual values can be obtained from the manufacturer's data sheet.

Is My Power Meter Broken?

After years of watching students test various fiber-optic links with an OLTS, I have seen many confused looks when the reading on the power meters drops to a very low value and never stabilizes. As discussed in this chapter, the manufacturer defines the optical input power range for the power meter. This value is typically between +3dBm and −50dBm.

The output power of an LED light source is typically −20dBm with a 62.5/125μm optical fiber and −23dBm with a 50/125μm optical fiber. A break in an optical fiber will cause significant attenuation, typically between 20 and 30dB. If a light source coupled −23dBm into the fiber-optic link under test and the break added 30dB of attenuation, the maximum light energy after the break would equal −53dBm, which is below the power range for the power meter. The actual power would be less because of the losses from the optical fiber and interconnections.

When the light energy into the power meter falls below the lowest value defined by the manufacturer, the power meter typically appears to be hunting. In other words, the numbers slowly increase and decrease; they never seem to stop changing. When this happens to you, remember that the power meter is not broken; however, the optical fiber is probably broken.

Single-Mode OLTS

The single-mode OLTS consists of a laser-based light source and an optical power meter. The typical light source consists of a 1310nm laser and a 1550nm laser. Some models have a separate output port for each laser, and others combine both wavelength lasers into a single output port. Another popular laser combination is 1550nm and 1625nm.

Unlike the LED multimode light source, the single-mode light source does not require a mandrel wrap. No mandrel wrap is required because the laser does not overfill the core of the optical fiber with high-order modes. The optical output power of the laser is typically around −5dBm into a 9/125μm single-mode optical fiber. Some single-mode light sources have the ability to attenuate the output of the laser. Most single-mode light sources are designed to operate between the temperatures of 0° to +40° C or 0° to +50° C. Within these temperature ranges, they should provide a stable optical output power that typically varies no more than ±0.1dB over an eight-hour period. To achieve this performance, the light source should be allowed to warm up before beginning testing. The manufacturer specifies warm-up times.

The single-mode optical power meter is typically designed to handle a range of wavelengths. It may measure common multimode optical fiber wavelengths and single-mode optical fiber wavelengths. A typical power meter will be selectable for 850nm, 1300nm, 1310nm, 1480nm, 1550nm, and 1625nm.

Single-mode optical power meters are typically designed to operate between the temperatures of 0° to +40° C or 0° to +50° C. When operated within this temperature range, they provide the best accuracy. Absolute accuracy is typically within ±0.25dB when measured at +25° C with an optical input power of −10dBm, relative accuracy within ±0.15dB, and repeatability within ±0.04dB. The values stated here are typical. Actual values can be obtained from the manufacturer's data sheet.

WHY IS THE LOSS FOR THIS CABLE ASSEMBLY DIFFERENT EACH TIME I MEASURE IT?

One thing I have always enjoyed is the friendly competition between the students during lab. As the students get to know each other and openly communicate, they compete for the lowest-loss cable assembly. Sometimes there is a substantial gap between the loss measurements and the winner is clearly defined. Other times the gap is not measurable with the equipment in the lab and the results vary slightly each time a cable assembly is measured.

In this chapter, the typical values for absolute accuracy, relative accuracy, and repeatability for a power meter were defined. The stability of the light source was also defined. Each of these parameters has a range, and absolute accuracy is typically the largest. What these parameter ranges mean is that insertion loss measurements will vary. There is no such thing as a perfect piece of test equipment.

Patch Cord

ANSI/TIA-568-C.3, section 6 sets the performance specifications for optical-fiber patch cords recognized in premises cabling standards. The optical-fiber *patch cord* is used at cross-connections to connect optical-fiber links. They are also used as equipment or work area cords to connect telecommunications equipment to horizontal or backbone cabling.

A patch cord is a length of optical-fiber cable with connectors on both ends. It uses the same connector type and optical-fiber type as the optical-fiber cabling that it is connected to. The patch cord must meet the cable transmission performance requirements and physical cable specifications of clauses 4.2 and 4.3.1 of ANSI/TIA-568-C.3. The patch cord must also meet the connector and adapter requirements of clause 5.2 of ANSI/TIA-568-C.3.

Optical-fiber patch cords used for either cross-connection or interconnection to equipment shall have a termination configuration defined in clause 6.4 of ANSI/TIA-568-C.3. Clause 6.4 describes configurations for simplex, duplex, and array patch cords. A simplex patch cord is a single-fiber cable with simplex connection terminations. A duplex patch cord is a two-fiber cable with duplex connectors. An array patch cord is a multifiber cable with array connectors on each end. Array connectors feature a single ferrule that contains multiple optical fibers arranged in a row or rows and columns. This connector is covered in detail in Chapter 26, "Connectors."

Test Jumper

The terms *patch cord* and *jumper* are often interchanged. A test jumper can be a single-fiber cable or a multifiber cable. This section focuses on multimode test jumpers as described in ANSI/TIA-526-14-A.

In the fiber-optic industry, the *test jumper* has several names. The test jumper connected to the light source is typically called the *transmit jumper*. The test jumper connected to the optical power meter is typically called the *receive jumper*. The U.S. Navy calls the test jumper a *measurement quality jumper (MQJ)*. You may also see the test jumper referred to as a *reference jumper*. Regardless of the name, the test jumper is a critical part of your optical power measurement equipment.

Test jumpers must have a core diameter and numerical aperture nominally equal to the optical fiber being tested. You cannot test a 50/125µm link with a 62.5/125µm test jumper, or vice versa. Per ANSI/TIA-526-14-A, jumpers shall be no less than 1 meter in length and no greater than 5 meters in length. The termination of the test jumper shall be compatible with the cable being tested.

Test jumpers should be cleaned and inspected prior to making measurements. The endface of each connector should be evaluated under a microscope. They should always meet or exceed the good criteria of ANSI/TIA-455-57-B. This means that there can be no scratches, notches, or chips in the endface of the optical fiber. You should also clean and inspect each connector of the cable under test. Mating a dirty or damaged connector with a test jumper connector can destroy the test jumper connector.

Test jumpers should be tested for insertion loss prior to performing any of the ANSI/TIA-526-14-A methods. They should be tested the same way a patch cord is tested. This test method is described later in this chapter. The maximum acceptable loss for a test jumper is 0.4dB. This value is not defined by the standard but serves as a good rule of thumb. Many test jumpers will have a measured loss of less than 0.2dB.

Test jumpers should be treated with care, especially multiple fiber measurement quality jumpers (MQJs) like the one shown in Figure 33.15, which can be very expensive. The measured loss of the test jumper will increase gradually as it is used. However, you can continue to use the test jumper as long as the loss does not exceed 0.4dB and the endface meets or exceeds the good criteria of ANSI/TIA-455-57-B.

FIGURE 33.15
Multiple-fiber MQJ

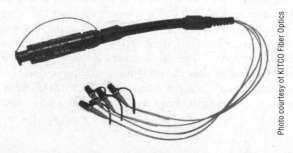

Photo courtesy of KITCO Fiber Optics

Mode Filter

The *mandrel wrap* or *mode filter* is required for light sources that do not meet the launch requirements of ANSI/TIA-455-78-B, method A. It is required because the attenuation in a short link of multimode optical fiber may be higher than calculated when making insertion loss measurements due to the power lost in the high-order modes. The high-order modes are caused by an LED light source that overfills the optical fiber.

When you're measuring the insertion loss of a multimode link with an LED source that overfills the optical fiber, a mandrel wrap must be used on the transmit test jumper. The transmit test jumper should have five nonoverlapping turns around a smooth mandrel. The diameter of the mandrel depends on the cable diameter and optical-fiber core diameter, as shown in Table 33.1. Figure 33.16 shows a transmit test jumper with a mandrel wrap attached to an LED light source.

TABLE 33.1   Mandrel Diameters for Multimode Optical Fiber

FIBER CORE SIZE (µM)	MANDREL DIAMETER FOR 900µM BUFFERED OPTICAL FIBER (MM)	MANDREL DIAMETER FOR 3MM JACKETED CABLE
50	25	22
62.5	20	17

An unfiltered LED light source overfills an optical fiber by launching high- and low-order modes into the core and cladding, as shown in Figure 33.17. The high-order modes experience more attenuation than the low-order modes at interconnections, splices, and bends. This causes higher-than-expected loss in short multimode links. The mandrel wrap takes advantage of the high loss of high-order modes caused by excessive bending of the optical fiber. The mandrel wrap removes the high-order modes by inserting a series of macrobends in the transmit test jumper, as shown in Figure 33.18.

FIGURE 33.16
Transmit test jumper with mandrel wrap

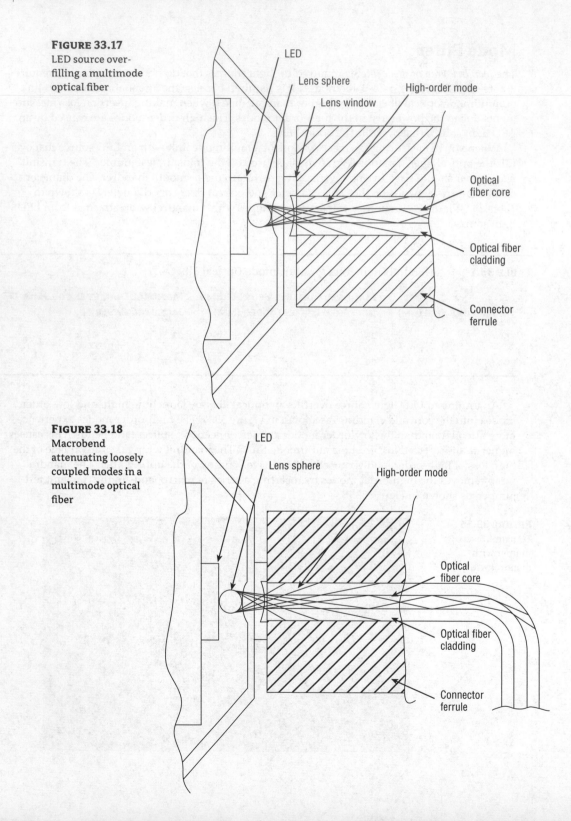

FIGURE 33.17
LED source over-filling a multimode optical fiber

LED
Lens sphere
Lens window
High-order mode
Optical fiber core
Optical fiber cladding
Connector ferrule

FIGURE 33.18
Macrobend attenuating loosely coupled modes in a multimode optical fiber

LED
Lens sphere
High-order mode
Optical fiber core
Optical fiber cladding
Connector ferrule

The five small-radius nonoverlapping loops around the mandrel wrap significantly attenuate the high-order modes with minimal attenuation of the low-order modes. The mandrel wrap is actually a low-order or low-pass mode filter. It allows the low-order modes to pass with very little attenuation while greatly attenuating the high-order modes. The mandrel wrap should always be on the transmit test jumper at every stage of testing when making insertion loss measurements.

ANSI/TIA-526-14-A Optical Loss Measurement

ANSI/TIA-526-14-A provides three methods for testing the *cable plant*. The difference between the three methods is how the reference power measurement is taken. Each method uses a different number of test jumpers. Method A uses two test jumpers to obtain the reference power measurement, method B uses one test jumper, and method C uses three test jumpers. Each of these methods is described in detail in this section.

Test methods A, B, and C each measure the optical loss of the cable plant at different points, as shown in Figure 33.19. Test method A measures the loss of the cable plant plus one connection loss. Test method B measures the loss of the cable plant plus two connection losses. Test method C measures the loss of the cable plant only. Test method B is the only method that can be used when testing to ANSI/TIA-568-B.

FIGURE 33.19
Cable plant measured values for methods A, B, and C

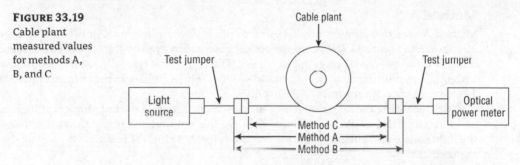

The first step before performing any of the optical power loss measurements is to clean and inspect the test jumpers and connectors at the ends of the cable plant to be tested. Verify that your test jumpers have the same optical-fiber type as the cable plant you are going to test.

While you are cleaning and inspecting the test jumpers and cable plant connectors, turn on the light source and optical power meter. Set both of them to the test wavelength. Allow sufficient time for the light source and optical power meter to warm up and stabilize. Warm-up information can be found in the manufacturer's data sheet.

After sufficient time has passed for the light source and optical power meter to stabilize, you need to measure the loss of the test jumpers. Connect test jumper 1 as shown in Figure 33.20. Record the optical power displayed on the optical power meter. This number is the reference power measurement. This number is typically around –20dBm with a 62.5/125µm multimode optical fiber and –23dBm with a 50/125µm multimode optical fiber. These numbers can vary from OLTS to OLTS.

FIGURE 33.20
Test jumper 1
optical reference
power measure-
ment

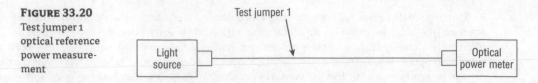

Connect test jumper 2 as shown in Figure 33.21. Record the optical power displayed on the optical power meter. The difference between the reference power measurement and this value is the connection loss for test jumpers 1 and 2. The difference between this value and the reference power measurement should be less than or equal to 0.4dB. If this number is greater than 0.4dB, clean the connectors again and retest. If the loss still exceeds 0.4dB, replace test jumper 2. Repeat this test for test jumper 3 if method C is being used.

FIGURE 33.21
Test jumper 2
optical reference
power measure-
ment

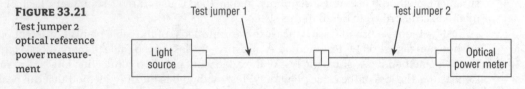

Method A

Method A uses two jumpers to establish the optical power reference. Connect test jumpers 1 and 2 as shown in Figure 33.21. Record the optical power displayed by the optical power meter. This number is the reference power measurement. This number is typically around –20dBm with a 62.5/125μm multimode optical fiber and –23dBm with a 50/125μm multimode optical fiber. These numbers can vary from OLTS to OLTS.

After recording the reference power measurement, separate the test jumpers at their point of interconnection, as shown in Figure 33.22. Do this without disturbing their attachment to the light source and optical power meter. Attach the test jumpers to the cable plant as shown in Figure 33.23.

FIGURE 33.22
Test jumpers 1 and
2 separated

FIGURE 33.23
Test jumpers 1 and
2 connected to the
cable plant

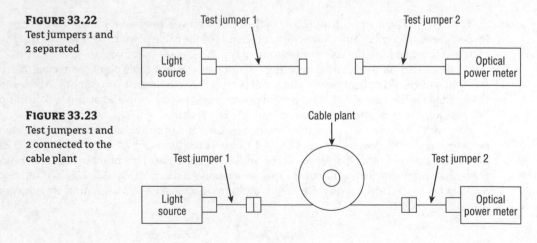

With the test jumpers attached, record the optical power displayed on the optical power meter. This is your test power measurement. The loss for the cable plant is the difference between the reference power measurement and the test power measurement.

To obtain the optical power loss for the cable plant, subtract the test power measurement from the reference power measurement. If the reference power measurement was –20.4dBm and the test power measurement was –21.6dBm, the optical power loss for the cable plant would be 1.2dB.

NOTE   Remember the mandrel wrap is always installed on the test jumper connected to the light source when an unfiltered light source is used.

Method B

Method B uses one jumper to establish the optical power reference. However, two test jumpers are required to perform the test. This method is required by ANSI/TIA-568-B to measure link attenuation. Per ANSI/TIA-568-B.1, section 11.3.3.1, horizontal links shall be tested at 850nm and 1300nm in one direction. Per section 11.3.3.2, multimode backbone links shall be tested in one direction at both 850nm and 1300nm. Per section 11.3.3.3, centralized links shall be tested at 850nm or 1300nm in one direction; 850nm is recommended. Link attenuation using this method is equal to cable attenuation plus connector insertion loss plus splice insertion loss.

NOTE   Single-mode backbone links should be tested at 1310nm and 1550nm in accordance with ANSI/TIA-526-7.

Connect test jumper 1 as shown earlier in Figure 33.20. Record the optical power displayed by the optical power meter. This number is the reference power measurement. This number is typically around –20dBm with a 62.5/125µm multimode optical fiber and –23.5dBm with a 50/125µm multimode optical fiber. These numbers can vary from OLTS to OLTS.

After recording the reference power measurement, disconnect the test jumper from the optical power meter. Do not disturb the jumper's attachment to the light source. Attach test jumper 2 to the optical power meter as shown in Figure 33.22. Attach the test jumpers to the cable plant as shown in Figure 33.23.

With the test jumpers attached, record the optical power displayed on the optical power meter. This is your test power measurement. The loss for the cable plant is the difference between the reference power measurement and the test power measurement.

To obtain the optical power loss for the cable plant, subtract the test power measurement from the reference power measurement. If the reference power measurement was –20.7dBm and the test power measurement was –23.6dBm, the optical power loss for the cable plant would be 2.9dB.

Method C

Method C uses three jumpers to establish the optical power reference. Connect test jumpers 1, 2, and 3 as shown in Figure 33.24. Record the optical power displayed by the optical power meter. This number is the reference power measurement.. This number is typically around –20dBm with a 62.5/125µm multimode optical fiber and –23.5dBm with a 50/125µm multimode optical fiber. These numbers can vary from OLTS to OLTS.

FIGURE 33.24
Method C reference power measurement

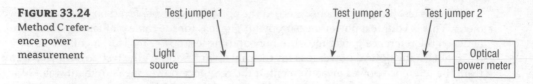

After recording the reference power measurement, separate the test jumpers at their point of interconnection, as shown in Figure 33.25. Do this without disturbing their attachment to the light source and optical power meter. Attach the test jumpers to the cable plant as shown earlier in Figure 33.23.

FIGURE 33.25
Removal of test jumper 3 from method C reference power measurement

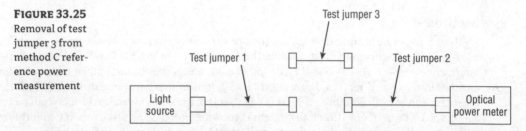

NOTE   Test jumper 3 is not used during testing. It is used only to establish reference power.

With the test jumpers attached, record the optical power displayed on the optical power meter. This is your test power measurement. The loss for the cable plant is the difference between the reference power measurement and the test power measurement.

To obtain the optical power loss for the cable plant, subtract the test power measurement from the reference power measurement. If the reference power measurement was –20.2dBm and the test power measurement was –21.6dBm, the optical power loss for the cable plant would be 1.4dB.

Patch Cord Optical Power Loss Measurement

Multimode patch cord optical loss power measurement is performed using the steps described in ANSI/TIA-526-14-A, method A. The patch cord is substituted for the cable plant. Because patch cords are typically no longer than 5m, the loss for the optical fiber is negligible and testing can be performed at 850nm or 1300nm. The loss measured in this test is the loss for the patch cord connector pair. ANSI/TIA-568-C states that the maximum loss for a connector pair is 0.75dB.

After setting up the test equipment as described in ANSI/TIA-526-14-A, method A, clean and inspect the connectors at the ends of the patch cords to be tested. Verify that your test jumpers have the same optical-fiber type and connectors as the patch cords you are going to test. Ensure that there are no sharp bends in the test jumpers or patch cords during testing.

Because both patch cord connectors are easily accessible, optical power loss should be measured in both directions. The loss for the patch cord is the average of the two measurements. If the loss for the patch cord exceeds 0.75dB in either direction, the patch cord needs to be repaired or replaced.

Optical Time-Domain Reflectometer

So far in this chapter, you have learned about tools and test equipment that can be used to test a fiber-optic link or cable. Of all the tools and test equipment discussed, none provides more information about the fiber-optic cable or link than the *optical time-domain reflectometer (OTDR)*. The OTDR can be used to evaluate the loss and reflectance of interconnections and splices. It will measure the attenuation rate of an optical fiber and locate faults.

This section of the chapter focuses on basic OTDR theory, setup, and testing. The OTDR is a complex piece of test equipment with many variables that must be programmed correctly before testing can be performed. Proper OTDR setup and cable preparation will ensure accurate test results.

OTDR Theory

The OTDR is nothing more than a device that launches a pulse or pulses of light into one end of an optical fiber and records the amount of light energy that is reflected back. Unlike all the test equipment discussed up to this point, the OTDR provides a graphical representation of what is happening in the fiber-optic link or cable under test. With the OTDR, the fiber-optic link or cable is no longer a black box. The OTDR shows how light passes through every segment of the fiber-optic link.

Light reflecting back in an optical fiber is the result of a reflection or backscatter. Reflections are when the light traveling through the optical fiber encounters changes in the refractive index. These reflections are called Fresnel reflections. Backscatter, or Rayleigh scattering, results from evenly distributed compositional and density variations in the optical fiber. Photons are scattered along the length of the optical fiber. The photons that travel back toward the OTDR, as shown in Figure 33.26, are considered backscatter.

FIGURE 33.26
Backscattered
photons

OTDRs come in many different shapes and sizes, as shown in Figure 33.27. The newer models are almost pocket-sized while others require a shoulder strap. The small OTDR is lighter and easier to carry; however, the small screen permits only a limited viewing area.

FIGURE 33.27
Large and small
OTDRs

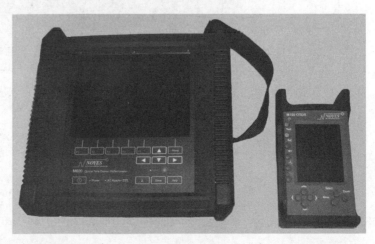

The OTDR is typically a battery-powered device. It is battery powered because many places where the OTDR is used have no electrical power available. It is a good idea to bring a charged spare battery with you when you are performing testing.

Many OTDRs can be configured to test both multimode and single-mode optical fibers. A typical OTDR may hold up to three light source modules, like the OTDR shown in Figure 33.28. Modules can be added or removed as testing requirements change. Some OTDRs even contain a visual fault locator like the one shown in Figure 33.29.

FIGURE 33.28
Light source module locations on an OTDR

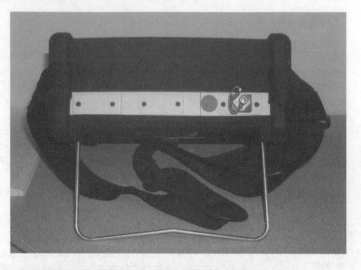

FIGURE 33.29
The visual fault locator built into an OTDR

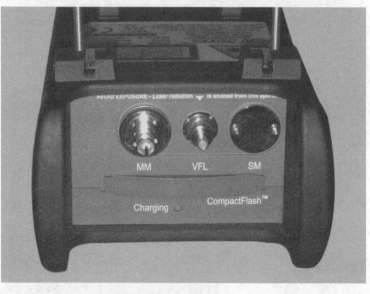

A typical OTDR can be broken up into eight basic components: the *directional coupler*, laser, *timing circuit*, *single-board computer*, *digital signal processor (DSP)*, *analog-to-digital converter*, *sample-and-hold circuit*, and *avalanche photodiode*. Figure 33.30 is a block diagram of the OTDR showing how light is launched from the laser through the directional coupler into the optical fiber. The directional coupler channels light returned by the optical fiber to the avalanche photodiode.

FIGURE 33.30
OTDR block
diagram

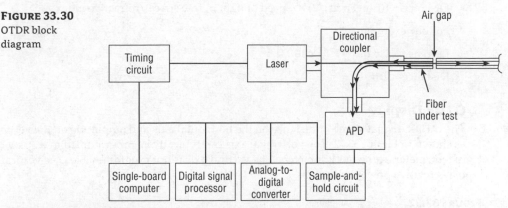

The avalanche photodiode converts the light energy into electrical energy. The electrical energy is sampled at a very high rate by the sample-and-hold circuit. The sample-and-hold circuit maintains the instantaneous voltage level of each sample long enough for the analog-to-digital converter to convert the electrical value to a numerical value. The numerical value from the analog-to-digital converter is processed by the DSP and the result is sent to the single-board computer to be stored in memory and displayed on the screen. This entire process is typically repeated many times during a single test of an optical fiber and coordinated by the timing circuit.

For the OTDR to produce accurate results, the refractive index of the optical fiber under test must be known. The refractive index of an optical fiber is different for each wavelength tested. The operator must enter the correct refractive index into the OTDR for each wavelength. The correct refractive index for a fiber-optic cable can be obtained from the manufacturer.

The OTDR samples light energy from the optical fiber at precise intervals. Each sample taken by the OTDR represents the round-trip time for light energy in the optical fiber. Let's assume that the OTDR is taking 500 million samples per second, or one sample every two nanoseconds. If the refractive index for the optical fiber under test were equal to 1.5, every five samples would represent the distance of 1 meter, as shown in Figure 33.31.

FIGURE 33.31
OTDR sampling at
2ns rate

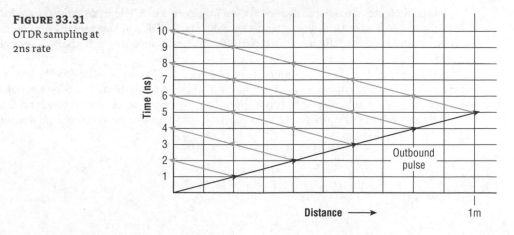

The following formula can be used to find distance based on time and refractive index. In this formula, the speed of light is rounded up to 3×10^8 m/s:

Distance = ((time in ns)/2) × (speed of light in free space)/refractive index

For our example:

Distance = $(10 \text{ ns}/2) \times (3 \times 10^8)/1.5$

Distance = 1m

OTDR Display

The OTDR displays time or distance on the horizontal axis and amplitude on the vertical axis, as shown in Figure 33.32. The horizontal axis can typically be programmed to display distance in feet, meters, miles, or kilometers. The vertical axis is not programmable. The vertical axis displays relative power in decibels.

FIGURE 33.32
OTDR display

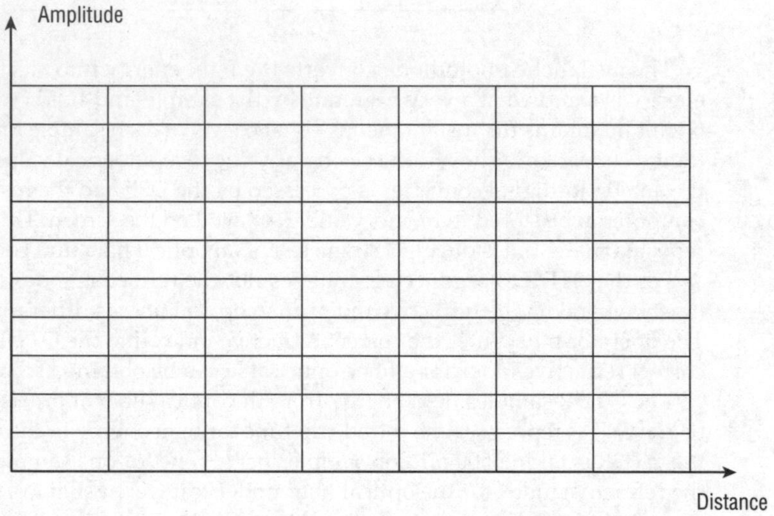

Amplitude

Distance

The OTDR creates a trace like the one shown in Figure 33.33. The trace shows event loss, event reflectance, and optical-fiber attenuation rate. The OTDR can horizontally and vertically zoom in on any section of the trace. This permits a more detailed inspection of the optical fiber or event.

The light pulses from OTDR produce a blind spot or dead zone, like the one shown in Figure 33.34. The dead zone is an area where the avalanche photodiode has been saturated by the reflectance of a mechanical interconnection. The size of the dead zone depends on the length of each light pulse and the amount of light reflected back toward the avalanche photodiode.

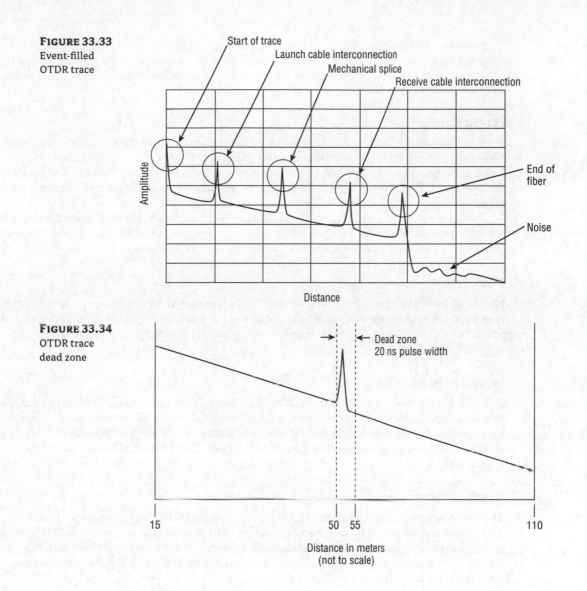

FIGURE 33.33
Event-filled
OTDR trace

FIGURE 33.34
OTDR trace
dead zone

An interconnection that has very little reflectance will not saturate the avalanche photodiode. This type of interconnection will produce the smallest dead zone. Dividing the pulse length in nanoseconds by 10 can approximate the length of the light pulse as seen by the OTDR in meters. A pulse length of 20ns will yield an ideal dead zone of 2m.

The ratio of the dead zone to the pulse length depends on the amount of light energy reflected back toward the OTDR. Poor interconnections or interconnections with an air gap typically saturate the avalanche photodiode and produce a greater-than-ideal dead zone, as much as 10 times greater than ideal.

OTDR Setup

Prior to testing, the OTDR needs to be set up correctly to provide the most accurate readings. When setting up the OTDR, you need to select the correct fiber type, wavelength or wavelengths, range and resolution, pulse width, averages, refractive index, thresholds, and backscatter coefficient. This process takes only a couple of minutes and ensures the most accurate results possible.

There are many different OTDRs on the market, and it is impossible to describe the setup for each. This section focuses on general setup parameters. Some OTDRs may have additional parameters, and some may not include all of the parameters discussed.

FIBER TYPE

Each light source or light source module in an OTDR is designed for one or several specific optical-fiber types. A multimode module should not be used to test a single-mode optical fiber, and vice versa. Before heading for the test site, ensure that your OTDR has the correct module for the optical fiber to be tested.

WAVELENGTH

The wavelength that your OTDR can test with depends on the light source module or modules in your OTDR. Some light source modules contain a single laser while others contain two different wavelength lasers. A light source with two lasers allows testing of the optical fiber at two wavelengths without disconnecting the cable under test. This simplifies testing and reduces testing time.

RANGE AND RESOLUTION

The distance range of an unzoomed trace displayed on the OTDR and the distance between data points is determined by range and resolution. As a general rule of thumb, the OTDR range should be set to 1.5 times the length of the fiber-optic link. If the range is set too short, the results will be unpredictable and the entire link may not be displayed. If the range is set too long, the trace will fill only a small portion of the display area.

Increasing the range automatically increases the distance between the data points. When the distance between the data points is increased, resolution is reduced. When setting range, choose the first range value that exceeds your fiber-optic link length. Selecting a 2km range for a 1.3km link will yield more accurate results than selecting a 20km range.

PULSE WIDTH

The pulse width determines the size of the dead zone and the maximum length optical fiber that can be tested. A short pulse width produces a small dead zone. However, a short pulse width reduces the length of optical fiber that can be tested.

The pulse width should be selected so that the trace never disappears into the noise floor. If the pulse width is set properly, the trace will stay smooth until the end of the fiber-optic link. If the pulse is set too wide, events may be lost in the dead zone.

AVERAGES

When setting the averages parameter on an OTDR, select the number of averages that produce a smooth trace in the least amount of time. If too few averages are taken, the trace will appear noisy because the noise floor is too high. If too many samples are taken, the trace will be smooth; however, valuable testing time will be wasted.

REFRACTIVE INDEX

As mentioned earlier in the chapter, the manufacturer should specify the refractive index or the group index of refraction for an optical fiber. The refractive index of similar optical fibers does not vary much from manufacturer to manufacturer. Most values are typically within 1 percent of each other. If the value for the optical fiber you are testing is not known, use the values shown in Table 33.2. Entering a low refractive index will produce measurements that are too long, and entering a high refractive index will produce measurements that are too short.

TABLE 33.2   Refractive Index Default Values

WAVELENGTH	REFRACTIVE INDEX
850nm	1.4960
1300nm	1.4870
1310nm	1.4675
1550nm	1.4681

THRESHOLDS

Thresholds can typically be set for end of fiber, event loss, and reflectance. Many OTDRs have a default value preset for each of these. Depending on the testing you are performing, the default values may be too sensitive or not sensitive enough.

Most OTDRs will generate event tables automatically based on the threshold settings. To capture a majority of the events in the event table, set the thresholds on the sensitive side. A good starting place would be to set the values as shown in Table 33.3. If these values are too sensitive, you can always increase them.

TABLE 33.3   Threshold Default Values

EVENT	DEFAULT VALUE
End of fiber	6.0dB
Event loss	0.05dB
Reflectance	−65dB

BACKSCATTER COEFFICIENT

The OTDR uses the backscatter coefficient to calculate reflectance. As with the refractive index, the optical-fiber manufacturer specifies backscatter coefficient. Backscatter coefficient does vary from manufacturer to manufacturer. However, the variation is typically not that great, and the default values shown in Table 33.4 can be used with good results when the manufacturer's specified value is not known.

TABLE 33.4   Backscatter Coefficient Default Values

WAVELENGTH	BACKSCATTER COEFFICIENT
850nm	68.00
1300nm	76.00
1310nm	80.00
1550nm	83.00

Cable Plant Test Setup

Now that the OTDR has been configured with the correct light source modules and the setup parameters have been entered, the OTDR can be connected to the fiber-optic link. There are many different ways that the OTDR can be used to test a single optical-fiber cable or fiber-optic link. In this chapter, the test setup will include a *launch cable* and *receive cable*. Launch and receive cables may also be referred to as *pulse suppression cables*.

Launch and receive cables allow the OTDR to measure the insertion loss at the near and far ends of the fiber-optic link. Like test jumpers, launch and receive cables should be constructed of optical fiber similar to the optical fiber under test. Launch and receive cables are available from many different manufacturers and come in all shapes and sizes. Some launch and receive cables are designed for rugged environments, like the cable in Figure 33.35. Others are designed to be compact, almost pocket-sized, like the one in Figure 33.36.

FIGURE 33.35
Rugged launch/
receive cable

FIGURE 33.36
Compact launch/
receive cable

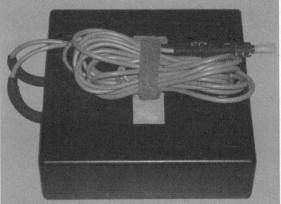

You should select the length of the launch and receive cables based on the pulse width you are using to test the optical fiber. Launch and receive cables are typically the same length. A good rule of thumb is to have at least 1 meter of launch or receive cable for each nanosecond of pulse width, with the minimum length cable being 100m. A 2km link being tested with a 20ns pulse width would require at least a 100m launch and receive cable. A 10km link being tested with a 200ns pulse would require at least a 200m launch and receive cable.

After the launch and receive cables have been selected, they can be attached to the fiber-optic link, as shown in Figure 33.37. Remember to clean and inspect the connector endfaces on the cable under test and on the launch and receive cables. Dirty or damaged endfaces on the launch and receive cables can damage the endface of the connectors that you are going to test. A dirty or damaged endface on a launch or receive cable typically results in a high-loss interconnection.

FIGURE 33.37
OTDR connected to fiber-optic link under test with launch and receive cables

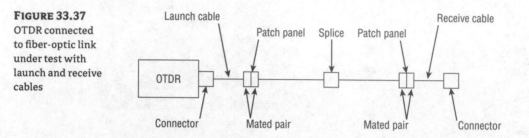

IF YOUR PULSE SUPPRESSION CABLES DON'T MATCH, YOU MAY SEE A GHOST

One day I was watching some students test a reel of multifiber cable with the OTDR. The Army had recently used this cable in the desert. Each fiber in the cable had been terminated on both ends.

After cleaning and visually inspecting the connectors, the pulse suppression cables were connected to both ends of the cable. The OTDR was connected to the launch cable and each optical fiber was tested. Looking at the OTDR trace, the students concluded that each optical fiber had multiple breaks because the OTDR trace had multiple back reflections every 100 meters.

I asked the students to test the continuity of each optical fiber with the continuity tester. Each optical fiber had good continuity. I then asked the students the core diameter of the optical fiber they were testing and the pulse suppression cables.

It turned out that the repetitive pulses or ghosts on the OTDR trace were caused by the core diameter mismatch between the pulse suppression cables and the optical fiber under testing. The students were testing a 50/125µm optical fiber with 62.5/125µm pulse suppression cables.

Testing and Trace Analysis

Testing the cable plant or a fiber-optic link with the OTDR should be done in both directions. Multimode optical fiber should be tested at both 850nm and 1300nm. Single-mode optical fiber should be tested at both 1310nm and 1550nm. Event loss and optical-fiber attenuation is typically the average of the bidirectional values.

There are many standards on how to test both multimode and single-mode optical fiber with an OTDR. This section describes several techniques that will allow you to measure the distances and losses in a typical cable plant or fiber-optic link. We will focus on short links—the type you would typically find in a commercial building.

All measurement techniques discussed in this chapter utilize the 2-point method. All OTDRs can measure loss using the 2-point method, but not all OTDRs can automatically perform 2-point subtraction. This section of the chapter assumes that your OTDR can perform 2-point subtraction. If your OTDR can't, you will have to perform the calculations manually.

Some OTDRs are capable of *least-squares averaging (LSA)*. LSA can be used to measure attenuation. This section of the chapter does not address LSA measurements. If your OTDR can perform LSA measurements, consult the operator's manual for how to perform them.

WARNING   When testing with the OTDR, always observe the manufacturer's precautions. Never directly view the end of an optical fiber being tested with an OTDR. Viewing the end of an optical fiber being tested with an OTDR directly or with a microscope may cause eye damage.

BASELINE TRACE

The first step in trace analysis is to generate the baseline trace. You should not press the test button on the OTDR to generate a baseline trace until you ensure that:

◆   All connectors have been cleaned and inspected and are undamaged.

◆   Launch and receive cables have optical fiber similar to the optical fiber under test.

◆   Launch and receive cables are the correct length.

◆   Launch and receive cables are properly connected to each end of the fiber-optic link under test.

◆   The correct fiber type, wavelength, range and resolution, pulse width, averages, refractive index, and backscatter coefficient have been entered into the OTDR.

Figure 33.38 is a drawing of a baseline trace as presented on the OTDR screen. This baseline trace contains a launch cable, horizontal segment, and receive cable. The launch and receive cables are 100m in length. The horizontal segment is 85m in length.

FIGURE 33.38
Baseline trace of horizontal segment

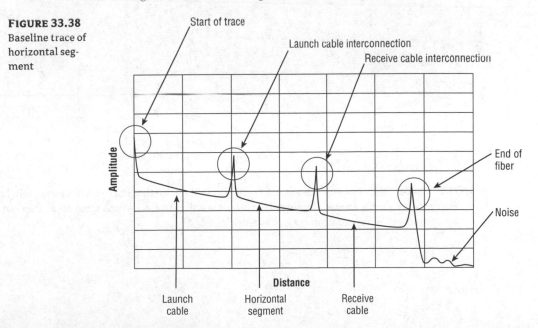

Looking at the trace from left to right, you will notice there is a large back reflection at the input to the launch cable. Because a 20ns pulse width was selected, the trace is smooth within 10m. The smooth trace slopes gradually to the back reflection caused by the connector pair where the launch cable and horizontal segment are connected together.

The trace becomes smooth again 10m after the interconnection back reflection. The trace remains smooth up to the back reflection caused by the connector pair where the receive cable and horizontal segment interconnect. The trace again becomes smooth 10m after the interconnection back reflection until a large back reflection is generated by the end of the receive cable. The receive cable back reflection is followed by a large reduction in amplitude, and then the trace disappears into the noise floor.

Measuring the Attenuation of a Partial Length of Optical Fiber

When using the 2-point method, the first thing you should do is measure the attenuation for a partial segment of the cable under test. This should be done for each wavelength that you are testing the optical fiber with. It has to be done only one time for each cable type being tested. The data should be recorded—this information will be needed to accurately measure interconnection loss.

After taking the baseline trace, position the two cursors on a smooth section of the optical fiber. The longer the section, the more accurate your results will be, because noise will have less impact on the overall measurement. The trace in Figure 33.39 has the A and B cursors 50m apart on the smooth section of the horizontal segment of the trace. The loss for this 50m segment at a wavelength of 850nm is approximately 0.14dB.

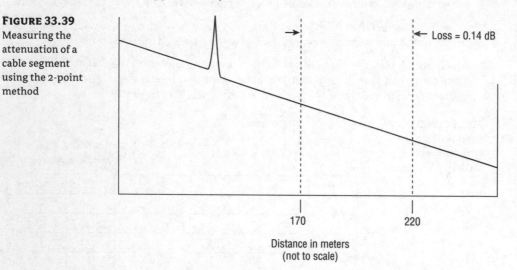

FIGURE 33.39
Measuring the attenuation of a cable segment using the 2-point method

Loss = 0.14 dB

170   220

Distance in meters
(not to scale)

NOTE Because of ripple on the smooth trace, you will notice that the A-B loss changes slightly every time a cursor moves. Ripple caused by noise will cause variations from measurement to measurement. This variation may be as much as 0.05dB.

MEASURING THE DISTANCE TO THE END OF THE OPTICAL FIBER

A break in an optical fiber in a fiber-optic cable looks like the end of the optical fiber on the OTDR display because the OTDR cannot see beyond the break. Regardless of the number of optical fiber breaks in a cable, the OTDR can only see the one closest to it. Because a break looks like the end of the optical fiber on the OTDR, the same method is employed to measure the distance to a break in an optical fiber or measure the distance to the end of an optical fiber.

Light from the OTDR exiting the optical fiber under test typically produces a strong back reflection. The strong back reflection is caused by the Fresnel reflection when the light from the optical fiber hits the air. However, a break in an optical fiber is not always exposed to the air. The break may be exposed to index matching gel from a mechanical splice, gel from a loose buffer, or water.

The same method can be used to find the distance to the end of an optical fiber whether or not a Fresnel reflection happens. Figure 33.40 has two OTDR traces. The top trace is the end of an optical fiber with a Fresnel reflection. The bottom trace is the end of an optical fiber without a Fresnel reflection.

FIGURE 33.40
Measuring the distance to the end of an optical fiber using the 2-point method

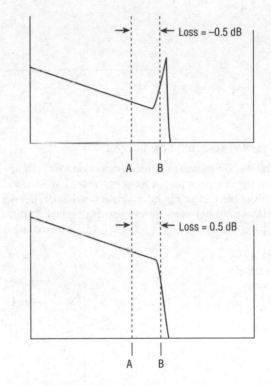

To measure the distance to the end of the optical fiber after zooming in on the back reflection, place the A cursor on a smooth section of the trace just in front of the back reflection or the drop in the trace. Move the B cursor toward the A cursor until it is in the leading edge of the back reflection. Keeping moving the B cursor toward the A cursor until the A-B loss is ±0.5dB. It should be 0.5dB of loss for the nonreflective trace and 0.5dB of gain for the reflective trace. The length for the entire span is the distance for the B cursor.

MEASURING THE LENGTH OF A CABLE SEGMENT

The first step in measuring the length of a cable segment is to horizontally zoom in on the cable segment, as shown in Figure 33.41. Place the cursors in the leading edge of the reflective events for that segment. The cursors should intersect the leading edge of the reflective event at the same vertical height above the smooth part of the trace, as shown in Figure 33.41. The distance between cursors is the length of the cable segment.

FIGURE 33.41
Measuring the length of a cable segment

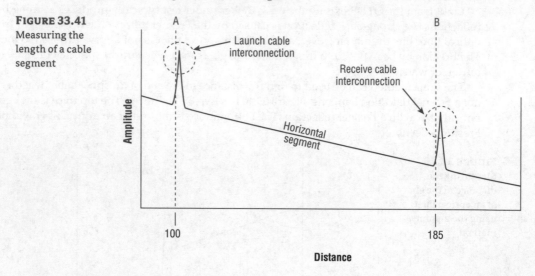

MEASURING INTERCONNECTION LOSS

The first step in measuring interconnection loss is to horizontally zoom in on the interconnection. Position the A cursor just in front of the back reflection; then position the B cursor on a smooth area on the trace after the interconnection back reflection. The loss for the interconnection and the optical fiber between the cursors is displayed on the OTDR screen. The loss for the inter-connection shown in Figure 33.42 is 0.4dB. The distance between the A and B cursors is 50m.

FIGURE 33.42
Measuring inter-connection loss with the OTDR

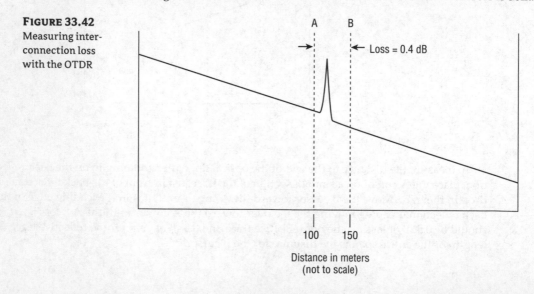

NOTE   The B cursor should always be in a smooth section of the trace. When a long pulse width is used, the dead zone may be very large. The distance between the A and B cursors may need to be several hundred meters. In that case, you measure the attenuation of a partial length of optical fiber no less than several hundred meters.

To find the loss for only the interconnection, the loss for the optical fiber between cursors A and B needs to be subtracted from the A-B loss displayed by the OTDR. This loss was previously at 0.14dB for 50m. To find the actual interconnection loss, subtract 0.14dB from the 0.4dB A-B loss. The loss for only the interconnection is approximately 0.26dB.

MEASURING THE LOSS OF A FUSION SPLICE OR MACROBEND

A fusion splice or macrobend does not produce a back reflection. A macrobend will always appear as a loss in the form of a dip in the smooth trace. A fusion splice may appear as a dip or a bump in the trace. The fusion splice will appear as a dip in the trace when tested in both directions when the optical fibers fusion-spliced together have the same backscatter coefficient.

When optical fibers with different backscatter coefficients are fusion-spliced together, the splice may appear as a loss in one direction and a gain in the form of a bump in the trace when tested in the other direction. This is referred to as a *gainer*. To find the loss for the fusion splice, the splice must be tested in both directions and the results averaged together. The losses or loss and gain should be added together and the sum divided by 2. This is the loss for the fusion splice.

To find the loss of a fusion splice or macrobend, horizontally zoom in on the event. The loss from a fusion splice or macrobend is typically very small and will require vertical zoom in addition to horizontal zoom. Place the A cursor on the smooth part of the trace before the dip in the trace. Place the B cursor on the smooth part of the trace after the dip, as shown in Figure 33.43. The loss for this event is 0.25dB. The loss for the event includes the loss for the fusion splice or macrobend plus the 50m of optical fiber between the cursors. Subtract the loss for the 50m of optical fiber that was previously measured at 0.14dB from the event loss. The loss for this fusion splice or macrobend is 0.11dB.

FIGURE 33.43
Measuring the loss of a fusion splice or macrobend

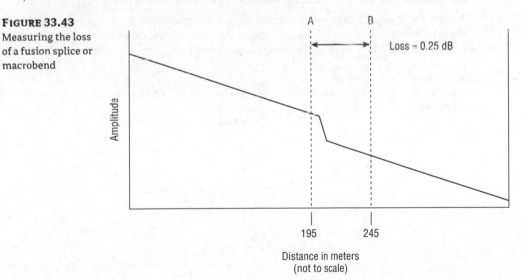

To find the gain of a fusion splice, horizontally zoom in on the event. The gain from a fusion splice is typically very small and will require vertical zoom in addition to horizontal zoom. Place the A cursor on the smooth part of the trace before the bump in the trace. Place the B cursor on the smooth part of the trace after the bump, as shown in Figure 33.44. The gain for this event is 0.15dB. The gain for the event includes the gain for the fusion splice plus the 50m of optical fiber between the cursors. Add the value for the loss for the 50m of optical fiber that was previously measured at 0.14dB to the event gain. The gain for this fusion splice is 0.29dB.

FIGURE 33.44
Measuring the gain
of a fusion splice

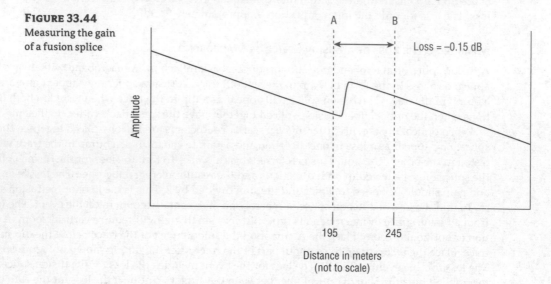

MEASURING THE LOSS OF A CABLE SEGMENT AND INTERCONNECTIONS

To find the loss for a cable segment plus the interconnections, the exact length of the segment must be known. This can be done by using the method described to measure the length of a cable segment. The length of the cable segment in this example, as shown in Figure 33.45, is 85m.

FIGURE 33.45
Measuring the
loss of a cable
segment and
interconnections

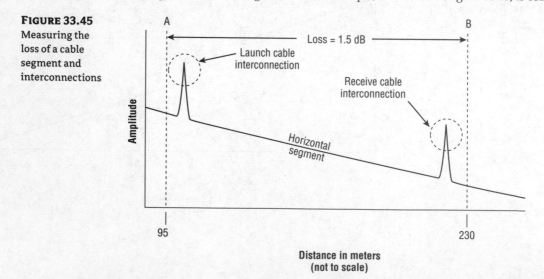

The first step is to horizontally zoom in on the cable segment. After zooming in on the cable segment, place the A cursor on a smooth section of the trace just in front of the leftmost cable segment interconnection back reflection. Place the B cursor on the smooth part of the trace after the rightmost cable segment interconnection back reflection. Looking at the A-B distance on the OTDR, move the B cursor to the right until the distance equals the length of the cable segment plus 50m, as shown in Figure 33.45. The A-B loss is 1.5dB.

To find the loss for the cable segment plus the interconnections, subtract the loss for the 50m of optical fiber from the 1.5dB loss for the cable segment and interconnections. The previously measured loss for 50m of optical fiber at 850nm is 0.14dB. The loss for the cable segment and the interconnections is 1.36dB.

Documentation of OTDR Testing

As you test the cable plant with the OTDR, you need to properly document the results of the test. The documentation should include at a minimum the following:

◆ The date of the testing

◆ The number or identification of fiber or cable tested

◆ The test procedure number

◆ OTDR make, model, serial number, and calibration dates

◆ Pulse width

◆ Range and resolution

◆ Number of averages

◆ Wavelength(s)

◆ Launch conditions

The Bottom Line

Perform optical loss measurement testing of a cable plant.   Remember ANSI/TIA-526-14-A provides three methods for testing the cable plant. The difference between the three methods is how the reference power measurement is taken. Each method uses a different number of test jumpers. Method A uses two test jumpers to obtain the reference power measurement, method B uses one test jumper, and method C uses three test jumpers.

> **Master It**   Prior to testing the cable plant using method B, you obtained your reference power measurement of –23.5dBm. Your test power measurement was –27.6dBm. What was the loss for the cable plant?

Determine the distance to a break in the optical fiber with an OTDR.   Remember a break in an optical fiber in a fiber-optic cable is not the end of the optical fiber. However, a break in an optical fiber looks like the end of the optical fiber on the OTDR display. The same method is employed to measure the distance to a break in an optical fiber and to measure the distance to the end of an optical fiber.

Master It   You are troubleshooting fiber-optic cables with the OTDR. One optical fiber appears to be broken. Refer to the following graphic and determine the distance to the break in the optical fiber.

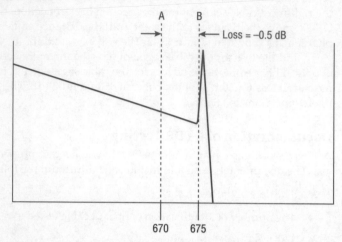

Measure the loss of a cable segment with an OTDR.   Remember that to find the loss for a cable segment plus the interconnections, the exact length of the segment must be known.

Master It   You are testing a new fiber-optic cable installation with the OTDR and the customer has requested the loss information for a cable segment including the interconnections. Refer to the following graphic and determine the cable segment.

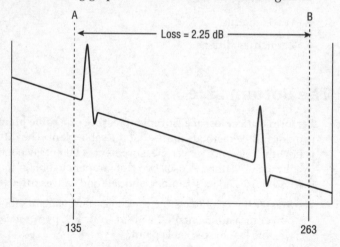

Measure the loss of an interconnection with an OTDR.   When measuring the loss for an interconnection with the OTDR, you must remember to subtract the loss for the optical fiber.

Master It   You are testing a fiber-optic cable installation with the OTDR and the customer has requested the loss information for each interconnection. Refer to the following graphic and determine the cable segment.

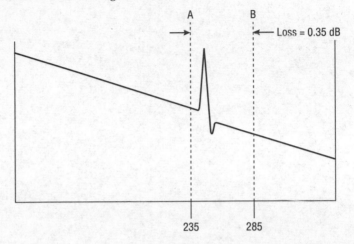

Measure the loss of a fusion splice or macrobend with an OTDR.   Remember fusion splices or macrobends do not produce a back reflection. A macrobend will always appear as a loss in the form of a dip in the smooth trace. A fusion splice may appear as a dip or a bump in the trace.

Master It   You are testing a fiber-optic cable installation with the OTDR and see a dip in the trace. Since there are no splices in this cable, you determine that this event must be a macrobend. Refer to the following graphic and determine the loss for the macrobend.

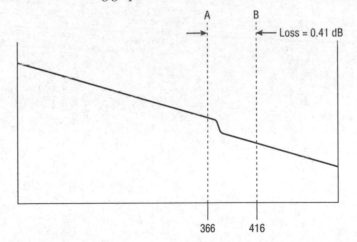

Chapter 34

Troubleshooting and Restoration

So far you have learned about fiber-optic theory, cables, connectors, splices, transmitters, receivers, and many different passive components. We have shown you how to predict the performance of a link and test that link to industry standards using various pieces of test equipment. However, we have not discussed what to do when the link doesn't work.

This chapter takes the knowledge and skills you have learned up to this point and applies them to troubleshooting a fiber-optic link or cable. You will learn basic techniques that will allow you to quickly analyze and determine the fault. You will learn fault location techniques with devices as simple as a flashlight and as complicated as an OTDR.

In this chapter, you will learn to:

◆ Identify a high-loss interconnection with an OTDR

◆ Identify a high-loss mechanical splice with an OTDR

◆ Identify a high-loss fusion splice or macrobend with an OTDR

Inspection and Evaluation

Good inspection and evaluation skills are essential for anyone attempting to troubleshoot a fiber-optic link or cable. Often the cause of a problem is basic and can be discovered with a thorough inspection. For many troubleshooting scenarios, expensive test equipment is not necessary.

This section of the chapter focuses on the visual inspection of the connector and the receptacle or mating sleeve.

Connector Inspection

The weakest link in any fiber-optic installation is likely to be the connector. When you look at a fiber-optic connector, you don't see anything exciting. Fiber-optic connectors look simple. They have very few moving parts and don't require electricity. How hard can it be to properly install a fiber-optic connector?

As you learned in Chapter 26, "Connectors," the fiber-optic connector provides a way to connect an optical fiber to a transmitter, receiver, or other fiber-optic device. The optical fiber is physically much smaller than the connector and virtually invisible to the naked eye. The naked eye can't tell the difference between a perfectly polished fiber-optic connector and a fiber-optic connector in which the optical fiber is shattered.

Fiber-optic connectors can have poor performance for the many reasons that are described in detail in Chapter 26. However, in our experience, poor surface finishes and dirt are the primary causes. We can't stress enough how important it is to make every effort possible to ensure that the connector is kept clean.

Dirt particles on the connector endface are not the only cause of poor connector performance. Dirt anywhere on the connector ferrule or inside the connector receptacle can also cause poor performance. Occasionally epoxy that runs onto the side of the connector ferrule during oven curing goes unnoticed, preventing the connector ferrule from aligning properly in the connector receptacle or mating sleeve. This can result in a high-loss interconnection.

It is a good idea to have an eye loupe like the one shown in Figure 34.1 in your tool bag. The eye loupe will allow you to spot small dirt or epoxy particles that may be on the connector ferrule. Excess epoxy needs to be removed from the connector ferrule before the connector is inserted into a receptacle. The endface of the connector should be cleaned following the procedures in Chapter 26. After the connector has been properly cleaned, it should be evaluated with an inspection microscope.

FIGURE 34.1
Eye loupe

Photo courtesy of KITCO Fiber Optics

Connector Endface Evaluation

The endface of a fiber-optic connector can only be properly evaluated with an inspection microscope. As you know from Chapter 26, many different inspection microscopes are available. Typically, a multimode connector can be evaluated with a 100X microscope, while a single-mode connector requires a minimum of 200X magnification. A 400X microscope works even better for both multimode and single-mode. Figure 34.2 shows two handheld microscopes. The smaller microscope is a 100X and the larger is a 400X.

ANSI/TIA-455-57-B provides a guideline for examination of an optical-fiber endface. Figure 34.3 contains possible core and cladding fiber endface results. Each endface in Figure 34.3 is labeled with a number and a letter. The number and letter combinations are listed in Table 34.1. This table is used to identify always acceptable, usually acceptable, and often acceptable endface results.

FIGURE 34.2
400X and 100X microscopes

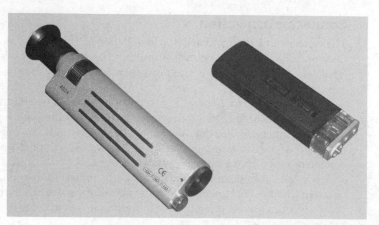

FIGURE 34.3
ANSI/TIA-455-57-B core and cladding endface results

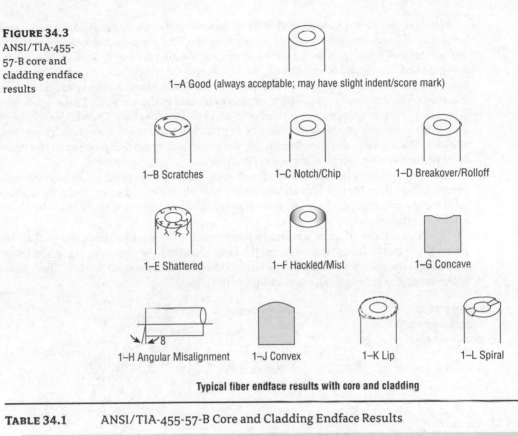

1–A Good (always acceptable; may have slight indent/score mark)

1–B Scratches   1–C Notch/Chip   1–D Breakover/Rolloff

1–E Shattered   1–F Hackled/Mist   1–G Concave

1–H Angular Misalignment   1–J Convex   1–K Lip   1–L Spiral

Typical fiber endface results with core and cladding

TABLE 34.1   ANSI/TIA-455-57-B Core and Cladding Endface Results

| | APPEARANCE ACCEPTABILITY | | |
Figure	Always Acceptable	Usually Acceptable[1]	Often Acceptable[1]
34.3	1-A	1-B, 1-C, 1-D, 1-G, 1-H[2]	1-E, 1-F, 1-J, 1-K, 1-L

1. Tighter or looser limits may be specified by the fiber-optic test procedure (FOTP).
2. Unless defect extends into the core.

TIP   Remember that a dirt particle invisible to the naked eye can damage the optical fiber in a fiber-optic connector beyond repair.

Looking at Table 34.1, you will see that one endface result is always acceptable. This is endface finish 1-A. Note that 1-A is drawn to have no imperfections, or to be "cosmetically perfect." However, the note next to the figure states that a slight indent or score mark is acceptable.

When you are evaluating the endface of a connector with a microscope, the magnification level of the scope has everything to do with the detection of imperfections. Earlier in the text, we mentioned that most multimode fiber-optic connectors could be properly evaluated with a 100X inspection microscope. Viewing a multimode endface with this magnification level will reveal only gross defects—small scratches from the polishing abrasive are typically undetectable.

The same connector viewed with a 400X inspection microscope may show many small scratches that were caused by the polishing abrasive. The endface that looked cosmetically perfect or defect-free with 100X magnification might look really ugly with 400x magnification. Imagine how that endface would look with 600X or 800X magnification. Typically, a multimode fiber-optic connector endface does not have to be cosmetically perfect to provide a low-loss interconnection. In my classes, I (Bill) demonstrate this to the students. I take a patch cord with cosmetically perfect connectors on each end and measure the loss. Then I take another patch cord with connector endfaces that have been polished with only a 1μm abrasive and measure the loss. The students are surprised when the cosmetically perfect patch cord has more loss than the less-than-cosmetically-perfect patch cord.

For most applications, a multimode connector endface does not need to be polished with an abrasive finer than 1μm. A 1μm abrasive will leave visible scratches in the endface when viewed with a 400X microscope. However, this same endface may appear scratch-free when viewed with a 100X microscope.

To evaluate the endface of a multimode connector, you need to divide the endface into three parts: the core, the inner cladding, and the outer cladding. As shown in Figure 34.4, the outer cladding is the area from the center of the cladding to the connector ferrule. The inner cladding is the area from the center of the cladding to the core.

FIGURE 34.4
Sections of a multi-mode connector endface

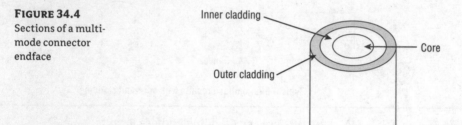

Field-terminated multimode connectors are typically polished with an aluminum oxide abrasive. As you learned in Chapter 26, aluminum oxide is harder than the optical fiber but softer than the ceramic ferrule. Because the aluminum oxide does not polish down the ceramic ferrule during the polishing process, chipping around the epoxy ring extends into the outer cladding. Sometimes this chipping may extend in excess of 10μm from the epoxy ring. This is typical, which means that this is acceptable.

TIP A good rule of thumb when evaluating a multimode connector is to accept all endface finishes that have no defects in the core and inner cladding. The connector endface should be defect-free when viewed with a 100X microscope. Minor scratches and pitting are acceptable when viewed with a 200X microscope or a 400X microscope.

Figure 34.5 shows a multimode endface polished with a 1μm aluminum oxide abrasive viewed with a 400X video microscope. Note the minor chipping along the epoxy ring and the minor scratches on the inner cladding and core. This is an acceptable multimode endface. This connector endface viewed with a 100X microscope would look cosmetically perfect.

FIGURE 34.5
Multimode con-
nector endface pol-
ished with a 1μm
aluminum oxide
abrasive viewed
with a 400X
microscope

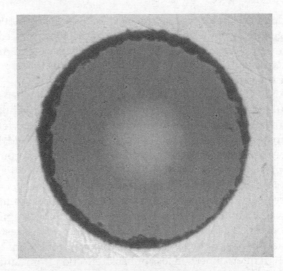

Single-mode connector evaluation requires 400X magnification. The endface of a single-mode connector should look cosmetically perfect like the endface shown in Figure 34.6. Remember from Chapter 26 that a single-mode connector endface is typically finished with a diamond abrasive. A diamond abrasive polishes down the ferrule and the glass.

FIGURE 34.6
Single-mode con-
nector endface
with no defects
viewed with a
400X microscope

Photo courtesy of Mic oCare

A single-mode endface polished with diamond abrasive should have a very thin epoxy ring in comparison to an endface polished with aluminum oxide. There should be no pitting or chipping of the optical fiber along the epoxy ring; however, the epoxy ring may appear to have pitting or chipping. The inner cladding, outer cladding, and core should be defect-free when viewed with a 400X microscope.

NOTE A multimode connector endface that has been polished with a diamond abrasive should not have any pitting or chipping on the optical fiber along the epoxy ring.

In some applications, single-mode terminations are performed in the field with an aluminum oxide abrasive. These terminations may have endface finishes similar to a field-terminated multimode endface. The endface should be evaluated as you would evaluate a multimode endface.

Up to this point, we have focused only on the cosmetics of the endface. We have not covered the geometry of the endface. When a connector is viewed with a microscope, you see only a two-dimensional view—you can't see depth or height. As you learned in Chapter 26, virtually all connector ferrules have a radius. This means that a properly polished optical fiber should have the same radius as the ferrule.

ANSI/TIA-455-57-B does not provide guidance when it comes to measuring or evaluating ferrule or endface radius. Ferrule or endface radius cannot be measured with a microscope; it can only be measured with an interferometer. An interferometer is an expensive piece of test equipment that is not typically carried out into the field. However, some portable video microscopes are now incorporating interferometers. Interferometers are covered in greater detail in Chapter 26.

In the early stages of troubleshooting, a visual inspection of the connector endface is all that is necessary. If the connector endface meets the requirements established by ANSI/TIA-455-57-B you can go to the next step, which would be to check optical-fiber continuity. You don't need to be concerned about connector endface radius, unless testing with the OTDR reveals a high interconnection loss or high interconnection back reflection.

High interconnection loss or back reflection from a mated connector pair with acceptable endface finishes is usually caused by a change in the refractive index. This may be the result of an air gap or contamination. An air gap has several causes:

♦ Dirt that is not allowing the ferrule to properly seat in the receptacle

♦ The connector not being inserted properly into the receptacle

♦ One or both connectors having a concave or flat as opposed to convex endface finish

Contamination is often caused from handling. If you touch the connector endface, oil from your skin will be deposited on the endface. The oil on the endface of the connector does not disappear when the connector is mated with another connector. It creates a thin film between the two optical fibers. This thin oil film, as shown in Figure 34.7, has a different refractive index than the optical fiber, which causes a Fresnel reflection.

FIGURE 34.7
Single-mode con-
nector endface
with skin oil
viewed with a
200X microscope

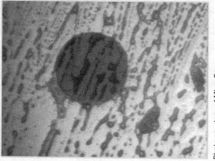

Photo courtesy of MicroCare

Receptacle/Mating Sleeve Inspection and Cleaning

As discussed in Chapter 27, "Fiber-Optic Light Sources," and Chapter 28, "Fiber-Optic Detectors and Receivers," many fiber-optic transceivers feature receptacles that mate with the connectors. Sometimes the problem is as simple as a connector not being latched or fully inserted into the receptacle. The transceiver shown in Figure 34.8 has ST connector receptacles. If you look closely at the photograph, you will see that connector on the right has not been fully inserted and latched. However, the connector on the left has.

FIGURE 34.8
The ST connector on the right is not fully mated with the receptacle.

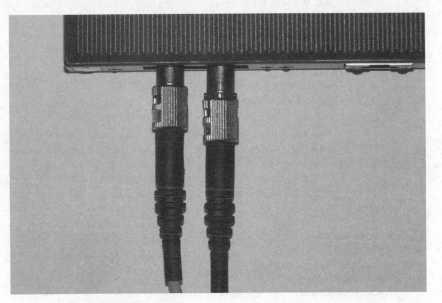

If the connector fails to properly mate with the transmitter or the receiver, problems may occur. The same is true at the mating sleeve when two connectors do not mate properly. Figure 34.9 shows two SC connectors in a mating sleeve. The connector on the right has not been fully inserted into the mating sleeve. This results in an air gap that produces a Fresnel reflection.

FIGURE 34.9
SC-to-SC mating sleeve with connector on right not fully mated

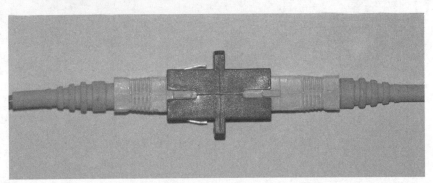

Receptacles and mating sleeves get dirty. If a dirty connector is plugged into a transceiver receptacle, the dirt from the endface of the connector will transfer to the mating surface inside the receptacle. This could cause a problem and affect the operation of the transmitter or receiver. It is difficult to examine the mating surface inside a receptacle. If the endface of the connector you just removed was dirty, the receptacle should be cleaned.

There are different cleaning products available to clean the contact area inside the receptacle. These same products can also be used to clean the endface of a connector in a mating sleeve. In this section we will examine two different approaches to cleaning these hard-to-reach areas.

CLEANING AN ENDFACE WITH A CLEANING STICK

The connector cleaning stick is a simple and inexpensive tool that can be used to clean the endface of a connector in a mating sleeve or the contact area in a receptacle. When using a cleaning stick, you should select a stick that has a cleaning tip the same diameter as the connector ferrule the receptacle or mating sleeve was designed to accept. Ferrule diameters are listed in Chapter 26.

Without touching the tip, lightly moisten the tip of the cleaning stick with an optical quality cleaning fluid like the fluid shown in Figure 34.10. Unlike isopropyl alcohol, an optical quality cleaning fluid will not leave a residue. Insert the tip in the receptacle or mating sleeve until it makes contact with the mating surface or connector endface, as shown in Figure 34.11. Rotate the cleaning stick in a clockwise direction 10 revolutions, applying varying pressure to create a gentle pumping action. Remove and dispose of the stick; never use a cleaning stick twice.

FIGURE 34.10
Optical quality cleaning fluid engineered for cleaning connector endfaces

Photo courtesy of MicroCare

FIGURE 34.11
Using a cleaning
stick to clean
the contact area
in a connector
receptacle

CLEANING AN ENDFACE WITH A TAPE TYPE CLEANER

A tape type cleaner is also a simple and inexpensive cleaning tool that can reach the contact area in a receptacle or the endface to a connector in a mating sleeve. Unlike the cleaning stick, the tape type cleaner is used without a cleaning fluid. Multiple manufacturers offer these cleaners; some are refillable and some are disposable.

The tape type cleaner uses a small cleaning cloth wrapped around a spool stored in the cleaner. The cleaning cloth is exposed at the tip of the cleaner, as shown in Figure 34.12. To clean a connector endface or contact area inside a receptacle, insert the tip of the cleaner, as shown in Figure 34.13. When using this cleaner, gently push the handle toward the contact surface. This will cause the cleaning cloth to move across the endface of the connector. Stop pushing when you hear a click. This type of cleaner is typically good for 500 uses before it needs to be replaced.

FIGURE 34.12
Cleaning cloth
exposed at the tip
of the cleaner

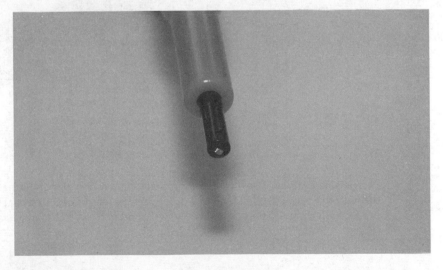

FIGURE 34.13
Cleaner tip
inserted in ST-to-
SC mating sleeve

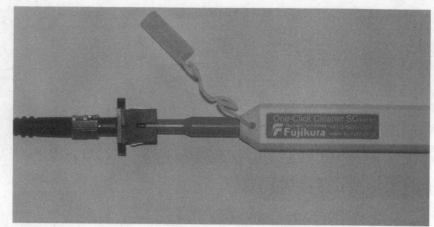

Continuity Tester Fault Location Techniques

As you learned in Chapter 33, "Test Equipment and Link/Cable Testing," the continuity tester is really no more than a flashlight. There are many different continuity testers on the market. Some use red LED light sources; others use incandescent lights. If you don't have a continuity tester, you can just use a flashlight. Figure 34.14 shows a rugged incandescent continuity tester. As you can see, this is a modified flashlight.

FIGURE 34.14
Rugged and focus-
able incandescent
continuity tester.

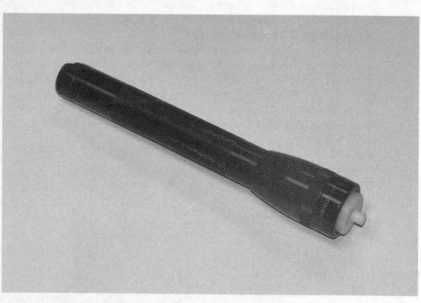

Photo courtesy of KITCO Fiber Optics

The key to using a flashlight is alignment. Remember, only light that enters an optical fiber through the cone of acceptance will propagate the length of the optical fiber. You may have to practice aligning the flashlight on a known good patch cord before testing a longer fiber-optic cable. It is much easier to use a flashlight to test multimode optical fiber than single-mode optical fiber.

The only drawback to using a flashlight, besides manual alignment, is the fact that you have to hold the connector and flashlight together. This could be a real challenge when you are working by yourself and the other end of the fiber-optic cable is in another building. If that is the case, you may have to ask your boss to break down and spend the $30 on a continuity tester.

As you learned in Chapter 33, the optical output power of an LED or incandescent continuity tester is relatively low when compared to an LED or laser transmitter. Most of the LED continuity testers I have examined output somewhere between –30dBm and –40dBm when measured at the end of one meter of 62.5/125μm optical fiber. The output of a focusable incandescent continuity tester when measured at the end of one meter of 62.5/125μm optical fiber can exceed –30dBm.

The rugged continuity tester shown in Figure 34.13 uses a mirror to focus the light. If focused properly, this continuity tester will provide an output greater than –25dBm when measured at the end of one meter of 62.5/125μm optical fiber. Regardless of the light source, all continuity testers should be made eye safe so that you can directly view the connector endface emitting the light energy.

So what can you do with a continuity tester? Basically, you can check the continuity of an optical fiber—that is, check to see if an optical fiber allows light to pass from one end to the other. A continuity tester tests only for breaks in the optical fiber. Macrobends or microbends can't be identified with a continuity tester.

Let's work through an example and take a step-by-step approach to troubleshooting a fiber-optic link with a continuity tester. As you already know and as has been stated in this book many times, the first step in troubleshooting a fiber-optic link is to clean the connector endface. If you have forgotten how to clean a connector endface, refer to Chapter 26.

With a clean connector in hand, the next step is to examine the connector endface with an inspection microscope. As we discussed earlier in this chapter, a 100X microscope is typically adequate for multimode applications. Single-mode applications require a 400X microscope. If your examination of the endface of each connector on opposite ends of the fiber-optic cable shows no defects, you can go to the next step. However, if your examination reveals a defect, there is no sense in testing continuity until the connector is repaired or replaced.

At this point, the connectors at each end of the fiber-optic cable are clean and you are ready to test cable continuity. Remember, before attaching the connector to the continuity tester, verify that the continuity tester works. To verify continuity tester operation, turn it on and see if it emits light. This seems like beginner-level stuff, but how many times have you heard the story about the technician who shows up to fix a computer only to find that it's simply not plugged in?

CAN A FLASHLIGHT KEEP YOUR COMPANY FROM GOING OUT OF BUSINESS?

Over the years, many students have shared stories about their experiences in the field. The story that really applies to this chapter is about a small company that made its livelihood doing aerial copper cable installations.

One day, this company took a job doing an aerial installation of a multifiber cable. Because this company had never installed a fiber-optic cable, they handled it like a copper cable. After several days of hard work, the cable was in place and the installation company called a fiber-optic contractor to complete the installation.

The first thing the fiber-optic contractor did was test the cable with the OTDR. The first optical fiber tested was broken several hundred meters from the OTDR. The next optical fiber tested was also broken. Testing from the opposite end of the cable revealed at least two breaks in each optical fiber tested.

This small company had to install a new fiber-optic cable and absorb the cost of the cable and the labor. Needless to say, this small company almost went out of business because of that. Eventually they determined that the fiber-optic cable had minor damage from what could have been a forklift.

The company never wanted to make this mistake again, so they sent their sharpest employee to our fiber-optic installer course. You can imagine the look on this person's face when we demonstrated how to use a continuity tester to look for a break in an optical fiber.

The moral of the story? Always test your fiber-optic cable before installation. If your company doesn't have an OTDR, use a continuity tester. If you don't have a continuity tester, use a flashlight.

Now that you know the continuity tester works, plug in or attach a connector at one end of the fiber-optic cable. With the continuity tester on, observe the connector at the other end of the fiber-optic cable. If the connector at the other end of the fiber-optic cable is not emitting light, the optical fiber in the fiber-optic cable is broken. If the connector at the other end of the fiber-optic cable is emitting light, you have continuity and know that the optical fiber in not broken.

Figure 34.15 shows a portion of a patch panel with two ST type connector receptacles. Each connector receptacle is populated on the back side. In other words, a fiber-optic cable is connectorized and plugged into the back side of each receptacle. On the other end of each fiber-optic cable, a continuity tester is attached and turned on. Receptacle A shows good continuity. No light is emitted from receptacle B, indicating a broken optical fiber.

FIGURE 34.15
Continuity test
of two fiber-optic
cables

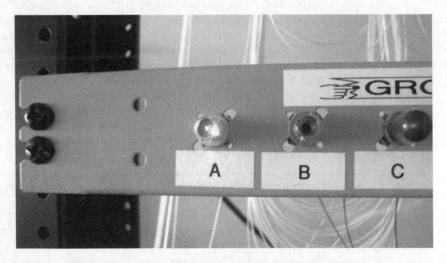

So far you have learned how to test the continuity of the optical fiber in a fiber-optic cable. Now let's look at the limitations of the continuity tester. Receptacle A is emitting light, so you know that there is no break in the optical fiber. Unfortunately, that is all you can tell about the optical fiber in this cable.

Our eyes respond to light in a *logarithmic* manner, which means that we can't detect small changes in optical power. A change of 0.5dB or 1.0dB, for example, is not detectable. This means that we can't look at the light being emitted and tell if there is a macrobend or microbend in the optical fiber. The continuity tester is just that, a continuity tester. It can only tell you that the optical fiber is not broken.

Another drawback to the continuity tester is the fact that it uses visible light. You may remember from Chapter 22, "Optical Fiber Characteristics," that visible light is a poor choice for fiber-optic transmission because of attenuation. The high attenuation of visible light in an optical fiber limits the use of the continuity tester to cables no longer than 2km.

Remember that the continuity tester is basically a flashlight with no lens system to direct the light energy into the core of the optical fiber. Back in Chapter 27, we explained that a 62.5μm core accepts more light energy than a 50μm core when plugged into the same transmitter. This means that the continuity tester will not direct much light energy into the core of a single-mode optical fiber.

The continuity tester can be used with single-mode optical fiber. However, the light emitted from the connector at the opposite end of the cable will be dim in comparison to the multimode optical fiber. This is one application where the performance of the multimode optical fiber literally outshines the performance of the single-mode optical fiber.

When you are using the continuity tester with single-mode optical fiber, you may want to test a short cable to see how well it works before testing a longer cable. Dimming the lights can help you see the light emitted from the connector endface. If you cannot dim the lights, shroud the connector with your hand as shown in Figure 34.16 to block ambient light.

FIGURE 34.16
Shroud the connector with your hands to reduce ambient light during a continuity test.

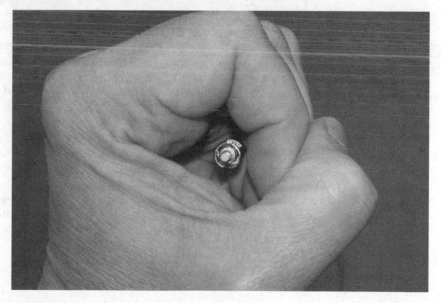

Visual Fault Locator

The continuity tester is a valuable piece of test equipment. However, due to its low output power, it is limited to just testing the continuity of the optical fiber. The visual fault locator (VFL) is similar to the continuity tester; however, it uses a visible laser source instead of an LED or incandescent lamp. The powerful laser in the VFL allows you to visibly identify breaks or macrobends in the optical fiber.

VFLs come in all shapes and sizes. Some are the size and shape of a large pen, like the one shown in Figure 34.17, while others may have the size and shape of a digital multimeter like the one shown in Figure 34.18. Regardless of the design, all of them perform the same function. VFLs are also incorporated into some OTDRs. Incorporating the VFL into the OTDR reduces the number of pieces of test equipment that you need to carry to the job site.

The laser light sources used in most VFLs typically output 1mW of optical energy somewhere in the 650nm range. The output of the VFL is roughly 1,000 to 10,000 times greater than the output of an LED continuity tester. This means that you should never look directly at the endface of a connector emitting light from the VFL. The VFL is an eye hazard, and proper safety precautions need to be taken when operating the VFL. You can refer to Chapter 23, "Safety," to find out detailed information on light source classification and safety.

FIGURE 34.17
Pen-type VFL

Photo courtesy of KITCO Fiber Optics

Now let's take a look at how to employ the VFL to locate faults in a fiber-optic cable. As with the continuity tester, the first thing you will need to do is clean the connector endface and inspect it with a microscope. If the endface finish is acceptable, the VFL can be connected to a connector at one end of the fiber-optic cable. The connector at the other end of the fiber-optic cable should not be viewed directly during this testing.

The VFL is designed to fill the core of an optical fiber with visible light. Depending on the cable type, the VFL may illuminate faults in an optical fiber through the buffer, strength member, and even jacket. However, the VFL performs best with tight-buffered optical fiber.

Figure 34.19 shows a broken tight-buffered optical fiber. You can see in the photograph how the VFL illuminates the break. The light energy from the VFL penetrates the buffer, allowing the person performing the test to quickly identify exactly where the fault is located.

FIGURE 34.18
Multimeter-
style VFL

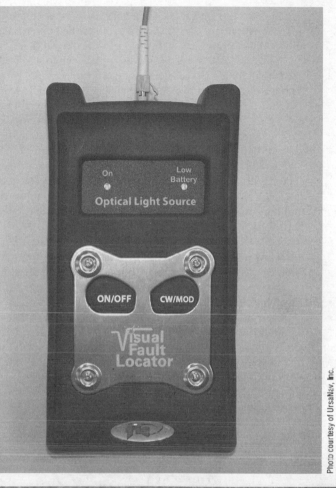

Photo courtesy of UrsaNav, Inc.

FIGURE 34.19
Broken tight-buff-
ered optical fiber

The VFL and the OTDR work hand in hand with each other when it comes to locating breaks in an optical fiber. The OTDR can provide the operator with the distance to the break. The VFL allows the operator to see the break in the optical fiber.

Fiber-optic cables are not the only place where the optical fiber may break. The optical fiber may break inside the connector or connector ferrule. Unless the optical fiber is broken at the endface of the connector, it is not visible with a microscope.

Often, students connect cables that look great when viewed with the microscope but fail continuity testing. When this happens, the hardest part is determining which connector contains the break in the optical fiber. Without a VFL in the classroom, students would have to cut the cable in half and use the continuity tester to identify the bad connection.

The VFL will often identify the bad termination or connector. Figure 34.20 shows an ST connector with a broken optical fiber in the ferrule of the connector. Looking at the photograph, you can see VFL illuminating the break in the optical fiber. The output of the VFL is so powerful that it penetrates the ceramic ferrule. Figure 34.21 shows an ST connector without a break in the optical fiber.

The VFL can also be used to locate a macrobend in an optical fiber. However, macrobends do not allow nearly as much light to penetrate the buffer and jacket as does a break in the optical fiber. Locating a macrobend with the VFL may require darkening the room. Figure 34.22 is a photograph of a severe macrobend in a tight-buffered optical fiber. You can see from the photograph how the VFL illuminates the macrobend.

Macrobends and high-loss fusion splices appear the same on an OTDR trace. The VFL allows the identification of a high-loss fusion splice. Figure 34.23 shows three fusion splices and three mechanical splices inside a splice enclosure. The splice illuminated by the VFL is a high-loss fusion splice. Believe it or not, the loss from this splice was not great enough to impact system performance. The loss for this splice was only 0.6dB.

FIGURE 34.20
Broken optical fiber in ST connector ferrule being tested with VFL

FIGURE 34.21
Good ST connector
being tested
with VFL

FIGURE 34.22
Macrobend in a
tight-buffered
optical fiber

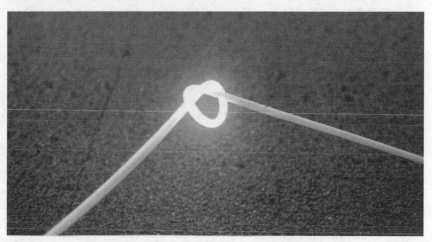

FIGURE 34.23
High-loss fusion
splice being tested
with VFL

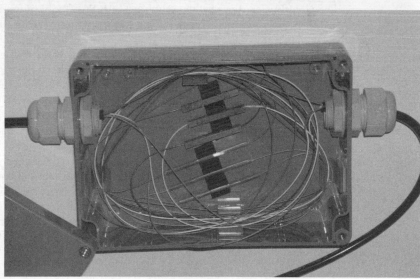

Fiber Identifier

The fiber identifier is a piece of test equipment that allows you to see through the jacket, strength member, and buffer of the fiber-optic cable. As you learned in Chapter 33, "Test Equipment and Link/Cable Testing," the fiber identifier is designed to place a macrobend in the fiber-optic cable under test.

Photodiodes in the fiber identifier detect light penetrating through the fiber-optic cable. The electronics in the fiber identifier measure the detected light energy and display the direction of light travel through the optical fiber.

The fiber identifier is used very much like the VFL when it comes to troubleshooting. One key difference is that the fiber identifier replaces your eyes. Another difference is that fiber-optic cable under test typically does not have to be disconnected from an active circuit—it can remain plugged into the transmitter and receiver. The infrared light traveling through the optical fiber during normal operation is often enough to perform most tests. However, sometimes an additional infrared light source is required to adequately troubleshoot.

Up to this point, we have only discussed troubleshooting with test equipment that emits visible light. You may have noticed that we keep mentioning that the fiber identifier works with infrared light. We have not mentioned using visible light with the fiber identifier. Visible light can be used with the fiber identifier; however, most fiber identifiers perform best with longer wavelength (800–1700nm) light sources.

Figure 34.24 shows a fiber identifier clamped around a tight-buffered optical fiber. This tight-buffered optical fiber is connected between a 1310nm laser transmitter and receiver. You can see in the photograph that the arrow is pointing to the direction the light is traveling through the optical fiber. This is an example of a fiber identifier being used to detect traffic on an optical fiber.

The fiber identifier can also be used with an external light source. Often the external light source is an OTDR. Many OTDR manufacturers build or program in a pulsed output function. When set for a pulsed output, the OTDR emits a continuous pulse train at a predetermined frequency. The electronics in the fiber identifier can detect preset frequencies and illuminate the corresponding LED. This feature can be very helpful when you are trying to identify an unmarked tight-buffered optical fiber within a bundle of tight-buffered optical fibers. This feature can also be helpful when you are trying to approximate the location of a break in the optical fiber.

The fiber identifier can be used with the OTDR to narrow down the location of a break in an optical fiber when a VFL is not available or when the light from the VFL is not visible through the jacket of the fiber-optic cable. If the index of refraction is correct, the OTDR should provide an accurate distance to the fault. Remember from Chapter 33 that the OTDR measures the length of optical fiber to the fault, not the length of fiber-optic cable. The cable length may be shorter than the optical fiber length. This is especially true if loose buffer cable exists in the fiber-optic link.

Once you have found the approximate location of the fault with the OTDR, set the OTDR or infrared light source to pulse at a predetermined frequency. Clamp the fiber identifier on the faulted fiber-optic cable several meters before the approximate location of the fault. Check the fiber identifier for the predetermined frequency. If the fiber identifier does not detect the predetermined frequency, move the fiber identifier several meters closer to the OTDR or infrared light source and recheck for the predetermined pulse. If you have selected the correct fiber-optic cable to test and you're confident about the distance to the fault, you should detect the predetermined frequency. If you still do not detect the frequency, double-check everything and retest. If you still do not detect the predetermined frequency, there may not be enough optical energy for the fiber identifier to function properly.

FIGURE 34.24
Fiber identifier

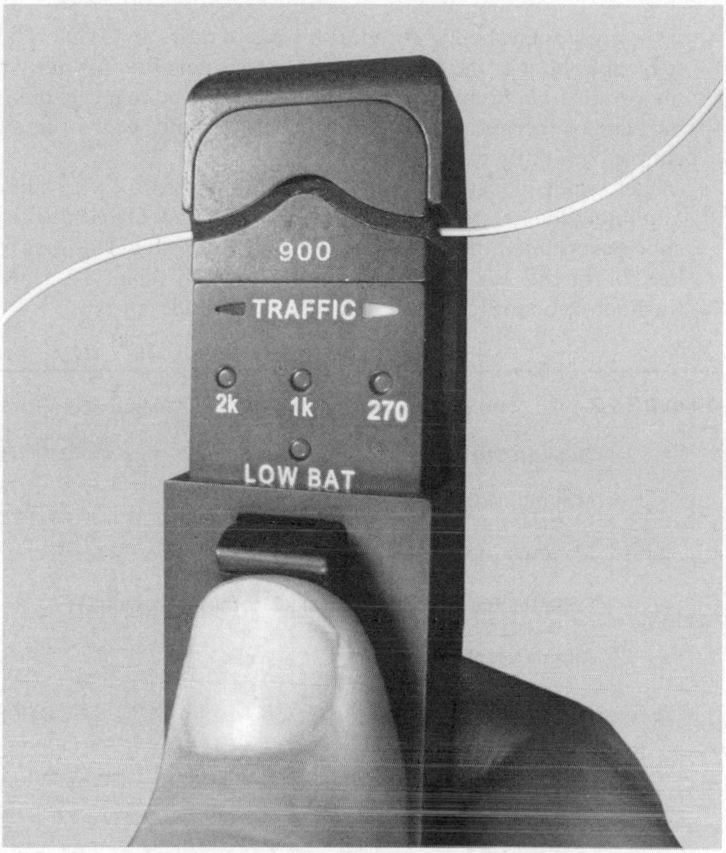

If you are able to detect the predetermined frequency, move the fiber identifier down the fiber-optic cable away from the OTDR or infrared light source in one-meter increments. Continue to do this until the fiber identifier no longer detects the predetermined pulse. You now know within one meter where the break in the optical fiber is located. At this point, you may want to disconnect the OTDR or infrared light source and connect the visible fault locator. The visible fault locator may illuminate the exact location of the fault. If the visible fault locator does not illuminate and conditions permit, darken the area around the fault. This may allow you to see the illuminated fault.

OTDR Fault Location Techniques

This section will show you how to use the OTDR to troubleshoot or locate faults. Fiber-optic links exist in virtually every place imaginable. This section does not attempt to address every possible scenario; rather, it discusses how to look for and find typical faults in a fiber-optic link.

NOTE   This section is not designed to teach you how to operate the OTDR or use an OTDR to test a fiber-optic link or cable. OTDR operation and OTDR fiber-optic link and cable testing are covered in Chapter 33.

Many troubleshooting scenarios never require an OTDR. The OTDR may be used exclusively to troubleshoot or the OTDR may be brought in after it is determined that there is a fault in a fiber-optic link. Remember that the OTDR is used to locate the fault in a fiber-optic cable or to evaluate connector or splice performance. A faulty fiber-optic cable can be identified with a simple continuity test.

Let's say you just finished testing a fiber-optic link with a fiber-optic source and power meter. You know the length of the link and you have calculated the loss for the link, connectors, and splice as outlined in Chapter 32, "Fiber-Optic System Design Considerations." Your calculated loss for the link is 2.5dB @ 1300nm, as shown in Table 34.2. However, your measured loss with the fiber-optic source and power meter is 3.75dB.

TABLE 34.2   Completed Multimode ANSI/TIA-568-C.3 Worst-Case Loss Calculation

#	DESCRIPTION	VALUE
1	Minimum EOL optical output power.	
2	Maximum optical input power.	
3	Subtract line 2 from line 1 to calculate the power budget.	
4	Kilometers of optical fiber.	0.467
5	Number of interconnections.	2.0
6	Number of splices.	1.0
7	Multiply line 4 × 3.5.	1.64
8	Multiply line 4 × 1.5.	0.7
9	Multiply line 5 × 0.75.	1.5
10	Multiply line 6 × 0.3.	0.3
11	Add lines 7, 9, and 10 for total link loss at 850nm.	3.44 dB
12	Add lines 8, 9, and 10 for total link loss at 1300nm.	2.5 dB
13	Subtract line 11 from line 3 for headroom at 850nm.	
14	Subtract line 12 from line 3 for headroom at 1300nm.	

Remember from Chapter 32 that there are typically three different methodologies that can be used to develop a loss budget or power budget: the worst-case method, statistical method, and numerical method. The worst-case method is commonly viewed as the most conservative engineering approach. This method uses the worst-case maximum value loss for each passive component. This is the method we are going to use to calculate the loss for the link under test.

The minimum transmitter output power and minimum receiver sensitivity are not required to calculate the worst-case loss for the link. These values are not used because we are concerned only with the losses from the optical fiber, connectors, and or splices. The worst-case loss should be calculated at both wavelengths and the link should be tested at both wavelengths, as described in Chapter 33.

As mentioned several times in this chapter, the first thing you should do is clean and perform a visual inspection of the connectors. You should also clean and inspect the receptacle that each connector is plugged into. After cleaning, retest with the fiber-optic source and power meter. If the cleaning does not improve link performance to the point where it meets or exceeds the worst-case calculated loss, it's time to start troubleshooting.

You learned earlier in this chapter that a cosmetically perfect connector may not necessarily result in a low-loss connection. A connector with a great endface finish can have a distorted geometry, which could increase the loss in an interconnection. Remember that you can't see height or depth with a two-dimensional microscope. But with the OTDR, you can quickly measure the loss for a mated connector pair and look at the light energy reflected back toward the OTDR.

Chapter 33 taught you how to test a fiber-optic link with the OTDR. You probably remember that launch and receive cables must be attached to both ends of the fiber-optic link under test. Ensure that these cables are long enough for your application, as described in Chapter 33.

With the launch and receive cables attached, test the fiber-optic link as described in Chapter 33. After the link has been tested, the high-loss component or components need to be identified. The measured loss for this link with the fiber-optic source and power meter was 1.25dB greater than the worst-case calculated value. Remember that the calculated value is the worst-case scenario. A well-constructed fiber-optic link should test well under the calculated value.

The next step is to evaluate the OTDR trace. There are many possible causes for the loss. The only thing that testing has revealed so far is that there are no breaks in the optical fiber in the link. If there was a break, the loss for the link when testing with the light source and power meter would typically be in excess of 30dB.

You typically approach any troubleshooting scenario looking for a single fault. However, at times you will discover more than one fault. This troubleshooting scenario is going to be approached as if there were a single fault.

Based on the light source and power meter testing, this link suffers from higher than acceptable loss. This loss could be caused by an interconnection, splice, or macrobend. It's the OTDR operator's job to be able to identify which of these three possible causes is the problem.

In Chapter 33, you learned how to measure the loss for interconnections and splices using the OTDR. To find the loss for the interconnection or splice, you had to subtract the loss for the length of optical fiber between cursors A and B. In this section, you will learn to look at the OTDR trace and quickly notice an event. This event could be a high-loss interconnection, splice, or macrobend.

When an event occurs that involves significant loss, there is no need to bring the cursors to the event and measure the exact loss. The vertical scale on the OTDR is there to help the operator visually approximate the amount of loss. A good OTDR operator can detect an event in the trace and use the vertical scale to quickly determine if that event exceeds the loss limitations defined in ANSI/TIA-568-C.3.

The operator has to learn to visualize the trace beyond the event. In other words, the operator has to know what the trace would look like if the event had very low loss versus the high loss shown on the OTDR display. Doing this will allow the operator to quickly identify high-loss events without having to move the cursors.

Let's say that a bad interconnection is the problem. A bad interconnection would look like Figure 34.25 on the OTDR. This bad interconnection can be quickly identified by following the sloping line as it enters and exits the interconnection back reflection. Visualize the sloping line traveling through the back reflection and exiting like the dashed line in Figure 34.25. The dashed line represents how a low loss interconnection would appear.

FIGURE 34.25
Bad interconnection

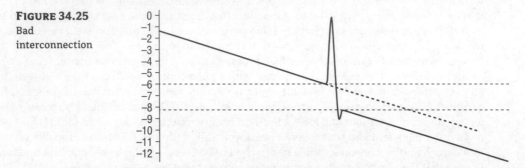

Remember from Chapter 33 that the OTDR display has a vertical and horizontal axis. The vertical axis shows relative power in decibels. Look at the amount of energy entering the interconnection and exiting the interconnection. Use the vertical scale on the left to approximate the loss without having to move the A and B cursors. This interconnection has a loss of approximately 2.25dB, as shown by the horizontal lines extending from the vertical axis in Figure 34.25.

Remember that the maximum allowable loss for an interconnection, according to ANSI/TIA-568-C.3, is 0.75dB. After this interconnection is repaired, the measured loss for the link should improve by no less than 1.5dB (2.25dB − 0.75dB = 1.5dB).

The next possible problem is a bad splice. This link contains one mechanical splice and does not contain a fusion splice. A poor fusion splice looks very similar to a macrobend and will be covered in the next paragraph. A bad mechanical splice looks just like a bad interconnection. The bad mechanical splice trace is shown in Figure 34.26. You should immediately notice that there is virtually no difference between the bad interconnection trace and this trace. The only difference is the amount of loss. When you look at the energy entering the splice and the energy leaving the splice, you'll see that a loss of 1.75dB is realized. This loss is 1.45dB greater than the maximum amount allowable by ANSI/TIA-568-C.3.

FIGURE 34.26
Bad mechanical splice

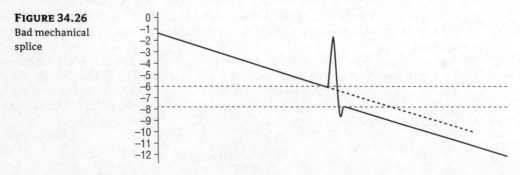

The last possible problem with this fiber-optic link is a macrobend. Although 1.25dB or greater is a significant loss and not a very typical macrobend loss, this amount of loss is possible from a macrobend. Macrobends and bad fusion splices look the same on the OTDR. Neither the fusion splice nor the macrobend produces a back reflection. Remember that back reflections are produced only from mechanical interconnections or breaks in the optical fiber. As with the previous two examples, look at the energy entering the macrobend and the energy exiting the macrobend, as shown in Figure 34.27. The loss for this macrobend is 1.8dB. When this macrobend is repaired, the measured loss for the fiber-optic link should decrease by 1.8dB.

FIGURE 34.27
Macrobend

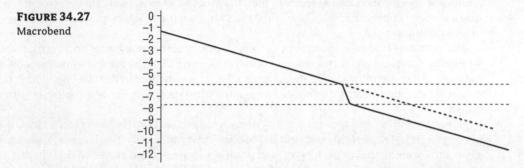

We have not yet addressed using the OTDR to find a break in the optical fiber. A break in an optical fiber can be detected with a continuity tester or visible fault locator. However, the OTDR can indicate the exact location of the break. You find the break in an optical fiber with the OTDR just as you find the length of an optical fiber. Chapter 33 describes in detail how to measure the length of an optical fiber.

Restoration Practices

So far in this chapter, you have learned about some of the tools available to troubleshoot a fiber-optic link. In Chapter 27 and Chapter 28, you learned how to understand the operating specifications of a fiber-optic transmitter and receiver. This section will outline a logical approach to troubleshooting a single fault to restore a fiber-optic link that has stopped working. We will proceed under the assumption that you have access to all of the tools described in this chapter.

The first step in any restoration is to ask the customer three questions:

◆ What do they believe the problem is?

◆ When do they believe the problem first occurred?

◆ What was the last thing done to the system?

The best technicians know that many problems are caused by operator error or customer error. For example, the problem may have been caused by the customer changing the connections at the patch panel and making an improper connection. Or the customer might have just rearranged the furniture in their office and damaged a fiber-optic cable in the process.

After speaking with the customer, you should begin to assess the situation and attempt to identify the fault. Fiber-optic cables do not just break, splices do not just go bad, and interconnections do not just fail unless someone has handled them. However, electronics do go bad and electronics do fail. Once you have identified the fiber-optic link that has the problem, the next step is to make sure that all cables are connected properly.

If all the cables are connected properly, the next step is to check to see if the electronics are functioning properly. Most fiber-optic transmitters output light energy as soon as they are powered up. In other words, the transmitter will output a series of pulses, typically at a constant frequency, when power is applied. The first step is to look for this light energy at the receiver end of the fiber-optic link and measure it. The measured energy should be within the input specification range for the receiver, which is covered in detail in Chapter 28. If the measured energy is within the specification for the receiver, then the problem is most likely the receiver. If the measured energy is not within the range of the receiver, then the problem may be in the fiber-optic link or the transmitter.

The next step is to measure the optical output power of the transmitter and verify that it is within the specification, as discussed in detail in Chapter 27. If the measured output power of the transmitter is within specification, the problem is somewhere in the fiber-optic link. If the measured output power of the transmitter is not within specification, the problem is with the transmitter.

If the transmitter and receiver are functioning to specification, the next place to look is the fiber-optic link. The first things to test in the fiber-optic link are the patch cords. You should clean and examine the connectors on each patch cord as described earlier in this chapter, and test the patch cords for loss as described in Chapter 33. You should also clean and inspect the connectors on the back side of the patch panel. Faulty patch cords should be replaced and bad connections on the main cable span should be repaired. If any repairs are made, the system should be retested.

After you inspect the patch cords and connectors, if no faults have been located, the next step is to test the entire cable span. Fiber-optic links should be well documented, though that is not always the case. The best scenario would be having that documentation for a comparison to your testing. However, if the customer can't provide documentation, you should still be able to analyze the link and locate faults.

The next step is to connect the OTDR as outlined in Chapter 33 and test the cable plant. If documentation exists, compare the OTDR traces and repair the faulty cable or splice. If documentation does not exist, evaluate the OTDR trace as outlined in Chapter 33 and earlier in this chapter and repair the faulty cable or splice. You may need to employ the VFL or fiber identifier to help you locate the exact physical location of the fault.

The Bottom Line

Identify a high-loss interconnection with an OTDR.   Remember a bad interconnection can be quickly identified by following the sloping line as it enters and exits the interconnection back reflection. Visualize the sloping line traveling through the back reflection and exiting.

Master It   You are troubleshooting a new fiber-optic cable installation with the OTDR and the trace shown in the following graphic. The back reflection is the result of an interconnection. Using only the relative decibel information on the vertical scale, approximate the loss for the interconnection and determine whether this interconnection meets the loss requirements defined by ANSI/TIA-568-C.3.

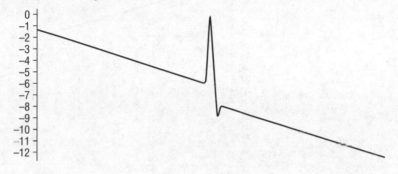

Identify a high-loss mechanical splice with an OTDR.   Remember a high-loss mechanical splice can be quickly identified by following the sloping line as it enters and exits the splice back reflection. Visualize the sloping line traveling through the back reflection and exiting.

Master It   You are troubleshooting an older fiber-optic cable installation with the OTDR and the trace shown in the following graphic. The back reflection is the result of a mechanical splice. Using only the relative decibel information on the vertical scale, approximate the loss for the splice and determine whether or not this splice meets the loss requirements defined by ANSI/TIA-568-C.3.

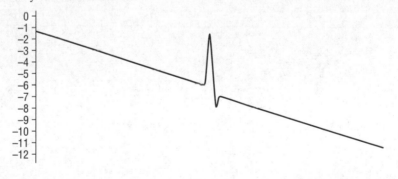

Identify a high-loss fusion splice or macrobend with an OTDR. Remember a high-loss fusion splice or macrobend can be quickly identified by the dip in the trace. These events do not produce a back reflection.

Master It You are troubleshooting a new fiber-optic cable installation with the OTDR and the trace shown in the following graphic. The dip in the trace is the result of a macrobend. Using only the relative decibel information on the vertical scale, approximate the loss for the macrobend.

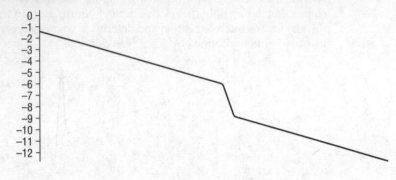

Appendices

Appendix A

The Bottom Line Solutions

Each of The Bottom Line sections in the chapters challenges your understanding of the chapters' concepts and improves your skills.

Chapter 1: Introduction to Data Cabling

Identify the key industry standards necessary to specify, install, and test network cabling. Early cabling systems were unstructured and proprietary, and often worked only with a specific vendor's equipment. Frequently, vendor-specific cabling caused problems due to lack of flexibility. More important, with so many options, it was difficult to utilize a standard approach to the design, installation, and testing of network cabling systems. This often led to poorly planned and poorly implemented cabling systems that did not support the intended application with the appropriate cable type.

Master It In your new position as a product specification specialist, it is your responsibility to review the end users' requirements and specify products that will support them. Since you must ensure that the product is specified per recognized U.S. industry standards, you will be careful to identify that the customer request references the appropriate application and cabling standards. What industry standards body and standards series numbers do you need to reference for Ethernet applications and cabling?

Solution For Ethernet standards, you would reference the IEEE standards body and the 802.3 series of standards. For cabling standards, you would reference the ANSI/TIA standards body and the ANSI/TIA-568-C series of standards.

Understand the different types of unshielded twisted-pair (UTP) cabling. Standards evolve with time to support the need for higher bandwidth and networking speeds. As a result, there have been many types of UTP cabling standardized over the years. It is important to know the differences among these cable types in order to ensure that you are using the correct cable for a given speed.

Master It An end user is interested in ensuring that the network cabling they install today for their 1000Base-T network will be able to support future speeds such as 10Gbps to a maximum of 100 meters. They have heard that Category 6 is their best option. Being well versed in the ANSI/TIA-568-C standard, you have a different opinion. What are the different types of Category 6 cable and what should be recommended for this network?

Solution There are two types of Category 6 cable: Standard Category 6 and Category 6A (augmented Category 6). Standard Category 6 has a limited bandwidth of 250MHz, while Category 6A has a bandwidth of 500MHz. Since 10Gbps operation is required, Category 6A should be recommended to the end user.

Understand the different types of shielded twisted-pair cabling. Shielded twisted-pair cabling is becoming more popular in the United States for situations where shielding the cable from external factors (such as EMI) is critical to the reliability of the network. In reviewing vendor catalogs, you will see many options. It is important to know the differences.

Master It Your customer is installing communications cabling in a factory full of stray EMI. UTP is not an option and a shielded cable is necessary. The customer wants to ensure capability to operate at 10GBase-T. What cable would you recommend to offer the best shielding performance?

Solution S/STP cabling, also known as screened fully shielded twisted-pair (S/FTP), should be recommended. This type of cabling offers the best protection from interference from external sources. Category 6A STP or ScTP cables will perform extremely well in this application. Category 7 is an S/STP cable standardized in ISO 11801 Ed. 2. Category 7 offers a usable bandwidth to 1000MHz. It can also be used for 10 Gigabit Ethernet, 10GBase-T applications.

Determine the uses of plenum- and riser-rated cabling. There are two main types of UTP cable designs: plenum and riser. The cost difference between them is substantial. Therefore, it's critical to understand the differences between the two.

Master It Your customer is building a traditional star network. They plan to route cable for horizontal links through the same space that is used for air circulation and HVAC systems. They plan to run cable vertically from their main equipment room to their telecommunications rooms on each floor of the building. What type of cable would you use for:

The horizontal spaces

The vertical links

Solution Plenum cable is necessary in this space since it shares the space with HVAC systems.

Plenum or riser cable can be used. Although plenum is more than enough, riser cable can be used since it does not share the space with HVAC systems.

Identify the key test parameters for communications cables. As you begin to work with UTP cable installation, you will need to perform a battery of testing to ensure that the cabling system was installed properly and meets the channel requirements for the intended applications and cable grades. If you find faults, you will need to identify the likely culprits and fix it.

Master It Crosstalk is one of the key electrical phenomena that can interfere with the signal. There are various types of crosstalk: NEXT, FEXT, AXT, among others. This amount of crosstalk can be caused in various ways. What would you look for in trying to find fault if you had the following failures:

NEXT and FEXT problems in 1Gbps links

Difficulty meeting 10Gbps performance requirements

Solution a)   In 1Gbps links there is likely no issue with AXT. Therefore, you should pay attention to problems within the cable of a given link. The first thing to do is to inspect the ends of the connectors and ensure that the cable has not been untwisted too much.

b)   Since AXT is the main type of crosstalk affecting 10Gbps performance, make sure the cables are widely spaced apart and that the use of cable ties is limited. And if ties must be used, use only Velcro ties.

Chapter 2: Cabling Specifications and Standards

Identify the key elements of the ANSI/TIA-568-C Commercial Building Telecommunications Cabling Standard.   The ANSI/TIA-568-C Commercial Building Telecommunications Cabling Standard is the cornerstone design requirement for designing, installing and testing a commercial cabling system. It is important to obtain a copy of this standard and to keep up with revisions. The 2009 standard is divided into several sections, each catering to various aspects.

Master It

1. Which subsection of the ANSI/TIA-568-C standard would you reference for UTP cabling performance parameters?

2. Which subsection of the ANSI/TIA-568-C standard would you reference for optical fiber cabling and component performance parameters?

3. What is the recommended multimode fiber type per the ANSI/TIA-568-C.1 standard for backbone cabling?

4. Which is typically a more expensive total optical fiber system solution: single-mode or multimode?

5. The ANSI/TIA-568-C.1 standard breaks structured cabling into six areas. What are these areas?

Solution

1. ANSI/TIA-568-C.2

2. ANSI/TIA-568-C.3

3. 850nm laser-optimized 50/125-micron multimode fiber

4. Single-mode

5. The areas are the horizontal cabling, the backbone cabling, the work area, telecommunications rooms and enclosures, equipment rooms, and the entrance facility (the building entrance).

Identify other ANSI/TIA/EIA standards required to properly design the pathways and spaces and grounding of a cabling system.   The ANSI/TIA-568-C Commercial Building Telecommunications Cabling Standard is necessary for designing, installing, and testing a cabling system, but there are specific attributes of the pathways and spaces of the cabling systems in ANSI/TIA-568-C that must be considered. In addition, these systems need to follow specific grounding regulations. It is just as important to obtain copies of these standards.

Master It Which other TIA standards need to be followed for:

1. Pathways and spaces?

2. Grounding and bonding?

3. Data centers?

Solution

1. TIA-569-B, Commercial Building Standard for Telecommunications Pathways and Spaces

2. ANSI/TIA-607-B, Telecommunications Grounding (Earthing) and Bonding for Customer Premises, as well as the NEC code

3. ANSI/TIA-942, Telecommunications Infrastructure Standard for Data Centers

Identify key elements of the ISO/IEC 11801 Generic Cabling for Customer Premises Standard. The International Organization for Standardization (ISO) and the International Electrotechnical Commission (IEC) publish the ISO/IEC 11801 standard predominantly used in Europe. It is very similar to ANSI/TIA-568-C, but they use different terminology for the copper cabling. It is important to know the differences if you will be doing any work internationally.

Master It What are the ISO/IEC 11801 copper media classifications and their bandwidths (frequency)?

Solution

Class A: 100kHz

Class B: 1MHz

Class C: 16MHz

Class D: 100MHz

Class E: 250MHz

Class E_A: 500MHz

Class F: 600MHz

Class F_A: 1000MHz

Chapter 3: Choosing the Correct Cabling

Identify important network topologies for commercial buildings. Over the years, various network topologies have been created: bus, ring, and star. From the perspective of cabling, the hierarchical star topology is now almost universal. It is also the easiest to cable. The ANSI/TIA-568-C and ISO/IEC 11801 Ed. 2 standards assume that the network architecture uses a hierarchical star topology as its physical configuration. This is the configuration you will most likely be involved in as you begin cabling commercial buildings. The ANSI/TIA-568-C standard specifies the use of the hierarchical star network; however, there are several ways of implementing this.

Master It

1. What is the most typical implementation of the hierarchical star? Specifically, where are the horizontal cross-connection and workgroup switches typically placed?

2. To reduce the cost of a hierarchical star, the network designer has the option to locate network elements closer to equipment outlets. What is this called and what are the benefits?

3. In certain situations it makes sense to install all network equipment in a central location. What is this implementation of the hierarchical star topology called and what are the pros and cons?

Solution

1. The typical implementation of the hierarchical star topology involves placing the horizontal cross-connection and work group switches in a telecommunications room(s) on each floor of a building. This allows equipment to be located in a room separate from the equipment outlets. Typically, the utilization of switch ports is low and can lead to higher costs compared to other implementations of the hierarchical star.

2. Placing horizontal cross-connections and work group switches closer to equipment outlets is called FTTE (fiber-to-the-telecommunications enclosure). The benefits include savings of 25 percent or more through better port utilization and lower building costs of telecommunications rooms.

3. This is called centralized cabling, or FTTD (fiber-to-the-desk). On one hand this provides the greatest port utilization and reduces the size and load of telecommunications rooms. On the other hand it can be more expensive since it requires the use of media converters or optical NICs near the equipment outlets to convert from optical to electrical. For examples of situations where cost savings were obtained, visit TIA's Fiber Optics LAN Section: www.fols.org.

Understand the basic differences between UTP and optical fiber cabling and their place in future-proofing networks. As network applications are evolving, better UTP and optical fiber cabling media are required to keep up with bandwidth demand. As you will see from standards, the end user has many options within a media category. There are many types of UTP and optical fiber cabling. Standards will continue to evolve, but it's always a good idea to install the best grade of cabling since the cost of the structured cabling systems (excluding installation cost) is usually only 5–10 percent of the total project cost. Therefore, making the right decisions today can greatly future-proof the network.

Master It

1. What is the preferred UTP cabling media to support 10Gbps network speeds?

2. What is the preferred optical fiber cabling media to support low-cost transmission at 10Gbps network speeds?

Solution

1. Category 6A cabling

2. 850nm laser-optimized 50/125 micron multimode fiber (a.k.a. OM3)

Identify key network applications and the preferred cabling media for each. In your network design planning and installation you will most likely be running Ethernet-based network applications. There is an "alphabet soup" full of jargon associated with naming the application for a given speed, the reach range, and media type. This chapter has provided a good starting point.

Master It You are asked to design a commercial building network. The owner wants to ensure that the desktops are able to operate using UTP with a maximum of 1Gbps capability. The distance between the workgroup switches located in the telecommunications rooms and the equipment room is 250 meters. What would you recommend for the following?

1. What Ethernet network application ports/modules should you use for the horizontal links?

2. What are your horizontal cabling options?

3. What backbone speed and network application should you use to obtain the lowest cost assuming a 10:1 rule?

4. What backbone cabling should you use?

Solution

1. 1000Base-T

2. Category 5e or Category 6

3. 10GBase-SR

4. OM3, 850nm laser-optimized 50/125 micron multimode fiber

Chapter 4: Cable System and Infrastructure Constraints

Identify the key industry codes necessary to install a safe network cabling system. In the United States, governing bodies issue codes for minimum safety requirements to protect life, health, and property. Building, construction, and communications codes originate from a number of different sources. Usually, these codes originate nationally rather than at the local city or county level.

Master It What industry code is essential in ensuring that electrical and communications systems are safely installed in which an industry listing body is typically used to test and certify the performance of key system elements?

Solution The National Electrical Code (NEC) under the National Fire Protection Association (NFPA) is the code necessary to ensure compliance to safety regulations. The Underwriters Laboratories (UL) organization is typically used to test and certify products.

Understand the organization of the National Electrical Code. The NEC is divided into chapters, articles, and sections. It is updated every three years. As of the time of this writing, the NEC is in its 2008 edition. This is a comprehensive code book, and it is critical to know where to look for information.

Master It

1. What section describes the organization of the NEC?

2. What is the pertinent chapter for communications systems?

Solution

1. Section 90-3

2. Chapter 8

Identify useful resources to make knowing and following codes easier. It can take years to properly understand the details of the NEC. Luckily there are some useful tools and Internet websites that can make this a lot easier.

Master It

1. Where can you view the NEC online?

2. What is one of the best references on the Internet for the NEC?

3. Where should you go for UL information?

4. Where can you go to buy standards?

Solution

1. www.nfpa.org

2. www.mikeholt.com

3. http://ulstandardsinfonet.ul.com

4. http://global.ihs.com

Chapter 5: Cabling System Components

Differentiate between the various types of network cable and their application in the network. ANSI/TIA-568-C.1 defines the various types of network cabling typically used in an office building environment. Much like major arteries and capillaries in the human body, backbone and horizontal cabling serve different functions within a network. Each of these cable types can use either fiber-optic or copper media. In specifying the network cable it is important to understand which type of cable to use and the proper media to use within it.

Master It You are asked to design a network for a company fully occupying a three-story office building using a traditional star network topology. The equipment room, containing the company's server, is located on the first floor, and telecommunications rooms are distributed among each of the floors of the building. The shortest run between the equipment room and the telecommunications rooms is 200 meters. What type of network cable would you use to connect the telecommunications rooms to the equipment room and what type of media would you specify inside the cable?

Solution   Since you are connecting a telecommunications room to an equipment room, you would use a *backbone cable*. Since the minimum distance is 200 meters, you would use one of the fiber-optic media instead of copper media since copper is limited to a 90 meter distance.

Estimate the appropriate size of a telecommunications room based on the number of workstations per office building floor.   The equipment room is like the "heart" (and brain) of the office building LAN. This is where servers, storage equipment, and the main cross-connection are located, and it also typically serves as the entrance of an outside line to the wide area network. It is important that the size of the equipment room be large enough to contain the key network devices.

Master It   In the previous example, you are asked to design a network for a company fully occupying a three-story office building using a traditional star network topology. The company presently has 300 workstations, but they expect to grow to 500 workstations in 1–2 years. Which standard should you reference for the answer and what is the estimate for the minimum size (in square feet) of the equipment room?

Solution   You should refer to TIA-569-B to determine the minimum size of the equipment room. As Table 5.1 shows, you should size the room to a minimum of 800 square feet.

Identify the industry standard required to properly design the pathways and spaces of your cabling system.   Providing the proper pathways and spaces for your cabling system is critical to ensuring that you obtain the level of performance the media is rated to. Too many bends, sharp bends, or twists and turns can significantly impair the bandwidth of the media and could prevent your network from operating

Master It   Which of these industry standards provides you with the recommendations you must follow to ensure a properly working network?

A. J-STD-607-A

B. ANSI/TIA/EIA-606-A

C. TIA-569-B

D. ANSI/TIA-568-C.1

Solution   The answer is C, TIA-569-B.

Chapter 6: Tools of the Trade

Select common cabling tools required for cabling system installation.   Don't start any cabling job without the proper tools. Having the proper tools not only makes the installation quicker and easier, but can also improve the quality of the finished product, making the testing portion go with minimal faults. Therefore, picking tools specific for the applications will make life much easier.

Master It   What are three basic tools required to install a plug for UTP wire?

Solution   A wire stripper, wire cutter, and cable crimper.

Identify useful tools for basic cable testing.   You will be installing fiber-optic cable in your next infrastructure project; however, a separate group will be performing the final testing. At a minimum, you want to make sure the fiber-optic cable was installed without any fault and that enough light is being transmitted through the fiber.

Master It   What type of fiber-optic tester should you purchase to make sure the fiber-optic cable was installed without any fault and that enough light is being transmitted through the fiber, and what are the three key attributes to know before you purchase a tester?

Solution   Since you want to know whether there is enough light power transmitting through the fiber, you should purchase an attenuation tester. The tester should include capability for the correct fiber connectors, support the correct type of fiber (single-mode vs. multimode), and be able to test at the appropriate wavelength.

Identify cabling supplies to make cable installations easy.   Copper and fiber-optic cables are sensitive to the tension applied during the installation of these cables over long lengths through conduits and cabling trays. In fact, industry standards have maximum pulling loads that can be applied.

Master It   Which of the following supplies enables the reduction of friction on the cables while being pulled?

1. Cable pulling tool

2. Cable pulley

3. Wire-pulling lubricant

4. Gopher Pole

Solution   The wire-pulling lubricant (option 3).

Chapter 7: Copper Cable Media

Recognize types of copper cabling.   Pick up any larger cabling catalog, and you will find a myriad of copper cable types. However, many of these are unsuitable for data or voice communications. It's important to understand which copper cables are recognized by key standards.

Master It   What are the recognized copper cables in ANSI/TIA-568-C? Which cable is necessary to support 100 meters (328´) for 10GBase-T?

Solution   Categories 3, 5e, 6 and 6A are the recognized copper cables in ANSI/TIA-568-C. Category 6A is required to support 100 meters (328´) for 10GBase-T.

Understand best practices for copper installation. This discipline is loaded with standards that define key requirements for copper cable installation, but it is important to know and follow some key "best practices."

Master It What are the three primary classes of best practices for copper installation?

Solution (1) Follow the ANSI/TIA-568-C standard, (2) make sure the cables do not exceed the distance limits for the rated application, and (3) use good installation techniques for pulling and placing cable.

Perform key aspects of testing. Every cable run must go through extensive testing to certify the installation to support the intended applications. The ANSI/TIA-568-C standard specifies they key elements. It's important to understand the basic types of testers required.

Master It What are the cable testers that you can use to perform testing?

Solution The cable testers you should consider at a minimum are tone generators and amplifier probes, continuity testers, wire-map testers, and cable-certification testers.

Chapter 8: Fiber-Optic Media

Understand the basic aspects of fiber-optic transmission. Optical-fiber based systems work differently than copper-based systems. Optical-fiber systems are based on transmitting a digital signal by modulating light pulses on and off. Since information is transmitted optically, you must understand certain factors.

Master It

A. What are the common wavelengths of operation for multimode fiber?

B. What are the common operating wavelengths of single-mode fiber?

Solution

A. 850 and 1300nm

B. 1300, 1490, 1550nm

Identify the advantages and disadvantages of fiber-optic cabling. There are pros and cons to fiber-optic cabling. However, with fiber-based systems, the list of pros continues to increase and the list of cons decreases with time.

Master It

A. What are the key advantages of fiber-based systems?

B. Based on these advantages, in which applications would you use fiber instead of UTP copper?

Solution

A. The typical advantages of fiber-based systems are immunity to electromagnetic interference (EMI), higher data rates, longer maximum distances, and better security.

B. Fiber should be used instead of copper when the application requires an application speed to deliver a signal over a distance greater than 100 meters and when better security and resistance to EMI are critical operating factors.

Understand the types and composition of fiber-optic cables.   Optical fiber comes in many flavors. There are key features of optical fiber that dictate a specific choice of ancillary components, such as optical transmitters and connectors. When choosing an optical cabling system, look at the full system cost.

Master It   In most short-distance applications of 100–550 meters using optical fiber, multimode fiber is the preferred choice over single-mode fiber. This is because single-mode–based systems are more expensive than multimode. Although single-mode cable is less expensive than multimode cable, what factors lead to single-mode systems being more expensive?

Solution   A total optical system includes the cable, connectors, connectivity apparatus, and electronics. Since single-mode fiber has a much smaller core size, the manufacturing tolerances of lasers, connectors, and connectivity apparatus is much tighter than multimode parts. As a result, these parts are more expensive and lead to the overall single-mode system being more expensive.

Identify the key performance factors of fiber optics.   You have learned several performance factors that affect the ability of optical fiber to transport signals. Since multimode fibers are the preferred optical media for commercial buildings, you must understand key performance factors affecting these types of optical fiber.

Master It   What is the primary performance factor affecting multimode fiber transmission capability and how can you minimize this?

Solution   The main performance factor affecting multimode transmission is modal dispersion (also called DMD). Modal dispersion can be minimized by specifying a 50 micron multimode fiber instead of a 62.5 micron fiber because 50 micron fiber transmits fewer modes and thereby has lower inherent modal dispersion. Once 50 micron fiber is specified, modal dispersion can be further minimized by picking a fiber with very low DMD.

Chapter 9: Wall Plates

Identify key wall plate design and installation aspects.   The wall plate is the "gateway" for the workstation (or network device) to make a physical and electrical/optical connection to the network cabling system. There are important aspects to consider when designing and installing wall plates.

Master It   What are the main aspects that must be addressed when designing and installing wall plates?

Solution Manufacturer system, wall plate location, wall plate mounting system, and whether to use a fixed design or modular plate.

Identify the industry standards required to ensure proper design and installation of wall plates. You'll have to make certain choices about how best to conform to the relevant industry standards when designing and installing wall plates.

Master It Which TIA standards should you refer to for ensuring proper design and installation of wall plates?

Solution You should refer to ANSI/TIA-570-B (for residential) and ANSI/TIA-568-C.1 (for commercial installation).

Understand the different types of wall plates and their benefits and suggested uses. There are two basic types of wall plate designs: fixed-design and modular wall plates. Each of these has some advantages and disadvantages and recommended uses.

Master It Your commercial office building environment is expected to change over the next few years as some of the network devices that are presently installed with coax connections need to be converted to RJ-45 and some network devices that are currently connected with RJ-45 connections may need to be converted to fiber-optic connections. In addition, the number of connections for each wall plate may increase. Which type of wall plate would provide you with the most flexibility and why?

Solution Modular wall plates will provide you with the greatest flexibility because these designs have individual components that can be installed in varying configurations depending on your cabling system needs today and tomorrow. As opposed to a fixed wall plate, the individual jacks in the wall plate can be replaced.

Chapter 10: Connectors

Identify key twisted-pair connectors and associated wiring patterns. Table 10.1 showed the common modular jack designations for copper cabling. As business systems, phones, and cameras continue to converge to IP/Ethernet-based applications, the types of jacks and plugs will narrow down quickly.

Master It What is the most common modular jack designation for IP/Ethernet-based systems (for example, 100Base-TX)? What are your wiring pattern options? Which of these wiring patterns is necessary for government installations?

Solution The most common modular jack designation for Ethernet applications is the 8-pin RJ-45. These jacks can be wired per either the T568A or T568B wiring pattern. The U.S. government recommends the use of the T568A wiring pattern.

Identify coaxial cable connectors. Although coaxial connectors are not commonly used in Ethernet-based operations of 100Base-T or above, there is still a need to use coaxial connectors for coaxial cables connecting security cameras to the building security systems.

Master It What is the common coaxial connector type to support security-camera service?

Solution   The F-series connector is the common connector used to support security camera cabling.

Identify types of fiber-optic connectors and basic installation techniques.   You will be using optical-fiber connectors more and more in years to come. It's important to understand the differences and the basic installation options to mount the connectors. An understanding of both of these aspects will enable savings in time and money as well as provide a more space-efficient installation.

Master It

A. What class of optical-fiber connector would you use to increase the density of fiber connections?

B. Which one of these types would you use for single fiber connections?

C. Which one of these types would you use for 12 or more fiber connections in one footprint?

D. What are the three types of adhesives that can be used to glue fiber into position of a connector housing, and which one provides the quickest cure?

Solution

A. SFF (small form factor) connectors are used to increase the density (more connections per space) of fiber connections.

B. You should use an LC connector for single fiber connections.

C. You should use an MPO connector to provide multiples of 12-fiber connections in one footprint.

D. Your adhesive options are (1) heat-cured adhesives, (2) UV-cured adhesives, or (3) anaerobic-cured adhesives. The anaerobic-cured adhesive takes the shortest amount of time to set in place.

Chapter 11: Network Equipment

Identify the basic active components of a hierarchical star network for commercial buildings and networks.   Active network equipment is connected by the structured cabling system to support the topology and applications required for the network. More simply, the active equipment is what sends, aggregates, directs and receives actual data. If you are responsible for specifying and procuring the active equipment for a commercial building network, it's important t understand the differences between equipment and their primary functions.

Master It

1. What is the active component that a patch cord is plugged into when it is attached to a network connected device, such as a desktop computer?

2. What type of switch, and over what type of cabling, are workstation connections funneled into in a hierarchical star network?

3. What type of switch, and over what type of cabling, are workgroup switches connected?

Solution

1. Patch cords are plugged into workstation ports. These are either part of the motherboard or part of a network interface card.

2. All workstation connections are connected to workgroup switches using horizontal cabling.

3. Workgroup switches are connected to core switches using backbone cabling.

Identify differences between various types of transceiver modules. The emergence of 1 and 10Gbps Ethernet speeds brought a host of different transceiver modules. At times, there appeared to be a competition as to who could develop the smallest and fastest module. These modules involve an alphabet soup of terminology, and it's important to be able to navigate your way through this.

Master It

1. You are asked to specify a pluggable module for 850nm transmission over multimode fiber at 1Gbps with an LC connector interface. What module do you need to order?

2. You are asked to specify a pluggable module for 1310nm transmission over single-mode fiber at 10Gbps from one building to another. The IT manager wants you to order the smallest possible form factor. What module do you need to order?

Solution

1. 1000Base-SX SFP mini-GBIC supports operation at 1Gbps over multimode fiber with an LC connection.

2. 10GBase-LR SFP+ module would provide you the smallest form factor transceiver for 1310nm operation over single-mode fiber.

Determine if your workgroup switching system is blocking or nonblocking. Certain situations may require you to design your workgroup switching system to support the maximum potential bandwidth to and from your end users' workstation ports. This is called a nonblocking configuration.

Master It You have nine users who have 100Base-T NIC cards on their computers. The 12-port workgroup switch they are connected to has a 1000Base-SX uplink connection. What is the maximum throughput per user that this switch is capable of? Is this presently a blocking or nonblocking configuration? How many additional users could be added before this situation changes?

Solution   This switch has the ability to support a maximum throughput of 83Mbps per port for each of the possible 12 connections (1000Mbps divided by 12 ports). Since there are only nine users, each with a 100Mbps maximum potential, the total demand on the switch is 900Mbps. Since the switch is capable of 1000Mbps, this is a nonblocking situation. The addition of two more users, each with a 100Mbps connection, would create a blocking situation, since the maximum potential demand on the switch would be greater than the switch capability of 1000Mbps.

Chapter 12: Wireless Networks

Understand how infrared wireless systems works.   The use of "free-space-optic" infrared wireless systems to communicate between buildings in large campuses is getting more attention. It's important to understand how these work and their key advantages and disadvantages.

Master It   Point-to-point systems use tightly focused beams of infrared radiation to send information between transmitters and receivers located some distance apart. What is the key parameter in how these individual devices are installed? What are some of the advantages and disadvantages of these systems?

Solution   Since these devices use tightly focused beams of infrared radiation, alignment of these devices is the critical installation parameter. They are relatively inexpensive, no FCC license is required, and they provide ease of installation and portability. Some disadvantages are weather effects and the need for unobstructed paths to support line-of-sight.

Know the types of RF wireless networks.   The use of wireless networks in LAN applications has grown substantially since 1999. The IEEE 802.11 standard was created to support standardization and interoperability of wireless equipment used for this purpose. It's important to understand this standard.

Master It   What are the three existing 802.11 standard subsets and which one would provide the highest range and speed?

Solution   The IEEE 802.11 standard is broken down into the following three subsets: 802.11a, 802.11b, and 802.11g. The 802.11g provides the highest range and speed.

Understand how microwave communication works and its advantages and disadvantages. Microwave networks are not typically used for LAN systems; however, you might come across them for satellite TV or broadband systems.

Master It   What frequency range is used for microwave systems? What are the two main types of microwave systems? What are some of the general advantages and disadvantages of microwave systems?

Solution   Microwave systems use the lower gigahertz frequencies of the electromagnetic spectrum. The two main types of microwave systems are terrestrial microwave and satellite microwave. Microwave systems offer relatively high bandwidth of 100Mbps, and signals can be point-to-point or broadcast. Some disadvantages are expensive equipment and atmospheric attenuation.

Chapter 13: Cabling System Design and Installation

Identify and understand the elements of a successful cabling installation. These may seem basic, but most networks succeed or fail based on how much care was taken in identifying the key elements of a successful installation and understanding how they are performed. If at the end of a network installation you find that the network does not operate as planned, chances are that one of these was not addressed carefully.

Master It

1. What are the key elements of a successful cabling installation?

2. What are the key aspects of a proper design?

Solution

1. Proper design, quality materials and good workmanship.

2. Desired standards and performance characteristics must be met. Other aspects are flexibility, longevity, ease of administration, and economic assessment.

Identify the pros and cons of network topologies. At this point, most commercial building network designs and installations will follow a hierarchical star network topology per the ANSI/TIA-568-C standard. There are many reasons why this is the only recognized cabling topology in this standard.

Master It What are the key reasons why the hierarchical star topology is the only recognized topology in the ANSI/TIA-568-C standard?

Solution The most important reason is that a single cable failure will not bring down the entire network. Other advantages include the ability to centralize electronics when running fiber-to-the-desk, and the use of telecommunications enclosures in addition to telecommunications rooms to reduce costs.

Understand cabling installation procedures. Many of the "microscopic" components necessary in a cabling network have been discussed in this chapter, from choice of media, telecommunications rooms, and cabling management devices to means of ensuring security. However, it is first important to understand the cable installation from a macro-level.

Master It

1. What are the five basic steps of the cabling installation?

2. What are some useful cabling tools?

3. Which TIA standard should you use to ensure that your labeling is done in a standardized fashion?

Solution

1. Design the cabling system, schedule the installation, install the cables, terminate the cables, and test the installation.

2. Pen and paper, hand tools, cable spool racks, fish tape, pull string, cable-pulling lubricant, two-way radio, labeling materials, tennis ball.

3. ANSI/TIA/EIA-606-A

Chapter 14: Cable Connector Installation

Install twisted-pair cable connectors. This chapter covered detailed installation procedures for twisted-pair cable connectors. Although these are pretty straightforward, it is important to understand some key aspects that will make this easier.

Master It

1. When cutting a cable to meet a certain length, what is the recommended additional length of cable you should cut from the master reel? For example, if you are trying to connectorize a 5″ patch cable, how much should you cut?

2. In step 5 of the process, what is the recommended total length of exposed conductor?

Solution

1. It's recommended that you always cut 3^2 of additional cable. For example, if the final patch cable should be 5″, cut 5″-3^2.

2. The recommended length of exposed conductor should be $\frac{1}{2}^2$ to $\frac{5}{8}^2$.

Install coaxial cable connectors. Although less popular than either twisted-pair or fiber-optic cable connectors, you may still encounter the need for installing coaxial connectors for modern video systems. There are several types of connector mounting systems, and they have their advantages and disadvantages.

Master It Basically, two types of connectors exist: *crimp-on* and *screw-on* (also known as *threaded*). The crimp-on connectors require that you strip the cable, insert the cable into the connector, and then crimp the connector onto the jacket of the cable to secure it. Most BNC connectors used for LAN applications use this installation method. Screw-on connectors, on the other hand, have threads inside the connector that allow the connector to be screwed onto the jacket of the coaxial cable. These threads cut into the jacket and keep the connector from coming loose. Which one of these methods is considered less reliable and why? Which is the recommended connector type?

Solution Screw-on connectors are generally unreliable because they can be pulled off with relative ease. Whenever possible, use crimp-on connectors.

Install fiber-optic cable connectors. This chapter covered detailed installation procedures for fiber-optic cable connectors. Similar to twisted-pair connectors, it is important to understand some key aspects that will make this easier.

Master It

1. What color surface should you use to perform the connector mounting and why?

2. Fiber-optic cables typically have a specific yarn that will need to be cut back. What is this made of and what types of scissors are required?

3. What is the proper size of the bead of epoxy to expel into the connector?

4. Polishing is a key step when using epoxy connectors. What is the first step of the polishing process?

Solution

1. You should use a black surface because it makes it easier to see the fiber.

2. This yarn is typically aramid. Use scissors capable of cutting aramid yarn to cut the yarn.

3. The bead of epoxy should be approximately half the diameter of the inside of the ferrule.

4. Air-polishing to remove burrs and sharp edges is the first step of polishing a fiber-optic cable.

Chapter 15: Cable System Testing and Troubleshooting

Identify important cable plant certification guidelines.   The tests you perform and the results you receive enable you to certify your cable installation as complying with a specific standard of performance. Luckily the ANSI/TIA-568-C standards provide a very detailed set of requirements that will be likely be specified on behalf of the building owner. Although many high-quality cable testers on the market perform their various tests automatically and provide you with a list of pass/fail results, it is important to know not only what is being tested but also what results the tester is programmed to evaluate as passes and failures.

Master It

1. What is the difference between the permanent link and channel testing?

2. If you have installed a Category 6A cable, what is the maximum NEXT that can be supported for a 500MHz channel?

3. What is the insertion loss budget for a 300 meter link intended to operate at 10GBase-SR using OM3 fiber?

Solution

1. The *permanent link* refers to the permanently installed cable connection that typically runs from a wall plate at the equipment site to a patch panel in a wiring closet or data center. The *channel link* refers to the complete end-to-end cable run including the basic link and the patch cables used to connect the equipment to the wall plate and the patch-panel jack to the hub or other device.

2. 26.1dB is the maximum NEXT that can be tolerated for a Category 6A cable operating in a 500MHz channel.

3. 2.6dB.

Identify cable testing tools. Cable-testing tools can range from simple, inexpensive, mechanical devices to elaborate electronic testers that automatically supply you with a litany of test results in an easy-to-read pass/fail format. It's important to identify these and understand their benefits.

> **Master It** You are asked to install a fiber-optic link that requires multiple connections and splice points. Based on the cable length, number of connections, and splice points, you have estimated the insertion loss to be very close to the insertion loss budget for the intended application. After installing the link, you find that the insertion loss is higher than the budget. It's now time to go and find the points in the link that could be causing unusually high loss. You suspect that it may be due to the splices in the system. Which measurement tool would you use and why?

> **Solution** You should consider using an OTDR since it enables the ability to test splice points and connector losses for a continuous link.

Troubleshoot cabling problems. Cabling problems account for a substantial number of network support calls. Since network uptime is a critical condition for maximizing an organization's business, troubleshooting problems quickly is very important. Understanding some basic steps will help you improve the speed of your process and repair of a customer's network.

> **Master It** What are the basic recommended steps for troubleshooting a cable problem?

> **Solution**

> 1. Split the system into its logical elements.

> 2. Locate the element that is most likely the cause of the problem.

> 3. Test the element or install a substitute to verify it is the cause of the problem.

> 4. If the suspected element is not the cause, move on to the next likely element.

> 5. After locating the cause of the problem, repair or replace it.

Chapter 16: Creating a Request for Proposal

Determine key aspects of writing an RFP. The RFP is essential to the success of your telecommunications-infrastructure project. It is often referred to as a combination of a guidebook, map and rulebook. The guidebook quality ensures that the proper instructions are given for the design, installation, and testing phases of the project. The rulebook quality ensures that the roles and responsibilities of the relevant parties are established along with each of their liabilities. The map quality ensures that a direction has been set. This is a very important quality of the RFP and will help ensure the long-term success of the project and many years of successful service after the project is completed.

> **Master It** What analysis must be performed as the first part of the development of the RFP in order to define the customer's current and future needs, and key restrictions and constraints?

> **Solution** The Needs Analysis.

Determine the components of the cabling infrastructure that should be included in the RFP. The design phase is an integral part of the RFP process. This may seem a bit intimidating, but even the largest and most intimidating projects become manageable when you divide the project into smaller sections.

Master It Identify the four major subsystems of the cabling infrastructure that should be included in the RFP.

Solution

◆ Telecommunications rooms

◆ Backbone cabling

◆ Horizontal cabling

◆ Work area

Understand what to do after the RFP is written. The steps following the writing of the RFP are just as important as the creation of the RFP. Once the guidebook, map, and rulebook are written, it is time to select the team and initiate the plan.

Master It What two steps should be taken after writing the RFP and obtaining approval by the customer in order to get started implementing the project?

Solution

◆ Distribute the RFP to prospective vendors.

◆ Select the vendor.

Chapter 17: Cabling @ Work: Experience from the Field

Employ useful hints for planning the installation. A well-thought-out and documented plan is critical for ensuring a well-executed job. Taking the time up-front to carefully plan the project will save a lot of time, materials, labor, and equipment.

Master It What are the recommended steps for developing a realistic plan?

Solution

1. Get the complete specification.

2. Perform a job walk.

3. Clarify inconsistencies and ambiguities.

4. Calculate the lengths of network runs.

5. Plan for service loops and waste.

6. Evaluate your labor and skill level.

Understand tips for working safely. Safety should be your top priority. Projects may need to be expedited or hurried if delayed. We will often need to work faster; but this may compromise our attention to safety. Please take the extra time to educate all of your employees of the importance and top priority of safety.

Master It What are some important aspects of safety training?

Solution Your safety training should include knowing proper safety methods for handling ladders, wearing safety helmets, using dust masks, dealing with electricity, and ensuring proper eye safety.

Plan for contingencies. Projects typically do not go as planned. No matter how good the plan is, there will likely be some things we may not have predicted. That's why every good plan has a set of contingency plans established in the event that things go wrong.

Master It What are the typical contingencies you should plan for?

Solution

1. Leave service loops.

2. Plan for additional drops.

3. Plan for extra time.

4. Plan for labor shortages.

5. Plan for equipment and material shortages.

Chapter 18: History of Fiber Optics

Recognize the refraction of light. Refraction is the bending of light as it passes from one material into another.

Master It You are fishing with a friend and in front of your boat; you see a large fish. You place your bait in front of the fish. Why does the fish not take the bait?

Solution Remember refraction is easily observed by placing a stick into a glass of water. When viewed from above, the stick appears to bend because light travels more slowly through the water than through the air. The same is true for the fish. The fish was viewed from above the water and refraction bent the light so the fish appeared to be in a different location then it actually was.

Identify total internal reflection. In 1840, Colladon and Babinet demonstrated that bright light could be guided through jets of water through the principle of total internal reflection.

Master It You are swimming in a pool at night with a waterproof flashlight. When you point the light beam at the surface of the water, you see it bounce off the surface of the water and project on the bottom of the pool. Why does this happen?

Solution Remember that Colladon and Babinet demonstrated that bright light could be guided through jets of water through the principle of total internal reflection. In their demonstration, light from an arc lamp was used to illuminate a container of water. You essentially recreated their experiment except on a larger scale. Instead of an arc lamp and a small barrel of water, you had a flashlight and a large pool of water. The light reflected off the surface of the pool because of total internal reflection.

Detect cross-talk between two optical fibers. Crosstalk or optical coupling is the result of light leaking out of one fiber into another fiber.

Master It You have bundled two small glass rods together in an effort to make a fiber-optic scope that will allow you to look into the combustion chamber of a car engine when the spark plug is removed. After the engine cools, you insert your fiber-optic scope into the combustion chamber and are disappointed with the image you see. What is one possible cause for the poor performance of your fiber-optic scope?

Solution Remember that in 1930, Lamm bundled commercially produced fibers and managed to transmit a rough image through a short stretch of the first fiber-optic cable. The process had several problems, including the fact that the fiber ends were not arranged exactly, and they were not properly cut and polished. Another issue was to prove more daunting. The image quality suffered from the fact that the quartz fibers were bundled against each other. This meant that the individual fibers were no longer surrounded by a medium with a lower index of refraction. Much of the light from the image was lost to crosstalk. Crosstalk or optical coupling is the result of light leaking out of one fiber into another fiber.

Chapter 19: Principles of Fiber-Optic Transmission

Calculate the decibel value of a gain or loss in power. The decibel is used to express gain or loss relative to a known value. In fiber optics, the decibel is most commonly used to describe signal loss through the link after the light has left the transmitter.

Master It The output power of a transmitter is 10mW and the power measured at the input to the receiver is 3mW. Calculate the loss in dB.

Solution To calculate the decibel value of a gain or loss in signal power, use the following equation:

$dB = 10 Log_{10} (P_{out} \div P_{in})$

$dB = 10 Log_{10} (3 \div 10)$

$dB = 10 Log_{10} 0.3$

$dB = -5.23$

$Loss = 5.23 dB$

Calculate the gain or loss in power from a known decibel value. If the gain or loss in power is described in dB, the gain or loss can be calculated from the dB value. If the input power is known, the output power can be calculated, and vice versa.

Master It The loss for a length of optical fiber is 11dB. Calculate the loss in power from the optical fiber and the output power of the optical fiber for an input power of 3mW.

Solution First solve for the loss, then for the output power. The equation would be written as follows:

$(P_{out} \div P_{in}) = antilog (dB \div 10)$

$(P_{out} \div 3mW) = antilog (-11 \div 10)$

$(P_{out} \div 3mW) = \text{antilog} -1.1$

$(P_{out} \div 3mW) = 0.0794$

$P_{out} = 0.0794 \times 3mW$

$P_{out} = 0.2382mW$

Convert dBm to a power measurement.   In fiber optics, dBm is referenced to 1mW of power.

Master It   A fiber-optic transmitter has an output power of –22dBm. Calculate the output power in watts.

Solution   To calculate the power in watts, the equation would be written as:

$-22 = 10\text{Log}_{10}(P \div 1mW)$

$-2.2 = \text{Log}_{10}(P \div 1mW)$

$0.0063 = P \div 1mW$

$P = 0.0063 \times 1mW$

$P = 0.0063mW$

Chapter 20: Basic Principles of Light

Convert various wavelengths to corresponding frequencies.   The relationship between wavelength and frequency can be expressed with the formula $\lambda = v \div f$.

Master It   If an electromagnetic wave has a frequency of 800kHz, what is its wavelength?

Solution   By dividing the speed of the electromagnetic wave, 300,000km/s, by the frequency, 800kHz, we arrive at the wavelength, 375 meters.

Calculate the amount of energy in a photon using Planck's constant.   The amount of energy in each photon depends on the electromagnetic energy's frequency: the higher the frequency, the more energy in the particle. To express the amount of energy in a photon, we use the equation $E = hf$.

Master It   Calculate the energy of a photon at a frequency of 375.25THz.

Solution   Using Planck's constant, 6.626×10^{-34} joule-seconds, and the formula $E = hf$, you can solve for the energy as shown in this equation:

$(6.626 \times 10^{-34}) \times (3.7525 \times 10^{14}) = 2.486 \times 10^{-19} \, W$

Calculate the speed of light through a transparent medium using its refractive index. The index of refraction is a relative value, and it is based on the speed of light in a vacuum, which has an index of refraction of 1. The index of refraction can be calculated using the equation $n = c/v$.

Master It   If light is passing through a medium with a refractive index of 1.25, what will the velocity of the light through that medium be?

Solution   To find the speed of light in a medium with a known refractive index, we divide the speed of light, 300,000km/s, by the medium's refractive index, in this case, 1.25. The result is a speed of 240,000km/s.

Use Snell's law to calculate the critical angle of incidence.   The incident angle required to produce a refracted angle of 90° is called the critical angle. To find the critical angle of two materials, we can use a modified version of Snell's equation:

$\theta_c = \arcsin(n_2 \div n_1)$

Master It   Calculate the critical angle for two materials where n_2 is 1.48 and n_1 is 1.52.

Solution   The critical angle can be found by dividing n_2 by n_1, then finding the arcsin of the result as shown in the equation below.

$\theta_c = \arcsin(1.48 \div 1.52) = 76.826°$

Calculate the loss in decibels from a Fresnel reflection.   Augustin-Jean Fresnel determined how to calculate the amount of light lost from a Fresnel reflection using the equation $\rho = ((n_1 - n_2) \div (n_1 + n_2))^2$.

Master It   Calculate the loss in decibels from a Fresnel reflection that is the result of light passing from a refractive index of 1.48 into air.

Solution   To calculate the loss in decibels from a Fresnel reflection, we need to use two equations shown here. While the refractive index of air was not given in the problem, it can be obtained from Table 20.1.

$\rho = ((1.48 - 1) \div (1.45 + 1))^2 = 0.2^2 = 0.037$

To calculate the loss in decibels, we use the equation

$dB = 10Log_{10} (1 - 0.037)$

plugging in the Fresnel reflection value from above,

$dB = 10Log_{10} 0.963 = -0.163$

This gives us a loss of 0.163dB.

Chapter 21: Optical Fiber Construction and Theory

Calculate the numerical aperture of an optical fiber.   To find the numerical aperture of an optical fiber, you need to know the refractive index of the core and the cladding.

Master It   The core of an optical fiber has a refractive index of 1.49 and the cladding has a refractive index of 1.47. Calculate the numerical aperture.

Solution   Using the formula shown here, square the value for the refractive index of the core and then do the same for the cladding. Subtract the cladding value from the core value and take the square root of that value. Your answer should be 0.243.

$$NA = \sqrt{1.49^2 - 1.47^2} = 0.243$$

Calculate the numbers of modes in an optical fiber.   To find the number of modes in an optical fiber, you need to know the diameter of the core, the wavelength of the light source, and the numerical aperture.

Master It   Calculate the number of possible modes in an optical fiber with the following specifications:

Core diameter = 62.5µm

Core refractive index = 1.50

Cladding refractive index = 1.46

Wavelength of the light source = 1300nm

Solution   First, determine the numerical aperture of the fiber's core using this equation:

$$NA = \sqrt{1.50^2 - 1.46^2} = 0.344$$

Next, use the equation $M = (2D \times \pi \times NA/\lambda)^2/2$ to find the number of modes.

$M = (125 \times 10^{-6} \times \pi \times 0.344/1300 \times 10^{-9})^2/2 = 5399.08$ modes.

Since you cannot have a fraction of a mode, the number of possible modes is 5399.

Select the proper optical fiber coating for a harsh environment.   The coating is the true protective layer of the optical fiber; it absorbs the shocks, nicks, scrapes, and even moisture that could damage the cladding. Without the coating, the optical fiber is very fragile. The coating found on an optical fiber is selected for a specific type of performance or environment.

Master It   Choose a coating for an optical fiber that must operate in temperatures as high as 250° C.

Solution   Coatings found on optical fibers include acrylate, silicone, carbon, and polyimide. Of these four, acrylate has the lowest operating temperature while carbon and polyimide have the highest. Silicone has a much higher operating temperature than acrylate; however, it is still limited to 200° C. Polyimide can operate at temperatures up to 350° C and carbon cannot be used by itself. Whether or not the optical fiber is carbon coated, it will require a polyimide coating.

Chapter 22: Optical Fiber Characteristics

Calculate the attenuation in dB for a length of optical fiber.   The attenuation values for a length of optical-fiber cable can be calculated using the attenuation coefficient for a specific type of optical-fiber cable. This information can be found in the manufacturer's data sheet or in a standard.

Master It   Calculate the maximum attenuation for a 15km ITU-T G.657 optical-fiber cable at 1550nm.

Solution   First, locate the attenuation coefficient for an ITU-T G.657 optical-fiber cable at 1550nm in Tables 22.5 and 22.6. You will notice that they are identical, 0.3dB/km. Then multiply the attenuation coefficient by the length as shown here:

Attenuation for 15km of optical fiber = $15 \times 0.3 = 4.5$

The maximum attenuation for 15km of ITU-T G.657 optical fiber at 1550nm is 4.5dB.

Calculate the usable bandwidth for a length of optical fiber.   The product of bandwidth and length (MHz km) expresses the information carrying capacity of a multimode optical fiber. Bandwidth is measured in megahertz (MHz) and the length is measured in kilometers. The MHz km figure expresses how much bandwidth the fiber can carry per kilometer of its length. The fiber's designation must always be greater than or equal to the product of the bandwidth and the length of the optical fiber.

Master It   Refer to Table 22.2 and calculate the usable bandwidth for 700 meters of 50/125μm multimode optical fiber at a wavelength of 850nm.

Solution   The minimum bandwidth-length product for a 50/125μm multimode optical fiber at 850nm and 1300nm is 500MHz km. To find the usable bandwidth for this optical fiber at a length of 700 meters, you must change the equation around and solve as shown here:

Bandwidth-length product = 500MHz km

Optical fiber length = 0.7km

Bandwidth in MHz = $500 \div 0.7$

Bandwidth = 714.29MHz

Calculate the macrobending loss in a single-mode optical fiber.   The macrobending loss is defined for a number of turns around a mandrel at a specified radius and wavelength.

Master It   A single-mode ITU-T G.657 class B bend-insensitive fiber-optic cable is installed in a home with three small radius bends. Two bends have a radius of 7.5mm and one bend has a radius of 10mm. The wavelength of the light source is 1625nm.

Solution   Referencing Table 22.6 we see that the loss for each 7.5mm bend is 1.0dB and the loss for the 10mm bend is 0.2dB. The total macrobending loss (MBL) will be the sum of the three losses as shown in this equation:

MBL 1 + MBL 2 + MBL 3 = Total MBL

1.0dB + 1.0dB + 0.2dB = 2.2dB

Total macrobending loss = 2.2dB

Calculate the acceptance angle for an optical fiber.   The acceptance angle defines the acceptance cone. Light entering the core of the optical fiber at an angle greater than the acceptance angle will not propagate the length of the optical fiber. For light to propagate the length of the optical fiber, it must be at an angle that does not exceed the acceptance angle.

Master It   Determine the acceptance angle of an optical fiber with a core refractive index of 1.48 and a cladding refractive index of 1.45.

Solution   To determine the acceptance angle of an optical fiber with a core refractive index of 1.48 and a cladding refractive index of 1.45, we must determine the NA of the fiber using this formula:

$$NA = \sqrt{[(n_1)^2 - (n_2)^2]}$$

The acceptance angle is the value of θ.

NA = sinθ

So

θ = arcsinNA = arcsin 0.2965 = 17.25

The acceptance angle is 17.25°.

Chapter 23: Safety

Classify a light source based on optical output power and wavelength.   The OSHA Technical Manual describes the Optical Fiber Service Group (SG) Designations based on ANSI Z136.2. These SG designations relate to the potential for ocular hazards to occur only when the OFCS is being serviced. OSHA also classifies standard laser systems.

Master It   An OFCS does not fall into any of the SG designations defined in the OSHA Technical Manual because during servicing it is possible to be exposed to laser emissions greater than 550mW. What classification would be assigned to this OFCS based on the OSHA Technical Manual?

Solution   If the total power for an OFCS is at or above 0.5W, it does not meet the criteria for an optical fiber SG designation and should be treated as a standard laser system. A Class IV laser hazard as defined in the OSHA Technical Manual has a continuous wave output above 500mW, or +27dBm. This OFCS would be Class IV.

Identify the symptoms of exposure to solvents.   Solvents can cause organ damage if inhaled or ingested. The molecules that make up most solvents can take the place of oxygen in the bloodstream and find their way to the brain and other organs. As the organs are starved of oxygen, they can become permanently damaged.

Master It   Your co-worker spilled a bottle of liquid several minutes ago and now appears impaired. What has your co-worker potentially been exposed to?

Solution   Because solvent molecules can displace oxygen molecules in the blood, they can starve the brain of oxygen and give the appearance of intoxication. Your co-worker may have been exposed to a solvent.

Calculate the proper distance the base of a ladder should be from a wall.   Remember non-self-supporting ladders, such as extension ladders, require a firm vertical support near the work area and a stable, nonslip surface on which to rest the ladder. When setting up a non-self-supporting ladder, it is important to place it at an angle of 75½° to support your weight and be stable.

Master It   A non-self-supporting ladder is 16′ in length. How far should the feet of the ladder be from the wall?

Solution   A good rule for finding the proper angle is to divide the working height of the ladder (the length of the ladder from the feet to the top support) by 4, and place the feet of the ladder that distance from the wall. This ladder was 16′ in length so the ladder's feet should be 4′ from the wall.

Chapter 24: Fiber-Optic Cables

Determine the cable type from the NEC markings.   Article 770 of the NEC states that optical-fiber cables installed within buildings shall be listed as being resistant to the spread of fire. These cables are also required to be marked.

Master It   You have been asked to install the two cables shown here in a plenum space. From the markings on the cables, determine the cable types and determine if each cable can be installed in a plenum space.

Solution   The cable marked OFNR is a nonconductive riser cable and the cable marked OFNP is a nonconductive plenum cable. Only the cable marked OFNP can be installed in the plenum space.

Identify the fiber number from the color code.   TIA-598-C defines a color code for optical fibers within a cable assembly.

Master It   You have been asked to identify the first and last tight-buffered optical fibers in the 12-fiber cable shown here. There are no numbers on the tight buffers; however, each buffer is colored as defined in TIA-598-C. What are the colors of the first and last tight-buffered optical fibers?

Solution   Table 24.2 list the optical fiber color codes as defined in TIA-598-C. The color for fiber number one is blue and the color for fiber number 12 is aqua.

Identify the optical-fiber type from the color code.   Premises cable jacket colors are defined in TIA-598-C.

Master It   It is your first day on the job and your supervisor has asked you to get some laser-optimized multimode optical fiber from the back of the van. The van contains several different cables, each with a different jacket color. What is the jacket color of the laser-optimized cable?

Solution   Table 24.3 list the premises jacket colors as defined in TIA-598-C. The color for the laser-optimized cable is aqua.

Determine the cable length using sequential markings.   Many markings are found on a cable. Sequential markings typically appear every 2 feet or every 1m.

Master It   You have been asked to determine the length of fiber-optic cable coiled up on the floor. The sequential markings for this cable are meters. The marking on one end of the cable is 1,235 and the marking on the other end is 2,485. How long is this cable?

Solution   To find the length of the cable, determine the distance between the markings as shown here:

2,485m – 1,235m = 1,250m

Chapter 25: Splicing

Determine if the splice loss is from an intrinsic factor.   Even when fibers are manufactured within specified tolerances, there are still slight variations from one optical fiber to another. These variations can affect the performance of the splice even though the optical fibers are perfectly aligned when mated.

Master It   Two optical fibers are about to be fusion spliced together, as shown here. Determine what type of problem exists with the optical fibers about to be spliced.

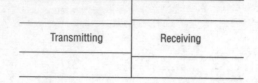

Solution   A close inspection of the graphic reveals that the core diameter of the transmitting optical fiber is slightly larger than the receiving optical fiber. This core diameter mismatch is the result of an intrinsic factor.

Determine the splice loss is from an extrinsic factor.   Extrinsic factors that affect optical fiber splice performance are factors related to the condition of the splice itself, external to the optical fiber.

Master It   Two optical fibers are about to be fusion spliced together, as shown here. Determine what type of problem exists with the optical fibers about to be spliced.

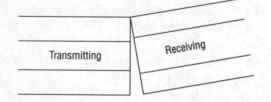

Solution   A close inspection of the graphic reveals that there is some angular misalignment. The angular misalignment is the result of an extrinsic factor.

Calculate the potential splice loss from a core diameter mismatch.   A core diameter mismatch loss results when the core diameter of the transmitting optical fiber is greater than the core diameter of the receiving optical fiber

Master It   Calculate the worst-case loss for a splice where the transmitting optical fiber has a core diameter of 51μm and the receiving optical fiber has a core diameter of 49.5μm.

Solution   We can calculate the worst-case loss percentage for a splice that joins a 51μm core and 49.5μm core using the following formula:

$$\text{Loss} = [(d_1)^2 - (d_2)^2] / (d_2)^2$$

where d_1 is the diameter of the transmitting core and d_2 is the diameter of the receiving core.

$51^2 - 49.5^2 = 150.75$

$150.75/49.5^2 = 0.0615$ or 6.15%

Using this formula, the decibel loss can be calculated:

$dB = 10Log_{10}(P_{out} / P_{in})$

Note that the decibel formula is a ratio, where the power output is divided by the power input. We know that the receiving optical fiber will not accept 6.15 percent of the light from the transmitting optical fiber. Subtracting 6.15 percent from 100 percent, as shown here, results in a difference of 93.85 percent, which can be written as 0.9385.

$100\% - 6.15\% = 93.85\%$ or 0.9385

$10Log_{10} (0.9385 / 1) = -0.276dB$

Worst-case loss = 0.276dB

Chapter 26: Connectors

Evaluate connector endface polishing.   To evaluate the endface of a fiber-optic connector, you need a microscope designed for this application.

Master It   You are inspecting the multimode endface shown here with a 200X inspection microscope. Is there anything wrong with this connector endface?

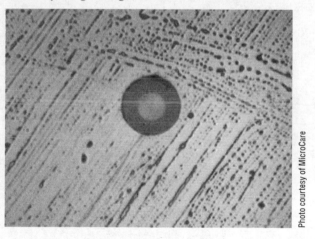

Photo courtesy of MicroCare

Solution   This connector has oil from a fingertip on the endface; however, the endface has a good polish. After cleaning and re-inspection, this connector will be ready to be put to use.

Evaluate connector endface cleaning. A connector that has been cleaned properly will be free of any dirt or contamination.

Master It You are inspecting the connector endfaces shown in these two graphics. Which of these connectors has oil on the endface?

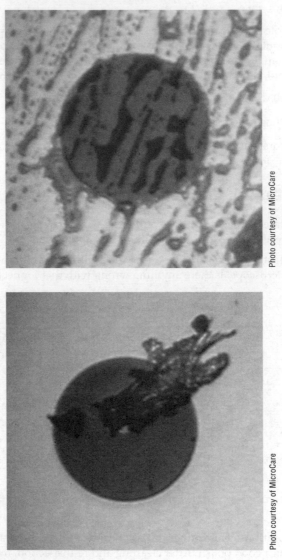

Photo courtesy of MicroCare

Photo courtesy of MicroCare

Solution The connector endface shown in the first graphic has oil on it, and the connector endface shown in the second graphic has debris from a cotton shirt on it. Both of these connectors need to be cleaned and re-inspected.

Evaluate connector endface geometry. The key measurements provided by the interferometer are radius of curvature, apex offset, and fiber undercut or protrusion. These critical parameters are required by Telcordia GR-326 to evaluate connector endface geometry.

Master It You are evaluating a machine polished endface with the interferometer. The radius of curvature measured by the interferometer is 25mm. Does this endface radius of curvature fall within the range specified by Telcordia GR-326?

Solution The minimum and maximum radius of curvature values as defined in Telcordia GR-326 are 7 and 22mm. Values below or above this range increase the risk of optical fiber damage; 25mm exceeds the maximum acceptable value.

Identify an optical fiber type from the color of the connector strain relief. Multimode and single-mode connectors and adapters can be identified using the color code in ANSI/TIA-568-C.3.

Master It You need to mate two 850nm laser-optimized 50/125μm fiber-optic cables together with a jumper at a patch panel. What color strain reliefs would the correct jumper have?

Solution 850nm laser-optimized 50/125μm optical fiber should have an aqua-colored strain relief.

Chapter 27: Fiber-Optic Light Sources

Determine the minimum optical output power for an LED transmitter. The minimum optical output power of an LED transmitter should be defined in the manufacturer's data sheet.

Master It Refer to the following table to determine the BOL and EOL values for the minimum optical output power of a 50/125μm optical fiber.

Parameter	Symbol	Min.	Typ.	Max.	Unit
Optical Output Power: BOL62.5/125μm, NA = 0.275	P_0	−22	−19.8	−17	dBm avg.
Fiber: EOL			23		
Optical Output Power: BOL50/125μm, NA = 0.20	P_0	−25.5	−23.3	−17	dBm avg.
Fiber: EOL		−26.5			
Optical Extinction Ratio			0.001	0.03	%
			−50	−35	dB
Optical Output Power at Logic "0" State	P_0 ("0")			−45	dBm avg.
Center Wavelength	λ_c	1270	1308	1380	nm

Parameter	Symbol	Min.	Typ.	Max.	Unit
Spectral Width—FWHM	Δl		137	170	nm
Optical Rise Time	t_r	0.6	1.0	3.0	ns
Optical Fall Time	t_f	0.6	2.1	3.0	ns
Duty Cycle Distortion Contributed by the Transmitter	DCD		0.02	0.6	$ns_{p\text{-}p}$
Data-Dependent Jitter Contributed by the Transmitter	DDJ		0.02	0.6	$ns_{p\text{-}p}$

Solution   To find the minimum optical output power for a specific optical fiber, locate the Optical Output Power row for that optical fiber in the table. In the Min. column, there is a BOL and EOL value. The higher number, –25.5 dBm, is the BOL value and the lower number, –26.5 dBm, is the EOL value.

Determine the maximum optical output power for an LED transmitter.   The manufacturer's data sheet should contain the maximum value for the optical output power of an LED transmitter.

Master It   Refer to the following table to determine the maximum optical output power of a 62.5/125µm optical fiber.

Parameter	Symbol	Min.	Typ.	Max.	Unit
Optical Output Power: BOL62.5/125µm NA = 0.275 Fiber: EOL	P_0	–24 –25	–21.8	–19	dBm avg.
Optical Output Power: BOL50/125µm, NA = 0.20 Fiber: EOL	P_0	–27.5 –28.5	–25.3	–19	dBm avg.
			0.001 –50	0.03 –35	% dB
Optical Output Power at Logic "0" State	P_0 ("0")			–45	dBm avg.
Center Wavelength	λ_c	1270	1308	1380	nm
Spectral Width—FWHM	Δl		137	170	nm
Optical Rise Time	t_r	0.6	1.0	3.0	ns
Optical Fall Time	t_f	0.6	2.1	3.0	ns
Duty Cycle Distortion Contributed by the Transmitter	DCD		0.02	0.6	$ns_{p\text{-}p}$
Data-Dependent Jitter Contributed by the Transmitter	DDJ		0.02	0.6	$ns_{p\text{-}p}$

Solution   To find the maximum optical output power for a specific optical fiber, locate the Optical Output Power row for that optical fiber in the table. The number in the Max. column, –19 dBm, is the value for the maximum optical output power.

Determine the minimum optical output power for a laser transmitter.   The manufacturer's data sheet should contain the minimum, maximum, and typical value for the optical output power of a laser transmitter.

Master It   Refer to the following table to determine the BOL and EOL values for the minimum optical output power.

Parameter	Symbol	Min.	Typ.	Max.	Unit
Optical Output Power 9μm SMF	P_{OUT}	–6	0	+3	dBm
Center Wavelength	λ_c	1260		1360	nm
Spectral Width—rms	ΔLOW		1.8	4	nm rms
Optical Rise Time	t_r		30	70	ps
Optical Fall Time	t_f		150	225	ps
Extinction Ratio	E_R	8.2	12		dB
Optical Output Eye		Compliant with eye mask Telecordia GR-253-CORE			
Back-Reflection Sensitivity				–8.5	dB
Jitter Generation	pk to pk			70	mUI
	RMS			7	mUI

Solution   To find the minimum optical output power, locate the Optical Output Power row in the table. In the Min. column, there is a value; since BOL and EOL are not stated, assume this value is the BOL value. To approximate the EOL value, subtract 1.5dB. The BOL value is –6dBm and the EOL value is –7.5dBm.

Determine the maximum optical output power for a laser transmitter.   The manufacturer's data sheet should contain the maximum value for the optical output power of a laser transmitter.

Master It   Refer to the following table to determine the maximum optical output power.

Parameter	Symbol	Min.	Typ.	Max.	Unit
Optical Output Power 9μm SMF	P_{OUT}	–12	–8	–5	dBm
Center Wavelength	λ_c	1260		1360	nm
Spectral Width—rms	ΔLOW		1.8	4	nm rms

Parameter	Symbol	Min.	Typ.	Max.	Unit
Optical Rise Time	t_r		30	70	ps
Optical Fall Time	t_f		150	225	ps
Extinction Ratio	E_R	8.2	12		dB
Optical Output Eye	Compliant with eye mask Telecordia GR-253-CORE				
Back-Reflection Sensitivity				−8.5	dB
Jitter Generation	pk to pk			70	mUI
	RMS			7	mUI

Solution   To find the maximum optical output power, locate the Optical Output Power row in the table. The number in the Max. column is the value for the maximum optical output power. There is no BOL or EOL value for the maximum optical output power. The maximum value is −5dBm.

Chapter 28: Fiber-Optic Detectors and Receivers

Calculate the dynamic range for an LED receiver.   The dynamic range of the LED receiver is the difference between the minimum value for the maximum optical input power and the maximum value for the minimum optical input.

Master It   Refer to the following table and calculate the dynamic range for the LED receiver.

Parameter	Symbol	Min.	Typ.	Max.	Unit
Optical input power minimum at window edge	$P_{IN\,Min.}$ (W)		−30.5	−28	dBm avg.
Optical input power maximum	$P_{IN\,Max.}$	−16	−13.8		dBm avg.
Operating wavelength	λ	1270		1380	nm

Solution   To calculate the dynamic range, subtract the minimum value for the maximum optical input power ($P_{IN\,Max}$) from the maximum value for the minimum optical input power ($P_{IN\,Min}$) as shown in this formula:

$P_{IN\,Max}$ − $P_{IN\,Min}$ = dynamic range

−16dBm − −28dBm = 12dB

Dynamic range = 12dB

Calculate the dynamic range for a laser receiver. The dynamic range of the laser receiver is the difference between the minimum value for the maximum optical input power and the maximum value for the minimum optical input.

Master It Refer to the following table and calculate the dynamic range for the laser receiver.

Parameter	Symbol	Min.	Typ.	Max.	Unit
Receiver sensitivity	$P_{IN\,Min}$		–20	–16	dBm avg.
Receiver overload	$P_{IN\,Max}$	–1	+3		dBm avg.
Input operating wavelength	λ	1260		1570	nm
Signal detect—asserted	P_A		–24.3	–16.5	dBm avg.
Signal detect—deasserted	P_D	–32	–25.7		dBm avg.
Signal detect—hysteresis	P_H	0.5	1.4	4	dB
Reflectance			–35	–27	dB

Solution To calculate the dynamic range, subtract the minimum value for the maximum optical input power ($P_{IN\,Max}$) from the maximum value for the minimum optical input power ($P_{IN\,Min}$) as shown in this formula:

$P_{IN\,Max} - P_{IN\,Min}$ = dynamic range

–1dBm – –16dBm = 15dB

Dynamic range = 15dB

Chapter 29: Passive Components and Multiplexers

Calculate the output power of a real tee coupler. Remember that manufacturers may or may not provide the insertion loss combined with the splitting ratio.

Master It The manufacturer provided the insertion loss combined with the loss from the splitting ratio for a 90:10 tee coupler. Calculate the power available at each output port.

Solution The manufacturer states that the insertion loss in decibels is 0.63/11. This means that the power at the 90 percent port will be 0.63dB less than the input power to the coupler and the power at the 10 percent port will be 11dB less than the input power to the coupler.

If the input power to the coupler is –10dBm, the power at each port can be calculated by subtracting the insertion loss for each port as defined by the manufacturer from –10dBm as shown here:

–10dBm – 0.63dB = –10.63dBm

–10dBm – 11dB = –21dBm

The output power at the 10 percent port is –10.63dBm and the output power at the 90 percent port is –21dBm. Remember this calculation did not take into account the interconnection losses and uniformity.

Calculate the output power of a real star coupler.   Remember that unlike the tee coupler, a manufacturer typically only provides the insertion loss for the coupler and does not take into account the loss for the splitting ratio.

Master It   Calculate the output power at each port of a star coupler with one input port and 10 output ports. Take into account the interconnection insertion losses. Assume the insertion loss for the coupler is 1.7dB and each interconnection insertion loss is 0.25dB. The input power to the coupler is –5dBm.

Solution   The total loss is the sum of the insertion losses for the interconnections (LI) and the insertion loss for the coupler (LC).

0.25dB × 11 = 2.75dB

2.75dB + 1.7dB = 4.45dB

Then account for the star coupler insertion loss and the interconnection losses. This is done by subtracting 4.45dB from the input power of –5dBm. The remaining power available to the ports is –9.45dBm, as shown here:

–5dBm – 4.45dB = –9.45dBm

Then split the remaining power using the decibel rules of thumb covered in Chapter 19. Since this has 10 output ports, each port will receive 10 percent of the energy because the energy is distributed evenly between the output ports. A loss of 90 percent is a change of 10dB. The output power at each port will equal –9.45dBm minus 10dB. Each output port will have a power output of –19.45dBm, as shown here:

–9.45dBm – 10dB = –e19.45dBm

Keep in mind that couplers are not perfect and the amount of power actually available will vary depending on the uniformity.

Calculate attenuator values.   Remember that if the maximum optical input power for a fiber-optic receiver is exceeded, the receiver will not operate properly.

Master It   Calculate the minimum attenuation required to prevent the receiver from being overloaded. Assume the transmitter power is –3dBm, the total losses are 6dB, and the maximum optical input power the receiver can tolerate is –14dBm.

Solution   To calculate the minimum attenuation required we need to subtract all the known losses from the output power of the transmitter, as shown here:

Transmitter power (TP) = –3dBm

Receiver maximum optical input power (MP) = –14dBm

Total losses (TL) = 6dB

Minimum attenuation required = MP + TL − TP

−14dBm + 6dB − (−3dBm) = −5dB

At a minimum, a 5dB attenuator is required. However, an attenuator with a larger value could be used as long as it did not over-attenuate the signal. Refer to Chapter 28 to determine the dynamic range of a receiver.

Chapter 30: Passive Optical Networks

Identify the different PON configurations.   FTTX can be used to describe any optical fiber network that replaces all or part of a copper network.

Master It   In this PON, the optical signals are converted right in front of the house or some distance down the block. What type of PON is this?

Solution   In an FTTC PON, optical fiber runs from the central office and stops at the curb. The curb may be right in front of the house or some distance down the block. Where the optical fiber stops is where the optical signal from the optical fiber is converted into electrical signals.

Identify the cables used in a PON.   Several different types of cables are employed in an FTTX PON.

Master It   This cable type runs from the central switching point to the local convergence point. What type of cable is it?

Solution   Feeder cables run from the central switching point to the local convergence point.

Identify different access points in a PON.   Cables connect the different access points in the PON.

Master It   What do you call the point where a distribution cable is broken out into multiple drop cables?

Solution   The network access point (NAP) is located close to the homes or buildings it services. This is the point where a distribution cable is broken out into multiple drop cables.

Chapter 31: Cable Installation and Hardware

Determine the minimum cable bend radius.   Remember that cable manufacturers typically define the minimum bend radius when the cable is stressed and unstressed. These numbers are typically in compliance with ANSI/TIA-568-C.3.

Master It   Refer to the following table and determine the minimum installation bend radius and the minimum operational bend radius for the 24-fiber cable.

Cable Type	Diameter	Weight	Minimum Bend Radius: Short Term	Minimum Bend Radius: Long Term	Maximum Load: Short Term	Maximum Load: Long Term
12-fiber cable	0.26″, 6.6mm	27lb/kft, 40kg/km	3.9″, 9.9cm	2.6″, 6.6cm	300lb., 1334N	90lb., 396N
24-fiber cable	0.375″, 9.5mm	38.2lb/kft, 56.6kg/km	5.5″ 14cm	3.7″, 9.4cm	424lb., 1887N	127lb., 560N

Solution   The short-term minimum bend radius in the table is the installation bend radius. The minimum installation bend radius for the 24-fiber cable is 5.5^2.

The long-term minimum bend radius in the table is the operational bend radius. The minimum operational bend radius for the 24-fiber cable is 3.7^2.

Determine the maximum cable tensile rating.   Remember that cable manufacturers typically define a dynamic and static pull load.

Master It   Refer to the following table and determine the maximum dynamic tensile load and the maximum static tensile load for the 12-fiber cable.

Cable Type	Diameter	Weight	Minimum Bend Radius: Short Term	Minimum Bend Radius: Long Term	Maximum Load: Short Term	Maximum Load: Long Term
12-fiber cable	0.26″, 6.6mm	27lb/kft, 40kg/km	3.9″, 9.9cm	2.6″, 6.6cm	300lb., 1334N	90lb., 396N
18-fiber cable	0.325″, 8.3mm	33.8lb/kft, 50kg/km	4.9″, 12.4cm	3.25″, 8.3cm	375lb., 1668N	112lb., 495N

Solution   The short-term maximum load in the table is the dynamic tensile load. The dynamic tensile load for the 12-fiber cable is 300lb.

The long-term maximum load in the table is the static tensile load. The static tensile load for the 12-fiber cable is 90lb.

Determine the fill ratio for a multiple-cable installation.   When preparing for a multiple cable conduit installation, you must be aware of the fill ratio for the conduit. If you are installing cable in existing conduit already occupied by one or more cables that contain electrical conductors, you will need to determine the fill ratio of the conduit prior to selecting your cable.

Master It   You need to install one cable into a conduit that has three existing electrical cables. The inside diameter of the conduit is 1^2. The cable to be installed has an outside diameter of 0.25^2. The outside diameters of the existing cables are 0.28^2, 0.32^2, and 0.38^2. Determine the fill ratio of the conduit before and after the fiber-optic cable is installed and verify that this installation will not overfill the conduit as define by the NEC.

Solution   The NEC specifies the following fill ratios by cross-sectional area for conduit:

1 cable: 53 percent

2 cables: 31 percent

3 or more cables: 40 percent

The first step is to determine the conduit fill ratio with the existing cables. To determine that fill ratio, use the following formula:

Fill Ratio = $(OD^2\text{cable}_1 + OD^2\text{cable}_2 + OD^2\text{cable}_3) / ID^2\text{conduit}$

Fill Ratio = $(0.28^2 + 0.32^2 + 0.38^2) / 1^2$

Fill Ratio = $(0.0784 + 0.1024 + 0.1444) / 1$

Fill Ratio = $0.3252 / 1$

Fill Ratio = 0.3252 or 32.52%

The next step is to determine the conduit fill ratio with the addition of the fiber-optic cable. To determine that fill ratio, use the following formula:

Fill Ratio = $(OD^2\text{cable}_1 + OD^2\text{cable}_2 + OD^2\text{cable}_3 + OD^2\text{cable}_4) / ID^2\text{conduit}$

Fill Ratio = $(0.28^2 + 0.32^2 + 0.38^2 + 0.25^2) / 1^2$

Fill Ratio = $(0.0784 + 0.1024 + 0.1444 + 0.0625) / 1$

Fill Ratio = $0.3877 / 1$

Fill Ratio = 0.3877 or 38.77%

The calculated fill ratio is less than 40 percent so the conduit is large enough as defined by the NEC.

Chapter 32: Fiber-Optic System Design Considerations

Calculate the bandwidth for a length of optical fiber.   The minimum effective bandwidth-length product in MHz km is in TIA-568-C.3.

Master It   Refer to the following table and determine the minimum bandwidth for 700 meters of 850nm laser-optimized 50/125µm multimode optical fiber.

Optical Fiber Cable Type	Wavelength (nm)	Maximum Attenuation (dB/km)	Minimum Overfilled Modal Bandwidth-Length Product (MHz · km)	Minimum Effective Modal Bandwidth-Length Product (MHz · km)
62.5/125µm Multimode	850 1300	3.5 1.5	200 500	Not required
50/125µm Multimode	850 1300	3.5 1.5	500 500	Not required
850nm laser-optimized 50/125µm multimode	850 1300	3.5 1.5	1500 500	2000 Not required

Optical Fiber Cable Type	Wavelength (nm)	Maximum Attenuation (dB/km)	Minimum Overfilled Modal Bandwidth-Length Product (MHz · km)	Minimum Effective Modal Bandwidth-Length Product (MHz · km)
Single-mode indoor-outdoor	1310	0.5	N/A	N/A
	1550	0.5	N/A	N/A
Single-mode inside plant cable	1310	1.0	N/A	N/A
	1550	1.0	N/A	N/A
Single-mode outside plant cable	1310	0.5	N/A	N/A
	1550	0.5	N/A	N/A

Solution The steps to calculate the minimum bandwidth for 700m of 850nm laser-optimized 50/125µm multimode optical fiber are shown here:

Bandwidth-length product = 2,000 MHz · km

Optical fiber length = 0.7km

Bandwidth in MHz = 2,000 ÷ 0.7

Bandwidth = 2,857.14MHz

Calculate the attenuation for a length of optical fiber. The maximum allowable attenuation in an optical fiber is defined in section 4.2 of ANSI/TIA-568-C.3.

Master It Refer to the following table and determine the maximum allowable attenuation for 5.6km of outside plant single-mode optical fiber.

Optical Fiber Cable Type	Wavelength (nm)	Maximum Attenuation (dB/km)
50/125µm multimode	850	3.5
	1300	1.5
62.5/125µm multimode	850	3.5
	1300	1.5
Single-mode inside plant cable	1310	1.0
	1550	1.0
Single-mode outside plant cable	1310	0.5
	1550	0.5

Solution The maximum attenuation for outside plant single-mode optical fiber is the same for both wavelengths. The steps to calculate the maximum attenuation are as follows:

Attenuation coefficient at 1310 or 1550nm = 0.5dB/km

Optical fiber length = 5.6km

Attenuation for the length of optical fiber = 5.6 × 0.5 = 2.8

Maximum attenuation for 5.6m of outside plant single-mode optical fiber is 2.8dB

Calculate the power budget and headroom for a multimode fiber-optic link. Before we can analyze the link and do a power budget, we need to know the optical performance characteristics for transmitter and receiver.

Master It   Refer to the following graphic and two tables to calculate the power budget and headroom for the multimode fiber-optic link with 50/125μm multimode optical fiber.

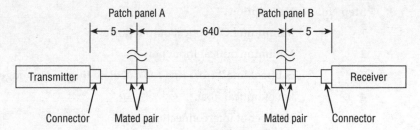

Distance in meters (not to scale)

Parameter		Symbol	Min.	Typ.	Max.	Unit
Optical output power	BOL	P_0	−19	−16.8	−14	dBm avg.
62.5/125μm, NA = 0.275 fiber	EOL		−20			
Optical output power	BOL	P_0	−22.5	−20.3	−14	dBm avg.
50/125μm, NA = 0.20 fiber	EOL		−23.5			
Optical extinction ratio				0.001	0.03	%
				−50	−35	dB
Optical output power at logic "0" state		P_0 ("0")			−45	dBm avg.
Center wavelength		λ_c	1270	1300	1380	nm
Spectral width—FWHM		Δ		137	170	nm
Optical rise time		t_r	0.6	1.0	3.0	ns
Optical fall time		t_f	0.6	2.1	3.0	ns
Duty cycle distortion contributed by the transmitter		DCD		0.02	0.6	ns_{p-p}
Duty cycle jitter contributed by the transmitter		DDJ		0.02	0.6	ns_{p-p}

Parameter	Symbol	Min.	Typ.	Max.	Unit
Optical input power minimum at window edge	$P_{IN\,Min.}$ (W)		−33.5	−31	dBm avg.
Optical input power maximum	$P_{IN\,Max.}$	−14	−11.8		dBm avg.
Operating wavelength	λ_c	1270		1380	nm

Solution   Using the minimum EOL optical output power from the first table and the maximum value for minimum optical input power from the second table, calculate the power budget as shown in the table that follows. Next, calculate all the link losses for optical fiber, interconnections, and splices, as shown in the following table. Next, calculate the total link loss and finally the headroom as shown in that table.

Step No.	Description	Value
1	Minimum EOL optical output power.	–23.5dBm
2	Minimum optical input power.	–31dBm
3	Subtract line 2 from line 1 to calculate the power budget.	7.5dB
4	km of optical fiber.	.65
5	Number of interconnections.	2.0
6	Number of splices.	0
7	Multiply line 4 × 3.5.	
8	Multiply line 4 × 1.5.	0.975
9	Multiply line 5 × 0.75.	1.5
10	Multiply line 6 × 0.3.	0
11	Add lines 7, 9, and 10 for total link loss at 850nm.	
12	Add lines 8, 9, and 10 for total link loss at 1300nm.	2.475dB
13	Subtract line 11 from line 3 for headroom at 850nm.	
14	Subtract line 12 from line 3 for headroom at 1300nm.	5.025dB

Calculate the power budget and headroom for a single-mode fiber-optic link.   Before we can analyze the link and do a power budget, we need to know the optical performance characteristics for transmitter and receiver.

Master It   Refer to the following graphic and two tables to calculate the power budget and headroom for the single-mode fiber-optic link.

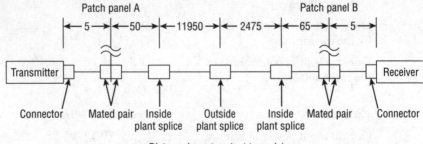

Distance in meters (not to scale)

Parameter	Symbol	Min.	Typ.	Max.	Unit
Optical output Power 9μm SMF, EOL	P_{OUT}	–5	–2	0	dBm
Center wavelength	λ_c	1260		1360	nm
Spectral width—rms	Δ_c		1.8	4	nm rms
Optical rise time	t_r		30	70	ps
Optical fall time	t_f		150	225	ps
Extinction ratio	E_R	8.2	12		dB
Optical output eye	Compliant with eye mask Telecordia GR-253-CORE				
Back-reflection Sensitivity				–8.5	dB
Jitter generation	pk to pk			70	mUI
	RMS			7	mUI

Parameter	Symbol	Min.	Typ.	Max.	Unit
Receiver sensitivity	P_{IN} MIN		–23	–19	dBm avg.
Receiver overload	P_{IN} MAX	0	+1		dBm avg.
Input operating wavelength	λ	1260		1550	nm
Signal detect — asserted	P_A		–27.3	–19.5	dBm avg.
Signal detect — deasserted	P_D	–35	–28.7		dBm avg.
Signal detect — hysteresis	P_H	0.5	1.4	4	dB
Reflectance			–35	–27	dB

Solution Using the minimum EOL optical output power from the first table and the maximum value for minimum optical input power from the second table, calculate the power budget as shown in the following table. Next, calculate all the link losses for optical fiber, interconnections, and splices as shown in the table that follows. Next, calculate the total link loss and then the headroom, as shown in the table that follows.

Step No.	Description	Value
1	Minimum EOL optical output power.	–5.0dBm
2	Minimum optical input power.	–19.0dBm
3	Subtract line 2 from line 1 to calculate the power budget.	14.0dB
4	km of inside plant optical fiber.	0.125
5	km of outside plant optical fiber.	14.425

6	Number of interconnections.	2
7	Number of inside plant splices.	2
8	Number of outside plant splices.	1
9	Multiply line 4 × 1.0.	0.125
10	Multiply line 5 × 0.5.	7.2125
11	Multiply line 6 × 0.75.	1.5
12	Multiply line 7 × 0.3.	0.6
13	Multiply line 8 × 0.1.	0.1
14	Add lines 9, 10, 11, 12, and 13 for total link loss.	9.5375dB
15	Subtract line 14 from line 3 for headroom.	4.4625dB

Chapter 33: Test Equipment and Link/Cable Testing

Perform optical loss measurement testing of a cable plant.   Remember ANSI/TIA-526-14-A provides three methods for testing the cable plant. The difference between the three methods is how the reference power measurement is taken. Each method uses a different number of test jumpers. Method A uses two test jumpers to obtain the reference power measurement, method B uses one test jumper, and method C uses three test jumpers.

Master It   Prior to testing the cable plant using method B, you obtained your reference power measurement of –23.5dBm. Your test power measurement was –27.6dBm. What was the loss for the cable plant?

Solution   To obtain the optical power loss for the cable plant, subtract the test power measurement from the reference power measurement. In this case, the reference power measurement was –23.5dBm and the test power measurement was –27.6dBm, so the loss for the cable plant would be 4.1dB.

Determine the distance to a break in the optical fiber with an OTDR.   Remember a break in an optical fiber in a fiber-optic cable is not the end of the optical fiber. However, a break in an optical fiber looks like the end of the optical fiber on the OTDR display. The same method is employed to measure the distance to a break in an optical fiber and to measure the distance to the end of an optical fiber.

Master It   You are troubleshooting fiber-optic cables with the OTDR. One optical fiber appears to be broken. Refer to the following graphic and determine the distance to the break in the optical fiber.

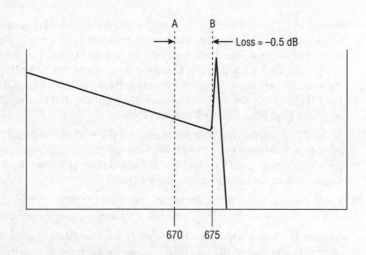

Solution To measure the distance to the break in the optical fiber, zoom in on the back reflection, and place the A cursor on a smooth section of the trace just in front of the back reflection. Move the B cursor toward the A cursor until it is in the leading edge of the back reflection. Keeping moving the B cursor toward the A cursor until the A-B loss is −0.5dB. The distance to the break is the distance for the B cursor, which is 675m.

Measure the loss of a cable segment with an OTDR. Remember that to find the loss for a cable segment plus the interconnections, the exact length of the segment must be known.

Master It You are testing a new fiber-optic cable installation with the OTDR and the customer has requested the loss information for a cable segment including the interconnections. Refer to the following graphic and determine the cable segment.

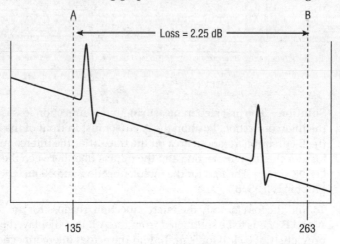

Solution   The first step in finding the loss for a cable segment with interconnections is to measure the length of the cable segment. After zooming in on the cable segment, place the A cursor on a smooth section of the trace just in front of the leftmost cable segment interconnection back reflection. Place the B cursor on the smooth part of the trace after the rightmost cable segment interconnection back reflection. Looking at the A-B distance on the OTDR, move the B cursor to the right until the distance equals the length of the cable segment plus 50m. The A-B loss is 2.25dB.

To find the loss for the cable segment plus the interconnections, subtract the loss for the 50m of optical fiber from the 2.25dB loss for the cable segment and interconnections. The previously measured loss for 50m of optical fiber at 850nm is 0.14dB. The loss for the cable segment and the interconnections is 2.11dB.

Measure the loss of an interconnection with an OTDR.   When measuring the loss for an interconnection with the OTDR, you must remember to subtract the loss for the optical fiber.

Master It   You are testing a fiber-optic cable installation with the OTDR and the customer has requested the loss information for each interconnection. Refer to the following graphic and determine the cable segment.

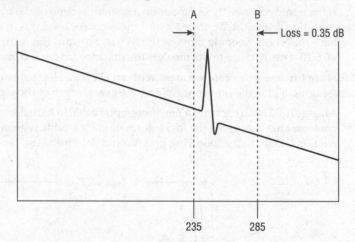

Solution   The first step in measuring interconnection loss is to horizontally zoom in on the interconnection. Position the A cursor just in front of the back reflection, then position the B cursor on a smooth area on the trace after the interconnection back reflection. The loss for the interconnection and the optical fiber between the cursors is displayed on the OTDR screen. The loss for the interconnection and 50 meters of optical fiber shown in the graphic is 0.35dB.

To find the loss for only the interconnection, the loss for the optical fiber between cursors A and B needs to be subtracted from the A-B loss displayed by the OTDR. This loss was previously at 0.14dB for 50m. To find the actual interconnection loss, subtract 0.14dB from the 0.35dB A-B loss. The loss for only the interconnection is approximately 0.21dB.

Measure the loss of a fusion splice or macrobend with an OTDR.   Remember fusion splices or macrobends do not produce a back reflection. A macrobend will always appear as a loss in the form of a dip in the smooth trace. A fusion splice may appear as a dip or a bump in the trace.

Master It You are testing a fiber-optic cable installation with the OTDR and see a dip in the trace. Since there are no splices in this cable, you determine that this event must be a macrobend. Refer to the following graphic and determine the loss for the macrobend.

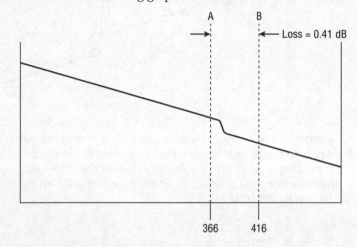

Solution To find the loss of a macrobend, horizontally zoom in on the event. The loss from a macrobend may be very small and require vertical zoom in addition to horizontal zoom. Place the A cursor on the smooth part of the trace before the dip in the trace. Place the B cursor on the smooth part of the trace after the dip, as shown in the graphic. The loss for this event is 0.41dB. The loss for the event includes the loss for the macrobend plus the 50m of optical fiber between the cursors. Subtract the loss for the 50m of optical fiber that was previously measured at 0.14dB from the event loss. The loss for this fusion splice or macrobend is 0.27dB.

Chapter 34: Troubleshooting and Restoration

Identify a high-loss interconnection with an OTDR. Remember a bad interconnection can be quickly identified by following the sloping line as it enters and exits the interconnection back reflection. Visualize the sloping line traveling through the back reflection and exiting.

Master It You are troubleshooting a new fiber-optic cable installation with the OTDR and the trace shown in the following graphic. The back reflection is the result of an interconnection. Using only the relative decibel information on the vertical scale, approximate the loss for the interconnection and determine whether this interconnection meets the loss requirements defined by ANSI/TIA-568-C.3.

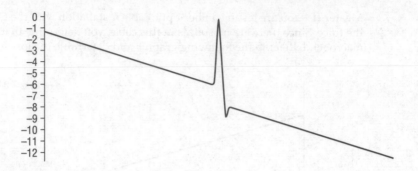

Solution The vertical axis shows relative power in decibels. Look at the amount of energy entering the interconnection and exiting the interconnection. Using the vertical scale on the left, approximate the loss as shown in the following graphic. The approximate loss for this interconnection is 2dB, which exceeds the 0.7 dB maximum interconnection loss allowed by ANSI/TIA-568-C.3.

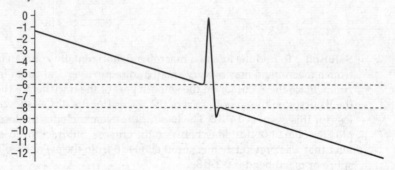

Identify a high-loss mechanical splice with an OTDR. Remember a high-loss mechanical splice can be quickly identified by following the sloping line as it enters and exits the splice back reflection. Visualize the sloping line traveling through the back reflection and exiting.

Master It You are troubleshooting an older fiber-optic cable installation with the OTDR and the trace shown in the following graphic. The back reflection is the result of a mechanical splice. Using only the relative decibel information on the vertical scale, approximate the loss for the splice and determine whether or not this splice meets the loss requirements defined by ANSI/TIA-568-C.3.

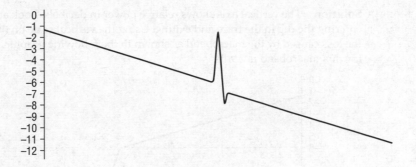

Solution The vertical axis shows relative power in decibels. Look at the amount of energy entering the splice and exiting the splice. Using the vertical scale on the left, approximate the loss as shown in the graphic. The approximate loss for this mechanical splice is 1dB, which exceeds the 0.3dB maximum loss allowed by ANSI/TIA-568-C.3

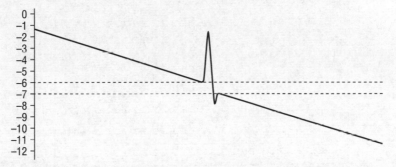

Identify a high-loss fusion splice or macrobend with an OTDR. Remember a high-loss fusion splice or macrobend can be quickly identified by the dip in the trace. These events do not produce a back reflection.

Master It You are troubleshooting a new fiber-optic cable installation with the OTDR and the trace shown in the following graphic. The dip in the trace is the result of a macrobend. Using only the relative decibel information on the vertical scale, approximate the loss for the macrobend.

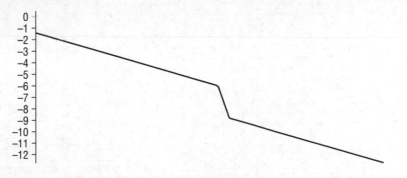

Solution The vertical axis shows relative power in decibels. Look at the amount of energy entering the dip in the trace and exiting. Using the vertical scale on the left, approximate the loss caused by the macrobend as shown in the following graphic. The approximate loss for this macrobend is 2.9dB.

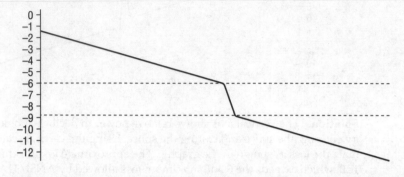

Appendix B

Cabling Resources

This appendix contains information about vendor resources, Internet sites, books, publications, and other tools that may be useful for learning more about cabling. The resources are generally listed in order of what we find most useful.

Informational Internet Resources

The Internet is wonderful because it enables speedy communication and access to valuable information. Without the Internet, the time required to complete projects such as this book would be much longer.

Information is changing very quickly and even though many of these resources were current the last time we visited them, they may not be up-to-date when you read this book—especially those websites that are maintained by one or two individuals.

Also note whether the website is for-profit or not-for-profit. A for-profit website may have a particular bias with regard to cable or connectivity selection and that might not always correspond to what you really need.

comp.dcom.cabling

If you have newsreader software (such as Outlook Express), point your newsreader to the Usenet newsgroup comp.dcom.cabling. This group can also be accessed through Google groups at groups.google.com. This interactive forum has a plethora of information. You can post your own questions, respond to others' questions, or just read the existing postings and learn from them. This particular forum has a number of dedicated and knowledgeable individuals who monitor it and try to assist everyone who posts queries. A word to the wise: because this forum is not moderated, any self-proclaimed expert can post responses to questions. Use common sense, and if an answer doesn't seem right, get a second opinion.

Whatis

Whatis is one of our favorite reference sites on the Internet; its URL is http://whatis.techtarget .com. It contains more than 2,000 commonly used computer and telecommunications terms and seems to grow every day. Also included at this reference site is the word of the day, information for the beginner, concepts, and book recommendations.

Wikipedia

By far the best maintained and most current website is Wikipedia (www.wikipedia.org), a website that has content contributors from all over the world and donates any proceeds to charity. At the end of every article there are Notes, References, and External Links for accessing additional information. Unlike many other noncommercial websites, it is not maintained by one or two individuals, but by experts in their respective fields from the entire Internet community.

TIA Online

The Telecommunications Industry Association's website is found at www.tiaonline.org. This site is the place to go for updated information on the TIA committee meetings, current proposals and standards, and current events; it also includes a tremendous glossary.

Fiber Optics LAN Section (FOLS)

The Fiber Optics LAN Section (FOLS) of the Telecommunications Industry Association is a consortium of leading optical fiber cable, components, and electronics manufacturers dedicated to increasing the use of optical fiber technologies in customer-owned networks. They provide cases studies and are well known for an interactive cost model that compares copper to fiber-optic networks for a variety of network architectures. Their website is found at www.fols.org.

TechFest

Is everything you ever wanted to know about networking here? Well, not quite, but it is pretty close. A huge amount of information is available on LANs, WANs, cabling, protocols, switching, networking standards, ATM, and more at www.techfest.com/networking/.

TechEncyclopedia

CMP's TechWeb sponsors the TechEncyclopedia. This thorough listing of more than 13,000 computer- and technology-related terms can be found at www.techweb.com/encyclopedia/.

National Electrical Code Internet Connection

This site is operated by Mike Holt (better known among electrical and data cabling professionals as Mr. Code). Mike Holt is *the* expert on the National Electrical Code, and he gives excellent seminars around the world. Visit his site for more information on the National Electrical Code, some really interesting stories from people in the field, his free email newsletter, and more. Holt has also written a number of books on how to interpret and work with the National Electrical Code; professional electricians and data communications designers should own his book on the 2002 NEC. The site can be found at www.mikeholt.com.

Charles Spurgeon's Ethernet Website

This site, at www.ethermanage.com/ethernet/, is the first place on the Internet we go for information on Ethernet, Fast Ethernet, and Gigabit Ethernet. Extensive information can be found here about various Ethernet technologies, planning cabling for Ethernet, and Ethernet analyzing software, as well as FAQs, technical papers, a history of Ethernet, and troubleshooting information.

On the home page is a neat drawing of Ethernet done by Bob Metcalfe, the original designer of Ethernet.

ATIS Telecom Glossary

This is one of the most thorough Internet sites you can use for finding technology-related terms. The standard was prepared by T1A1, the Technical Subcommittee on Performance and Signal Processing, and can be found at www.atis.org/glossary/.

Protocols.com

Though it's not specifically related to cabling, we found a site for learning more about networking and communications protocols. The site, at www.protocols.com, includes protocol references and information on physical network interfaces.

Webopedia: Online Computer Dictionary for Internet Terms and Technical Support

Another valuable site for finding technology-related terms is Webopedia. This online dictionary, www.webopedia.com, has a thorough computer dictionary, technology-related news, and a listing of the top terms.

Books, Publications, and Videos

You may wish to own a number of the following books and videos for your professional library. In addition, two publications that we highly recommend are listed in this section.

Cabling Business Magazine

Cabling Business Magazine covers copper and fiber-optic cabling for voice, data, and imaging. The monthly has features written by some of the leaders in the industry, a stock-market watch, how-to columns, and more. *Cabling Business Magazine* also offers seminars and classes on a variety of telecommunications topics. For more information on the magazine or information on a free subscription to qualified subscribers, check out the publication on the Web at www.cablingbusiness.com.

Cabling Installation and Maintenance Magazine

Cabling Installation and Maintenance is a monthly magazine published by PennWell that has columns (including Q&A), a Standards Watch, and more. Its website, at http://cim.pennnet.com, has links to contractors, a buyers' guide, a calendar of events, and an article archive. Online subscriptions are free to qualified print subscribers.

The Fiber Optic Association (FOA)

The Fiber Optic Association (www.thefoa.org) is an international nonprofit professional society for the fiber-optic industry. Its charter is to develop educational programs, certify fiber-optic technicians, approve fiber-optic training courses, participate in standards-making processes, and generally promote fiber optics. The FOA has certified over 22,000 CFOTs (Certified Fiber-Optic Technicians) through over 180 approved training organizations worldwide.

Newton's Telecom Dictionary

The 24th edition of *Newton's Telecom Dictionary* (Flatiron Publishing, 2008), by Harry Newton, is a 1,120-page guide to the world of modern data and voice telecommunications terms. It's not easy to keep up with all the terms, acronyms, and concepts today, but this book is a great start. Anyone who owns one will tell you it is indispensable. It is available through most bookstores.

Premises Network

This virtual community is designed for people who work with premises cabling. It includes a buyers' guide and allows you to submit an RFQ/RFP, respond to an RFQ/RFP, search for jobs, and purchase used equipment and products in its online marketplace. The site can be found at www.premisesnetworks.com.

BICSI's *Telecommunications Distribution Methods Manual* and *Information Transport Systems Installation Methods Manual*

The Building Industry Consulting Services International's *Telecommunications Distribution Methods Manual* (TDMM) and *Information Transport Systems Installation Methods Manual* are great study guides for the RCDD certification and excellent resources for professional cable installers. BICSI members can get discounts to these publications on the BICSI website (www.bicsi.org).

ANSI/TIA-568-C Commercial Building Telecommunication Cabling Standard

The ANSI/TIA-568-C standard is the definitive guide to commercial building cabling. All professional cable designers and installers should own or have access to this standards document. It can be purchased online through Global Engineering Documents at http://global.ihs.com. The company has a great deal on a CD containing the entire structured wiring collection of standards, which includes free updates for a year.

Manufacturers

To say that many manufacturers and vendors handle telecommunications and cabling products would be a bit of an understatement. The following is a list of vendors and manufacturers that we have found to provide not only good products but also good service and information.

The Siemon Company

Telecommunications vendor The Siemon Company has an informative website, at www.siemon.com, which includes technical references, standards information, an online catalog, white papers, and frequently asked questions. If you visit the site, order the company's catalog—it is one of the best telecommunications catalogs in the industry.

MilesTek, Inc.

MilesTek has one of the neatest sites and is one of the easiest companies on the Internet to work with for purchasing cabling supplies and tools. It also has a good catalog and helpful people on the ordering side if you are not quite ready for online commerce. The company sells cabling supplies, tools, cable, components, connectors, and more. Its site, at www.milestek.com, also has good information on standards, cabling, and telecommunications.

IDEAL Industries, Inc.

IDEAL Industries, Inc., is a leading supplier of cabling and wiring tools and supplies. If you have handled any cabling tools, you have probably used one of IDEAL's tools. Its website, at www.idealindustries.com, has much useful information about the tools that it sells, as well as tips and tricks for premises cabling and electrical wiring. The company's customer service is good.

Leviton

Part of a 100-year-old company, Leviton's Network Solutions product line was created in 1988 to meet the growing need for telecommunications and high-speed data technologies. Today, the division provides complete copper, fiber, and power network connectivity for enterprise, data center, service provider, and residential applications. They can be found on the Web at www.leviton.com.

Ortronics

Manufacturer Ortronics maintains one of our favorite websites and catalogs. Ortronics also offers training and a certified installer program. Its catalog is easy to understand and follow. You can find the company on the Internet at www.ortronics.com.

Superior Essex

Superior Essex is one of the largest manufacturers of premises, outside plant, and fiber-optic cable in the world. Its website, at www.superioressex.com, has excellent technical information pertaining to copper and fiber cabling as well as performance specifications for its products.

CommScope

CommScope is one of the world's largest manufacturers of premises and outside plant copper and fiber-optic cable and connectivity for the broadband/CATV, enterprise, and wireless markets. They can be found on the Web at www.commscope.com. CommScope includes the SYSTIMAX division, formerly of Avaya and Lucent Technologies.

Jensen Tools

Jensen Tools come in a huge variety of tools, toolkits, and other products for the computer and telecommunications professional. They can be purchased from Stanley Tools. Visit their website at www.stanleysupplyservices.com and order their catalog.

Labor Saving Devices, Inc.

This company has one of the best selections we've found of cabling installation tools for those who need to retrofit existing structures. In addition to its impressive product offering, Labor Saving Devices, Inc. offers online instructions and an interactive Q&A forum. You can find LSD Inc.'s website at www.lsdinc.com.

OFS

OFS is a world-leading designer, manufacturer, and provider of optical fiber, optical fiber cable, FTTX, optical connectivity, and specialty photonics products. OFS's corporate lineage dates back to 1876 and included technology powerhouses such as AT&T and Lucent Technologies (now Alcatel-Lucent). Today, OFS is owned by Furukawa Electric Co., Ltd., a multibillion-dollar global leader in optical communications. Visit their website at www.ofsoptics.com.

Erico

Erico is a leading manufacturer of electrical products, including the CADDY system of fastening and fixing solutions. These products help you to better organize and run cabling. Visit Erico on the Internet at www.erico.com and check out the CADDY Fixings, Fasteners, and Support sections of its website.

Berk-Tek

Cable manufacturer and supplier Berk-Tek, part of the Nexans company, has a great website that includes good technical sections. You can find it at www.berktek.com.

Fluke

Fluke is a premier manufacturer of electronic test equipment. Its DSP series of LAN testers are widely used by the professional installer community. Fluke's website address is www.fluke.com.

Panduit

Panduit is a leading supplier of telecommunications products and tools. Its website, at www.panduit.com, offers online ordering and information about its products.

Anixter

Telecommunications product vendor Anixter has a website that includes one of the most extensive technical libraries on cabling and standards found on the Internet. It can be found at www.anixter.com.

Graybar

Graybar is one of the leading distributors of network cabling. It can be found at www.Graybar.com.

Communications Supply Corporation

Communications Supply Corporation offers structured cabling systems from major manufacturers. They are part of the Wesco Distribution family, and their website can be found at www.gocsc.com.

Appendix C

Registered Communications Distribution Designer (RCDD) Certification

Certification programs are all the rage in the technology industry. The cabling business is no exception. As with all industry certifications, an organization has to be responsible for the certification program, manage the testing, and set the quality requirements. BICSI (Building Industry Consulting Services International) is a professional organization devoted to promoting standards for the design and installation of communications systems and is instrumental in the provision of guidelines, training, and professional certification of knowledge and experience related to the communication infrastructure.

The breakup and divestiture of the Bell system in 1984 created a void of expertise where communication system design and deployment was concerned. Not only was a single national entity no longer governing standard practices, but also many of the old-timers (the ones who really knew what worked—*and why*) left the baby Bells during, and shortly following, the divestiture period.

BICSI recognized the need to acknowledge and certify cabling professionals; it filled the void with a two-pronged effort. First, BICSI developed and published the *TDMM* (*Telecommunications Distribution Methods Manual*), a compendium of guidelines, standards, and best practices for the engineering of a communications system. Second, it initiated the RCDD (Registered Communications Distribution Designer) accreditation program to establish a benchmark of expertise for the industry.

BICSI membership exceeds 23,000 members worldwide, representing more than 90 countries. More than 6,400 members have achieved the professional accreditation of Registered Communications Distribution Designer. The designation of RCDD is not awarded lightly. It takes knowledge, experience, and much hard work to qualify for those four letters that follow your name on your business card and stationery.

Today, the RCDD is a recognized standard of excellence and professionalism within the communications industry. Increasingly, architects and building-management consortia specify that an RCDD must perform the design of a building project's low-voltage systems: video, security, and communications. Often, the contractor awarded the installation work is required to have an RCDD either on staff or as a direct supervisor of the work performed. In fact, BICSI estimates that RCDD participation is a requirement for approximately 60 percent of structured wiring RFQs. In some companies, obtaining your RCDD is a condition of employment if you are in technical sales or technical support. As a result, RCDDs are very much in demand in the job market. These reasons are pretty compelling ones to become an RCDD if you are serious about working in the field of communications infrastructure.

How do you get there? You will go through three stages:

1. Apply and be accepted as a candidate for the designation of RCDD.

2. Successfully pass the stringent RCDD exam.

3. Maintain your accreditation through continuing membership and education.

Apply and Be Accepted as a Candidate for the Designation of RCDD

You must be a member of BICSI before you can apply to become an RCDD. Membership requires a nominal annual fee, currently $150 for an individual. (BICSI is a nonprofit organization, and all its fees are very reasonable compared to other professional organizations of the same caliber.)

NOTE You can find out more about membership in BICSI by visiting its website at www.bicsi.org.

Next, you must submit your qualifications to become an RCDD. BICSI doesn't let just anyone sit for the exam; you must be a bona fide member of the industry. The requirement is part of the quality control that elevates the RCDD designation in the industry. BICSI requires that you have a minimum of two years of system design experience. You show your experience by doing the following:

◆ Filling out an application, much as if you were applying for employment. You list work experience, educational background, and any awards, accomplishments, or other professional credentials you may have. The application includes a Communications Distribution Design Experience list that you must complete by ranking your experience in a number of different distribution design areas.

◆ Supplying three letters of reference:

  ◆ One letter from your current employer detailing your involvement in design activities

  ◆ One from a client for whom you have performed design work

  ◆ One personal reference touting you as an all-around swell person

◆ Sending a nonrefundable application fee of $295 for members and $495 for nonmembers, and waiting a couple of weeks to see if you are accepted as a candidate to sit for the exam.

NOTE BICSI members, RCDD candidates, and those who have written letters of reference for colleagues and employees will attest to the fact that BICSI diligently follows up these letters of reference to establish their legitimacy.

Successfully Pass the Stringent RCDD Exam

Okay, you've been accepted, now what? Well, next is the hard (some would say *grueling*) part. You have to study for and take the RCDD examination, which consists of 280 multiple-choice questions drawn from BICSI's *TDMM*. Each copy of the exam is unique. It is given during a strictly proctored 3½-hour session. To be admitted to the test room, you must show a photo ID.

The testing procedure includes using sealed envelopes and a personal test-code number that's assigned to you. The exam is a closed-book test, and no calculators or reference materials are permitted in the test room. You won't be told your score—only that you either passed or failed. These procedures may sound hokey and unnecessary, but in fact, they ensure the integrity of the RCDD title by eliminating the possibility of cheating or favoritism and putting all RCDDs on level ground.

The exam is difficult. It takes a grade of 78 percent to pass, meaning you can miss only 61 of the 280 questions. Only 30 to 40 percent of those who take the test pass the first time. BICSI allows you to retake the test up to three times within one year. If, after that third attempt, you still don't pass, you must wait a full year and then begin the application process again.

The potential test questions all come directly from the *TDMM*. This is a very important point, and you will fail if you don't take it to heart. *Do not* answer questions from your own experience or based on reference material other than the *TDMM*. The *TDMM* is the bible as far as the RCDD exam is concerned, and you should *only* respond to test questions with answers from the *TDMM*, *regardless of whether or not you agree*. RCDDs are convinced that two reasons account for the 60 to 70 percent first-time failure rate. The first is that people come to the test believing their own experience and know-how should supersede what BICSI teaches in the *TDMM*; the second is that they simply don't study the *TDMM* enough.

For an RCDD candidate with two years of experience and heavy specialization in a few areas of communication-system distribution design, BICSI recommends the following study regimen prior to sitting for the exam:

- Study the *TDMM* for 125 hours.
- Attend the six-day BICSI DD102 distribution system design course.
- Attend the four-day BICSI DD200 distribution system review course immediately prior to the exam.

NOTE   More information on the BICSI classes can be found on BICSI's website at www.bicsi.org.

More extensive experience over a broader range of subjects reduces the number of hours required in study and can eliminate the need for the distribution design courses. However, we recommend the courses to anyone who even slightly doubts his or her ability to pass the exam. Who knows—even skilled professionals may learn something in the process.

Here's the regimen successful candidates have used in the past. Almost everyone who followed this course of study passed the RCDD test on the first try, which is a pretty good endorsement:

- Study the *TDMM* as much as possible prior to any course work.
- Attend the BICSI DD102 class. (Note: In general, first-time test takers who attend this class as administered by BICSI raise their chances of passing significantly. Some instructor teams have 70 percent or better first-time pass rates for students in their classes.)

♦ Participate in study groups in the evenings during the DD102 class. Studying and quizzing each other using the *TDMM* or flash cards greatly reinforces the learning and memorization required to pass the exam.

♦ Purchase and use one of the RCDD practice test packages available from third-party vendors. One such third-party practice exam is from NetCBT (formerly Clark Technology Group); it is computerized test software. You can find the company at www.netcbt.com. Purchase at least two practice tests. Then, begin testing yourself with one version. Use it over and over, interweaving it with study periods. When you are consistently achieving above 90 percent correct on the first practice test, switch to the other and follow the same procedure. These tests are also a great adjunct to the study group if you use them interactively and answer by panel instead of by individual. Discuss your answers with the group until you all agree why the answer is right or wrong.

♦ Take the RCDD exam immediately following the DD102 course. There is some controversy over this approach. Some think it is better to wait, let the course material sink in, then take the DD200 as a refresher. Many successful candidates have gone full steam ahead right into the exam and been successful.

TIP Assuming you're going for broke directly from DD102 into the exam, your brain will be packed to bursting with information from the *TDMM*. The most fragile items are the numerous tables that you've memorized, on information like the number and size of conduits required to service a building of *x* square feet. These tables don't lend themselves to logical sequence or mnemonic clues. The morning of the exam, you won't want to talk with anyone for fear that you'll lose concentration and these delicate matrices will collapse like a house of cards. Get your test packet, tear it open, and dump your brain onto the back of the test. Then, start the exam.

The exam is offered at each BICSI conference. Three of these are held each year in the United States, with several others held in other parts of the world. The exam is also offered immediately following BICSI courses such as the DD200 and DD102. In addition, the exam is scheduled periodically in various locations across the United States. Finally, special proctoring of the exam can be scheduled for a special-purpose group, such as when a company processes a large number of RCDD candidates at the same time. Usually, such exams are held in conjunction with specially scheduled design courses.

One interesting fact about the exam: BICSI recognizes that neither the *TDMM* nor the exam is absolutely infallible. The *TDMM* has different sections written by different individuals. In a few cases, contradictory information is given. A few buggy questions are also in the question bank (although BICSI works hard to weed them out).

If you think you've encountered one of these contradictory or erroneous questions, answer it as best you're able but make a note by the question to refer to the back of the test. There, you are allowed to challenge the question by providing an explanation of why you believe the question to be faulty. It's best if you can quote chapter and paragraph numbers for the conflicting or bad information (doing so is not far-fetched, considering the preparation you'll have done). If you really know your stuff, you will probably catch a buggy question. Though one question probably won't make a difference if you are ready to take the test, it could swing the balance if your score is marginal.

The test costs $100 each time you re-take it. This is in addition to the membership fee and RCDD application fee you will have already paid.

TIP Don't rely completely on the practice tests to memorize potential types of questions. Learn the material and concepts and then use the tests to ensure that you are ready. Memorizing test-type questions will only be a disservice to you and the industry.

Maintain Your Accreditation Through Continuing Membership and Education

When you've passed (congratulations!), you can't just sit back and coast. The world of communications changes very rapidly. BICSI recognizes that for RCDDs to be effective in servicing clients, they must keep up with ever-changing standards and best practices.

Your RCDD designation is awarded for a three-year period. During those three years, you must maintain your BICSI membership. *Don't let your membership lapse, or you'll have to sit for the exam again.* (Yikes!)

You must also accumulate a minimum of 45 Continuing Education Credits (CECs). You can accomplish that by attending more BICSI or third-party education courses (check with the BICSI office to see if the third-party courses are sanctioned for CECs) or by attending BICSI's scheduled technical conferences.

A BICSI conference is worth 15 CECs. You must attend at least one during your three-year period as an RCDD. By attending one each year, you accumulate all the CECs you need for renewal of your RCDD accreditation.

A BICSI conference is packed with presentations, workshops, or seminars that will keep you up-to-date on what's happening in your industry. Virtually everyone will have an update on what's happening in the industry-standard development committees, such as the committee that works on ANSI/TIA-568-C. New developments in areas such as cable performance, connectors, electronics, fiber versus twisted pair, and changes to the NEC are all topics of regular discussion. In addition, information on how to better serve your customers and run your business is often provided. Nightly receptions allow you to network with other industry professionals and review the offerings of vendors that are mainstays of the communications market.

Check Out BICSI and the RCDD Program for Yourself

Visit the BICSI website at www.bicsi.org for additional information on the organization and the RCDD program. You can download applications, view membership lists, review presentations made at prior BICSI conferences, read status reports from the TIA working groups involved in cabling standards, and much more.

Appendix D

Home Cabling: Wiring Your Home for Now and the Future

In case you haven't noticed, you're standing knee-deep in the future of data interchange. High-speed, inexpensive computers, high-bandwidth network technologies, IP-telephony, Internet-connected entertainment devices, HDTV, and the usefulness of the Internet as a whole are advancing faster than most can assimilate and react to.

Corporations, to a large degree, are keeping up when it comes to providing LAN and Internet functionality for their employees. The home, however, is the last bastion of low-end technology for data exchange. Even the home you just built may only have voice-grade wiring in it—and it may not even support more than one phone line. Category 5e wiring may be the buzz in new houses above a certain price point, but few contractors know how to properly install and terminate it.

If cable is improperly pulled or terminated, it will be useless for data communications. You can quickly convert Category 5e to Category 1 by stretching it, crimping it, daisy-chaining it, bridge tapping, or using low-quality connectors.

Many homes constructed today simply do not have enough outlets to meet the demands of a modern family. If your home builder wired your home for a typical number of outlets—that is, one in the master bedroom, one in the kitchen, maybe one in the family room—you won't have enough connection points to take advantage of a number of in-home systems that are rapidly being adopted.

Increasingly, more people run businesses out of the home or simply use them as a second office, requiring them to install multiple personal computers as well as home-automation appliances. A home that is wired to support voice, video, and data networking is becoming a valuable asset. The Multiple Listing Service is considering a technology rating for homes. One that is properly wired will score well on this rating. Realtors estimate that future-wiring a home can increase value by 5 to 10 percent.

KEY TERM *SOHO (small office, home office)* A *SOHO* is a small office or home office. The trend toward having offices at home is driving a need for more sophisticated cabling.

Home Cabling Facts and Trends

The growth of the home computer in the United States over the past 15 years is staggering. The growth of the Internet and the use of home computers have far outpaced similar growth patterns of telephone or DVD usage. Consider too the explosive growth of other "connected" devices and services. In addition to the ubiquitous cell phone, the PDA has transitioned to a "smart phone" that not only keeps your life organized and make phone calls but can also act as a GPS and give you access to the Internet through both the wireless phone company's network and WiFi networks. Apple's revolutionary iPhone is a good example of this.

NOTE Parks Associates conducts extensive market research in the area of communications. Read about trends for residential structured cabling at www.parksassociates.com. Another excellent resource for both home and building automation is the Continental Automated Buildings Association; they can be found at www.caba.org.

Imagine buying a house and then discovering that you can't install an air conditioner or that the wiring isn't adequate for running a dishwasher. A similar situation exists for the communication wiring in many homes in the United States today when it comes to having a network, multiple phone lines, or even connecting to the Internet at decent speeds via a modem. A home is expected to last years—at least as long as the term of its mortgage. Based on the aforementioned trends, many new homes have communication wiring that will be obsolete in fewer months than the loan term for a new car.

Structured Residential Cabling

The FCC recognized some of the problems associated with residential cabling and, on February 15, 2000, put into effect a requirement that voice and data wiring in new residential structures should be a minimum of Category 3 UTP. Enforcement of this ruling began in July 2000. (Note that because the requirement is not a safety-related issue, enforcement by most local code authorities is minimal.) The FCC ruling only addresses part of the problem and doesn't address the need for flexible, centrally controlled access to the wiring in your home.

What's the rest of the answer for residences? Just as in the commercial, corporate world, the answer is structured cabling. A cabling system distributed throughout the building, with every connection point leading back to a central location where systems and outside services are connected, is what works.

Enter ANSI/TIA-570-B, or the Residential Telecommunications Infrastructure Standard. It details the requirements for low-voltage cabling systems in single and multitenant residences. Included in the ANSI/TIA-570-B standard are definitions for the two grades of residential cabling installations, which are shown in Table D.1.

TABLE D.1   Grades of Residential Cabling

ANSI/TIA-570-A Grades	Cable Types		
	Four-Pair UTP	**75 Ohm Coax**	**Fiber Optic**
1	Category 5e[1]	RG-6	Not included
2	Category 5e[2]	RG-6[3]	Two-fiber multimode or single mode[4]

[1]Category 6 is recommended.
[2]Category 6 is recommended.
[3]Two cables are recommended.
[4]Optional.

For both grades of installation, the following guidelines shall apply to cabling and communication-services providers:

- Providers, such as a phone company or CATV company, bring service to the exterior of the house. This is the demarcation point, or the location at which ownership of the infrastructure (wiring and connections) transfers to you, the homeowner.

- From the demarcation point, a cable brings the service into a central distribution device installed inside the residence. The distribution device consists of a cross-connect device of some sort so that incoming service can be transferred to the horizontal wiring that runs to wall outlets. The distribution device provides a flexible method of distributing service to desired locations throughout the house.

- From the distribution device, cables are run to each wall outlet on a one-to-one basis. Outlets are never connected in series (daisy-chained). The one-to-one method is referred to as *star*, or *home-run*, wiring.

- At a minimum, one outlet is required in the kitchen, each bedroom, family/living room, and den/study. In addition, one outlet is required in any wall with an unbroken space of 12′ or more. Additional outlets should be added so that no point on the floor of the room is more than 25′ from an outlet. Note that these minimum-outlet requirements refer to UTP, coax, and fiber-optic cables (if installed).

- Connecting hardware (plugs, jacks, patch panels, cross-connects, patch cords) shall be at least the same grade as the cable to which they connect—for example, Category 5e cable requires Category 5e or better connectors.

- UTP plugs and jacks shall be eight-position (RJ-45 type), configured to the T568-A wiring scheme. Coaxial plugs and jacks shall be F-type connectors, properly sized for the grade and shielding of the coaxial cable used. Fiber-optic connectors shall be SC-type in either simplex or duplex configuration.

- Proper labeling of the cables is required, along with documentation that decodes the labeling.

For a Grade 1 installation, you should populate each wall plate with just one cable of each type. For a Grade 2 installation, two cables of each type should be run to each wall plate. We recognize that Grade 2 may seem like overkill *at the present*. But again, break your old ways of thinking. One coax with signals going *to* your entertainment center and one coax with signals *from* your entertainment center means that you can watch a movie in a completely different room from the one in which your DVD player is playing (from your computer monitor, for instance). Or, you could use a centrally located video switch to route satellite, DVD, CATV, or antenna inbound signals around your house to wherever a TV is located. Likewise, say you've got ADSL coming to your computer via UTP for direct access to the Internet, but you still want to use the modem in your computer for faxing. You need two UTP outlets right there.

Following installation, the cabling system should be tested. At a minimum, continuity tests for opens, shorts, and crosses should be performed. You may want to insist on the UTP and optical fiber cable (if terminated) being tested to transmission performance requirements; further, you may want to assure yourself that your Category 5e or better components will actually deliver Category 5e or better performance.

This appendix is just a short summary of what is contained in the ANSI/TIA 570-B standard. The standard specifies a great deal of additional detail about cables, installation techniques, electrical grounding and bonding, etc. If you want to read about these details (perhaps you're an insomniac), you can purchase a copy from Global Engineering Documents at http://global.ihs.com.

Picking Cabling Equipment for Home Cabling

So you're going to move forward with your home-cabling project. One of the decisions you will need to make is what type of equipment you will have to purchase and install. Although you can use any commercial structured cabling components in your home, many vendors are now building cabinets, panels, and wall plates that are specifically designed for home use. Manufacturers Ortronics (www.ortronics.com) and Leviton (www.leviton.com) build a line of cabling products that are designed with the small or home office in mind. Figure D.1 and Figure D.2 show two cabinets that can hold specially designed Category 5e patch panels, a hub, and a video patch panel.

Pick a centrally located but out-of-the-way location for your residential cabling cabinet; you probably don't want it on the wall in your living room. Common locations for residential wiring cabinets include the utility room, laundry room, and mechanical room in the basement. The cabinet should ideally be placed close to the termination point of the telephone service, cable-TV service, and a power source.

FIGURE D.1
Ortronics'
compact-sized In
House cabinet

Photo courtesy of Ortronics

FIGURE D.2
Ortronics' large-sized In House cabinet

Photo courtesy of Ortronics

In addition to the cabinets, both Leviton and Ortronics make a specially designed series of modular faceplates for residential structured cabling systems. The faceplates (pictured in Figure D.3) offer a variety of modular outlets for RJ-45 connectors, coaxial video, RCA audio/video, S-video, and RJ-25C for voice.

FIGURE D.3
Different configurations of Ortronics' In House modular faceplate

Photo courtesy of Ortronics

A Word About Wireless

When the first edition of this book was written, wireless networking was a poor choice compared to a cabled network. Data-throughput rates were slow, the equipment was relatively expensive, and it was not so easy to set up and use. That's all changed, and the growth of wireless networking in homes has outstripped most analysts' predictions.

Your greatest data throughput is still achieved using a Category 5e or better structured cabling system, and wireless networks still display some warts, such as the occasional disruption to your network because a family member is talking on a cordless phone. However, wireless networking has developed into a viable alternative to retrofitting your existing home with structured wiring.

Thinking Forward

Maybe you have looked at the ANSI/TIA-570-B standard and are now asking why you may need to put coaxial cable and data cable in every room in the house. After all, you may only have one TV in the family room. But think about other possible services, such as Internet service via cable modem. What about networking using coax? What about delivering DVD output from your entertainment center in the family room to your computer in the study or home office?

It's time to break your old habits of thinking about wiring only being used in traditional ways. The modern idea is to put a universal wiring infrastructure in place so that you can reconfigure at will.

Builders, as a rule, don't add costs willingly, unless it will clearly allow differentiation or some additional premium to be added to the price of the home. Smart, forward-thinking builders are beginning to install higher-grade wiring, and some are installing structured cabling systems as well. However, many builders forego the extra cost unless the prospective homeowner insists on properly installed structured wiring. It will be up to the consumer to take the trend into the mainstream.

TIP For the latest info on electronic gadgetry for communications and entertainment in the home, check out www.electronichouse.com.

Appendix E

Overview of IEEE 1394 and USB Networking

In 1986, Apple Computer conceived and developed a computer-to-computer and computer-to-peripheral interface system to transmit large volumes of data between devices very, very quickly. Apple turned the interface over to IEEE (the Institute of Electrical and Electronics Engineers) in 1990 to develop as a multimedia serial-bus standard. Now correctly called IEEE 1394, or just 1394, the system is also marketed under the trade names FireWire by Apple and i.LINK by Sony. The names are used interchangeably because the technology and functionality are identical. IEEE 1394 can be either a small network, a connecting interface for peripheral devices, or a combination of both.

USB, which stands for Universal Serial Bus, is a corresponding technology to IEEE 1394. It shares a number of characteristics with 1394 but is more limited in bandwidth and distance. Its niche is the connection of peripheral devices such as mice, printers, keyboards, scanners, and PDAs directly to computers.

Both technologies are specifically designed to allow consumer electronics, such as entertainment devices (TVs, CD players, DVD players), video cameras, and digital cameras, to connect and interact with computers.

TRADE ASSOCIATIONS

The 1394 Trade Association, found on the Web at www.1394ta.com, was formed to advance the technology behind IEEE 1394 and to promote its acceptance by the marketplace. Apple is a staunch supporter of IEEE 1394, as are Sony, Intel, Microsoft, JVC, IBM, Matsushita, Compaq, NEC, Philips, Samsung, and a host of other companies in the computer and consumer-electronics market.

USB-IF (IF stands for Implementers Forum) is an organization that supports USB and is similar to the 1394 Trade Association. A consortium of computing industry giants formed and supports this organization—Compaq, Hewlett-Packard, Intel, Microsoft, NEC, and Philips all share positions on the board of directors of the USB-IF. USB-IF has a website at www.usb.org.

The websites of 1394 Trade Association and USB-IF are chock-full of technical and marketing information on their respective technologies, and both were relied on heavily to research this appendix.

Both IEEE 1394 and USB interfaces are growing rapidly in popularity and deployment. Among the PCs recommended to users by *Consumer Reports* in its December 2008 edition, 75 percent featured FireWire, including brands from Sony, Acer, Dell, Lenovo, and Hewlett Packard. USB is growing at an even faster rate. Some 10 billion PCs (*PC World*, January 6, 2009) and peripherals incorporate USB connectivity. USB technology is already built into 99 percent of PCs, 54 percent of video cameras, and 38 percent of scanners sold, and it is well on its way to wholesale replacement of serial- and parallel-port interfaces for printers.

Although IEEE 1394 and USB are not networking technologies on the scale of Ethernet, they are still important in the overall scheme of networking PCs and peripherals. In addition, recent revisions to the IEEE specification now allow IEEE 1394 to compete with Ethernet over the structured wiring installations described in this book.

Among the features shared by IEEE 1394 and USB technologies are the following:

◆ High data throughput compared to many network technologies.

◆ Short distances between devices and for the network overall.

◆ Hot-swap technology. Devices can be plugged in and unplugged without taking down the PC or network.

◆ Self-configuring (plug-and-play) devices.

◆ Similar cable constructions.

◆ Robust jacks and plugs with small footprints.

◆ Daisy-chain or "tree" network topology. No loops are allowed.

NOTE   IEEE-1394-2008 and USB 3.0 are documents defining the respective interfaces. They are specifications written by experts in the field, but they have not been sanctioned by ANSI as national standards. They have become, however, de facto standards.

IEEE 1394

After Apple turned the technology over to IEEE to develop as a specification, IEEE published IEEE 1394-1995, a specification defining the technology. The original specification defined data-transfer rates of 100, 200, and 400Mbps. It also defined the cables, plugs, and jacks required for connecting devices. Cable-length limits were set to 4.5 meters (about 15′) from node to node. Network architectures, signaling protocols, and electrical requirements were all defined as well.

The specification was then revised and published as IEEE 1394a-2000, which enhanced parts of the specification but did not alter the original cable, plug, and distance requirements. IEEE 1394b was published in early 2002 and allows throughput speeds of up to 3.2Gbps and distances of up to 100 meters using fiber-optic cables. (Specifically, 1394b supports 100 meters with glass fiber-optic cables and 50 meters with plastic fiber-optic cables.) It also allows the interface to transmit 100Mbps over Category 5e or better cabling at up to 100 meters distance between nodes.

IEEE 1394c-2006 was published on June 8, 2007. It supported 800Mbps over the same RJ45 connectors with Category 5e cable along with a corresponding automatic negotiation that allows the same port to connect to either IEEE 1394 or IEEE 802.3 Ethernet devices.

IEEE 1394-2008 was published on July 9, 2008, and updates and revises all prior 1394 standards dating back to the original 1394-1995 version, including 1394a, 1394b, 1394c, enhanced UTP, and the 1394 beta plus PHY-Link interface. It also incorporates the complete specifications for S1600 (1.6Gbps bandwidth) and for S3200, which provides 3.2Gbps speeds.

NOTE The current version of the IEEE 1394 specification may be purchased from Global Engineering Documents at http://global.ihs.com.

Up to 63 network nodes (devices) can be daisy-chained on an IEEE 1394 network bus. The implementation of the specification using fiber-optic cables or the installed base of Category 5e or better UTP removes the length limitations of earlier standards, so IEEE 1394 can be a powerfully viable competitor to Ethernet.

Unless you are using fiber-optic or UTP in a structured wiring environment, the cable for 1394 is shielded twisted pair (STP). It contains two 28 AWG twisted pairs for data transmission, each enclosed in a metallic foil/braid combination shield. It also contains one unshielded, 22 AWG, twisted pair for carrying power to devices. These components make up the core of the cable. A metallic foil/braid combination shield is applied over the core. Finally, the shielded core is enclosed in a plastic jacket. Conductors must be stranded for flexibility.

NOTE IEEE 1394 cables can be purchased with (six position) and without (four position) the power-carrying pair, depending on the device being connected. The four-position cables are usually for manufacturer-specific configurations, as the IEEE 1394 document specifies a power pair. Check your device before buying a cable.

As discussed in Chapter 1, "Introduction to Data Cabling," conductor resistance is a major factor in signal attenuation, or decay of the signal. The smaller the conductor, the greater the resistance and the more decay of the signal. The 28 AWG conductors in an IEEE 1394 cable have much greater resistance, and therefore much greater attenuation, than does a 24 AWG UTP cable. Although all the shielding helps protect the signal from noise, the conductor size limits how far the signal can travel before a receiver can't detect it. That is the primary reason why the 4.5 meter length limitation existed for the earlier 1394-type cables, although electronics and protocol have built-in limitations as well. Larger conductor sizes in the data pairs, such as in Category 5e or better UTP, are what allow distances between nodes to increase in the structured wiring environment of an IEEE 1394 network implementation.

IEEE 1394 plugs and connectors are simple male/female designs, keyed so that they can only be plugged in the proper way. The devices have the female socket, so most cables are male-to-male, as shown in Figure E.1.

FIGURE E.1
Male-end plugs on a six-position IEEE 1394 cable

Photo from MilesTek.com

Most cable/plug assemblies are purchased. Although you could make your own IEEE 1394 cables and terminate the plugs on the ends, the complexity of the plug design, combined with the necessity for correctly tying in the shielding components, make this ill advised. You should buy ready-made cables, with molded-on plugs, for your 1394 connections.

IEEE 1394 transmits data in both *isochronous* and *asynchronous* modes. Isochronous data transfer is a real-time, constant delivery technique. It is perfect for delivery of multimedia from device to device; audio and video streams utilize isochronous transfer. Asynchronous transfer is typically used for data in a networking environment. Instead of a steady stream of information being delivered, data is sent in packets and acknowledged by the receiver before more is transmitted.

IEEE 1394 peripheral and networking abilities make it extremely well suited for storage-device connections, multimedia, and networks. As its popularity grows, we envision it making major inroads in the residential networking arena. What a perfect place for this technology to be implemented! Home networking, tied directly to the entertainment devices of the home using IEEE 1394 technology, will facilitate the convergence of traditional computing, the Internet, and home entertainment. Those homeowners with the foresight to install structured wiring systems are positioned to take advantage of 1394 connectivity immediately—without needing room-to-room cabling installed.

USB

The USB interface can be thought of as a younger sibling of IEEE 1394. It uses similar cable and connectors and shares the ability to transfer large chunks of data in a small amount of time.

The current specification defining USB interfacing is USB 3.0. Like the IEEE 1394 specification, USB 3.0 defines the architecture, signaling protocol, electrical requirements, cable types, and connector configuration for the technology.

NOTE USB 3.0 is available for free download at www.usb.org.

USB 1.1 allowed data-transfer speeds of 1.5 and 12Mbps. USB 2.0 allowed data-transfer speeds of 480Mbps. USB 3.0 has increased this to 5Gbps.

USB is not a networking technology in the same sense as is IEEE 1394. Peripheral devices, such as printers, must be connected to a PC directly or through a USB hub. The PC must then connect to a traditional network such as Ethernet, or even an IEEE 1394 network, for the peripheral to be used as a shared network device.

USB 1.1 allowed up to 127 devices to be connected to a single PC using USB hubs. USB 2.0 allows up to five hubs. USB devices, from digital cameras to pocket-watch-sized hard drives, are sprouting like toadstools (in some cases almost literally, as in the USB Mushroom Lamp based on the Super Mario Brothers video game). USB is fully supported by Microsoft OS platforms from Windows NT on and by Macintosh OS 8.5 and higher. Its true Plug-and-Play capabilities make it a dream for PC users who are inexperienced at adding equipment to their computers.

The cable construction specified for USB is screened twisted pair (ScTP). The cable contains one 28 AWG unshielded data pair and a power-conducting pair that can range from 28 AWG to 20 AWG in size (depending on the power requirements of the device for which the cable is designed). Conductors must be stranded for flexibility. These core components are enclosed in an overall metallic foil/braid combination shield. A plastic jacket covers this outer shield. This construction is mandated for the full speed USB 1.1 (12Mbps) and USB. 2.0 high-speed (480Mbps) implementations. For low-speed devices (1.5Mbps), the specification does not require that the data pair be twisted.

As with IEEE 1394, the 28 AWG data conductors necessitate short distances between devices if the data rates are to be maintained. For low-speed devices, a 3 meter (10′) cable-length limit is imposed. Full-speed and high-speed devices can be connected using a 5 meter (16′) cable. The specification allows up to five hubs between devices, so from PC to the farthest device, a 30 meter (98′) distance can be achieved. PC-to-PC connections are possible using a USB bridge.

Connectors for USB cables have *A* ends and *B* ends. The A end goes upstream to the host (PC or hub), and the B end goes downstream to the device.

Figure E.2 shows various USB male and female connectors.

WARNING The USB-IF is adamant that only A-to-B cables be used. Some cable vendors are selling A-to-A cables in the mistaken idea that two PCs or devices can be connected together directly. Doing so will short the power supplies and/or create a shock hazard. Beware.

FIGURE E.2
USB connector configurations

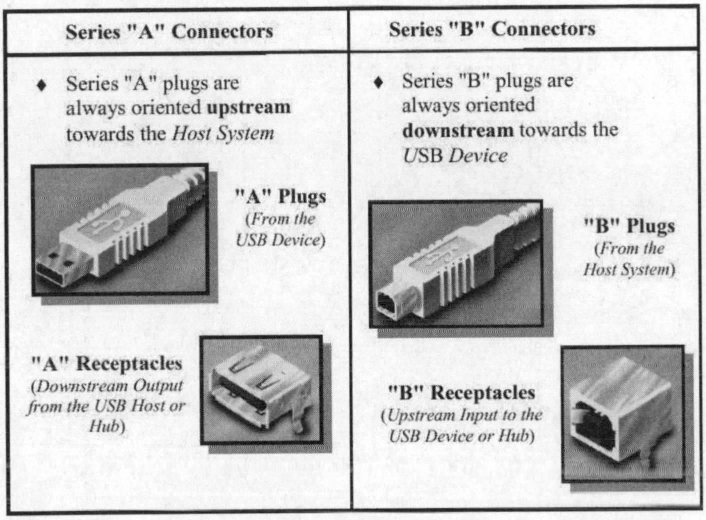

Series "A" Connectors	Series "B" Connectors
◆ Series "A" plugs are always oriented **upstream** towards the *Host System*	◆ Series "B" plugs are always oriented **downstream** towards the *USB Device*
"A" Plugs (*From the USB Device*)	**"B" Plugs** (*From the Host System*)
"A" Receptacles (*Downstream Output from the USB Host or Hub*)	**"B" Receptacles** (*Upstream Input to the USB Device or Hub*)

Graphics from the USB 2.0 specification published by USB-IF, Inc.

You should purchase ready-made cables with molded-on connectors.

USB is a virtually ubiquitous technology on all new PCs, and more devices will utilize this interface. Conceivably, USB could replace all other PC interfaces except for ultra-fast IDE or 1394 transfer for mass-storage devices. Increasingly more consumer electronics will be shipped with USB ports. In conjunction with IEEE 1394, USB technology is helping to drive the convergence of information and entertainment technologies.

Appendix F

The Electronics Technicians Association, International (ETA) Certifications

The Electronics Technicians Association, International (ETA) is a nonprofit professional trade association established in 1978 "by technicians, for technicians" to promote the electronics-service industry and to better benchmark the knowledge and skills of individual electronics technicians. The ETA currently has a membership of thousands of individuals, primarily from the electronics installation, maintenance, and repair occupations. It provides over 30 different certifications in various electronics areas. This appendix lists the competencies for ETA's top three certifications:

1. Cabling Installer Certification (DCIC)
2. Fiber Optics Installer (FOI)
3. Fiber Optics Technician (FOT)

Data Cabling Installer Certification (DCIC) 2007 Competency Requirements

One of the ETA's top certifications is its Data Cabling Installer Certification (DCIC). The program was designed to test the knowledge and technical skills of technicians working with fiber- and copper-cabling media.

Data cabling installers are expected to obtain knowledge of basic concepts of copper cabling installation and service, which are then applicable to all the functions required to safely and competently install communications cabling. Once a CDCI has acquired these skills, abilities, and knowledge, he or she should be able to enter employment in the telecommunications cabling field. With minimal training in areas unique to the special requirements of individual products or systems designs, the data cabling installer should become a profitable and efficient part of the workforce.

Data cabling installers must be knowledgeable and have abilities in the following technical areas.

1.0 Basic Electricity

1.1   Describe the relationships between voltage, current, resistance and power

1.2   Identify components called resistors and also noncomponent types of resistance in cabling technology

1.3   Use ohms law to calculate power usage and power losses in cabling circuits

1.4   Explain how noise may be generated onto communications cabling and components

1.5   Define impedance and compare impedance with resistance

1.6. Explain signal-to-noise ratio

1.7   Explain the difference between inductance and inductive reactance; capacitance and capacitive reactance

1.8   Explain the importance of grounding cabling and electronics communications products

1.9   Identify wire sizes needed for grounding

1.10 Describe the types of conductor insulation used for communications wiring

1.11 Explain the difference between AC and DC circuits

2.0 Data Communications Basics

2.1   Provide a brief history of telephone and wireless communications

2.2   Draw a simple diagram of a basic telephone system

2.3   Describe the differences between analog and digital communications signals

2.4   Define *transfer mode* and explain *transfer speeds*

3.0 Definitions, Symbols, And Abbreviations

3.1   Define audio and radio or RF frequencies

3.2   Explain the term *bandwidth*

3.3   Explain the difference between frequency, bit rate, and baud

3.4   Trace the history of the BEL and decibel and explain how and why these terms are used

3.5   Convert signals from voltage levels to their corresponding decibel equivalents and decibel levels to their corresponding voltage or current levels

3.6   Convert signal gains or losses to comparative decibel readings

3.7   Define *attenuation*

3.8   Define *crosstalk* and explain how it occurs in communications cabling

3.9   Define *basic link, UTP, NEXT,* and other common Telco terms

4.0 Cable Construction

4.1   Explain the differences between electrical power wiring; 22 AWG twisted-pair telephone wiring; coaxial and fiber optics cables; 150-ohm shielded (STP); and 4-pair 100-ohm UTP

4.2   Describe coaxial cable and explain where it is normally the choice

4.3   Describe the differences between CAT 1, 2, 3, 4, 5, 5e, and 6 telephone-data cables

5.0 Cable Performance Characteristics

5.1   Explain the electrical characteristics of twisted pair cabling

5.2   Explain the electrical characteristics of coaxial cable

5.3   Explain the performance characteristics of twisted pair and coaxial cables

6.0 Cabling Standards

6.1   Explain the purpose and basic requirements of the following standards:

6.1.1   ANSI/TIA/EIA-568-A

6.1.2   ANSI/TIA/EIA-569-A

6.1.3   ANSI/TIA/EIA-607; also CSA T527 and C22.1; also NFPA 70

6.1.4   ANSI/TIA/EIA-570-A (Residential Telecom Cabling Standard)

6.1.5   ISO/IEC 11801

7.0 Basic Network Topologies

7.1   Draw a block diagram of a computer network using STAR topology

7.2   Draw a block diagram of a computer network using BUS topology

7.3   Draw a block diagram of a computer network using RING topology

7.4   Explain an advantage of each type of topology

8.0 Basic Network Architectures

8.1   Describe a network using Ethernet

8.2   Describe how a Token Ring network operates

8.3   Define *ATM*

8.4   Describe 100VG-Any LAN

9.0 National Electric Code – NEC and UL Requirements

9.1   Explain the purpose and requirements for the following NEC code chapters

      9.1.1   Chapter 1—General Requirements

      9.1.2   Chapter 2—Wiring Protection

      9.1.3   Chapter 3—Wiring Methods and Materials

      9.1.4   Chapter 5—Special Occupancy Requirements

      9.1.5   Chapter 7—Special Conditions Requirements

      9.1.6   Chapter 8—Communications Systems Wiring

      9.1.7   Explain UL 1863; explain the TIA/EIA-606 Standard

10.0 Cabling System Components

10.1   Describe horizontal and backbone cables

10.2   Explain why patch cords are used and describe them

10.3   Explain the differences between the various segments of cabling pathways

10.4   Describe the purpose, construction, and usage of communications wiring closets

10.5   Define MDF and IDF. Define *punch down block*

10.6   Compare power sources for telephone-data cabling equipment

11.0 DCIC Installation Tools

11.1   Explain the purpose and proper usage of wire strippers

11.2   Show how wire cutters and cable prep tools are used

11.3   Demonstrate the proper methods of using cable crimpers (TP and coaxial)

11.4   Describe a punch-down tool; show where it is used and how it is used

11.5   Explain the purpose and proper use of fish tape and pull devices

12.0 Connectors and Outlets

12.1   List the proper identification number for twisted pair plugs and jacks

12.2   List the proper identification numbers for coaxial cable plugs, splices and jacks

12.3   Describe fixed design wall plates and explain where they are used

12.4   Describe a modular wall plate and why and where it is used

12.5   Describe a biscuit jack and why it is used; describe a floor telecom outlet

12.6   Describe a cable splitter

12.7   Describe a cable junction box

13.0 Cabling System Design

13.1   Draw a block diagram showing a cabling topology

13.2   Describe how the telecommunications closet is wired

13.3   Explain the concept of "cabling management" and proper outlet placement and requirements

14.0 Cabling Installation

14.1   Describe the steps used in installing communications cabling

14.2   Explain cable stress and the precautions for aerial construction; underground and ducts and plenum installation; define *pulling tension* and *bend radius*

14.3   Describe cabling dressing and methods of securing cabling

14.4   Explain proper labeling of cables and what a wire map is

14.5   Demonstrate proper cable stripping

14.6   Explain the purpose of and requirements for OSHA height regulations

14.7   Explain safety precautions for underground construction

14.8   Explain what a CAM splice is

15.0 Connector Installation

15.1   Demonstrate proper installation of twisted pair connectors

15.2   Demonstrate proper installation of coaxial cable connectors

15.3   Describe the color code for telecom cabling and the pin/pair assignments

15.4   Explain how ducts are used for cabling installations

16.0 Cabling Testing and Certification

16.1   Explain the purpose of installation testing

16.2   Describe the purpose and methods of certifying the cable plant

16.3   Show the proper selection and use of cable testing tools and equipment

17.0 Cabling Troubleshooting

17.1 Explain how to establish a baseline for testing or repairing a cabling system

17.2 Demonstrate a method of locating a cabling defect or problem

17.3 Describe commonly encountered cable problems and the methods used to resolve them

17.4 Define *crosstalk* and *florescent lighting interference*

17.5 Explain loop-testing

17.6 Describe a radio detector and how it is used to locate interference sources

18.0 Documentation

18.1 Explain the purpose of documenting a cabling installation

18.2 Explain the required ingredients of the installation documents

18.3 Prepare a sample cable documentation record that meets industry standards

Certified Fiber Optics Installer (CFOI) 2009 Competency Requirements

Fiber-optics installers are expected to obtain knowledge of basic concepts of fiber optics installation and services, which are applicable to all the functions required to safely and competently install fiber optics communications cabling. Once a CFOI has acquired these skills, abilities, and knowledge, he or she should be able to enter employment in the telecommunications cabling field. With minimal training in areas unique to the special requirements of individual products or systems designs, the fiber optics installer should become a profitable and efficient part of the workforce.

Fiber optics installers must be knowledgeable and have abilities in the following technical areas.

1.0 History of Fiber Optic Cabling

1.1 Trace the evolution of light in communication

1.2 Summarize the evolution of optical fiber manufacturing technology

1.3 Track the evolution of optical fiber integration and application

2.0 Principles of Fiber Optic Transmission

2.1 Describe the basic parts of a fiber-optic link

2.2 Describe the basic operation of a fiber-optic transmitter

2.3 Describe the basic operation of a fiber-optic receiver

2.4   Demonstrate how to express gain or loss using dB

2.5   Demonstrate how to express optical power in dBm

3.0 Basic Principles of Light

3.1   Describe the following:

    3.1.1.   Light as electromagnetic energy

    3.1.2   Light as particles and waves

    3.1.3   The electromagnetic spectrum and locate light frequencies within the spectrum in relation to radio and microwave communication frequencies

    3.1.4   The refraction of light (Snell's Law)

    3.1.5   Describe total internal reflection

3.2   Explain the following:

    3.2.1   How the index of refraction is used to express the speed of light through a transparent medium

    3.2.2   Reflection, to include angle of incidence, critical angle, and angle of refraction

    3.2.3   Fresnel reflections and how they can impact the performance of a fiber optic communication system

4.0 Optical Fiber Construction and Theory

4.1   Describe the following:

    4.1.1   The basic parts of an optical fiber

    4.1.2   The different materials that can be used to construct an optical fiber

    4.1.3   Optical fiber manufacturing techniques

    4.1.4   The tensile strength of an optical fiber

    4.1.5   A mode in an optical fiber

    4.1.6   The three refractive index profiles commonly found in optical fiber

4.2   Explain the propagation of light through:

    4.2.1   Multimode step index optical fiber

    4.2.2   Multimode graded index optical fiber

    4.2.3   Legacy 62.5/125 multimode optical fiber

    4.2.4   Laser optimized 50/125 multimode optical fiber

    4.2.5   Single-mode optical fiber

4.3 Describe the TIA/EIA-568-B.3 recognized:

4.3.1 Multimode optical fibers

4.3.2 Single-mode optical fibers

4.4 Describe the ITU-T-G.652 recognized single-mode optical fibers

4.5 Define the ITU-T-G.655 recognized single-mode optical fibers

4.6 Cite commercially available:

4.6.1 PCS and HCS optical

4.6.2 Plastic optical fiber

5.0 Optical Fiber Characteristics

5.1 Explain dispersion in an optical fiber

5.1.1 Define modal dispersion and its effects on the bandwidth of an optical fiber

5.1.2 Describe material dispersion and its effects on the bandwidth of an optical fiber

5.2 Outline the causes of attenuation in an optical fiber

5.2.1 Compare attenuation versus wavelength in a multimode optical fiber

5.2.2 Compare attenuation versus wavelength in a single-mode optical fiber

5.3 Explain the numerical aperture of an optical fiber

5.4 Relate how the number of modes in an optical fiber is defined by core diameter and wavelength

5.5 Describe microbends in an optical fiber

5.6 Explain macrobends in an optical fiber

6.0 Fiber Optic Cabling Safety

6.1 Explain how to safely handle and dispose of fiber optic cable, optical fiber chips, and debris

6.2 List the safety classifications of fiber optic light sources

6.3 Discuss the potential chemical hazards in the fiber optic environment and the purpose of the material safety data sheet (MSDS)

6.4 Cite potential electrical hazards in the fiber optic installation environment

6.5 Outline typical workplace hazards in the fiber optic environment

7.0 Fiber Optic Cables

7.1 Draw a cross section of a fiber optic cable and explain the purposes of each segment

7.2 Identify why and where loose tube fiber optic cable is used

7.3 Describe tight buffer fiber optic cable

7.4 Relate common strength members found in fiber optic cables

7.5 Name common jacket materials found in fiber optic cables

7.6 Describe simplex and duplex cordage and explain the difference between cordage and cable

7.7 Describe:

　7.7.1 Distribution cable

　7.7.2 Breakout cable

　7.7.3 Armored cable

　7.7.4 Messenger cable

　7.7.5 Ribbon cable

　7.7.6 Submarine cable

　7.7.7 Hybrid cable

　7.7.8 Composite cable

7.8 Discuss fiber optic cable duty specifications

7.9 Review how and when a fan out kit is used

7.10 Identify how and when a breakout kit is used

7.11 List the National Electrical Code (NEC) optical fiber cable types

7.12 Describe the NEC listing requirements for:

　7.12.1 Optical fiber cables

　7.12.2 Optical fiber raceways

7.13 Outline the TIA/EIA-598-C color code and cable markings

7.14 List the TIA/EIA-568-B.3 bend radius specifications for inside and outside plant fiber optic cables

7.15 Cite the ITU-T G.657 and the Telcordia GR-20 bend radius specification for outside plant fiber optic cables

8.0 Splicing

8.1   Explain the intrinsic factors that affect splice performance

8.2   Relate the extrinsic factors that affect splice performance

8.3   Recall the basic parts of a mechanical splicer

8.4   Discuss how to perform a mechanical splice

8.5   Specify the operation of a fusion splicer

8.6   Indicate how to perform a fusion splice

8.7   List TIA/EIA-568-B.3 inside plant splice performance requirements

8.8   Cite TIA/EIA-758 outside plant splice performance requirements

9.0 Connectors

9.1   Indicate the basic parts of a fiber optic connector

9.2   Discuss these connectors:

    9.2.1   Flat finish connectors

    9.2.2   PC finish connectors

    9.2.3   APC finish connectors

9.3   Distinguish between:

    9.3.1   TIA/EIA-568-B.3 recognized connectors

    9.3.2   Small form factor (SFF) connectors

    9.3.3   Multi-fiber connectors

    9.3.4   An optical fiber pigtail

9.4   Review common connector ferrule materials

9.5   Specify the intrinsic factors that affect connector performance

9.6   Explain the extrinsic factors that affect connector performance

9.7   Describe return reflections in an interconnection

9.8   Identify the steps involved in an anaerobic epoxy connector termination and polish

9.9   Cite the steps involved in a UV epoxy connector termination and polish

9.10 List the steps involved in an oven cured epoxy connector termination and polish

9.11 Relate the steps involved in a pre-load epoxy connector termination and polish

9.12 Describe how to construct a no-polish, no-epoxy connector termination

9.13 Explain how to properly clean a connector

9.14 Relate how to examine the endface of a connector per TIA/EIA-455-57B

9.15 List the TIA/EIA-568-B.3 connector performance requirements

10.0 Fiber Optic Light Sources

10.1 Explain why an LED is best used for multimode fiber applications

10.2 Identify why a VCSEL light source is used with multimode optical fiber as opposed to an LED for gigabit or higher transmission rates

10.3 Specify the benefit of using a laser light source in fiber optic communication systems

10.4 Identify which fiber type is best used for communications systems with VCSEL light sources

10.5 Explain the differences between an overfilled launch condition and a restricted mode launch

11.0 Fiber Optic Detectors and Receivers

11.1 Summarize the basic operation of a photodiode

11.2 Explain why an optical attenuator is occasionally used in a communication system

12.0 Cable Installation and Testing

12.1 Cite manufacturer installation cable specifications

12.2 Explain the static and dynamic loading on a fiber optic cable during installation

12.3 List commonly used installation hardware

12.4 Describe these types of installation:

12.4.1 Tray and duct installation

12.4.2 Conduit installation

12.4.3 Direct burial installation

12.4.4 Aerial installation

12.4.5 Blown fiber installation

12.4.6 Wall outlet installation

12.5 Explain cable grounding and bonding per NEC Article 250

12.6 Summarize these types of preparation:

12.6.1 Patch panel

12.6.2 Racks and cable

12.6.3 Splice enclosure

12.7 Describe the process of labeling a cable in accordance with TIA/EIA-606A

13.0 Fiber Optic System Design Considerations

13.1 Compare the following advantages of optical fiber over twisted pair and coaxial cables:

13.1.1 Weigh the bandwidth advantages of optical fiber over twisted pair and coaxial copper cables

13.1.2 Relate the attenuation advantages of optical fiber over twisted pair and coaxial copper cables

13.1.3 Explain the electromagnetic immunity advantages of fiber optic cable over copper cable

13.1.4 Compare the weight saving advantages of fiber optic cable over copper cable

13.1.5 Describe the size advantages of fiber optic cable over copper cable

13.1.6 Identify the security advantages of fiber optic cable over copper cable

13.1.7 Define the safety advantages of fiber optic cable over copper cable

14.0 Test Equipment and Link/Cable Testing

14.1 Describe the types of fiber optic test equipment that can be used to test for continuity

14.2 Explain the limitations of a visual laser when troubleshooting a fiber span

14.3 Describe the basic operation of a single-mode and multimode light source and optical power meter

14.4 Review the difference between a test jumper and a patch cord

14.5 Define the purpose of a mode filter

14.6 Describe how to measure the optical loss in a patch cord with a light source and optical power meter

14.7 Summarize the basic operation of an optical time domain reflectometer (OTDR)

Fiber Optic Technician (FOT) 2007 Competency Requirements

The following are the competency requirements for the Fiber Optic Technician certification by ETS. This certification will provide the knowledge and skills needed to certify and troubleshoot fiber-optic cable networks accurately and efficiently. Trainees are typically required to certify or troubleshoot different cable systems with a combination of fibers, connectors, and splices.

1.0 Principles of Fiber Optic Transmission

1.1 Describe the basic parts of a fiber-optic link

1.2 Describe the basic operation of a transmitter

1.3 Describe the basic operation of a receiver

1.4 Explain amplitude modulation

1.5 Compare digital data transmission with analog

1.6 Graphically explain how analog to digital conversion (A/D) is accomplished

1.7 Graphically explain how digital to analog conversion (D/A) circuitry works

1.8 Explain the difference between pulse coded modulation and amplitude modulation

1.9 List the benefits of multiplexing signals

1.10 Explain the purpose of decibels (dBs) and convert power levels to and from decibel equivalents

1.11 Explain how optical power is measured (dBm), express optical power levels in dBm, and compare power gains and losses

2.0 Basic Principles of Light

2.1 Describe the electromagnetic spectrum and locate light frequencies within the spectrum in relation to communications frequencies

2.2 Convert various wavelengths to corresponding frequencies

2.3 Explain the theories of light particles and light waves

2.4 Describe how the index of refraction is calculated

2.5 Define *Fresnel reflection loss*

2.6 Explain the effects of refraction and Snell's Law

2.7 Calculate the Index of Refraction in two different mediums

3.0 Optical Fiber Construction and Theory

3.1   Name the materials out of which fiber optic cable core is manufactured

3.2   Describe the purpose of cladding and how it is applied to the core of fiber optics cables

3.3   Name the materials from which the FO coating is manufactured

3.4   Explain why multimode fiber optics cable may be selected over single mode

3.5   Describe single mode optical fiber and how it differs from multimode

3.6   List common classifications for optical fiber

3.7   Describe refractive index profiles and their purpose

3.8   Describe the basics of optical fiber manufacturing

4.0 Optical Fiber Characteristics

4.1   Explain modal dispersion and its importance to fiber optics

4.2   Define *material dispersion*

4.3   List bandwidth limitations for common communications cable types

4.4   Measure the attenuation loss in fiber optics cables

4.5   Describe the effects of a microbending fiber cables

4.6   Define macrobend and explain its importance in fiber cabling

4.7   Explain how the numerical aperture (NA) is arrived at in fiber connectors

4.8   Identify the cone of acceptance as used in fiber cabling

4.9   Compare the tensile strength of fiber and other communications cables

5.0 Advantages of Fiber Over Copper

5.1   List bandwidth limitations of various copper cables and fiber-optic cables

5.2   Describe attenuation in copper and fiber-optic cables

5.3   Explain why electromagnetic immunity is superior in fiber-optic cabling

5.4   List weight differences for telephone, coaxial and fiber-optic cables

5.5   Make a comparison between the size of fiber-optic cables versus copper cables

5.6   Explain why fiber-optic cables present fewer risks of safety concerns than copper cabling

5.7   Compare security features of fiber-optic cables versus electromagnetic cables

6.0 Fiber Optic Cables

6.1   Draw a cross section of a fiber-optic cable and explain the purposes of each segment

6.2   Explain why and where loose tube cable is used

6.3   Describe tight buffered optical fiber

6.4   Identify the strength member in a fiber-optic cable

6.5   Specify the cable jacket material used in common types of fiber-optic cables

6.6   Define *duty specifications*

6.7   Explain the differences between simplex cordage and duplex cordage

6.8   List locations where duplex cordage is preferred over simplex

6.9   Define *distribution cable*

6.10 Explain how and where breakout cable is used

6.11 List reasons for utilizing armored fiber-optic cables

6.12 Explain what a messenger cable is and how it is used

6.13 Describe ribbon cable

6.14 Explain what hybrid cables are and where they are ordinarily used

6.15 Define *composite cables*

6.16 Explain where the TIA/EIA-598 color code is used and how the colors are used to identify individual cables

6.17 Describe cable markings and how they are used

6.18 Define *tensile strength of a fiber-optic cable* and explain the reasons an installer would need to know the strength of various cables

6.19 Describe the dynamic load of a fiber-optic cable

6.20 Define *static load* as it refers to fiber-optic cabling

6.21 Demonstrate the detrimental effects of exceeding the minimum dynamic bend radius of a fiber cable

6.22 Define *static bend radius*

7.0 Sources

7.1   Explain the safety classifications for light sources used in fiber-optic communications

7.2   Describe the effects of laser exposure to the human body

7.3   Explain the differences between light emitting diodes and laser diodes

7.4 List the common wavelengths used in fiber-optic communications and the advantages and disadvantages of each

7.5 Measure the output power of a light generator

7.6 Define spectral width of the output of a light generator

7.7 Compare the speed of operation of one light source versus another

7.8 Compare ease of operation of the types of light sources commonly used in fiber signal generation

7.9 Present historical data and future estimates for the lifetime of fiber cables and fiber light generators

7.10 Compare the source characteristics of LEDs and lasers

8.0 Detectors

8.1 Describe the optical and electrical theory of operation of a photodiode

8.2 Compare common detector designs

8.3 Explain the use for PIN photodiodes and theory of operation

8.4 Describe the action of an avalanche photodiode (APD)

8.5 Locate the performance characteristic- responsivity specifications for various optical detectors

8.6 Explain bit error rate (BER) and how it is utilized in fiber optics communications equipment

9.0 Connectors

9.1 Identify TIA/EIA-568-A Standard connector types

9.2 List connector requirements for various types

9.3 Describe ferrule materials used with fiber-optic connectors

9.4 Explain intrinsic factors applicable to fiber-optic connectors

9.5 Explain extrinsic factors

9.6 Measure interconnection losses using common measuring equipment

9.7 Measure return reflections in a completed cable

9.8 Define *PC finish*

9.9 Explain *APC finish*

9.10 Describe how and where pigtails are used in fiber cabling

9.11 Demonstrate the proper use of anaerobic epoxy

9.12 Explain how and where UV epoxy is used

9.13 Demonstrate the use of two-part epoxy

9.14 Describe preload epoxy

9.15 Explain why and where epoxyless connectors are best used

9.16 Demonstrate proper fiber preparation techniques

9.17 Properly assemble common types of connectors

9.18 List steps taken to properly perform a visual inspection of a connector

9.19 List ways to properly care for fiber-optic connectors

9.20 Explain and demonstrate TIA/EIA 568-A performance testing

10.0 Passive Components

10.1 Explain the uses and benefits as well as disadvantages of using fiber-optic couplers

10.2 Describe where a tee coupler is used

10.3 Describe where a star coupler is used

10.4 Explain the operation of a reflective star coupler

10.5 Define *transmissive star coupler*

10.6 Explain modulation and how wavelength division multiplexing (WDM) is used in fiber-optic distribution systems

10.7 Explain the difference between dense wavelength division multiplexing (DWDM) and WDM

11.0 Types of Splicing

11.1 Mechanical Splicing

11.1.1 Explain the differences between intrinsic factors and extrinsic factors when splicing fiber-optic cables

11.1.2 List extrinsic factors important in fiber splicing

11.1.3 Demonstrate correct fiber preparation

11.1.4 Define index matching gel

11.1.5 Explain performance characteristics of different gel compositions

11.1.6 Perform TIA/EIA-568-A performance testing procedures

11.2 Fusion Splicing

    11.2.1 List fusion splice requirements

    11.2.2 Demonstrate proper fiber preparation

    11.2.3 Explain the use of the splice closure

    11.2.4 Perform TIA/EIA-568-A performance testing of the fusion splice

12.0 Cable Installation and Hardware

12.1 Explain dynamic tensile loading

12.2 Explain static tensile loading

12.3 Compare the dynamic bend radius minimums for common fiber cables

12.4 Describe the effects of exceeding static bend radius minimums

12.5 Describe the proper use of pulling tape

12.6 Define *pulling grip*

12.7 Explain where conduit should be installed to enclose fiber cables

12.8 Describe the requirements for tray and duct installation of fiber cabling

12.9 Explain the National Electric Code (Article 770) rules

13.0 Fiber Optic Link

13.1 List the considerations for basic fiber-optic system design

13.2 Prepare a basic optical link power budget

14.0 Optical Fiber Measurement and Testing

14.1 Explain Fiber Optic Cable Measurement Standards

14.2 Demonstrate the use of optical loss test sets

14.3 List common fiber optic sources (FOS)

14.4 Demonstrate proper and safe use of the fiber optic light meter (FOM)

14.5 Locate a fault using an optical time domain reflectometer (OTDR) and point out the aberrations in the display presentation and describe their likely causes

14.6 Describe the proper use of a continuity tester

14.7 Describe the fiber identifier

15.0 Link and Cable Testing

15.1   FOS/FOM link testing

15.2   FOS/FOM patch cable testing

15.3   OTDR loss per unit length testing

15.4   OTDR connector and splice evaluation

15.5   OTDR fault location

15.6   Acceptance testing documentation

ETA CONTACT INFORMATION

For more information about the ETA and its certification program, see the following information:

Address and Contact Numbers

ETA, International
5 Depot Street
Greencastle, Indiana 46135
Phone: (800) 288-3824 or (765) 653-8262
Fax: (765) 653-4287

Mail and Web

eta@eta-i.net

www.etainternational.org

Glossary

Numbers

2B+D

Shortcut for describing basic ISDN service (2B+D = two bearer channels and one data channel, which indicates two 64Kbps channels and one 16Kbps channel used for link management).

4B/5B

Signal encoding method used in 100Base-TX/FX Fast Ethernet and FDDI standards. Four-bit binary values are encoded into 5-bit symbols.

5-4-3 Rule

A rule that mandates that between any two nodes on the network, there can only be a maximum of five segments, connected through four repeaters, or concentrators, and only three of the five segments may contain user connections.

6-around-1

A configuration used to test alien crosstalk in the lab whereby six disturber cables completely surround and are in direct contact with a central disturbed cable. Required testing for all CAT-6A cables by ANSI/TIA-568-C.2.

8B/10B

Signal encoding method used by the 1000Base-X Gigabit Ethernet standards.

8B/6T

Signal encoding method used in 100Base-T4 Fast Ethernet standard.

50-pin connector

Commonly referred to as a telco, CHAMP, or blue ribbon connector. Commonly found on telephone switches, 66-blocks, 110-blocks, and 10Base-T Ethernet hubs and used as an alternate twisted-pair segment connection method. The 50-pin connector connects to 25-pair cables, which are frequently used in telephone wiring systems and typically meet Category 3. Some manufacturers also make Category 5e–rated cables and connectors.

66-type connecting block

Connecting block used by voice-grade telephone installations to terminate twisted pairs. Not recommended for LAN use.

110-block

A connecting block that is designed to accommodate higher densities of connectors and to support higher-frequency applications. The 110-blocks are found on patch panels and cross-connect blocks for data applications.

A

abrasion mark

A flaw on an optical surface usually caused by an improperly polished termination end.

absorption

The loss of power (signal) in an optical fiber resulting from conversion of optical power (specific wavelengths of light energy into heat. Caused principally by impurities, such as water, ions of copper or chromium (transition metals), and hydroxyl ions, and by exposure to nuclear radiation. Expressed in dB/km (decibels per kilometer). Absorption and scattering are the main causes of attenuation (loss of signal) of an optical waveguide during transmission through optical fiber.

abstract syntax notation (ASN) 1

Used to describe the language interface standards for interconnection of operating systems, network elements, workstations, and alarm functions.

acceptance angle

With respect to optical-fiber cable, the angle within which light can enter an optical fiber core of a given numerical aperture and still reflect off the boundary layer between the core and the cladding. Light entering at the acceptance angle will be guided along the core rather than reflected off the surface or lost through the cladding. Often expressed as the half angle of the cone and measured from the axis. Generally measured as numerical aperture (NA); it is equal to the arcsine. The acceptance angle is also known as the cone of acceptance, or acceptance cone. *See also* numerical aperture.

acceptance cone

The cross section of an optical fiber is circular; the light waves accepted by the core are expressed as a cone. The larger the acceptance cone, the larger the numerical aperture (NA); this means that the fiber is able to accept and propagate more light.

acceptance pattern

The amount of light transmitted from a fiber represented as a curve over a range of launch angles. *See also* acceptance angle.

access coupler

An optical device to insert or withdraw a signal from a fiber from between two ends. Many couplers require connectors on either end, and for many applications they must be APC (angled physical contact) connectors. The most popular access coupler is made by the fused biconic taper process, wherein two fibers are heated to the softening point and stretched so that the mode fields are brought into intimate contact, thus allowing a controlled portion of light to move from one core to the other.

access method

Rules by which a network peripheral accesses the network medium to transmit data on the network. All network technologies use some type of access method; common approaches include Carrier Sense Multiple Access/Collision Detection (CSMA/CD), token passing, and demand priority.

acknowledgment (ACK)

A message confirming that a data packet was received.

active branching device

Converts an optical input into two or more optical outputs without gain or regeneration.

active coupler

A device similar to a repeater that includes a receiver and one or more transmitters. The idea is to regenerate the input signal and then send them on. These are used in optical fiber networks.

active laser medium

Lasers are defined by their medium; laser mediums such as gas, (CO2, helium, neon) crystal (ruby) semiconductors, and liquids are used. Almost all lasers create coherent light on the basis of a medium being activated electronically. The stimulation can be electronic or even more vigorous, such as exciting molecular transitions from higher to lower energy states, which results in the emissions of coherent light.

active monitor

In Token Ring networks, the active monitor is a designated machine and procedure that prevents data frames from roaming the ring unchecked. If a Token Ring frame passes the active monitor too many times, it is removed from the ring. The active monitor also ensures that a token is always circulating the ring.

active splicing

A process performed with an alignment device, using the light in the core of one fiber to measure the transmittance to the other. Ensures optimal alignment before splicing is completed. The active splicing device allows fusion splicing to perform much better with respect to insertion losses when compared to most connectors and splicing methods. A splicing technician skilled at the use of an active splicing device can reliably splice with an upper limit of 0.03dB of loss.

adapter

With respect to optical fiber, a passive device used to join two connectors and fiber cores together. The adapter is defined by connector type, such as SC, FC, ST, LC, MT-RJ, and FDDI. Hybrid adapters can be used to join dissimilar connectors together, such as SC to FC. The adapter's key element is a "split sleeve," preferably made from zirconia and having a specific resistance force to insertion and withdrawal of a ferrule. This resistance, typically between 4 and 7 grams, ensures axial alignment of the cores.

address

An identifier that uniquely identifies nodes or network segments on a network.

adjustable attenuator

An attenuator in which the level of attenuation is varied with an internal adjustment. Also known as variable attenuator.

administration

With respect to structured cabling, the procedures and standards used to accurately keep track of the various circuits and connections, as well as records pertaining to them. The ANSI/TIA/EIA-606 Administration Standard for Telecommunications Infrastructure of Commercial Buildings defines specifications for this purpose.

ad hoc RF network

An RF network that exists only for the duration of the communication. Ad hoc networks are continually set up and torn down as needed.

Advanced Intelligent Network (AIN)

Developed by Bell Communications Research, a telephone network architecture that separates the service logic from the switching equipment. This allows the rapid deployment of new services with major equipment upgrades.

aerial cable

Telecommunications cable installed on aerial supporting structures such as poles, towers, sides of buildings, and other structures.

air polishing
Polishing a connector tip in a figure-eight motion without using a backing for the polishing film.

Alco Wipes
A popular brand of medicated wipes premoistened with alcohol.

alien crosstalk (AXT)
The crosstalk noise experienced by two or more adjacent twisted pair cables; measured in the lab by performing a 6-around-one test where six disturber cables impose crosstalk on the cable in the center (disturbed cable).

all-dielectric self-supporting cable (ADSS)
An optical fiber cable with dielectric strength elements that can span long distances between supports without the aid of other strength elements.

alternating current (AC)
An electric current that cyclically reverses polarity.

American Standard Code for Information Interchange (ASCII)
A means of encoding information. ASCII is the method used by Microsoft to encode characters in text files in their operating systems.

American Wire Gauge (AWG)
Standard measuring gauge for nonferrous conductors (i.e., noniron and nonsteel). Gauge is a measure of the diameter of the conductor. The higher the AWG number, the smaller the diameter of the wire. See Chapter 1 for more information.

ampere
A unit of measure of electrical current.

amplifier
Any device that intensifies a signal without distorting the shape of the wave. In fiber optics, a device used to increase the power of an optical signal.

amplitude
The difference between high and low points of a wavelength cycle. The greater the amplitude, the stronger the signal.

amplitude modulation (AM)
A method of signal transmission in which the amplitude of the carrier is varied in accordance with the signal. With respect to optical fiber cabling, the modulation is done by varying the amplitude of a light wave, common in analog/RF applications.

analog
A signal that varies continuously through time in response to an input. A mercury thermometer, which gives a variable range of temperature readings, is an example of an analog instrument. Analog electrical signals are measured in hertz (Hz). Analog is the opposite of digital.

analog signal
An electrical signal that varies continuously (is infinitely variable within a specified range) without having discrete values (as opposed to a digital signal).

analog-to-digital (A/D) converter
A device used to convert analog signals to digital signals for storage or transmission.

ANDing
To determine whether a destination host is local or remote, a computer will perform a simple mathematical computation referred to as an AND operation.

angle of incidence
With respect to fiber optics, the angle of a ray of light striking a surface or boundary as measured from a line drawn perpendicular to the surface. Also called incident angle.

angle of reflection
With respect to fiber optics, the angle formed between the normal and a reflected beam. The angle of reflection equals the angle of incidence.

angle of refraction
With respect to fiber optics, the angle formed between the normal (a line drawn perpendicular to the interface) and a refracted beam.

angled end
An optical fiber whose end is polished to an angle to reduce reflectance.

angled physical contact (APC) connector
A single-mode optical-fiber connector whose angled end-face helps to ensure low mated reflectance and low unmated reflectance. The ferrule is polished at an angle to ensure physical contact with the ferrule of another APC connector.

angular misalignment loss
The loss of optical power caused by deviation from optimum alignment of fiber-to-fiber. Loss at a connector due to fiber angles being misaligned because the fiber ends meet at an angle.

antireflection (AR)
Coating used on optical fiber cable to reduce light reflection.

AppleTalk
Apple Computer's networking protocol and networking scheme, integrated into most Apple system software, that allows Apple computing systems to participate in peer-to-peer computer networks and to access the services of AppleTalk servers. AppleTalk operates over Ethernet (EtherTalk), Token Ring (TokenTalk), and FDDI (FDDITalk) networks. *See also* LocalTalk.

application
(1) A program running on a computer. (2) A system, the transmission method of which is supported by telecommunications cabling, such as 100Base-TX Ethernet, or digital voice.

aramid strength member
The generic name for Kevlar® or Twaron® found in fiber optic cables. A yarn used in fiber-optic cable that provides additional tensile strength, resistance to bending, and support to the fiber bundle. It is not used for data transmission.

ARCnet (Attached Resource Computer network)
Developed by Datapoint, a relatively low-speed form of LAN data link technology (2.5Mbps, in which all systems are attached to a common coaxial cable or an active or passive hub). ARCnet uses a token-bus form of medium access control; only the system that has the token can transmit.

armoring
Provides additional protection for cables against severe, usually outdoor, environments. Usually consists of plastic-coated steel corrugated for flexibility; *see also* interlock armor.

ARP (Address Resolution Protocol)
The method for finding a host's Link layer (hardware) address when only its Internet layer or some other Network layer address is known.

ARP table
A table used by the ARP protocol on TCP/IP-based network nodes that contains known TCP/IP addresses and their associated MAC (media access control) addresses. The table is cached in memory so that ARP lookups do not have to be performed for frequently accessed TCP/IP and MAC addresses.

array connector
A connector designed for use with multiple optical fibers; *see also* MPT and MPO connectors.

Asymmetric Digital Subscriber Line (ADSL)
Sometimes called Universal ADSL, G.Lite, or simply DSL. ADSL is a digital communications method that allows high-speed connections between a central office (CO) and telephone subscriber over a regular pair of phone wires. It uses different speeds for uploading and downloading (hence the Asymmetric in ADSL) and is most often used for Internet connections to homes or businesses.

asynchronous
Transmission where sending and receiving devices are not synchronized (without a clock signal). Data must carry markers to indicate data division.

asynchronous transfer mode (ATM)
In networking terms, asynchronous transfer mode is a connection-oriented networking technology that uses a form of very fast packet switching in which data is carried in fixed-length units. These fixed-length units are called cells; each cell is 53 bytes in length, with 5 bytes used as a header in each cell. Because the cell size does not change, the cells can be switched very quickly. ATM networks can transfer data at extremely high speeds. ATM employs mechanisms that can be used to set up virtual circuits between users, in which a pair of users appear to have a dedicated circuit between them. ATM is defined in specifications from the ITU and ATM Forum. For more information, see the ATM Forum's website at www.atmforum.com.

attachment unit interface (AUI) port
A 15-pin connector found on older network interface cards (NICs). This port allowed connecting the NIC to different media types by using an external transceiver. The cable that connected to this port when used with older 10Base5 media was known as a transceiver cable or a drop cable.

attenuation
A general term indicating a decrease in power (loss of signal) from one point to another. This loss can be a loss of electrical signal or light strength. In optical fibers, it is measured in decibels per kilometer (dB/km) at a specified wavelength. The loss is measured as a ratio of input power to output power. Attenuation is caused by poor-quality connections, defects in the cable, and loss due to heat. The lower the attenuation value, the better. Attenuation is the opposite of gain. See Chapter 1 for additional information on attenuation and the use of decibels.

attenuation-to-crosstalk ratio (ACR)

A copper cabling measurement, the ratio between attenuation and near-end crosstalk, measured in decibels, at a given frequency. While it is not a requirement of ANSI/TIA-568-C.2, it is used by every manufacturer as a figure of merit in denoting signal to noise ratio

attenuation-to-crosstalk ratio, far end (ACRF)

The ratio between attenuation and far-end crosstalk, measured in decibels, at a given frequency; a requirement of ANSI/TIA-568-C.2 for twisted-pair cables; formerly referred to as equal-level far end crosstalk (ELFEXT) in ANSI/TIA/EIA-568-B.2.

attenuation-limited operation

In a fiber-optic link, the condition when operation is limited by the power of the received signal.

attenuator

A passive device that intentionally reduces the strength of a signal by inducing loss.

audio

Used to describe the range of frequencies within range of human hearing; approximately 20 to 20,000Hz.

auxiliary AC power

A standard 110V AC power outlet found in an equipment area for operation of test equipment or computer equipment.

avalanche photodiode (APD)

With respect to optical fiber equipment, a specialized diode designed to use the avalanche multiplication of photocurrent. The photodiode multiplies the effect of the photons it absorbs, acting as an amplifier.

average picture level (APL)

A video parameter that indicates the average level of video signal, usually relative to blank and a white level.

average power

The energy per pulse, measured in joules, times the pulse repetition rate, measured in hertz (Hz). This product is expressed as watts.

average wavelength (l)

The average of the two wavelengths for which the peak optical power has dropped to half.

axial ray

In fiber-optic transmissions, a light ray that travels along the axis of a fiber-optic filament.

B

backbone

A cable connection between telecommunications or wiring closets, floor distribution terminals, entrance facilities, and equipment rooms either within or between buildings. This cable can service voice communications or data communications. In star-topology data networks, the backbone cable interconnects hubs and similar devices, as opposed to cables running between hub and station. In a bus topology data network, it is the bus cable. Backbone is also called riser cable, vertical cable, or trunk cable.

backbone wiring

The physical/electrical interconnections between telecommunications rooms and equipment rooms. *See also* backbone.

backscatter

Usually a very small portion of an overall optical signal, backscatter occurs when a portion of scattered light returns to the input end of the fiber; the scattering of light in the direction opposite to its original propagation. Light that propagates back toward the transmitter. Also known as back reflection or backscattering.

bait

The mold or form on which silicon dioxide soot is deposited to create the optical fiber preform.

balance

An indication of signal voltage equality and phase polarity on a conductor pair. Perfect balance occurs when the signals across a twisted-pair cable are equal in magnitude and opposite in phase with respect to ground.

balanced cable

A cable that has pairs made up of two identical conductors that carry voltages of opposite polarities and equal magnitude with respect to ground. The conductors are twisted to maintain balance over a distance.

balanced coupler

A coupler whose output has balanced splits; for example, one by two is 50/50, or one by four is 25/25/25/25.

balanced signal transmission

Two voltages, commonly referred to as tip and ring, equal and opposite in phase with respect to each other across the conductors of a twisted-pair cable.

balun

A device that is generally used to connect balanced twisted-pair cabling with unbalanced coaxial cabling. The balun is an impedance-matching transformer that converts the impedance of one transmission media to the impedance of another transmission media. For example, a balun would be required to connect 100 ohm UTP to 120 ohm STP. Balun is short for balanced/unbalanced.

bandpass
A range of wavelengths over which a component will meet specifications.

bandwidth
Indicates the transmission capacity of media. For copper cables, bandwidth is defined using signal frequency and specified in hertz (Hz). For optical fiber, wavelength in nanometers (nm) defines bandwidth. Also refers to the amount of data that can be sent through a given channel and is measured in bits per second.

bandwidth-limited operation
Systems can be limited by power output or bandwidth; bandwidth-limited operation is condition when the total system bandwidth is the limiting factor (as opposed to signal amplitude).

barrier layer
A layer of glass deposited on the optical core to prevent diffusion of impurities into the core.

baseband
A method of communication in which the entire bandwidth of the transmission medium is used to transmit a single digital signal. The signal is driven directly onto the transmission medium without modulation of any kind. Baseband uses the entire bandwidth of the carrier, whereas broadband only uses part of the bandwidth. Baseband is simpler, cheaper, and less sophisticated than broadband.

basic rate interface (BRI)
As defined by ISDN, consists of two 64Kbps B-channels used for data and one 16Kbps D-channel (used primarily for signaling). Thus, a basic rate user can have up to 128Kbps service.

battery distribution fuse bay (BDFB)
A type of DC "power patch panel" for telecommunication equipment where power feeder cables are connected to a box of fused connections. Distribution cables run from this device to the equipment.

baud
The number of signal level transitions per second. Commonly confused with bits per second, the baud rate does not necessarily transmit an equal number of bits per second. In some encoding schemes, baud will equal bits per second, but in others it will not. For example, a signal with four voltage levels may be used to transfer two bits of information for every baud.

beacon
A special frame in Token Ring systems indicating a serious problem with the ring such as a break. Any station on the ring can detect a problem and begin beaconing.

beamsplitter
An optical device, such as a partially reflecting mirror, that splits a beam of light into two or more beams. Used in fiber optics for directional couplers.

beamwidth
For a round light beam, the diameter of a beam, measured across the width of the beam. Often specified in nanometers (nm) or millimeters (mm).

bearer channel (B-channel)
On an ISDN network, carries the data. Each bearer channel typically has a bandwidth of 64Kbps.

bel
Named for Alexander Graham Bell, this unit represents the logarithm of the ratio of two levels. See Chapter 1 for an explanation of bel and decibels.

bend loss
A form of increased attenuation in a fiber where light is escaping from bent fiber. Bend loss is caused by bending a fiber around a restrictive curvature (a macrobend) or from minute distortions in the fiber (microbend). The attenuation may be permanent if fractures caused by the bend continue to affect transmission of the light signal.

bend radius (minimum)
The smallest bend a cable can withstand before the transmission is affected. UTP copper cabling usually has a bend radius that is four times the diameter of the cable; optical fiber is usually 10 times the diameter of the cable. Bending a cable any more than this can cause transmission problems or cable damage. Also referred to as cable bend radius.

biconic connector
A fiber-optic termination connector that is cone-shaped and designed for multiple connects and disconnects. The biconic connector was developed by AT&T but is not commonly used.

bidirectional attenuation testing
Refers to measuring the attenuation of a link using an OTDR in each direction.

bidirectional couplers
Couplers that operate in both directions and function in the same way in both directions.

bifurcated contact prongs
Contacts in a 66- or 110-block that are split in two so that the wire can be held better.

binary
A value that can only be expressed as one of two states. Binary values may be "1" or "0"; "high" or "low" voltage or energy; or "on" or "off" actions of a switch or light signal. Binary signals are used in digital data transmission.

binder
A tape or thread used to hold assembled cable components in place.

binder group
A group of 25 pairs of wires within a twisted-pair cable with more than 25 total pairs. The binder group has a strip of colored plastic around it to differentiate it from other binder groups in the cable.

biscuit jacks
A surface mount jack used for telephone wiring.

bistable optics
Optical devices with two stable transmission states.

bit
A binary digit, the smallest element of information in the binary number system, and thus the smallest piece of data possible in a digital communication system. A 1 or 0 of binary data.

bit error
An error in data transmission that occurs when a bit is a different value when it is received than the value at which it was transmitted.

bit error rate (BER)
In digital applications, a measure of data integrity. It is the ratio of bits received in error to bits originally sent or the ratio of incorrectly transmitted bits to total transmitted bits. BERs of one error per billion bits sent are typical.

bit error rate tester (BERT)
A device that tests the bit error rate across a circuit. One common device that does is this is called a T-BERD because it is designed to test T-1 and leased line error rates.

bit rate
The actual number of light pulses per second being transmitted through a fiber-optic link. Bit rates are usually measured in megabits per second (Mbps) or gigabits per second (Gbps).

bit stream
A continuous transfer of bits over some medium.

bit stuffing
A method of breaking up continuous strings of 0 bits by inserting a 1 bit. The 1 bit is removed at the receiver.

bit time
The number of transmission clock cycles that are used to represent one bit.

bits per second (bps)
The number of binary digits (bits) passing a given point in a transmission medium in one second.

BL
Blue. Refers to blue cable pair color in UTP twisted-pair cabling.

black body
A body or material that, in equilibrium, will absorb 100 percent of the energy incident upon it, meaning it will not reflect the energy in the same form. It will radiate nearly 100 percent of this energy, usually as heat and/or IR (infrared).

blown fiber
A method for installing optical fiber in which fibers are blown through pre-installed conduit or tube using air pressure to pull the fiber through the conduit.

BNC connector
Bayonet Neill-Concelman connector (Neill and Concelman were the inventors). A coaxial connector that uses a "bayonet" style turn-and-lock mating method. Historically used with RG-58 or smaller coaxial cable or with 10Base-2 Ethernet thin coaxial cable.

bonding
(1) The method of permanently joining metallic parts to form an electrical contact that will ensure electrical continuity and the capacity to safely conduct any current likely to be imposed on it. (2) Grounding bars and straps used to bond equipment to the building ground. (3) Combining more than one ISDN B-channel using ISDN hardware.

bounded medium
A network medium that is used at the physical layer where the signal travels over a cable of some kind, as opposed to an unbounded medium such as wireless networking.

BR

Brown. Refers to brown cable pair color in UTP twisted-pair cabling.

braid

A group of textile or metallic filaments interwoven to form a tubular flexible structure that may be applied over one or more wires or flattened to form a strap. Designed to give a cable more flexibility or to provide grounding or shielding from EMI.

breakout cable

Multifiber cables composed of simplex interconnect cables where each fiber has additional protection by using additional jackets and strength elements, such as aramid yarn.

breakout kit

A collection of components used to add tight buffers, strength members, and jackets to individual fibers from a loose tube buffer cable. Breakout kits are used to build up the outer diameter of fiber cable when connectors are being installed and designed to allow individual fibers to be terminated with standard connectors.

bridge

A network device, operating at the Data Link layer of the OSI model, that logically separates a single network into segments but lets the multiple segments appear to be one network to higher-layer protocols.

bridged tap

Multiple appearances of the same cable pair at several distribution points, usually made by splicing into a cable. Also known as parallel connections. Bridge taps were commonly used in coaxial cable networks and still appear in residential phone wiring installations. Their use is not allowed in any structured cabling environment.

broadband

A transmission facility that has the ability to handle a variety of signals using a wide range of channels simultaneously. Broadband transmission medium has a bandwidth sufficient to carry multiple voice, video, or data channels simultaneously. Each channel occupies (is modulated to) a different frequency bandwidth on the transmission medium and is demodulated to its original frequency at the receiving end. Channels are separated by "guard bands" (empty spaces) to ensure that each channel will not interfere with its neighboring channels. For example, this technique is used to provide many CATV channels on one coaxial cable.

broadband ISDN

An expansion of ISDN digital technology that allows it to compete with analog broadband systems using ATM or SDH.

broadcast

Communicating to more than one receiving device simultaneously.

brouter

A device that combines the functionality of a bridge and a router but can't be distinctly classified as either. Most routers on the market incorporate features of bridges into their feature set. Also called a hybrid router.

buffer (buffer coating)

A protective layer or tube applied to a fiber-optic strand. This layer of material, usually thermoplastic or acrylic polymer, is applied in addition to the optical-fiber coating, which provides protection from stress and handling. Fabrication techniques include tight or loose-tube buffering as well as multiple buffer layers. In tight-buffer constructions, the thermoplastic is extruded directly over the coated fiber. In loose-tube buffer constructions, the coated fiber "floats" within a buffer tube that is usually filled with a nonhygroscopic gel. See Chapter 10 for more information.

buffer tube

Used to provide protection and isolation for optical fiber cable. Usually hard plastic tubes, with an inside diameter several times that of a fiber, which holds one or more fibers.

building cable

Cable in a traditional cabling system that is inside a building and that will not withstand exposure to the elements. Also referred to as premises cable.

building distributor (BD)

An ISO/IEC 11801 term that describes a location where the building backbone cable terminates and where connections to the campus backbone cable may be made.

building entrance

The location in a building where a trunk cable between buildings is typically terminated and service is distributed through the building. Also the location where services enter the building from the phone company and antennas.

Building Industry Consulting Service International (BICSI)

A nonprofit association concerned with promoting correct methods for all aspects of the installation of communications wiring. More information can be found on their website at www.bicsi.org.

bundle (fiber)
A group of individual fibers packaged or manufactured together within a single jacket or tube. Also a group of buffered fibers distinguished from another group in the same cable core.

bundled cable
An assembly of two or more cables continuously bound together to form a single unit prior to installation.

bus topology
In general, a physical or logical layout of network devices in which all devices must share a common medium to transfer data.

busy token
A data frame header in transit.

bypass
The ability of a station to isolate itself optically from a network while maintaining the continuity of the cable plant.

byte
A binary "word" or group of eight bits.

C

c
The symbol for the speed of light in a vacuum.

C
The symbol for both capacitance and Celsius. In the case of Celsius, it is preceded by the degree symbol (°).

cable
(1) Copper: A group of insulated conductors enclosed within a common jacket. (2) Fiber: One or more optical fibers enclosed within a protective covering and material to provide strength.

cable assembly
A cable that has connectors installed on one or both ends. If connectors are attached to only one end of the cable, it is known as a pigtail. If its connectors are attached to both ends, it is known as a jumper. General use of these cable assemblies includes the interconnection of multimode and single-mode fiber optical cable systems and optical electronic equipment.

cable duct
A single pathway (typically a conduit, pipe, or tube) that contains cabling.

cable entrance conduits
Holes or pipes through the building foundation, walls, floors, or ceilings through which cables enter into the cable entrance facility (CEF).

cable entrance facility (CEF)
The location where the various telecommunications cables enter a building. Typically this location has some kind of framing structure (19-inch racks, or plywood sheet) with which the cables and their associated equipment can be organized.

cable management system
A system of tools, hold-downs, and apparatus used to precisely place cables and bundles of cables so that the entire cabling system is neat and orderly and that growth can be easily managed.

cable modem
A modem that transmits and receives signals usually through copper coaxial cable. Connects to your CATV connection and provides you with a 10Base-T connection for your computer. All of the cable modems attached to a cable TV company line communicate with a cable modem termination system (CMTS) at the local CATV office. Cable modems can receive and send signals to and from the CMTS only and not to other cable modems on the line. Some services have the upstream signals returned by telephone rather than cable, in which case the cable modem is known as a telco-return cable modem; these require the additional use of a phone line. The theoretical data rate of a CATV line is up to 27Mbps on the download path and about 2.5Mbps of bandwidth for upload. The overall speed of the Internet and the speed of the cable provider's access pipe to the Internet restrict the actual amount of throughput available to cable modem users. However, even at the lower end of the possible data rates, the throughput is many times faster than traditional modem connections to the Internet.

cable plant
Consists of all the copper and optical elements, including patch panels, patch cables, fiber connectors, and splices, between a transmitter and a receiver.

cable pulling lubricant
A lubricant used to reduce the friction against cables as they are pulled through conduits.

cable rearrangement facility (CRF)
A special splice cabinet used to vertically organize cables so that they can be spliced easier.

cable sheath

A covering over the core assembly that may include one or more metallic members, strength members, or jackets.

cable tray

A shallow tray used to support and route cables through building spaces and above network racks.

cabling map

A map of how cabling is run through a building network

campus

The buildings and grounds of a complex, such as a university, college, industrial park, or military establishment.

campus backbone

Cabling between buildings that share data and telecommunications facilities.

campus distributor (CD)

The ISO/IEC 11801 term for the main cross-connect; this is the distributor from which the campus backbone cable emanates.

capacitance

The ability of a dielectric material between conductors to store electricity when a difference of potential exists between the conductors. The unit of measurement is the farad, which is the capacitance value that will store a charge of a one coulomb when a one-volt potential difference exists between the conductors. In AC, one farad is the capacitance value, which will permit one ampere of current when the voltage across the capacitor changes at a rate of one volt per second.

carrier

An electrical signal of a set frequency that can be modulated in order to carry voice, video, and/or data.

carrier detect (CD)

Equipment or a circuit that detects the presence of a carrier signal on a digital or analog network.

carrier sense

With Ethernet, a method of detecting the presence of signal activity on a common channel.

Carrier Sense Multiple Access/Collision Avoidance (CSMA/CA)

A network media access method that sends a request to send (RTS) packet and waits to receive a clear to send (CTS) packet before sending. Once the CTS is received, the sender sends the packet of information. This method is in contrast to CSMA/CD, which merely checks to see if any other stations are currently using the media.

Carrier Sense Multiple Access/Collision Detect (CSMA/CD)

A network media access method employed by Ethernet. CSMA/CD network stations listen for traffic before transmitting. If two stations transmit simultaneously, a collision is detected and each station waits a brief (and random) amount of time before attempting to transmit again.

Category 1

Also called CAT-1. Unshielded twisted pair used for transmission of audio frequencies up to 1MHz. Used as speaker wire, doorbell wire, alarm cable, etc. Category 1 cable is not suitable for networking applications or digital voice applications. See Chapter 1 for more information. Not a recognized media by ANSI/TIA-568-C.

Category 2

Also called CAT-2. Unshielded twisted pair used for transmission at frequencies up to 4MHz. Used in analog and digital telephone applications. Category 2 cable is not suitable for networking applications. See Chapter 1 for more information. Not a recognized medium by ANSI/TIA-568.C.

Category 3

Also called CAT-3. Unshielded twisted pair with 100 ohm impedance and electrical characteristics supporting transmission at frequencies up to 16MHz. Used for 10Base-T Ethernet and digital voice applications. Recognized by the ANSI/TIA-568-C standard. See Chapters 1 or 7 for more information.

Category 4

Also called CAT-4. Unshielded twisted pair with 100 ohm impedance and electrical characteristics supporting transmission at frequencies up to 20MHz. Not used and not supported by the standards. See Chapter 1 for more information.

Category 5

Also called CAT-5. Unshielded twisted pair with 100 ohm impedance and electrical characteristics supporting transmission at frequencies up to 100MHz. Category 5 is not a recognized cable type for new installations by the ANSI/TIA-568-C standard, but its requirements are included in the standard to support legacy installations of Category 5 cable. See Chapters 1 or 7 for more information.

Category 5e

Also called CAT-5e or Enhanced CAT-5. Recognized in ANSI/TIA-568-C. Category 5e has improved specifications for NEXT, PSNEXT, Return Loss, ACRF (ELFEXT), PSACRF (PS-ELFEXT), and attenuation as compared to Category 5. Like Category 5, it consists of unshielded twisted pair with 100 ohm impedance and electrical characteristics supporting transmission at frequencies up to 100MHz. See Chapters 1 or 7 for more information.

Category 6

Also called CAT-6. Recognized in ANSI/TIA-568-C, Category 6 supports transmission at frequencies up to 250MHz over 100 ohm twisted pair. See Chapters 1 or 7 for more information.

Category 6A

Also called CAT-6A. Recognized in ANSI/TIA-568-C, Category 6A supports transmission at frequencies up to 500MHz over 100 ohm twisted pair. This is the only category cable required to meet alien crosstalk specifications. See Chapters 1 or 7 for more information.

CCIR

Consultative Committee for International Radio. Replaced by the International Telecommunications Union – Radio (ITU-R).

center wavelength (laser)

The nominal value central operating wavelength defined by a peak mode measurement where the effective optical power resides.

center wavelength (LED)

The average of two wavelengths measured at the half amplitude points of the power spectrum.

central member

The center component of a cable, which serves as an antibuckling element to resist temperature-induced stresses. Sometimes serves as a strength element. The central member is composed of steel, fiberglass, or glass-reinforced plastic.

central office (CO)

The telephone company building where subscribers' lines are joined to telephone company switching equipment that serves to interconnect those lines. Also known as an exchange center or head end. Some places call this a public exchange.

central office ground bus

Essentially a large ground bar used in a central office to provide a centralized grounding for all the equipment in that CO (or even just a floor in that CO).

channel

The end-to-end transmission or communications path over which application-specific equipment is connected. Through multiplexing several channels, voice channels can be transmitted over an optical channel.

channel bank

Equipment that combines a number of voice and sometimes digital channels into a digital signal; in the case of a T-1 channel bank, it converts 24 separate voice channels into a single digital signal.

channel insertion loss

The static signal loss of a link between a transmitter and receiver for both copper and fiber systems. It includes the signal loss of the cable, connectors/splices and patch cords.

channel service unit/digital service unit (CSU/DSU)

A hardware device that is similar to a modem that connects routers' or bridges' WAN interfaces (V.35, RS-232, etc.) to wide area network connections (Fractional-T1, T-1, Frame Relay, etc.). The device converts the data from the router or bridge to frames that can be used by the WAN. Some routers will have the CSU/DSU built directly into the router hardware, while other arrangements require a separate unit.

characteristic impedance

The impedance that an infinitely long transmission line would have at its input terminal. If a transmission line is terminated in its characteristic impedance, it will appear (electrically) to be infinitely long, thus minimizing signal reflections from the end of the line.

cheapernet

A nickname for thin Ethernet (thinnet) or 10Base-2 Ethernet systems.

cheater bracket

A bracket used for making cable installation easier.

chromatic dispersion

The spreading of a particular light pulse because of the varying refraction rates of the different colored wavelengths. Different wavelengths travel along an optical medium at different speeds. Wavelengths reach the end of the medium at different times, causing the light pulse to spread. This chromatic dispersion is expressed in picoseconds per kilometer per nanometer (of bandwidth). It is the sum of material and waveguide dispersions. *See also* waveguide dispersion, material dispersion.

churn

Cabling slang for the continual rearrangement of the various connections in a data connection frame. Office environments where network equipment and phones are frequently moved will experience a high churn rate.

circuit

A communications path between two pieces of associated equipment.

cladding

Name for the material (usually glass, sometimes plastic) that is put around the core of an optical fiber during manufacture. The cladding is not designed to carry light, but it has a slightly lower index of refraction than the core, which causes the transmitted light to travel down the core. The interface between the core and the cladding creates the mode field diameter, wherein the light is actually held reflectively captive within the core.

cladding mode

A mode of light that propagates through and is confined to the cladding.

Class A

(1) ISO/IEC 11801 channel designation utilizing twisted-pair cabling rated to 100KHz. Used in voice and low-frequency applications. Comparable to TIA/EIA Category 1 cabling; not suitable for networking applications. (2) IP addresses that have a range of numbers from 1 through 127 in the first octet.

Class B

(1) ISO/IEC 11801 channel designation utilizing twisted-pair cabling rated to 1MHz. Used in medium bit-rate applications. Comparable to TIA/EIA Category 2 cabling; not suitable for networking applications. (2) IP addresses that have a range of numbers from 128 through 191 in the first octet.

Class C

(1) ISO/IEC 11801 channel designation utilizing twisted-pair cabling rated to 16MHz. Used in high bit-rate applications. Corresponds to TIA/EIA Category 3 cabling. (2) IP addresses that have a range of numbers from 192 through 223 in the first octet.

Class D

(1) ISO/IEC 11801 channel designation utilizing twisted-pair cabling rated to 100MHz. Used in very high bit-rate applications. Corresponds to TIA/EIA Category 5 cabling. (2) IP addresses used for multicast applications that have a range of numbers from 224 through 239 in the first octet.

Class E

(1) ISO/IEC 11801 channel utilizing twisted-pair cabling rated to 250MHz. Corresponds to the TIA/EIA Category 6 cabling Standard. (2) IP addresses used for experimental purposes that have a range of numbers from 240 through 255 in the first octet.

Class Ea

ISO/IEC 11801 proposed channel utilizing twisted-pair cabling rated to 500MHz. Corresponds closely to the TIA/EIA Category 6A cabling standard, but there are some differences in some internal performance requirements. (2) IP addresses used for experimental purposes that have a range of numbers from 240 through 255 in the first octet.

Class F

ISO/IEC 11801 channel utilizing twisted-pair cabling rated to 600MHz.

Class Fa

ISO/IEC 11801 proposed channel utilizing twisted-pair cabling rated to 1000MHz.

cleaving

The process of separating an unbuffered optical fiber by scoring the outside and pulling off the end. Cleaving creates a controlled fracture of the glass for the purpose of obtaining a fiber end that is flat, smooth, and perpendicular to the fiber axis. This is done prior to splicing or terminating the fiber.

clock

The timing signal used to control digital data transmission.

closet

An enclosed space for housing telecommunications and networking equipment, cable terminations, and cross-connect cabling. It contains the horizontal cross-connect where the backbone cable cross-connects with the horizontal cable. Called a telecommunications room by the TIA/EIA standards; sometimes referred to as a wiring closet.

CMG

A cable rated for general purpose by the NEC.

CMR

A cable rated for riser applications by the NEC.

coating

The first true protective layer of an optical fiber. The plastic-like coating surrounds the glass cladding of a fiber and put on a fiber immediately after the fiber is drawn to protect it from the environment. Do not confuse the coating with the buffer.

coaxial cable
Also called coax. Coaxial cable was invented in 1929 and was in common use by the phone company by the 1940s. Today it is commonly used for cable TV and by older Ethernet; twisted-pair cabling has become the desirable way to install Ethernet networks. It is called coaxial because it has a single conductor surrounded by insulation and then a layer of shielding (which is also a conductor) so the two conductors share a single axis; hence "co"-axial. The outer shielding serves as a second conductor, ground, and to reduce the effects of EMI. Can be used at high bandwidths over long distances.

code division multiple access (CDMA)
In time division multiplexing (TDM), one pulse at a time is taken from several signals and combined into a single bit stream.

coherence
In light forms, characterized as a consistent, fixed relationship between points on the wave. In each case, there is an area perpendicular to the direction of the light's propagation in which the light may be highly coherent.

coherence length or time
The distance time over which a light form may be considered coherent. Influenced by a number of factors, including medium, interfaces, and launch condition. Time, all things being equal, is calculated by the coherence length divided by the phase velocity of light in the medium.

coefficient of thermal expansion
The measure of a material's change in size in response to temperature variation.

coherent communications
Where the light from a laser oscillator is mixed with the received signal, and the difference frequency is detected and amplified.

coherent light
Light in which photons have a fixed or predictable relationship; that is, all parameters are predictable and correlated at any point in time or space, particularly over an area in a lane perpendicular to the direction of propagation or over time at a particular point in space. Coherent light is typically emitted from lasers.

collision
The network error condition that occurs when electrical signals from two or more devices sharing a common data transfer medium crash into one another. This commonly happens on Ethernet-type systems.

committed information rate (CIR)
A commitment from your service provider stating the minimum bandwidth you will get on a frame relay network averaged over time.

common mode transmission
A transmission scheme where voltages appear equal in magnitude and phase across a conductor pair with respect to ground. May also be referred to as longitudinal mode.

community antenna television (CATV)
More commonly known as cable TV, a broadband transmission facility that generally uses a 75 ohm coaxial cable to carry numerous frequency-divided TV channels simultaneously. CATV carries high-speed Internet service in many parts of the world.

compliance
A wiring device that meets all characteristics of a standard is said to be in compliance with that standard. For example, a data jack that meets all of the physical, electrical, and transmission standards for ANSI/TIA/EIA-568-B Category 5e is compliant with that standard.

composite cable
A cable that typically has at least two different types of transmitting media inside the same jacket; for example, UTP and fiber.

concatenation
The process of joining several fibers together end to end.

concatenation gamma
The coefficient used to scale bandwidth when several fibers are joined together.

concentrator
A multiport repeater or hub.

concentric
Sharing a common geometric center.

conductivity
The ability of a material to allow the flow of electrical current; the reciprocal of resistivity. Measured in "mhos" (the word ohm spelled backward).

conductor
A material or substance (usually copper wire) that offers low resistance (opposition) to the flow of electrical current.

conduit
A rigid or flexible metallic or nonmetallic raceway of circular cross section in which cables are housed for protection and to prevent burning cable from spreading flames or smoke in the event of a fire.

cone of acceptance

A cone-shaped region extending outward from an optical fiber core and defined by the core's *acceptance angle*.

Conference of European Postal and Telecommunications Administrations (CEPT)

A set of standards adopted by European and other countries, particularly defining interface standards for digital signals.

connecting block

A basic part of any telecommunications distribution system. Also called a terminal block, a punch-down block, a quick-connect block, and a cross-connect block, a connecting block is a plastic block containing metal wiring terminals to establish connections from one group of wires to another. Usually each wire can be connected to several other wires in a bus or common arrangement. There are several types of connecting blocks, such as 66 clip, BIX, Krone, or 110. A connecting block has insulation displacement connections (IDCs), which means you don't have to remove insulation from around the wire conductor before you punch it down or terminate it.

connectionless protocol

A communications protocol that does not create a virtual connection between sending and receiving stations.

connection-oriented protocol

A communications protocol that uses acknowledgments and responses to establish a virtual connection between sending and receiving stations.

connector

With respect to cabling, a device attached to the end of a cable, receiver, or light source that joins it with another cable, device or fiber. A connector is a mechanical device used to align and join two conductors or fibers together to provide a means for attaching and decoupling it to a transmitter, receiver, or another fiber. Commonly used connectors include the RJ-11, RJ-45, BNC, FC, ST, LC, MT-RJ, FDDI, Biconic, and SMA connectors.

connector-induced optical-fiber loss

With respect to fiber optics, the part of connector insertion loss due to impurities or structural changes to the optical fiber caused by the termination within the connector.

connector plug

With respect to fiber optics, a device used to terminate an optical fiber.

connector receptacle

With respect to fiber optics, the fixed or stationary half of a connection that is mounted on a patch panel or bulkhead.

connector variation

With respect to fiber optics, the maximum value in decibels of the difference in insertion loss between mating optical connectors (e.g., with remating, temperature cycling, etc.). Also known as optical connector variation.

consolidation point (CP)

A location defined by the ANSI/TIA/EIA-568-B standard for interconnection between horizontal cables that extends from building pathways and horizontal cables that extend into work area pathways. Often an entry point into modular furniture for voice and data cables. ISO/IEC 11801 defines this as a transition point (TP).

consumables kit

Resupply material for splicing or terminating fiber optics.

continuity

An uninterrupted pathway for electrical or optical signals.

controlled environmental vault (CEV)

A cable termination point in a below-ground vault whose humidity and temperature are controlled.

Copper Distributed Data Interface (CDDI)

A version of FDDI that uses copper wire media instead of fiber-optic cable and operates at 100Mbps. *See also* twisted-pair physical media dependent (TP-PMD).

cordage

Fiber-optic cable designed for use as a patch cord or jumper. Cordage may be either single-fiber (simplex) or double fiber (duplex).

core

The central part of a single optical fiber in which the light signal is transmitted. Common core sizes are 8.3 microns, 50 microns, and 62.5 microns. The core is surrounded by a cladding that has a higher refractive index that keeps the light inside the core. The core is typically made of glass or plastic.

core eccentricity

A measurement that indicates how far off the center of the core of an optical fiber is from the center of that fiber's cladding.

counter-rotating

An arrangement whereby two signal paths, one in each direction, exist in a ring topology.

coupler

With respect to optical fiber, a passive, multiport device that connects three or more fiber ends, dividing the input between several outputs or combining several inputs into one output.

coupling

The transfer of energy between two or more cables or components of a circuit. *See also* crosstalk.

coupling efficiency

How effective a coupling method is at delivering the required signal without loss.

coupling loss

The amount of signal loss that occurs at a connection because of the connection's design.

coupling ratio/loss

The ratio/loss of optical power from one output port to the total output power, expressed as a percent.

cover plate mounting bracket

A bracket within a rack that allows a plate to cover some type of hardware, such as a patch panel.

crimper, crimping, crimp-on

A device that is used to install a crimp-on connector. Crimping involves the act of using the crimping tool to install the connector.

critical angle

The smallest angle of incidence at which total internal reflection occurs; that is, at which light passing through a material of a higher refractive index will be reflected off the boundary with a material of a lower refractive index. At lower angles, the light is refracted through the cladding and lost. Due to the fact that the angle of reflection equals the angle of incidence, total internal reflection assures that the wave will be propagated down the length of the fiber. The angle is measured from a line perpendicular to the boundary between the two materials, known as normal.

cross-connect

A facility enabling the termination of cables as well as their interconnection or cross-connection with other cabling or equipment. Also known as a punch-down or distributor. In a copper-based system, jumper wires or patch cables are used to make connections. In a fiber-optic system, fiber-optic jumper cables are used.

cross-connection

A connection scheme between cabling runs, subsystems, and equipment using patch cords or jumpers that attach to connecting hardware at each end.

crossover

A conductor that connects to a different pin number at each end. *See also* crossover cable.

crossover cable

A twisted-pair patch cable wired in such a way as to route the transmit signals from one piece of equipment to the receive signals of another piece of equipment, and vice versa. Crossover cables are often used with 10Base-T or 100Base-TX Ethernet cards to connect two Ethernet devices together "back-to-back" or to connect two hubs together if the hubs do not have crossover or uplink ports. See Chapter 9 for more information on Ethernet crossover cables.

crosstalk

The coupling or transfer of unwanted signals from one pair within a cable to another pair. Crosstalk can be measured at the same (near) end or far end with respect to the signal source. Crosstalk is considered noise or interference and is expressed in decibels. Chapter 1 has an in-depth discussion of crosstalk.

crush impact

A test typically conducted using a press that is fitted with compression plates of a specified cross sectional area. The test sample is placed between the press plates and a specified force is applied to the test specimen. Cable performance is evaluated while the cable is under compression and/or after removal of load depending on the test standard specifications.

current

The flow of electrons in a conductor. *See also* alternating current and direct current.

curvature loss

The macro-bending loss of signal in an optical fiber.

customer premises

The buildings, offices, and other structures under the control of an end-user or customer.

cutback

A technique or method for measuring the optical attenuation or bandwidth in a fiber by measuring first from the end and then from a shorter length and comparing the difference. Usually one is at the full length of the fiber-optic cable and the other is within a few meters of the input.

cutoff wavelength

For a single-mode fiber, the wavelength above which the operation switches from multimode to single-mode propagation. This is the longest wavelength at which a single-mode fiber can transmit two modes. At shorter wavelengths the fiber fails to function as a single-mode fiber.

cut-through resistance

A material's ability to withstand mechanical pressure (such as a cutting blade or physical pressure) without damage.

cycles per second

The count of oscillations in a wave. One cycle per second equals a hertz.

cyclic redundancy check (CRC)

An error-checking technique used to ensure the accuracy of transmitting digital code over a communications channel after the data is transmitted. Transmitted messages are divided into predetermined lengths that are divided by a fixed divisor. The result of this calculation is appended onto the message and sent with it. At the receiving end, the computer performs this calculation again. If the value that arrived with the data does not match the value the receiver calculated, an error has occurred during transmission.

D

daisy chain

In telecommunications, a wiring method where each telephone jack in a building is wired in series from the previous jack. Daisy chaining is not the preferred wiring method, since a break in the wiring would disable all jacks "downstream" from the break. Attenuation and crosstalk are also higher in a daisy-chained cable. *See also* home run cable.

dark current

The external current that, under specified biasing conditions, flows in a photo detector when there is no incident radiation.

dark fiber

An unused fiber; a fiber carrying no light. Common when extra fiber capacity is installed.

data communication equipment (DCE)

With respect to data transmission, the interface that is used by a modem to communicate with a computer.

data-grade

A term used for twisted-pair cable that is used in networks to carry data signals. Data-grade media has a higher frequency rating than voice-grade media used in telephone wiring does. Data-grade cable is considered Category 3 or higher cable.

data packet

The smallest unit of data sent over a network. A packet includes a header with addressing information, and the data itself.

data rate

The maximum rate (in bits per second or some multiple thereof) at which data is transmitted in a data transmission link. The data rate may or may not be equal to the baud rate.

data-terminal equipment (DTE)

(1) The interface that electronic equipment uses to communicate with a modem or other serial device. This port is often called the computer's RS-232 port or the serial port. (2) Any piece of equipment at which a communications path begins or ends.

datagram

A unit of data larger than or equal to a packet. Generally it is self-contained and its delivery is not guaranteed.

DB-9

Standard 9-pin connector used with Token Ring and serial connections.

DB-15

Standard 15-pin connector used with Ethernet transceiver cables.

DB-25

Standard 25-pin connector used with serial and parallel ports.

DC loop resistance

The total resistance of a conductor from the near end to the far end and back. For a single conductor, it is just the one-way measurement doubled. For a pair of conductors, it is the resistance from the near end to the far end on one conductor and from the far end to the near end on the other.

D-channel

Delta channel. On ISDN networks, the channel that carries control and signaling information at 16Kbps for BRI ISDN services or 64Kbps for PRI ISDN services.

decibel (dB)

A standard unit used to express a relative measurement of signal strength or to express gain or loss in optical or electrical power. A unit of measure of signal strength, usually the relation between a received signal and a standard signal source. The decibel scale is a logarithmic scale, expressed as the logarithmic ratio of the strength of a received signal to the strength of the originally transmitted signal. For example, every 3dB equals 50 percent of signal strength, so therefore a 6dB loss equals a loss of 75 percent of total signal strength. See Chapter 1 for more information.

degenerate waveguides

A set of waveguides having the same propagation constant for all specified frequencies.

delay skew

The difference in propagation delay between the fastest and slowest pair in a cable or cabling system. See Chapter 1 for more information on delay skew.

delta

In fiber optics, equal to the difference between the indexes of refraction of the core and the cladding divided by the index of the core.

demand priority

A network access method used by Hewlett-Packard's 100VG-AnyLAN. The hub arbitrates requests for network access received from stations and assigns access based on priority and traffic loads.

demarcation point

A point where the operational control or ownership changes, such as the point of interconnection between telephone company facilities and a user's building or residence.

demodulate

To retrieve a signal from a carrier and convert it into a usable form.

demultiplex

The process of separating channels that were previously joined using a multiplexer.

demultiplexer

A device that separates signals of different wavelengths for distribution to their proper receivers.

dense wavelength division multiplexing (DWDM)

A form of multiplexing that separates channels by transmitting them on different wavelengths at intervals of about 0.8nm (100GHz).

detector

(1) A device that detects the presence or absence of an optical signal and produces a coordinating electrical response signal. (2) An optoelectric transducer used in fiber optics to convert optical power to electrical current. In fiber optics, the detector is usually a photodiode.

detector noise limited operations

Occur when the detector is unable to make an intelligent decision on the presence or absence of a pulse due to the losses that render the amplitude of the pulse too small to be detected or due to excessive noise caused by the detector itself.

diameter mismatch loss

The loss of power that occurs when one fiber transmits to another and the transmitting fiber has a diameter greater than the receiving fiber. It can occur at any type of coupling where the fiber/coupling sizes are mismatched: fiber-to-fiber, fiber-to-device, fiber-to-detector, or source-to-fiber. Fiber-optic cables and connectors should closely match the size of fiber required by the equipment.

dichroic filter

Selectively transmits or reflects light according to selected wavelengths. Also referred to as dichromatic mirror.

dielectric

Material that does not conduct electricity, such as non-metallic materials that are used for cable insulation and jackets. Optical fiber cables are made of dielectric material.

dielectric constant

The property of a dielectric material that determines the amount of electrostatic energy that can be stored by the material when a given voltage is applied to it. The ratio of the capacitance of a capacitor using the dielectric to the capacitance of an identical capacitor using a vacuum as a dielectric. Also called permittivity.

dielectric loss

The power dissipated in a dielectric material as the result of the friction produced by molecular motion when an alternating electric field is applied.

dielectric nonmetallic

Refers to materials used within a fiber-optic cable.

differential group delay (DGD)

The total difference in travel time between the two polarization states of light traveling through an optical fiber. This time is usually measured in picoseconds, and can differ depending on specific conditions within an optical fiber and the polarization state of the light passing through it.

differential mode attenuation
A variation in attenuation in and among modes carried in an optical fiber.

differential mode transmission
A transmission scheme where voltages appear equal in magnitude and opposite in phase across a twisted-pair cable with respect to ground. Differential mode transmission may also be referred to as balanced mode. Twisted-pair cable used for Category 1 and above is considered differential mode transmission media or balanced mode cable.

differential modal dispersion (DMD)
The test method used to determine the variation in arrival time of modes in a pulse of light in an optical fiber (typically for multimode fiber).

diffraction
The deviation of a wave front from the path predicted by geometric optics when a wave front is restricted by an edge or an opening of an object. Diffraction is most significant when the aperture is equal to the order of the wavelength.

diffraction grating
An array of fine, parallel, equally spaced reflecting or transmitting lines that mutually enhance the effects of diffraction to concentrate the diffracted light in a few directions determined by the spacing of the lines and by the wavelength of the light.

diffuse infrared
It enables the use of infrared optical emissions without the need for line-of-sight between the transmitting and receiving communication entities.

digital
Refers to transmission, processing, and storage of data by representing the data in binary values (two states: on or off). On is represented by the number 1 and off by the number 0. Data transmitted or stored with digital technology is expressed as a string of 0s and 1s. Digital signals are used to communicate information between computers or computer-controlled hardware.

digital loop carrier
A carrier system used for pair gain and providing multiple next generation digital services over traditional copper loop in local loop applications.

digital signal (DS)
A representation of a digital signal carrier in the TDM hierarchy. Each DS level is made up of multiple 64Kbps channels (generally thought of as the equivalent to a voice channel) known as DS-0 circuits. A DS-1 circuit (1.544Mbps) is made up of 24 individual DS-0 circuits. DS rates are specified by ANSI, CEPT, and the ITU.

digital signal cross-connect (DSX)
A piece of equipment that serves as a connection point for a particular digital signal rate. Each DSX equipment piece is rated for the DS circuit it services—for example, a DSX-1 is used for DS-1 signals, DSX-3 for DS-3 signals, and so on.

digital signal processor (DSP)
A device that manipulates or processes data that has been converted from analog to digital form.

digital subscriber line (DSL)
A technology for delivering high bandwidth to homes and businesses using standard telephone lines. Though many experts believed that standard copper phone cabling would never be able to support high data rates, local phone companies are deploying equipment that is capable of supporting up to theoretical rates of 8.4Mbps. Typical throughput downstream (from the provider to the customer) are rates from 256Kbps to 1.544Mbps. DSL lines are capable of supporting voice and data simultaneously. There are many types, including HDSL (high bit-rate DSL) and VDSL (very high bit-rate DSL).

diode
A device that allows a current to move in only one direction. Some examples of diodes are light-emitting diodes (LEDs), laser diodes, and photodiodes.

direct current (DC)
An electric current that flows in one direction and does not reverse direction, unlike alternating current (AC). Direct current also means a current whose polarity never changes.

direct frequency modulation
Directly feeding a message into a voltage-controlled oscillator.

direct inside wire (DIW)
Twisted-pair wire used inside a building, usually two- or four-pair AWG 26.

directional coupler
A device that samples or tests data traveling in one direction only.

dispersion
A broadening or spreading of light along the propagation path due to one or more factors within the medium (such as optical fiber) through which the light

is traveling. There are three major types of dispersion: modal, material, and waveguide. Modal dispersion is caused by differential optical path lengths in a multi-mode fiber. Material dispersion is caused by a delay of various wavelengths of light in a waveguide material. Waveguide dispersion is caused by light traveling in both the core and cladding materials in single-mode fibers and interfering with the transmission of the signal in the core. If dispersion becomes too great, individual signal components can overlap one another or degrade the quality of the optical signal. Dispersion is one of the most common factors limiting the amount of data that can be carried in optical fiber and the distance the signal can travel while still being usable; therefore, dispersion is one of the limits on bandwidth on fiber-optic cables. It is also called pulse spreading because dispersion causes a broadening of the input pulses along the length of the fiber.

dispersion flattened fiber
A single-mode optical fiber that has a low chromatic dispersion throughout the range between 1300nm and 1600nm.

dispersion limited operation
Describes cases where the dispersion of a pulse rather than loss of amplitude limits the distance an optical signal can be carried in the fiber. If this is the case, the receiving system may not be able to receive the signal.

dispersion shifted fiber
A single-mode fiber that has its zero dispersion wavelength at 1550nm, which corresponds to one of the fiber's points of low attenuation. Dispersion shifted fibers are made so that optimum attenuation and bandwidth are at 1550nm.

dispersion unshifted fiber
A single-mode optical-fiber cable that has its zero dispersion wavelength at 1300nm. Often called conventional or unshifted fiber.

distortion
Any undesired change in a waveform or signal.

distortion-limited operation
In fiber-optic cable, the limiting of performance because of the distortion of a signal.

distributed feedback (DFB) laser
A semiconductor laser specially designed to produce a narrow spectral output for long-distance fiber optic communications.

distribution subsystem
A basic element of a structured distribution system. The distribution subsystem is responsible for terminating equipment and running the cables between equipment and cross connects. Also called distribution frame subsystem.

distribution cable
An optical fiber cable used "behind-the-shelf" of optical fiber patch panels; typically composed of 900 micron tight buffered optical fibers supported by aramid and/or glass-reinforced plastic (GRP).

disturbed pair
A pair in a UTP cable that has been disturbed by some source of noise during cable testing.

DIX Connector
Digital, Intel, Xerox (DIX) connector used to connect thinnet cables and networks.

DoD Networking Model
The Department of Defense's four-layer conceptual model describing how communications should take place between computer systems. From the bottom up, the four layers are network access, Internet, transport (host-to-host), and application.

dopants
Impurities that are deliberately introduced into the materials used to make optical fibers. Dopants are used to control the refractive index of the material for use in the core or the cladding.

D-ring
An item of hardware, usually a metal ring shaped like the letter D. It may be used at the end of a leather or fabric strap, or may be secured to a surface with a metal or fabric strap.

drain wire
An uninsulated wire in contact with a shielded cable's shield throughout its length. It is used for terminating the shield. If a drain wire is present, it should be terminated.

drop
A single length of cable installed between two points.

dry nonpolish connector (DNP)
Optical fiber connector used for POF (plastic optical fiber).

DS-1
Digital Service level 1. Digital service that provides 24 separate 64Kbps digital channels.

DSX bay

A combination of the various pieces of DSX apparatus and its supporting mechanism, including its frame, rack, or other mounting devices.

DSX complexes

Any number of DSX lineups that are connected together to provide DSX functionality.

DSX lineup

Multiple contiguous DSX bays connected together to provide DSX functions.

D-type connector

A type of connector that connects computer peripherals. It contains rows of pins or sockets shaped like a sideways D. Common connectors are the DB-9 and DB-25.

DU connector

A fiber-optic connector developed by the Nippon Electric Group in Japan.

dual-attachment concentrator (DAC)

An FDDI concentrator that offers two attachments to the FDDI network that are capable of accommodating a dual (counter-rotating) ring.

dual-attachment station (DAS)

A term used with FDDI networks to denote a station that attaches to both the primary and secondary rings; this makes it capable of serving the dual (counter-rotating) ring. A dual-attachment station has two separate FDDI connectors.

dual ring

A pair of counter-rotating logical rings.

dual-tone multifrequency (DTMF)

The signal that a touch-tone phone generates when you press a key on it. Each key generates two separate tones, one in a high-frequency group of tones and one from a low-frequency group of tones.

dual-window fiber

An optical fiber cable manufactured to be used at two different wavelengths. Single-mode fiber cable that is usable at 1300nm and 1550nm is dual-window fiber. Multimode fiber cable is optimized for 850nm and 1300nm operations and is also dual-window fiber. Also known as double-window fiber.

duct

(1) A single enclosed raceway for wires or cable. (2) An enclosure in which air is moved.

duplex

(1) A link or circuit that can carry a signal in two directions for transmitting and receiving data. (2) An optical fiber cable or cord carrying two fibers.

duplex cable

A two-fiber cable suitable for duplex transmission. Usually two fiber strands surrounded by a common jacket.

duplex transmission

Data transmission in both directions, either one direction at a time (half duplex) or both directions simultaneously (full duplex).

duty cycle

With respect to a digital transmission, the product of a signal's repetition frequency and its duration.

dynamic loads

Loads, such as tension or pressure, that change over time, usually within a short period.

dynamic range

The difference between the maximum and minimum optical input power that an optical receiver can accept.

E

E-1

The European version of T-1 data circuits. Runs at 2.048Mbps.

E-3

The European version of T-3 data circuits. Runs at 34.368Mbps.

earth

A term for zero reference ground (not to mention the planet most of us live on).

edge-emitting LED

An LED that produces light through an etched opening in the edge of the LED.

effective modal bandwidth (EMB)

The bandwidth of a particular optical fiber cable, connections and splices, and transmitter combination. This can be measured in a complete channel.

effective modal bandwidth, calculated (EMBc)

The calculation of the EMB from the optical fiber's DMD and specific weighting functions of VCSELs. This is also known as CMB (calculated modal bandwidth).

effective modal bandwidth, minimum (min EMB)
The value of bandwidth used in the IEEE 802.3ae model that corresponds to the fiber specification.

EIA rack
A rack with a standard dimension per EIA. *See also* 19-inch racks.

electromagnetic compatibility (EMC)
The ability of a system to minimize radiated emissions and maximize immunity from external noise sources.

electromagnetic field
The combined electric and magnetic field caused by electron motion in conductors.

electromagnetic immunity
Protection from the interfering or damaging effects of electromagnetic radiation such as radio waves or microwaves.

electromagnetic interference (EMI)
Electrical noise generated in copper conductors when electromagnetic fields induce currents. Copper cables, motors, machinery, and other equipment that uses electricity may generate EMI. Copper-based network cabling and equipment are susceptible to EMI and also emit EMI, which results in degradation of the signal. Fiber-optic cables are desirable in environments that have high EMI because they are not susceptible to the effects of EMI.

Electronic Industries Alliance (EIA)
An alliance of manufacturers and users that establishes standards and publishes test methodologies. The EIA (with the TIA and ANSI) helped to publish the ANSI/TIA/EIA-568-B cabling standard.

electrostatic coupling
The transfer of energy by means of a varying electrostatic field. Also referred to as capacitive coupling.

electrostatic discharge (ESD)
A problem that exists when two items with dissimilar static electrical charges are brought together. The static electricity jumps to the item with lower electrical charge, which causes ESD; ESD can damage electrical and computer components.

emitter
A source of optical power.

encoding
Converting analog data into digital data by combining timing and data information into a synchronized stream of signals. Encoding is accomplished by representing digital 1s and 0s through combining high- and low-signal voltage or light states.

end finish, endface finish
The quality of a fiber's end surface—specifically, the condition of the end of a connector ferrule. Endface finish is one of the factors affecting connector performance.

end separation
(1) The distance between the ends of two joined fibers. The end separation is important because the degree of separation causes an extrinsic loss, depending on the configuration of the connection. (2) The separation of optical fiber ends, usually taking place in a mechanical splice or between two connectors.

end-to-end loss
The optical signal loss experienced between the transmitter and the detector due to fiber quality, splices, connectors, and bends.

energy density
For radiation. Expressed in joules per square meter. Sometimes called irradiance.

entrance facility (EF)
A room in a building where antenna, public, and private network service cables can enter the building and be consolidated. Should be located as close as possible to the entrance point. Entrance facilities are often used to house electrical protection equipment and connecting hardware for the transition between outdoor and indoor cable. Also called an entrance room.

entrance point
The location where telecommunications enter a building through an exterior wall, a concrete floor slab, a rigid metal conduit, or an intermediate metal conduit.

epoxy
A glue that cures when mixed with a catalyzing agent.

epoxy-less connector
A connector that does not require the use of epoxy to hold the fiber in position.

equal level far-end crosstalk (ELFEXT)
ELFEXT is the name for the crosstalk signal that is measured at the receiving end and equalized by the attenuation of the cable.

equipment cable
Cable or cable assembly used to connect telecommunications equipment to horizontal or backbone cabling systems in the telecommunications room and equipment room. Equipment cables are considered to be outside the scope of cabling standards.

equipment cabling subsystem

Part of the cabling structure, typically between the distribution frame and the equipment.

equipment room (ER)

A centralized space for telecommunications equipment that serves the occupants of the building or multiple buildings in a campus environment. Usually considered distinct from a telecommunications closet because it is considered to serve a building or campus; the telecommunications closet serves only a single floor. The equipment room is also considered distinct because of the nature of complexity of the equipment that is contained in it.

erbium doped fiber amplifier (EDFA)

An optical amplifier that uses a length of optical fiber doped with erbium and energized with a pump laser to inject energy into a signal.

error detection

The process of detecting errors during data transmission or reception. Some of the error checking methods including CRC, parity, and bipolar variations.

error rate

The frequency of errors detected in a data service line, usually expressed as a decimal.

Ethernet

A local area network (LAN) architecture developed by Xerox that is defined in the IEEE 802.3 standard. Ethernet nodes access the network using the Carrier Sense Multiple Access/Collision Detect (CSMA/CD) access method.

European Computer Manufacturers Association (ECMA)

A European trade organization that issues its own specifications and is a member of the ISO.

excess loss

(1) In a fiber-optic coupler, the optical loss from that portion of light that does not emerge from the nominally operational pods of the device. (2) The ratio of the total output power of a passive component with respect to the input power.

exchange center

Any telephone building where switch systems are located. Also called exchange office or central office.

extrinsic loss

When describing a fiber-optic connection, the portion of loss that is not intrinsic to the fiber but is related to imperfect joining, which may be caused by the connector or splice, as opposed to conditions in the optical fiber itself. The conditions causing this type of loss are referred to as extrinsic factors. These factors are defects and imperfections that cause the loss to exceed the theoretical minimum loss due to the fiber itself (which is called intrinsic loss).

F

f

See frequency.

Fabry-Pérot

A type of laser used in fiber-optic transmission. The Fabry-Pérot laser typically has a spectral width of about 5nm.

faceplate

A plate used in front of a telecommunications outlet. *See also* wall plate.

fanout cable

A multifiber cable that is designed for easy connectorization. These cables are sometimes sold with installed connectors or as part of a splice pigtail, with one end carrying many connectors and the other installed in a splice cabinet or panel ready for splicing or patching.

fanout kit

A collection of components used to add tight buffers to optical fibers from a loose tube buffer cable. A fanout kit typically consists of a furcation unit and measured lengths of tight buffer material. It is used to build up the outer diameter of fiber cable for connectorization.

farad

A unit of capacitance that stores one coulomb of electrical charge when one volt of electrical pressure is applied.

faraday effect

A phenomenon that causes some materials to rotate the polarization of light in the presence of a magnetic field.

far-end

The end of a twisted pair cable that is at the far end with respect to the transmitter, i.e., at the receiver end.

far-end crosstalk (FEXT)

Crosstalk that is measured on the nontransmitting wires and measured at the opposite end from the source. See Chapter 1 for more information on various types of crosstalk.

Fast Ethernet

Ethernet standard supporting 100Mbps operation. Also known as 100Base-TX or 100Base-FX (depending on media)

F-series connector
A type of coaxial RF connector used with coaxial cable.

FC connector
A threaded optical fiber connector that was developed by Nippon Telephone and Telegraph in Japan. The FC connector is good for single-mode or multimode fiber and applications requiring low back reflection. The FC is a screw type and is prone to vibration loosening.

Federal Communications Commission (FCC)
The federal agency responsible for regulating broadcast and electronic communications in the United States.

feeder cable
A voice backbone cable that runs from the equipment room cross-connect to the telecommunications cross-connect. A feeder cable may also be the cable running from a central office to a remote terminal, hub, head end, or node.

ferrule
A small alignment tube attached to the end of the fiber and used in connectors. These are made of stainless steel, aluminum, zirconia, or plastic. The ferrule is used to confine and align the stripped end of a fiber so that it can be positioned accurately.

fiber
A single, separate optical transmission element characterized by core and cladding. The fiber is the material that guides light or waveguides.

fiber channel
A gigabit interconnect technology that, through the 8B/10B encoding method, allows concurrent communications among workstations, mainframes, servers, data storage systems, and other peripherals using SCSI and IP protocols.

fiber curl
Occurs when there is misalignment in a mass or ribbon splicing joint. The fiber or fibers curl away from the joint to take up the slack or stress caused by misalignment of fiber lengths at the joint.

Fiber Distributed Data Interface (FDDI)
ANSI Standard X3T9.5, Fiber Distributed Data Interface (FDDI)—This standard details the requirements for all attachment devices concerning the 100Mbps fiber-optic network interface. It uses a counter-rotating, token-passing ring topology. FDDI is typically known as a backbone LAN because it is used for joining file servers together and for joining other LANs together.

fiber distribution frame (FDF)
Any fiber-optic connection system (cross-connect or interconnect) that uses fiber-optic jumpers and cables. *See also* horizontal distribution frame.

fiber identifier
A testing device that displays the direction of travel of light within an optical fiber by introducing a macrobend and analyzing the light that escapes the fiber.

fiber illumination kit
Used to visually inspect continuity in fiber systems and to inspect fiber connector end-faces for cleanliness and light quality.

fiber-in-the-loop (FITL)
Indicates deployment of fiber-optic feeder and distribution facilities.

fiber loss
The attenuation of light in an optical-fiber transmission.

fiber-optic attenuator
An active component that is installed in a fiber-optic transmission system that is designed to reduce the power in the optical signal. It is used to limit the optical power received by the photodetector to within the limits of the optical receiver.

fiber-optic cable
Cable containing one or more optical fibers.

fiber-optic communication system
Involves the transfer of modulated or unmodulated optical energy (light) through optical-fiber media.

fiber-optic connector panel
A patch panel where fiber-optic connectors are mounted.

fiber-optic inter-repeater link (FOIRL)
An Ethernet fiber-optic connection method intended for connection of repeaters.

fiber-optic pigtail
Used to splice outside plant cable to the backside of a fiber-optic patch panel.

fiber-optic test procedures (FOTP)
Test procedures outlined in the EIA-RS-455 standards.

fiber-optic transmission
A communications scheme whereby electrical data is converted to light energy and transmitted through optical fibers.

fiber-optic waveguide
A long, thin strand of transparent material (glass or plastic), which can convey electromagnetic energy in the optical waveform longitudinally by means of internal refraction and reflection.

fiber optics
The optical technology in which communication signals in the form of modulated light beams are transmitted over a glass or plastic fiber transmission medium. Fiber optics offers high bandwidth and protection from electromagnetic interference and radioactivity; it also has small space needs.

fiber protection system (FPS)
A rack or enclosure system designed to protect fiber-optic cables from excessive bending or impact.

fiber test equipment
Diagnostic equipment used for the testing, maintenance, restoration, and inspection of fiber systems. This equipment includes optical attenuation meters and optical time-domain reflectometers (OTDRs).

fillers
Nonconducting components cabled with insulated conductors or optical fibers to impart flexibility, tensile strength, roundness, or a combination of all three.

figure-8
A fiber-optic cable with a strong supporting member incorporated for use in aerial installations. *See also* messenger cable.

filter
A device that blocks certain wavelengths to permit selective transmission of optical signals.

firestop
Material, device, or collection of parts installed in a cable pathway (such as a conduit or riser) at a fire-rated wall or floor to prevent passage of flame, smoke, or gases through the rated barrier.

fish tape
Also called fish cord. A tool used by electricians to route new wiring through walls and electrical conduit.

flex life
The average number of times a particular cable or type of cable can bend before breaking.

floating
A floating circuit is one that has no ground connection.

floor distributor (FD)
The ISO/IEC 11801 term for horizontal cross-connect. The floor distributor is used to connect between the horizontal cable and other cabling subsystems or equipment.

fluorinated ethylene-propylene (FEP)
A thermoplastic with excellent dielectric properties that is often used as insulation in plenum-rated cables. FEP has good electrical-insulating properties and chemical and heat resistance and is an excellent alternative to PTFE (Teflon®). FEP is the most common material used for wire insulation in Category 5 and better cables that are rated for use in plenums.

four-wave mixing
The creation of new light wavelengths from the interaction of two or more wavelengths being transmitted at the same time within a few nanometers of each other. Four-wave mixing is named for the fact that two wavelengths interacting with each other will produce two new wavelengths, causing distortion in the signals being transmitted. As more wavelengths interact, the number of new wavelengths increases exponentially.

frame check sequence (FCS)
A special field used to hold error correction data in Ethernet (IEEE 802.3) frames.

frame relay
A packet-switched, wide area networking (WAN) technology based on the public telephone infrastructure. Frame relay is based on the older, analog, X.25 networking technologies.

free space optics transmission
The transmission of optical information over air using specialized optical transmitters and receivers.

free token
In Token Ring networks, a free token is a token bit set to 0.

frequency (f)
The number of cycles per second at which a waveform alternates—that is, at which corresponding parts of successive waves pass the same point. Frequency is expressed in hertz (Hz); one hertz equals one cycle per second.

frequency division multiplexing (FDM)
A technique for combining many signals onto a single circuit by dividing the available transmission bandwidth by frequency into narrower bands; each band is used for a separate communication channel. FDM can be used with any and all of the sources created by wavelength division multiplexing (WDM).

frequency hopping
Frequency hopping is the technique of improving the signal to noise ratio in a link by adding frequency diversity.

frequency modulation (FM)
A method of adding information to a sine wave signal in which its frequency is varied to impose information on it. Information is sent by varying the frequency of an optical or electrical carrier. Other methods include amplitude modulation (AM) and phase modulation (PM).

frequency response
The range of frequencies over which a device operates as expected.

Fresnel diffraction pattern
The near-field diffraction pattern.

Fresnel loss
The loss at a joint due to a portion of the light being reflected.

Fresnel reflection
Reflection of a small amount of light passing from a medium of one refractive index into a medium of another refractive index. In optical fibers, reflection occurs at the air/glass interfaces at entrance and exit ends.

Fresnel reflection method
A method for measuring the index profile of an optical fiber by measuring reflectance as a function of position on the end-face.

full-duplex
A system in which signals may be transmitted in two directions at the same time.

full-duplex transmission
Data transmission over a circuit capable of transmitting in both directions simultaneously.

fundamental mode
The lowest number mode of a particular waveguide.

furcation unit
A component used in breakout kits and fanout kits for separating individual optical fibers from a cable and securing tight buffers and/or jackets around the fibers.

fusion splicing
A splicing method accomplished by the application of localized heat sufficient to fuse or melt the ends of the optical fiber, forming continuous single strand of fiber. As the glass is heated it becomes softer, and it is possible to use the glass's "liquid" properties to bond glass surfaces permanently.

G

G
Green. Used when referring to color-coding of cables.

gain
An increase in power.

gainer
A fusion splice displayed in an OTDR trace that appears to have gain instead of loss.

gamma
The coefficient used to scale bandwidth with fiber length.

gap loss
The loss that results when two axially aligned fibers are separated by an air gap. This loss is often most significant in reflectance. The light must launch from one medium to another (glass to air to glass) through the waveguide capabilities of the fiber.

gigahertz (GHz)
A billion hertz or cycles per second.

glass-reinforced plastic (GRP)
A relatively stiff dielectric, nonconducting strength element used in optical fiber cables. It is composed of continuous lengths of glass yarn formed together with thermally or UV cured high-strength epoxy. It exhibits not only excellent tensile strength, but compressive strength as well.

graded-index fiber
An optical fiber cable design in which the index of refraction of the core is lower toward the outside of the core and progressively increases toward the center of the core, thereby reducing modal dispersion of the signal. Light rays are refracted within the core rather than reflected as they are in step-index fibers. Graded-index

fibers were developed to lessen the modal dispersion effects found in multimode fibers with the intent of increasing bandwidth.

ground

A common point of zero potential such as a metal chassis or ground rod that grounds a building to the earth. The ANSI/TIA/EIA-607 Commercial Building Grounding and Bonding Requirements for Telecommunications Standard is the standard that should be followed for grounding requirements for telecommunications. Grounding should never be undertaken without consulting with a professional licensed electrician.

ground loop

A condition where an unintended connection to ground is made through an interfering electrical conductor that causes electromagnetic interference. *See also* ground loop noise.

ground loop noise

Electromagnetic interference that is created when equipment is grounded at ground points having different potentials, thereby creating an unintended current path. Equipment should always be grounded to a single ground point.

guided ray

A ray that is completely confined to the fiber core.

H

half-duplex

A system in which signals may be sent in two directions, but not at the same time. In a half-duplex system, one end of the link must finish transmitting before the other end may begin.

half-duplex transmission

Data transmission over a circuit capable of transmitting in either direction. Transmission can be bidirectional but not simultaneously.

halogen

One of the following elements: chlorine, fluorine, bromine, astatine, or iodine.

hard-clad silica fiber

An optical fiber with a hard plastic cladding surrounding a step index silica core.

hardware address

The address is represented by six sections of two hexadecimal addresses; for example, 00-03-fe-e7-18-54. This number is hard-coded into networking hardware by manufacturers. Also called the physical address; *see also* MAC address.

hardware loopback

Connects the transmission pins directly to the receiving pins, allowing diagnostic software to test whether a device can successfully transmit and receive.

head end

(1) The central facility where signals are combined and distributed in a cable television system or a public telephone system. See central office.

header

The section of a packet (usually the first part of the packet) where the layer-specific information resides.

headroom

The number of decibels by which a system exceeds the minimum defined requirements. The benefit of more headroom is that it reduces the bit-error rate (BER) and provides a performance safety net to help ensure that current and future high-speed applications will run at peak accuracy, efficiency, and throughput. Also called overhead or margin.

hertz (Hz)

A measurement of frequency defined as cycles per second.

heterojunction structure

An LED design in which the pn (P-type and N-type semiconductor) junction is formed from similar materials that have different refractive indices. This design is used to guide the light for directional output.

hicap service

A high-capacity communications circuit service such as a private line T-1 or T-3.

home-run cable

A cable run that connects a user outlet directly with the telecommunications or wiring closet. This cable has no intermediate splices, bridges, taps, or other connections. Every cable radiates out from the central equipment or wiring closet. This configuration is also known as star topology and is the opposite of a daisy-chained cable that may have taps or splices along its length. Home-run cable is the recommended installation method for horizontal cabling in a structured cabling system.

homojunction structure

An LED design in which the pn (P-type and N-type semiconductor) junction is formed from a single semiconductor material.

hop

A connection. In routing terminology, each router a packet passes through is counted as a hop.

horizontal cabling

The cabling between and including the telecommunications outlet and the horizontal cross-connect. Horizontal cabling is considered the permanent portion of a link; may also be called horizontal wiring.

horizontal cross-connect (HC)

A cross connect that connects the cabling of the work area outlets to any other cabling system (like that for LAN equipment, voice equipment).

hub

A device that contains multiple independent but connected modules of network and internetworking equipment that form the center of a hub-and-spoke topology. Hubs that repeat the signals that are sent to them are called active hubs. Hubs that do not repeat the signal and merely split the signal sent to them are called passive hubs. In some cases, hub may also refer to a repeater, bridge, switch, or any combination of these.

hybrid adapter

A mating or alignment sleeve that mates two different connector types such as an SC and ST.

hybrid cable

A cable that contains fiber, coaxial, and/or twisted-pair conductors bundled in a common jacket. May also refer to a fiber-optic cable that has strands of both single-mode and multimode optical fiber.

hybrid connector

A connector containing both fiber and electrical connectivity.

hybrid mesh network

Hybrid networks use a combination of any two or more topologies in such a way that the resulting network does not exhibit one of the standard topologies (e.g., bus, star, or ring). A hybrid topology is always produced when two different basic network topologies are connected.

hydrogen loss

Optical signal loss (attenuation) resulting from hydrogen found in the optical fiber. Hydrogen in glass absorbs light and turns it into heat and thus attenuates the light. For this reason, glass manufacturers serving the fiber-optic industry must keep water and hydrogen out of the glass and deliver it to guaranteed specifications in this regard. In addition, they must protect the glass with a cladding that will preclude the absorption of water and hydrogen into the glass.

Hypalon

A DuPont trade name for a synthetic rubber (chlorosulfonated polyethylene) that is used as insulating and jacketing material for cabling.

I

I

Symbol used to designate current.

IBM data connector

Used to connect IBM Token Ring stations using Type 1 shielded twisted-pair 150 ohm cable. This connector has both male and female components, so every IBM data connector can connect to any other IBM data connector.

IEEE 802.1 LAN/MAN Management

The IEEE standard that specifies network management, internetworking, and other issues that are common across networking technologies.

IEEE 802.2 Logical Link Control

The IEEE standard that provides specifications for the operation of the Logical Link Control (LLC) sublayer of the OSI data link layer. The LLC sublayer provides an interface between the MAC sublayer and the Network layer.

IEEE 802.3 CSMA/CD Networking

The IEEE standard that specifies a network that uses a logical bus topology, baseband signaling, and a CSMA/CD network access method. This is the standard that defines Ethernet networks. *See also* Carrier Sense Multiple Access/Collision Detect.

IEEE 802.4 Token Bus

The IEEE standard that specifies a physical and logical bus topology that uses coaxial or fiber-optic cable and the token-passing media access method.

IEEE 802.5 Token Ring

The IEEE standard that specifies a logical ring, physical star, and token-passing media access method based on IBM's Token Ring.

IEEE 802.6 Distributed Queue Dual Bus (DQDB) Metropolitan Area Network

The IEEE standard that provides a definition and criteria for a metropolitan area network (MAN), also known as a Distributed Queue Dual Bus (DQDB).

IEEE 802.7 Broadband Local Area Networks
The IEEE standard for developing local area networks (LANs) using broadband cabling technology.

IEEE 802.8 Fiber-Optic LANs and MANs
The IEEE standard that contains guidelines for the use of fiber optics on networks, including FDDI and Ethernet over fiber-optic cable.

IEEE 802.9 Integrated Services (IS) LAN Interface
The IEEE standard that contains guidelines for the integration of voice and data over the same cable.

IEEE 802.10 LAN/MAN Security
The IEEE standard that provides a series of guidelines dealing with various aspects of network security.

IEEE 802.11 Wireless LAN
The IEEE standard that provides guidelines for implementing wireless technologies such as infrared and spread-spectrum radio.

IEEE 802.12 Demand Priority Access Method
The IEEE standard that defines the concepts of a demand priority network such as HP's 100VG-AnyLAN network architecture.

impact test
Used to determine a cable's susceptibility to damage when subjected to short-duration crushing forces. Rate of impact, the shape of the striking device, and the force of the impact all are used to define the impact test procedures.

impedance
The total opposition (resistance and reactance) a circuit offers to the flow of alternating current. It is measured in ohms and designated by the symbol Z.

impedance match
A condition where the impedance of a particular cable or component is the same as the impedance of the circuit, cable, or device to which it is connected.

impedance mismatch
A condition where the impedance of a particular cable or component is different than the impedance of the device to which it is connected.

impedance matching transformer
A transformer designed to match the impedance of one circuit to another.

incident angle
The angle between the subject light wave and a plane perpendicular to the subject optical surface.

incoherent light
Light in which the electric and magnetic fields of photons are completely random in orientation. Incoherent light is typically emitted from lightbulbs and LEDs.

index matching gel
A clear gel or fluid used between optical fibers that are likely to have their ends separated by a small amount of air space. The gel matches the refractive index of the optical fiber, reducing light loss due to Fresnel reflection.

index matching material
A material in liquid, paste, gel, or film form whose refractive index is nearly equal to the core index; it is used to reduce Fresnel reflections from a fiber end-face. Liquid forms of this are also called index matching gel.

index of refraction
(1) The ratio of the speed of light in a vacuum to the speed of light in a given transmission medium. This is usually abbreviated n. (2) The value given to a medium to indicate the velocity of light passing through it relative to the speed of light in a vacuum.

index profile
The curve of the refractive index over the cross section of an optical waveguide.

Infiniband
A standard for a switched fabric communications link primarily used in high-performance computing.

infrared
The infrared spectrum consists of wavelengths that are longer than 700nm but shorter than 1mm. Humans cannot see infrared radiation, but we feel it as heat. The commonly used wavelengths for transmission through optical fibers are in the infrared at wavelengths between 1100nm and 1600nm.

infrared transceiver
A transmitting and receiving device that emits and receives infrared energy.

infrared laser
A laser that transmits infrared signals.

infrared transmission
Transmission using infrared transceivers or lasers.

injection laser diode (ILD)
A laser diode in which the lasing takes place within the actual semiconductor junction and the light is emitted from the edge of the diode.

innerduct
A separate duct running within a larger duct to carry fiber-optic cables.

insertion loss
Light or signal energy that is lost as the signal passes through the optical fiber end in the connector and is inserted into another connector or piece of hardware. A good connector minimizes insertion loss to allow the greatest amount of light energy through. A critical measurement for optical fiber connections. Insertion loss is measured by determining the output of a system before and after the device is inserted into the system. Loss in an optical fiber can be due to absorption, dispersion, scattering, microbending, diffusion, and the methods of coupling the fiber to the power. Usually measured in dB per item; for example, a coupler, connector, splice, or fiber. Most commonly used to describe the power lost at the entrance to a waveguide (an optical fiber is a waveguide) due to axial misalignment, lateral displacement, or reflection that is most applicable to connectors.

inside plant (IP)
(1) Cables that are the portion of the cable network that is inside buildings, where cable lengths are usually shorter than 100 meters. This is the opposite of outside-the-plant (OP or OSP) cables. (2) A telecommunications infrastructure designed for installation within a structure.

installation load
The short term tensile load that may be placed on a fiber-optic cable.

Institute of Electrical and Electronics Engineers (IEEE)
A publishing and standards-making body responsible for many standards used in LANs, including the 802 series of standards.

insulation
A material with good dielectric properties that is used to separate electrical components close to one another, such as cable conductors and circuit components. Having good dielectric properties means that the material is non-conductive to the flow of electrical current. In the case of copper communication cables, good dielectric properties also refer to enhanced signal-transfer properties.

insulation displacement connection (IDC)
A type of wire termination in which the wire is punched down into a metal holder that cuts into the insulation wire and makes contact with the conductor, thus causing the electrical connection to be made. These connectors are found on 66-blocks, 110-blocks, and telecommunications outlets.

integrated optical circuit
An optical circuit that is used for coupling between optoelectronic devices and providing signal processing functions. It is composed of both active and passive components.

integrated optics
Optical devices that perform two or more functions and are integrated on a single substrate; analogous to integrated electronic circuits.

integrated optoelectronics
Similar in concept to integrated optics except that one of the integrated devices on the semiconductor chip is optical and the other electronic.

Integrated Services Digital Network (ISDN)
A telecommunications standard that is used to digitally send voice, data, and video signals over the same lines. This is a network in which a single digital bit stream can carry a great variety of services. For the Internet it serves much better than analog systems on POTS (plain old telephone service), which is limited to 53Kbps. *See also* basic rate interface and primary rate interface.

intelligence
A signal that is transmitted by imposing it on a carrier such as a beam of light to change the amplitude of the carrier.

intelligent hub
A hub that performs bridging, routing, or switching functions. Intelligent hubs are found in collapsed backbone environments.

intelligent network
A network that is capable of carrying overhead signaling information and services.

intensity
The square of the electric field amplitude of a light wave. Intensity is proportional to irradiance and may be used in place of that term if relative values are considered.

interbuilding backbone
A telecommunications cable that is part of the campus subsystem that connects one building to another.

interconnect
A circuit administration point, other than a cross-connect or an information outlet, that provides capability for routing and rerouting circuits. It does not use patch cords or jumper wires and is typically a jack-and-plug device that is used in smaller distribution arrangements or connects circuits in large cables to those in smaller cables.

interconnect cabinet

Cabinets containing connector panels, patch panels, connectors, and patch cords to interface from inside the plant to outside the plant. The interconnect cabinet is used as an access point for testing and rearranging routes and connections.

interconnection

A connection scheme that provides direct access to the cabling infrastructure and the capability to make cabling system changes using patch cords.

interface

The boundary layer between two media of different refractive indexes.

interference

(1) Fiber optic: the interaction of two or more beams of coherent or partially coherent light. (2) Electromagnetic: interaction that produces undesirable signals that interfere with the normal operation of electronic equipment or electronic transmission.

interlock armor

A helical armor applied around a cable where the metal tape sides interlock giving a flexible, yet mechanically robust cable; specifically adds to the compression resistance in optical fiber cables and acts as an EMI/RFI shield in twisted-pair copper cables. The interlock armor can be made of either aluminum (more common) or steel.

intermediate cross-connect (ICC)

A cross-connect between first-level and second-level backbone cabling. This secondary cross-connect in the backbone cabling is used to mechanically terminate and administer backbone cabling between the main cross-connect and horizontal cross-connect (station cables).

intermediate distribution frame (IDF)

A metal rack (or frame) designed to hold the cables that connect interbuilding and intrabuilding cabling. The IDF is typically located in an equipment room or telecommunications room. Typically, a permanent connection exists between the IDF and the MDF.

International Organization for Standardization (ISO)

The standards organization that developed the OSI model. This model provides a guideline for how communications occur between computers. See www.iso.org for more information.

International Telecommunication Union (ITU)

The branch of the United Nations that develops communications specifications.

International Telephone and Telegraph Consultative Committee (CCiTT)

International standards committee that develops standards for interface and signal formats. Currently exists as the ITU-T.

Internet Architecture Board (IAB)

The committee that oversees management of the Internet, which is made up of several subcommittees including the Internet Engineering Task Force (IETF), the Internet Assigned Numbers Authority (IANA, which is now known as ICANN), and the Internet Research Task Force (IRTF). See www.iab.org for more information.

Internet Engineering Task Force (IETF)

An international organization that works under the Internet Architecture Board to establish specifications and protocols relating to the Internet. See www.ietf.org for more information.

Internet Research Task Force (IRTF)

An international organization that works under the Internet Architecture Board to research new Internet technologies. See www.irtf.org for more information.

interoffice facility(IOF)

A communication channel of copper, fiber, or wireless media between two central offices. Often, IOF refers to telephone channels that can transport voice and/or data.

internetworking, internetwork

Involves connecting two or more computer networks via gateways using a common routing technology. The result is called an internetwork (often shortened to internet).

Inter-repeater link (fiber-optic)

Defined in IEEE 802.3 and implemented over two fiber links, transmit and receive, this medium may be up to 500m and 1 kilometer long depending on the number of repeaters in the network.

intersymbol interference (ISI)

Form of distortion in digital communications systems. ISI is caused by multipath propagation and nonlinear frequency response of a channel resulting from not having enough bandwidth. ISI leads to BER and causes the loss of the channel to increase.

intersymbol loss (ISL)

Channel loss resulting from intersymbol interference.

intrabuilding backbone

Telecommunications cables that are part of the building subsystem that connect one telecommunications closet to another or a telecommunications room to the equipment room.

intrinsic factors
When describing an optical fiber connection, factors contributing to attenuation that are determined by the condition of the optical fiber itself. See also extrinsic loss.

intrinsic joint loss
The theoretical minimum loss that a given joint or device will have as a function of its nature. Intrinsic joint loss may also be used to describe the given theoretical minimum loss that a splice joint, coupler, or splitter may achieve.

intrinsic splice loss
The optical signal loss arising from differences in the fibers being spliced.

intrinsic performance factor (IPF)
Performance specification whereby total optical channel performance is specified, rather than performance of individual components.

intumescent
An intumescent is a substance that swells as a result of heat exposure, thus increasing in volume and decreasing in density.

inverter
A device that inverts the phase of a signal.

ion exchange techniques
A method for making and doping glass by ion exchange.

irradiance
The measure of power density at a surface upon which radiation is directed. The normal unit is watts per square centimeter.

ISDN terminal adapter
The device used on ISDN networks to connect a local network or single machine to an ISDN network (or any non-ISDN compliant device). The ISDN terminal adapter provides line power and translates data from the LAN or individual computer for transmission on the ISDN line.

isochronous
Signals that are dependent on some uniform timing or carry their own timing information embedded as part of the signal.

isolated ground
A separate ground conductor that is insulated from the equipment or building ground.

isolation
The ability of a circuit or component to reject interference.

isolator
In fiber optics, a device that permits only forward transmission of light and blocks any reflected light.

J

jabber
A term used with Ethernet to describe the act of continuously sending data or sending Ethernet frames with a frame size greater than 1,518 bytes. When a station is jabbering, its network adapter circuitry or logic has failed, and it has locked up a network channel with its erroneous transmissions.

jack
A receptacle used in conjunction with a plug to make electrical contact between communication circuits. A variety of jacks and their associated plugs are used to connect hardware applications, including cross-connects, interconnects, information outlets, and equipment connections. Jacks are also used to connect cords or lines to telephone systems. A jack is the female component of a plug/jack connector system and may be standard, modified, or keyed.

jacket
The outer protective covering of a cable, usually made of some type of plastic or polymer.

jacketed ribbon cable
A cable carrying optical fiber in a ribbon arrangement with an elongated jacket that fits over the ribbon.

jitter
A slight movement of a transmission signal in time or phase that can introduce errors and loss of synchronization. More jitter is introduced when cable runs are longer than the network-topology specification recommends. Other causes of jitter include cables with high attenuation and signals at high frequencies. Also called phase jitter, timing distortion, or intersymbol interference.

joint
Any joining or mating of a fiber by splicing (by fusion splicing or physical contact of fibers) or connecting.

jumper

(1) A small, manually placed connector that connects two conductors to create a circuit (usually temporary). (2) In fiber optics, a simplex cable assembly that is typically made from simplex cordage.

jumper wire

A cable of twisted wires without connectors used for jumpering.

junction laser

A semiconductor diode laser.

K

Kevlar

A strong, synthetic material developed and trademarked by DuPont; the preferred strength element in cable. Also used as a material in body armor and parts for military equipment. Also known by the generic name aramid; *see also* aramid yarn.

keying

A mechanical feature of a connector system that guarantees correct orientation of a connection. The key prevents connection to a jack or an optical-fiber adapter that was intended for another purpose.

kHz

Kilohertz; 1,000 hertz.

KPSI

KiloPSI. A unit of tensile strength expressed in thousands of pounds per square inch.

L

L

Symbol used to designate inductance.

ladder rack

A rack used to hold cabling that looks like a ladder.

large core fiber

Usually a fiber with a core of 200 microns or more. This type of fiber is not common in structured cabling systems.

laser

Acronym for light amplification by stimulated emission of radiation. The laser produces a coherent source of light with a narrow beam and a narrow spectral bandwidth (about 2cm). Lasers in fiber optics are usually solid-state semiconductor types. Lasers are used to provide the high-powered, tightly controlled light wavelengths necessary for high-speed, long-distance optical fiber transmissions.

laser diode (LD)

A special semiconductor that emits laser light when a specific amount of current is applied. Laser diodes are typically used in higher speed applications (622Mbps to 10Gbps) such as ATM, 1000Base-LX, and SONET. The mode is usually ellipse shaped and therefore requires a lens to make the light symmetrical with the mode of the fiber, which is usually round.

lasing threshold

The energy level that, when reached, allows the laser to produce mostly stimulated emissions rather than spontaneous emissions.

lateral displacement loss

The loss of signal power that results from lateral displacement from optimum alignment between two fibers or between a fiber and an active device.

launch angle

In fiber-optic transmissions, the launch angle is defined as the difference between the incoming direction of the transmitting light and the alignment of the optical fiber.

launch cable

Used to connect fiber-optic test equipment to the fiber system.

launch fiber

An optical fiber used to introduce light from an optical source into another optical fiber. Also referred to as launching fiber.

lay

The axial distance required for one cabled conductor or conductor strand to complete one revolution around the axis around which it is cabled.

lay direction

The direction of the progressing spiral twist of twisted-pair wires while looking along the axis of the cable away from the observer. The lay direction can be either left or right.

Layer 2 switch

A network device that operates at the Data Link layer. The switch builds a table of MAC addresses of all connected stations and uses the table to intelligently forward data to the intended recipients.

Layer 3 switch
A network device that can route LAN traffic (Layer 3) at a speed that is nearly as quick as a Layer 2 switch device. Layer 3 switches typically perform multiport, virtual LAN, and data pipelining functions of a standard Layer 2 switch and can also perform routing functions between virtual LANs.

lbf
Abbreviation for pounds force.

LC
A small-form-factor optical fiber connector originally designed and manufactured by AT&T Bell Labs; closely resembles the RJ-11 connector.

leakage
An undesirable passage of current over the surface of or through a connector.

leased line
A private telephone line (usually a digital line) rented for the exclusive use of a leasing customer without interchange switching arrangements.

least-squares averaging
A method of measuring attenuation in a fiber-optic signal that reduces the effect of high-frequency noise on the measurement.

light
The electromagnetic radiation visible to the human eye between 400nm and 700nm. The term is also applied to electromagnetic radiation with properties similar to visible light; this includes the invisible near-infrared radiation in most fiber-optic communication systems.

light-emitting diode (LED)
A semiconductor device used in a transmitter to convert information from electric to optical form. The LED typically has a large spectral width (that is, it produces incoherent light); LED devices are usually used on low-speed (100–256Mbps) fiber-optic communication systems, such as 100Base-FX and FDDI, that do not require long distances or high data rates.

line build-out (LBO)
A device that amplifies a received power level to ensure that it is within proper specs.

line conditioner
A device used to protect against power surges and spikes. Line conditioners use several electronic methods to clean all power coming into the line conditioner so that clean, steady power is put out by the line conditioner.

line-of-sight transmission
The transmission of free-space-optics devices where the transmitter and receiver are in the same plane.

link
An end-to-end transmission path provided by the cabling infrastructure. Cabling links include all cables and connecting hardware that compose the horizontal or backbone subsystems. Equipment and work-area cables are not included as part of a link.

link light
A small light-emitting diode (LED) that is found on both the NIC and the hub and is usually green and labeled "Link." A link light indicates that the NIC and the hub are making a Data Link layer connection.

listed
Equipment included on a list published by an organization, acceptable to the authority having jurisdiction, that maintains periodic inspection of production of listed equipment, and whose listing states either that the equipment or material meets appropriate standards, or that it has been tested and found suitable for use in a specified manner. In the United States, electrical and data communications equipment is typically listed with Underwriters Laboratories (UL).

lobe
An arm of a Token Ring that extends from a multistation access unit (MSAU) to a workstation adapter.

local area network (LAN)
A network connecting multiple nodes within a defined area, usually within a building. The linking can be done by cable that carries optical fiber or copper. These are usually high bandwidth (4Mbps or greater) and connect many nodes within a few thousand meters. LANs can, however, operate at lower data rates (less than 1Mbps) and connect nodes over only a few meters.

local exchange carrier (LEC)
The local regulated provider of public switched telecommunications services. The LEC is regulated by the local Public Utilities Commission.

local loop
The loop or circuit between receivers (and, in two-way systems, receivers and senders), who are normally the customers or subscribers to the service's products, and the terminating equipment at the central office.

LocalTalk
A low-speed form of LAN data link technology developed by Apple Computer. It was designed to transport Apple's AppleTalk networking scheme; it uses a Carrier

Sense Multiple Access/Collision Avoidance (CSMA/CA) form of media access control. Supports transmission at 230Kbps.

logical network addressing
The addressing scheme used by protocols at the OSI Network layer.

logical ring topology
A network topology in which all network signals travel from one station to another, being read and forwarded by each station. A Token Ring network is an example of a logical ring topology.

logical topology
Describes the way the information flows. The types of logical topologies are the same as the physical topologies, except that the information flow specifies the type of topology.

long wavelength
Light whose wavelength is greater than 1000nm (longer than one micron).

longitudinal conversion loss (LCL)
A measurement (in decibels) of the differential voltage induced on a conductor pair as a result of subjecting that pair to longitudinal voltage. This is considered to be a measurement of circuit balance.

longitudinal conversion transfer loss (LCTL)
Measures cable balance by the comparison of the signal appearing across the pair to the signal between ground and the pair, where the applied signal is at the opposite end of the cable from the location at which the across-pair signal is measured. LCTL is also called far-end unbalance attenuation.

longitudinal modes
Oscillation modes of a laser along the length of its cavity. Each longitudinal mode contains only a very narrow range of wavelengths; a laser emitting a single longitudinal mode has a very narrow bandwidth. The oscillation of light along the length of the laser's cavity is normally such that two times the length of the cavity will equal an integral number of wavelengths. Longitudinal modes are distinct from transverse modes.

loop
(1) A complete electrical circuit. (2) The pair of wires that winds its way from the central office to the telephone set or system at the customer's office, home, or factory.

loopback
A type of diagnostic test in which a transmitted signal is returned to the sending device after passing through a data communications link or network. This test allows the comparison of a returned signal with the transmitted signal to determine if the signal is making its way through the communications link and how much signal it is losing upon its total trip.

loose tube
A protective tube loosely surrounding an optical fiber, often filled with gel used as a protective coating. Loose-tube cable designs are usually found in outdoor cables, not inside buildings.

loose tube buffer
Optical fiber that is carried loosely in a buffer many times the diameter of the fiber. Loose-buffered fiber is typically terminated with a breakout kit or a fanout kit and connected to a patch panel. Also known as *loose buffer*.

loss
The attenuation of optical or electrical signal, normally measured in decibels (dB). With respect to fiber-optic cables, there are two key measurements of loss: insertion loss and return loss. Both are measured in decibels. The higher the decibel number, the more loss there is. Some copper-based and optical fiber–based materials are lossy and absorb electromagnetic radiation in one form and emit it in another; for example, heat. Some optical fiber materials are reflective and return electromagnetic radiation in the same form as it is received, usually with little or no power loss. Still others are transparent or translucent, meaning they are "window" materials; loss is the portion of energy applied to a system that is dissipated and performs no useful work. *See also* attenuation.

loss budget
A calculation and allowance for total attenuation in a system that is required in order to ensure that the detectors and receivers can make intelligent decisions about the pulses they receive.

lossy
Describes a connection having poor efficiency with respect to loss of signal.

M

MAC
(1) See media access control. (2) Abbreviation for moves, adds, and changes.

macrobend

A major bend in an optical fiber with a radius small enough to change the angle of incidence and allow light to pass through the interface between the core and the cladding rather than reflect off it. Macrobends can cause signal attenuation or loss by allowing light to leave the optical fiber core.

macrobending loss

Optical power loss due to large bends in the fiber.

main cross-connect

A cross-connect for first-level backbone cables, entrance cables, and equipment cables. The main cross-connect is at the top level of the premises cabling tree.

main distribution frame (MDF)

A wiring arrangement that connects the telephone lines coming from outside on one side and the internal lines on the other. The MDF may be a central connection point for data communications equipment in addition to voice communications. An MDF may also carry protective devices or function as a central testing point.

Manchester coding

A method of encoding a LAN in which each bit time that represents a data bit has a transition in the middle of the bit time. Manchester coding is used with 10Mbps Ethernet (10Base-2, 10Base-5, 10Base-F, and 10Base-T) LANs.

mandrel wrap

A device used to remove high-order modes caused by overfilling in a length of optical fiber for insertion loss measurements.

margin

The allowance for attenuation in addition to that explicitly accounted for in system design.

mass splicing

The concurrent and simultaneous splicing of multiple fibers at one time. Currently mass splicing is done on ribbon cable, and the standard seems to be ribbon cable with 12 fibers. Special splice protectors are made for this purpose, as well as special equipment for splicing.

material dispersion

A pulse dispersion that results from each wavelength traveling at a speed different from other wavelengths through an optical fiber. *See also* chromatic dispersion.

maximum tensile rating

A manufacturer's specified limit on the amount of tension, or pulling force, that may be applied to a fiber-optic cable.

Mean time between failures (MTBF)

A measurement of how reliable a hardware component is. Usually measured in thousands of hours.

mechanical splice

With respect to fiber-optic cables, a splice in which fibers are joined mechanically (e.g., glued, crimped, or otherwise held in place) but not fused together using heat. Mechanical splice is the opposite of a fusion splice in which the two fiber ends are butted and then joined by permanently bonding the glass end-faces through the softening of the glass, which is fused together.

media

Wire, cable, or conductors used for transmission of signals.

media access

The process of vying for transmission time on the network media.

media access control (MAC)

A sublayer of the OSI Data Link layer (Layer 2) that controls the way multiple devices use the same media channel. It controls which devices can transmit and when they can transmit. For most network architectures, each device has a unique address that is sometimes referred to as the MAC address. *See also* media access control (MAC) address.

media access control (MAC) address

Network-adapter cards such as Ethernet, Token Ring, and FDDI cards are assigned addresses when they are built. A MAC address is 48 bits represented by six sections of two hexadecimal digits. No two cards have the same MAC address. The IEEE helps to achieve the unique addresses by assigning the first half to manufacturers so that no two manufacturers have the same first three bytes in their MAC address. The MAC address is also called the hardware address.

media filter

An impedance-matching device used to change the impedance of the cable to the expected impedance of the connected device. For example, media filters can be used in Token Ring networks to transform the 100 ohm impedance of UTP cabling to the 150 ohm impedance of media interface connections.

media interface connector (MIC)

A pair of fiber-optic connectors that links the fiber media to the FDDI network card or concentrator. The MIC consists of both the MIC plug termination of an optical cable and the MIC receptacle that is joined with the FDDI node.

medium attachment unit (MAU)

When referring to Ethernet LANs, the transceiver in Ethernet networks.

medium dependent interface (MDI)

Used with Ethernet systems; it is the connector used to make the mechanical and electrical interface between a transceiver and a media segment. An 8-pin RJ-45 connector is the MDI for Ethernet implemented using UTP.

medium independent interface (MII)

Used with 100Mbps Ethernet systems to attach MAC-level hardware to a variety of physical media systems. Similar to the AUI interface used with 10Mbps Ethernet systems. The MII is a 40-pin connection to outboard transceivers or PHY devices.

medium interface connector (MIC)

A connector that is used to link electronics and fiber-optic transmission systems.

meridian plane

Any plane that includes or contains the optical axis.

meridional ray

A light ray that passes through the axis of an optical fiber.

messenger cable

A cable with a strong supporting member attached to it for use in aerial installations. *See also* figure-8.

metropolitan area network (MAN)

An interconnected group of local area networks (LANs) that encompasses an entire city or metropolitan area.

microbend

A small radius bend in the optical fiber that changes the angle of incidence, allowing light to pass through the interface rather than reflect off it.

microbending loss

The optical power loss due to microscopic bends in the fiber.

microfarad

One millionth of a farad. Abbreviated µF and, less commonly, µfd, mf, or mfd.

micrometer

Also referred to as a micron; one millionth of a meter, often abbreviated with the symbol µ.

midsplit broadband

A broadcast network configuration in which the cable is divided into two channels, each using a different range of frequencies. One channel is used to transmit signals and the other is used to receive.

minimum bend radius

A fiber-optic cable manufacturer's specified limit on the amount of bending that the cable can withstand without damage.

misalignment loss

The loss of optical power resulting from angular misalignment, lateral displacement, or end separation.

modal bandwidth

The bandwidth-limiting characteristic of multimode fiber systems caused by the variable arrival times of various modes.

modal dispersion

A type of dispersion or spreading that arises from differences in the amount of time that different modes take to travel through multimode fibers. Modal dispersion potentially can cause parts of a signal to arrive in a different order from the one in which they were transmitted. *See also* DMD.

modal noise

A disturbance often measured in multimode fiber-optic transmissions that are fed by diode lasers. The higher quality the laser light feeding the fiber, the less modal noise will be measured.

mode

A single wave traveling in an optical fiber or in a light path through a fiber. Light has modes in optical-fiber cable. A high-order mode is a path that results in numerous reflections off the core/cladding interface. A low-order mode results in fewer reflections. A zero-order mode is a path that goes through the fiber without reflecting off the interface at all. In a single-mode fiber, only one mode (the fundamental mode) can propagate through the fiber. Multimode fiber has several hundred modes that differ in field pattern and propagation velocity. The number of modes in an optical fiber is determined by the diameter of the core, the wavelength of the light passing through it, and the refractive indexes of the core and cladding. The number of modes increases as the core diameter increases, the wavelength decreases, or the difference between refractive indexes increases.

mode field diameter (MFD)

The actual diameter of the light beam traveling through the core and part of the cladding and across the end-face in a single-mode fiber-optic cable. Since the mode field diameter is usually greater than the core diameter, the mode field diameter replaces the core diameter as a practical parameter.

mode filter
A device that can select, attenuate, or reject a specific mode. Mode filters are used to remove high-order modes from a fiber and thereby simulate EMD. *See also* mandrel wrap.

mode mixing
Coupling multiple single modes into a single multimode strand by mixing the different signals and varying their modal conditions.

mode stripper
A device that removes high-order modes in a multimode fiber to give standard measurement conditions.

modem
A device that implements modulator-demodulator functions to convert between digital data and analog signals.

modified modular jack (MMJ)
A six-wire modular jack used by the DEC wiring system. The MMJ has a locking tab that is shifted to the right-hand side.

modular
Equipment is said to be modular when it is made of plug-in units that can be added together to make the system larger, improve the capabilities, or expand its size. Faceplates made for use with structured cabling systems are often modular and permit the use of multiple types of telecommunications outlets or modular jacks such as RJ-45, coaxial, audio, or fiber.

modular jack
A female telecommunications interface connector. Modular jacks are typically mounted in a fixed location and may have four, six, or eight contact positions, though most typical standards-based cabling systems will have an eight-position jack. Not all positions need be equipped with contacts. The modular jack may be keyed or unkeyed so as to permit only certain types of plugs to be inserted into the jack.

modulate
(1) To convert data into a signal that can be transmitted by a carrier. (2) To control.

modulation
(1) Coding of information onto the carrier frequency. Types of modulation include amplitude modulation (AM), frequency modulation (FM), and phase modulation (PM). (2) When light is emitted by a medium, it is coherent, meaning that it is in a fixed-phase relationship within fixed points of the light wave. The light is used because it is a continuous, or sinusoidal, wave (a white or blank form) upon which a signal can be superimposed by modulation of that form. The modulation is a variation imposed upon this white form, a variation of amplitude, frequency, or phase of the light. There are two basic forms of this modulation: one by an analog form, another by a digital signal. This signal is created in the form of the "intelligence" and superimposed upon the light wave. It is then demodulation by a photodetector and converted into electrical energy.

monochromatic
Light having only one color, or more accurately only one wavelength.

Motion Pictures Experts Group (MPEG)
A standards group operating under the ISO that develops standards for digital video and audio compression.

MT-RJ connector
A duplex fiber-optic connector that looks similar to the RJ-45-type connector.

multifiber jumpers
Used to interconnect fiber-optic patch panels from point to point.

multimedia
An application that communicates to more than one of the human sensory receptors such as audio and video components.

multimode
Transmission of multiple modes of light. *See also* mode.

multimode distortion
The signal distortion in an optical waveguide resulting from the superposition of modes with differing delays.

multimode fiber
Optical-fiber cable whose core is between 8 and 200 microns with a refractive index that is graded or stepped; multimode fiber has a core diameter large enough to allow light to take more than one possible path through it. It allows the use of inexpensive LED light sources, and connector alignment and coupling is less critical than with single-mode fiber. Distances of transmission and transmission bandwidth are less than with single-mode fiber due to dispersion. The ANSI/TIA/EIA-568-B standard recognizes the use of 62.5/125-micron and 50/125-micron multimode fiber for horizontal cabling.

multimode laser

A laser that produces emissions in two or more longitudinal modes.

multiple reflection noise (MRN)

The noise at the receiver caused by the interface of delayed signals from two or more reflection points in an optical fiber span.

multiplex

The combination of two or more signals to be transmitted along a single communications channel.

multiplexer

A device that combines two or more discrete signals into a single output. Many types of multiplexing exist, including time-division multiplexing and wavelength-division multiplexing.

multiplexing

Transmitting multiple data channels in the same signal.

multipoint RF network

An RF network where the RF transmitters can access more than one receiver.

multistation access unit (MAU or MSAU)

Used in Token Ring LANs, a wiring concentrator that allows terminals, PCs, printers, and other devices to be connected in a star-based configuration to Token Ring LANs. MAU hardware can be either active or passive and is not considered to be part of the cabling infrastructure.

multiuser telecommunications outlet assembly (MuTOA)

A connector that has several telecommunications/outlet connectors in it. These are often used in a single area that will have several computers and telephones.

mutual capacitance

The capacitance (the ability to store a charge) between two conductors when they are brought adjacent to each other.

Mylar

The DuPont trademark for biaxially-oriented polyethylene terephthalate (polyester) film.

MZI

Mach-Zehnder interferometer. A device used to measure the optical phase shift of various materials.

N

National Electrical Code (NEC)

The U.S. electrical wiring code that specifies safety standards for copper and fiber-optic cable used inside buildings. See Chapter 4 for more information.

National Security Agency (NSA)

The U.S. government agency responsible for protecting U.S. communications and producing foreign intelligence information. It was established by presidential directive in 1952 as a separately organized agency within the Department of Defense.

N-connector

A coaxial cable connector used for Ethernet 10Base-5 thick coaxial segments.

near-end

Defined in copper twisted pair cables as the end of the cable where the transmitter is located.

near-end crosstalk (NEXT)

Crosstalk noise between two twisted pairs measured at the near end of the cable. Near is defined as the end of the cable where the transmission originated. See Chapter 1 for more information.

near-field radiation pattern

The distribution of the irradiance over an emitting surface (over the cross section of an optical waveguide).

near infrared

The part of the infrared spectrum near the visible spectrum, typically 700nm to 1500nm or 2000nm; it is not rigidly defined.

network

Ties things together. Computer networks connect all types of computers and computer-related peripherals—terminals, printers, modems, door entry sensors, temperature monitors, and so forth. The networks we're most familiar with are long-distance ones, such as phone or train networks. Local area networks (LANs) connect computer equipment within a building or campus.

NFPA 262

The fire test method that measures flame spread, peak smoke optical density, and average smoke optical density. Formerly referred to as UL 910. Cables are required to pass this test and be listed by a nationally recognized test laboratory (e.g., UL or ETL) for the cables to be allowed to be placed in plenum spaces.

network media
The physical cables that link computers in a network; also known as physical media.

NIC card
A circuit board installed in a computing device that is used to attach the device to a network. A NIC performs the hardware functions that are required to provide a computing device physical communications capabilities with a network.

NIC diagnostics
Software utilities that verify that the NIC is functioning correctly and that test every aspect of NIC operation, including connectivity to other nodes on the network.

node
Endpoint of a network connection. Nodes include any device connected to a network such as file servers, printers, or workstations.

noise
In a cable or circuit, any extraneous signal (electromagnetic energy) that interferes with the desired signal normally present in or passing through the system.

noise equivalent power (NEP)
The optical input power to a detector needed to generate an electrical signal equal to the inherent electrical noise.

Nomex
A DuPont trademark for a temperature-resistant, flame-retardant nylon.

nominal velocity of propagation (NVP)
The speed that a signal propagates through a cable expressed as a decimal fraction of the speed of light in a vacuum. Typical copper cables have an NVP value of between 0.6c and 0.9c.

non-return to zero (NRZ)
A digital code in which the signal level is low for a 0 bit and high for a 1 bit and which does not return to zero volts between successive 1 bits or between successive 0 bits.

non-zero-dispersion-shifted fiber
A type of single-mode optical fiber designed to reduce the effects of chromatic dispersion while minimizing four-wave mixing.

normal
A path drawn perpendicular to the *interface*, or boundary layer between two media, that is used to determine the *angle of incidence* of light reaching the interface.

normal angle
The angle that is perpendicular to a surface.

NT-1
Used to terminate ISDN at the customer premises. It converts a two-wire ISDN U interface to a four-wire S/T interface.

numerical aperture (NA)
The light-gathering ability of a fiber, defining the maximum angle to the fiber axis at which light will be accepted and propagated through the fiber. The numerical aperture is determined by the refractive indexes of the core and cladding. The numerical aperture is also used to determine the fiber's acceptance angle.

Nyquist Minimum
The calculated minimum effective sampling rate for a given analog signal based on its highest expected frequency. The Nyquist Minimum requires sampling to take place at a minimum of twice the expected highest frequency of an analog signal. For example, if an analog signal's highest frequency is expected to be 10kHz, it must be sampled at a rate of at least 20kHz.

O

O
Orange, when used in conjunction with color-coding for twisted-pair cabling.

OC-1
Optical carrier level one, equal to 51.84Mbps. This is a SONET channel, whose format measures 90 bytes and is composed of the transport overhead and the synchronous payload envelope.

OC-3
SONET Channel of 155.52Mbps.

OC-12
SONET channel of 622.08Mbps.

OC-48
SONET channel of 2.5Gbps.

OC-192
SONET channel of 10Gbps, currently the highest level now available.

octet
Eight bits (also called a byte).

OFCP

An optical fiber cable that has conducting (metal) elements in its construction and that meets the plenum test requirements of NFPA 262; examples of conducting elements in the cable include the copper wire or interlock aluminum armoring

OFCR

An optical fiber cable that has conducting (metal) elements in its construction and that meets the riser test requirements of UL 1666; examples of conducting elements in the cable include the copper wire or interlock aluminum armoring

OFNP

An optical fiber cable that has no conducting (metal) elements and meets the plenum test requirements of NFPA 262 (UL910).

OFNR

An optical fiber cable that has no conducting (metal) elements and meets the riser test requirements of UL 1666.

off-hook

The handset's state of being lifted from its cradle. The term originated from when the early handsets were actually suspended from a metal hook on the phone. With modern telephones, when the handset is removed from its hook or cradle, it completes the electrical loop, thus signaling the central office to provide dial tone. Opposite of on-hook.

office principle ground point (OPGP)

The main grounding point in a central office. Usually connects directly to an earth ground like a water pipe.

ohm

A unit of electrical resistance. The value of resistance through which a potential of one volt will maintain a current of one ampere.

on-hook

The telephone handset's state of resting in its cradle. The phone is not connected to any particular line. Only the bell is active—that is, it will ring if a call comes in. Opposite of off-hook.

open circuit

An incomplete circuit. It can be either a break in a cable or a switch that's turned off.

open fault

A break in the continuity of a circuit. This means that the circuit is not complete or the cable/fiber is broken. This condition is also called unmated, open, or unterminated.

Open Systems Interconnection (OSI)

A model defined by the ISO to categorize the process of communication between computers in terms of seven layers. *See also* International Organization for Standardization.

operating wavelength

The wavelength at which a fiber-optic receiver is designed to operate. Typically, an operating wavelength includes a range of wavelengths above and below the stated wavelength.

operations, administration, maintenance, and provisioning (OAM&P)

A telecommunications term for the support functions of a network.

optical amplifier

Increases the power of an optical signal without converting any of the signals from optical to electrical energy and then back to optical so that the amplification processes the optical signal wholly within optical amplification equipment. The two most common optical amplifiers are semiconductor laser amplifiers and those made from doped fiber, such as the EDFA (erbium doped fiber amplifier), which amplifies with a laser pump diode and a section of erbium doped fiber.

optical attenuator

Reduces the intensity of light waves, usually so that the power is within the capacity of the detector. There are three basic forms of attenuators: fixed optical attenuators, stepwise variable optical attenuators, and continuous variable optical attenuators. Attenuation is normally achieved either by a doped fiber or an offset or core misalignment. *See also* attenuator.

optical bandpass

The range of optical wavelengths that can be transmitted through a component.

optical carrier n

Optical signal standards. The *n* indicates the level where the respective data rate is exactly *n* times the first level OC-1. OC-1 has a data rate of 51.84Mbps. OC-3 is three times that rate, or 155.52Mbps, and so on. Associated with SONET. OC levels are medium-dependant on fiber.

optical combiner

A device used to combine fiber-optic signals.

optical continuous wave reflectometer (OCWR)

A device that measures optical return loss, or the loss of optical signals due to reflection back toward the transmitter.

optical detector
A transducer that generates an electronic signal when excited by an optical power source.

optical directional coupler (ODC)
A directional coupler used to combine or separate optical power.

optical-fiber cable
An assembly consisting of one or more optical fibers. These optical fibers are thin glass or plastic filaments used for the transmission of information via light signals. The individual optical fibers are the signal carrying part of a fiber-optic cable. *See also* single-mode fiber and multimode fiber.

optical-fiber duplex adapter
A mechanical media termination device designed to align and join two duplex connectors.

optical-fiber duplex connector
A mechanical media termination device designed to transfer optical power between two pairs of optical fibers.

optical isolator
A component used to block out reflected and other unwanted light.

optical loss test set
An optical power meter and a light source calibrated for use together to detect and measure loss of signal on an optical cable.

optical polarization
A term used to describe the orientation in space of a time varying field vector of an optical signal.

optical receiver
An optoelectronic circuit that converts an incoming signal to an electronic signal. The optical receiver will include a transducer in the form of a detector, which might be a photodiode or other device. When irradiated by an optical power device, it will be able to translate the optical signal into an electronic signal.

optical reference plane
Defines the optical boundary between the MIC (media interface connector) plug and the MIC receptacle.

optical repeater
An optoelectronic device, which could include an amplifier, that receives a signal and amplifies it, especially in the case of analog signals. In the case of a digital signal, the optical repeater reshapes or retimes the signal and then retransmits it.

optical return loss (ORL)
ORL is a ratio expressed in decibels. The reflection is caused by a component or an assembly.

optical spectrum
Starts with red, then orange, yellow, green, blue, indigo, and finally violet. Each color represents a wavelength or frequency of electromagnetic energy; the spectrum is between 400nm and 700nm. 400nm is the ultraviolet portion of the spectrum and 700nm is the infrared portion of the spectrum.

optical splitter
A device used to split fiber-optic signals.

optical subassembly
The portion of a fiber-optic receiver that guides light from the optical fiber to the photodiode.

optical time-domain reflectometer (OTDR)
A device used to test a fiber-optic link, including the optical fiber and connectors, by launching an optical signal (light pulse) through the link and measuring the amount of energy that is reflected back. The OTDR is a troubleshooting device that can pinpoint faults throughout a fiber-optic link.

optical transmitter
An optoelectronic circuit that converts an electronic signal into an optical signal.

optical waveguide
Any structure that can guide light; the optical waveguide is a nonconductive material with a central core of optically transparent material (usually silica glass) surrounded by a transparent cladding material that has a lower refractive index than the core.

optoelectronic
Any device that uses or responds to optical power in its internal operation.

outlet box
A metallic or nonmetallic box mounted within a wall, floor, or ceiling used to hold outlet, connector, or transition devices.

output
The useful signal or power delivered by a circuit or device.

outside plant
A telecommunications infrastructure designed for installation outside of any structure.

outside plant (OSP) cables
Typically used outside of the wire center but also may be routed into the CEF. Since OSP cables are more flammable than premises (indoor) cables, the distance of penetration into the building must be limited.

oversampling
A method of synchronous bit synchronization. The receiver samples the signal at a much faster rate than the data rate. This permits the use of an encoding method that does not add clocking transitions.

over-voltage threshold
The level of over-voltage that will trip the circuit breaker in a surge protector.

P

P region
The area in a semiconductor that is doped to have an abundance of electron acceptors in which vacancies in the valence electron level are the dominant current carriers.

packet
Bits grouped serially in a defined format containing a command or data message sent over a network. The packet is the major structure of data sent over a network.

packet switching
The process of breaking messages into packets. Each packet is then routed optimally across the network. Packet sequence numbers are used at the destination node to reassemble packets.

packing fraction
At a cut end, the fraction of the face surface area of a fiber-optic bundle that is fiber core.

PAM5x5
The signal-encoding technique used in the Ethernet 100Base-T2 and 1000Base-T media systems.

Part 68 requirements
Specifications established by the FCC as the minimum acceptable protection that communications equipment must provide to the telephone network.

passive branching device
A device that divides an optical input into two or more optical outputs.

passive coupler
Divides light without generating new light.

patch cable
Any flexible piece of cable that connects one network device to the main cable run or to a patch panel that in turn connects to the main cable run; also called *patch cord*. Used for interconnecting circuits on a patch panel or cross-connect. Patch cables are short distance, usually have connectors preinstalled on both ends, are used to connect equipment, and are generally between 3 and 6 meters long.

patch panel
A connecting hardware that typically provides means to connect horizontal or backbone cables to an arrangement of fixed connectors that may be accessed using patch cords or equipment cords to form cross-connections or interconnections. Patch panels may connect either copper or optical fiber cables.

patching
A means of connecting circuits via cords and connectors that can be easily disconnected and reconnected at another point. May be accomplished by using modular patch cords connected between jack fields or by patch cord assemblies that plug onto connecting blocks.

pathway
A facility (e.g., conduit, cable tray, raceway, ducting, or plenum) for the placement and protection of telecommunications cables.

peak
The maximum instantaneous value of a varying current or voltage.

peak wavelength
The optical wavelength at which the power output of a source is at its maximum level.

pedestal
A device, usually mounted on the floor or ground, which is used to house voice/data jacks or power outlets at the point of use. Also commonly referred to as a monument, tombstone, above-floor fitting, or doghouse.

periodicity
Uniformly spaced variations in the insulation diameter of a transmission cable that result in reflections of a signal.

permanent virtual circuit (PVC)
Technology used by frame relay (as well as other technologies like X.25 and ATM) that allows virtual data circuits to be set up between the sender and receiver over a packet-switched network.

phase

An angular relationship between waves or the position of a wave in its oscillation cycle.

phase modulation (PM)

One of three basic methods of adding information to a sine wave signal in which its phase is varied to impose information on it. *See also* amplitude modulation and frequency modulation.

phase shift

A change in the phase relationship between two alternating quantities.

photo-bleaching

A reduction in added loss that occurs when a fiber is exposed to light. Ionizing radiation causes added loss. This loss can be reduced by transmitting light through the fiber during normal operation or by exposing the fiber to sunlight.

photodetector

An optoelectronic transducer, such as a pin photodiode or avalanche photodiode, that acts as a light detector.

photodiode

A component that converts light energy into electrical energy. The photodiode is used as the receiving end of a fiber-optic link.

photon

A basic unit of light; the smallest quantum particulate component of light.

photonic

A term coined to describe devices using photons, analogous to electronic, describing devices working with electrons.

physical bus topology

A network that uses one network cable that runs from one end of the network to the other. Workstations connect at various points along this cable. These networks are easy to run cable for, but they are typically not as reliable as a star topology. 10Base-2 Ethernet is a good example of a network architecture that uses a physical bus topology.

physical contact (PC)

Description for a connector that places an optical fiber end in direct physical contact with the optical fiber end of another connector.

physical mesh topology

A network configuration that specifies a link between each device in the network. A physical mesh topology requires a lot of cabling and is difficult to reconfigure.

physical ring topology

A network topology that is set up in a circular fashion. Data travels around the ring in one direction, and each device on the ring acts as a repeater to keep the signal strong as it travels. Each device incorporates a receiver for the incoming signal and a transmitter to send the data on to the next device in the ring. The network is dependent on the ability of the signal to travel around the ring. Cabling a physical ring topology is difficult because of the amount of cable that must be run. FDDI is an example of a network that can be wired to use a physical ring topology.

physical star topology

A network in which a cable runs from each network device to a central device called a hub. The hub allows all devices to communicate as if they were directly connected. The network may logically follow another type of topology such as bus or ring topology, but the wiring is still a star topology.

physical topology

The physical layout of a network, such as bus, star, ring, or mesh.

picofarad

One millionth of one millionth of a farad. Abbreviated pf.

picosecond (PS)

One trillionth of a second.

pigtail

(1) A short length of fiber with a permanently attached device, usually a connector, on one end. (2) A fiber-optic cable assembly consisting of a connector or hardware device (such as a light source package installed by a manufacturer) on one end and an unterminated fiber at the other end. Normally found in applications wherein a splice is convenient for terminating a device with a connector. Also used when the loss characteristics of the connector must be known precisely. For instance, a splice of .03dB might be reliably predicted and controlled, but the variability of most commercially available terminations is unacceptable, so a precharacterized cable assembly is cut into a pigtail and attached to the device through splicing.

pinout scheme

The pinout scheme is the pattern that identifies which wires in a UTP cable are connected to each pin of a connector.

plain old telephone service (POTS)
The basic service that supplies standard single-line telephones, telephone lines, and access to the public switched network; it only receives and places calls and has no added features like call waiting or call forwarding.

planar waveguide
A waveguide fabricated in a flat material such as a thin film.

plastic-clad silica (PCS) fiber
A step-index multimode fiber that has a silica core and is surrounded by a lower index plastic cladding.

plastic fiber
Optical fiber having a plastic core and plastic cladding rather than using glass.

plasticizer
A chemical added to plastics to make them softer and more flexible.

plenum
The air-handling space between the walls, under structural floors, and above drop ceilings when used to circulate and otherwise handle air in a building. Plenum-grade cable can be run through these spaces if local building codes permit it.

plenum cable
Cable whose flammability and smoke characteristics allow it to be routed in plenum spaces without being enclosed in a conduit; all cables with this rating must pass NFPA 262 (formerly UL 910).

plug
The male component of a plug/jack connector system. In premises cabling, a plug provides the means for a user to connect communications equipment to the communications outlet.

point-to-point transmission
Carrying a signal between two endpoints without branching to other points.

polarity
Identifies which side of an electrical circuit is positive and which is negative.

polarization
(1) Alignment of the electric and magnetic fields that make up an electromagnetic wave. Normally refers to the electric field. If all light waves have the same alignment, the light is polarized. (2) The direction of vibration of the photons in the light wave.

polarization maintaining fiber
Optical fiber that maintains the polarization of light that enters it.

polarization mode dispersion
Spreading of the light wave caused by imperfections in a single-mode optical fiber that slow down a polarization mode of the signal. When one polarization mode lags behind another, the signal spreads out and can become distorted.

polarization stability
The degree of variation in insertion loss as the polarization state of the input light is varied.

polling
A media access control method that uses a central device called a controller, which polls each device in turn and asks if it has data to transmit. 100VG-AnyLAN hubs poll nodes to see if they have data to transmit.

polybutadiene
A type of synthetic rubber often blended with other synthetic rubbers to improve their dielectric properties.

polyethylene (PE)
A thermoplastic material with excellent electrical properties. PE is used as an insulating material and as jacket material where flame-resistance requirements allow.

polyimide
A polymer that is used to coat optical fiber for harsh environment applications.

polymer
A substance made of repeating chemical units or molecules. The term is often used as a synonym for plastic, rubber, or elastomer.

polypropylene
A thermoplastic material that is similar to polyethylene, which is somewhat stiffer and has a higher softening point (temperature), with comparable electrical properties.

polyurethane (PUR)
A broad class of polymers that are noted for good abrasion and solvent resistance. Not as common as PVC (polyvinyl chloride).

polyvinyl chloride (PVC)
A general-purpose thermoplastic used for wire and cable insulation and jackets.

potting
The process of sealing by filling with a substance to exclude moisture.

power brownout
Occurs when power drops below normal levels for several seconds or longer.

power level
The ratio between the total power delivered to a circuit, cable, or device and the power delivered by that device to a load.

power overage
Occurs when too much power is coming into a piece of equipment. *See also* power spike and power surge.

power ratio
The ratio of power appearing at the load to the input power. Expressed in decibels.

power sag
Occurs when the power level drops below normal and rises to normal in less than one second.

power spike
Occurs when the power level rises above normal and drops back to normal for less than a second. *See also* power overage and power surge.

power sum
A test method for cables with multiple pairs of wire whereby the mathematical sum of pair-to-pair crosstalk from a reference wire pair is measured while all other wire pairs are carrying signals. Power-sum tests are necessary on cables that will be carrying bidirectional signals on more than two pairs. Most commonly measured on four-pair cables.

power-sum attenuation-to-crosstalk ratio (PSACR)
The ratio of attenuation to power-sum near-end crosstalk; while not a requirement of ANSI/TIA-568-C.2, it is figure of merit used by all manufacturers to denote the signal to power-sum noise ratio in the near end.

power-sum attenuation-to-crosstalk ratio, far end (PSACRF)
The ratio of attenuation to power-sum far-end crosstalk; a requirement of ANSI/TIA-568-C.2; formerly known as power-sum equal-level far-end crosstalk (PSELFEXT)

power-sum alien near-end crosstalk (PSANEXT)
The power-sum measurement of alien crosstalk in the near end (at the transmitter); a newly established requirement in ANSI/TIA-568-C.2 for Category 6A cables.

power-sum alien far-end crosstalk (PSAFEXT)
The power-sum measurement of alien crosstalk in the far end (away from the transmitter); the result from this test is used with the attenuation to calculate power-sum alien attenuation-to-crosstalk ratio (far end) (PSAACRF), a requirement of ANSI/TIA-568-C.2 for Category 6A cables.

power-sum alien attenuation-to-crosstalk ratio, far end (PSAACRF)
The ratio of the attenuation (signal) to the power-sum measurement of alien crosstalk in the far end (noise); a requirement of ANSI/TIA-568-C.2 for Category 6A cables.

power surge
Occurs when the power level rises above normal and stays there for longer than a second. *See also* power overage and power spike.

power underage
Occurs when the power level drops below the standard level. Opposite of power overage.

preform
A short, thick glass rod that forms the basis for an optical fiber during the manufacturing process. The preform is created first, and then melted and drawn under constant tension to form the long, thin optical fiber.

prefusing
Fusing the end of a fiber-optic cable with a low current to clean the end; it precedes fusion splicing.

preload
A connector with built-in adhesive that must be preheated before the fiber can be installed.

premises
A telecommunications term for the space occupied by a customer or an authorized/joint user in a building on continuous or contiguous property that is not separated by a public road or highway.

premises wiring system
The entire wiring system on a user's premises, especially the supporting wiring that connects the communications outlets to the network interface jack.

prewiring
Wiring that is installed before walls and ceilings are enclosed. Prewiring is usually easier than waiting until the walls are built to install wire.

primary coating

The specialized coating applied to the surface of the fiber cladding during manufacture.

primary rate interface (PRI)

As defined by the ISDN standard, consists of 23 B-channels (64Kbps each) and one 64Kpbs D-channel (delta channel) in the United States, or 30 B-channels and one D-channel in Europe.

private branch exchange (PBX)

A telephone switching system servicing a single customer, usually located on that customer's premises. It switches calls both inside a building and outside to the telephone network, and it can sometimes provide access to a computer from a data terminal. Now used interchangeably with PABX (private automatic branch exchange).

profile alignment system (PAS)

A fiber splicing technique for using non-electro-optical linked access technology for aligning fibers for splicing.

propagation delay

The difference in time between when a signal is transmitted and when it is received.

protector

A device that limits damaging voltages on metallic conductors by protecting them against surges and transients.

protocol

A set of predefined, agreed-upon rules and message formats for exchanging information among devices on a network.

protocol analyzer

A software and hardware troubleshooting tool used to decode protocol information to try to determine the source of a network problem and to establish baselines.

public data network

A network established and operated for the specific purpose of providing data transmission services to the public. *See also* public switched network.

public switched network

A network provided by a common carrier that provides circuit switching between public users, such as the public telephone network, telex, Sprint's Telenet, or MCI's Execunet.

public switched telephone network (PSTN)

The basic phone service provided by the phone company. *See also* plain old telephone service.

puck

In fiber optics, metal disc that holds a connector in the proper position against an abrasive for polishing the connector endface.

pull strength

The pulling force that can be applied to a cable without damaging a cable or affecting the specified characteristics of the cable. Also called pull tension.

pull string

A string that is tied to a cable and is used to pull cables through conduits or over racks. Similar to fish tape.

pulse

A current or voltage that changes abruptly from one value to another and back to the original value in a finite length of time.

pulse code modulation (PCM)

The most common method of converting an analog signal, such as speech, to a digital signal by sampling at a regular rate and converting each sample to an equivalent digital code. The digital data is transmitted sequentially and returned to analog format after it is received.

pulse dispersion

The dispersion of pulses as they travel along an optical fiber.

pulse spreading

The dispersion of an optical signal with time as it propagates through an optical fiber.

punch-down

A generic name for any cross-connect block where the individual wires in UTP are placed into a terminal groove and "punched down" with a special tool. The groove pierces the insulation and makes contact with the inner conductor. The punch-down operation may also trim the wire as it terminates. Punch-downs are performed on telecommunications outlets, 66-blocks, and 110-blocks. Also called cut down.

Q

Quadrature amplitude modulation (QAM)

The modulation of two separate signals onto carriers at a single frequency and kept separate by having the two signals 90 degrees out of phase.

quality of service (QoS)
Data prioritization at the Network layer of the OSI model. QoS results in guaranteed throughput rates.

quantum
A basic unit, usually used in reference to energy. A quantum of light is called a photon.

quantizer IC
An integrated circuit (IC) that measures received optical energy and interprets each voltage pulse as a binary 1 or 0.

quantizing error
Inaccuracies in analog to digital conversion caused by the inability of a digital value to match an analog value precisely. Quantizing error is reduced as the number of bits used in a digital sample increases, since more bits allow greater detail in expressing a value.

quartet signaling
The encoding method used by 100VG-AnyLAN, in which the 100Mbps signal is divided into four 25Mbps channels and then transmitted over different pairs of a cable. Category 3 cable transmits one channel on each of four pairs.

R

R
Symbol for resistance.

rack
A frame-like structure where patch panels, switches, and other network equipment are installed. The typical dimension is 19 inches.

raceway
Any channel or structure used in a building to support and guide electrical and optical fiber wires or cables. Raceways may be metallic or nonmetallic and may totally or partially enclose the wiring (e.g., conduit, cable trough, cellular floor, electrical metallic tubing, sleeves, slots, under-floor raceways, surface raceways, lighting fixture raceways, wireways, busways, auxiliary gutters, and ventilated flexible cableways). See also pathway.

radial refractive index profile
The refractive index measured in a fiber as a function of the distance from the axial core or center.

radiant flux
Radiant flux is the measured amount of energy on a surface per unit time.

radiation (rad) hardened
Used to describe material that is not sensitive to the effects of nuclear radiation; such material is typically used for military applications.

radio frequency (RF)
The frequencies in the electromagnetic spectrum that are used for radio communications.

radio frequency interference (RFI)
The interference on copper cabling systems caused by radio frequencies.

Raman amplification
An amplification method using a pump laser to donate energy to a signal to amplify it without using a doped length of fiber.

ray
A beam of light in a single direction. Usually a representation of light traveling in a particular direction through a particular medium.

Rayleigh scattering
The redirection of light caused by atomic structures and particles along the light's path. Rayleigh scattering is responsible for some attenuation in optical fiber, because the scattered light is typically absorbed when it passes into the cladding.

reactance
A measure of the combined effects of capacitance and inductance on an alternating current. The amount of such opposition varies with the frequency of the current. The reactance of a capacitor decreases with an increase in frequency. The opposite occurs with an inductance.

receiver
A device whose purpose is to capture transmitted signal energy and convert that energy for useful functions. In fiber-optic systems, an electronic component that converts light energy to electrical energy.

receiver sensitivity
In fiber optics, the amount of optical power required by a particular receiver in order to transmit a signal with few errors. Can be considered a measure of the overall quality of receiver. The more sensitive the receiver, the better its quality.

receptacle
The part of a fiber-optic receiver that accepts a connector and aligns the ferrule for proper optical transmission.

reflectance

A percentage that represents the amount of light that is reflected back along the path of transmission from the coupling region, the connector, or a terminated fiber.

reflection

(1) A return of electromagnetic energy that occurs at an impedance mismatch in a transmission line, such as a LAN cable. *See also* return loss. (2) The immediate and opposite change in direction that happens to a light beam when it strikes a reflective surface. Reflection causes several spectral problems, including high optical distortion and enhanced intensity noise.

refraction

The bending of a beam of light as it enters a medium of different density. Refraction occurs as the velocity of the light changes between materials of two different refractive indexes.

refractive index gradient

The change in refractive index with respect to the distance from the axis of an optical fiber.

refractive index profile

A graphical description of the relationship between the refractive indexes of the core and the cladding in an optical fiber.

regenerator

A receiver-transmitter pair that detects a weak signal, cleans it up, then sends the regenerated signal through another length of fiber.

Registered Communications Distribution Designer (RCDD)

A professional accreditation granted by BICSI (the Building Industry Consulting Service International). RCDDs have demonstrated a superior level of knowledge of the telecommunications wiring industry and associated disciplines.

registered jack (RJ)

Telephone and data jacks/applications that are registered with the FCC. Numbers such as RJ-11 and RJ-45 are widely misused in the telecommunications industry—the RJ abbreviation was originally used to identify a type of service and wiring pattern to be installed, not a specific jack type. A much more precise way to identify a jack is to specify the number of positions (width of opening) and number of conductors. Examples include the eight-position, eight-conductor jack and the six-position, four-conductor jack.

remodel box

An electrical box that is designed to clamp to the wallboard, as opposed to a "new construction" box, which is nailed to a wall stud.

repeater

(1) A device that receives, amplifies (and reshapes), and retransmits a signal. It is used to overcome attenuation by boosting signal levels, thus extending the distance over which a signal can be transmitted. Repeaters can physically extend the distance of a LAN or connect two LAN segments.

resistance

In DC (direct current) circuits, the opposition a material offers to current flow, measured in ohms. In AC (alternating current) circuits, resistance is the real component of impedance and may be higher than the value measured at DC.

resistance unbalance

A measure of the inequality of the resistance of the two conductors of a transmission line.

responsivity

(1) The ratio of a detector's output to input, usually measured in units of amperes per watt (or microamperes per microwatt). (2) The measure of how well a photodiode converts a wavelength or range of wavelengths of optical energy into electrical current.

restricted mode launch (RML)

A launch that limits a laser's launch condition so that fewer modes are used.

retermination

The process of disconnecting, then reconnecting a cable to a termination point (possibly moving the cable in the process).

retractile cord

A cord with a specially treated insulation or jacket that causes it to retract like a spring. Retractile cords are commonly used between a telephone and a telephone handset.

return loss

The ratio of reflected power to inserted power. Return loss is a measure of the signal reflections occurring along a channel or basic link and is related to various electrical mismatches along the cabling. This ratio, expressed in decibels, describes the ratio of optical power reflected by a component, for instance a connector, to the optical power introduced to that component.

return reflection

Light energy that is reflected from the end of a fiber through Fresnel reflection.

return to zero (RZ)

A digital coding scheme where the signal level is low for a 0 bit and high for a 1 bit during the first half of a bit interval; in either case, the bit returns to zero volts for the second half of the interval.

reversed pair

A wiring error in twisted-pair cabling where the conductors of a pair are reversed between connector pins at each end of a cable. A cabling tester can detect a reversed pair.

RF access point

A wireless receiver and transmitter used in a wireless network. This point accesses incoming RF signals and transmits RF signals.

RG-58

The type designation for the coaxial cable used in thin Ethernet (10Base-2). It has a 50 ohm impedance rating and uses BNC connectors.

RG-62

The type designation for the coaxial cable used in ARCnet networks. It has a 93 ohm impedance and uses BNC connectors.

RG/U

Radio grade/universal. RG is the common military designation for coaxial cable.

ribbon

Multiple conductors or optical fibers clad in a single, flat, ribbon-like cable.

ring

(1) A polarity designation of one wire of a pair indicating that the wire is that of the secondary color of a five-pair cable (which is not commonly used anymore) group (e.g., the blue wire of the blue/white pair). (2) A wiring contact to which the ring wire is attached. (3) The negative wiring polarity (*see also* tip). (4) Two or more stations in which data is passed sequentially between active stations, each in turn examining or copying the information before finally returning it to the source. *See also* ring topology

ring conductor

A telephony term used to describe one of the two conductors that is in a cable pair used to provide telephone service. This term was originally coined from its posi-

tion as the second (ring) conductor of a tip-ring-sleeve switchboard plug. *See also* ring.

ring topology

A network topology in which terminals are connected in a point-to-point serial fashion in an unbroken circular configuration. Many logical ring topologies such as Token Ring are wired as a star for greater reliability.

ripcord

A length of string built into optical fiber cables that is pulled to split the outer jacket of the cable without using a blade.

riser

(1) A designation for a type of cable run between floors Fire-code rating for indoor cable that is certified to pass through the vertical shaft from floor to floor. (2) A space for indoor cables that allow cables to pass between floors, normally a vertical shaft or space.

riser cable

A type of cable used in vertical building shafts, such as telecommunications and utility shafts. Riser cable typically has more mechanical strength than general use cable and has an intermediate fire protection rating.

RJ-45

A USOC code identifying an eight-pin modular plug or jack used with unshielded twisted-pair cable. Officially, an RJ-45 connector is a telephone connector designed for voice-grade circuits. Only RJ-45-type connectors with better signal handling characteristics are called eight-pin connectors in most standards documents, though most people continue to use the RJ-45 name for all eight-pin connectors.

RJ-connector

A modular connection mechanism that allows for as many as eight copper wires (four pairs). Commonly found in phone (RJ-11) or 10Base-T (RJ-45) connections.

rope strand

A conductor composed of groups of twisted strands.

router

A device that connects two networks and allows packets to be transmitted and received between them. A router may also determine the best path for data packets from source to destination. Routers primarily operate on Layer 3 (the Network layer) of the OSI model.

routing

A function of the network layer that involves moving data throughout a network. Data passes through several network segments using routers that can select the path the data takes.

RS-232C
The EIA's registered standard that defines an interface that computers use to talk to modems and other serial devices such as printers or plotters.

S

sample-and-hold circuit
A circuit that samples an analog signal such as a voltage level and then holds the voltage level long enough for the analog-to-digital (A/D) converter to change the level to a numerical value.

SC connector
An optical-fiber connector made from molded plastic using push-pull mechanics for joining to a fiber adapter. The SC connector has a 2.5mm ferrule push-pull latching mechanism and can be snapped together to form duplex and multifiber connectors. SC connectors are the preferred fiber-optic cable for premises cabling and are recognized by the ANSI/TIA/EIA-568-B standard for structured cabling.

scanner
A cable-testing device that uses TDR methods to detect cable transmission anomalies and error conditions.

scattering
(1) A property of glass that causes light to deflect from the fiber and contributes to losses. (2) The redirection of light caused by atomic structures and particles along the light's path. *See also* Rayleigh scattering.

screened twisted-pair (ScTP) cable
A balanced four-pair UTP with a single foil or braided screen surrounding all four pairs in order to minimize EMI radiation or susceptibility. Screened twisted-pair is also sometimes called foil twisted-pair (FTP). ScTP is a shielded version of Category 3, 5, 5e, and 6 UTP cables; they are less susceptible to EMI than UTP cables but are more susceptible than STP cables.

scribe
A tool used to mark an object prior to some type of drilling or cutting.

segment
A portion of a network that uses the same length of cable (electrically contiguous). Also the portion of a network that shares a common hub or set of interconnected hubs.

SELFOC Lens
A trade name used by the Nippon Sheet Glass Company for a graded-index fiber lens. A segment of graded index fiber made to serve as a lens.

semiconductor
In wire industry terminology, a material possessing electrical conductivity that falls somewhere between that of conductors and insulators. Usually made by adding carbon particles to an insulator. This is not necessarily the same as semiconductor materials such as silicon or germanium.

semiconductor laser
A laser in which the injection of current into a semiconductor diode produces light by recombination of holes and electrons at the junction between p- and n-doped regions. Also called a semiconductor diode laser.

semiconductor optical amplifier (SOA)
A laser diode with optical fibers at each end instead of mirrors. The light from the optical fiber at either end is amplified by the diode and transmitted from the opposite end.

sensitivity
For a fiber-optic receiver, the minimum optical power required to achieve a specified level of performance, such as BER.

separator
Pertaining to wire and cable, a layer of insulating material such as textile, paper, or Mylar, which is placed between a conductor and its dielectric, between a cable jacket and the components it covers, or between various components of a multiple conductor cable. It can be utilized to improve stripping qualities or flexibility, or it can offer additional mechanical or electrical protection to the components it separates.

sequential markings
Markings on the outside of a fiber-optic cable to aid in measuring the length of the cable.

series wiring
See daisy chain.

service loop
A loop or slack left in a cable when the cable is installed and terminated. This loop allows future trimming of the cable or movement of equipment if necessary.

service profile identifier (SPID)
The ISDN identification number issued by the phone company that identifies the ISDN terminal equipment attached to an ISDN line.

sheath
An outer protective layer of a fiber-optic or copper cable that includes the cable jacket, strength members, and shielding.

shield
A metallic foil or multiwire screen mesh that is used to reduce electromagnetic fields from penetrating or exiting a transmission cable. Also referred to as a screen.

shield coverage
The physical area of a cable that is actually covered by shielding material, often expressed as a percentage.

shield effectiveness
The relative ability of a shield to screen out undesirable interference. Frequently confused with the term shield coverage.

shielded twisted pair (STP)
A type of twisted-pair cable in which the pairs are enclosed in an outer braided shield, although individual pairs may also be shielded. STP most often refers to the 150 ohm IBM Type 1, 2, 6, 8, and 9 cables used with Token Ring networks. Unlike UTP cabling, the pairs in STP cable have an individual shield, and the individual shielded cables are wrapped in an overall shield. The primary advantages of STP cable are that it has less attenuation at higher frequencies and is less susceptible to EMI. Since the advent of standards-based structured wiring, STP cable is rarely used in the United States.

short wavelength
In reference to light, a wavelength shorter than 1000nm.

SI units
The standard international system of metric units.

signal
The information conveyed through a communication system.

signal encoding
The process whereby a protocol at the Physical layer receives information from the upper layers and translates all the data into signals that can be transmitted on a transmission medium.

signaling
The process of transmitting data across the medium. Types of signaling include digital and analog, baseband and broadband.

signal-to-noise ratio (SNR or S/N)
The ratio of received signal level to received noise level, expressed in decibels and abbreviated SNR or S/N. A higher SNR ratio indicates better channel performance. The relationship between the usable intended signal and the extraneous noise present. If the SNR limit is exceeded, the signal transmitted will be unusable.

silica glass
Glass made mostly of silicon dioxide used in conventional optical glass that is used commonly in optical fiber cables.

Silicone
A General Electric trademark for a material made from silicon and oxygen. Can be in thermosetting elastomer or liquid form. The thermosetting elastomer form is noted for high heat resistance.

silver satin cable
The silver-gray voice-grade patch cable used to connect a telephone to a wall jack such as that used by home telephones. Silver satin cables are unsuitable for use in LAN applications because they do not have twisted pairs, and this results in high levels of crosstalk and capacitance.

simplex
(1) A link that can only carry a signal in one direction. (2) A fiber-optic cable or cord carrying a single fiber. Simplex cordage is mainly used for patch cords and temporary installations.

simplex cable
A term sometimes used to describe a single fiber cable.

simplex transmission
Data transmission over a circuit capable of transmitting in only one direction.

single attachment station (SAS)
With FDDI networks, denotes a station that attaches to only one of two rings in a dual-ring environment.

single-board computer
A circuit board containing the components needed for a computer that performs a prescribed task. A single-board computer is often the basis of a larger piece of equipment that contains it.

single-ended line

An unbalanced circuit or transmission line, such as a coaxial cable (see also balanced line and unbalanced line).

single-frequency laser

A laser that emits a range of wavelengths small enough to be considered a single frequency.

single-mode fiber (SMF)

Optical-fiber cable with a small core, usually between two and nine microns, which can support only one wavelength. It requires a laser source for the input because the acceptance cone is so small. The small core radius approaches the wavelength of the source. Single-mode optical-fiber cable is typically used for backbones and to transmit data over long distances.

single polarization fibers

Optical fibers capable of carrying light in only one polarization.

sintering

A process in optical fiber manufacturing in which the soot created by heating silicon dioxide is compressed into glass to make the fiber preform.

sinusoidal

A signal that varies over time in proportion to the sine of an angle. Alternating current (AC) is sinusoidal.

skew ray

A light ray that does not intersect the fiber axis and generally enters the fiber at a very high angle.

skin effect

The tendency of alternating current to travel on the surface of a conductor as the frequency increases.

small form factor (SFF) connector

A type of optical fiber connector that is designed to take up less physical space than a standard-sized connector. An SFF connector provides support for two strands of optical fiber in a connector enclosure that is similar to an RJ-45. There is currently no standard for SFF connectors; types include the LC and the MT-RJ connectors.

sneak current

A low-level current that is of insufficient strength to trigger electrical surge protectors and thus may be able to pass between these protectors undetected. The sneak current may result from contact between communications lines and AC power circuits or from power induction. This current can cause equipment damage unless secondary protection is used.

solar cell

A device used to convert light energy into electrical current.

solid-state laser

A laser whose active medium is a glass or crystal.

soliton

A special type of light pulse used in fiber-optic communications in combination with optical amplifiers to help carry a signal longer distances.

source

In fiber optics, the device that converts the information carried by an electrical signal to an optical signal for transmission over an optical fiber. A fiber-optic source may be a light-emitting diode or laser diode.

source address

The address of the station that sent a packet, usually found in the source area of a packet header. In the case of LAN technologies such as Ethernet and Token Ring, the source address is the MAC (media access control) address of the sending host.

spectral bandwidth

(1) The difference between wavelengths at which the radiant intensity of illumination is half its peak intensity. (2) Radiance per unit wavelength interval.

spectral width

A measure of the extent of a spectrum. The range of wavelengths within a light source. For a source, the width of wavelengths contained in the output at one half of the wavelength of peak power. Typical spectral widths are between 20nm and 170nm for an LED and between 1nm and 5nm for a laser diode.

spectrum

Frequencies that exist in a continuous range and have a common characteristic. A spectrum may be inclusive of many spectrums; the electromagnetic radiation spectrum includes the light spectrum, the radio spectrum, and the infrared spectrum.

speed of light (c)

In a vacuum, light travels 299,800,000 meters per second. This is used as a reference for calculating the index of refraction.

splice

(1) A permanent joint between two optical waveguides. (2) Fusing or mechanical means for joining two fiber ends.

splice enclosure

A cabinet used to organize and protect splice trays.

splice tray

A container used to organize and protect spliced fibers.

split pair

A wiring error in twisted-pair cabling where one of a pair's wires is interchanged with one of another pair's wires. Split pair conditions may be determined with simple cable testing tools (simple continuity tests will not reveal the error because the correct pin-to-pin continuity exists between ends). The error may result in impedance mismatch, excessive crosstalk, susceptibility to interference, and signal radiation.

splitting ratio

The ratio of power emerging from two output ports of a coupler.

spontaneous emission

The emission of random photons (incoherent light) at the junction of the p and n regions in a light-emitting diode when current flows through it. Occurs when a semiconductor accumulates spurious electrons. Spontaneous emission interferes with coherent transmission

S/T interface

The four-wire interface of an ISDN terminal adapter. The S/T interface is a reference point in ISDN.

ST connector

A fiber-optic connector with a bayonet housing; it was developed by AT&T but is not in favor as much as SC or FC connectors. It is used with older Ethernet 10Base-FL and fiber-optic inter-repeater links (FIORLs).

stabilized light source

An LED or laser diode that emits light with a controlled and constant spectral width, central wavelength, and peak power with respect to time and temperature.

standards

Mutually agreed-upon principles of protocol or procedure. Standards are set by committees working under various trade and international organizations.

star coupler

A fiber-optic coupler in which power at any input port is distributed to all output ports. Star couplers may have up to 64 input and output ports.

star network

A network in which all stations are connected through a single point such as a hub.

star topology

(1) A method of cabling each telecommunications outlet/connector directly to a cross-connect in a horizontal cabling subsystem. (2) A method of cabling each cross-connect to the main cross-connect in a backbone cabling subsystem. (3) A topology in which each outlet/connector is wired directly to the hub or distribution device.

static charge

An electrical charge that is bound to an object.

static load

Load such as tension or pressure that remains constant over time, such as the weight of a fiber-optic cable in a vertical run.

station

A unique, addressable device on a network.

stay cord

A component of a cable, usually of high tensile strength, used to anchor the cable ends at their points of termination and keep any pull on the cable from being transferred to the electrical conductors.

step index fiber

An optical fiber cable, usually multimode, that has a uniform refractive index in the core with a sharp decrease in index at the core/cladding interface. The light is reflected down the path of the fiber rather than refracted as in graded-index fibers. Step-index multimode fibers generally have lower bandwidths than graded-index multimode fibers.

step-index single-mode fiber

A fiber with a small core that is capable of carrying light in only one mode. Sometimes referred to as single-mode optical fiber cable.

step insulated

A process of applying insulation to a cable in two layers. Typically used in shielded networking cables so that the outer layer of insulation can be removed and the remaining conductor and insulation can be terminated in a connector.

stimulated emission

The process in which a photon interacting with an electron triggers the emission of a second photon with the same phase and direction as the first. Stimulated emission is the basis of a laser..

stitching

The process of terminating multiconductor cables on a punch-down block such as a 66-block or 110-block.

STP-A

Refers to the enhanced IBM Cabling System specifications with the Type A suffix. The enhanced Type 1A, 2A, 6A, and 9A cable specifications were designed to support operation of 100Mbps FDDI signals over copper. See Type 1A, Type 2A, Type 6A, or Type 9A.

strength member

The part of a fiber-optic cable composed of Kevlar aramid yarn, steel strands, or fiberglass filaments that increases the tensile strength of the cable.

structural return loss (SRL)

A measurement of the impedance uniformity of a cable. It measures energy reflected due to structural variations in the cable. The higher the SRL number, the better the performance; this means more uniformity and lower reflections.

structured cabling system

Telecommunications cabling that is organized into a hierarchy of wiring termination and interconnection structures. The concept of structured wiring is used in the common standards from the TIA and EIA. See Chapter 1 for more information on structured cabling and Chapter 2 for more information on structured cabling standards.

stud cavity

A type of frame used in building construction.

submarine cable

A cable designed to be laid underwater.

subminiature D-connector

The subminiature D-connector is a family of multipin data connectors available in 9-, 15-, 25- and 37-pin configurations. Sometimes referred to as DB9, DB15, DB25, and DB37 connectors, respectively.

subnetwork

A network that is part of another network. The connection is made through a router.

subnet mask

Subnet masks consist of 32 bits, usually a block of 1s followed by a block of 0s. The last block of 0s designate that part as being the host identifier. This allows a classful network to be broken down into subnets.

supertrunk

A cable that carries several video channels between the facilities of a cable television company.

surface-emitting LED

An LED in which incoherent light is emitted at all points along the pn junction.

surface light emitting diode (SLED)

A light-emitting diode (LED) that emits light from its flat surface rather than its side. These are simple and inexpensive and provide emission spread over a wide range.

surface mount assembly (SMA) connector

An optical fiber cable connector that is a threaded type connector. The SMA 905 version is a straight ferrule design, whereas the SMA 906 is a stepped ferrule design.

surface-mount box

A telecommunications outlet that is placed in front of the wall or pole.

surface-mount panel

A panel for electrical or data communications connections placed on a wall or pole.

surge

A rapid rise in current or voltage, usually followed by a fall back to a normal level. Also referred to as a transient.

surge protector

A device that contains a special electronic circuit that monitors the incoming voltage level and then trips a circuit breaker when an over-voltage reaches a certain level, called the over-voltage threshold.

surge suppression

The process by which transient voltage surges are prevented from reaching sensitive electronic equipment. *See also* surge protector.

switch

A mechanical, optical, or optomechanical device that completes or breaks an optical path or routes an optical signal.

switched network

A network that routes signals to their destinations by switching circuits or packets. Two different packets of information may not take the same route to get to the same destination in a packet switched network.

synchronous

Transmission in which the data is transmitted at a fixed rate with the transmitter and receiver being synchronized.

Synchronous Optical Network (SONET)

The underlying architecture in most systems, which uses cells of fixed length.

T

T-1

A standard for digital transmission in North America. A digital transmission link with a capacity of 1.544Mbps (1,619,001 bits per second), T-1 lines are used for connecting networks across remote distances. Bridges and routers are used to connect LANs over T-1 networks.

T-3

A 44.736Mbps multichannel digital transmission system for voice or data provided by long-distance carriers. T-3C operates at 90Mbps.

tap

(1) A device for extracting a portion of the optical fiber. (2) On Ethernet 10Base-5 thick coaxial cable, a method of connecting a transceiver to the cable by drilling a hole in the cable, inserting a contact to the center conductor, and clamping the transceiver onto the cable at the tap. These taps are referred to as vampire taps.

tap loss

In a fiber-optic coupler, the ratio of power at the tap port to the power at the input port. This represents the loss of signal as a result of tapping.

tapered fiber

An optical fiber whose transverse dimensions vary monotonically with length.

T-carrier

A carrier that is operating at one of the standard levels in the North American Digital Hierarchy, such as T-1 (1.544Mbps) or T-3 (44.736Mbps).

T-coupler

A coupler having three ports.

tee coupler

A device used for splitting optical power from one input port to two output ports.

Teflon

DuPont Company trademark for fluorocarbon resins. *See also* fluorinated ethylene propylene and tetrafluoroethylene.

telco

An abbreviation for telephone company.

telecommunications

Any transmission, emission, or reception of signs, signals, writings, images, sounds, or information of any nature by cable, radio, visual, optical, or other electromagnetic systems.

telecommunications bus bar

Refers to thick strips of copper or aluminum that conduct electricity within a switchboard, distribution board, substation, or other electrical apparatus in a telecommunications network.

Telecommunications Industry Association (TIA)

The standards body that helped to author the ANSI/TIA/EIA-568-B Commercial Building Telecommunications Cabling Standard in conjunction with EIA and that continues to update it, along with standards for pathways, spaces, grounding, bonding, administration, field testing, and other aspects of the telecommunications industry. See Chapter 2 for more information.

telecommunications infrastructure

A collection of telecommunications components that together provide the basic support for the distribution of all information within a building or campus. This excludes equipment such as PCs, hubs, switches, routers, phones, PBXs, and other devices attached to the telecommunications infrastructure.

telecommunications outlet

A fixed connecting device where the horizontal cable terminates that provides the interface to the work area cabling. Typically found on the floor or in the wall. Sometimes referred to as a telecommunications outlet/connector or a wall plate.

telecommunications room

An enclosed space for housing telecommunications equipment, cable terminations, and cross-connect cabling used to serve work areas located on the same floor. The telecommunications room is the typical location of the horizontal cross-connect and is distinct from an equipment room because it is considered to be a floor-serving (as opposed to building- or campus-serving) facility.

Telecommunications Systems Bulletin (TSB)

A document released by the TIA to provide guidance or recommendations for a specific TIA standard.

tensile strength

Resistance to pulling or stretching forces.

terminal

(1) A point at which information may enter or leave a communications network. (2) A device by means of which wires may be connected to each other.

terminal adapters

ISDN customer-premise equipment that is used to connect non-ISDN equipment (computers and phones) to an ISDN interface.

terminate

(1) To connect a wire conductor to something, typically a piece of equipment like patch panel, cross-connect, or telecommunications outlet. (2) To add a component such as a connector or a hardware connection to a bare optical fiber end.

terminator

A device used on coaxial cable networks that prevents a signal from bouncing off the end of the network cable, which would cause interference with other signals. Its function is to absorb signals on the line, thereby keeping them from bouncing back and being received again by the network or colliding with other signals.

test jumper

A single- or multi-fiber cable used for connections between an optical fiber and test equipment.

tetrafluoroethylene (TFE)

A thermoplastic material with good electrical insulating properties and chemical and heat resistance.

theoretical cutoff wavelength

The shortest wavelength at which a single light mode can be propagated in a single-mode fiber. Below the cutoff several modes will propagate; in this case, the fiber is no longer single-mode but multimode.

thermal rating

The temperature range in which a material will perform its function without undue degradation such as signal loss.

thermoplastic

A material that will soften, flow, or distort appreciably when subjected to sufficient heat and pressure. Examples are polyvinyl chloride and polyethylene, which are commonly used in telecommunications cable jackets.

thicknet

Denotes a coaxial cable type (similar to RG-8) that is commonly used with Ethernet (10Base-5) backbones. Originally, thicknet cabling was the only cabling type used with Ethernet, but it was replaced as backbone cabling by optical fiber cabling. Thicknet cable has an impedance of 50 ohms and is commonly about 0.4 inches in diameter.

thinnet

Denotes a coaxial cable type (RG-58) that was commonly used with Ethernet (10Base-2) local area networks. This coaxial cable has an impedance of 50 ohms and is 0.2 inches in diameter. It is also called cheapernet due to the fact that it was cheaper to purchase and install than the bulkier (and larger) thicknet Ethernet cabling. The maximum distance for a thinnet segment is 180 meters.

tight buffer

A type of optical fiber cable construction where each glass fiber is buffered tightly by a protective thermoplastic coating to a diameter of 900 microns. High tensile strength rating is achieved, providing durability and ease of handling and connectorization.

tight-buffered

An optical fiber with a buffer that matches the outside diameter of the optical fiber, forming a tight outer protective layer; typically 900 microns in diameter.

time division multiple access (TDMA)

A method used to divide individual channels in broadband communications into separate time slots, allowing more data to be carried at the same time.

time division multiplexing (TDM)

Digital multiplexing that takes one pulse at a time from separate signals and combines them in a single stream of bits.

time domain reflectometry (TDR)

A technique for measuring cable lengths by timing the period between a test pulse and the reflection of the pulse from an impedance discontinuity on the cable. The returned waveform reveals undesired cable conditions, including shorts, opens, and transmission anomalies due to excessive bends or crushing. The length to any anomaly, including the unterminated cable end or cable break, may be computed from the relative time of the wave return and nominal velocity of propagation of the pulse through the cable. For optical fiber cables, see also optical time-domain reflectometry.

timing circuit

In an optical time-domain reflectometer, the circuit used to coordinate and regulate the testing process.

tinsel

A type of electrical conductor composed of a number of tiny threads, each having a fine, flat ribbon of copper or other metal closely spiraled about it. Used for small-sized cables requiring limpness and extra-long flex life.

tip

(1) A polarity designation of one wire of a pair indicating that the wire is that of the primary (common) color of a five-pair cable (which is not commonly used anymore) group (e.g., the white/blue wire of the blue pair). (2) A wiring contact to which the tip wire is connected. (3) The positive wiring polarity. *See also* ring.

tip conductor

A telephony term used to describe the conductor of a pair that is grounded at the central office when the line is idle. This term was originally coined from its position as the first (tip) conductor of a tip-ring-sleeve switchboard plug. *See also* tip.

TNC

A threaded connector used to terminate coaxial cables. TNC is an acronym for Threaded Neill-Concelman (Neill and Concelman invented the connector).

token passing

A media access method in which a token (data packet) is passed around the ring in an orderly fashion from one device to the next. A station can transmit only when it has the token. The token continues around the network until the original sender receives the token again. If the host has more data to send, the process repeats. If not, the original sender modifies the token to indicate that the token is free for anyone else to use.

Token Ring

A ring topology for a local area network (LAN) in which a supervisory frame, or token, must be received by an attached terminal or workstation before that terminal or workstation can start transmitting. The workstation with the token then transmits and uses the entire bandwidth of whatever communications media the token ring network is using. The most common wiring scheme is called a star-wired ring. Only one data packet can be passed along the ring at a time. If the data packet goes around the ring without being claimed, it eventually makes its way back to the sender. The IEEE standard for Token Ring is 802.5.

tone dial

A push-button telephone dial that makes a different sound (in fact, a combination of two tones) for each number pushed. The technically correct name for tone dial is dual-tone multifrequency (DTMF) since there are two tones generated for each button pressed.

tone generator

A small electronic device used to test network cables for breaks and other problems that sends an electronic sig-nal down one set of UTP wires. Used with a tone locator or probe.

tone locator

A testing device or probe used to test network cables for breaks and other problems; designed to sense the signal sent by the tone generator and emit a tone when the signal is detected in a particular set of wires.

topology

The geometric physical or electrical configuration describing a local communication network, as in network topology; the shape or arrangement of a system. The most common topologies are bus, ring, and star.

total attenuation

The loss of light energy due to the combined effects of scattering and absorption.

total internal reflection

The reflection of light in a medium of a given refractive index off of the interface with a material of a lower refractive index at an angle at or above the critical angle. Total internal reflection occurs at the core/cladding interface within an optical fiber cable.

tracer

The contrasting color-coding stripe along an insulated conductor of a wire pair.

tractor

A machine used to pull a preform into an optical fiber using constant tension.

transceiver

The set of electronics that sends and receives signals on the Ethernet media system. Transceivers may be small outboard devices or they may be built into an Ethernet port. Transceiver is a combination of the words transmitter and receiver.

transducer

A device for converting energy from one form to another, such as optical energy to electrical energy.

transfer impedance

For a specified cable length, relates to a current on one surface of a shield to the voltage drop generated by this current on the opposite surface of the shield. The transfer impedance is used to determine shield effectiveness against both ingress and egress of interfering signals. Shields with lower transfer impedance are more effective than shields with higher transfer impedance.

transient

A high-voltage burst of electrical current. If the transient is powerful enough, it can damage data transmission equipment and devices that are connected to the transmission equipment.

transimpedance amplifier

In a fiber-optic receiver, a device that receives electrical current from the photodiode and amplifies it before sending it to the *quantizer IC*.

transition point (TP)

ISO/IEC 11801 term that defines a location in the horizontal cabling subsystem where flat under-carpet cabling connects to round cabling or where cable is distributed to modular furniture. The ANSI/TIA/EIA-568-B equivalent of this term is consolidation point.

transmission line

An arrangement of two or more conductors or an optical waveguide used to transfer a signal from one location to another.

transmission loss

The total amount of signal loss that happens during data transmission.

transmission media

Anything such as wire, coaxial cable, fiber optics, air, or vacuum that is used to carry a signal.

transmitter

With respect to optical fiber cabling, a device that changes electrical signals to optical signals using a laser and associated electronic equipment such as modulators. Among various types of light transmitters are light-emitting diodes (LEDs), which are used in lower speed (100 to 256Mbps) applications such as FDDI, and laser diodes, which are used in higher speed applications (622Mbps to 10Gbps) such as ATM and SONET.

transverse modes

In the case of optical fiber cable, light modes across the width of the waveguide.

tree coupler

A passive fiber-optic coupler with one input port and three or more output ports, or with three or more input ports and one output port.

tree topology

A LAN topology similar to linear bus topology, except that tree networks can contain branches with multiple nodes.

triaxial cable

Coaxial cable with an additional outer copper braid insulated from signal carrying conductors. It has a core conductor and two concentric conductive shields. Also called triax.

triboelectric noise

Electromagnetic noise generated in a shielded cable due to variations in capacitance between the shield and conductor as the cable is flexed.

trunk

(1) A phone carrier facility such as phone lines between two switches. (2) A telephone communication path or channel between two points, one of them usually a telephone company facility.

trunk cable

The main cable used in thicknet Ethernet (10Base-5) implementations.

trunk line

A transmission line running between telephone switching offices.

T-series connections

A type of digital connection leased from the telephone company or other communications provider. Each T-series connection is rated with a number based on speed. T-1 and T-3 are the most popular.

TSI

Time slot interchanger. A device used in networking switches to provide non-port blocking.

turn-key agreement

A contractual arrangement in which one party designs and installs a system and "turns over the keys" to another party who will operate the system. A system may also be called a turn-key system if the system is self-contained or simple enough that all the customer has to do is "turn the key."

twinaxial cable

A type of communications cable consisting of two center conductors surrounded by an insulating spacer, which is in turn surrounded by a tubular outer conductor (usually a braid, foil, or both). The entire assembly is then covered with an insulating and protective outer layer. Twinaxial is often thought of as dual-coaxial cable. Twinaxial cable was commonly used with Wang VS terminals, IBM 5250 terminals on System/3x, and AS/400 minicomputers. Also called twinax.

twin lead

A transmission line used for television receiving antennas having two parallel conductors separated by insulating material. Line impedance is determined by the diameter and spacing of the conductors and the insulating material. The conductors are usually 300 ohms.

twisted pair

Two insulated copper wires twisted around each other to reduce induction (thus interference) from one wire to the other. The twists, or lays, are varied in length from pair to pair to reduce the potential for signal interference between pairs. Several sets of twisted-pair wires may be enclosed in a single cable. In cables greater than 25 pairs, the twisted pairs are grouped and bound together in groups of 25 pairs.

twisted-pair–physical media dependent (TP-PMD)

Technology developed by the ANSI X3T9.5 working group that allows 100Mbps transmission over twisted-pair cable.

Type 1

150-ohm shielded twisted-pair (STP) cabling conforming to the IBM Cabling System specifications. Two twisted pairs of 22 AWG solid conductors for data communications are enclosed in a braided shield covered with a sheath. Type 1 cable has been tested for operation up to 16MHz. Available in plenum, nonplenum, riser, and outdoor versions.

Type 1A

An enhanced version of IBM Type 1 cable rated for transmission frequencies up to 300MHz.

Type 2

150 ohm shielded twisted-pair (STP) cabling conforming to the IBM Cabling System specifications. Type 2 cable is popular with those who insist on following the IBM Cabling System because there are two twisted pairs of 22 AWG solid conductors for data communications that are enclosed in a braided shield. In addition to the shielded pairs, there are four pairs of 22 AWG solid conductors for telephones that are included in the cable jacket but outside the braided shield. Tested for transmission frequencies up to 16MHz. Available in plenum and nonplenum versions.

Type 2A

An enhanced version of IBM Type 2 cable rated for transmission speeds up to 300MHz.

Type 3

100 ohm unshielded twisted-pair (UTP) cabling similar to ANSI/TIA/EIA 568-B Category 3 cabling. 22 AWG or 24 AWG conductors with a minimum of two twists per linear foot. Typically four twisted pairs enclosed within cable jacket.

Type 5

100/140-micron optical fiber cable conforming to the IBM Cabling System specifications. Type 5 cable has two optical fibers that are surrounded by strength members and a polyurethane jacket. There is also an IBM Type 5J that is a 50/125-micron version defined for use in Japan.

Type 6

150 ohm shielded twisted-pair (STP) cabling that conforms to the IBM Cabling System specifications. Two twisted pairs of 26 AWG stranded conductors for data communications. Flexible for use in making patch cables. Tested for operation up to 16MHz. Available in nonplenum version only.

Type 6A

An enhanced version of IBM Type 6 cable rated for transmission speeds up to 300MHz.

Type 8

150 ohm under-carpet cable conforming to the IBM Cabling System specifications. Two individually shielded parallel pairs of 26 AWG solid conductors for data communications. The cable includes "ramped wings" to make it less visible when installed under carpeting. Tested for transmission speeds up to 16MHz. Type 8 cable is not very commonly used.

Type 9

150 ohm shielded twisted-pair (STP) cabling that conforms to the IBM Cabling System specifications. A plenum-rated cable with two twisted pairs of 26 AWG solid or stranded conductors for data communications enclosed in a braided shield covered with a sheath. Tested for transmission speeds up to 16MHz.

Type 9A

An enhanced version of IBM Type 9 cable rated for transmission speeds up to 300MHz.

U

UL Listed

If a product carries this mark, it means UL found that representative samples of this product met UL's safety requirements.

UL Recognized

If a product carries this mark, it means UL found it acceptable for use in a complete UL Listed product.

ultraviolet

The electromagnetic waves invisible to the human eye with wavelengths between 10nm and 400nm.

unbalanced line

A transmission line in which voltages on the two conductors are unequal with respect to ground; one of the conductors is generally connected to a ground point. An example of an unbalanced line is a coaxial cable. This is the opposite of a balanced line or balanced cable.

underground cable

Cable that is designed to be placed beneath the surface of the ground in ducts or conduit. Underground cable is not necessarily intended for direct burial in the ground.

Underwriters Laboratories, Inc. (UL)

A privately owned company that tests to make sure that products meet safety standards. UL also administers a program for the certification of category-rated cable with respect to flame ratings. See Chapter 4 for more information on Underwriters Laboratories.

unified messaging

The integration of different streams of communications (email, SMS, fax, voice, video, etc.) into a single interface, accessible from a variety of different devices. Unified messaging is a subset of a fully integrated u system.

uniformity

The degree of insertion loss difference between ports of a coupler.

uninterruptible power supply (UPS)

A natural line conditioner that uses a battery and power inverter to run the computer equipment that plugs into it. The battery charger continuously charges the battery. The battery charger is the only thing that runs off line voltage. During a power problem, the battery charger stops operating, and the equipment continues to run off the battery.

Universal Service Order Code (USOC)

Developed by AT&T/the Bell System, the Universal Service Order Code (pronounced "U-sock") identifies a particular service, device, or connector wiring pattern. Often used to refer to an old cable color-coding scheme that was current when USOC codes were in use. USOC is not used in wiring LAN connections and is not supported by current standards due to the fact that high crosstalk is exhibited at higher frequencies. See Chapter 9 for more information.

unmated

Optical fiber connectors in a system whose end-faces are not in contact with another connector, resulting in a fiber that is launching light from the surface of the glass into air. Also called unterminated or open.

unshielded twisted pair (UTP)

(1) A pair of copper wires twisted together with no electromagnetic shielding around them. (2) A cable containing multiple pairs of UTP wire. Each wire pair is twisted many times per foot (higher grade UTP cable can have more than 20 twists per foot). The twists serve to cancel out electromagnetic interference that the transmission of electrical signal through the pairs generates. An unshielded jacket made of some type of plastic then surrounds the individual twisted pairs. Twisted-pair cabling includes no shielding. UTP most often refers to the 100 ohm Categories 3, 5e, and 6 cables specified in the ANSI/TIA/EIA-568-B standard.

uptime

The portion of time a network is running.

V

velocity of propagation

The transmission speed of electrical energy in a length of cable compared to speed in free space. Usually expressed as a percentage. Test devices use velocity of propagation to measure a signal's transit time and thereby calculate the cable's length. *See also* nominal velocity of propagation (NVP).

vertical-cavity surface-emitting laser (VCSEL)

A type of laser that emits coherent energy along an axis perpendicular to the pn junction.

very high frequency (VHF)

Frequency band extending from 30MHz to 300MHz.

very low frequency (VLF)

Frequency band extending from 10KHz to 30KHz.

video

A signal that contains visual information, such as a picture in a television system.

videoconferencing

The act of conducting conferences via a video telecommunications system, local area network, or wide area network.

videophone

A telephone-like device that provides a picture as well as sound.

visual fault locator (VFL)

A testing device consisting of a red laser that fills the fiber core with light, allowing a technician to find problems such as breaks and macrobends by observing the light through the buffer, and sometimes the jacket, of the cable.

visible light

Electromagnetic radiation visible to the human eye at wavelengths between 400nm and 700nm.

voice circuit
A telephone company circuit capable of carrying one telephone conversation. The voice circuit is the standard unit in which telecommunications capacity is counted. The U.S. analog equivalent is 4KHz. The digital equivalent is 56Kbps in the U.S. and 64Kbps in Europe. In the U.S., the Federal Communications Commission restricts the maximum data rate on a voice circuit to 53Kbps.

voice-grade
A term used for twisted-pair cable used in telephone systems to carry voice signals. Usually Category 3 or lower cable, though voice signals can be carried on cables that are higher than Category 3.

volt
A unit of expression for electrical potential or potential difference. Abbreviated as V.

voltage drop
The voltage developed across a component by the current flow through the resistance of the component.

volt ampere (VA)
A designation of power in terms of voltage and current.

W

W
(1) Symbol for watt or wattage. (2) Abbreviation for white when used in conjunction with twisted-pair cable color codes; may also be Wh.

wall plate
A wall plate is a flat plastic or metal that usually mounts in or on a wall. Wall plates include one or more jacks.

watt
A unit of electrical power.

waveform
A graphical representation of the amplitude of a signal over time.

waveguide
A structure that guides electromagnetic waves along their length. The core fiber in an optical-fiber cable is an optical waveguide.

waveguide couplers
A connection in which light is transferred between waveguides.

waveguide dispersion
That part of the chromatic dispersion (spreading) that occurs in a single-mode fiber as some of the light passes through the cladding and travels at a higher velocity than the signal in the core, due to the cladding's lower refractive index. For the most part, this deals with the fiber as a waveguide structure. Waveguide dispersion is one component of chromatic dispersion.

waveguide scattering
The variations caused by subtle differences in the geometry and fiber index profile of an optical fiber.

wavelength
(1) The distance between two corresponding points in a series of waves. (2) With respect to optical fiber communications, the distance an electromagnetic wave travels in the time it takes to oscillate through a complete cycle. Wavelengths of light are measured in nanometers or micrometers. Wavelength is preferred over the term frequency when describing light.

wavelength division multiplexing (WDM)
A method of carrying multiple channels through a fiber at the same time (multiplexing) whereby signals within a small spectral range are transmitted at different wavelengths through the same optical-fiber cable. *See also* frequency division multiplexing.

wavelength isolation
A wave division multiplexer's isolation of a light signal from the unwanted optical channels in the desired optical channel.

wavelength variance
The variation in an optical parameter caused by a change in the operating wavelength.

wide area network (WAN)
A network that crosses local, regional, and international boundaries. Some types of internetwork technology such as a leased-line, ATM, or frame relay connect local area networks (LANs) together to form WANs.

window
In optical transmission, a wavelength at which attenuation is low, allowing light to travel greater distances through the fiber before requiring a repeater.

wire center
(1) Another name for a wiring or telecommunications closet. (2) A telephone company building where all local telephone cables converge for service by telephone switching systems. Also called central office or exchange center.

wire cross-connect

A piece of equipment or location at which twisted-pair cabling is terminated to permit reconnection, testing, and rearrangement. Cross-connects are usually located in equipment rooms and telecommunications closets and are used to connect horizontal cable to backbone cable. Wire cross-connects typically use a 66- or 110-block. These blocks use jumpers to connect the horizontal portion of the block to the backbone portion of the block.

wire fault

A break in a segment or cable that causes an error. A wire fault might also be caused by a break in the cable's shield.

wireless bridge

A wireless bridge is a hardware component used to connect two or more network segments (LANs or parts of a LAN) that are physically and logically (by protocol) separated. It does not necessarily always need to be a hardware device, as some operating systems (such as Windows, Linux, Mac OS X, and FreeBSD) provide software to bridge different protocols.

work area

The area where horizontal cabling is connected to the work area equipment by means of a telecommunications outlet. A telecommunications outlet serves a station or desk. *See also* work area telecommunications outlet.

work area cable

A cable used to connect equipment to the telecommunications outlet in the user work area. Sometimes called a patch cable or patch cord.

work area telecommunications outlet

Sometimes called a wall plate, a connecting device located in a work or user area where the horizontal cabling is terminated. A work-area telecommunications outlet provides connectivity for work area patch cables, which in turn connect to end-user equipment such as computers or telephones. The telecommunications outlet can be recessed in the wall, mounted on the wall or floorboard, or recessed in the floor or a floor monument.

workgroup

A collection of workstations and servers on a LAN that are designated to communicate and exchange data with one another.

workstation

A computer connected to a network at which users interact with software stored on the network. Also called a PC (personal computer), network node, or host.

X

X

(1) Symbol for reactance. (2) Symbol often used on wiring diagrams to represent a cross-connect.

xDSL

A generic description for the different DSL technologies such as ADSL, HDSL, RADSL. *See also* digital subscriber line.

XTC

An optical-fiber connector developed by OFTI; not in general use.

Y

Y-coupler

In fiber optics, a variation on the T-coupler, where input light is split between two channels that branch out like a "Y" from the input.

Z

z

Symbol for impedance.

zero dispersion slope

In single-mode fiber, the chromatic dispersion slope at the fiber's zero dispersion wavelength.

zero-dispersion point

In an optical fiber of a given refractive index, the narrow range of wavelengths within which all wavelengths travel at approximately the same speed. The zero-dispersion point is useful in reducing chromatic dispersion in single-mode fiber.

zero dispersion wavelength

In single-mode fiber, the wavelength where waveguide dispersion cancels out material dispersion and total chromatic dispersion is zero.

zip cord

Duplex fiber optic interconnect cable consisting of two tight-buffered fibers with outer jackets bonded together, resembling electrical wiring used in lamps and small appliances.

Index

Note to the reader: Throughout this index **boldfaced** page numbers indicate primary discussions of a topic. *Italicized* page numbers indicate illustrations.

tree coupler, 746
trenches, safety and, 589
troubleshooting problems, **449–453**
  attenuation, 452
  baseline determination, **449–450**
  communication failure, hardware vs. software, 737
  crosstalk, 452–453
  excessive length, 451
  inspection and evaluation, **871–879**
    connector endface, **872–876**
    of connectors, **871–872**
    receptacle/mating sleeve, **877–879**
  locating problems, **450–451**
  logical approach to, **893–894**
  noise, 453
  opens and shorts, 452
  using power budget, 816
  wire-map fault resolution, 451
trunk line, 365
TSB (Telecommunications Systems Bulletin)-95, 11
TSBs (Telecommunications Systems Bulletins), 67
TV coax cable connector, 398, *399*
TWCSTAC (Telecommunications Wire and Cable
  Standards Technical Advisory Committee), 62
tweezers, 673, *673*
twisted-pair cable connectors, **291–307**
  conductor arrangement, **395–396**
  crimping procedure, **396–402**
    installing connector, *398–401*, 398–402
    prerequisites, **397**
    testing connectorization, 398
  crossover cables, **305–306**
  installation, **393–402**
  modular jacks and plugs, *293*, **293–306**
    pin assignments, **296–297**
    tips for terminating, 302
    wiring schemes, 297–298. *See also* wiring
     schemes
  patch-panel terminations, **291**
  shielded, *306*, **306–307**
  single horizontal cable run for 10Base-T
    connections, **304–305**
  solid vs. stranded conductor cables, **292**
  types, **393–394**, *394*
twisted-pair cabling, 7, 8–14
  balanced signal transmission, 46
  vs. coaxial cable, *320*
  nomenclature for, 12
twisted-pair continuity tester, **192**, *192*
twisted-pair strippers, *179*, **179**
twisting, 453
twists, **31**, 46
  impact of untwisted, 48
two-way radio, 382
Tyndall, John, 498

U

U/FTP, 12
UL (Underwriters Laboratories), 27, **63**, **132–133**
  website, 133
UL 444 test, 133
UL 1581 standard, 133
UL 1666 test, 133
UL Listing Mark, 132
UL Recognized Component Mark, 132
(UL) symbol, 26
ultraviolet light, 527
unbalances, 46
underground circuits, cables entering building
  from, 145
Understanding the NEC (Holt), 132
underwater hazards, for cables, 604
Underwriters Laboratories (UL), 27, **63**, **132–133**
  website, 133
unfiltered LED light sources, 841
unidirectional WDM multiplexers, 761
unified messaging, 369
uniformity, 745
  of passive devices, 741
uninterruptible power supply (UPS), **375**
  RFP on, 462
U. S. Food and Drug Administration, Center for
  Devices and Radiological Health (CDRH), 579
U. S. National Electrical Code (NEC), 20, 62,
  135–150, 273
  cable markers, 610, *611*, 796
  communications systems, **144–150**
  fiber-optic cable requirements, 795
  fire resistance, **609–611**, 796
  flame ratings, 20–21
  flame-test requirements, 20
  general requirements, **135–136**
  importance of knowing, 150
  special conditions, **141–144**
  special occupancy, **141**
  standards for fiber-optic cables and raceways,
    609–612
  wiring and protection, **136–140**
  wiring methods and materials, **140–141**
U. S. Navy, fiber-optic telephone link, 502
universal data connector, 306
Universal Service Order Code (USOC), 294
  color code order for wiring, 396
unregulated frequency, 342
unshielded twisted-pair (UTP), 8, **9–10**, *10*, *25*, *209*
  ANSI/TIA-568-C standard requirements, **84–85**
  application-specific pair assignments for, 76
  applications using, 219
  Category 5 cable as minimum, 466
  in hierarchical topology, 107